Sir Ronald Fisher (1890-1962)

Fisher is best known in statistics for the development of the concepts of the design of statistical experiments to draw inductive inferences from data. In 1925, he published the book, *Statistical Methods for Research Workers,* which for many years was considered to be the "bible" for researchers using statistics. Among his other important contributions were the concepts of consistency and efficiency, the maximum likelihood estimation criterion and the F-statistic. Fisher also made significant contributions in quantitative genetics. In 1933, he became the Galton Professor of Eugenics at University College, London and later moved to Cambridge University where he was elected to the Balfour Chair of Genetics. He was knighted in 1952.

H. O. Hartley (1912-1980)

Professor Hartley is the holder of three doctorate degrees: Ph.D.-math, University of Berlin; Ph.D.-math statistics, Cambridge and Doctor of Science, University of London. Born in Germany, he developed an interest in statistics upon moving to England where he became in 1946 a lecturer in statistics at University College, London under E. S. Pearson, the son of Karl Pearson. He is widely recognized for his unique talent for coupling mathematics and statistics to the solution of real world problems. He collaborated with E. S. Pearson in the publication of the *Biometrika Tables for Statisticians.* In 1973, he was awarded the S. S. Wilks Memorial Medal in recognition of his diverse and significant contributions in the theory and practice of statistics, and in the teaching of statistics.

Statistical methods
for business and economics

The Irwin Series in
Quantitative Analysis for Business
Consulting Editor   **Robert B. Fetter**   *Yale University*

Roger C. Pfaffenberger
Texas Christian University

James H. Patterson
University of Missouri—Columbia

Statistical methods

for business and economics

REVISED EDITION

 1981

RICHARD D. IRWIN, INC.   Homewood, Illinois   60430

ISBN 0-256-02350-6
Library of Congress Catalog Card No. 80–85122
Printed in the United States of America

3 4 5 6 7 8 9 0 D 8 7 6 5 4 3

Preface

☐ A note to the instructor

In most disciplines, including business and economics, there is an increasing awareness of the need to comprehend basic statistical methodology for successful careers in industry, government, and self-employed positions. With the widespread use of computers today, the collection and analysis of data for decision making are, for many enterprises, standard operating procedure. In a typical organization, perhaps only a few individuals are directly involved in the collection, tabulation, and analysis of data, but many others must have a sufficient understanding of statistics to *interpret* the analyses for making decisions.

Because of the widespread coverage of statistical methods in non-mathematics curricula, it is no longer possible or, indeed, proper to assume that individuals who take the typical introductory statistics course are well trained mathematically. The demand for statistical training has increased so dramatically in the past several years and has affected so many disciplines that most introductory courses *cannot* require more than a knowledge of algebra on the part of the student. We believe the modest mathematics prerequisites to be not only a pragmatic necessity, but also appropriate for the needs of the majority of students taking an introductory course. Most of these students do not need the level of knowledge that would enable them to pursue training for a career in statistics. Rather, they need to understand the fundamental ideas of statistical analysis presented in a competent, but not necessarily mathematically rigorous, fashion. This is the spirit of this text. Among our goals is to motivate students by suggesting the potential importance that statistical analysis has in pursuing varied careers. Furthermore, we have emphasized the development of the logic of statistical procedures without resorting to mathematical derivations or proofs. We firmly believe that the extensive use of examples, the careful discussion of assumptions required to implement statistical procedures, and the proper interpretation of results can competently and effectively replace a mathematically rigorous development of statistics in an introductory course.

The text is designed to be quite flexible. A typical one-semester course would cover Chapters 1–10, the core of classical statistical inference, and perhaps two or three additional chapters that may be selected to satisfy specific course objectives. The accompanying chart indicates the interdependence of the chapters.

Chapters 1-10				
Additional inference	Linear models	Nonparametric statistics	Decision theory	Index numbers
11	13	19	21	18
12	14	20	22	
	15		23	
	16			
	17			

In a two-semester sequence, we suggest Chapters 1–10, the linear models chapters (Chapters 13–17), including analysis of variance (Chapter 12), and a selection of two or three additional chapters. We have included what we feel are the most important special statistical methods in this text. Nonparametric statistical methods (Chapters 19 and 20) are becoming increasingly important in business and social science applications, where, all too often, the data do not support the appropriate classical statistics assumptions. Forecasting (Chapter 17) also has become an increasingly used statistical tool in recent years. Most financial institutions depend heavily on the accuracy of forecasts for their success. Statistical decision theory (Chapters 21–23) is finding an important place in business applications of statistics, where prior information relevant to a current problem is available and may be combined with current information for decision-making purposes.

We have profited a great deal from the feedback we have received from users of the first edition, faculty and students alike. In particular, a survey was undertaken in the summer of 1978 to learn what those who used the first edition of the text liked most about it and what they liked the least! Of 104 survey instruments mailed, 72 responded indicating potential changes they would like to see in the revised edition. The questionnaire was fairly simple to complete, and this undoubtedly was a factor in the relatively high response rate (69 percent) achieved (take note questionnaire designers!). The information provided by this survey was most helpful in assisting us in the revision of the text, and to those who responded, we offer our sincerest thanks.

Basically, users of the first edition told us they most wanted a change from the discrete distribution examples in Chapter 10 (hypothesis testing) to continuous distribution examples; inclusion of two-way analysis of variance in the ANOVA chapter; better integration of the simple and multiple linear regression and correlation material; continuation of the emphasis on random variables in the text; and a discussion of p-values or probability values. We listened. The hypothesis-testing chapter (Chapter 10) uses a continuous distribution to develop this important statistical topic. The analysis of variance chapter (Chapter 12) now includes two-way analysis of variance as well as a discussion of factorial experiments. The correlation and regression material begins with point estimation of the regression and correlation parameters (Chapter 13), and continues (Chapter 14) with interval estimation and hypothesis-testing decision rules for these same quantities. Multiple linear regression and correlation are then covered in Chapter 15, and Chapters 16 and 17 extend these concepts into time series analysis and forecasting. The forecasting chapter now includes a very brief introduction to the Box-Jenkins forecasting methodology, another common request.

Notationally, we further emphasize the concept of a random variable in this revised edition by distinguishing among a population parameter [usually denoted by a Greek character (μ, σ, π, β, etc.)], a random variable function or an *estimator* of the population parameter [denoted in boldface (**X, S, P, b,** etc.)], and the *estimate* of the parameter usually obtained in a sample [denoted by lowercase letters (x, s, p, b, etc.)]. This notation emphasizes the notion of a random variable being a *function,* and that what is obtained after sampling is an estimate, which is a *value* of the random variable or the function. This distinction is used throughout the text after its development in Chapter 5. For those who do not wish to emphasize random variables in the teaching of statistics, this distinction, for the most part, can be ignored. But we have found that several instructors desire to highlight this distinction, so we have developed the notation to facilitate it. Finally, p-values or probability values are used throughout the text after their introduction in Chapter 10—this is especially true with regard to the regression and correlation materials.

An instructor's manual is available, which gives complete solutions to all exercises at the end of the chapters and provides helpful guidance to the instructor on how to present the text material. The instructor is encouraged to use a statistical software package to minimize the computational burden on the student while adding realism to the analysis of data sets. We recommend MINITAB, a statistical programming language that has proved over time to be easy for students to use and very effective as an aid in illustrating statistical concepts, or SAS (Statistical Analysis System), SPSS (Statistical

Procedures for the Social Sciences), or any of the other fine software statistical systems available.

☐ A note to the student

The competent interpretation of statistical data and analyses is possible only with an understanding of basic statistical concepts. All *research* that involves the collection of data inevitably leads to a statistical analysis of these data. A well-known research effort that has involved considerable statistical analysis in recent years is the study conducted under the authority of the Surgeon General of the United States to assess the effect of smoking on the health of an individual. While this study has received notoriety because of various interpretations of the results reported, it is in no way unique in the sense of using *statistical analysis to interpret data*. Today, the applications and reports of statistical data abound, and we are frequently called upon to draw *our own* conclusions based on an analysis of the reported results. It is not surprising, therefore, to find that many disciplines have added introductory statistics courses to their undergraduate curricula in recent years.

It is the authors' view that statistics is best learned on first exposure by presenting an example of the application of the statistical concept immediately after its introduction. This should solidify an understanding of the statistical concept involved. Accordingly, most discussions of statistical techniques and concepts are followed by at least one example (with solution) of an application of the statistical technique discussed. The inclusion of these example problems and solutions accounts for the text being somewhat longer than several introductory statistics texts with which the student may be familiar. We feel that this increased size is warranted by the exposure to actual problem solutions. The student is encouraged in reviewing the text material to attempt to solve several of the examples given in the various chapters and then compare his or her results with those presented. This "self-test" feature of the book should provide rapid feedback on the understanding of the material presented and should lessen the time required for mastery of the concepts involved. We have also highlighted important definitions, theorems, rules, and so on, and, where possible, have provided summary computational formulas at the end of selected chapters.

To assist the student in doing the exercises at the end of each chapter, we have provided the answers to selected end-of-chapter problems in the back of the text. Consistent with our belief that statistics can be better learned "by doing," the student is encouraged to work as many problems at the end of each chapter as possible.

☐ Acknowledgments

The material in this text was class-tested over a number of years. Much of it has been revised in the second edition based on student and instructor response to the first edition. We are indebted to those students and instructors who used the first edition of the text—their suggestions about revising the material have been most helpful in preparing the second edition.

We particularly wish to thank Warren Boe and Bruce Bowerman, who spent a considerable amount of time reviewing and reacting to changes we made in the first edition. Their thoughtful criticisms and suggestions have been most helpful to us in the revision of the first-edition material. Deborah Dabbs typed much of the revised material, as well as helped to verify the solutions to many of the end-of-chapter problems. Anita Gehr read with us through the typesetting portion of the text development, and we are indebted to her for locating errors that otherwise might have found their way into the text.

As in the first edition, we wish to acknowledge the contribution made by Tim Kaage and staff in preparing the art work for the second edition. By the clever and inspirational use of colors, much clarity has been added to the use of figures in illustrating key statistical concepts.

Roger Pfaffenberger
James Patterson

Contents

Introduction

1

■ 1.1 EXAMPLES OF STATISTICAL PROBLEMS

• The word *statistics* conveys a variety of meanings to people, many of which are inaccurate or, at the very least, misleading. To some, the word suggests only a plethora of mind-boggling tables, charts, and figures. Others consider statistics to be an imposing form of mathematics. The use of the word certainly had an inauspicious beginning, as might be suspected from a cursory study of the word, for it was originally a term used to denote a collection of figures, graphs, and the like that contained useful information for the state (primarily budget information such as taxation figures).

Used in the context of its original meaning(statistics generally refers to information about an activity or a process that is expressed in numbers listed in tables or illustrated in figures.)But, since its early connotation, statistics has grown to encompass a larger role than presenting us with charts, graphs, and tables or figures. In a modern setting, statistics refers to the science of collecting, presenting, and analyzing data. A statistician is a person who engages in one or more of the following tasks: (1) the clerical activities of tabulating, summarizing, and displaying statistical data; (2) analyzing data by using statistical methods, usually for the purposes of decision making; or (3) advancing the science of statistics by developing new and better analysis methods.)The level of expertise required by statisticians ranges from mastering simple clerical operations with data to advanced training in applied mathematics, and statisticians are needed at all levels.

The use of statistics has permeated almost every facet of our lives. The daily newspapers and the televised news reports supply us with numerous summaries of data such as stock market reports, financial summaries, and

crime statistics—and with the results of statistical analyses—we
casts, political election outcome predictions, and so on.

Governments, businesses, and individuals collect statistical data re
to carry out their activities efficiently and effectively. The rate at wi
statistical data are being collected is staggering and is primarily due to ti
realization that better decisions are possible with more information and
perhaps more importantly, to technological advances that have enabled the
efficient collection and analysis of large bodies of data. The most important
technological advance in this area has, of course, been the development of
the electronic digital computer. Statistical concepts and methods, and the
use of computers in statistical analyses, have affected virtually all
disciplines—physics, engineering, economics, sociology, psychology, business, and others. In business and economics, the development and application of statistical methods have led to greater production efficiency, better
forecasting techniques, and better management practices. It is becoming
increasingly apparent that some knowledge of statistics and computers is
essential for careers in economics, business, administration, and many other
fields as well. To gain an appreciation for the breadth of applications of
statistics to business and economic problems in particular, let us consider
five examples.

Example 1.1 Most major city newspapers provide in the Sunday edition a
summary of the New York Stock Exchange transactions during the previous
week. For example, on a particular Sunday, we may find that 2,174,000
shares of IBM (International Business Machines) were traded during the
previous week, with a weekly high of 59⅛ and a low of 56¼ (dollars per
share). The value of a share at the end of the week is listed as 58⅞, with a
change of +1¾ from the closing value at the end of the previous week. This
information is illustrated as an excerpt from the Sunday edition of the *Dallas
Morning News* on June 15, 1980, in Figure 1.1. Since it would not be feasible
to list the values of *all* 2,174,000 shares traded during the week, values of the
weekly high, low, closing price, and net change *statistics* are used to summarize the entire set of transactions. The values of the statistics presented in

FIGURE 1.1 Excerpt from a summary of weekly New York Stock Exchange
transactions

1980				Weekly			
High	Low	Stock	Volume (in 100s)	High	Low	Weekly closing	Change from last week
33¾	22½	IntrLak	203	27½	25⅞	26¼	−1⅛
15⅞	9½	IntlAlum	119	11⅞	11⅜	11¾	+ ¼
76¾	50⅜	IBM Cp.	21740	59⅛	56¼	58⅞	+1¾
23⅜	16⅝	IntlFlav	1431	22⅞	21¾	22¼	+ ⅝
45½	23	IntHarv	2662	27	25½	26¾	+ ⅛

Source: *Dallas Morning News*, June 15, 1980.

these tables in the Sunday newspapers provide valuable information for investment decisions.

Example 1.2 Annually, the Bureau of the Census publishes the *Statistical Abstract of the United States,* which contains economic and business data on a variety of subjects. The data are compiled from the U.S. census of the population (conducted every ten years), annual samples drawn from the U.S. population by the bureau, surveys conducted by other government agencies, and government-supported research projects conducted by private organizations.

In Table 1.1, extracted from the 1979 *Statistical Abstract,* statistics relating to the receipts and outlays of the federal government are presented. Although the sources and functions are listed only by major sources, the table contains valuable information for business decision making. For example,

TABLE 1.1 Federal budget receipts, by major source, and outlays, by function—percent distribution: 1960 to 1979 (for years ending June 30)

Major source	1960	1965	1970	1972	1973	1974	1975	1976	1977	1978	1979 est.
Total receipts, by source	100.0	100.0	100.0	100.0	100.0	100.0	100.0	100.0	100.0	100.0	100.0
Individual income taxes	44.0	41.8	46.7	45.4	44.4	44.9	43.6	43.9	44.0	45.0	44.6
Corporation income taxes	23.2	21.8	16.9	15.4	15.6	14.6	14.5	13.8	15.5	14.9	15.4
Social insurance taxes and contributions	15.9	19.1	23.4	25.8	27.8	28.9	30.7	31.0	30.4	30.7	31.1
Employment taxes and contributions	12.1	14.9	20.1	22.1	23.6	24.9	26.8	26.6	25.8	25.8	26.2
Excise taxes	12.6	12.5	8.1	7.4	7.0	6.4	5.9	5.7	4.9	4.6	4.0
Customs, estate, and gift taxes	2.9	3.6	3.1	4.2	3.5	3.2	2.9	3.1	3.5	3.0	2.9
Miscellaneous receipts	1.3	1.4	1.8	1.7	1.7	2.0	2.4	2.7	1.8	1.8	1.9

Function	1960	1965	1970	1972	1973	1974	1975	1976	1977	1978	1979
Total outlays	100.0	100.0	100.0	100.0	100.0	100.0	100.0	100.0	100.0	100.0	100.0
National defense	49.0	40.1	40.0	33.0	30.1	28.9	26.2	24.4	24.2	23.3	23.2
Income security	19.8	21.7	21.9	27.5	29.5	31.3	33.3	34.8	34.2	32.4	32.2
Health	0.9	1.5	6.7	7.5	7.6	8.2	8.5	9.1	9.6	9.7	10.0
Veterans benefits and services	5.9	4.8	4.4	4.6	4.9	5.0	5.1	5.0	4.5	4.2	4.1
Education, training, employment	1.1	1.8	4.4	5.4	5.1	4.6	4.8	5.1	5.2	5.9	6.2
Commerce and housing credit	1.7	0.9	1.1	0.9	0.4	1.4	1.7	1.0	0.0	0.7	0.6
Transportation	4.4	4.9	3.6	3.6	3.7	3.4	3.2	3.7	3.6	3.4	3.5
Natural resources and environment	1.7	2.1	1.5	1.8	1.9	2.1	2.2	2.2	2.5	2.4	2.3
Energy	0.5	0.6	0.5	0.6	0.5	0.3	0.7	0.8	1.0	1.3	1.7
Community and regional development	0.2	0.9	1.2	1.5	1.9	1.5	1.1	1.3	1.6	2.4	1.8
Agriculture	2.8	3.3	2.6	2.3	2.0	0.8	0.5	0.7	1.4	1.7	1.3
Net interest	7.5	7.3	7.3	6.7	7.0	8.0	7.1	7.3	7.5	7.9	8.7
Revenue sharing	0.2	0.2	0.3	0.3	3.0	2.6	2.2	2.0	2.4	2.1	1.8
International affairs	3.3	4.4	2.2	2.0	1.6	2.1	2.1	1.5	1.2	1.3	1.5
General science, space, and technology	0.7	4.9	2.3	1.8	1.6	1.5	1.2	1.2	1.2	1.0	1.1
General government	1.1	1.3	1.0	1.1	1.1	1.2	1.0	0.8	0.8	0.8	0.9
Administration of justice	0.4	0.4	0.5	0.7	0.8	0.9	0.9	0.9	0.9	0.8	0.9
Undistributed offsetting receipts	−1.3	−1.2	−1.3	−1.3	−2.8	−3.7	−2.0	−1.9	−1.7	−1.6	−1.8

Source: U.S. Bureau of the Census, *Statistical Abstract of the United States: 1979,* 100th ed. (Washington, D.C., 1979).

the increase in health percentage outlay and the decrease in defense percentage outlay may dictate a company's future strategies for new product and services development. The negative percentages recorded as outlays reflect the failure of the federal budget to balance in recent years.

Caution should be exercised in utilizing the values of statistics such as those in Table 1.1. Although the *percentage* outlay for national defense decreased from 1960 to 1979, the *total* expenditure on defense did not necessarily decrease. Indeed, the total federal outlay for defense increased from approximately $45 billion to $114.5 billion during this period. Thus, while the percentage expended for defense decreased, the total defense outlay increased from 1960 to 1979. Of course, much of the increase was due to inflation and the corresponding devaluation of the dollar during this period.

Example 1.3 In operations management, a primary concern is controlling the quality of the items being produced. If the product is a transistor radio battery, for example, we may be concerned with the longevity of the batteries. Suppose it is desired that at least 95 percent of the batteries last through at least 20 hours of continuous use. The actual percentage of batteries lasting more than 20 hours could be determined by inserting every battery produced into a transistor radio and recording its time to failure, but then there would be no batteries to sell. Rather, a manager may decide in a day's production to pull every 100th battery off the production line, insert the sample batteries in electric test circuits and record their times to failure. The percentage of these batteries lasting through more than 20 hours of continuous use could be used to estimate the percentage of all batteries produced during that day that will last more than 20 hours. Moreover, if this estimated percentage drops much below 95 percent (say, to 80 percent), the manager may wish to stop the production line until he can determine why the percentage of bad batteries appears to be greater than the tolerated 5 percent. The manager is using the value of a percentage statistic computed from a sample of all batteries produced to arrive at a decision regarding the quality of the set of all batteries produced on a given day.

This example illustrates a common phenomenon in quality control: destructive sampling. It is impossible to test the quality (longevity) of each battery produced because the test for longevity will ordinarily involve its destruction. The manager has little recourse but to sacrifice a small number of batteries (the *sample*) in order to gain information about the entire set of batteries comprising the daily production (the *population*).

Example 1.4 Determining the salability of a new product is a constant problem posed to many marketing research groups. To determine whether a new kitchen-ware product will sell, the marketers might conduct a house-to-house survey of 1,000 households selected randomly in the product target areas, during which they present the product to the housewife for evaluation. The percentage of the housewives willing to buy the product at its listed price, together with other information obtained from the interviews, could be used to decide whether or not the new item should undergo full-scale production.

Example 1.5 Politicians and their supporters are immensely interested in knowing their prospects of winning an election as the campaign heads toward final balloting. By sampling 1,000 registered voters prior to the election, the percentage who claim they will vote for a given candidate may be used to estimate the percentage of the votes the candidate will receive in the election. The estimated percentage could be used to decide, for example, whether a greater campaign effort (more money) is required to assure the candidate's election.

There are many more examples in business and other areas that might be cited, but the preceding five should indicate the many ways in which statistics can be used. In the first two examples, statistics is used to *describe* large bodies of data. In this application, the word *statistic* is used to describe the value of a specific numerical measure such as an average or a total, and the collection of the values of the statistics is used to summarize or condense a large set of numbers. These compiled values of the statistics may, in turn, be used to assist in decision making. In the last three examples, statistics may be interpreted in a much broader sense—namely, the process of drawing conclusions about an entire *population* or collection of things based on a *sample,* a subset of the population or collection.

Most students probably view statistics in the context of the first two examples—that is, as tables of figures, charts, and graphs (batting averages, pie charts illustrating the sources of government revenue, and so on). This concept is called *descriptive statistics* and was at one time the principal use of statistics in business. Currently, there is an increasing interest in the methods and uses of *inferential statistics*—the process of drawing inferences about the whole (the *population*) from a subset of it (the *sample*), as exemplified in Examples 1.3–1.5. Schematically, the process of drawing inferences about an unknown population numerical quantity (the proportion of defectives in a production lot, mean income of a class of laborers, etc.) is illustrated in Figure 1.2. Units are selected from the population to form the sample, which in turn is used to draw inferences about the population characteristic of interest. Much of this text is devoted to the study of statistical inference. In

FIGURE 1.2 The statistical inference process

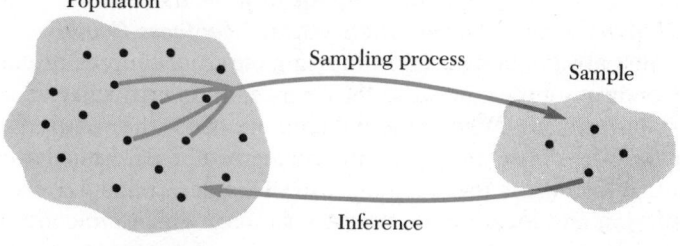

subsequent sections of this chapter, we will focus attention on the sources of data, the methods of obtaining data, and data measurement considerations.

1.2 SOURCES OF DATA

In many applications of statistics, businesses use *internal* data—that is, data arising from bookkeeping practices, standard operating business procedures, or planned experiments by research divisions within the company. Examples are profit and loss statements, employee salary information, production data and economic forecasts. Occasionally, it may be necessary or desirable to use sources of *external* data. By external, we mean sources of data outside the firm. External data may be of two types: *primary* data and *secondary* data. By primary data, we mean data obtained from the organization that originally collected them. An example is the population data collected by and available from the U.S. Bureau of the Census. Secondary data come from a source other than the one that originally collected them. The federal budget information used in Example 1.2 was published by the Bureau of the Census, but originated from the Bureau of the Budget. Ordinarily, if external data must be used, it is recommended that primary data be sought since it will not have undergone any "refining" by the secondary source.

In the election survey in Example 1.5, the *Statistical Abstract* provides numerous tables of both primary and secondary data, such as past voting records in districts and numbers of registered Democrats and Republicans, which may supply important information in conjunction with the internal sampled data on estimating a candidate's probability of being elected.

There are many excellent sources of published (primary and secondary) data compiled by the state and federal government, by business and economic associations, and by commercial sources (periodicals). Some examples are: *The Statistical Abstract of the United States* (published annually by the Bureau of the Census), *Survey of Current Business* (published by the Department of Commerce), *Monthly Labor Review* (published by the Bureau of Labor Statistics), *Harvard Business Review* (periodical), *Business Week* (periodical), *The Wall Street Journal* (periodical), *Dun's Statistical Review* (periodical), and *The Journal of Management Science* (association journal). Additional sources of external data that are available in most reference libraries are *The Economic Almanac, Federal Reserve Bulletin, Life Insurance Fact Book, International Financial Statistics,* and *Business Conditions Digest.*

Caution must always be exercised in using external sources of data, particularly secondary sources, because they may contain errors in transcription from the primary source. When external data are used, the conditions under which the data were collected and summarized must be determined to assure that they are relevant for the intended use. This determination usually requires identifying and locating the primary source, which typically will discuss any restrictions placed on the data due to the process of their collection.

Thus, while secondary sources of data are convenient, it usually is prudent to seek out and use primary sources of external data.

■ 1.3 STATISTICAL DATA TERMINOLOGY

When statistical data are collected and analyzed, it is usually in the context of *populations* and their *characteristics.*

> **Definition 1.1**
> Population and population characteristic
> A *population* is the totality of units under study. A *population characteristic* is an attribute of a population unit.

We may be interested, for example, in the salaries of workers in a particular industry. If so, the population is the totality of these workers and the characteristic of interest is each worker's salary. In collecting the salary data, we may be interested in other population characteristics as well, including sex, age, educational level, and other information. In general, a population unit may have one or more characteristics of interest in a particular study.

As another illustration of a population, a firm may be interested in the proportion of defective units of a certain product that it has produced in a large lot stored in a warehouse. The population is the totality of units in the warehouse, and the characteristic is the acceptability of each unit of the product—it is either defective or nondefective.

Suppose we are interested in information about a population characteristic. There are two ways we can collect data to study the population characteristic of interest: inspect all the population units or inspect a portion of the population units. When the data are produced by measuring the population characteristic for *each and every* unit in the population, we say that a *census* of the population has been taken.

> **Definition 1.2**
> A population census
> A *population census* is the evaluation of each and every unit in the population under study.

In some situations, it is possible to take a complete census of the population. This rarely occurs in business unless the population size is very small, due to cost and time considerations. A census of the U.S. population is undertaken every ten years and it is truly a Herculean effort, paid for, naturally, by the taxpayers. The U.S. census produces a wealth of data of considerable importance to the federal government and to firms and institutions, many of whom view the census as an important source of external data.

In most instances, it is not possible to take a census of a population. It may be too costly or too time-consuming, or the evaluation process may destroy the population unit as in Example 1.3.

Definition 1.3
A sample

(A *sample* is a part of a population in which the population characteristic is studied so that inferences may be made from the sample study about the entire population.)

(A classic example of a situation in which samples must be used rather than a census taken is *destructive sampling,* in which the process of evaluating a unit of the population destroys or irrevocably damages that unit.) For example, suppose a tire manufacturer wishes to claim its new radial tire will last 40,000 miles or more. To support this claim, a sample of all tires produced (the population) is selected for testing to determine how many miles the tires will last. Since testing destroys the tires, a complete census of the population is impossible.

(The advantages of sampling over taking a census are rather obvious. A sample is less expensive than a census, it can produce data more quickly, and the data are often more reliable because more time can be spent studying each sampled unit. But there clearly is a price to be paid as well. By looking at only a portion of the population, we are subject to committing errors because the sample may not be representative of the whole population.)

Definition 1.4
Sampling error

(*Sampling error* is the difference between studying a sample and inferring a *result* about a population characteristic, and determining the *result* by taking a census of the population and evaluating the population characteristic.)

(As an illustration of sampling error, suppose we are interested in the average salary (the characteristic) of unionized workers in a specific industry and we know from union membership lists that there are 1,000 workers in the industry (the population). If we had taken a census, we may have found that the average salary is, let us say, $25,000 (the result by taking the population census). Based on a sample of, say, 100 workers, we may find that the sample average salary is $27,200 (the result by taking the sample). The difference between these two figures—$2,200—is the sampling error, if nonsampling errors have not occurred.)

Errors in acquiring and tabulating statistical data can arise in other ways as well, and these errors are called *nonsampling errors*.

Definition 1.5
Nonsampling error

Nonsampling errors are errors that occur in acquiring, recording, or tabulating statistical data that cannot be ascribed to sampling error. They may arise in either a census or a sample.

Nonsampling errors are usually more difficult to control and detect than sampling errors. Suppose we are acquiring data on the 1,000 unionized workers mentioned earlier. If we approached a particular worker and asked for his or her income, we could be lied to—a troublesome and frequent source of nonsampling error when a sensitive question is asked directly of a person. (What is your grade point average?) In some instances, a person may give a false response out of ignorance rather than by design. Another source of nonsampling error is in recording that data. A "7" may be written as a "9," the decimal point may be incorrectly placed, and so on. Errors may also occur in tabulating the data—keypunching errors in preparing computer cards and typing errors in transcribing data, for instance. It is always necessary to edit data carefully to minimize the chance of nonsampling errors adversely affecting the statistical analysis of the data.

The identification of the units in a population under study can often be a surprisingly difficult task. We refer to a listing of population units as a *frame*.

Definition 1.6
Population frame

The listing of all units in the population under study is called the *population frame*.

If the population is a production lot of units stored in a warehouse, production records will give us a listing of the serial numbers of the units from which each unit may be identified. If the population is the 1,000 unionized workers in a specific industry, union membership records may serve as a frame. But, what about a frame for all persons who will vote in a particular election? A listing of registered voters is not appropriate, because in many elections less than 50 percent of the registered voters actually vote. Some classic errors have been made in identifying the frame. In the 1948 Dewey-Truman presidential race, one survey prior to the election used telephone directories as the frame. Random telephone numbers were selected and called; the person was asked whether he or she was a registered voter and intended to vote on Election Day. As it turned out, not everyone owned a telephone in 1948 (incredible as that may seem today), and a preponderance of those who did were Republicans. The survey came to the incorrect conclusion that Dewey would be elected president of the United States.

1.4 THE ACQUISITION OF DATA: SURVEYS AND EXPERIMENTS

When internal or external data are not readily available or are incomplete in a study attempting to answer questions about a population characteristic, a *survey* or an *experiment* may be conducted to provide the required information.

Definition 1.7
Statistical survey

A *survey* is a process of collecting data from existing population units, with no particular control over factors that may affect the population characteristics of interest in the study.

Most of us are very familiar with surveys. As students, we are asked about our opinions regarding dining hall food, impending tuition hikes, teaching effectiveness, and so on. Filling out survey questionnaires or answering an interviewer's questions has become a routine occurrence in most of our lives. To better understand our definition of a survey, suppose we are interested in acquiring data on the salaries of 1,000 unionized workers in a specific industry. The population characteristic "salary" may be affected by a host of factors—age, race, sex, educational level, etc. As we elicit a particular worker's salary by a survey, we have no control over educational level, age, and so on—these are existing attributes of the worker.

In contrast to a survey is a statistical *experiment* in which we do exercise control over factors that may affect the population characteristics of interest.

Definition 1.8
Statistical experiment

An *experiment* is a process of collecting data about population characteristics when control is exercised over some or all factors that may affect the characteristics of interest in the study.

We may be interested, for example, in the yield of a chemical process that is affected by temperature and pressure. A variety of settings for temperature and pressure could be selected, and the chemical process run for each setting to determine the yield. In this way, the joint effect of temperature and pressure on yield is studied in a controlled manner.

In management, we may be interested in the effects of a training program on the first-year performance of new employees. A set of new employees may be split into two groups such that both groups are approximately alike in terms of age, sex, education, and other factors. The training program could be administered to one group and not to the other (the *control* group). At the end of the first year, performance characteristics could be measured to assess the effects of the training program, accounting for factors other than the training program that may affect performance.

Experiments almost always provide better information than do surveys, but both are extremely important and useful tools for acquiring data. Though an experiment is preferred to a survey, much of the data used in statistical analyses in business and economics are survey data. There are a number of reasons for this. First, most internal and external data are collected by surveys. Second, it is not always possible to conduct an experiment to acquire the needed information. An interesting example of this is the effect of smoking on health. Virtually all data on the relationship between smoking and health are survey data; other factors that may affect health, such as age, race, sex, and physiological properties, are not under the control of those collecting the data. To run an experiment in this case would involve controlling persons' lives. Some people in the experiment would be required to smoke while others would not. It is neither feasible nor desirable to approach the acquisition of the data for the study of the relationship between smoking and health in this way.

The planning of a survey or an experiment is essential to ensure that the resulting information will be useful. A good plan usually involves the following steps (these steps are applied to the quality control problem in Section 1.1):

1. A clear and detailed statement of the problem.

   The statement of the problem should clearly indicate that we are interested in determining whether or not the percentage of good batteries (those lasting through 20 hours) exceeds a specified number (95 percent). The population is composed of all batteries produced during a chosen period of time (a day or a week), and the characteristic in the population of interest is the number of hours the battery will continuously operate before failure.

2. A decision to survey or to experiment.

   In this problem, it is possible to answer the question about the population by experimentation. We may test each selected battery under the set of conditions in which it was designed to operate.

3. A decision to take a census or a sample.

   This is a case of destructive sampling. In determining the proportion of batteries that will last 20 hours or more, the tested batteries are "spent." We must, therefore, take a sample of batteries.

4. Designing the survey or experiment.

   The experiment must be designed so that we isolate the characteristic of interest—the lifetime of the battery. Test circuits must be constructed and carefully monitored when the selected batteries are inserted for testing.

5. Collecting and analyzing the data.

For each battery, the time to failure is recorded and the proportion of batteries lasting 20 hours or more is calculated.

6. Reaching conclusions about the population characteristics.

The sample proportion of batteries surviving 20 or more hours is used as an estimate of the population proportion that survives 20 or more hours.

7. Reporting the results.

The report should include a thorough description of the problem, the sampling design, the testing method, and the inferences. Sufficient monies should be allocated for a competent writing of the report. Indeed, many companies employ technical writers to put into "laymen's" words the experimental results.

The manner in which a sample is drawn, the methods of analyzing statistical data, and the kinds of inferences that may be drawn from the analysis (steps 4, 5, and 6) are major topics in this text. It is important not to minimize the other steps, particularly steps 1 and 7. A clear and detailed statement of the problem is essential in planning a survey or an experiment. And the best analysis of survey or experimental results is meaningless unless the analysis can be accurately and understandably reported.

1.5 OBTAINING DATA

Once it has been determined that a survey or experiment is required, there are a variety of methods that may be used. The most difficult problems arise when gathering information from people in surveys, and the methods most relevant to this situation will be emphasized.

1.5.1 Self-enumeration

Self-enumeration is probably the most common method of acquiring data from people in a survey or experiment. Questionnaires are usually distributed to selected individuals by mail, although the distribution mechanism depends to a large extent on the purpose and nature of the questionnaire. For example, if the purpose of the questionnaire is to survey the attitudes of those using public transportation, the questionnaire may be distributed to persons while commuting to and from work on buses, subways, and trains.

The use of questionnaires suffers from two serious drawbacks. First, if the respondent has difficulty in interpreting the questions, no one is available for assistance. If this situation arises, the information received may contain a high degree of nonsampling error, or the respondent may become frustrated and not bother to complete or return the questionnaire. Furthermore, if a questionnaire is mailed to a household, it is often not clear who in the household responded to it. Second, questionnaires typically have an extremely poor response rate. It is not uncommon to have less than 30 percent returned on the first mailing of a questionnaire. The principal advantage of a questionnaire is the low cost relative to the other means of obtaining information. Most mail questionnaires may be bulk mailed at a reasonable rate. But it is almost always necessary to contact nonrespondents to the first mailing by subsequent mailings, telephone calls, or personal interviews, and these costs must be planned for in a well-designed self-enumeration survey or experiment. In most instances, those who do respond to the first mailing of a questionnaire are not representative of the entire population. To use only their responses would tend to bias the analytical results. Some self-enumeration questionnaires do enjoy high initial response rates. Examples are questions asked on warranty cards that must be returned to the manufacturer for warranty coverage of a new product and Census Bureau questions asked on federal income tax returns.

The federal government, particularly the Census Bureau, frequently uses mail questionnaires to acquire information necessary for economic forecasts and planning. The response rate to government questionnaires tends to be much greater than to those distributed by the private sector.

☐ 1.5.2 Personal interview

In most situations, the best method of eliciting information from individuals is by a personal interview. The interviewer personally contacts individuals who are selected to participate in the survey or experiment. Responses are recorded on a *schedule* (a questionnaire form filled out by the interviewer).

The personal interview method produces a higher response rate than self-enumeration questionnaires and allows the interviewer to clear up any misunderstandings about any questions on the schedule. But personal interviews are generally very expensive. Interviewers must be carefully selected and trained, and sufficient remuneration must be provided to ensure that the interviewer is competent and dedicated to the chore. It is always prudent in a personal interview survey to call a selected set of respondents to ensure that they were in fact contacted (as opposed to the interviewer filling fake responses), to ascertain if the interviewer's demeanor was appropriate, and to determine whether the interviewer may have biased responses by making gestures or comments when asking the questions or recording the responses.

Overall, the personal interview method of conducting an experiment or survey is the best way to acquire data when the population units are people, when the process is properly planned and executed, and if it can be afforded.

☐ 1.5.3 Telephone interview

Occasionally, it is possible to conduct an interview over the telephone with the interviewer working from a schedule as in a personal interview.

Polls to determine the most popular programs on television are frequently conducted in this manner. Telephone interviews are usually less expensive than personal interviews, but the response rate is lower and fewer questions may be asked before the respondent tires of the proceedings. And, not everyone owns a phone—even today.

■ 1.6 CONSTRUCTING QUESTIONNAIRES AND SCHEDULES

There are three steps in constructing and analyzing a questionnaire or a schedule: (1) designing the instrument, (2) conducting the pretest, and (3) editing the results. The construction of a questionnaire or schedule instrument is time-consuming and difficult. There is a natural tendency to rush through the construction of the instrument so that the data collection process can commence. But time spent in this stage of a well-planned survey or experiment is invariably found to be extremely valuable in retrospect.

The proper construction of questionnaires is a skill that is generally developed only by experience in the use of research methodology or by on-the-job training. We will discuss only some of the basic concepts concerning the construction of a questionnaire. For further information on this subject, see the Hansen et al. and Kish references listed at the end of this chapter.

□ 1.6.1 The design

There are basically three kinds of questions that may be asked: dichotomous, multiple choice, or free answer. In the *dichotomous* question, the respondent is asked to select one of two responses, usually yes and no. For example, in a transportation study, a worker may be asked:

Did you drive a car to work this morning?
Yes ( )   No ( )

The dichotomous question is simple and straightforward, and perhaps comes closest to decisions that respondents are used to making.

In the multiple-choice question, the respondent is asked to select one of a number of responses:

What is the likelihood of your using the following services for preventive health care purposes in the next two years? (*a*) Dental checkup, (*b*) eye exam, (*c*) general physical.

	a	b	c
Extremely unlikely	( )	( )	( )
Unlikely	( )	( )	( )
Slightly unlikely	( )	( )	( )
Not certain	( )	( )	( )
Slightly likely	( )	( )	( )
Likely	( )	( )	( )
Extremely likely	( )	( )	( )

(The multiple-choice question gives the respondent a greater range of responses to choose from, but it may also request a more qualified response than the respondent is prepared to make.) For instance, a respondent may answer yes to the dichotomous question, "Will you have a physical this year?" when it is not a certain event; good intentions are not always realized. Yet, the respondent may not be able to conceptualize properly the assignment of a *likelihood* (slightly likely, likely, extremely likely) to the event, "I will have a physical this year." All too often, responses in situations like this lead to "end-loading"—selecting the response that most closely approaches a simple yes-no response. In this case, the respondent would select the "extremely likely" response in place of yes if he or she is given the multiple-choice response format, for instance.

(In the free answer form, the respondent is asked to answer a question in his or her own words in essay form.) For example:

What is your opinion of the dining hall food and service?

(The difficulty with the free answer question is in classifying the responses. This not only may be difficult and somewhat arbitrary, but it is also time-consuming.)

In most instruments, it usually is necessary to use all three types of questions to elicit the information required.

(The order of the questions in the instrument can be extremely important. The questionnaire or survey should begin slowly with easily answered questions to develop rapport with the respondent. Respondents tend to "tie" questions together, and one particular ordering of questions may produce a different set of responses than another set for this reason.)

(The degree of directness of the questions is also important. If sensitive questions are asked directly, respondents may distort their answers. This invariably happens when a person is asked for his income. To elicit information about sensitive questions, indirect questions may be used.) For example, we may ask the respondent to indicate his salary range among a set of ranges. Later, we may ask what proportion of his monthly income is spent on food, and much later, what his average monthly expenditure for food is. We may be able to determine a person's salary indirectly in this way better than by directly asking for income. At the very least, we have a consistency check to determine how reliable the responses are.

(It is important that the questions are stated clearly and do not bias the results. Ideally, the question should have the same meaning to every respondent in the survey or experiment. And the questions should be relatively short. Bias may arise when leading questions are used,) such as:

The food in the dining hall is rotten.
    Agree ( )   Uncertain ( )   Disagree ( )

Given to the typical college student, the response will invariably be, "Agree." A less biased question might be, "The food in the dining hall is of acceptable quality."

□ 1.6.2 The pretest

The pretest is an essential step in constructing a questionnaire or schedule instrument. The instrument is given to a small number of respondents to determine whether there are any problems with it. Almost always there are. There may be ambiguous questions, the ordering may require changing, and some questions may have to be asked in different forms. The time to identify difficulties with the instrument is before the full-scale survey or experiment is conducted—not after.

Furthermore, the information gathered during the pretest phase may be used to estimate quantities required for the proper planning of the statistical design of the experiment. We will return to the discussion of this aspect of the pretest in Chapter 9.

□ 1.6.3 Editing

The completed questionnaires or schedules must be carefully checked and edited for errors. Often it is possible to design questions that represent internal consistency checks for the respondent's answers. Finding recording, transcription, or clerical errors can be very tedious work, but it is necessary if the data are going to be of value in decision making.

Today, the computer is used extensively to edit data. Various computer-assisted techniques have been developed to identify "outliers"—responses that are greatly different from the majority of the responses. Many outliers result from recording, transcription, or clerical errors, or from false information provided by the respondent.

■ 1.7 VARIABLES AND SCALES OF MEASUREMENT

The characteristic of the population under study is called a *variable* if it can take on two or more different values among the population units. For instance, if we are interested in the incomes of workers in a particular industry, we may record other characteristics about the worker as well, for example, age, race, level of education, and sex. In this instance, the five characteristics—income, age, race, level of education, and sex—are variables in the survey or experiment.

Furthermore, we would call income a *dependent variable* and the other four *independent variables* if we are concerned with how sex, age, level of education, and race affect income. Income is the basic variable of interest, and our interest in the other variables is in their influence on income.

If we are measuring a set of variables from a population, determining which are dependent and independent variables is a function of the purpose of the survey or experiment. An independent variable in one study may be a dependent variable in another.

A *quantitative* variable is one that can be measured numerically, such as income and age. A *qualitative* variable is one that is nonnumeric, such as sex, race, and level of education (high school, college, graduate school, etc.). In preparing data for analysis, we must be familiar with the four numerical scales of measurement: *nominal, ordinal, interval,* and *ratio.* The *nominal* scale applies whenever we use numbers only to categorize values of a variable. For instance, we could let a male be 1 and a female be 0, but this numerical assignment is clearly arbitrary; a female could be assigned 100 and a male 0. The *ordinal* scale differs from the nominal scale in that the ordering of the numbers has meaning. An example is the responses to a multiple-response question:

Strongly agree	Agree	Uncertain	Disagree	Strongly disagree
−2	−1	0	+1	+2

The numerical assignments of −2, −1, 0, 1, and 2 indicate the degree of agreement, but they could just as easily have been 0, 10, 100, 200, and 500, respectively. The key here is that while a −2 indicates stronger agreement than a −1, the difference in the strength of the respondent's conviction between −2 and −1 may not be the same as between 0 and +1. In the *interval* scale, the relative order of the numbers is important, but so is the difference between them. This scale uses the concept of unit distance such that the difference between any two numbers may be expressed as some number of units. The interval scale requires a zero point, but its location may be arbitrary. Good examples of interval scales are the Fahrenheit and Celsius temperature scales. Both have different zero points and unit distances. The principle of an interval scale is not violated by a change in scale or location or both. The *ratio* scale is used when the interval size is important and also the ratio between two numbers has meaning. By this, we mean it is appropriate to speak of one number being, say, twice as big as another. This is clearly not possible with an interval scale, where, for instance, 80°F is not twice as "hot" as 40°F; measured on the Celsius scale, these two temperatures are 27°C and 4°C, respectively, and 27°C is not twice 4°C. Examples of instances when ratio scales are appropriate are measurements of heights, weights, and age.

Section 20.2 in Chapter 20 contains additional information about and examples of the four numerical scales of measurement.

Most of the statistical methods we will develop in this book require that the variable be measured at least on the interval scale.

■ 1.8   TEXT ORIENTATION

In most practical applications of statistics, it is not feasible to take a census of the population under study. Therefore we will emphasize the development and proper use of statistical methods that allow us to infer results about population characteristics from *sample* information. This process is called *inferential statistics*. In effect, we are defining inductive reasoning, for it is reasoning from the part to the whole. This is the very essence of statistics as it is used today in research and decision making. Considerable space is devoted in this text to describing the inductive reasoning process of inferring from the part (the *sample*) to the whole (the *population*).

■ 1.9 SUMMARY

The science of statistics encompasses the methodologies that are used for the collection, presentation, and analysis of quantitative data, and the interpretations of these data. Descriptive statistics refers to methods used to describe data numerically or graphically, whereas inferential statistics refers to the methods of inferring results about one or more characteristics in the population from a sample. A sample is a subset of the population units, and a census is a complete enumeration and use of all population units in a survey or experiment. A listing of all population units is called a *frame*. In a survey, factors that may affect the population characteristics of interest are not controlled, while in an experiment they are. When the population units are people, personal interviews using schedules are usually the most effective method of acquiring data, although self-enumeration questionnaires and telephone interviews are more commonly used and are less expensive than interviews. In constructing a questionnaire, there are three important steps: the design, the pretest, and editing the responses.

A population characteristic under study is called a *variable* if it can take on two or more values among the population units. If the value of a variable is numeric, it is called a *quantitative* variable; otherwise, it is called a *qualitative* variable. There are four scales of measurement: nominal, ordinal, interval, and ratio.

With the extensive use of inferential statistics in business operations and the power of modern computers to do great amounts of computation in a very short time, there is a tendency for students to assume that their training in statistics will provide them with the *only* tools needed for successful decision making. Nothing could be further from the truth. Appropriately applied, inferential statistics will provide the effective manager with *one* source of information which, in conjunction with other sources (the manager's intuition for one), enables him to make a good decision. Inferential statistics should always be used with common sense; indeed, it *is* the use of common sense by definition (inductive reasoning).

■ REFERENCES

Hansen, M. H., Hurwitz, N. W., and Madow, W. G. *Sample Survey Methods and* *Theory*. Vols. 1 and 2. New York: John Wiley & Sons, Inc., 1953.

Kish, L. *Survey Sampling*. New York: John Wiley & Sons, Inc., 1965.

Ostle, B., and Mensing, R. W. *Statistics in Research*. 3d ed. Ames: Iowa State University Press, 1976.

Snedecor, G., and Cochran, W. *Statistical Methods*. 7th ed. Ames: Iowa State University Press, 1980.

Tanur, J., ed. *Statistics: A Guide to the Unknown*. San Francisco: Holden-Day, 1972.

■ **PROBLEMS**

1.1. Briefly describe each of the following terms:
   a. A statistic.
   b. Descriptive statistics.
   c. Inferential statistics.
   d. Population.
   e. Population characteristic.
   f. Census.
   g. Sample.
   h. Sampling error.
   i. Nonsampling error.
   j. Frame.
   k. Schedule.
   l. Questionnaire.
   m. Survey.
   n. Experiment.
   o. Variable.
   p. Quantitative variable.
   q. Qualitative variable.
   r. Dependent variable.
   s. Independent variable.

1.2. Distinguish between a *schedule* and a *questionnaire*. What is each used for?

1.3. Distinguish between a *survey* and an *experiment*. Which is preferred and why?

1.4. Distinguish between *primary* and *secondary* data. Which are the most reliable? Why?

1.5. There are three kinds of questions that may be used in a schedule or questionnaire. Describe each, and discuss its advantages and disadvantages.

1.6. In constructing a schedule or questionnaire, there are three primary steps: *design, pretest,* and *editing*. Describe each step.

1.7. There are four measurement scales: *nominal, ordinal, interval,* and *ratio*. Describe each, and give an example of a survey question that may use measurements of each type.

1.8. For each of these listed sources of statistical data, determine the publisher, frequency of publication, and nature of the data—monthly, quarterly, semiannually, or yearly. Cite two statistics from each source. Example: The number of persons in the United States in 1969 who were 20 years of age or less was: _____.
   a. *Statistical Abstract of the United States*.
   b. *Survey of Current Business*.
   c. *Monthly Labor Review*.
   d. *Federal Reserve Bulletin*.
   e. *Life Insurance Fact Book*.
   f. *International Financial Statistics*.
   g. *Business Conditions Digest*.
   h. *Dun's Statistical Review*.
   i. *The Economic Almanac*.
   j. *Automobile Facts and Figures*.

1.9. For each of the following, indicate the scale of measurement:
   a. Red (1), Blue (0), Yellow (−1)
   b. Extremely likely (5), Likely (4), Indifferent (3), Unlikely (2), and Extremely unlikely (1).
   c. Pressure in pounds per square inch; from 0 to ∞.
   d. Volume in cubic centimeters; from 0 to ∞.
   e. Age in years; 0 to ?
   f. Salary in dollars, 0 to ?

g. Rank of a state in population; 1 to 50.

1.10. For each of the following, indicate whether it is a *quantitative* or *qualitative* variable.

a. Hair color.

b. Sales volume of an automotive firm.

c. Sex of an individual.

d. Number of persons unemployed in the United States.

1.11. Distinguish between *sampling* and *nonsampling* error. Which can occur in a census? Which can occur in a sample?

1.12. A manufacturer buys electronic parts from a supplier with the understanding that 1 percent or fewer of the parts are defective. In a particular shipment of 5,000 parts, the supplier finds in a sample of 100 parts that none is defective. The manufacturer decides to check the parts as well and, in another sample of 100 parts, finds that four are defective. On this basis, the manufacturer decides to reject the lot.

a. How is it possible that one sample produced 0 percent defectives while another produced 4 percent defectives?

b. Is it possible that the manufacturer is making a mistake by not accepting the shipment? If you were the supplier, how would you defend the shipment as possibly being acceptable (1 percent or less defectives)?

1.13. How can sampling error be minimized? Discuss. How can nonsampling error be reduced? Discuss. Is it possible for sample results to be more accurate than census results? Explain.

1.14. A corner drugstore is losing business and wishes to determine the causes for the loss. Customers entering the store are given a questionnaire that inquires about their satisfaction with the store, the service, and so on, and asks for suggestions on improvements.

a. What is the population in this survey?

b. Is the sample biased? That is, are the respondents representative of the population?

c. How would you suggest collecting data to determine the causes for the loss in business?

1.15. There are 5,000 members in a community group health association, and 500 are selected for a mail survey on level of satisfaction with the services provided by the association. Thirty percent reply to the initial mailing of the questionnaire.

a. Is the sampled population the same as the target population (the 5,000 members)? Explain why the sample is likely to be biased in this instance.

b. If 90 percent responded in the sample of 500, would you be less concerned about bias? Why?

1.16. In the following situations, indicate whether a sample or a census should be taken and explain why.

a. A firm that employs 500 persons wishes to determine the acceptability of subscribing to a new employee insurance program.

b. A car manufacturer wishes to obtain information on customer preferences with respect to size of cars.

c. The Internal Revenue Service wishes to obtain data on the proportion of income tax returns that contain arithmetic mistakes.

1.17. Is it possible to "solve" the potential problem of sample bias due to a low response rate by simply sending out a sufficiently large number of questionnaires to produce a sizable sample? Explain.

1.18. In the following set of questions, find at least one fault in each. Also, suggest an improved rewording of the question.

a. Do you agree that too much money is being spent on national defense?

b. How many tubes of toothpaste did you purchase in the past year?

c. Does the name "Frontier" come to mind when hi-fidelity equipment is mentioned?

d. It is a waste of money to send men into outer space.
(Strongly agree, Agree, Undecided, Disagree, Strongly disagree)

1.19. In the following situations, which method of data collection—self-enumeration, personal interview, or telephone interview—would you select and why? Keep in mind the cost of the method, the response rate, and the time necessary to obtain the information, as well as other relevant factors.

a. Consumer acceptance of a new camera model before it is placed on the market.

b. National survey to determine, in the public's view, the best way to fund the social security system in future years.

c. Data on preventive health care behavior of persons in the United States.

d. The determination of the national ranking of a television show in a specific time slot.

e. Data on electricity rates nationally.

1.20. In the following situations, indicate whether a survey or an experiment is more suitable, and explain why.

a. Collecting data on the gas mileage of a specific model and make of car.

b. Collecting data on the number of sick days per month used by workers in a large company.

c. Collecting data on the interest paid for home mortgages in a large metropolitan area.

1.21. Athletic teams generally wear numbers on the backs of their uniforms. These numbers represent which measurement scale? Explain.

1.22. The *Seal Point Daily Tribune* conducted a telephone survey asking the question, "Should the proposed nuclear power plant at Seal Point be built?" Of the 514 people contacted, 183 said yes, 307 said no, and the rest said they had no opinion on the matter.

a. What is the target population of the survey?

b. What is the sampled population?

c. The survey produces three statistics. What are they?

d. On what measurement scale are the three statistics measured?

e. What are the values of the statistics?

1.23. For each of the following, indicate an appropriate scale of measurement.

a. Temperature (in degrees Centigrade).

b. Stock prices per share.

c. Ranking of the horses finishing a horse race (first, second, etc.).

d. Bond ratings (AAA, AA, A, etc.).

1.24. Give an example of each of the following.

a. Descriptive statistics.

b. Inferential statistics.

1.25. In each of the following statements, indicate whether descriptive or inferential statistics is being used.

a. Last year, the inflation rate rose by 5 percent.

b. Based on a sample of 233 students at Tech University, 72 percent of all university students at least occasionally drink alcoholic beverages.

c. Based on an average of 50 selected stocks, the stock market's average value of a share rose by 5.2 points at the end of the trading today.

d. The average grade point average of all business students at Tech University is 3.12, based on current enrollment data supplied by the registrar's office.

1.26. Criticize or defend the following statements.

a. Only 23 percent of the Tech University students who had flu shots last winter caught the flu. This provides evidence that the flu shots were effective in preventing a flu attack.

b. The responses to an interviewer's questions are free from nonsampling error if they are accurate and honest.

c. When mail questionnaires are used, it must be kept in mind that more responses tend to be from persons with special interests in the outcome of the survey than from others.

1.27. Discuss the following statement: "A census will always give better information than does a sample."

Descriptive statistics

2

When a survey or an experiment has produced a body of data, the original state of the data will not generally convey much information about the characteristics of interest. Typically, there will simply be too many observations to give one an insight into the nature of the data. Whether a data set represents a sample or a population, it is necessary to organize and reduce the data into meaningful forms such as graphs and charts or numerical quantities such as averages, totals, and percentages. The resulting statistical summaries of the data can be used as a convenient and meaningful framework for data analysis and interpretation.

There are basically two methods to describe data: *graphical* methods and *numerical* methods. We shall discuss both in this chapter. As the methods are being discussed, it is important to keep in mind that they may be applied to either population or sample data sets. If it is possible to take a complete census, then these methods may be used to summarize information about the population characteristics. An excellent example of this case is the U.S. population census taken every ten years. Without summarizing and reducing the data, it would be impossible to extract much meaning from the millions of numbers. If the data set represents a sample, the statistical summaries, particularly the numerical measures, may be used to draw inferences about the population characteristics.

It would be remiss not to mention the impact of the computer on our ability to analyze data. Whereas it may take many hours to compute averages and totals, and to form charts from a large data set, the computer can typically produce the summaries in only minutes, once the data have been

23

encoded on computer cards. This "computing power" has made it possible to undertake larger experiments and surveys than before the advent of the computer. It has also considerably diminished the time it takes to put the resulting data in a suitable form to aid decision making.

2.2 FREQUENCY DISTRIBUTIONS

As a means of explaining the concepts in this chapter, we shall make extensive use of the data in the following example.

Example 2.1 At a small state college in upper Suburbia, the problem of illegally parked automobiles on campus has constantly plagued both students (presumably the bearers of most parking tickets) and the faculty and staff whose parking places are occasionally "appropriated." As a means of gaining insight into the magnitude of the problem, it is decided to take a census of the students who were enrolled full time at Suburbia during the past term. Data on each student are recorded in computer files, including the number of parking tickets assessed during each term. From the computer files, it is found that there were 2,200 full-time students enrolled during the past term.

As a means of efficiently summarizing the data, it is decided to form six classes for the characteristic "number of parking tickets received" and then record the number of students who received the number of tickets in each class. These data are illustrated in Table 2.1.

TABLE 2.1 Frequency distribution of the numbers of tickets assessed to students during the past term

Class number	Class limits	Frequency	Relative frequency
1	0–2	1053	1,053/2,200 = 0.4786
2	3–5	592	592/2,200 = 0.2691
3	6–8	305	305/2,200 = 0.1386
4	9–11	195	195/2,200 = 0.0886
5	12–14	36	36/2,200 = 0.0164
6	15–17	19	19/2,200 = 0.0087
		2,200	1.0000

Source: Computer files, Suburbia State College.

The grouping of data in Table 2.1 is called a *frequency distribution.* (A frequency distribution is a convenient way of grouping data so that the important aspects of the raw data are more readily apparent.) In this instance, the 2,200 tickets comprising the census data have been reduced to six "numbers"—the frequencies in each class, which have been tallied and tabulated in Table 2.1. The relative frequency in each class gives the ratio of the number of students in each class to the total number of students. Here, for instance, the proportion 0.4786 of the students received from 0 to 2 tickets during the past term.

When data are grouped into a frequency distribution, some information is lost. We cannot determine from Table 2.1 the number of students who received no tickets, for example. If we had been interested in this number in particular, the first class in the frequency distribution could have been "0," with the second class having the limits 1–2. As this discussion suggests, selecting the number of classes and the class limits in constructing a frequency distribution is somewhat arbitrary. A useful series of steps for forming a frequency distribution is given here. [In brackets, these steps are applied to Example 2.1.]

1. Determine the *range* of the ungrouped numbers by finding the difference between the largest and smallest number.[1]

$$[17 - 0 = 17]$$

2. Select the number of *classes* into which the range will be divided. As a rule of thumb, the number of classes should be between 4 and 20. In general, the number of classes will depend on the size and nature of the data set.

$$[6]$$

3. Divide the number of classes into the range and round the result to the next larger integer. This number represents the *class width* of each class.

$$[17/6 = 2.83 \rightarrow 3]$$

4. Usually, select the *class limits* by beginning with the smallest number and constructing classes with the width determined in step 3 and stopping when the range of numbers has been covered.

$$[0–2, 3–5, 6–8, 9–11, 12–14, 15–17]$$

5. Form column headings for the class number, the class limits, the tally, and the frequency. Read off the numbers in the ungrouped data, and for each number, record a mark in the tally column for the appropriate class.

6. Write the sum of the marks for each class in the frequency column. The sum of the frequencies should equal the total number of observations.

7. Form the column of relative frequencies by dividing the frequency in each class by the total number of observations.

A few comments are in order regarding these steps. If at all possible, the widths of all classes, possibly excluding the first and last classes, should be

---

[1] For convenience of illustration, it is assumed in this example that the most tickets acquired by a student was 17.

equal. The class limits should be chosen so that each number in the census falls into one and only one class. For example, we could not select the first and second class limits to be 0–2 and 2–4. The number 2 could be placed either in class one or in class two in this case. The class limits do not have to be integers. If, for example, the data were composed of numbers specified to two decimal places, we might have to select class limits such as 0–2.99, 3–5.99, etc., to ensure that each number can be placed in one and only one class. Alternatively, the classes could be defined as 0 or greater but less than 3, 3 or greater but less than 6, etc.

If the first or last class is *open-ended* (for example, the last class in Example 2.1 could have been "15 or more"), then the rules for constructing a frequency distribution must be modified. For example, suppose that we had one student who received 35 tickets. In step 1, the range of the numbers would be $35 - 0 = 35$. With six classes, the class widths would be $35/6 = 5.83 \to 6$, producing classes 0–5, 6–11, 12–17, 18–23, 24–29, 30–35. Then, 74.77 percent of the data would fall in the first class and none in classes 18–23 and 24–29. In a case such as this, it is best to determine a "largest number" so that the frequency distribution does not become too elongated. If one or two classes contain almost all the data and the remaining classes are essentially empty, important information about the nature of the data set is lost. By selecting the largest number to be 17, the range in step 1 would be $17 - 0 = 17$, and we could then add the class "more than 17" to pick up the one student with 35 tickets.

The construction of a frequency distribution from ungrouped data is somewhat of an art, because there are many possible frequency distributions that could be formed from one set of data. It is up to the statistician to select the number of classes, the class widths, and the limits. For further examples of constructing frequency distributions, the reader is directed to the references at the end of this chapter.

There are additional aspects of a frequency distribution, three of which we shall now discuss.

1. Associated with each class are its *boundaries*. The class boundaries are computed by determining the midpoint between the upper and lower limits of adjacent classes. For example, the boundaries for Class 2 in Table 2.1 are $(2 + 3)/2 = 2.5$ and $(5 + 6)/2 = 5.5$. The width of any class can be determined by computing the difference between its upper and lower boundaries. In Class 2, for example, the class width is 3 units: $5.5 - 2.5 = 3$.

2. The *class mark* is the midpoint of each class and may be determined by averaging either the class limits or the class boundaries. In Class 2, for example, the class mark is 4: $(3 + 5)/2 = 8/2 = 4$, *or* $(2.5 + 5.5)/2 = 8/2 = 4$.

3. The *cumulative relative frequency* for each class is determined by summing the relative frequencies for the class and all prior classes. For

example, the cumulative relative frequency for Class 3 is: 0.4786 + 0.2691 + 0.1386 = 0.8863.

The complete frequency distribution for the data given in Table 2.1 (Example 2.1) is presented in Table 2.2.

TABLE 2.2 Complete frequency distribution for the student ticket data in Example 2.1

Class number	Class limit	Class boundary	Class mark	Frequency	Relative frequency	Cumulative relative frequency
1	0–2	−0.5 to 2.5	1	1,053	0.4786	0.4786
2	3–5	2.5 to 5.5	4	592	0.2691	0.7477
3	6–8	5.5 to 8.5	7	305	0.1386	0.8863
4	9–11	8.5 to 11.5	10	195	0.0886	0.9749
5	12–14	11.5 to 14.5	13	36	0.0164	0.9913
6	15–17	14.5 to 17.5	16	19	0.0087	1.0000
		Total number of observations N =		2,200	1.0000	

2.3 GRAPHICAL METHODS OF DATA PRESENTATION

The frequency distribution discussed in Section 2.2 conveniently places the data in a form for a graphical display called a *histogram*. The following steps are required in constructing a histogram:

1. Mark off the x (horizontal) axis with the class boundaries taken from the frequency distribution.
2. Construct rectangles over the class boundaries so that the area of each rectangle is proportional to the class frequency with a constant coefficient of proportionality for all classes. If the class widths are equal, choosing the height proportional to class frequency is equivalent to choosing area proportional to class frequency.

In applying these steps to the frequency distribution in Table 2.2, we see that the height of each rectangle may be taken as the class frequency, because each class is of equal width. The resulting histogram is shown in Figure 2.1. The histogram clearly indicates that the preponderance of the Suburbia State College students have received only a few tickets in the past term.

From a frequency distribution, it is possible to construct numerous other forms for graphically displaying data. A *frequency polygon* is constructed by plotting the class frequency and the class mark for each class and joining the resulting points by straight lines. A frequency polygon for the data given in Table 2.2 is shown in Figure 2.2. Notice that the polygon is "tied down" at the ends by starting at 0 and ending with a frequency of 0 at 19, the class mark for the "next" class. This gives the polygon a neater appearance than if

FIGURE 2.1 Histogram for the Suburbia State College ticket data constructed from the frequency distribution in Table 2.2.

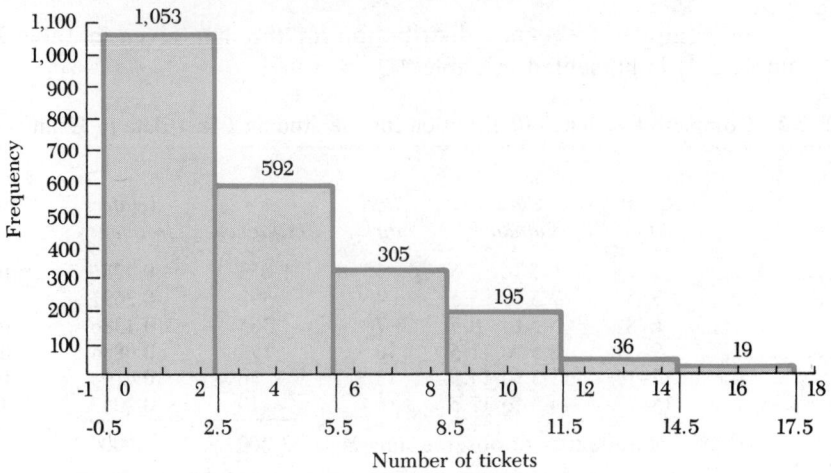

FIGURE 2.2 Frequency polygon for the data in Table 2.2

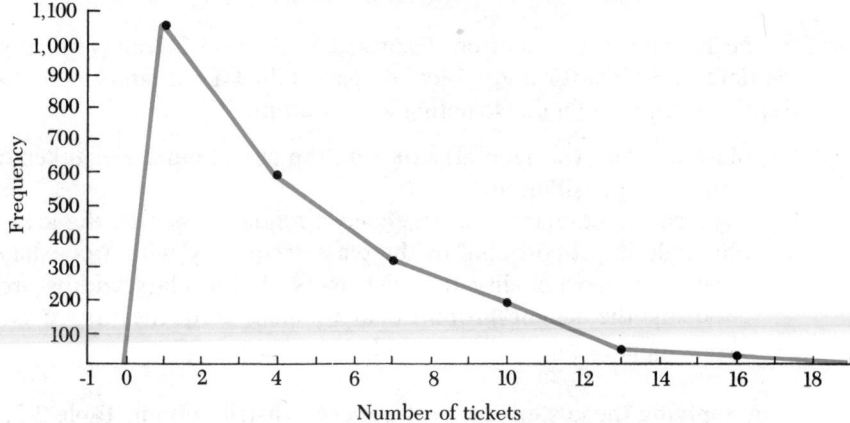

we had just connected the dots corresponding to the class marks in the frequency distribution alone. The frequency polygon is another means of shedding light on the population distribution. As with the histogram, the polygon clearly indicates that most students acquire only a few tickets and relatively few students acquire many.

The cumulative relative or absolute frequencies may also be used to construct another graphical display of the data. A frequency polygon constructed from the cumulative frequencies is called an *ogive*. A "less than or equal to" cumulative relative frequency polygon (ogive) for the frequency distribution in Table 2.2 is illustrated in Figure 2.3. Notice that the ogive

FIGURE 2.3 A "less than or equal to" ogive for the frequency distribution in Table 2.2

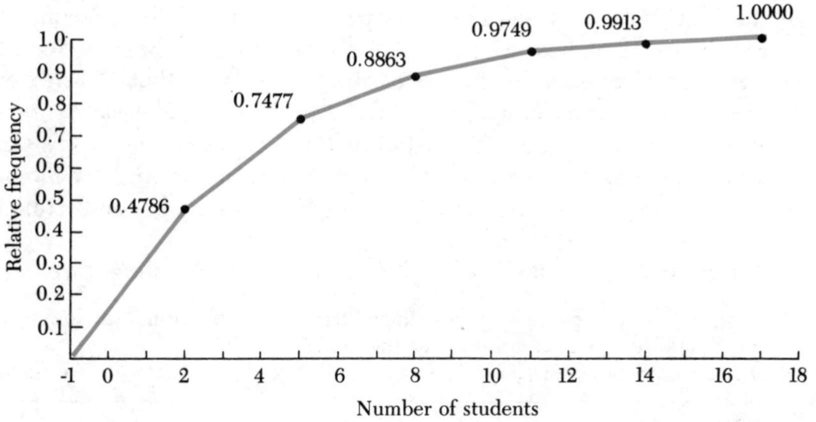

Relative frequency

1.0 —
0.9 —
0.8 —
0.7 —
0.6 —
0.5 —
0.4 —
0.3 —
0.2 —
0.1 —

0.4786
0.7477
0.8863
0.9749
0.9913
1.0000

-1 0   2   4   6   8   10   12   14   16   18

Number of students

begins at $(-1, 0)$. The value -1 represents the upper limit of the class preceding the "first class" with limits 0–2. It is also possible to construct a "greater than or equal to" ogive for either the relative or absolute frequencies.

There are several forms of graphical displays that are not usually associated with a frequency distribution. Examples are *bar charts, line charts,* and *pie charts.* In Figure 2.4, pie charts summarize the 1979 distribution of sales and earnings for the Black and Decker Manufacturing Company.[2] In a pie

FIGURE 2.4 Distribution of sales and earnings of the Black and Decker Manufacturing Company, 1979

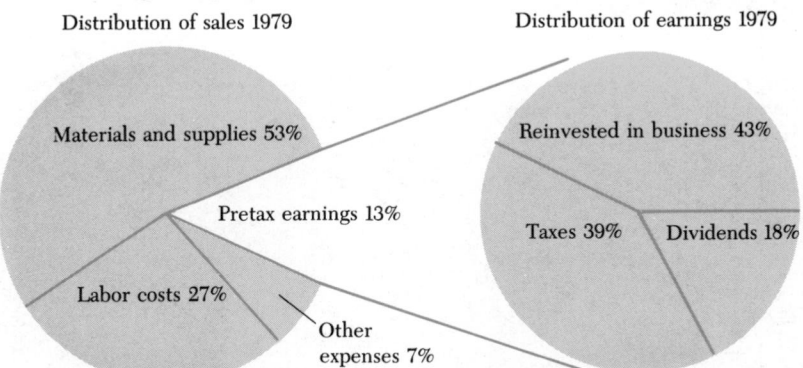

Distribution of sales 1979   Distribution of earnings 1979

Materials and supplies 53%   Reinvested in business 43%

Pretax earnings 13%

Taxes 39%   Dividends 18%

Labor costs 27%

Other expenses 7%

Source: Black and Decker Manufacturing Company, *1979 Annual Report.*

[2] *1979 Annual Report* (Towson, Md.: The Black and Decker Manufacturing Company).

chart, an area or "slice" of the pie is allocated to each class or grouping in proportion to the relative frequency or percentage of the total. For example, in the distribution of sales pie, 53 percent of the sales revenue went for materials and supplies, 27 percent for labor costs, 7 percent for other expenses, and 13 percent for pretax earnings. For the "slice" pretax earnings, the distribution of earnings pie indicates that 43 percent was reinvested, 39 percent was paid in taxes, and 18 percent was paid in dividends to shareholders. Pie charts can be used quite effectively to illustrate how monies were distributed (revenues, profits, or costs) by a company in conducting its business.

In Figure 2.5, bar and line charts are combined to show three economic

FIGURE 2.5 Stock price monthly ranges, earnings, and annual dividend rate for the Black and Decker Manufacturing Company, 1979

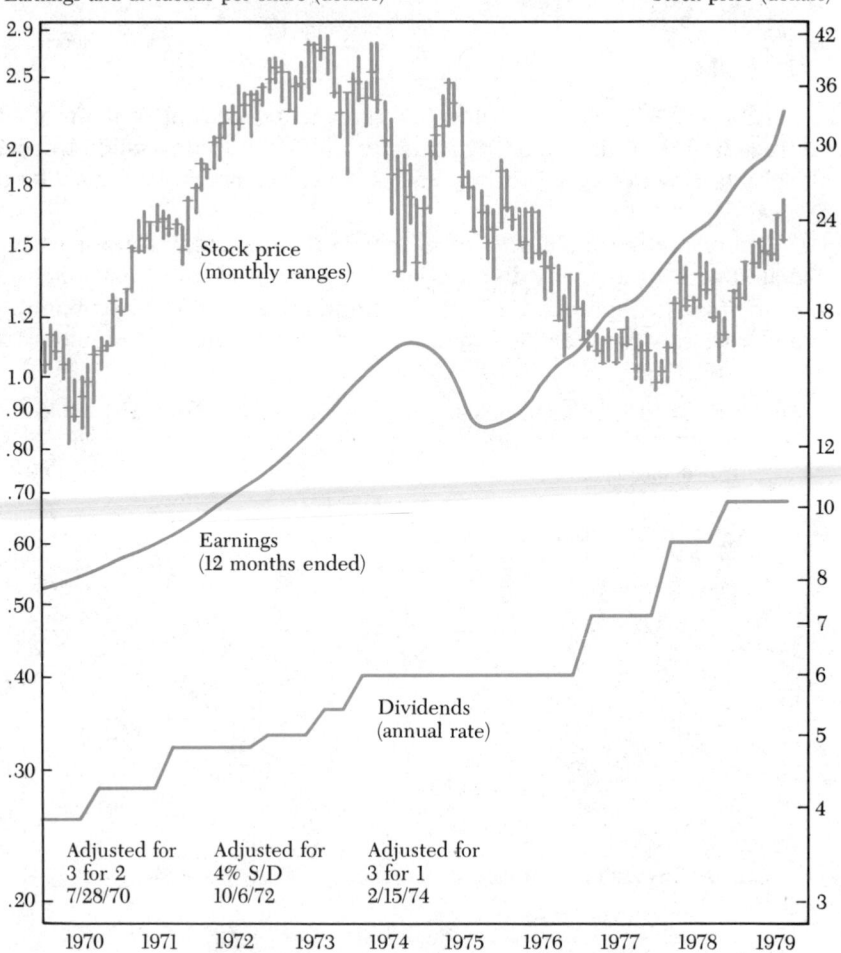

Source: Black and Decker Manufacturing Company, *1979 Annual Report.*

indicators for the Black and Decker Manufacturing Company. The bar chart at the top of the figure gives the monthly stock price ranges from 1970 to 1979. The horizontal "crosshatch" gives the monthly closing prices, and the vertical bar gives the monthly range of the stock prices. Two line charts illustrate the earnings and annual dividend rates for 1970 to 1979. Most companies use charts extensively in their annual reports to summarize important economic measures for their stockholders.

For the interested reader, additional information on graphical methods is available in many of the references listed at the end of this chapter. Histograms, polygons, bar charts, line charts, and pie charts provide easily understood pictorial representations of data. These graphical methods can be of considerable value in providing insights about the nature and properties of a data set. It is the task of the statistician to select the form of display that most clearly and accurately depicts the information in the data set.

■ 2.4   REPRESENTATIVE MEASURES

The frequency distribution and its graphical display, the histogram, can provide considerable insights into the nature of the distribution of a data set. From the histogram, we can quickly tell the form or shape of the distribution—the degree of spread in the observations, their tendency to cluster about a "middle" value, the existence of one or few very large or very small observations, and so forth. It is also possible to characterize this information by computing *numerical measures* such as averages or quantities which measure the spread of the observations. Although these numbers may not contain as much information as a frequency distribution or its histogram, when they are taken collectively, the statistician can readily construct a visual image of how the frequency distribution must look. More importantly, these numerical measures are easy to work with mathematically and are used extensively in the inference-making process if they have been calculated from sample data.

In this chapter, we will assume that the data set represents a population of N observations as opposed to a sample set of n observations taken from a population. Throughout the text, we will use greek letters (μ, σ, ρ, etc.) to denote *population* numerical measures. In Chapter 8, we will develop the notation for numerical measures calculated from sample data sets.

The first numerical measures we will consider are called *representative measures*. By this we mean measures that are "representative" of all the observations. These measures may attempt to locate the "center" or "middle" value of the distribution of observations. But in some instances it may not make much sense to talk about a central value, because of a wide spread in the observations or concentrations in the ends of the distribution rather than in the middle.

The most commonly used representative measure is the simple arithmetic mean.

> **Definition 2.1**
> The arithmetic mean
>
> The *arithmetic mean*, denoted by μ, of a set of N measurements x_1, x_2, . . . , x_N is given by the formula:
>
> $$\mu = \frac{\sum_{i=1}^{N} x_i}{N} \qquad (2.1)$$

Very simply put, the mean μ of a set of N measurements is their sum divided by N. It is perhaps better known as the *simple average* of a set of numbers.

Example 2.2   Find the arithmetic mean of the following five measurements:

$$x_1 = 10 \quad x_2 = 15 \quad x_3 = 6 \quad x_4 = 12 \quad x_5 = 11$$

Solution

$$\mu = \frac{\sum_{i=1}^{5} x_i}{5} = \frac{10 + 15 + 6 + 12 + 11}{5} = \frac{54}{5} = 10.8$$

The arithmetic mean gives the "center of gravity" of a set of numbers as indicated in Figure 2.6 for the data set of Example 2.2. The mean of 10.8 is the "balancing point" of the five numbers.

FIGURE 2.6   Arithmetic mean as the center of gravity

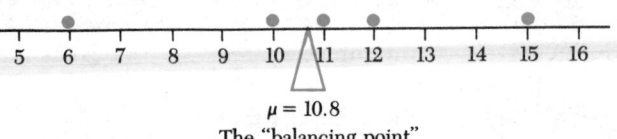

$\mu = 10.8$

The "balancing point"

Example 2.3   The following eight observations represent the gain or loss in the daily closing price of a commodity on eight consecutive market days:

$$3.375 \quad -1 \quad 0 \quad 0 \quad 2.25 \quad 2.5 \quad 1.75 \quad -1.875$$

Find the average loss or gain in the commodity price.

Solution

$$\mu = \frac{\sum_{i=1}^{8} x_i}{8} = \frac{3.375 + (-1) + 0 + 0 + 2.25 + 2.5 + 1.75 + (-1.875)}{8}$$

$$= 0.875 \quad \text{(the average gain)}$$

> **Definition 2.2**
> **Median**
>
> The *median*, denoted by M, of a set of N measurements $x_1, x_2, \ldots, x_N$ is the value x such that it falls in the middle of the array of N values when they have been ordered from the least numerical value to the greatest value.

Example 2.4　Find the median of the five measurements given in Example 2.2.

Solution　We must first order the numbers from the smallest to the largest: 6, 10, 11, 12, and 15. The "middle" value is 11. Thus, $M = 11$.

Example 2.5　Find the median of the gains and losses in Example 2.3.

Solution　The numbers must first be ordered from the smallest to the largest: -1.875, -1, 0, 0, 1.75, 2.25, 2.5, and 3.375.

In this case, there is no "middle" number. When the number of items in the data set is even, we define the median M to be the value halfway between the two "middle" numbers in the ordered set. Thus, in this example,

$$M = \frac{0 + 1.75}{2} = 0.875$$

> **Definition 2.3**
> **Mode**
>
> The *mode*, denoted by MO, of a set of N measurements $x_1, x_2, \ldots, x_N$ is the value x that occurs with the greatest frequency.

Example 2.6　Find the mode of the following set of six measurements:

$$x_1 = 3 \quad x_2 = 4 \quad x_3 = 3 \quad x_4 = 5 \quad x_5 = 2$$

Solution　$MO = 3$ since the value 3 occurs with the greatest frequency (twice).

Example 2.7　Find the mode of the following set of numbers:

$$2 \quad 3 \quad 2 \quad 5 \quad -1 \quad -2 \quad -1 \quad 3 \quad 2 \quad -1 \quad 5$$

Solution　Both -1 and 2 occur three times. Thus, there are two modes: $MO = -1$ and $MO = 2$. In this case, we say that the set of numbers is *bimodal*.

Example 2.8　Find the mode of the following set of numbers:

$$1 \quad 3 \quad -1 \quad -2 \quad 4 \quad 0 \quad 5$$

Solution　Each number occurs once. In this case, we say that the set of numbers does not have a mode.

A reasonable question to raise at this point is: Which statistic, the mean, the median, or the mode, should be used as the representative measure of a data set? The arithmetic mean is most often used and for some very good reasons. All three measures are functions of all the observations, but the mean seems intuitively to take all the observations into account "better" than does the median or the mode. More important, the mean lends itself "best" among the representative measures for the purpose of statistical inference when it has been computed over a sample data set.

In most instances, all three representative measures should be computed for a data set. If the three measures differ in value, this information has implications regarding the shape of the distribution of the observations. The arithmetic mean is the center of gravity of the distribution. If its value for a specific data set is different from the value of the median, then the distribution of the observations must be *skewed*—the distribution has a longer "tail" either to the right or to the left. For example, if a distribution is skewed to the right, most values are clustered about the left end of the range of values, with a few large values occurring at the upper end of the range. A distribution that is skewed to the right will appear to have a rather long tail on the right end of the range of values. Further, the mean will be greater than the median for the data set. The few large values in the upper tail tend to pull the center of gravity to the right, away from the center of the distribution as illustrated in Figure 2.7.

FIGURE 2.7   Distribution skewed to the right

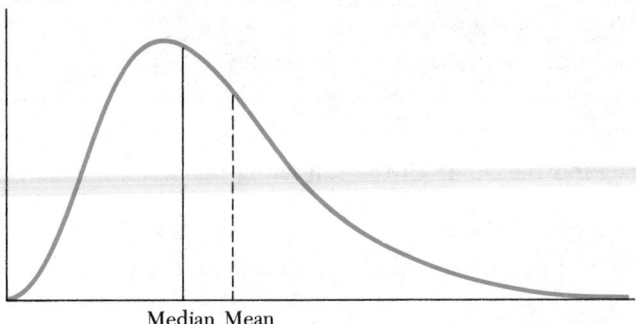

Median Mean

A distribution is called *symmetric* if there is a value such that the portion of the distribution to the left of this value is a mirror image of the portion of the distribution to the right of this value. In Figure 2.8, examples of symmetric, skewed right, and skewed left distributions are illustrated. If the distribution is unimodal and symmetric, the three measures of central tendency have identical numerical values. In Section 2.6, numerical measures for determining the degree of skewness will be discussed. However, as the above discussion suggests, the three representative measures frequently provide a clue as to the symmetry or skewness of the distribution.

FIGURE 2.8   Examples of symmetric and skewed distributions

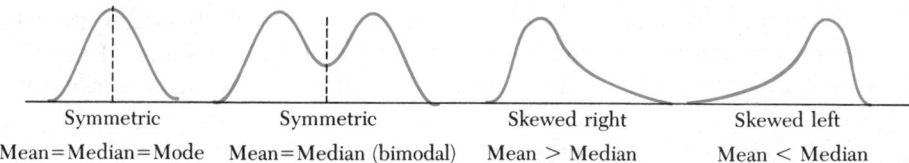

Symmetric	Symmetric	Skewed right	Skewed left
Mean=Median=Mode	Mean=Median (bimodal)	Mean > Median	Mean < Median

When the distribution is skewed, there is a problem in selecting the representative measure that best represents the data set. Unfortunately, in some cases the decision is made on the basis of which measure best supports an individual's argument, and this is a clear abuse of statistics. For example, suppose a basketball team of ten players has the annual salaries listed in Table 2.3.

TABLE 2.3   Annual salaries of ten basketball players

Player	Annual salary	Player	Annual salary
Joe Dribble	$100,000	"Hands" Slat	$ 250,000
"Touch" Phelps	100,000	Allie Hoops	175,000
Sly Slack	200,000	"Radar" Slater	275,000
Rally Cheer	125,000	Ron Rebound	150,000
Joe Owner, Jr.	225,000	"Sky" Hook	1,400,000

The salaries of nine of the players are less than or equal to $275,000 per year, while "Sky" Hook is making $1.4 million per year. For this data set, the average (mean) salary is $300,000, the median is $187,500 [($200,000 + $175,000)/2], and the mode is $100,000. In arguing against increases in player salaries, the owner is likely to quote an average salary of $300,000 (the mean), but the median in this case is arguably the better representative measure, at least from the viewpoint of nine of the players. A "fair" owner would quote all three measures, recognizing that the distribution is severely skewed to the right due to Hook's annual salary.

In business applications, the mode has been commonly used in demand analysis. A retail shoe store might be concerned, for example, about the modal sizes of shoes that are demanded by customers. The median has been used extensively in the social sciences and is a favorite government statistic as a representative measure in skewed distributions (median income, for instance). However, for the purposes of inference making, the mean will be used extensively in this text as a measure of a representative value for a data set. Its advantages outweigh its disadvantages. Additionally, in many practical applications of statistical inference, the distribution of the values is nearly symmetric, so that the three representative measures are approximately equal numerically in the sample selected.

■ **2.5 MEASURES OF VARIABILITY**

Although representative measures provide certain information about a distribution, more is needed before a clear picture of the shape of the distribution can be formulated. In Figure 2.9, for example, both distributions

FIGURE 2.9 Two dissimilar distributions with identical means

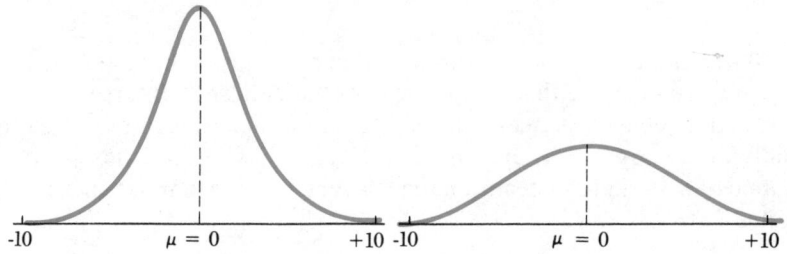

have the same mean value but obviously differ in another respect—the amount of *dispersion* or *variability* of the values. The concept of variability is very important in statistics. For example, in production management, a major concern is the variability of the quality of a product being produced or the variability of a crucial measurement of a product such as a bearing diameter. More importantly, in statistical inference, we use the concept of variability to determine how good our inferences are. For the moment, it is sufficient to recognize the need for measuring the variability of a set of values to get a better idea of the shape of the distribution of the values.

The first and simplest measure of variability is the *range*.

Definition 2.4
Range

The range of a set of N measurements $x_1, x_2, \ldots, x_N$ is the algebraic difference between the largest and smallest values.

Example 2.9 Given the following six numbers, determine their range.

$$x_1 = 0 \quad x_2 = 5 \quad x_3 = 6 \quad x_4 = 2 \quad x_5 = -1 \quad x_6 = 10$$

Solution The largest number is 10, while the smallest is -1. Thus, the range is $10 - (-1) = 11$.

Although the range is easy to compute, it is not a very satisfactory measure of variability, as Figure 2.10 illustrates. The distribution on the left clearly is less variable than the distribution on the right, yet the ranges for the two distributions are identical ($10 - 0 = 10$).

FIGURE 2.10 Two distributions with equal ranges

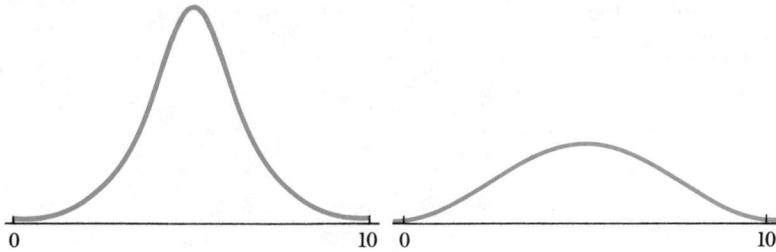

A better measure of variability is provided by using the *deviations* of the measurements from their mean or average value. For example, suppose we have a set of six measurements:

$$x_1 = 1 \quad x_2 = 4 \quad x_3 = 6 \quad x_4 = 7 \quad x_5 = 4 \quad x_6 = 8$$

The mean of these six measurements is:

$$\mu = \frac{\sum_{i=1}^{6} x_i}{6} = \frac{1 + 4 + 6 + 7 + 4 + 8}{6} = \frac{30}{6} = 5$$

In Figure 2.11, the six values are placed on the x axis and the deviation $(x_i - \mu)$ of each value from the mean is indicated by a horizontal line. The

FIGURE 2.11 Deviations of six values from their mean

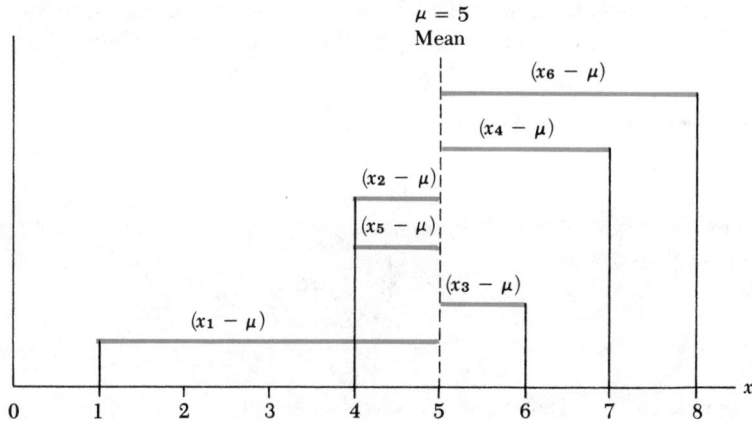

"lengths" of these horizontal lines contain valuable information about the dispersion of the values. However, if we attempt to form a numerical measure based on the sum of these deviations, we find that it is not very helpful in measuring dispersion—the sum is 0 as the computations in

TABLE 2.4 Computation of $\sum_{i=1}^{N} (x_i - \mu)$ for the data associated with Figure 2.11

x_i	$x_i - \mu$
1............	$1 - 5 = -4$
4............	$4 - 5 = -1$
6............	$6 - 5 = 1$
7............	$7 - 5 = 2$
4............	$4 - 5 = -1$
8............	$8 - 5 = \underline{3}$
	0

Table 2.4 verify. Indeed, it can be shown that for *any* set of N measurements, $\sum_{i=1}^{N} (x_i - \mu) = 0$. We will not prove this result algebraically. Rather, the result can be appreciated on intuitive grounds. Since μ is the center of gravity, the positive and negative deviations about μ must balance out. Thus, the sum of the positive deviations corresponding to the x_i values that are greater than μ equal the sum of the negative deviations corresponding to the x_i values that are less than μ.

One way to use the deviations of the x_i about their mean μ is to take their absolute values—the absolute values will eliminate any negative signs. Although the sum of the absolute values of the deviations has been used in statistics to measure dispersion, it is not a favored statistic because of the mathematical problems involved in working with the absolute value function. A more attractive statistic is the "average" of the squared deviations.

Definition 2.5
Variance

The *variance* of a set of N measurements $x_1, x_2, \ldots, x_N$, denoted by σ^2, is given by the formula

$$\sigma^2 = \frac{\sum_{i=1}^{N} (x_i - \mu)^2}{N} \tag{2.2}$$

where μ is the mean of the set of N measurements.

Example 2.10 Compute the variance for the six measurements 1, 4, 6, 7, 4, and 8.

Solution We can compute σ^2 by forming a table of deviations and their squares, as in Table 2.5.

TABLE 2.5 Computation of σ^2 for data in Example 2.10

x_i	$(x_i - \mu)$	$(x_i - \mu)^2$
1	$(1 - 5) = -4$	$(-4)^2 = 16$
4	$(4 - 5) = -1$	$(-1)^2 = 1$
6	$(6 - 5) = 1$	$(1)^2 = 1$
7	$(7 - 5) = 2$	$(2)^2 = 4$
4	$(4 - 5) = -1$	$(-1)^2 = 1$
8	$(8 - 5) = 3$	$(3)^2 = 9$
		$\overline{32}$

$$\sigma^2 = \frac{32}{6} = 5.33$$

Example 2.11 Compute the variance for the six numbers 0, -10, 20, 30, -15, and 5.

Solution

$$\mu = \frac{\sum\limits_{i=1}^{6} x_i}{6} = \frac{0 - 10 + 20 + 30 - 15 + 5}{6} = \frac{30}{6} = 5$$

$$\sigma^2 = \frac{(0 - 5)^2 + (-10 - 5)^2 + (20 - 5)^2 + (30 - 5)^2 + (-15 - 5)^2 + (5 - 5)^2}{6}$$

$$= \frac{(-5)^2 + (-15)^2 + (15)^2 + (25)^2 + (-20)^2 + (0)^2}{6}$$

$$= \frac{25 + 225 + 225 + 625 + 400 + 0}{6} = \frac{1,500}{6} = 250$$

Examples 2.10 and 2.11 indicate that the larger σ^2 is, the more variable the distributions of values are. In these examples, both sets of measurements have the same mean value, but the larger σ^2 in the second example clearly indicates that the second set is more variable.

The variance σ^2 has more important uses than as a relative measure to deduce which of two sets of data has more variability. In Section 2.8, we will discuss the information that σ^2 supplies about the distribution of a set of data. Specifically, we shall make use of the square root of σ^2, which leads us to the definition of another measure of variability.

> **Definition 2.6**
> **Standard deviation**
>
> The *standard deviation* of a set of N measurements $x_1, x_2, \ldots, x_N$, denoted by σ, is the *positive* square root of the variance, σ^2.

For example, the standard deviation of the set of values in Example 2.11 is $\sigma = \sqrt{250} = 15.81$.

It is usually more convenient to work with the standard deviation than the variance since σ is in the same units (inches, for example) as the measurements, while σ^2 is measured in the original units squared (inches^2). Furthermore, interpretations about the amount of variability in a set of data are more easily drawn using σ rather than σ^2, as we shall see in Section 2.8.

There is an easier way to calculate σ^2 (and, consequently σ) than the preceding method. We may use the following *computing formula* for the variance:

$$\sigma^2 = \frac{\sum_{i=1}^{N} x_i^2 - \dfrac{\left(\sum_{i=1}^{N} x_i\right)^2}{N}}{N} \tag{2.3}$$

The formula directs us to square each value and sum these squared quantities ($\sum_{i=1}^{N} x_i^2$); subtract from this the square of the sum of the values divided by N, $[(\Sigma x_i)^2/N]$, and divide the difference by N.

Example 2.12 Compute σ^2 using the formula in equation (2.3) for the set of values in Example 2.11.

Solution First, form a table to sum the values and their squares (see Table 2.6).

TABLE 2.6
Calculations required for σ^2 (data from Example 2.11)

x_i	x_i^2
0	0
-10	100
20	400
30	900
-15	225
5	25
30	1,650

$$\sigma^2 = \frac{\sum_{i=1}^{6} x_i^2 - \dfrac{\left(\sum_{i=1}^{6} x_i\right)^2}{6}}{6} = \frac{1,650 - \dfrac{(30)^2}{6}}{6}$$

$$= \frac{1,650 - \dfrac{900}{6}}{6} = \frac{1,650 - 150}{6} = \frac{1,500}{6} = 250$$

At this stage, your primary concern should be to compute the variance and to verify that the two formulas give the same result. The interpretation and use of σ^2 are exceedingly important, but these considerations will be reserved for later sections.

2.6 OTHER NUMERICAL MEASURES

There are numerous representative measures other than the mean, median, and mode. The *geometric mean* is commonly used in business problems to describe the "average" of ratios. Although the geometric mean is not as important as the three principal representative measures (mean, median, and mode), we shall introduce it now while reserving its application until the discussion of Fisher's ideal price index in Chapter 18.

☐ 2.6.1 Geometric mean

Definition 2.7
Geometric mean

The *geometric mean* of a set of N measurements $x_1, x_2, \ldots, x_N$, denoted by GM, is defined by

$$GM = (x_1 \cdot x_2 \cdot x_3 \cdot \cdots \cdot x_N)^{1/N} \tag{2.4}$$

The geometric mean is the Nth root of the product of all the measurements. It is not as easily computed as the arithmetic mean—the computation is eased somewhat by taking logarithms of both sides of equation (2.4):

$$\log(GM) = \frac{1}{N} (\log x_1 + \log x_2 + \cdots + \log x_N) \tag{2.5}$$

From equation (2.5), it is apparent that the geometric mean can be computed by taking the antilog of the arithmetic mean of the logs of the measurements.

Example 2.13 Determine the geometric mean of the three measurements:

$$x_1 = 2 \quad x_2 = 4 \quad x_3 = 8$$

Solution

$$GM = \sqrt[3]{(2)(4)(8)} = \sqrt[3]{64} = 4$$

Example 2.14 Find the arithmetic and geometric mean of 100, 100, 100, and 1,000.

Solution Arithmetic mean:

$$\mu = \frac{100 + 100 + 100 + 1,000}{4} = 325$$

Geometric mean:

$$\log(GM) = \frac{1}{4} (\log 100 + \log 100 + \log 100 + \log 1,000)$$

$$= \frac{1}{4} (2 + 2 + 2 + 3) = \frac{1}{4} (9) = 2.25$$

$$GM = 10^{2.25} = 177.8$$

This example illustrates that the geometric mean is less affected by one (or two) extremely large (or small) values than is the arithmetic mean. Unfortunately, the geometric mean is neither easy to compute nor amenable to use for statistical inferences. It is very useful, however, in averaging ratios—a process that frequently arises in computing the cost of living or other index numbers.

☐ 2.6.2 Quartiles and percentiles

The mean, median, and mode can be thought of as measures of location—they attempt to locate the most representative value. Other measures of location are *quartiles* and *percentiles*.

Definition 2.8
100*p*th Percentile

The *100pth percentile* of a set of N measurements $x_1, x_2, \ldots, x_N$ that have been ordered from the smallest to the largest is a value such that at least 100p percent of the measurements are at or to the left of (below) this value *and* at least $100 (1 - p)$ percent of the measurements are at or to the right of (above) this value.

As a matter of convention, we will choose as the value a measurement in the data set as the 100*p*th percentile *except* when two adjacent ordered measurements satisfy the definition, in which case their average is taken as the 100*p*th percentile. The averaging of the two adjacent ordered measurements follows the suggested computation of the median introduced in Section 2.4. In Example 2.5, the 50th percentile (the median) was determined by averaging 0 and 1.75. Note that both measurements satisfy the requirement that at least 50 percent of the eight measurements are at or to the left of each measurement and at least 50 percent of the eight measurements are at or to the right of each measurement. You should verify that this is the case in Example 2.5.

Example 2.15 Find the 20th percentile for the data set:

20 34 17 18 28 33 12 15 17 12 41
                   45 18 19 16 21 26 14 26 13 29

Solution Ordered from the smallest to the largest, the 21 measurements are:

12 12 13 14 15 16 17 17 18 18 19
                   20 21 26 26 28 29 33 34 41 45

To determine the 20th percentile, we look for a value so that at least $100(0.20) = 20$ percent of the measurements are at or to the left of this value and at least $100(1 - 0.20) = 80$ percent of the measurements are at or to the right of this value. The measurement, 15, satisfies these conditions: at least 20 percent of the measurements $[(0.20)(21) = 4.2]$ are at or to the left of 15 (exactly 5 measurements are at or to the left of 15), and at least 80 percent of the measurements $[(0.80)(21) = 16.8]$ are at or to the right of 15 (exactly 17 measurements are at or to the right of 15). Therefore, the 20th percentile is 15.

> **Definition 2.9**
> Lower, middle, and upper quartiles
>
> The *lower quartile* (Q_1) of a set of N measurements is the 25th percentile. The *middle quartile* (Q_2), the median, is the 50th percentile. The *upper quartile* (Q_3) is the 75th percentile.

Example 2.16　Find Q_1, Q_2, and Q_3 for the data set in Example 2.15.

Solution　For Q_1, at least 25 percent of the ordered measurements should be at or to the left of Q_1 and at least 75 percent of the ordered measurements should be at or to the right of Q_1. Thus, at least $(0.25)(21) = 5.25$ measurements are at or to the left of Q_1, and at least $(0.75)(21) = 15.75$ measurements are at or to the right of Q_1. Therefore, Q_1 is 16.

For Q_2, at least 50 percent of the ordered measurements should be at or to the left of Q_2 and at least 50 percent of the ordered measurements should be at or to the right of Q_2. Thus, at least $(0.50)(21) = 10.5$ measurements are at or to the left of Q_2, and at least $(0.50)(21) = 10.5$ measurements are at or to the right of Q_2. Therefore, Q_2, the median, is equal to 19.

For Q_3, at least 75 percent of the ordered measurements should be at or to the left of Q_3 and at least 25 percent of the ordered measurements should be at or to the right of Q_3. Thus, at least $(0.75)(21) = 15.75$ measurements are at or to the left of Q_3, and at least $(0.25)(21) = 5.25$ measurements are at or to the right of Q_3. Therefore, Q_3 is 28.

By calculating the difference between the lower and upper quartiles, we can get some feel for the variability of the data. The difference, $Q_3 - Q_1$, is called the *interquartile range*. The larger the interquartile range is for a data set, the more variable (spread out) the set of measurements is.

☐ 2.6.3　Skewness

As suggested earlier in this chapter, one possible measure of the *skewness* of a distribution of a set of measurements is the difference between its mean and its median. We use a function of this difference as a measure of skewness.

> **Definition 2.10**
> Skewness measure γ
> (Pearson's second coefficient of skewness)
>
> The *skewness measure* of a set of N measurements $x_1, x_2, \ldots, x_N$, denoted by γ, is defined as three times the difference between the mean and the median, divided by the standard deviation:
>
> $$\gamma = \frac{3(\mu - M)}{\sigma}$$

If the distribution is skewed right, the mean will be larger than the median and γ will be positive. If the distribution is skewed left, the mean will be smaller than the median and γ will be negative.

The effect of dividing by σ in γ is to produce a statistic that is not dependent on the unit of measurement. The mean, median, and standard deviation are all measured in the same units for a given data set.

The skewness measure γ can be used in two ways. First, the sign of γ indicates the direction of skewness: $+$, skewed right and $-$, skewed left. Second, if γ is larger in magnitude in one data set than in another, the first data set distribution is more skewed than the other. That is, γ can be used as a relative measure of the degree of skewness among data sets.

Example 2.17 Compute γ for the following five measurements:

$$x_1 = 10 \quad x_2 = 4 \quad x_3 = 4 \quad x_4 = 6 \quad x_5 = 1$$

Solution The mean, median, and standard deviation for this data set are $\mu = 5$, $M = 4$, and $\sigma = 2.97$. Therefore,

$$\gamma = \frac{3(5 - 4)}{2.97} = 1.01$$

Since γ is positive, the distribution of the five measurements is skewed to the right.

There are other properties of a set of measurements and their corresponding frequency distributions than representative measure, variance, and skewness. But these three numerical measures provide the foundation for almost all the basic methods of statistical inference. We will, therefore, discuss no other kinds of numerical measures than those presented here.

■ 2.7 COMPUTATION OF NUMERICAL MEASURES FOR GROUPED DATA

After data have been grouped into a frequency distribution, numerical measures such as the mean and variance may be computed *approximately* from a frequency distribution. Before the advent of the computer, forming a frequency distribution was an important computational device. For large data sets, the numerical measures could be approximated from the frequency distribution with considerably less computational effort than by determining the numerical measures exactly from ungrouped data. Today, computers allow us to find numerical measures for extremely large data sets (1 million or more numbers) in a short period of "computer time." Hence, this use of the frequency distribution is no longer important, except in the rare instance when the only source of data is a secondary one that provides only the frequency distribution and not the ungrouped data. We will, therefore, only indicate how the mean and variance can be approximated from the frequency distribution. For the approximation of other numerical measures from a frequency distribution, the interested reader is directed to the Mendenhall and Reinmuth and the Neter et al. references at the end of this chapter.

The arithmetic mean μ of N measurements $x_1, x_2, \ldots, x_N$ may be computed approximately from a frequency distribution of the N measurements by using the formula.[3]

$$\mu \doteq \frac{\sum\limits_{i=1}^{k} f_i m_i}{\sum\limits_{i=1}^{k} f_i} \tag{2.6}$$

where m_i and f_i are the ith class mark and the ith frequency, respectively, and k is the number of classes in the frequency distribution.

The variance can also be approximated from a frequency distribution by using the formula:

$$\sigma^2 \doteq \frac{\sum\limits_{i=1}^{k} f_i m_i^2 - \dfrac{\left(\sum\limits_{i=1}^{k} f_i m_i\right)^2}{N}}{N} \tag{2.7}$$

Example 2.18 Suppose that 20 measurements,

0 11 8 0 0 0 3 7 8 0 0 1 4 3 0 0 5 3 0 3

are placed in the frequency distribution given in Table 2.7. Using the frequency distribution, approximate the mean and the variance of the measurements.

TABLE 2.7 Frequency distribution (Example 2.18 data)

Class	Class limit	Frequency f_i	Class mark m_i	$f_i m_i$	$f_i m_i^2$
1	0–2	$f_1 = 10$	$m_1 = 1$	10	10
2	3–5	$f_2 = 6$	$m_2 = 4$	24	96
3	6–8	$f_3 = 3$	$m_3 = 7$	21	147
4	9–11	$f_4 = 1$	$m_4 = 10$	10	100
Totals		20		65	353

Solution Since there are only 20 measurements, we can easily compute μ and σ^2; they are 2.8 and 10.96, respectively. But, by computing these numerical measures from the frequency distribution, we can better appreciate the

---

[3] The symbol $\doteq$ indicates an approximate answer.

quality of the approximation and the efficiency of computation in extremely large data sets. Now, from equation (2.6),

$$\mu \doteq \frac{\sum\limits_{i=1}^{4} f_i m_i}{\sum\limits_{i=1}^{4} f_i} = \frac{65}{20} = 3.25$$

The discrepancy of 0.45 between the actual arithmetic mean (2.80) and the approximation from the frequency distribution (3.25) is due to the fact that, in each class, we are using the class mark as the representative value. For example, in the first class, 1 is used as the representative value of the ten values ranging from 0 to 2. The quantity $f_1 m_1 = 10$ is the approximated sum of these ten values when indeed their true sum is only 1 (nine zeros and a one). In this example, the approximated value (3.25) is perhaps surprisingly close to the true value of the mean (2.80). If, for the k classes in the frequency distribution, the average of the values in each class is close to the associated class mark, the calculation of μ by equation (2.6) will generally give a very good answer. If the average value within each class were identically equal to the class mark of that class, equation (2.6) will give the exact value of the mean of the data.

From equation (2.7),

$$\sigma^2 \doteq \frac{\sum\limits_{i=1}^{k} f_i m_i^2 - \frac{\left(\sum\limits_{i=1}^{k} f_i m_i\right)^2}{N}}{N} = \frac{353 - \frac{(65)^2}{20}}{20} = 7.09$$

The approximate variance (7.09) is also not very close to its exact numerical quantity (10.96). By placing the measurements into the frequency distribution, some of the variability is lost in this case—it is embedded in the class frequencies and is not readily apparent from observing the frequency distribution.

■ 2.8 USE OF THE STANDARD DEVIATION: THE CHEBYCHEF THEOREM

The representative measures convey an intuitive idea of a property of the distribution of a set of measurements—the location of a representative or "middle" value of the distribution. The variance and its positive square root, the standard deviation, are not so easily interpreted; they measure dispersion, but a numerical value of σ^2 does not help us in visualizing the distribution of the measurements. A theorem by the Russian mathematician Chebychef does provide a means of using the standard deviation σ to gain insight into the dispersion of a set of measurements.

The remarkable property of the Chebychef theorem is that it is applicable to *any* set of measurements. The result of the theorem is illustrated in Figure 2.12; in the interval $\mu - k\sigma$ to $\mu + k\sigma$, at least $(1 - 1/k^2)$ of the measure-

FIGURE 2.12 Illustration of the Chebychef theorem

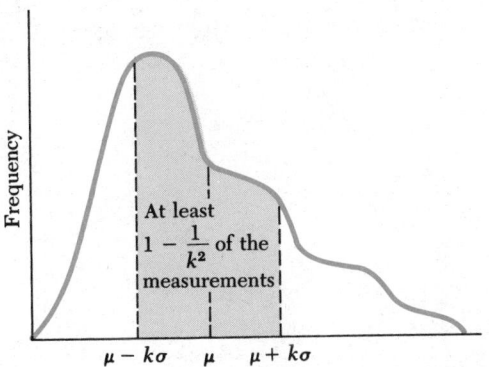

ments will reside. To see the effect of selecting different values of k, in Table 2.8 a few values of k are selected and the quantity $(1 - 1/k^2)$ calculated. When $k = 1$, at least 0 of the measurements lie in the interval from $\mu - \sigma$ to $\mu + \sigma$; thus, selecting $k = 1$ in the Chebychef theorem does not produce a useful result. But, say, when $k = 3$, at least $8/9$ of the measurements lie in the interval from $\mu - 3\sigma$ to $\mu + 3\sigma$.

TABLE 2.8 Values of k and $(1 - 1/k^2)$ for the Chebychef theorem

k	$1 - 1/k^2$
1	0
$3/2$	$5/9$
2	$3/4$
$5/2$	$21/25$
3	$8/9$

Example 2.19   Suppose a set of 100 measurements has a mean of 50 and a standard deviation of 10. Apply the Chebychef theorem to these population data to describe their dispersion.

Solution   In Table 2.9, three values of k ($^3/_2$, 2, and 3) are chosen and the corresponding intervals computed with $\mu = 50$ and $\sigma = 10$. These results are illustrated in Figure 2.13.

TABLE 2.9   Application of the Chebychef theorem to data from Example 2.19

		Interval
k	$1 - 1/k^2$	$\mu - k\sigma$ to $\mu + k\sigma$
$^3/_2$	$^5/_9$	$50 - (^3/_2)10$ to $50 + (^3/_2)10$; 35–65
2	$^3/_4$	$50 - (2)10$ to $50 + (2)10$; 30–70
3	$^8/_9$	$50 - (3)10$ to $50 + (3)10$; 20–80

FIGURE 2.13   Illustration of the Chebychef theorem applied to Example 2.19

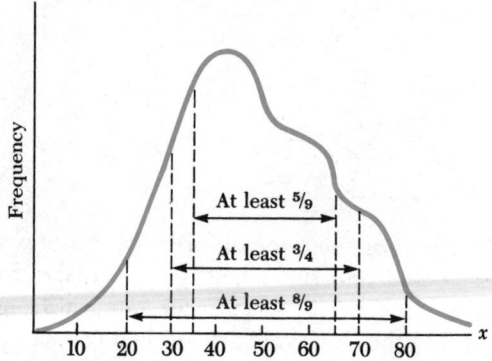

The mean of the measurements is 50. At least $^5/_9$ of the measurements fall between 35 and 65, at least $^3/_4$ between 30 and 70, and at least $^8/_9$ between 20 and 80. Thus, $^1/_9$ or less of the measurements are less than 20 and greater than 80. It is not necessarily true that half the $^1/_9$ of the measurements outside the interval are to the left of 20 and the other half are to the right of 80, because the distribution may be skewed either to the right or to the left.

2.9   SUMMARY

In this chapter, we have been concerned with describing sets of data by numerical and graphical methods. Both methods are extremely useful in describing either population or sample data sets. When the data set represents a sample, numerical measures lend themselves to inference making better than do graphical methods. As a consequence, numerical measures will be used extensively in the inference-making process, a major topic in this text.

Among the graphical methods, the frequency distribution histogram is very useful in summarizing a set of data, for it provides at a glance information about the shape of the data set. The other graphical methods can be useful if they supply an easily understood and accurate summary of the data.

Among the representative measures, the arithmetic mean is generally preferred in inference making, although the median, mode, geometric mean, quartiles, and percentiles can be helpful in determining a representative value for the data set and in providing information concerning the location of most values in the set.

The variance and the standard deviation are the preferred measures of variability, although determining differences of quartiles and percentiles may also be helpful in assessing the spread of the measurements. Used in conjunction with the Chebychef theorem, the standard deviation provides meaningful information about the dispersion of the data set.

The frequency distribution can be used to compute the numerical measures approximately, but the measures can easily be computed from ungrouped data by high-speed digital computers. The most important reason for constructing a frequency distribution is, therefore, to generate a histogram which gives a pictorial display of the distribution of the measurements.

■ REFERENCES

Croxton, F. E., and Cowden, D. J. *Practical Business Statistics*. Englewood Cliffs, N.J.: Prentice-Hall, Inc., 1960.

Croxton, F. E.; Cowden, D. J.; and Klein, S. *Applied General Statistics*. Englewood Cliffs, N.J.: Prentice-Hall, Inc., 1967.

Freund, J. E., and Williams, F. J. (revised by Perles, B., and Sullivan, C.). *Modern Business Statistics*. Englewood Cliffs, N.J.: Prentice-Hall, Inc., 1969.

Huff, D. *How to Lie with Statistics*. New York: W. W. Norton & Company, Inc., 1954.

Lewis, E. E. *Methods of Statistical Analysis in Economics and Business*. Boston: Houghton Mifflin Company, 1963.

Mendenhall, W., and Reinmuth, J. E. *Statistics for Management and Economics*. 3d ed. Belmont, Calif.: Duxbury Press, 1978.

Neiswanger, W. A. *Elementary Statistical Methods*. New York: Macmillan, Inc., 1956.

Neter, J.; Wasserman, W.; and Whitmore, G. *Applied Statistics*, Boston: Allyn & Bacon, Inc., 1978.

Ostle, B., and Mensing, R. W. *Statistics in Research*. 3d ed. Ames: Iowa State University Press, 1975.

Reichmann, W. J. *Use and Abuse of Statistics*. Baltimore: Penguin Books, Inc., 1971. (paperback)

Snedecor, G., and Cochran, W. *Statistical Methods*. 7th ed. Ames: Iowa State University Press, 1980.

Winkler, R. L., and Hays, W. L. *Statistics: Probability, Inference and Decision*. 2d ed. New York: Holt, Rinehart and Winston, Inc., 1975.

■ PROBLEMS

2.1. In the Complaints Department of a large department store on a given day, the lengths (in seconds) of the first 100 telephone calls were recorded (rounded to the nearest second), and the following frequency distribution was constructed:

Class	Class limit	Class boundary	Frequency
1	0–59	−0.5 to 59.5	5
2	60–119	59.5 to 119.5	20
3	120–179	119.5 to 179.5	40
4	180–239	179.5 to 239.5	25
5	240–299	239.5 to 299.5	10

   a. Construct a histogram from this frequency distribution.
   b. Construct a polygon from this frequency distribution.
   c. Compute the approximate mean length of the 100 telephone calls.
   d. Compute the approximate standard deviation of the 100 telephone calls.

2.2. The following 25 measurements represent the number of business trips taken annually by 25 claim adjusters of the Acme Insurance Company:

33 17 2 10 12 15 22 18 20
12 17 15 21 38 16 18 10 12
24 8 27 8 9 5 28

Construct a frequency distribution with five classes for these data. Give the relative frequencies and construct a histogram from the frequency distribution.

2.3. Compute the mean, median, mode, range, and variance for the following set of six measurements:

6 6 4 3 8 5

2.4. Compute the mean, median, mode, range, and variance for the following set of ten measurements:

−8 0 0 4 5 −10 6 10 4 2

2.5. Compute the mean, median, mode, range, and variance for the data given in Problem 2.2.

2.6. For the data in Problem 2.4, find the percentage of the values lying in the intervals $\mu \pm \sigma$, $\mu \pm 2\sigma$, and $\mu \pm 3\sigma$. Apply the Chebychef theorem to these data and compare the percentages of the values contained in these intervals.

2.7. For the data in the frequency distribution in Problem 2.1, determine by the Chebychef theorem, intervals that will contain *approximately* at least $3/4$ and $8/9$ of the values. Why must we add the word *approximately* in the preceding statement?

2.8. Use the measure of skewness, γ, to determine whether the data in Problem 2.4 are skewed left or right.

2.9. The average daily high temperature recorded at a weather station is 80°F during June. On six consecutive days, the difference between the recorded high and this average high were:

−10 −8 −2 6 4 10

Compute the average difference and the standard deviation.

2.10. Use the Chebychef theorem to find the fraction of measurements falling in the intervals $\mu \pm (3/4)\sigma$, $\mu \pm (5/2)\sigma$, and $\mu \pm 4\sigma$ for the data in Problem 2.9.

2.11. Can the geometric mean be computed for the data in Problem 2.9? Explain.

2.12. For the following five measurements, compute the arithmetic mean, geometric mean, median, and mode:

10 10 50 100 1,000

Which number(s) would you select as

the best representative measure, and why?

2.13. The Nitro-Fusion Company, in its annual report, states, "The following table supports our contention that shares of NF stock are widely held— more than 90 percent of the stockholders own 100 shares or less."

Number of shares held	Percent of total number of stockholders
1–20	48.76
21–50	25.44
51–100	18.53
101–500	5.81
Over 500	1.46

Do you agree with this statement? Discuss.

2.14. Under what conditions may the arithmetic average be a poor representative measure? Explain.

2.15. Can the mean be approximated from a frequency distribution with an open-end interval? Why or why not? Can the median be computed in this case? Why or why not?

2.16. An investment firm handles small investment portfolios for customers in a large metropolitan area. As a part of their annual review, they have sampled 60 customer accounts to determine the *net* (after taxes) return on investment for each account. The net returns are given as the following percentages:

5.81	6.39	5.67	5.96	6.54	5.85
4.94	4.98	6.30	6.14	5.04	5.01
6.02	4.06	5.58	5.98	5.85	5.26
4.84	5.96	5.30	3.96	5.96	5.24
5.92	4.55	4.98	6.02	5.28	4.20
4.46	5.08	5.71	5.27	4.89	4.92
4.34	5.62	6.18	4.30	4.64	6.18
6.14	6.08	6.18	4.18	6.01	5.92
6.04	5.70	5.87	5.74	5.66	6.13
3.75	6.37	6.12	5.04	4.72	5.84

a. Using the classes given in the table below, construct a frequency distribution giving both the frequency and the cumulative relative frequency for each class.

Net return	Frequency	Cumulative relative frequency
3.50–3.99		
4.00–4.49		
4.50–4.99		
5.00–5.49		
5.50–5.99		
6.00–6.49		
6.50–6.99		

b. Draw a histogram for this frequency distribution.

c. Draw a "less than or equal to" ogive based on the relative frequencies.

2.17. Compute the mean, median, mode, range, and variance for the following ten numbers:

$$-60 \quad 10 \quad 10 \quad 20 \quad -75$$
$$5 \quad 0 \quad 10 \quad 20 \quad 0$$

2.18. For the data in Problem 2.17, apply the Chebychef theorem to determine the minimum proportion of observations lying in the intervals $\mu \pm 2\sigma$ and $\mu \pm 3\sigma$. What proportions of the observations actually lie in these intervals?

2.19. Using the skewness measure, are the data in Problem 2.17 skewed? If so, in which direction?

2.20. It is known that the average annual salary among workers in a particular industry is $20,000 with a standard deviation of $2,000. The distribution of salaries is known to be skewed to the right. Using Chebychef's theorem, within what interval are contained at least 84 percent of the salaries?

2.21. For the following ten numbers, compute the arithmetic mean, the geometric mean, and the median. Is the distribution of the numbers skewed? Explain.

2 6 1 2 5 6 4 15 6 2

2.22. In its annual report to shareholders, Soybean International, Inc., provides the annual net earnings per common share for each year in the ten-year period from 1970 through 1979. The data are:

1970	1971	1972	1973	1974
$0.79	0.50	0.60	0.74	1.42

1975	1976	1977	1978	1979
4.73	3.50	2.20	1.26	1.86

Draw a line chart with the years on the horizontal axis and net earnings per common share on the vertical axis.

2.23. Polyglas, Inc., reports at its annual stockholders meeting that the net revenue for 1980 has been expended as follows:

Expenditure	Percent of total revenue
Raw materials	56.0
Supplies and services	28.7
Employee benefits	6.9
Depreciation	2.4
Taxes	1.8
Advertising and promotion	1.3
Other (interest, dividends, reinvestment)	2.9
Total	100.0

Form a pie chart to show how the net revenue was expended.

2.24. A coffee machine dispenses coffee in 8-ounce cups. The machine is supposed to fill the cups with 7.75 ounces of liquid. In 25 recent purchases from the machine, the amounts of the fills were:

7.2 7.2 7.6 7.2 7.0 7.7 7.3
7.2 7.6 7.3 7.0 7.8 7.6 7.6
7.1 7.2 7.5 7.4 7.2 7.8 7.1
7.4 7.2 7.5 7.4

a. Calculate the mean, mode, median, and standard deviation for these data.

b. Construct a frequency distribution using the following five classes: 7.00–7.19, 7.20–7.39, 7.40–7.59, 7.60–7.79, and 7.80–7.99.

c. Use the frequency distribution to approximate the mean, median, and standard deviation. Compare these answers with those calculated in part *a*.

2.25. The Painful Dentist Clinic, Inc., has a receptionist who makes appointments by telephone for patients who wish to see one of the four dentists in the clinic. If the phone is busy, the computerized system automatically answers up to seven calls, putting the callers on hold and providing them with musical entertainment. The computer also records how long each person must wait before being served by the receptionist (or before hanging up in disgust!). The average waiting time for service when the system is busy is 2.66 minutes with a standard deviation of 1.82 minutes.

a. Use the Chebychef theorem to determine an interval that contains at least 90 percent of the waiting times.

b. Since no waiting times can be negative, what problem does this present, if any, with the interpretation of the Chebychef interval in part *a?* Explain.

2.26. Most income measures are skewed to the right, which requires the use of the median as a representative measure rather than the mean in many instances. In a study of union salaries for garment workers in California, the average salary is found to be $18,642 with a standard deviation of $3,613. The median salary is $16,021.

  a. Calculate the skewness measure for these data. Is the distribution skewed, and if so, in which direction?

  b. Use the Chebychef theorem to explain the variability in these union salaries.

2.27. The average net profit on weekly sales invoices at the Maui Gift Shop is $11.62 and the standard deviation of the sales invoice profits is $6.81. Use the Chebychef theorem to describe the variability in the sales invoice net profits.

Introduction to probability

3

Probability plays an integral role in inferential statistics by building a "bridge" between the population and the sample taken from it. Our initial applications of probability in this connection will be to make deductions about a sample from a known population, whereas in Chapter 9 the "traffic" will flow in the opposite direction on this bridge. Thus, until Chapter 9, the use of probability is as indicated in Figure 3.1: probability reasons from the population to the sample, while statistical inferences are drawn about the population from the sample.

As an example of the use of probability in this context, consider the next national election for the office of president of the United States. Let us suppose that only two candidates are listed on the ballot for the presidency, the Democratic candidate (A) and the Republican candidate (B). Furthermore, suppose it is known in the population of registered persons who will vote on election day that 60 percent will vote for A and 40 percent will vote for B. If we now randomly sample one person from this population, what is the probability that he or she will vote for A? Since we know that 60 percent of the persons will vote for A, the probability that the one sampled person will vote for A is 0.60. Knowledge of the probabilities of the two possible outcomes of the experiment (voting for A or voting for B) enables us to deduce the probability of the outcome in our sample of one.

Indeed, by using this knowledge, we could deduce the probabilities of zero, one, or two persons voting for A in a sample of two, and so on for larger sample sizes. Thus, if the population is known in the sense that the prob-

FIGURE 3.1   Role of probability in the statistical inference process

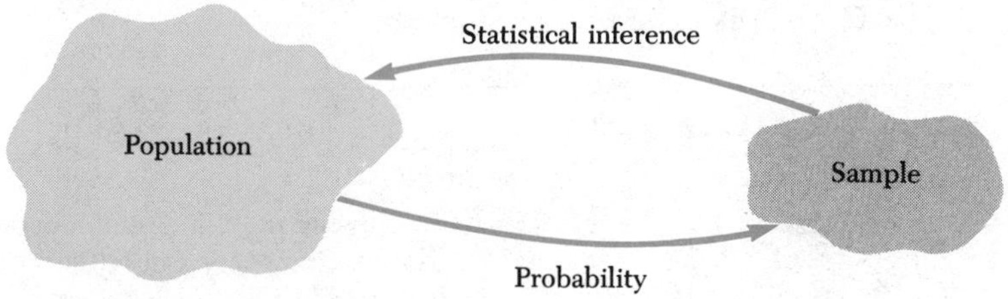

abilities associated with the values in it are known, then this knowledge can be used to deduce the probabilities of the outcomes in the sample.

To illustrate the use of probability in making inferences from a sample to the population, suppose candidate A conjectures *before* the election that 60 percent of the people in the voting population will vote for him. In order to check this conjecture, his campaign manager randomly samples ten individuals from the population and finds that all ten intend to vote for candidate B. If the probability of a randomly chosen person voting for A is really 0.60, it is extremely unlikely that ten randomly chosen persons would all vote for candidate B. It is more likely that the true percentage who will vote for A is considerably less than 60 percent. Hence, knowledge of this experiment (sample outcome) indicates to A that more resources (campaigning, etc.) may have to be used if he is to have a chance of winning the election. Candidate A is interested, of course, in testing whether this sample is indicative of the population characteristics (voting patterns), whether more sample information should be obtained, or whether the election is likely to go to B (in which case A would be wasting his time by campaigning further).

In practical situations, probability is used as a vehicle in drawing inferences about unknown population characteristics. Additionally, as we shall see later, probability concepts can be used to indicate how good these inferences are.

In this chapter, we will assume the population is known and compute the probability of the occurrence of various sample outcomes. In effect, we will be selecting a probability model depicting the outcomes in the population. In practical applications of statistics, we shall see that the selection of this model is an integral part of the statistical inference process.

▪ 3.2   THE SAMPLE AND EVENT SPACES

In the presidential election example discussed in the previous section, we defined a population consisting of registered persons who will vote on election day. Suppose we assign a 1 to an individual if he or she intends to vote for candidate A, and 0 if he or she intends to vote for candidate B. The population can then be thought of as a collection of 1s and 0s. How are these 1s and 0s generated? Each person in the population must be contacted and represented by a 1 or a 0. The process of contacting each person to determine the outcome (1 or 0) is called an *experiment*.

> **Definition 3.1**
> Experiment
> An *experiment* is a process that results in one and only one outcome of a set of disjoint outcomes, where the outcomes cannot be predicted with certainty.

In the voting example, there are only two possible outcomes, and they are *disjoint* (nonoverlapping): a 0 and a 1. With our previous assumption that only two candidates are listed on the ballot, each experiment results in one and only one of the two possible experimental outcomes. And we cannot predict with certainty the outcome before the experiment is conducted. Repeated trials of this experiment will generate the population of 0s and 1s. Other examples of experiments are:

1. A professor at a large university is selected and his salary is recorded.
2. A unit of a product is selected from an assembly line and is analyzed to determine whether it is defective.
3. A light bulb is randomly selected from the day's production and its time to failure measured.

By repeating an experiment many times, a population of outcomes can be generated. For example, if we repeated the experiment in 3 until each and every light bulb in the day's production run had been tested to failure, the population of all times to failure of this set of light bulbs would have been generated. In the process of doing this, however, the entire day's production of light bulbs (the population) would have been destroyed! We can also think of the sample being generated by repeated trials of an experiment. For example, if we wanted to sample ten light bulbs, we could repeat the experiment ten times.

> **Definition 3.2**
> Elementary outcome
> Each distinct outcome of an experiment is called an *elementary outcome* of the experiment.

The elementary outcomes of an experiment are sometimes referred to as the *simple events* of the experiment. We will, however, refer to the distinct outcomes of an experiment as the *elementary outcomes* throughout this chapter. The elementary outcomes of an experiment will be denoted by the capital letter E. A subscript is added to associate it with a distinct outcome of the experiment.

Example 3.1   A dice game is played by tossing a pair of dice and totaling the numbers of dots on the top two faces. There are 36 elementary outcomes when the pair of dice is tossed, and the numbers of dots appearing on the top two faces are recorded. The experiment in this case is the process of tossing the dice and recording the number of dots on the two top faces of the pair. The 36 elementary outcomes for this experiment are illustrated in Figure 3.2.

FIGURE 3.2   The 36 elementary outcomes E_{11}, E_{12}, . . . , E_{66} corresponding to the dice experiment

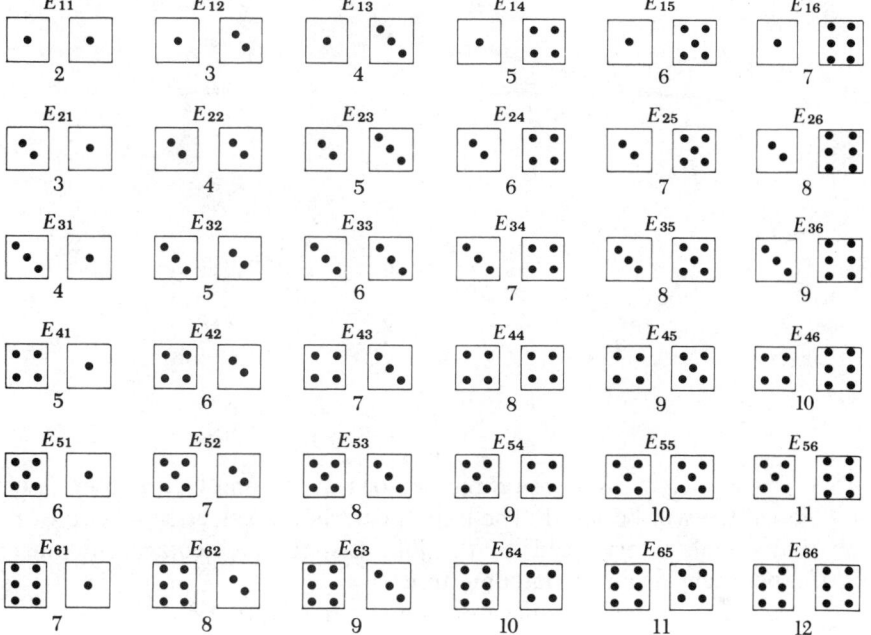

The elementary outcomes of the dice game experiment are denoted by the double-subscripted term E_{ij}, where i gives the number of dots on the top face of the first die and j gives the number of dots on the top face of the second die. Thus, the elementary outcomes of this experiment are labeled E_{11}, E_{12}, . . . , E_{66}; 36 elementary outcomes in all.

The 36 elementary outcomes of this experiment can be generated by an *outcome tree,* as illustrated in Figure 3.3. The first set of branches in the tree represents the possible outcomes for the first die. The second set of branches in the tree represents the possible outcomes for the second die. Thus, there are $(6)(6) = 36$ elementary outcomes of this experiment.

The outcome tree is a logical way to list the elementary outcomes of an experiment. It is very practical and efficient if the number of elementary outcomes is not too large.

FIGURE 3.3 Outcome tree for the dice experiment in Example 3.1

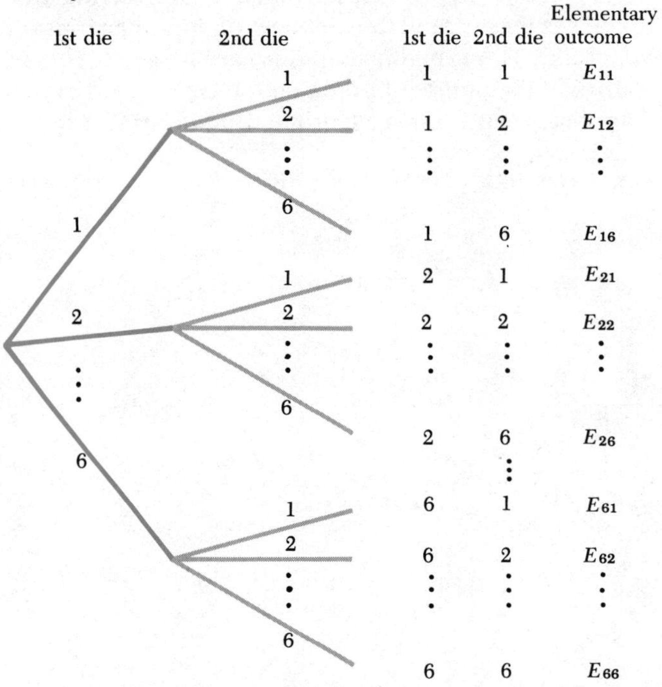

1st die	2nd die	1st die	2nd die	Elementary outcome
	1	1	1	E_{11}
	2	1	2	E_{12}
	⋮	⋮	⋮	⋮
	6	1	6	E_{16}
1	1	2	1	E_{21}
2	2	2	2	E_{22}
	⋮	⋮	⋮	⋮
⋮	6	2	6	E_{26}
6				⋮
	1	6	1	E_{61}
	2	6	2	E_{62}
	⋮	⋮	⋮	⋮
	6	6	6	E_{66}

Example 3.2 Suppose three persons, A, B, and C, are interviewing for a job. Two will be hired. The experiment is the selection of two of the three individuals interviewed for the job. List the elementary outcomes of this experiment using an outcome tree.

Solution The outcome tree is given in Figure 3.4.

FIGURE 3.4 Outcome tree for Example 3.2

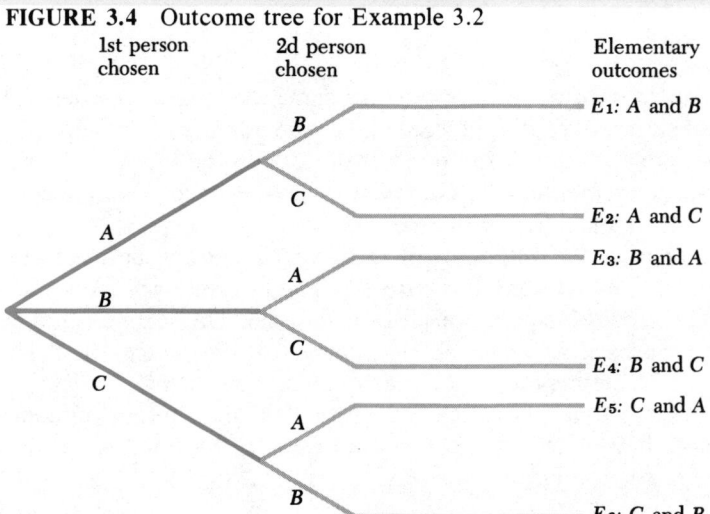

1st person chosen	2d person chosen	Elementary outcomes
A	B	E_1: A and B
A	C	E_2: A and C
B	A	E_3: B and A
B	C	E_4: B and C
C	A	E_5: C and A
C	B	E_6: C and B

Notice in Example 3.2 that the six elementary outcomes listed specify not only the two individuals selected, but also the *order* in which they are selected. If the order in which the two individuals are selected is not important, then we need not distinguish between E_1 and E_3, E_2 and E_5, and E_4 and E_6. In this case, a "simpler" set of outcomes would be:

E_1^*:   A and B are selected[1]
E_2^*:   A and C are selected
E_3^*:   B and C are selected

As suggested, it is often possible to define the elementary outcomes differently in the same experiment. To gain a better understanding of how to define the elementary outcomes of an experiment, consider the following example.

Example 3.3   Assume that Herman is to toss a "fair coin" twice. He informs you that there are three possible elementary outcomes associated with this experiment:

E_1^*:   Two heads   (no tails)
E_2^*:   One head   (one tail)
E_3^*:   No heads   (two tails)

He tells you that the probability of any one of the three elementary outcomes occurring in a single trial of the experiment is ⅓. He then wishes to wager with you that one of the elementary outcomes, say E_2^*, will occur in a single trial of the experiment. Construct an outcome tree for this experiment.

Solution   The outcome tree for this experiment is given in Figure 3.5.

FIGURE 3.5   Outcome tree for a coin-tossing experiment

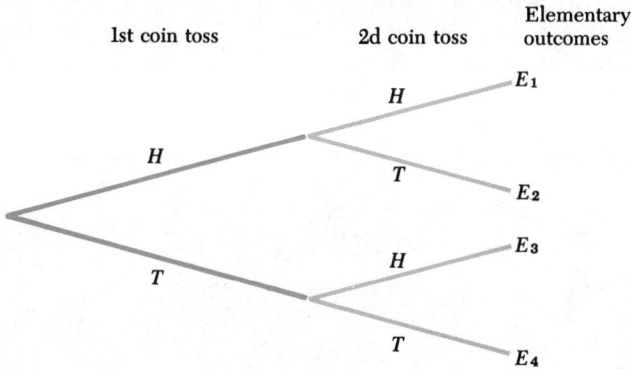

From the outcome tree, it is clear that it is possible to define another set of elementary outcomes for this experiment:

E_1: (H,H) E_3: (T,H)
E_2: (H,T) E_4: (T,T)

---

[1] The asterisk(*) differentiates between two sets of elementary outcomes, E_i and E_i^*, defined for the same experiment.

If the coin is "fair," then the probability of each of the elementary outcomes E_1, E_2, E_3, and E_4 in Figure 3.5 occurring is ¼.

In terms of the original three elementary outcomes, E_1^*, E_2^*, and E_3^*, it is clear that each does not have a ⅓ probability of occurring—the proper probabilities are:

$$E_1^*: \quad ¼ \quad [E_1^* = E_1 (H,H)]$$
$$E_2^*: \quad ²/4 \quad [E_2^* = E_2 (H,T) \text{ or } E_3 (T,H)]$$
$$E_3^*: \quad ¼ \quad [E_3^* = E_4 (T,T)]$$

You can use this new information to accept or reject the wager. For example, if Herman wagers that a single head will appear with probability ⅓ in each trial of the experiment, you know it will not be so from your outcome tree. He might suggest you pay him $1 if two heads ($E_1^*$) or two tails ($E_3^*$) occur in a trial of the experiment, and he will pay you $2 if one head and one tail (E_2^*) occurs. You certainly would be advised to accept his request to participate in this "game"—if the experiment is repeated many times, you can expect to receive $2 in 50 percent of the trials and lose $1 in the remaining 50 percent of the trials. For instance, if the experiment is repeated 100 times, you might win about $100 while losing only $50.

Example 3.3 suggests that there is more than one way to define a set of elementary outcomes of an experiment. Indeed, this is the case in most experiments. In Section 3.3, we will learn that probability questions formulated about an experiment can be correctly answered using two (or more) elementary outcome definition sets, provided that the probabilities have been correctly associated with the elementary outcomes of the experiment. In Example 3.3, if the probabilities of ¼, ½, and ¼ had been associated with the elementary outcomes E_1^*, E_2^*, and E_3^*, respectively, then Herman would not have proposed playing the game—he would have reached the same conclusion we did from the set of elementary outcomes E_1, E_2, E_3, and E_4.

After the elementary outcomes of an experiment have been defined, collections of the elementary outcomes, called *events,* may be defined.

Definition 3.3
Event
An *event* is a collection of one or more elementary outcomes of an experiment.

In Example 3.1, we might define the event A to be "a total of seven dots appear on the top two faces of the pair of dice." The event A is composed of the elementary outcomes E_{16}, E_{25}, E_{34}, E_{43}, E_{52}, and E_{61}. In each of these elementary outcomes, the total of the dots on the top faces of the dice is 7. Note that *each* of these elementary outcomes could be an event as well. For example, we may define the event F to be "a total of two dots on the top two faces of the dice." Event F is composed of the single elementary outcome E_{11}. Hence, E_{11} is both an elementary outcome and an event in this case.

Definition 3.4
Null event

A *null event* is an event that contains no elementary outcomes of the experiment. It is denoted by ϕ.

In Example 3.1, an example of a null event is "a total of one dot on the two top faces of the dice." It is impossible for this event to happen, because at least one dot appears on the top face of each die. In this instance, the event set is empty; it does not contain any of the elementary outcomes of the experiment.

The elementary outcomes of an experiment and events defined to be collections of one or more of these elementary outcomes can be illustrated graphically by a *Venn diagram*. The Venn diagram associated with the elementary outcomes in Example 3.1 is shown in Figure 3.6. Each elementary outcome in a Venn diagram is shown as a "point" with its corresponding

FIGURE 3.6 Venn diagram for the elementary outcomes in Example 3.1

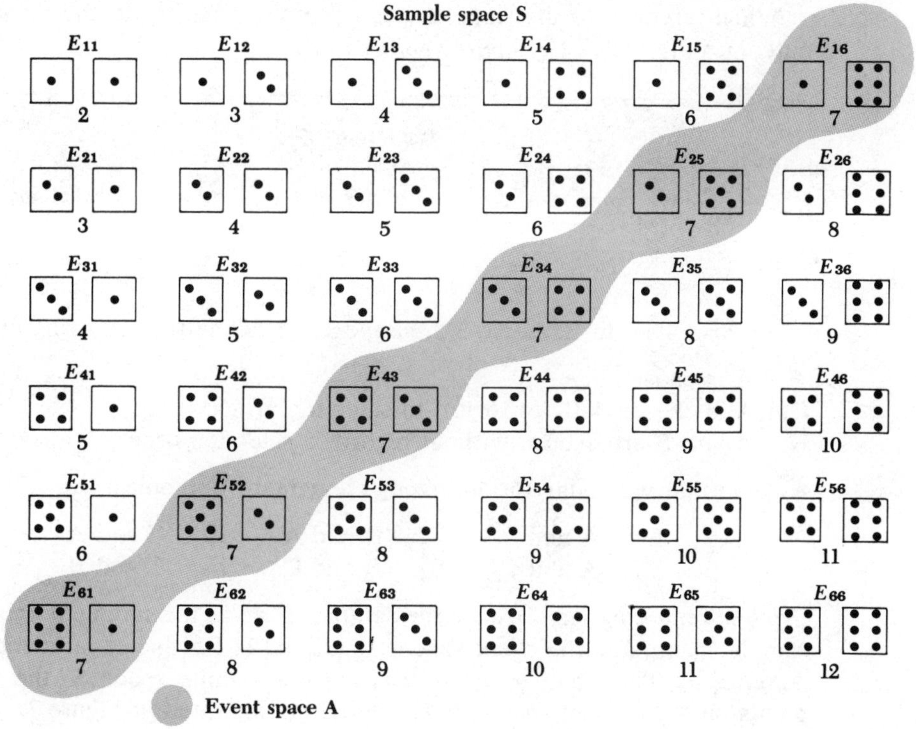

subscripted letter E. The collection of *all* elementary outcomes in an experiment is called the *sample space of the experiment*. The sample space of an experiment is denoted by S and is shown in the Venn diagram as the collection of all elementary outcomes. The event A, "a total of seven dots appearing on the top two faces of the dice," is illustrated in the Venn diagram by enclosing the elementary outcomes belonging to it as illustrated in Figure 3.6. The resulting enclosed region is called the *event space*, A.

Definition 3.5
Sample space

A *sample space* of an experiment is the collection of *all* elementary outcomes of the experiment.

Definition 3.6
Event space

An *event space* of an experiment is the collection of the elementary outcomes constituting the event.

When referring to the sample space S of an experiment, the elementary outcomes are called the *sample points* in S.

Definition 3.7
Sample point

A *sample point* in a sample space S of an experiment is an elementary outcome of the experiment.

Example 3.4 In Example 3.2, suppose we define the following events:

F: A is chosen first
G: A is chosen without regard to selection order
H: A and B are chosen without regard to selection order.

Assume that we define the following elementary outcomes:

E_1: A and B E_3: B and A E_5: C and A
E_2: A and C E_4: B and C E_6: C and B

Draw a Venn diagram showing the sample space S, the sample points in the sample space, and the event spaces F, G, and H for this experiment.

Solution The Venn diagram showing the sample space S, the sample points, and the event spaces F, G, and H is illustrated in Figure 3.7.

FIGURE 3.7   Venn diagram for the events defined in Example 3.4

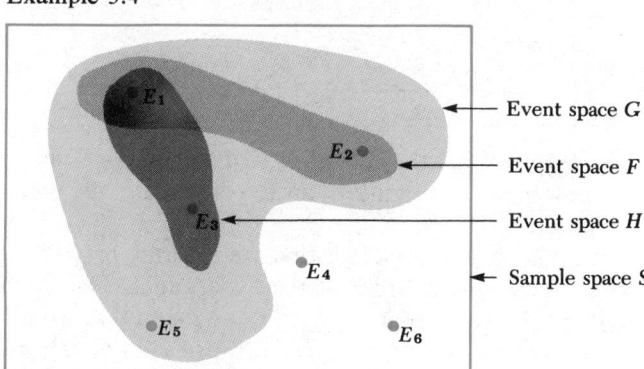

- Event space G
- Event space F
- Event space H
- Sample space S

■ 3.3   COMPUTING PROBABILITIES FROM THE SAMPLE SPACE

Suppose an event A is defined over a sample space S. To determine the probability that event A occurs in a single trial of the experiment, it is necessary to know the probability of each and every sample point in the sample space S occurring. Therefore, to answer the question, "What is the probability that an event, say event A, occurs in the experiment?", we first must assign probabilities to the elementary outcomes of the experiment (or equivalently, to the sample points in the sample space S of the experiment). We will denote by $P(E_i)$ the probability assigned to the elementary outcome E_i of an experiment. The assigned probabilities $P(E_1)$, $P(E_2)$, . . . , $P(E_N)$, where there are N elementary outcomes or sample points in the sample space S, must satisfy three probability axioms for experiments that have a finite number of outcomes.

Axiom 1
$$0 \leq P(E_i) \leq 1 \quad \text{for} \quad i = 1, 2, \ldots, N$$

The first axiom requires that every elementary outcome be assigned as its probability a nonnegative number between 0 and 1, inclusive.

Axiom 2
$$\sum_{i=1}^{N} P(E_i) = 1$$

The second axiom requires that the probabilities assigned to *all* the elementary outcomes in the experiment must total one.

> **Axiom 3**
> $P(E_1 \text{ or } E_2 \text{ or } E_3 \text{ or } \ldots) = P(E_1) + P(E_2) + P(E_3) + \cdots$

The third axiom requires that the probability of one or more members of a set of elementary outcomes occurring in an experiment is the sum of their respective probabilities.

Example 3.5 Suppose in Example 3.1 that we assign a probability of $1/36$ to each of the 36 elementary outcomes of the experiment. This would seem to be reasonable if the dice are "fair." By fair, we mean that each die, as a six-sided cube, will have any one of its faces upward with equal probability on each toss. If this is so for each die, then each of the 36 elementary outcomes in tossing the pair of dice should be equally likely. Thus, the assignment of a probability of $1/36$ to each elementary outcome is established.

Elementary outcome	Definition	Probability
E_{ij}	i dots on the top face of the first die j dots on the top face of the second die $i = 1, 2, 3, 4, 5, 6; \quad j = 1, 2, 3, 4, 5, 6$	$P(E_{ij}) = 1/36$

These probability assignments to the 36 elementary outcomes satisfy the three axioms of probability: (1) each probability is between 0 and 1, inclusive; (2) the probabilities sum to 1; and (3) the probability of one or more members of a set of elementary outcomes occurring in an experiment is the sum of their respective probabilities.

The satisfaction of the third axiom by these probability assignments may be argued as follows. By the definition of the elementary outcomes (E_{ij} values), it is obvious that one and only one of the elementary outcomes can occur in any single trial of the experiment. Thus, the probability of, say, either E_{11} or E_{12} occurring in a single trial of the experiment is simply the sum of the probability that E_{11} occurs and the probability that E_{12} occurs.

These axioms are intuitive; most of us understand them before taking any formal training in probability. If an event is certain to happen, its probability of occurrence is 1, and if an event is certain not to happen, its probability of occurrence is 0.

How do we, in fact, formally define probability? We will consider three ways.

> **Definition 3.8**
> Relative frequency definition of probability
>
> If an event E is defined in an experiment, the experiment is repeated a very large number of times, say, N, and the event E is observed to occur in n of these N experimental trials, then
>
> $$P(E) = \frac{n}{N}$$

The ratio n/N represents the proportion of the time that event E occurs in repeated experiments. Suppose we wish to assign a probability to the elementary outcome E_1: a head, in a coin-tossing experiment that consists of tossing a coin and observing whether it lands heads or tails. If we toss the coin a large number of times, say 10,000, and if the coin is fair, we would expect the proportion of heads that occurs in the 10,000 trials to be about ½. The larger the number of trials, the closer we should expect the proportion of heads to approach ½. The relative frequency definition of probability defines probability in a limiting sense; it is the limit of the ratio n/N as the total number of trials N approaches infinity.

> **Definition 3.9**
> Deductive logic definition of probability
>
> The probability of an event E is determined logically from symmetry or geometric considerations associated with the experiment.

An example of applying Definition 3.9 is the assignment of probability ½ to the elementary outcome E: a head, in a coin-tossing experiment, on the basis of the symmetry of a fair coin. Another example is the assignment of probability $1/36$ to each elementary outcome in the dice game in Example 3.5.

> **Definition 3.10**
> Subjective definition of probability
>
> The probability of an event E is determined subjectively, reflecting a person's "degree of belief" that an event will occur.

For example, a person who is knowledgeable about an experiment and its outcomes subjectively assigns probabilities to elementary outcomes. An example is football game prognostication. Let us suppose that Penn State is playing Texas in the "all-world college football bowl" on January 1. Before the game, "Jimmy the Greek," a well-known football game prognosticator, may quote a probability of $^1/_{10}$ that Penn State will win the game. The two possible elementary outcomes of the experiment are: E_1—Penn State wins and E_2—Penn State loses (we exclude the possibility of a tie here). In this example, the probabilities have been assigned subjectively as $P(E_1) = {}^1/_{10}$ and $P(E_2) = {}^9/_{10}$.

When probabilities are assessed in ways that are consistent with the relative frequency or the symmetry determination of probability, we call them *objective* probabilities. Objective and subjective probabilities are fundamentally different. If we asked a number of persons to determine the probability of an event objectively, they would all arrive at the same answer, provided they did the calculations properly. But, if we asked them to determine the probability subjectively, each individual would arrive at his or her own answer. As a consequence, not all probability theory and methods that can be applied to objective probabilities can be applied to subjective ones.

In this chapter, we shall assume that the probabilities of the elementary outcomes in an experiment have been determined objectively. In Chapter 4, we shall again discuss how one assigns objective probabilities to events. Subjective probabilities will be considered in Chapters 21, 22, and 23.

Now we will consider methods to determine the probability that an event occurs in an experiment.

Definition 3.11
Event occurrence

An event, defined over a sample space S in an experiment, *occurs* if one of the sample points in its event space occurs.

Once the probabilities have been assigned to the elementary outcomes of an experiment, we are able to compute the probabilities of events occurring that are defined over the sample space for the experiment by using the following theorem.

Theorem 3.1

Suppose event A is an event defined over a sample space. The probability of event A occurring is equal to the sum of the probabilities of the elementary outcomes comprising event A.

The following three examples illustrate the use of the sample space–event space framework for answering probability questions.

Example 3.6   In the dice game introduced in Example 3.1, suppose we have defined the following events:

A: sum of the dots appearing on the top two faces of the dice is seven
B: sum of the dots appearing on the top two faces of the dice is three or less

What are the probabilities of events A and B occurring?

Solution   The Venn diagram in Figure 3.8 shows the two event spaces, A and B. The elementary outcomes comprising event A are: E_{16}, E_{25}, E_{34}, E_{43}, E_{52}, and E_{61}. Thus,

$$P(A) = P(E_{16}) + P(E_{25}) + P(E_{34}) + P(E_{43}) + P(E_{52}) + P(E_{61})$$
$$= {}^1/_{36} + {}^1/_{36} + {}^1/_{36} + {}^1/_{36} + {}^1/_{36} + {}^1/_{36} = {}^6/_{36} = {}^1/_6$$

The elementary outcomes comprising event B are: E_{11}, E_{12}, and E_{21}. Thus,

$$P(B) = P(E_{11}) + P(E_{12}) + P(E_{21}) = {}^1/_{36} + {}^1/_{36} + {}^1/_{36} = {}^3/_{36} = {}^1/_{12}$$

FIGURE 3.8   Venn diagram of the events in Example 3.6

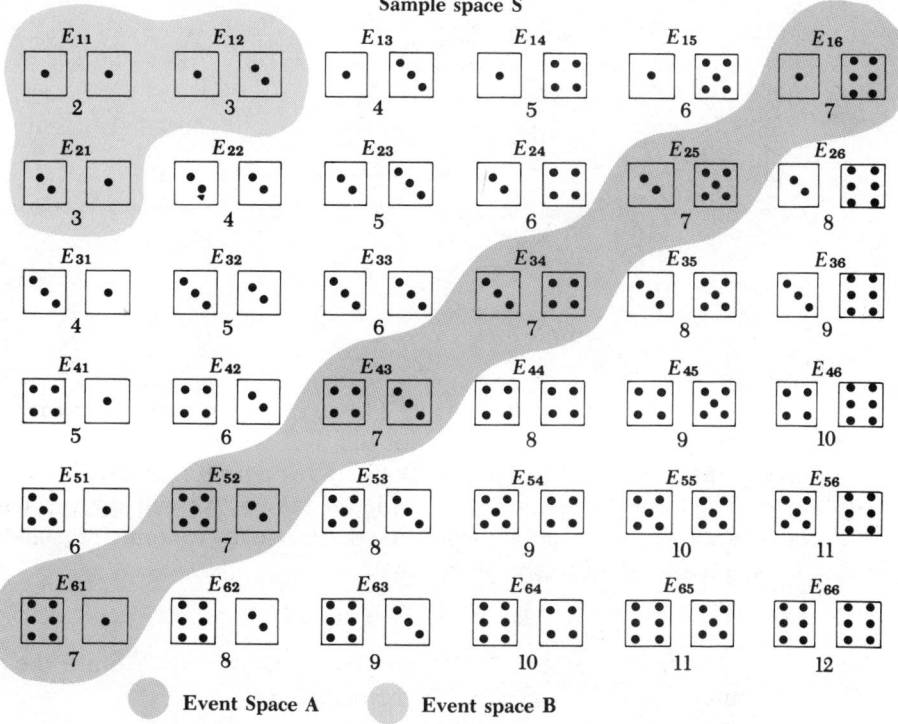

Event Space A   Event space B

Example 3.7   A construction manager is in charge of preparing bids on construction jobs for submission to a county commission, which awards contracts for these jobs. Let us suppose that during the current time period, there are three construction jobs available in the county for which the manager wishes to bid. What is the probability that the manager will win at least two of the three submitted bids if he has a 50 percent chance of winning each of the three bids?

Solution   The experiment consists of placing bids on three construction jobs. The elementary outcomes of this experiment are generated by using the outcome tree illustrated in Figure 3.9. Thus, there are eight points in the

FIGURE 3.9   Outcome tree for the experiment in Example 3.7

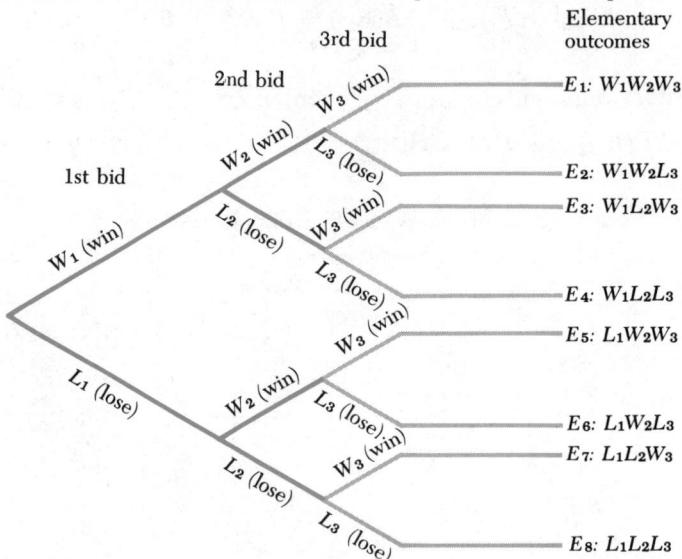

sample space for this experiment. The assumption that the manager has a probability of ½ of winning each of the bids will result in each of the eight elementary outcomes being equally likely to occur if the outcome of any one bid will not affect the outcomes of the other two. We will show in Chapter 4 how to arrive at this conclusion. For the moment, it seems reasonable, and we shall assume in this problem that

$$P(E_i) = \tfrac{1}{8}, \qquad i = 1, 2, \ldots, 8$$

Now, define the event A, "at least two bids are won." Since the elementary outcomes E_1, E_2, E_3, and E_5 belong to A,

$$P(A) = P(E_1) + P(E_2) + P(E_3) + P(E_5) = \tfrac{1}{8} + \tfrac{1}{8} + \tfrac{1}{8} + \tfrac{1}{8} = \tfrac{1}{2}$$

Example 3.8   Consider the problem introduced in Example 3.2: there are three persons, A, B, and C, interviewing for a job and two will be hired. If the recruiter views all three candidates as being equally qualified, what is the probability that A is among the two hired?

Solution   The elementary outcomes of the experiment are:

$$E_1: \quad \text{A and B are selected}$$
$$E_2: \quad \text{A and C are selected}$$
$$E_3: \quad \text{B and C are selected}$$

Since the recruiter considers all three to be equally qualified, each pair is equally likely to be chosen. Thus, $P(E_i) = \frac{1}{3}, i = 1, 2, 3$. Let the event D be "A is among the two hired." Since D is composed of E_1 and E_2,

$$P(D) = P(E_1) + P(E_2) = \frac{1}{3} + \frac{1}{3} = \frac{2}{3}$$

The rule for computing probabilities using the sample space–event space method can be modified if each sample point in the sample space is equally likely to occur.

Theorem 3.2

If each of the elementary outcomes in an experiment has the same probability of occurring, and A is an event in the sample space, then

$$P(A) = \frac{n}{N} \qquad \text{where } N = \text{number of points in sample space S}$$
$$n = \text{number of points in event space A}$$

Example 3.9   The sample space S for the dice game introduced in Example 3.1 together with the event space for the event A, "a total of seven dots on the top two faces of the dice," are shown in Figure 3.6. In Example 3.5, a probability of $\frac{1}{36}$ was assigned to each of the sample points (E_{ij}) in the sample space S. What is the probability of event A, "a total of seven dots appear on the top two faces," occurring?

Solution   Since each of the sample points in the sample space has the same probability of occurring, the probability that event A occurs in a single trial of the experiment is given by:

$$P(A) = \frac{\text{Number of points in event space A}}{\text{Number of points in sample space S}}$$

The number of points in the sample space S is 36 and the number of points in the event space A is 6 (E_{16}, E_{25}, E_{34}, E_{43}, E_{52}, and E_{61}). Thus, $P(A) = \frac{6}{36} = \frac{1}{6}$. The probability of tossing a 7 (the dots total seven on the two top faces) in a single roll of the dice is $\frac{1}{6}$. Therefore, rather than summing the probabilities of the elementary outcomes comprising A, we can simply *count* the number of sample points in the event space A and in the sample space S to compute $P(A)$.

Since in Examples 3.6–3.8, the elementary outcomes have equal probabilities assigned to them, Theorem 3.2 could have been used to answer the probability questions. In most problems, if Theorem 3.2 can be applied, it will usually be easier than summing probabilities of elementary outcomes to find the probability of an event occurring in a sample space. We will return to this point later, but first let's see how a sample space can be formed to take advantage of Theorem 3.2 in a probability problem.

Example 3.10   Herman purchases a new four-cylinder compact automobile. After one week, the car begins to run poorly. It is difficult to start, and the engine misses and quits at stoplights. Herman suspects that something is wrong. Indeed, unknown to Herman, two of the four spark plugs are defective. Herman takes the car to the garage and issues a complaint. On the basis of the symptoms Herman describes, the mechanic immediately suspects spark plug difficulties. He randomly selects two of the four spark plugs and pulls them out to be checked. What is the probability that he selected the two defective spark plugs?

Solution   A possible sample space is:

E_1^*:   Two defective spark plugs are selected
E_2^*:   One defective and one good spark plug are selected
E_3^*:   Two good spark plugs are selected

These three elementary outcomes are not equally likely to occur. There are more ways that E_2^* can occur than E_1^*. To see this, label the good spark plugs as G_1 and G_2 and the defective spark plugs as D_1 and D_2. The possible extracted pairs are:

$$E_1: \quad G_1G_2 \qquad E_3: \quad G_1D_2 \qquad E_5: \quad G_2D_2$$
$$E_2: \quad G_1D_1 \qquad E_4: \quad G_2D_1 \qquad E_6: \quad D_1D_2$$

These pairs are equally likely to occur, because the mechanic is selecting two spark plugs at random. Let A be the event "the two defectives are selected." Since $P(E_i) = 1/6$, $i = 1, \ldots, 6$, Theorem 3.2 may be applied:

$$P(A) = \frac{1}{6}$$

The sample space comprised of the points E_1^*, E_2^*, and E_3^* certainly could have been used to solve this problem. In this space, $P(E_1^*) = 1/6$, $P(E_2^*) = 4/6$, and $P(E_3^*) = 1/6$. Thus, by using Theorem 3.1,

$$P(A) = P(E_1^*) = \frac{1}{6}$$

It would have been very easy, for example, to mistakenly assign equal probabilities of $1/3$ to E_1^*, E_2^*, and E_3^*. That is, if Theorem 3.2 had been applied to the sample space composed of E_1^*, E_2^*, and E_3^*, the probability assigned to event A would have incorrectly been $1/3$—1 [the number of points in the event space (E_1^*)] divided by 3 [the number of points in the sample space (E_1^*, E_2^*, and E_3^*)]. Note the similarity between this problem and Example 3.3.

The reason for trying to construct a sample space so that the sample points have equal probabilities assigned to them becomes more pronounced as the number of sample points N in S increases. It is not uncommon for a sample space to contain thousands of points. Identifying them and adding the probabilities of a subset of the points comprising an event become rather troublesome and time-consuming. In the next section, we shall present various counting rules that enable us to compute probabilities using Theorem 3.2 when it becomes difficult to list the entire set of elementary outcomes corresponding to an experiment due to the sheer number of them.

■ 3.4 PERMUTATIONS, COMBINATIONS, AND OTHER COUNTING RULES

Numerous counting rules can be used to count the number of points in sample and event spaces. We shall consider four of the most important counting rules. Each will be presented without proof and followed by examples.

Rule 3.1
Product rule

Suppose there are m groups of objects. The first group contains k_1 distinguishable objects, the second group contains k_2 distinguishable objects, and so forth, with the mth and last group containing k_m distinguishable objects. If the experiment is to draw *one* object from *each* of the m groups, then there are $(k_1)(k_2) \cdots (k_m)$ elementary outcomes in the experiment.

Example 3.11   Herman has decided to purchase a new hi-fi system with the money he saved by buying a compact car instead of a large sedan. His hi-fi system will be composed of a receiver, a pair of speakers, a record changer, and a tape deck. In the store where he will make the purchase, there are ten different kinds of receivers, five kinds of speakers, four kinds of changers, and eight kinds of tape decks. How many systems can Herman choose from? Assume that all combinations are compatible.

Solution   There are $m = 4$ groups (types of equipment) with $k_1 = 10$ (receivers), $k_2 = 5$ (speakers), $k_3 = 4$ (changers), and $k_4 = 8$ (tape decks). Since he must select one object from each group, he can choose from $(10)(5)(4)(8) = 1,600$ possible systems.

The next two counting rules apply to a different type of experiment as indicated in the following example.

Example 3.12   Suppose three persons, A, B, and C, are competing for two job positions. How many ways can two people be selected for employment from the three?

Solution   In this problem, it is easy to list the possible outcomes of the experiment. There are three: AB, AC, and BC. However, we may be interested in the order of the selection as well as the content of the resulting pairs. If this is the case, there are six possible outcomes: AB, BA, AC, CA, BC, and CB.

In Example 3.12, there are two different ways to view the pairs formed by selecting two persons out of three. Suppose there are n distinguishable objects from which we are selecting a subset of size r. If we are concerned about the number of groups of size r that can be formed from the n where one group of size r is different from another if its content is different, then we want to determine the number of *combinations* of r things selected from n. If, on the other hand, we want to compute the number of groups of size r that can be formed from the n where one group of size r is different from another in terms of *both* its content and the order in which the r things were drawn, then we want to determine the number of *permutations* of r things drawn from n.

In Example 3.12, the number of combinations of two persons drawn from three is three, while the number of permutations of two persons drawn from three is six.

In conjunction with the rules for computing the number of permutations and combinations, the complete definitions follow.

Definition 3.12
Permutation

An ordered arrangement of r distinguishable objects is called a *permutation*. The number of permutations of r objects taken from n distinguishable objects is denoted by P_r^n.

Rule 3.2
Permutations

$$P_r^n = \frac{n!}{(n - r)!}$$

Definition 3.13
Combination

A set of r distinguishable objects is called a *combination*. The number of combinations of r objects taken from n distinguishable objects is denoted by C_r^n.

Rule 3.3
Combinations

$$C_r^n = \frac{n!}{r!(n - r)!} = \frac{1}{r!} P_r^n$$

The symbol $n!$ is called "n factorial"; $n! = n(n - 1)(n - 2) \cdots (2)(1)$. Thus, $4! = (4)(3)(2)(1) = 24$ and $6! = (6)(5)(4)(3)(2)(1) = 720$; $1! = 1$, and, by definition, $0! = 1$.

Example 3.13   A committee of three is to monitor the activities of the local lonely hearts club. The committee is to be formed by selecting three people from a group of five persons. How many different committees could be formed?

Solution   Nothing is mentioned about the order or arrangement of the three selected individuals. Thus, one committee will be different from another if it has one or more different people in it. We are only concerned about the *content* of each group, and therefore we want to determine the number of combinations of three things taken from five.

$$C_3^5 = \frac{5!}{3!(5-3)!} = \frac{5 \cdot 4 \cdot 3 \cdot 2 \cdot 1}{(3 \cdot 2 \cdot 1)(2 \cdot 1)} = 10$$

Thus, it is possible to develop ten different committees of three people selected from five. You can check this result by listing all possible groups of three drawn from five, where the five people are labeled A, B, C, D, and E.

Example 3.14   The lonely hearts club committee of three is to be formed by selecting three people from a group of five. One of the selected people will be chosen as chairman of the committee, another secretary, and the third person simply a member of the committee. How many different committees can be formed?

Solution   Suppose the three people (denoted by A, B, and C) have been chosen from among the five. Once we have this combination of three people, we must assign them to the three positions: chairman, secretary, and member. We can view this process as ordering the three persons. That is, let the first position in the ordering (or *permutation*) be the chairman, the second the secretary, and the third the member. The possible permutations are: ABC, ACB, BAC, BCA, CAB, and CBA. Each of these permutations is a different committee, although each contains the same three people, A, B, and C. Thus, we want to compute the number of *permutations of r* = 3 people chosen from *n* = 5:

$$P_3^5 = \frac{5!}{(5-3)!} = \frac{5 \cdot 4 \cdot 3 \cdot 2 \cdot 1}{2 \cdot 1} = 60$$

In this case, 60 different committees can be formed. Notice that this is six times as many committees as could be formed in Example 3.13, because for each combination in Example 3.13, there are six permutations in Example 3.14.

Rule 3.4
Hypergeometric rule

Suppose there are *n* objects in a group, and that n_1 are of one type and n_2 are of another type. The number of groups of *r* objects, where r_1 are of the first type and r_2 are of the second type, that can be formed by drawing *r* objects from the *n* is given by

$$C_{r_1}^{n_1} \cdot C_{r_2}^{n_2}, \text{ where } n_1 + n_2 = n; r_1 + r_2 = r$$

Example 3.15 A college recruiter for the Acme Corporation has set up interviews with ten college seniors. Six students are fraternity members and four are not. The recruiter plans to hire five of the ten interviewed students. How many possible groups of the five individuals hired will contain exactly three fraternity students?

Solution Let $n_1 = 6$ and $n_2 = 4$ in the hypergeometric rule. In the subgroup of size $r = 5$, we want $r_1 = 3$ fraternity students and $r_2 = 2$ nonfraternity students. The number of groups of size 5 with this property are:

$$C_3^6 \cdot C_2^4 = \frac{6!}{3!(6-3)!} \cdot \frac{4!}{2!(4-2)!} = \frac{6 \cdot 5 \cdot 4 \cdot 3 \cdot 2 \cdot 1}{3 \cdot 2 \cdot 1(3 \cdot 2 \cdot 1)} \cdot \frac{4 \cdot 3 \cdot 2 \cdot 1}{2 \cdot 1(2 \cdot 1)}$$
$$= (20)(6) = 120$$

Notice that the hypergeometric rule is "putting together" the product and the combination rules. In Example 3.15, we first compute the number of ways of selecting three fraternity students from six individuals [$C_3^6 = 20$], then we compute the number of ways of selecting two nonfraternity students from four individuals [$C_2^4 = 6$]. A particular set of three fraternity students and two nonfraternity students is formed by selecting one of the 20 groups of three fraternity students and one of the 6 groups of nonfraternity students. Thus, by the product rule, the number of "pairs" of groups is $(20)(6) = 120$. The reader is *not* encouraged to list all 120 elementary outcomes of this experiment!

Now that the rules have been introduced and applied to counting problems, we can use them in conjunction with Theorem 3.2 to solve probability problems using the sample space–event space method. The next three examples will illustrate this use of the counting rules.

Example 3.16 A corporation has expanded its business by opening an office in a new location. The office manager must select a cleaning service to care for the office and a delivery service to make local deliveries. In the Yellow Pages, she finds listed four cleaning services (A, B, C, and D) and three delivery services (I, II, and III). The services in each category charge similar rates, and all receive a good rating by the Better Business Bureau. Therefore, the manager decides to select one cleaning service and one delivery service at random. Unknown to her, cleaning service B and delivery service III are the poorest in each category. What is the probability that she will select the poorest service in each category?

Solution The number of elementary outcomes of the experiment can be determined from the product rule, because one service will be chosen from the first group of four cleaning services and one service will be selected from the second group of three delivery services. The selection is made at random, so each elementary outcome of the experiment will occur with the same probability as any other. Thus, Theorem 3.2 may be used. Let G denote the event "cleaning service B and delivery service III are selected." Then,

$$P(G) = \frac{\text{Number of sample points in event space G}}{\text{Number of sample points in sample space S}} = \frac{n}{N}$$

By using the product rule, the total number of points in the sample space S is $N = (4)(3) = 12$. The number of points in event space G is $n = 1$, since for event G to occur, the pair (B, III) must be chosen. Thus,

$$P(G) = \frac{n}{N} = \frac{1}{12}$$

Example 3.17   Suppose we take a well-shuffled deck of 52 playing cards and deal a five-card hand. That is, five cards are drawn randomly without replacement from the deck. What is the probability that our hand is a straight flush? A *straight flush* is five cards in one suit in order; A, K, Q, J, 10 of spades or 8, 7, 6, 5, 4 of hearts, for example. An ace can be counted either as a low card or as a high card.

Solution   Define A to be the event "a five-card straight flush is dealt." The number of points N in the sample space is $C_5^{52} = 2,598,960$—the total number of possible different five-card hands. The number of points n in the event space can be counted as follows. In any particular suit, there are ten possible straight flushes: (A, K, Q, J, 10), . . . , (6, 5, 4, 3, 2), (5, 4, 3, 2, A)—the ace may be used as the highest card or the lowest. There are four suits (hearts, diamonds, clubs, spades). Thus, there are $n = (4)(10) = 40$ possible straight flushes by using the product rule. Hence,

$$P(A) = \frac{n}{N} = \frac{40}{2,598,960} = 0.000015$$

A straight flush is very unlikely to occur, and a player holding this hand is in an ideal betting situation! The quality of a poker hand is determined by probability. Hand A is "better" than hand B if it is less likely to occur.

Notice that it would be virtually impossible to list all the points in the sample space in this problem. However, the combination rule makes it quite easy to count these points.

Example 3.18   The manager of a large accounting firm decides to send four employees in the tax division to a workshop on "New Ways to Shelter Income from Federal Income Taxes" presented by the local state university. There are ten employees in the tax division, three of whom are women. The manager decides to select the four employees at random from the ten. What is the probability that two of the three women are selected to attend the workshop?

Solution   Each possible sample point (set of four employees) in the experiment has the same probability of occurring as any other sample point.

Thus, we can use Theorem 3.2 to solve this problem. Let the event "two women and two men are selected" be denoted by A. The probability $P(A)$ is given by:

$$P(A) = \frac{\text{Number of points in event space A}}{\text{Number of points in the sample space S}} = \frac{n}{N}$$

Let us first consider the denominator in the probability calculation. The experiment is drawing four employees from the ten in the tax division. The set of elementary outcomes comprises the sample space S. The total number of points in S is the number of combinations of four things taken from ten, because the order in which the four employees are chosen is not important. Thus,

$$N = C_4^{10} = \frac{10 \cdot 9 \cdot 8 \cdot 7 \cdot 6 \cdot 5 \cdot 4 \cdot 3 \cdot 2 \cdot 1}{4 \cdot 3 \cdot 2 \cdot 1(6 \cdot 5 \cdot 4 \cdot 3 \cdot 2 \cdot 1)} = 210$$

Now, how many points are in the event space A? The ten employees are divided into two groups (male and female), so we can use the hypergeometric rule to determine the number of sample points in A. Let n_1 and r_1 denote the number of women in the division and group selected, respectively. Then, $n_1 = 3$, $r_1 = 2$, and $n_2 = 7$, $r_2 = 2$. Thus,

$$n = C_2^3 C_2^7 = \left[\frac{3 \cdot 2 \cdot 1}{2 \cdot 1(1)}\right] \left[\frac{7 \cdot 6 \cdot 5 \cdot 4 \cdot 3 \cdot 2 \cdot 1}{2 \cdot 1(5 \cdot 4 \cdot 3 \cdot 2 \cdot 1)}\right] = (3)(21) = 63$$

Therefore,

$$P(A) = \frac{C_2^3 C_2^7}{C_4^{10}} = \frac{63}{210} = 0.3$$

3.5 SUMMARY

In this chapter, we have introduced the basic concepts of probability—the experiment, elementary outcomes, events, the sample and event spaces, and the probability axioms. Three ways of determining the probability of an event were discussed—the relative frequency definition, probability deduced logically, and subjective probability. Counting rules—the product rule, permutations, combinations, and the hypergeometric rule—were shown to be helpful in determining the probability of an event occurring in an experiment.

REFERENCES

Feller, W. *An Introduction to Probability Theory and Its Applications*. 2d ed., New York: John Wiley & Sons, Inc., 1957.

Parzen, E. *Modern Probability Theory and Its Applications*. New York: John Wiley & Sons, Inc., 1960.

■ PROBLEMS ──────────────────────

3.1. Give the sample space of each of the following experiments in the form of a Venn diagram. Be certain to define the elementary outcomes corresponding to the sample points in the sample space.

 a. A fair coin is tossed three times.

 b. A coin and a die are tossed together.

 c. A student receives his score on a multiple-choice exam containing 20 questions.

 d. A student receives his grade on an exam.

 e. The number of telephone calls received at a switchboard during a five-minute interval is recorded.

 f. A child is selected in a first grade class, and his or her weight (to the nearest pound) is recorded.

3.2. A committee is composed of two men and two women. One member of the committee is selected to serve as chairman and another to serve as secretary.

 a. Define the elementary outcomes comprising the sample space of this experiment.

 Identify which sample points in the sample space belong to the following event spaces:

 b. The younger man is selected as chairman: event A.

 c. A man is selected as chairman: event B.

 d. A woman is selected as secretary: event C.

 e. Events A and C occur: event D.

 f. Events B or C or both occur: event E.

 g. Show the five event spaces in a Venn diagram.

3.3. Two college job recruiting officers, Herman and Bill, come to the University of Truth campus to fill positions in their organizations. Each officer is attempting to fill three positions. Three students qualify for the positions described, and each will be interviewed by the two officers. If a sample point is defined as a specific number of students hired by Herman or Bill, define the following events as specific collections of sample points. Note: There are six jobs and three students—three jobs (or possibly more!) will not be filled. (Hint: The sample space is two-dimensional.)

 a. The sample space S that consists of all elementary outcomes defining the number of students hired by each officer.

 b. Event A: Herman hires at least two students.

 c. Event B: All three students are hired by Bill.

 d. Event C: Exactly one student is hired by Bill.

 e. Event D: Bill hires two students and so does Herman.

 f. Event E: Events A or C or both occur.

 g. Event F: Events A and C occur.

3.4. In Problem 3.2, suppose each individual committee member has an equal chance of being selected as either chairman or secretary. Compute the probabilities of the events A, B, C, D, and E by summing the probabilities of the appropriate sample points.

3.5. In Problem 3.3, suppose all the sample points of the sample space S are equally likely to occur. Compute the probability of the events A, B, C, D, E, and F by summing the probabilities of the appropriate sample points.

3.6. A piece of equipment is assembled in three separate operations, which may be performed in any sequence. Herman must decide which order to use. He places three slips of paper marked #1, #2, and #3 in a hat and draws them out one at a time without looking.

a. Define the elementary events comprising the sample space of this experiment. For each event described below, identify its sample points and compute the probability of its occurrence. Assume all the sample points in the sample space are equally likely to occur.

b. The operation marked #1 comes first in the implemented production sequence.

c. The operation marked #2 comes either second or third in the implemented production sequence.

d. The operation marked #1 comes first *and* the operation marked #2 comes either second or third in the implemented production sequence.

3.7. A sales representative is traveling from New York to Los Angeles with a stop at Detroit. From New York to Detroit there are five possible flights, and from Detroit to Los Angeles there are six. In how many ways can the sales representative make the trip from New York to Los Angeles with a stopover in Detroit?

3.8. Suppose auto license plates in a particular state are designed so that the first three characters are letters of the alphabet and the second three characters are numbers (e.g., ABC 127). How many different license plates can be produced by following this convention?

3.9. A brief market research questionnaire requires the respondent to answer each of eight questions. The first four questions are to be answered either *yes* or *no,* while the remaining four questions have three possible answers (*always, sometimes, never*). The sequence of responses to all eight questions is defined as the respondent's "profile." How many different profiles are there?

3.10. Six brands of soft drinks are ranked in order of preference by independent taste testers.

a. How many distinct rankings (using all six brands and ruling out ties) are possible?

b. In how many of these possible rankings can brand A be ranked best?

3.11. A group is composed of six men and four women. How many committees of four members can be formed from this group if:

a. There are no restrictions placed on the composition of the committee?

b. The committee must contain two men and two women?

c. The committee must contain at least one woman?

3.12. Herman is taking four courses during the present term. In each course, he can receive one of five possible grades: A, B, C, D, and F.

a. At the end of the term, how many different combinations of grades can he receive in the four courses?

b. Suppose in his first course he knows he can make no better than a C. How many combinations of grades can he receive?

c. Suppose by midterm it is apparent that he will flunk botany (he quits going to class), but he knows he has A's in two other courses. How many grade combinations are possible?

3.13. A customer considering the purchase of a new car is told that the car comes in five body styles, six colors, two transmissions, and five engines. How many choices does the customer have?

3.14. A firm has ten sales representatives. In how many ways can they be assigned to two territories, with seven sales representatives in one territory and three in the other?

3.15. An investment specialist is to rate a bond issue. The rating is on an eight-point scale: A+, A, A−, B+, B, B−, C, F.

    a. Develop three alternative sample spaces for this experiment. [Hint: One possible space is E_1 = high rating (A+, A, A−), E_2 = Medium rating (B+, B, B−), E_3 = poor rating (C, F).]

    b. Is any one of the three sample spaces "best"? Discuss.

    c. Describe a situation where each sample space developed in part *a* is appropriate.

3.16. For each example below, discuss whether it may be viewed as an *objective* (frequency or deductive logic definition) probability or a *subjective* (personal) probability, or both.

    a. The probability that the 'Skins will beat the Cowboys in an upcoming football game.

    b. The probability that a new electric toaster will survive its warranty period (six months) without failure or defect.

    c. The probability that a newly purchased lot of 100 calculators contains no defectives.

    d. The probability that Herman will be fired from his job in the next year.

    e. The probability that Green will win the presidential nomination of her party.

3.17. Distinguish between objective and subjective probability. Is it possible to apply both definitions of probability to a specific event? Under what conditions is it impossible to determine an objective probability?

3.18. Define the experimental conditions under which each of the following counting rules is appropriate:

    a. The product rule.

    b. Combinations.

    c. Permutations.

    d. Hypergeometric rule.

3.19. In a certain region of the country, radio station call letters are restricted to either three-letter or four-letter sequences beginning with the letter W. How many possible call letters are there?

3.20. In a market survey, customers are asked to taste and rank three breakfast cereals from best to worst. The cereals are labeled A, B, and C.

    a. Develop a sample space for this experiment.

    b. If each elementary outcome in the sample space is equally likely to occur, what is the probability that either cereal A or cereal B ranks first?

3.21. A division minicomputer is used by four departments in a company. Jobs may be submitted as "priority" or "nonpriority" jobs. A job sent to the computer room is selected at random.

    a. Develop a sample space for this experiment.

    b. Is it likely that the events in this sample space would be equally likely to occur in a typical company setting? Explain.

3.22. A machine shop has six work orders that require lathe work and four work orders that require milling work. If the shop has three lathes and two milling machines, in how many different ways can the work orders be assigned to the machines?

3.23. How many different ways can five horses in a five-horse race finish if there are no ties?

3.24. How many ways can six people be seated in a row if:

    a. There are no restrictions on the seating arrangement?

    b. There are three married couples, and each couple must sit together?

3.25. A corporate information division of a large multinational company subscribes to national newspapers to keep up with news in the various regions of the country. To simplify matters, suppose the division divides the nation into four regions: West, North, South, and East. In the West, they have identified 10 potential papers for subscription, 6 in the North, 8 in the South, and 12 in the East. It is decided to subscribe to five papers in each region. How many choices are possible?

3.26. Five men are scheduled to participate in a 100-meter race at a track meet. What is the probability that a sportswriter picks the first-, second-, and third-place winners if he makes the selections at random?

3.27. In the dice game, a player wins on the first toss of the pair of dice if the total number of dots on the two top faces of the dice is either 7 or 11, and loses if the total number of dots on the two top faces is either 2 or 12.

a. What is the probability that the player wins on the first toss?

b. What is the probability that the player loses on the first toss?

Conditional probability, the probability of compound events, and Bayes theorem

4

INTRODUCTION

In Chapter 3, the basic concepts of probability were developed. In this chapter, we will continue the development of probability concepts by first introducing the compound event—an event that is the combination of other events. Second, we will develop the notion of conditional probability—the probability that an event occurs given (or conditioned upon) the knowledge that one or more other events have occurred. Finally, ways of describing the relationships between two events are presented and are used in stating probability laws concerning compound events.

■ 4.2 **COMPOUNDING EVENTS: UNIONS AND INTERSECTIONS**

In many experiments, it is convenient or necessary to combine or to *compound* events to answer probability questions formulated from the experiment. In many problems, the compounding of events may lead to a solution of a probability problem without the need of listing all points in the sample space.

The compounding of events can be manifested in two ways: *unions* and *intersections*.

Definition 4.1
Union of events

Suppose A and B are two events defined in a sample space S. The *union* of A and B, denoted by A ∪ B, is the set of all sample points in the event space A *or* in event space B *or* in both event spaces A and B.

Example 4.1   To illustrate the union, suppose there are six points in the sample space S; E_1, E_3, and E_6 belong to the event space A, and E_1, E_2, and E_6 belong to the event space B.

The union of events A and B is the set of sample points E_1, E_2, E_3 and E_6. The compound event A ∪ B may be illustrated in a Venn diagram as in Figure 4.1.

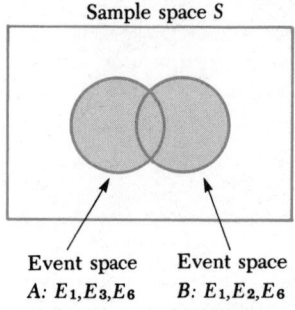

Sample space *S*

Event space
A: E_1, E_3, E_6 Event space
B: E_1, E_2, E_6

FIGURE 4.1   Venn diagram of the union of two events, A and B

Definition 4.2
Intersection of events

Suppose A and B are two events defined in a sample space S. The *intersection* of A and B, denoted by A ∩ B, is the set of sample points in the event space A *and* in the event space B.

Example 4.2   Suppose, as in the previous example, events A and B are defined as the following collections of elementary outcomes: A: E_1, E_3, and E_6, and B: E_1, E_2, and E_6. Then, A ∩ B is comprised of the events E_1 and E_6.

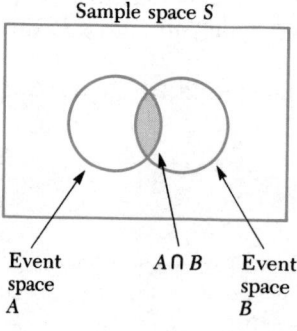

Sample space *S*

Event
space
A   A ∩ B   Event
space
B

FIGURE 4.2   Venn diagram of the intersection of two events, A and B

The intersection of two events, A and B, is illustrated in the shaded area in a Venn diagram in Figure 4.2. It may well be that the two events A and B do not overlap, as illustrated in Figure 4.3. In this instance, the compound event A ∩ B is empty—it contains no common elementary outcomes—and is denoted by ϕ, the null event.

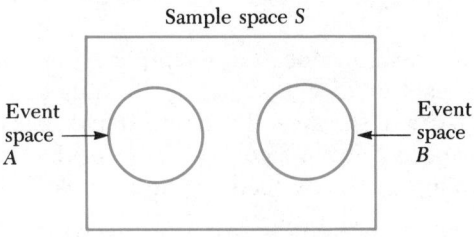

Sample space S

Event space A

Event space B

FIGURE 4.3 Venn diagram of the intersection of two non-overlapping events, A and B

Example 4.3 Recently, there has been considerable interest in computer literature searches. The method works as follows: a data base is formed of the titles or key words in articles appearing in magazines, journals, or government documents. If we are interested in doing research on, say, the effects of a recession on unemployment, we would instruct the computer literature search specialist to search the data base for papers, articles, or documents on this topic. The logic of the search usually uses the intersection logic of sets. For example, the specialist would enter into the computer on a terminal the request to search for and to print out the citations for all papers, articles, or documents that have the words *recession* and (intersection) *unemployment* in their titles or key words (many articles now contain a set of three to five key words that attempt to capture the main ideas in the article). Draw a Venn diagram illustrating the logic of the search process.

Solution Figure 4.4 shows the intersection logic used. Set A contains all citations that have the word *recession* in their titles or lists of key words, and set B contains all citations with the word *unemployment* in their titles or lists of key words. For example, there may be 512 citations in set A and 783 citations in set B, as illustrated in Figure 4.4. What we want is the *intersection*

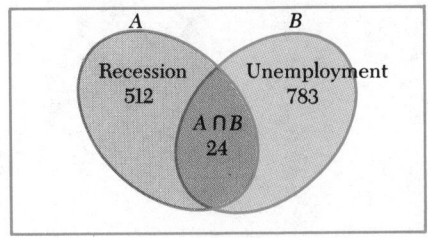

A

B

Recession
512

Unemployment
783

A ∩ B
24

FIGURE 4.4 The intersection of the title words or key words *recession* and *unemployment*

of the two sets, which contains, in this example, 24 citations. The computer search specialist would then instruct the computer to print out the citations for the 24 "hits."

Computer bibliographic literature searching is a relatively new field that is bound to grow dramatically in the next decade as more written work is stored in computer files. For anyone writing a term paper or a research paper, this innovation is significant, for it can save many hours of "searching the stacks" for references.

Example 4.4 The dice game was introduced in Example 3.1 in Chapter 3. The experiment is tossing a pair of dice and noting the number of dots appearing on each of the top faces of the dice. There are 36 possible elementary outcomes in the experiment and, if the dice are "fair," each elementary outcome has an equal probability of $1/36$ of occurring in a single trial of the experiment.

Suppose we define the following events in this experiment:

A: total of the dots on the top two faces of the dice is seven
B: total of the dots on the top two faces of the dice is even

FIGURE 4.5 Sample space and event spaces for Example 4.4

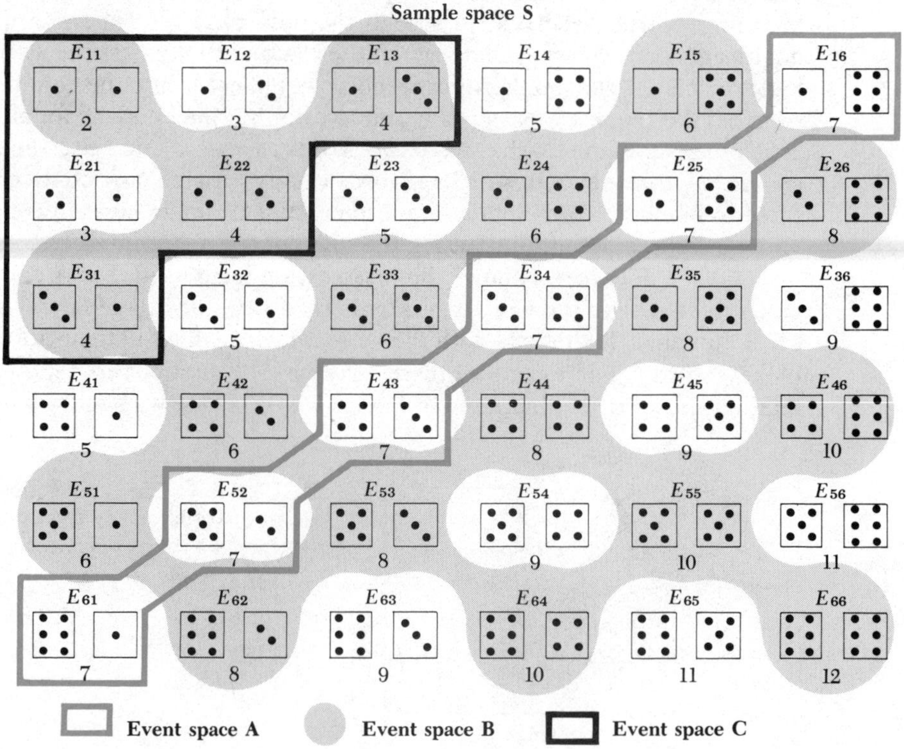

Sample space S

C: total of the dots on the top two faces of the dice is less than or equal to four

What are $P(A \cup B)$, $P(A \cup C)$, $P(B \cup C)$, $P(A \cap B)$, $P(A \cap C)$, and $P(B \cap C)$?

Solution   The event spaces A, B, and C and the sample space S are illustrated in Figure 4.5. Table 4.1 gives the elementary outcomes contained in each compound event. For the intersection compound events, $A \cap B$, $A \cap C$, and $B \cap C$, it is evident that A and B do not overlap (no intersection) and A and C do not overlap, as Figure 4.5 illustrates.

TABLE 4.1   Elementary outcomes contained in the compound events in Example 4.4

Compound event	*Elementary outcomes in compound events*	*Number of elementary outcomes*
A ∪ B	$E_{16}, E_{25}, E_{34}, E_{43}, E_{52}, E_{61},$ $E_{11}, E_{13}, E_{22}, E_{31}, E_{15}, E_{24},$ $E_{33}, E_{42}, E_{51}, E_{26}, E_{35}, E_{44},$ $E_{53}, E_{62}, E_{46}, E_{55}, E_{64}, E_{66}$	24
A ∪ C	$E_{16}, E_{25}, E_{34}, E_{43}, E_{52}, E_{61},$ $E_{11}, E_{12}, E_{21}, E_{13}, E_{22}, E_{31}$	12
B ∪ C	$E_{11}, E_{31}, E_{22}, E_{13}, E_{51}, E_{42}, E_{33},$ $E_{24}, E_{15}, E_{26}, E_{35}, E_{44}, E_{53}, E_{62},$ $E_{46}, E_{55}, E_{64}, E_{66}, E_{21}, E_{12}$	20
A ∩ B	ϕ	0
A ∩ C	ϕ	0
B ∩ C	$E_{11}, E_{31}, E_{22}, E_{13}$	4

Now, since each sample point in the sample space is equally likely to occur, Theorem 3.2 in Chapter 3 can be used:

$$P(\text{event}) = \frac{\text{Number of points in the event space}}{\text{Number of points in the sample space}}$$

Therefore,

$P(A \cup B) = {}^{24}/_{36} = {}^{2}/_{3}$ $P(A \cap C) = {}^{0}/_{36} = 0$ $P(A \cap B) = {}^{0}/_{36} = 0$
$P(B \cup C) = {}^{20}/_{36} = {}^{5}/_{9}$ $P(A \cup C) = {}^{12}/_{36} = {}^{1}/_{3}$ $P(B \cap C) = {}^{4}/_{36} = {}^{1}/_{9}$

■ **4.3   CONDITIONAL PROBABILITY**

Suppose in the dice game that the first die is black and the second die is white. The experiment, as before, is tossing the pair of dice and noting the total number of dots on the two top faces. The sample space is repeated in Figure 4.6. Suppose we define the following two events for this experiment:

A: dots on the top two faces total 11
B: six dots show on the top of the first (black) die

FIGURE 4.6 Reduced sample space and reduced event space for the dice game experiment

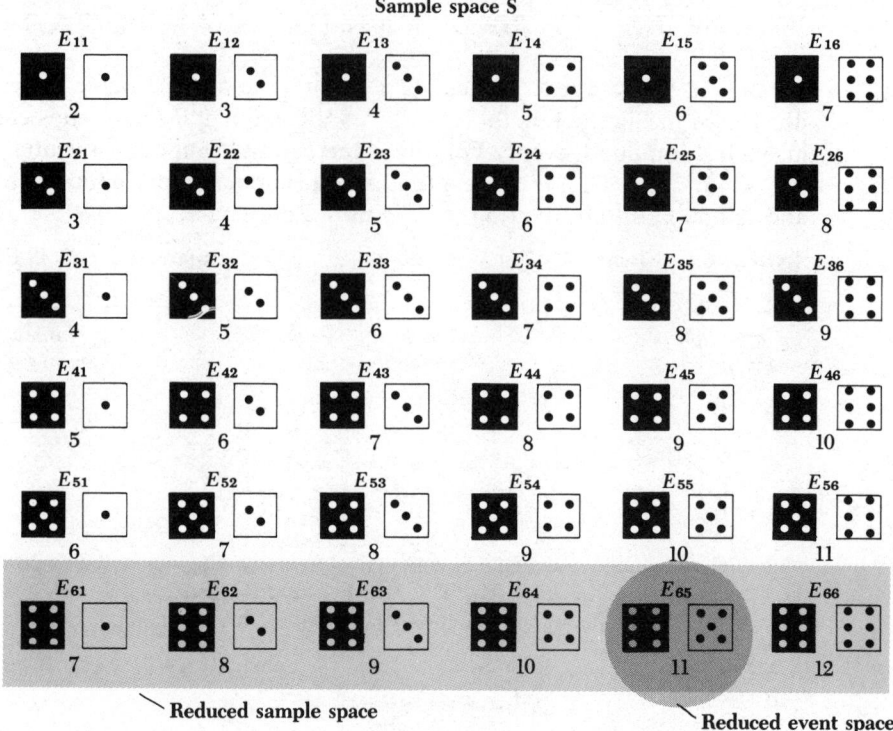

Sample space S

Since there are two points in the event space for A (E_{56} and E_{65}), $P(A) = {}^2/_{36} = {}^1/_{18}$. Now, suppose we have the following situation. The dice are tossed, but one of them rolls under a table so that its top face cannot be seen. The other die, the black one, shows six dots on its top face. Now, what is the probability that event A has occurred? This probability is called a *conditional probability* and is denoted by $P(A/B)$.

The notation $P(A/B)$ is read, "the probability that event A occurs, given (slash) that event B has occurred."[1] The conditional probability, $P(A/B)$, can be determined from the sample space. Since we know that event B has occurred (the black die shows six dots on its top face), the only possible elementary outcomes in the sample space are E_{61}, E_{62}, E_{63}, E_{64}, E_{65} and E_{66}. These elementary outcomes represent the shaded region in the sample space illustrated in Figure 4.6. These sample points comprise the *reduced sample space*—it is the portion of the sample space that still remains, given that event B has occurred (six dots on the black die). In the reduced sample space, the only sample point that produces a total of 11 dots on the top faces

---

[1] Notice that $P(A/B)$ does not mean $P(A) \div P(B)$.

of the dice is E_{65}. Since all the sample points are equally likely to occur, $P(A/B) = \frac{1}{6}$. In essence, we are merely reusing Theorem 3.2:

$$P(A/B) = \frac{\text{Number of points in the reduced event space}}{\text{Number of points in the reduced sample space}}$$

The number of points in the reduced sample space is six (E_{61}, E_{62}, E_{63}, E_{64}, E_{65}, and E_{66}). *Within* the reduced sample space, the reduced event space has one element, E_{65}. Thus, $P(A/B) = \frac{1}{6}$.

Notice that knowing that the first (black) die is showing six dots on its top face substantially changes the probability that the dots total 11 on the top faces of the pair of dice (event A). Without knowing that event B has occurred, $P(A) = \frac{2}{36} = \frac{1}{18}$, but if we know B has occurred, $P(A/B) = \frac{1}{6}$. Thus, (the knowledge that event B has occurred *triples* the probability that event A occurs in this example.)

Often two events are related in such a way that the probability of the occurrence of one event, say event A, is affected by the occurrence of another event, say event B, as in the dice game illustration. If we know that event B has occurred, then we can compute the *conditional probability* of A, *given* B.

Although it is possible to calculate conditional probabilities from the sample space by finding the appropriate reduced sample space and the appropriate reduced event space, there is a formula that may be used as well. It is given in Definition 4.3.

Definition 4.3
Conditional probability

(The *conditional probability* of event A, given event B has occurred, is the joint probability of events A and B occurring (the intersection event) divided by the probability that event B occurs:)

$$P(A/B) = \frac{P(A \cap B)}{P(B)} ; \quad \text{similarly,} \quad P(B/A) = \frac{P(A \cap B)}{P(A)}$$

Example 4.5 In the dice game illustrated earlier, events A and B were defined as follows:

   A: dots on the top faces of the pair of dice sum to 11
   B: dots on the top face of the first (black) die number 6

Now, from Definition 4.3,

$$P(A/B) = \frac{P(A \cap B)}{P(B)}$$

The intersection event space A ∩ B contains the sample point E_{65}. That is, E_{65} is the *only* sample point that has six dots showing on the top face of the first (black) die *and* a total of 11 dots on the top faces of the pair of dice. Thus, $P(A \cap B) = \frac{1}{36}$. The event space B contains the following sample points: $E_{61}, E_{62}, E_{63}, E_{64}, E_{65},$ and E_{66}. Thus, $P(B) = \frac{6}{36}$. Therefore,

$$P(A/B) = \frac{P(A \cap B)}{P(B)} = \frac{1/36}{6/36} = \frac{1}{6}$$

It should be apparent that these formulas need not always be used in computing conditional probabilities. The conditional probability $P(A/B)$, can be determined from the sample space associated with the experiment by reducing the space using the knowledge that event B has occurred, and then forming the ratio of the number of points in reduced event space A to the total number of points in the reduced sample space. This method of computing the conditional probability will work if each sample point in the original sample space is equally likely to occur.

The next example further illustrates the notion of conditional probability and the methods used to compute it.

Example 4.6 A study to determine the (possible) connection between smoking and lung cancer involved the survey of 100 patients at a local Veterans Administration hospital. Each patient was classified as a smoker or a nonsmoker and as having lung cancer or not having lung cancer. The results of this study are shown in the *contingency table* in Figure 4.7. For

FIGURE 4.7 Contingency table for Example 4.6

Cancer / Smoking	Lung cancer A	No lung cancer $\overline{A}$	Totals
Smokers B	15	25	40
Nonsmokers $\overline{B}$	5	55	60
Totals	20	80	100

example, it was found that 20 patients had lung cancer, 40 were smokers, and 15 had lung cancer *and* smoked. As indicated in Figure 4.7, the event "Lung cancer" is denoted by A, "No lung cancer" by $\overline{A}$, "Smoker" by B, and "Nonsmoker" by $\overline{B}$. The entries in the cells of the contingency table give the numbers of patients who belong to the intersections of the corresponding row and column events. The contingency table is a convenient way to summarize the elementary outcomes of an experiment and tabulate the sample points in the sample space S. The experiment in this problem consists of contacting each patient and recording whether or not he or she has lung cancer and/or smokes. The experiment was repeated 100 times, generating the 100 sample points summarized in the contingency table. There are, for example, 40 points in event space B and 80 points in event space $\overline{A}$.

Suppose that one patient is drawn at random from the 100 patients in this survey. Determine $P(B)$, $P(A \cap B)$, and $P(A/B)$.

Solution   Since each of the 100 sample points is equally likely to occur, we can use Theorem 3.2 (in Chapter 3) to determine each of these probabilities.

$$P(B) = \frac{\text{Number of points in event space B}}{\text{Number of points in sample space S}} = \frac{40}{100} = 0.4$$

$$P(A \cap B) = \frac{\text{Number of points in compound event space } (A \cap B)}{\text{Number of points in sample space S}}$$

$$= \frac{15}{100} = 0.15$$

$P(A/B)$

$$= \frac{\text{Number of points in reduced event space} A \text{ within reduced sample space} R}{\text{Number of points in reduced sample space } R}$$

$$= \frac{15}{40} = 0.375$$

The computation of $P(A/B)$ is illustrated in Figure 4.8.

FIGURE 4.8   Computation of $P(A/B)$

	A	$\overline{A}$	
B	15	25	40
$\overline{B}$	5	55	60
	20	80	100

Reduced sample space R (40 points)

$$P(A/B) = \frac{15}{40}$$

$$= 0.375$$

Given that event B has occurred, we know that the one patient we selected is a smoker. Therefore, he must be one of the 40 patients who smokes out of the total of 100 patients—the reduced sample space R has 40 points in it. Now, what is the probability that this patient has lung cancer? There are 15 sample points in the reduced space containing the 40 points that belong to event A "lung cancer." Thus,

$$P(A/B) = \frac{15}{40}$$

Notice that the conditional probability, $P(A/B)$, could also have been determined from the definitional formula:

$$P(A/B) = \frac{P(A \cap B)}{P(B)} = \frac{^{15}/_{100}}{^{40}/_{100}} = \frac{15}{40} = 0.375$$

Also, notice that the probability of a randomly sampled patient having lung cancer is 0.20 [$P(A) = {}^{20}/_{100}$], whereas if we know he smokes, the probability of his having lung cancer is 0.375 [$P(A/B) = {}^{15}/_{40}$]. Thus, the probability of lung cancer is much larger for patients in this group if we know that the patient smokes.

■ 4.4 EVENT RELATIONSHIPS

In most probability problems, the methods used in determining the probabilities associated with two or more events depend on the form of relationships among the events.

Definition 4.4
Independent events

Two events, A and B, are said to be *independent* if either

$$P(A/B) = P(A) \qquad \text{or} \qquad P(B/A) = P(B)$$

If either of these conditions is not satisfied, then the two events are said to be *dependent*.

The statement, $P(A/B) = P(A)$, says that the occurrence of event B does not affect the probability of the occurrence of event A. If this is true, it must necessarily follow that the second condition, $P(B/A) = P(B)$, must hold; that is, the occurrence of event A does not affect the probability of event B occurring.

Example 4.7 In Example 4.6, are the two events, lung cancer (A) and smoker (B), independent events?

Solution Since $P(A) = {}^{20}/_{100} = 0.20$, $P(A/B) = {}^{15}/_{40} = 0.375$, and $P(A) \neq P(A/B)$, the two events are *dependent*.

Example 4.8 For the two events, F and G, it is known that $P(F) = 0.6$, $P(G) = 0.4$, and $P(F \cap G) = 0.10$. Are the two events independent?

Solution Since $P(F \cap G) = 0.10$, $P(F/G) = P(F \cap G)/P(G) = 0.10/0.40 = ¼ = 0.25$. Since $P(F/G) = 0.25 \neq P(F) = 0.6$, the two events are *dependent*.

The fact that two events are dependent does not imply a cause-effect relationship. For instance, in Example 4.7, we can state that the occurrence of the event "smoking" affects the probability of the occurrence of the event "lung cancer." This does not mean that smoking *causes* lung cancer, but rather that the act of smoking is related to the *probability* of contracting lung cancer. The causal agent may be a physiological factor, for example, that increases a person's likelihood of both smoking and contracting lung cancer.

This is a rather salient point concerning the dependence of two events that will be dealt with in detail in later chapters.

> **Definition 4.5**
> **Mutually exclusive events**
> Two events, A and B, are said to be *mutually exclusive* if their intersection, A ∩ B, is the null set ϕ. Thus, if A and B are mutually exclusive events,
> $$P(A \cap B) = 0$$

In words, two events, A and B, are mutually exclusive if they cannot both occur in a single trial of the experiment. In Figure 4.9, two mutually exclusive events are illustrated; the event spaces A and B do not intersect.

Sample space

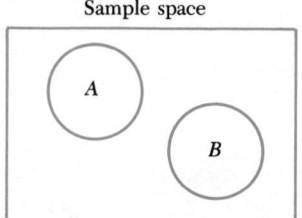

FIGURE 4.9 Venn diagram of two mutually exclusive events

Example 4.9 Suppose the experiment is flipping a coin once. Let events A and B be getting a head and a tail, respectively. In the single coin toss, are the two events mutually exclusive?

Solution Since, if a head occurs, a tail cannot, and vice versa, events A and B are mutually exclusive.

Example 4.10 In Example 4.6 where the experiment is randomly selecting one of the 100 hospital patients, are events A, "lung cancer," and B, "smoking," mutually exclusive?

Solution Since $P(A \cap B) = {}^{15}/_{100} \neq 0$, the two events are not mutually exclusive. They can occur jointly in a single trial of the experiment.

The two event relationships, *independence* and *mutual exclusivity,* are the most important ones in statistics! In most probability problems, the nature of the questions asked require determining the independence or dependence of events or whether the events are mutually exclusive. These determinations can be made separately by applying the definitions, but it is interesting to ask whether these two event relationships are related. Can a Venn diagram be used to show independent events as was done to illustrate mutually exclusive events. Let us suppose, for the moment, that two events, A and B, are mutually exclusive. This means that the occurrence of event A precludes the occurrence of event B, since by definition, the two events cannot occur jointly in a single trial of an experiment. This certainly ensures the *dependence* of the two events, because the occurrence of A, for example, means that $P(B/A) = 0 \neq P(B)$. If two events are not mutually exclusive, then they *may* be independent events, but it is not a certainty. In Example 4.6, the events A (lung cancer) and B (smoker) were not mutually exclusive but they were not independent either. The property of event independence may be illustrated by Venn diagrams as in Figure 4.10.

FIGURE 4.10 Venn diagrams of dependence and independence of two events

No intersection: Events
A and B are dependent

Intersection: Events A and B
may be independent

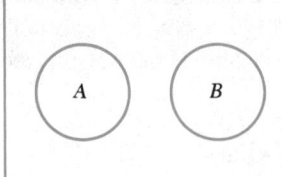
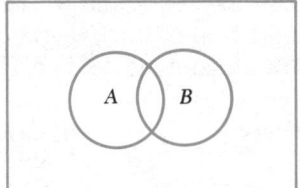

Definition 4.6
Mutally exclusive and collectively exhaustive events

A set of events, A_1, A_2, . . . , A_r is said to be *mutually exclusive* and *collectively exhaustive* if

$$\sum_{i=1}^{r} P(A_i) = 1 \quad \text{and} \quad A_1, A_2, \ldots, A_r \text{ are mutually exclusive events}$$

If a set of r events, A_1, A_2, . . . , A_r, are mutually exclusive and collectively exhaustive, then they absorb the entire sample space as illustrated in Figure 4.11. The implication is that one and only one of the events in the set of r events will occur in a single trial of the experiment.

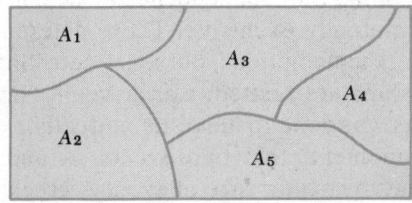

FIGURE 4.11 A set of five mutually exclusive and collectively exhaustive events

Example 4.11 An experiment consists of tossing a die a single time. Events are defined to be the number of dots that appear on the top face of the die. Is this set of events mutually exclusive and collectively exhaustive?

Solution The events are mutually exclusive, because if one occurs, none of the others may occur. If the die is fair, then each event has a probability of $1/6$ of occurring and the sum of the probabilities assigned to the individual events is 1. Thus, the set of events is mutually exclusive and collectively exhaustive.

Definition 4.7
Complement of an event

The *complement* of an event A is the collection of all points in the sample space that are *not* in the event space A. The complement of A is denoted by $\overline{A}$.

As illustrated by the Venn diagram in Figure 4.12, an event and its complement are simply a special case of a set of mutually exclusive and collectively exhaustive events.

From Figure 4.12, it should be apparent that

$$P(A) + P(\overline{A}) = 1 \quad \text{so that} \quad P(A) = 1 - P(\overline{A})$$

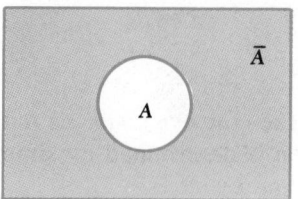

FIGURE 4.12 An event and its complement

The last statement frequently may be used to determine the probability of an event more easily than by computing $P(A)$ directly, as will be seen in the examples in Section 4.6.

4.5 ADDITIVE AND MULTIPLICATIVE PROBABILITY LAWS

In addition to answering probability questions by the sample space–event space approach, we can approach probability questions by defining events, determining event relationships, and then applying the following two probability laws.

> **Law 1**
> Additive law of probability
> The probability of the union of two events A and B (A ∪ B) is given by:
>
> $$P(A \cup B) = P(A) + P(B) - P(A \cap B)$$
>
> If A and B are mutually exclusive, then
>
> $$P(A \cup B) = P(A) + P(B)$$

The additive law can be used to compute the probability of the union of two events. If the two events are not mutually exclusive, then the law tells us to add the probability of event A to the probability of event B and subtract the probability of the intersection of A and B because it has been counted twice when $P(A)$ is added to $P(B)$, as Figure 4.13 illustrates.

FIGURE 4.13 Computation of $P(A \cup B)$

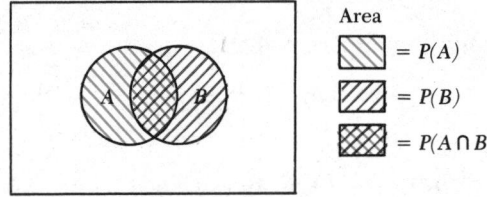

$$P(A \cup B) = P(A) + P(B) - P(A \cap B)$$

If the two events are mutually exclusive, then the event spaces A and B do not overlap, and the probability of the union is determined by simply summing the probabilities assigned to event A and event B.

Example 4.12 In Example 4.2, determine the probability that the randomly sampled patient is a smoker (B) or has lung cancer (A).

Solution This problem may be solved using the sample space–event space method by simply summing the number of sample points in the event space A ∪ B (5 + 15 + 25) and dividing this sum by the total number of points in the sample space (100); thus,

$$P(A \cup B) = \frac{5 + 15 + 25}{100} = \frac{45}{100} = 0.45$$

Alternatively, $P(A \cup B)$ may be determined by the event composition method and the additive probability law. Since $P(A) = 0.20$, $P(B) = 0.40$, $P(A \cap B) = 0.15$, and A and B are not mutually exclusive events,

$$P(A \cup B) = P(A) + P(B) - P(A \cap B) = 0.20 + 0.40 - 0.15 = 0.45$$

It should be apparent that both methods perform the same operations, but in slightly different ways. In many problems, the sample space may contain too many points to enumerate. In others, the only information we are given may already be in the form of the probabilities. Thus, we may have no choice but to use the event composition method.

> **Law 2**
> Multiplicative law of probability
>
> The probability of the intersection (A ∩ B) of two events A and B is given by:
>
> $$P(A \cap B) = P(A/B)P(B) = P(B/A)P(A)$$
>
> If A and B are independent events, then
>
> $$P(A \cap B) = P(A)P(B)$$

Example 4.13 An applicant for a quality control position in a manufacturing firm is given the following test: Twenty units of a product are placed in a large bin. Two of the units are defective. Inspect the units and select the two that are defective. What is the probability the applicant will select the two defective units if he randomly selects any pair of units from the bin?

Solution Define the two events A, the first unit selected is defective, and B, the second unit selected is defective. Then we wish to find the probability of the intersection of the two events; that is, $P(A \cap B)$. The events clearly are not independent. The probability that the second unit is defective depends on whether or not the first unit selected is defective. Thus, we must use the first form of the multiplicative law:

$$P(A \cap B) = P(A)P(B/A) = \left(\frac{2}{20}\right)\left(\frac{1}{19}\right) = \frac{1}{190} = 0.00526$$

The conditional probability $P(B/A)$ equals $1/19$, because if A occurs (a defective unit is selected in the first draw), there will be 19 units left to select from on the second draw, one of which is defective. Hence, it is very unlikely that the applicant would select the two defective units by chance alone.

Example 4.14 In the dice game, a total of seven dots on the top faces of the pair of dice wins. What is the probability that a player tosses two consecutive sevens?

Solution The experiment in this case is tossing the pair of dice twice. Let event A be "a total of seven dots appear on the top faces of the dice on the first toss," and let event B be "a total of seven dots appear on the top faces of the dice on the second toss." Then, we wish to find $P(A \cap B)$. It is reasonable to assume that the two tosses are independent; that is, the outcome in the second toss in no way is affected by or affects the outcome in the first toss. Thus, the second form of the multiplicative law may be used:

$$P(A \cap B) = P(A)P(B) = \left(\frac{6}{36}\right)\left(\frac{6}{36}\right) = \frac{1}{36}$$

The event composition approach using the probability laws to solve probability problems is more direct than the sample space–event space approach. Indeed, as suggested by Examples 4.13 and 4.14, it may not be feasible to use the latter approach due to the sheer number of points in the sample space. The sample space–event space method could have been used to solve the two example problems, but the work involved would be considerably more time-consuming. In Example 4.13, it would have been necessary to "list out" or count the 380 (20 × 19) points (pairs of units) in the sample space, and in Example 4.14 to work with the sample space containing 1,296 (36 × 36) points.

Although the event composition method of determining probabilities may be appealing in many problems, it takes considerable creativity and experience in using this method to select a proper set of event descriptions and to determine the proper relationships between events. The importance of event relationships should now be apparent; if we are finding $P(A \cap B)$, the determination that the events are independent will simplify computations considerably. Similarly, if we are finding $P(A \cup B)$, the determination that A and B are mutually exclusive will simplify computations.

In the next section, example problems are presented and solved using the event composition technique in conjunction with the additive and multiplicative probability laws. It will be seen that the use of probability trees can be particularly helpful in solving problems by the event composition method.

■ 4.6 COMPUTING PROBABILITIES: PROBABILITY LAWS AND PROBABILITY TREES

Example 4.15 The workers in a large assembly plant must check out and check in tools required during the day as the need for a particular tool arises. There is a centrally located tool bin operated by two employees who get the required tools for the workers and return them to the bin shelves when the workers are finished. The availability of one bin employee is assumed to be independent of the other. The probability that a specific employee in the bin is available to get a tool is 0.8. What is the probability that if a worker requires a tool, he can get it from the bin without waiting?

Solution Define the two events:

A: first bin employee is available for service
B: second bin employee is available for service

If the worker requiring a tool does not have to wait in line at the bin, then events A *or* B *or* both have occurred. Thus, we are interested in computing $P(A \cup B)$. Since we have determined that the question may be answered by using the additive probability law, the next step is to determine whether A and B are mutually exclusive events. Since both bin employees can be busy at the same time, they are clearly not mutually exclusive events. Then,

$$P(A \cup B) = P(A) + P(B) - P(A \cap B) = (0.8) + (0.8) - (0.8)(0.8)$$
$$= 1.6 - 0.64 = 0.96$$

Notice that $P(A \cap B)$ has been computed by the multiplicative law, using the fact that events A and B are independent (the availability of one bin employee is independent of the availability of the other employee). This problem can also be solved using a *probability tree* as illustrated in Figure 4.14. The first branch on the tree depicts the availability of employee A; he is

FIGURE 4.14 Probability tree for Example 4.15

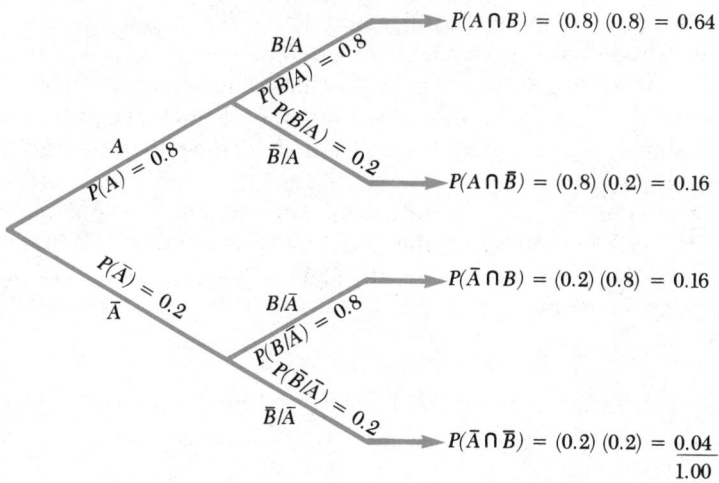

either available [with probability $P(A) = 0.8$] or not available [with probability $P(\bar{A}) = 0.2$]. On the upper branch A, we now branch on B. These two branches represent conditional events—employee B is available given that employee A is available (B/A) and employee B is not available given that employee A is available ($\bar{B}$/A). The conditional probabilities reflect that two events, A and B, are independent; that is, $P(B/A) = P(B) = 0.8$ and $P(\bar{B}/A) = P(\bar{B}) = 0.2$. By multiplying the probabilities along the branches, the probability of the intersection of events can be determined. For example, on the upper branch, $(0.8)(0.8) = 0.64$ is the probability of $(A \cap B)$. [By the multiplicative rule, $P(A)P(B/A) = P(A \cap B)$.] The probability tree gives us the required components to use in conjunction with the additive probability law to find $P(A \cup B)$: $P(A) = 0.8$, $P(B) = 0.8$, and $P(A \cap B) = 0.64$. Notice that we can determine $P(A \cup B)$ directly from the tree by summing the probabilities of the three mutually exclusive outcomes $A \cap B$, $A \cap \bar{B}$, and $\bar{A} \cap B$ $(0.64 + 0.16 + 0.16 = 0.96)$.

Further, notice that the "ends" of the probability tree list the four mutually exclusive and collectively exhaustive compound events associated with the problem—$(A \cap B)$, $(A \cap \bar{B})$, $(\bar{A} \cap B)$, and $(\bar{A} \cap \bar{B})$. The question could have been answered somewhat more easily had we recognized that at least one employee being available is the complement event to the event neither employee is available. Thus,

$$P(A \cup B) = P(A \cap B) + P(A \cap \bar{B}) + P(\bar{A} \cup B) = 1 - P(\bar{A} \cap \bar{B})$$
$$= 1 - 0.04 = 0.96$$

Example 4.16 Three fire detectors are placed in an office building. The probability that *each* will signal a fire when the temperature reaches or exceeds a specific value is 0.9. The signaling by one detector is independent of the signaling of the other two detectors when this minimum temperature is reached. If there is a fire in the office and the critical temperature is reached, what is the probability that at least one detector will signal a fire?

Solution If we define the event A, "at least one detector signals the fire," then $P(A) = 1 - P(\bar{A})$, where $\bar{A}$ is the complementary event, "none of the detectors signals a fire." Let F_1, F_2, and F_3 denote the events, "the first detector signals," "the second detector signals," and "the third detector signals" when the critical temperature has been reached, respectively. Then, $P(\bar{A}) = P(\bar{F}_1 \cap \bar{F}_2 \cap \bar{F}_3)$ since the probability that none will signal is the probability of the intersection $(\bar{F}_1 \cap \bar{F}_2 \cap \bar{F}_3)$. Since the signals of the detectors are independent events and $P(\bar{F}_i) = 1 - P(F_i) = 1 - 0.9 = 0.1$ for $i = 1$, 2, and 3,

$$P(\bar{F}_1 \cap \bar{F}_2 \cap \bar{F}_3) = P(\bar{F}_1)P(\bar{F}_2)P(\bar{F}_3) = (0.1)(0.1)(0.1) = 0.001$$

Thus,

$$P(A) = 1 - P(\bar{A}) = 1 - 0.001 = 0.999$$

Example 4.17 A local Ford dealer received a shipment of 50 new ½-ton pickup trucks and sold two of them to the Half-Fast Construction Company, which will use them on construction jobs. A week after the trucks were delivered, the dealer received a notice that 10 of the 50 trucks had faulty rear axles. What is the probability that the Half-Fast Company bought two defective trucks?

Solution Let event A be defined as "both trucks purchased by Half-Fast are defective." The event A can be written as the compound event, $A_1 \cap A_2$, where $A_i = i$th purchased truck is defective and $i = 1, 2$.

To stress that there are frequently many ways to approach a probability problem, this problem will be solved in two ways.

1. *Sample space–event space method.* Assuming that each of the total number of pairs of trucks that could have been selected by Half-Fast were equally likely to have been chosen, then $P(A)$ can be determined by counting the number of points in the event space A (n) and dividing this by the number of points in the sample space S (N).

Since two trucks are selected from the 50, the number of points in the sample space is $N = C_2^{50} = (50 \cdot 49)/2 = 1,225$. Since the set of 50 trucks from which the two were selected contains two types, defectives (faulty rear axles) and nondefectives, the hypergeometric counting rule is used to count the number of points in the event space A:

$$n = C_2^{10}C_0^{40} = \left(\frac{10 \cdot 9}{2}\right) (1) = 45$$

Thus,

$$P(A) = \frac{45}{1,225} = \frac{9}{245} = 0.037$$

2. *Multiplicative rule.* Since $P(A) = P(A_1 \cap A_2)$, the multiplicative rule can be used to determine $P(A)$. The two events A_1 and A_2 are not independent, because if the first truck purchased is defective (event A_1 occurs), this affects the probability that the second event A_2 occurs. Thus,

$$P(A) = P(A_1)P(A_2/A_1) = \frac{10}{50} \cdot \frac{9}{49} = \frac{9}{245} = 0.037$$

The conditional probability, $P(A_2/A_1)$, is determined by simple reasoning: if event A_1 occurs, then a defective truck was chosen as the first truck selected. This leaves 49 trucks, 9 of which are defective. Thus, the probability that the second purchased truck is defective given that the first one is defective is $9/49$.

■ 4.7 BAYES THEOREM

Bayes theorem provides another formula for computing conditional probability, $P(A/B)$.

Theorem 4.1
Bayes' theorem

Let A be an event. If B is another event such that P(B) is not 0, then

$$P(A/B) = \frac{P(A \cap B)}{P(B)} = \frac{P(B/A)P(A)}{P(B/A)P(A) + P(B/\bar{A})P(\bar{A})}$$

In applications of Bayes theorem, the computed probability $P(A/B)$ is often called the *posterior probability* of event A, given that event B has occurred. The probabilities $P(A)$ and $P(\bar{A})$ are called the *prior probabilities* of the event A and its complement $\bar{A}$. The prior probability $P(A)$ is thus *revised* using the information that event B has occurred. The probability of the intersection of events A and B, $[P(A \cap B)]$, is the *joint probability* of events A and B, and the divisor in Bayes theorem, $[P(B)]$, is called the *unconditional* or the *marginal probability* of event B. These relationships between events are developed in detail in Chapter 23.

The statement of Bayes theorem implies that

$$\frac{P(A \cap B)}{P(B)} = \frac{P(B/A)P(A)}{P(B/A)P(A) + P(B/\bar{A})P(\bar{A})}$$

From the multiplicative law of probability, the numerators of the two expressions above are equivalent: $P(A \cap B) = P(B/A)P(A)$. Thus, it must be true that the denominators are equal; that is,

$$P(B) = P(B/A)P(A) + P(B/\bar{A})P(\bar{A})$$

By the multiplicative law, $P(B/A)P(A) = P(A \cap B)$ and $P(B/\bar{A})P(\bar{A}) = P(\bar{A} \cap B)$. Thus,

$$P(B) = P(A \cap B) + P(\bar{A} \cap B)$$

It is easy to show that this statement does hold, using a Venn diagram as illustrated in Figure 4.15. The intersection $(A \cap B)$ plus the intersection $(\bar{A} \cap B)$ is the event space B.

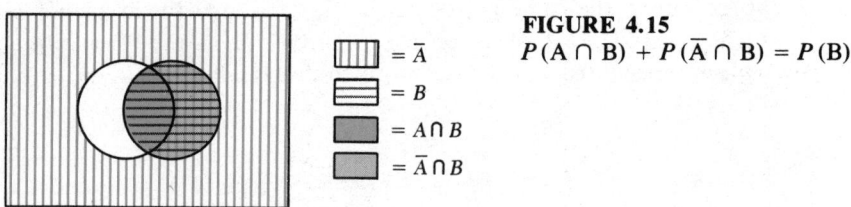

$\square = \bar{A}$
$\square = B$
$\square = A \cap B$
$\square = \bar{A} \cap B$

FIGURE 4.15
$P(A \cap B) + P(\bar{A} \cap B) = P(B)$

Thus, Bayes theorem decomposes the conditional probability $P(A/B) = P(A \cap B)/P(B)$ into smaller parts by writing $P(A \cap B)$ as $P(B/A)P(A)$ and $P(B)$ as $P(B/A)P(A) + P(B/\bar{A})P(\bar{A})$. For this reason, Bayes theorem is often interpreted as a simple restatement of conditional probability.

Bayes theorem can be very useful when we wish to determine a conditional probability, say $P(A/B)$, but we have the probabilities $P(A \cap B)$ and $P(B)$ given in decomposed forms, as the next example illustrates.

Example 4.18 The Superior Research Hospital has publicized the discovery of a new diagnostic test for cancer. If the person has cancer, the test will be *positive* (indicating the person has cancer) 99 percent of the time. If the person does not have cancer, the test is *negative* 95 percent of the time. From past medical histories and population census files, the probability that a person has cancer is 0.005. If the test is administered to a randomly selected individual in this population and it is positive, what is the probability that he has cancer?

Solution Define the events S, "test is positive," and C, "person has cancer." Using the given information, we can assign the following probabilities:

$$P(C) = 0.005 \qquad P(S/C) = 0.99 \qquad P(\overline{S}/\overline{C}) = 0.95$$

We wish to determine $P(C/S)$.

Bayes theorem may be used to determine $P(C/S)$. Associating the event C with A and S with B in Bayes theorem above,

$$P(C/S) = \frac{P(S/C)P(C)}{P(S/C)P(C) + P(S/\overline{C})P(\overline{C})}$$

In the statement of the problem, we have all the pieces required to calculate $P(C/S)$ by using the above formula except $P(S/\overline{C})$ and $P(\overline{C})$. But recall that

$$P(S/\overline{C}) + P(\overline{S}/\overline{C}) = 1$$

That is, given that the person does not have cancer, the test will be either positive or negative. The probabilities of these two conditional events $(S/\overline{C})$ and $(\overline{S}/\overline{C})$ must total 1. Since $P(\overline{S}/\overline{C}) = 0.95$, it follows that

$$P(S/\overline{C}) = 1 - P(\overline{S}/\overline{C}) = 1 - 0.95 = 0.05$$

Similarly, $P(\overline{C}) = 1 - P(C) = 1 - 0.005 = 0.995.$ Now.

$$P(C/S) = \frac{(0.99)(0.005)}{(0.99)(0.005) + (0.05)(0.995)} = \frac{0.00495}{0.00495 + 0.04975} = 0.091.$$

Thus, the probability that the person has cancer has been *revised* from 0.005 to 0.091 by knowing that the test for cancer was positive. Notice that the denominator $(0.00495 + 0.04975 = 0.0547)$ gives us the probability that the test will be positive $[P(S)]$.

This problem can also be solved by using a probability tree as illustrated in Figure 4.16. Notice that the first branching of the tree was performed on the given event C. The tree may be used to compute $P(C/S)$ from the definitional form, $P(C/S) = P(S \cap C)/P(S).$ To determine $P(S)$, add the joint prob-

FIGURE 4.16   Solution to Example 4.18 using a probability tree

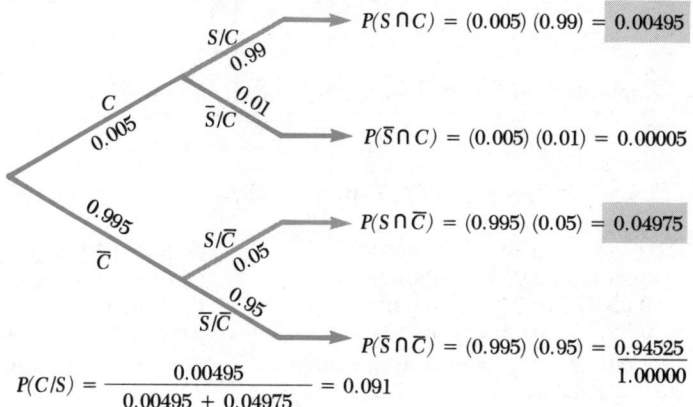

$$P(C/S) = \frac{0.00495}{0.00495 + 0.04975} = 0.091$$

abilities at the end of the tree with the event letter S in them—0.00495 + 0.04975 = $P(S)$.

Bayes theorem can be extended to more than two events that are conditioned by an event B. We will state the general form of Bayes theorem and then apply it in an example.

Theorem 4.2
Bayes' theorem—general form

Let $A_1, A_2, \ldots, A_K$ be a set of K mutually exclusive and collectively exhaustive events. If B is another event, such that $P(B)$ is not 0, then

$$P(A_i/B) = \frac{P(B/A_i)P(A_i)}{\sum_{i=1}^{K} P(B/A_i)P(A_i)}, \qquad i = 1, 2, \ldots, K.$$

If we let $K = 2$, $A_1 = A$, and $A_2 = \bar{A}$, Theorem 4.1 is a special case of the general form of Bayes theorem.

Example 4.19 A manufacturing firm produces units of a product in four plants. Define event A_i, " a unit is produced in plant i, where $i = 1, 2, 3, 4$," and event B, "a unit is defective." From past records of the proportions of defectives produced at each plant, the following conditional probabilities are set: $P(B/A_1) = 0.10$, $P(B/A_2) = 0.05$, $P(B/A_3) = 0.02$, and $P(B/A_4) = 0.15$. The first plant produces 40 percent of the units of the product, the second plant 20 percent, the third 5 percent, and the fourth 35 percent. A unit of the product made at one of these plants is tested and found to be defective. What is the probability that the unit was produced in plant 3?

Solution We wish to determine $P(A_3/B)$. From the general form of Bayes theorem this probability is given by

$$P(A_3/B) = \frac{P(B/A_3)P(A_3)}{\sum_{i=1}^{4} P(B/A_i)P(A_i)}$$

TABLE 4.2 Application of Bayes theorem to more than two events

Plant (event) i	$P(A_i)$	$P(B/A_i)$	$P(A_i)P(B/A_i)$	$P(A_i)P(B/A_i)/\sum_{i=1}^{4} P(A_i)P(B/A_i)$
1	0.40	0.10	0.0400	0.04/0.1035 = 0.3865
2	0.20	0.05	0.0100	0.01/0.1035 = 0.0966
3	0.05	0.02	0.0010	0.001/0.1035 = 0.0097
4	0.35	0.15	0.0525	0.0525/0.1035 = 0.5072
	1.00		0.1035 = $P(B)$	1.0000

The computation of $P(A_3/B)$ by Bayes theorem is eased somewhat by applying the tabular format as given in Table 4.2. From the third row, fifth column of this table, $P(A_3/B) = 0.0097$. The fourth column, when summed, gives the denominator of Bayes theorem. From Table 4.2, $P(B) = 0.1035$; that is, the probability that a defective part is produced by this firm is 0.1035. An advantage to using a table similar to Table 4.2 with Bayes theorem is that summing of probabilities in the last column to 1.0 (or close to 1.0, allowing for some numerical roundoff) gives some assurance that the required probability has been correctly determined.

4.8 ASSIGNMENT OF PROBABILITY

An often quoted advantage of following the relative frequency or the symmetry definition of probability is that all persons assessing probabilities of various events would arrive at the same probabilities (if calculations are performed correctly). For example, two different individuals would likely assess the probability of a head on the single toss of a fair coin at one half, or $P(H) = 0.50$. On the other hand, these same two individuals might assess the probability of Penn State beating Texas at the Cotton Bowl quite differently, depending on subjective beliefs concerning the outcome, which in turn are influenced by a host of other factors.

The interpretation of various statistical tests varies considerably, depending on how one uses subjective probabilities. Not all probability theory and procedures that can be applied to objective probabilities, for example, can be applied and interpreted similarly when using subjective probabilities. A statistician who uses subjective probabilities is called a *Bayesian* statistician. The topics of Bayesian statistics and Bayesian decision theory are developed in Chapters 21, 22, and 23.

4.9 SUMMARY

Probability measures the likelihood of an event occurring. It must be a number between 0 and 1, inclusive. If the probability of an event is 0, it means that the event is certain not to occur; if the probability is 1, then the event is certain to occur in a single trial of an experiment.

A convenient framework for computing probabilities is the use of sample spaces to list, at least conceptually, all the elementary outcomes that can occur in the experiment. The elementary outcomes should form a set of mutually exclusive and collectively exhaustive outcomes of the experiment. Additionally, they should be chosen in such a way that each elementary outcome is equally likely to occur in a single trial of the experiment.

The assignments of probabilities to the elementary outcomes may be accomplished in three ways: experimentally, using the frequency definition of probability; by symmetrical arguments; or subjectively. In any case, the assigned probabilities must all be numbers between 0 and 1, inclusive, and the sum of the probabilities associated with all the elementary outcomes in an experiment must be 1.

Once the elementary outcomes have been identified and probabilities assigned to them, the probabilities associated with more complex events in the experiment can be determined by one or more of the methods presented in this chapter, among which are: the sample space–event space method, the event composition method and the use of probability trees. The latter method is suggested for most problems, because it conveniently illustrates the events associated with an experiment and minimizes the risk of defining events incorrectly or misassigning probabilities from the wording of the problem.

Knowledge of the relationships among events is important in solving probability problems, because it indicates which probability law should be used to solve the problem. The two most important event relationships are *mutual exclusiveness* and *independence*.

■ REFERENCES

Feller, W. *An Introduction to Probability Theory and Its Applications*. 2d ed. New York: John Wiley & Sons, Inc., 1956.

Ferguson, T. S. *Mathematical Statistics: A Decision-Theoretic Approach*. New York: Academic Press, 1967.

Parzen, E. *Modern Probability Theory and Its Applications*. New York: John Wiley & Sons, Inc., 1960.

Winkler, R. L. *Introduction to Bayesian Inference and Its Applications*. New York: Holt, Rinehart and Winston, Inc., 1972.

■ PROBLEMS

4.1. Sales of an automobile dealer over the last month can be classified by method of payment and type of car sold, as indicated in the following table:

Type of car	Methods of payment	
	Cash	Installments
New	5	95
Used	25	25

A sales record is selected at random from last month's sales.

a. What is the probability that it represents a used car purchase?

b. If the purchase is of a new car, what is the probability that it was paid by installments?

c. Is the event, "a new car is purchased," independent of the event "installment payment"? Why or why not?

4.2. A poll taken among 50 employees of an industrial firm on the question of changing the work week from five days (8 A.M.–5 P.M.) to four days (7 A.M.–6 P.M.) yielded the following results:

Response	Classification of employees				
	Executive	Salesman	Office worker	Plant worker	Total
Favor change............	2	8	10	10	30
Oppose change	1	3	3	8	15
No opinion	0	2	2	1	5
Totals	3	13	15	19	50

One employee is selected at random from among the 50.

a. What is the probability that an employee is a plant worker and favors the change?

b. What is the probability that the employee will oppose the change given that he is a salesman?

c. If the employee is either an executive or a salesman, what is the probability that he expressed no opinion?

d. Given that the employee is *not* a plant worker, what is the probability that he favors the change?

e. What is the probability that the employee either favors the change or is an office worker?

f. Are the events "office worker" and "plant worker" statistically independent? Why or why not?

g. Are the events "office worker" and "favor change" statistically independent? Why or why not?

h. Are the events "executive" and "plant worker" mutually exclusive? Why or why not?

i. Are the events "executive" and "no opinion" mutually exclusive? Why or why not?

4.3 The data at the bottom of the page pertain to a survey taken of new car dealers in the United States. If a dealer is selected at random from this survey, determine:

a. $P(A_3)$

b. $P(A_1 \cap B_4)$

c. $P(A_4/B_1)$

d. $P(A_3 \cup B_4)$

e. $P[(B_3 \cup B_4)/A_1]$

f. $P[(A_1 \cup A_3)/B_5]$

g. $P(A_1 \cup A_2 \cup A_3)$

Event	Type of dealership	Region of dealer				
		A_1 Northeast	A_2 Northwest	A_3 Southeast	A_4 Southwest	Totals
B_1:	Fjord Motors	200	100	50	150	500
B_2:	Chysler Motors	150	50	30	70	300
B_3:	National Motors	50	50	20	30	150
B_4:	Chevylay	300	150	150	200	800
B_5:	Toysalot	10	40	10	90	150
	Totals	710	390	260	540	1,900

h. $P[B_1/(A_3 \cup A_4)]$
i. $P[A_3/(B_3 \cup B_5)]$
j. Are events A_1 and B_1 independent? Why or why not?

4.4 A department store surveyed 200 customer purchases to determine whether or not each purchase was charged, paid for in cash, or paid for by check. Define these events A_1: cash used, A_2: check used, A_3: charged; B_1: purchase less than $10, B_2: purchase more than $10 but less than $50, and B_3: purchase $50 or more. The survey results are:

Payment \ Purchase	B_1	B_2	B_3	Total
A_1	25	15	10	50
A_2	10	40	10	60
A_3	65	15	10	90
Totals	100	70	30	200

If one purchase is selected at random from this survey, determine:
a. $P(B_3)$
b. $P(A_3 \cap B_2)$
c. $P[(A_1 \cup A_2)/B_3]$
d. $P[(A_1 \cup A_3)/B_2]$
e. $P[A_1/(B_1 \cup B_2)]$
f. $P(A_1/B_3)$
g. Are events A_1 and B_1 independent? Why or why not?

4.5. The probability that an employee selected at random from a work force at a certain factory is a male is 0.65, the probability that the employee is married is 0.75, and the probability that the employee is both married and male is 0.50. If an employee in this group is chosen at random, calculate the probability that the employee is:

a. Female.
b. Single.
c. Single and female.
d. Female or married.

4.6. A town has both a morning and an evening newspaper. Of the families in the town, 50 percent buy the morning paper, 40 percent buy the evening paper, and 25 percent buy both papers. Define the events: A_1: buys a morning paper; A_2: does not buy a morning paper; B_1: buys an evening paper; and B_2: does not buy an evening paper.

a. List all pairs of events from the set that are mutually exclusive.
b. List all pairs of events that are statistically dependent.
c. Find $P(B_2/A_1)$.
d. Find the probability that a family buys either one paper or the other *but not both*.

4.7. A shipment of four automobile air conditioners is received by an automobile dealer. All four are identical in appearance, but one of them is defective. One unit is selected at random from the four and tested by mechanic A. If he does not find it defective, a second unit is randomly selected from the remaining three and tested by mechanic B. What is the probability that one of the two mechanics will discover the defective unit?

4.8. From company records it is known that 80 percent of all applicants pass the American Trust Company's Managerial Trainee Examination on the first attempt. An applicant who fails the exam on the first try has a probability of 0.90 of passing the exam on the second try.

a. Three candidates take the examination for the first time. What is the probability that two will pass? What is the probability that at least one will pass?

b. What is the probability that a candidate fails the exam twice?

c. If a candidate can only take the exam twice, what is the probability of passing the exam in one of the two trials?

4.9. A psychological experiment is designed to test whether intelligence of the type measured by IQ tests can be detected from photographs. Photographs of five individuals (whose IQs are known only by the experimenter) were presented to a subject. The subject was asked to arrange the photographs in order according to IQ. Suppose the subject could not detect IQ from photographs.

a. What is the probability that the subject would by chance achieve the correct ordering of the five pictures?

b. What is the probability that this subject would by chance succeed in placing the photograph of the individual having the highest IQ in one of the first two positions?

4.10. Suppose the last three male customers out of a restaurant have all lost their hat checks so that the checker hands back their hats in random order. What are the following probabilities:

a. No man will get the right hat.

b. Exactly one man will get the right hat.

c. Exactly two men will get the right hats.

4.11. Suppose two defective color TV sets have been included in a shipment of six. The buyer begins to test the six TVs one at a time.

a. What is the probability that the second defective is found on the fourth test?

b. What is the probability that no more than four TV sets need to be tested before locating both defectives?

4.12. An article in the newspaper claims that if the probability of destroying an attacking airplane were 0.3 at each of three independent defense barriers, and if the attacking plane had to pass all three barriers to reach its target, the probability that the plane would not reach its target is 0.90. Correct this statement.

4.13. A portable calculator salesperson finds that the probability of selling a unit to a prospective buyer on the first contact is 0.30 but improves to 0.60 on the second contact. The salesperson will not contact a prospective buyer more than twice. If the salesperson contacts Herman, a prospective client, determine:

a. The probability that Herman will buy a calculator.

b. The probability that Herman will not buy a calculator.

4.14. It is known that the probability is 0.10 that an item from a specific production line is defective. Three items are randomly selected from this line. Determine:

a. The probability that all three are defective.

b. The probability that at least one item is not defective.

4.15. Each of two packages of six flashlight batteries contains exactly two inoperable batteries. If two batteries are selected from each package, what is the probability that all four batteries will operate?

4.16. Herman awakens in the middle of the night with a headache and stumbles into the bathroom without turning on the light. In the dark, he grabs one of three bottles containing pills and takes a pill from the selected bottle. One hour later, he is feeling much worse and recalls that one of the three bottles contained tranquilizing pills and the other two aspirin. Herman quickly consults his handy medical text and finds that 80 percent of normal individuals show his symptoms after taking tranquilizing pills and only 5 percent have these symptoms after taking aspirin. Assume that each of the three bottles had an equal probability of being chosen in the dark.

   a. Ignoring Herman's symptoms, what is the probability that he took a tranquilizing pill?

   b. Find the probability that Herman took a tranquilizing pill, given his symptoms.

   c. Find the probability that Herman took an aspirin, given his symptoms.

4.17. Construct a probability tree to determine $P(A)$ in Example 4.16 in the text, where event A is defined to be, "At least one detector signals the fire."

4.18. Construct a probability tree to determine $P(A)$ in Example 4.17 in the text, where event A is defined to be, "Both trucks purchased by Half-Fast are defective."

4.19. Given events A and B, where $P(A) = 0.4$, $P(B) = 0.3$, and $P(B/A) = 0.5$, find:

   a. $P(A/B)$ *b.* $P(A \cup B)$

4.20. A company has three factories producing the same item. Factory A produces 20 percent of the total output, factory B, 50 percent of the output, and factory C, 30 percent of the total output. The proportion of defectives found at the three factories are 1 percent, 2 percent, and 3 percent, respectively. An item is selected at random from the combined output of the three factories and is found to be defective. What is the probability that the item came from factory B?

4.21. An oil-well drilling company must decide whether or not to drill at a particular site the company has under lease. On the basis of a geological survey, it is decided that there is a 0.45 probability that a formation of type 1 lies beneath the well site, a 0.30 probability that a formation of type 2 lies beneath the site, and a 0.25 probability of a type 3 formation. Records indicate that oil is discovered 30 percent of the time in type 1 formations, 40 percent of the time in type 2 formations, and 20 percent in type 3 formations. What are the following probabilities?

   a. If the well produces oil, it comes from a type 1 or type 2 formation.

   b. If the well is not productive, it comes from a type 3 formation.

4.22. If $P(B) = 0$, does the statement $P(A/B)$ have any meaning? Discuss.

4.23. Consider the statement: If two events are mutually exclusive, they must be dependent events. Is this statement true? Discuss. If two events are not mutually exclusive, does this imply that they are independent events? Discuss.

4.24. If three events, A_1, A_2, and A_3, are mutually exclusive and collectively exhaustive and B is not a null event, show that

$$P(B/A_1)P(A_1) + P(B/A_2)P(A_2) \\ + P(B/A_3)P(A_3) = P(B)$$

using a Venn diagram.

4.25. Using a Venn diagram, develop a general formula for $P(A \cup B \cup C)$.

4.26. Is $P(A \cup B)$ greater when $P(A \cap B) = 0$ or when $P(A \cap B) > 0$? Discuss.

4.27. If $P(A \cup B) = 0$, can the event $A \cap B$ occur? Discuss.

4.28. Given $P(A) = \frac{1}{3}$, $P(B) = \frac{3}{4}$, and $P(A \cap B) = \frac{1}{6}$.
  a. Find $P(A \cup B)$.
  b. Find $P(B/A)$.
  c. Are A and B independent? Explain.

4.29. An arts and crafts store decides to record, over its next 100 customer purchases, sales on a cash basis and on a charge basis as a function of the size of the sale. The results are shown in the table at the bottom of the page.

  a. Given that a purchase is charged, what is the probability that the size of the purchase is over $100?
  b. Given that a purchase is charged, what is the probability that the size of the purchase is under $25?
  c. Determine $P(A_1 \cup B_1)$.
  d. Determine $P(A_1 \cap \bar{B}_3)$.
  e. Are the compound events "type of purchase" and "size of purchase" independent? Explain.

4.30. A gambler is playing the following game: A coin is tossed and if it comes up heads, he wins $100. If the coin comes up tails, he loses $100. After five tosses of the coin, he has lost $500. Assume that the coin is fair.
  a. What is the probability of five straight tails?
  b. What is the probability that he will lose $100 on the next coin toss?

4.31. An insurance agency employs two salespersons. Sally contacts 75 percent of the prospective clients and signs a sales contract with 60 percent of the clients she contacts. Hal contacts 25 percent of the prospective clients and signs 40 percent of those contacted. If a sales contract has just been signed, what is the probability that it was signed by Sally?

4.32. A printing press makes at least one typesetting error on each page printed with probability 0.01. Assume a six-page brochure is printed.
  a. What is the probability that the brochure contains at least one page with at least one typesetting error?
  b. What is the probability that exactly three pages contain at least one typesetting error?

Size of purchase / Type of purchase	Less than $25 B_1	$25.01–$100 B_2	Over $100 B_3	Totals
A_1 (cash)	25	20	5	50
A_2 (charge)	5	20	25	50
Totals	30	40	30	100

Random variables and their distributions

5

In Chapters 3 and 4, the basic concepts of probability were developed in detail. Although the notion of chance events (e.g., head or tail in the flip of a coin, or winning or losing a bet) is familiar to most of us, the specific elements of probability theory, such as the concepts of independence, mutual exclusiveness, and a statistical experiment, are frequently difficult to grasp on first exposure. Indeed, the student may wonder why it is necessary to spend so much time on probability in an introductory statistics course. In this chapter, we shall answer this query. Probability provides a framework upon which is based the process of making inferences about a population characteristic from a sampled portion of the population.

■ 5.2 RANDOM VARIABLES

In a typical population, it is usually possible to identify more than one characteristic of the units that constitute it. For example, suppose the population is the *collection* of all full-time students registered at your university or college during the present academic term. In this instance, it is possible to identify numerous characteristics of the population unit—earned income (if any!), height, weight, sex, hair color, number of parking tickets accumulated during the term, grade point average, and so on. In a statistical study of the units in this population, we may be interested in just one characteristic or in a collection of such characteristics, such as sex and grade point average or earned income and grade point average and so on.

In Chapter 1, we referred to a population characteristic as a *variable* if it can assume one or more values in the population. We must now define *variable* more specifically when we use the word to mean a measure of a population characteristic.

Definition 5.1
Random variable[1]

A *random variable* is a numerically valued *function* with a value that is determined by a random experiment.

Suppose we consider the population of full-time students registered at your university or college. If we are interested in the population characteristic of grade point average, and we select one student at random from this population, then the characteristic may be viewed as a random variable. Its value is numerical and arises from a random experiment, and it is a function because it defines a correspondence between members of one set (the student population) and members of another set (the set of all possible grade point averages, from 0.00 to 4.00). For each student, the random variable defines one and only one grade point average, although more than one student may have the same grade point average.

To appreciate the concept of a random variable being a function, let us consider an example.

Example 5.1   Suppose the random experiment is tossing a coin twice and we define the random variable X as the number of heads in the two tosses. Give the correspondence between members of the experimental elementary outcomes and possible values of the random variable.

Solution   Figure 5.1 gives the correspondence between members of the experimental elementary outcomes and possible values of the random variable.

Notice that the random variable X is a function. To each member in the first set, the elementary outcomes in the experimental sample space, there

FIGURE 5.1   Random variable X = number of heads in two tosses of a coin

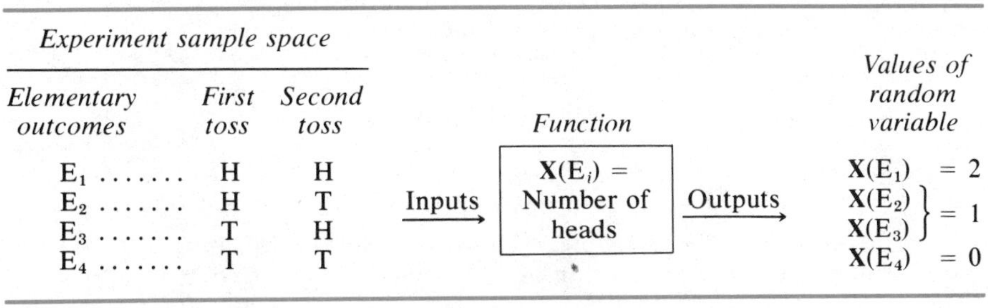

Experiment sample space				Values of random variable
Elementary outcomes	First toss	Second toss	Function	
E_1	H	H	$X(E_i) =$ Number of heads	$X(E_1)$ = 2
E_2	H	T	Inputs →	$X(E_2)$ ⎫ = 1
E_3	T	H	← Outputs	$X(E_3)$ ⎭
E_4	T	T		$X(E_4)$ = 0

---

[1] A more mathematically rigorous definition is: A random variable is a numerically valued function defined over a sample space.

corresponds one and only one member in the second set, the values of the random variable. Each value of the random variable may correspond to one or more elementary outcomes, however. Notice that we have written the random variable in functional notation in Figure 5.1 to emphasize its meaning.

Example 5.2   To appreciate more fully the notion that a random variable is a function (that the *value* of the random variable may correspond to one or more elementary outcomes of an experiment, but to each elementary outcome there corresponds only one value of the random variable), consider the die-tossing experiment of Example 3.1. Suppose we wish to establish a correspondence between the elementary outcomes of the experiment and the sum of the number of dots appearing on the top face of each die in the toss. Define the random variable in this experiment and list the elementary outcomes corresponding to the values the random variable can assume.

Solution   The random variable X is defined to be the *number* of dots that appear on the top faces of the two dice in a random roll. Thus, to each of the elementary outcomes of the experiment, we associate the number of dots on the two top faces. This is the random variable. Values of the random variable and elementary outcomes that produce the values of the random variable are given in Table 5.1.

TABLE 5.1   Values of the random variable and elementary outcomes corresponding to values of the random variable for the die-tossing experiment

Values of the random variable	Elementary outcomes
2	E_{11}
3	E_{12}, E_{21}
4	E_{13}, E_{22}, E_{31}
5	$E_{14}, E_{23}, E_{32}, E_{41}$
6	$E_{15}, E_{24}, E_{33}, E_{42}, E_{51}$
7	$E_{16}, E_{25}, E_{34}, E_{43}, E_{52}, E_{61}$
8	$E_{26}, E_{35}, E_{44}, E_{53}, E_{62}$
9	$E_{36}, E_{45}, E_{54}, E_{63}$
10	E_{46}, E_{55}, E_{64}
11	E_{56}, E_{65}
12	E_{66}

In most random experiments, we are interested in measuring more than one characteristic of the population. Suppose we randomly select two students from the population of full-time students at your university or college. We are interested in three population characteristics: grade point average, number of parking tickets received, and sex. If we define the three random variables as X = grade point average, Y = number of parking tickets received, and Z = sex (1 if male, 0 if female), then the measurements for the three students can be determined by using these functions as illustrated in Figure 5.2.

FIGURE 5.2   Experiment involving three random variables

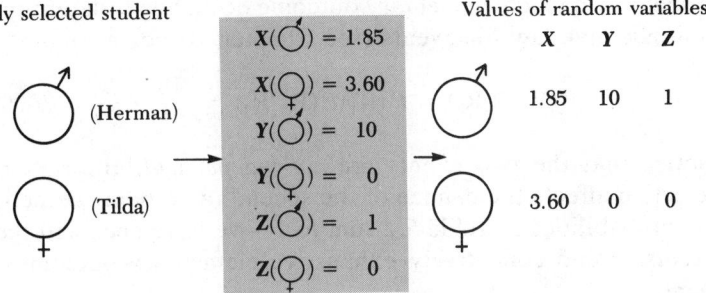

In this experiment, we can think of the sample space as being comprised of all possible pairs of students that could be selected from the population of full-time students at your college or university. The random variables **X, Y,** and **Z,** then, establish a functional correspondence between the sample space points (the two students) and the possible values of the three random variables (0.00 ≤ X ≤ 4.00, 0 ≤ Y ≤ ?, and Z = 0, 1). This may seem like a rather abstract way to say that Herman's grade point average is 1.85 (he is in trouble), the number of parking tickets he has received at the time of the experiment is 10 (he is in *big* trouble), and he is a male (what more can be said!). But, we will come to appreciate the need to establish a mathematical foundation for measuring population characteristics, as we have done here.

Note the treatment of the random variable **Z.** We could not let the values of this random variable be represented by "male" and "female" because these values are not numerical and a random variable is a *numerically* valued function. Whenever we are dealing with a qualitative population characteristic such as sex, we must assign to its outcomes numerical values if we are to treat the characteristic as a random variable.

When the probabilities of the elementary outcomes in the experimental sample space are known, it is possible to determine the probabilities of the values of the random variable by using the fact that the random variable is a function that establishes a correspondence between sample space points and the values of the random variable, as the next example illustrates.

Example 5.3   A production lot of 100 transistor radios contains 10 defectives. A retailer decides to select two of the radios randomly and, by extensive testing, determine how many are defective. If neither radio is defective, he will accept the lot. Define the random variable **X** = number of defective radios (**X** = 0, 1, or 2). Determine the probability that the random variable **X** assumes each of its three possible values.

Solution   In the experiment of selecting two radios at random, define the elementary outcomes:

R_1: first radio is defective $\overline{R}_1$: first radio is nondefective
R_2: second radio is defective $\overline{R}_2$: second radio is nondefective

The four elementary outcomes of the experiment are given in Table 5.2. The probability of each elementary outcome occurring is determined by using the multiplicative law for events (see Chapter 4). For example,

$$P(R_1 \cap R_2) = P(R_1)P(R_2/R_1) = \frac{10}{100} \cdot \frac{9}{99} = 0.0091$$

Notice that the two events are *not independent;* the outcome of the first selection affects the chance of the second radio being defective. Notice that the probabilities in Table 5.2 sum to 1—we have specified the four mutually exclusive and collectively exhaustive elementary outcomes of the experiment.

TABLE 5.2   Elementary outcomes and probabilities of the experiment described in Example 5.3

Elementary outcomes	Probability	Value of random variable X
1: $R_1 \cap R_2$.............	$(^{10}/_{100})\ (^{9}/_{99}) = 0.0091$	2
2: $R_1 \cap \overline{R}_2$.............	$(^{10}/_{100})\ (^{90}/_{99}) = 0.0909$	1
3: $\overline{R}_1 \cap R_2$.............	$(^{90}/_{100})\ (^{10}/_{99}) = 0.0909$	1
4: $\overline{R}_1 \cap \overline{R}_2$.............	$(^{90}/_{100})\ (^{89}/_{99}) = 0.8091$	0

Since the elementary outcomes are mutually exclusive and collectively exhaustive, $P(X = 2) = 0.0091$, $P(X = 1) = 0.0909 + 0.0909 = 0.1818$, and $P(X = 0) = 0.8091$. The values of the random variable X and its probabilities of occurrence are given in Table 5.3. This table represents the *probability distribution* of the random variable X—a list of each random variable value and its probability of occurrence.

In order not to confuse a random variable with values that the random variable might assume, we will adopt the notation that random variables are expressed as boldface capital letters, and *values* of the random variable are expressed as lowercase letters. Thus, for example, the random variable "X" will be denoted by X, and values that the random variable X may assume will be denoted by x. Similarly, $P(X)$ will denote the probability distribution of

TABLE 5.3   Probability distribution of X

x	$P(x)$
0.................	0.8091
1.................	0.1818
2.................	0.0091

the random variable **X**, while $P(x) = P(\mathbf{X} = x)$ will denote the probability that the random variable **X** assumes the particular value x. From Example 5.3, $P(1)$ represents the probability that the random variable **X** assumes the value of 1, or $P(1) = 0.1818$. The probability that the retailer will accept the lot is (from Table 5.3) $P(\mathbf{X} = 0) = P(0) = 0.8091$.

The probabilities in Table 5.3 are *relative frequency* probabilities. Since the population is composed of 100 distinguishable units, there are 4,950 possible pairs of different radios that could be drawn from this population[2]—the percentage of the total of 4,950 radio pairs in which both are nondefective is 80.91 percent. Thus, by experimentally forming every different pair of radios that could be randomly drawn from this population, we would find that the proportion of pairs that contained no defectives would be 0.8091.

Most statistical inference methods of inducing information about a population characteristic from a sample of units drawn from the population are based on defining a random variable and determining its probability distribution.

It is important to distinguish between two types of random variables—those that are discrete and those that are continuous. A *discrete* random variable is one for which the number of its values is finite (as in Example 5.3) or countably infinite. By *countably infinite*, we mean that the values of the random variable can be placed in one-to-one correspondence with the positive integers.[3] An example of a random variable that has a countably infinite number of values is a *counting variable*, a random variable with values of 0 and the positive integers 1, 2, . . . , ∞. There is an infinite number of values, but these values are restricted to specific real numbers—the set of positive integers plus 0. The counting random variable typically is appropriate when we are counting the number of occurrences of a particular event in a fixed interval of time, where there is no upper bound on the maximum number of occurrences of the event in the specified time interval.

Examples of discrete random variables are the number of parking tickets issued to a student during a term, the number of dots appearing on the top faces of a pair of dice, the number of heads occurring in a specified number of coin flips, and the sampling problem of Example 5.3. In each instance, there is only a finite number of possible outcomes in a simple statistical experiment (selection of a student to determine the number of tickets received, roll of a dice, flip of a coin, etc.). Some additional examples of discrete random variables are given in Table 5.4.

---

[2] The number of possible pairs of radios that can be selected from 100 is given by the combination rule of Chapter 3 (Rule 3.3). Since the order of selection is not important and we are selecting 2 items from 100, the number of possible different pairs selected is given by:

$$C_2^{100} = \frac{100!}{2!(100 - 2)!} = 4,950$$

[3] An alternative definition of a discrete random variable is: the set of values is such that there is at most a finite number of values in every finite interval on the real number line.

TABLE 5.4 Examples of discrete random variables

Random variable **X** definition	Values of the random variable	Number of values in the set
1. Let **X** be the number of correct answers by a student on a test containing ten questions	0, 1, 2, . . . , 10	Finite (11)
2. Let **X** be the number of tickets acquired by a student in a term— the student is expelled after the eighth ticket........................	0, 1, 2, . . . , 8	Finite (9)
3. Let **X** be the number of tickets acquired by a student in a term with no limit on the number received	0, 1, 2, . . .	Countably infinite
4. Let **X** be the number of customers entering Ed's Hamburger Joint during one day's business............	0, 1, 2, . . .	Countably infinite

In the last two examples in Table 5.4, the largest value of the random variables defined can be arbitrarily large, at least conceptually. Thus, the sets of values in these two examples are countably infinite. In practice, a countably infinite discrete random variable can be converted to a finite discrete random variable by placing an upper bound on the largest value that the random variable can assume. In fact, this is the case in the second example in Table 5.4—no student can receive more than eight tickets.

Definition 5.2
Discrete random variable

A random variable is *discrete* if its set of values is finite or countably infinite in number.

A *continuous* random variable assumes values that occur over an interval or on an intersection of intervals on the real number line. The number of values that a continuous random variable may assume is infinite. Examples of continuous random variables are the height and weight of individuals and the diameter of ball bearings produced by a certain machine. Although each variable is bounded (for example, the weight of individuals is bounded between 0 and, say, 500 pounds), the variable can assume any of an infinite number of values between these bounds. Other examples of continuous random variables are given in Table 5.5.

TABLE 5.5 Examples of continuous random variables

Random variable **X** definition	Range of values of **X**
1. Let **X** be the diameter of ½″ bolts produced by a machine in a machine shop	$X \geq 0$
2. Let **X** be the weight of a student in your class	$X \geq 0$
3. Let **X** be the amount of rainfall in inches recorded at a weather station on a given day	$X \geq 0$
4. Let **X** be the number of ounces of coffee contained in one fill from the school cafeteria vending machine, where the cup can contain at most 8 ounces before it overflows	$0 \leq X \leq 8$

Definition 5.3
Continuous random variable

A random variable is *continuous* if it can assume all the real number values in an interval on the real line (in theory).

In the following discussion of the probability distribution of a random variable, it is essential to keep in mind the difference between these two basic types of random variables. In the next two sections, our attention will be focused on *discrete* random variables. The discussion of the probability distribution of a *continuous* random variable will be deferred until Sections 5.5 and 5.6.

Before leaving this section, let us focus our attention on the distinction between quantitative and qualitative population characteristics, since qualitative characteristics present a problem when we attempt to associate them with a random variable.

A difficulty arises because a random variable is a *numerically* valued function, whereas qualitative characteristics by definition are nonnumerical. One way to circumvent this problem is to assign numbers to the categories of the qualitative characteristic. For example, we might let a 2 denote a male and a 1 denote a female if the qualitative characteristic is sex. However, this may create additional problems *unless* the qualitative characteristic can be quantified on a *meaningful* scale. To illustrate this point, consider the quantification of the qualitative characteristic sex: whereas $10,000 has twice the value of $5,000, being a male (2) does not have twice the value of being female (1).

This difficulty also arises, for example, in ratings. We may have four ratings of a new product—excellent, good, fair, and poor. To treat the rating characteristic as a random variable, we may assign the numbers 4, 3, 2, and 1 to the ratings excellent, good, fair, and poor, respectively. But is excellent four times the value of poor and twice the value of fair? Obviously, attention should be paid to these considerations before a qualitative characteristic is quantified. A more realistic quantification in this illustration might be 10, 8, 3, and 1 for the ratings excellent, good, fair, and poor, respectively.

In quantifying a qualitative characteristic, the problem is thus twofold: categories must be constructed that identify understandable and acceptable gradations of the qualitative characteristic, and a meaningful scale must be identified for its categories. For example, an acceptable gradation of a new product might be excellent, very good, good, fair, poor, and very poor as opposed to the gradations of excellent, good, fair, and poor. Note that this gradation difficulty presumably would not arise in assigning different gradations to the qualitative variable sex—one usually would use two gradations to represent this qualitative random variable! Although our concern is principally with *quantitative* characteristics of the population units, some methods of dealing with *qualitative* factors as random variables will be discussed in Chapters 19 and 20.

■ 5.3 DISCRETE RANDOM VARIABLES AND THEIR DISTRIBUTIONS

Consider the problem in Example 5.3: two radios are sampled from a population of 100 radios in which it is known that 10, or 10 percent of the radios are defective. As a radio is withdrawn from the population to be examined, it is not replaced, so the probability of a defective unit changes on the second and succeeding trials of the experiment. The experiment is to select randomly two radios and count the number of defectives. The sample space generated by this experiment contains four mutually exclusive and collectively exhaustive events. Table 5.6 summarizes the values that the random variable X can assume, the corresponding elementary outcomes of the sample space of the experiment, and the probabilities corresponding to the values of X.

TABLE 5.6 Probability distribution with corresponding elementary outcomes for Example 5.3 (random variable X, event $R_i = i$th radio is defective, event $\bar{R}_i = i$th radio is nondefective)

Value of random variable X	Corresponding elementary outcomes	Probability $P(X = x)$
2	$(R_1 \cap R_2)$	$P(2) = 0.0091$
1	$(R_1 \cap \bar{R}_2), (\bar{R}_1 \cap R_2)$	$P(1) = 0.0909 + 0.0909 = 0.1818$
0	$(\bar{R}_1 \cap \bar{R}_2)$	$P(0) = 0.8091$

Since the four elementary outcomes in the experiment form a set of mutually exclusive and collectively exhaustive events, we know that the probabilities corresponding to these events sum to 1. Thus, the probabilities associated with the values of the random variable—$P(0)$, $P(1)$, and $P(2)$—

must also sum to 1. Indeed, the probability distribution of a discrete random variable must have the property that the probabilities associated with the set of mutually exclusive and collectively exhaustive values of the random variable sum to 1.

The probability distribution can be described by a function $P(X)$, called a *probability mass function,* which assigns probabilities to the values of a discrete random variable.

Definition 5.4
Probability mass function (finite case)

Let the random variable **X** assume a finite number of values, r in total, and denote these values by $x_1, x_2, \ldots , x_r$, where $x_1 < x_2 < x_3 < \cdots < x_r$. Let $P(x_i)$ be the probability that the random variable **X** assumes the value x_i [i.e., $P(\mathbf{X} = x_i)$].

A *probability mass function* is a function that assigns probabilities to the values of a discrete random variable such that the following two conditions on the function $P(X)$ are satisfied:

$$1. \quad 0 \le P(x_i) \le 1, \qquad i = 1, 2, \ldots , r \qquad 2. \quad \sum_{i=1}^{r} P(x_i) = 1$$

The probability mass function, $P(X)$, of a discrete random variable can be stated in three ways. The first method is to list the values of the random variable **X** and the probabilities corresponding to these values $P(x)$ in tabular form as in Table 5.6. Second, it may be possible to write a function in the form of an equation to determine each value the random variable may assume. We give such a function in Table 5.7 for determining the probability of selecting 0, 1, or 2 defective radios from a population of 100 radios in which it is known that 10 of them are defective. It is *not* important at this time that you understand how the equation for determining $P(X)$ is derived in Table

TABLE 5.7   Three ways of presenting the distribution of the discrete random variable in Example 5.3.

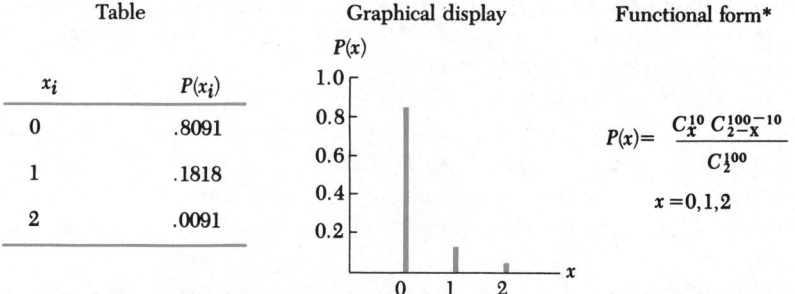

Table		Graphical display	Functional form*

x_i	$P(x_i)$
0	.8091
1	.1818
2	.0091

$$P(x) = \frac{C_x^{10} \, C_{2-x}^{100-10}}{C_2^{100}}$$

$$x = 0, 1, 2$$

* By substituting the values of 0, 1, and 2 into this equation, the probabilities given in the table on the left are determined. This probability mass function is the probability mass function of a discrete random variable discussed in Section 6.4.

5.7—indeed, it will be the topic of Section 6.4 to develop the probability mass function of this random variable. *It is important* at this time to recognize that it is possible to state the probability mass function of a discrete random variable in the form of an equation for many different discrete random variables. The functional form is an alternative to presenting the values of a discrete random variable along with their probabilities of occurrence in the form of a table or a graph, to be discussed next.

The third way to present a probability mass function is a graphical display called a *stick diagram*. In a stick diagram, the x-axis is used for the values of the random variable (X) and the y-axis is used as the probability scale (0 to 1). At each value of the random variable, say x_i, a "stick" of height $P(x_i)$ is drawn vertically, giving the probability of the occurrence of the value x_i.

In Table 5.7, the three ways of presenting a discrete random variable are illustrated for the random variable described in Example 5.3.

Certainly not all three methods of describing the distribution of a discrete random variable need to be used. In Example 5.3, the tabular form of presentation alone would normally be sufficient.

Example 5.4   Due to revelations made known in a congressional hearing on wiretapping, a senator (whom we shall call Herman) became intensely concerned about securing his office from intruders, "bugs," and the like. At his request, three intrusion-sensing devices were installed in his office (at taxpayers' expense!). The devices work independently of one another. Given an intrusion, the first device will detect it with probability 0.8, the second with probability 0.9, and the third with probability 0.7.

Suppose an intrusion occurs. Let X be the number of sensing devices that correctly signal the intrusion. Form the probability distribution of the random variable X.

Solution   The possible values of the random variable X are clearly 0, 1, 2, and 3. Define the events S_i: the ith device correctly triggers, and $\bar{S}_i$: the ith device fails, where $i = 1, 2,$ and 3. The elementary outcomes of the experiment are most easily enumerated by using an outcome tree as illustrated in Figure 5.3.

Since the three sensing devices work independently, the probabilities corresponding to the elementary outcomes are computed by using the corollary of the multiplicative law of probability; for example, $P(E_1) = P(S_1 \cap S_2 \cap S_3) = P(S_1)P(S_2)P(S_3) = (0.8)(0.9)(0.7) = 0.504$.

In Table 5.8, the probability distribution of X is formulated by accumulating the probabilities of corresponding elementary outcomes with the various values of the random variable X. From the probability distribution, it is clear that the probability that at least one device signals $[P(1) + P(2) + P(3) = 1 - P(0)]$ is 0.994—the senator should feel reasonably safe about discovering an intrusion.

The distribution of the random variable X in the form of a stick diagram is given in Figure 5.4A. In this instance, a reasonably simple functional form for $P(X)$ is not obvious. Can you formulate one?

FIGURE 5.3 Eight mutually exclusive and collectively exhaustive elementary outcomes of Example 5.4

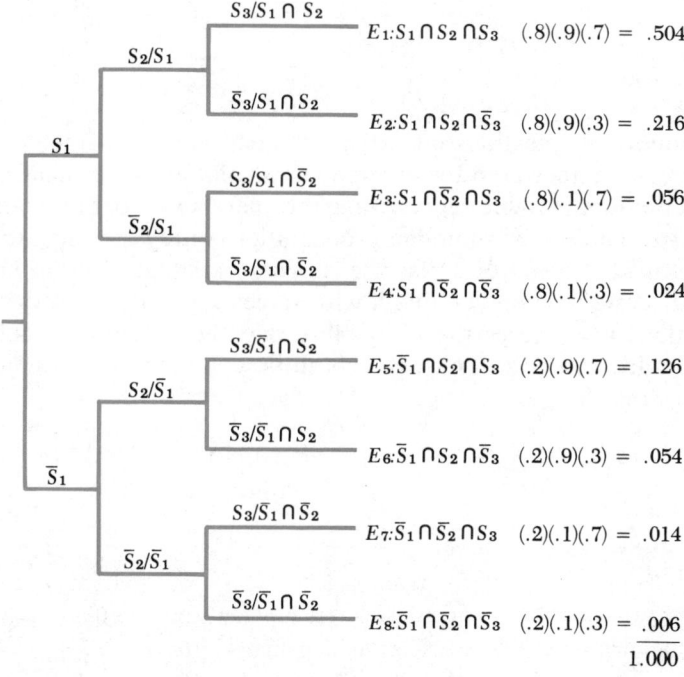

| $S_3/S_1 \cap S_2$ | $E_1: S_1 \cap S_2 \cap S_3$ | $(.8)(.9)(.7) = .504$ |

$$\overline{S}_3/S_1 \cap S_2 \qquad E_2: S_1 \cap S_2 \cap \overline{S}_3 \quad (.8)(.9)(.3) = .216$$

$$S_3/S_1 \cap \overline{S}_2 \qquad E_3: S_1 \cap \overline{S}_2 \cap S_3 \quad (.8)(.1)(.7) = .056$$

$$\overline{S}_3/S_1 \cap \overline{S}_2 \qquad E_4: S_1 \cap \overline{S}_2 \cap \overline{S}_3 \quad (.8)(.1)(.3) = .024$$

$$S_3/\overline{S}_1 \cap S_2 \qquad E_5: \overline{S}_1 \cap S_2 \cap S_3 \quad (.2)(.9)(.7) = .126$$

$$\overline{S}_3/\overline{S}_1 \cap S_2 \qquad E_6: \overline{S}_1 \cap S_2 \cap \overline{S}_3 \quad (.2)(.9)(.3) = .054$$

$$S_3/\overline{S}_1 \cap \overline{S}_2 \qquad E_7: \overline{S}_1 \cap \overline{S}_2 \cap S_3 \quad (.2)(.1)(.7) = .014$$

$$\overline{S}_3/\overline{S}_1 \cap \overline{S}_2 \qquad E_8: \overline{S}_1 \cap \overline{S}_2 \cap \overline{S}_3 \quad (.2)(.1)(.3) = \underline{.006}$$

$$1.000$$

TABLE 5.8 Probability distribution of the discrete random variable **X** in Example 5.4

Values of **X** x_i	Corresponding elementary outcomes E_i	Probabilities $P(x_i)$
0	E_8	0.006
1	E_4, E_6, and E_7	$0.024 + 0.054 + 0.014 = 0.092$
2	E_2, E_3, and E_5	$0.216 + 0.056 + 0.126 = 0.398$
3	E_1	0.504
		1.000

One final note regarding discrete random variables and their distributions: if **X** is a discrete random variable and we wish to determine the probability that **X** is greater than or equal to a but less than or equal to b, where $b \geq a$, we need only to sum the probabilities of the occurrence of the values a through b; that is

$$P(a \leq \mathbf{X} \leq b) = \sum_{x=a}^{x=b} P(x)$$

For instance, in Example 5.4, the probability that X, the number of intrusion devices that operate properly, is 1 or more but less than 3 is given by

$$P(1 \leq X < 3) = P(1 \leq X \leq 2) = \sum_{x=1}^{x=2} P(x)$$
$$= P(1) + P(2) = 0.092 + 0.398 = 0.490$$

Many important questions concerning a quantitative characteristic of a population can be answered by summing probabilities corresponding to the values of a random variable representing the characteristic. For example, an important special case of summing probabilities is the determination of the *cumulative mass function* of a discrete random variable, given in Definition 5.5. For a discrete random variable X with values $x_1 \leq x \leq x_r$, the cumulative mass function, $F(x)$, gives the probability that the random variable X assumes the value x *or less*. Thus, $F(x_i)$ can be determined by summing the probabilities that $X = x_1, x_2, x_3, \ldots, x_i$. Or,

$$F(x_i) = \sum_{x=x_1}^{x=x_i} P(x) = P(x_1) + P(x_2) + \cdots + P(x_i)$$

Definition 5.5
Cumulative mass function of a discrete random variable (finite case)

Let the random variable X assume a finite number of values $x_1, x_2, \ldots, x_r$ and have a probability mass function $P(X)$ as delineated in Definition 5.4. The *cumulative mass function* of the random variable X, denoted $F(X)$, is the probability that the random variable assumes a value less than or equal to $x = x_i$, where $1 \leq i \leq r$. Mathematically, it is given by:

$$F(X = x_i) = P(X \leq x_i) = \sum_{x=x_1}^{x=x_i} P(x)$$

We illustrate the computation of a cumulative mass function of a discrete random variable in the following example.

Example 5.5 Using the data in Example 5.4, determine the values of the cumulative mass function $F(x)$ where the random variable X is defined to be the number of devices that correctly signal an intrusion. Graph the resulting values.

Solution Using the data in Table 5.8, we can easily determine the required probabilities. A graph of the resulting cumulative mass function is given in Figure 5.4B, and a list of values of $F(x)$ is given in Table 5.9.

The use of the cumulative mass function of a discrete random variable is required in many of the nonparametric statistical tests described in Chapter 20.

We will not, in this text, develop the countably infinite discrete random variable case. We will, however, use an important discrete mass function in

TABLE 5.9 Cumulative mass function of the data in Example 5.4

Values of X		
x	P(x)	F(x) = P(X ≤ x)
0.......	0.006	0.006 = P(X ≤ 0)
1.......	0.092	0.098 = P(X ≤ 1)
2.......	0.398	0.496 = P(X ≤ 2)
3.......	0.504	1.000 = P(X ≤ 3)

TABLE 5.10 Expected value of the random variable in Example 5.3

x	P(x)	xP(x)
0	0.8091	0.0000
1	0.1818	0.1818
2	0.0091	0.0182
Total		0.2000

Chapter 6 that is based on a countably infinite discrete random variable. For the development of the countably infinite discrete random variable, the interested reader is directed to the Harnett reference at the end of this chapter.

FIGURE 5.4 Stick diagrams for the probability mass function and the cumulative mass function for the data in Examples 5.4 and 5.5

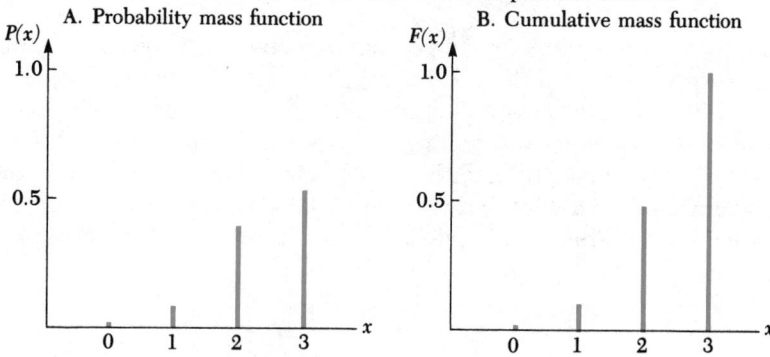

5.4 THE MEAN AND VARIANCE OF A DISCRETE RANDOM VARIABLE

In Example 5.3, we illustrated the connection between a random variable and a population of values. This relationship is shown in Figure 5.5 for the experiment in Example 5.3: selecting 2 radios at random from a production lot of 100 radios. There are 4,950 possible pairs of radios that can be randomly drawn from the population. In 4,005 of these pairs [(0.8091)(4,950)], neither radio is defective; in 900 of these pairs [(0.1818)(4,950)], one radio is defective; and in 45 of these pairs [(0.0091)(4,950)], both radios are defective. The mean of the population of values of the random variable X is, therefore,

$$\frac{(0)(4,005) + (1)(900) + (2)(45)}{4,950} = (0)\left(\frac{4,005}{4,950}\right) + (1)\left(\frac{900}{4,950}\right) + (2)\left(\frac{45}{4,950}\right) = 0.2$$

Thus, the average of the 0s, 1s, and 2s is 0.2—on the average, there are $0.2 = {}^2/_{10}$ radio defective in each pair. Notice that 0.2 is 10 percent of 2, the

FIGURE 5.5 Population set of values of the random variable in Example 5.3

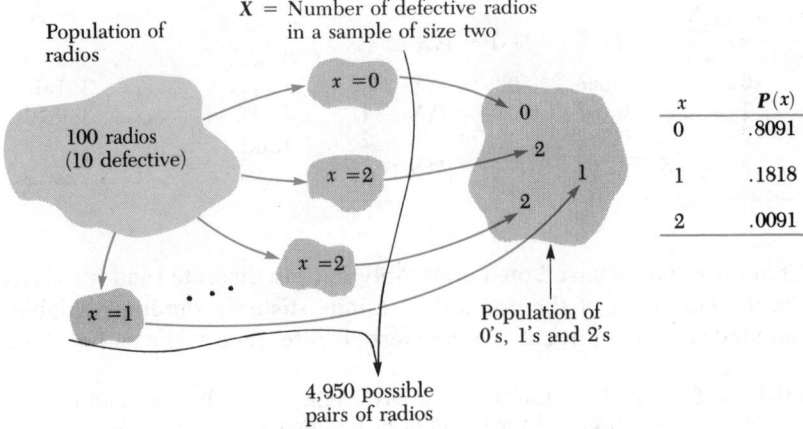

sample size. Since there are 10 percent defective radios in the population, it is not surprising to find that the average number of defectives in a sample of size 2 is 10 percent of 2, or 0.2. Also, notice that the average of the 0s, 1s, and 2s can be determined by summing the products of the values of the random variable **X** and their probabilities. This calculation is shown in Table 5.10. The average of the values of a random variable is called the *expected value* of the random variable and is given for a discrete random variable by the formula:

$$E(\mathbf{X}) = \sum_{\substack{\text{all values} \\ \text{of X}}} xP(x)$$

Definition 5.6
Expected value of a random variable
(discrete case)

Let **X** be a discrete random variable with a finite number of values denoted by $x_1, x_2, \ldots, x_r$. The *mean* or *expected value* of the random variable, denoted by $E(\mathbf{X})$, is given by

$$E(\mathbf{X}) = \sum_{i=1}^{r} x_i P(x_i)$$

We have shown that by using the formula for $E(\mathbf{X})$ we can determine the mean of the values of the random variable for the random variable given in Example 5.3. Indeed, this formula simply "weights" the 0s, 1s, and 2s with their relative frequencies of occurrence: 0.8091, 0.1818, and 0.0091, respectively. In an analogous manner, we can measure the variance of the values of a random variable—in this case, the variability of the 0s, 1s, and 2s.

Definition 5.7
Variance of a random variable
(discrete case)

Let **X** be a discrete random variable with a finite number of values denoted by $x_1, x_2, \ldots, x_r$. The *variance* of the random variable **X**, denoted by $V(\mathbf{X})$, is given by

$$V(\mathbf{X}) = \sum_{i=1}^{r} [x_i - E(\mathbf{X})]^2 P(x_i) = \sum_{i=1}^{r} x_i^2 P(x_i) - [E(\mathbf{X})]^2$$

The second form of $V(\mathbf{X})$ in Definition 5.7 is the "computing formula"—giving two expressions for $V(\mathbf{X})$ is similar to giving two expressions for the variance in Chapter 2: the first is a definitional form and the second is the computing form.

Example 5.6 Compute the variance of the random variable given in Example 5.3.

Solution The easiest way to compute the variance of a discrete random variable is by using a table similar to Table 5.11.

TABLE 5.11 Partial computation of the variance of the random variable **X** described in Example 5.3

x_i	$P(x_i)$	x_i^2	$x_i^2 P(x_i)$
0	0.8091	0	0.0000
1	0.1818	1	0.1818
2	0.0091	4	0.0364
Total			0.2182

From Table 5.10, $E(\mathbf{X}) = 0.2$. Thus,

$$V(\mathbf{X}) = \sum_{i=1}^{3} x_i^2 P(x_i) - [E(\mathbf{X})]^2$$

$$= 0.2182 - (0.2)^2 = 0.1782$$

The variance of **X** is 0.1782, and the standard deviation of **X** is $\sqrt{0.1782} = 0.422$. The population of 0s, 1s, and 2s has a standard deviation of 0.422.

Example 5.7 Compute the expected value, the variance, and the standard deviation of the random variable described in Example 5.4.

Solution In this problem, the random variable **X** is the number of sensing devices that correctly signal an intrusion into the senator's office. The computation of $E(\mathbf{X})$ and part of $V(\mathbf{X})$ is shown in Table 5.12. The expected value of **X** is $E(\mathbf{X}) = 2.4$, and the variance of **X** is

$$V(\mathbf{X}) = 6.220 - (2.4)^2 = 6.22 - 5.76 = 0.46$$

The standard deviation of **X** is thus $\sqrt{0.46} = 0.678$.

TABLE 5.12 Computation of $E(\mathbf{X})$ and $V(\mathbf{X})$ in Example 5.7

Value of $\mathbf{X}$ x_i	Probability $P(x_i)$	$x_i P(x_i)$	x_i^2	$x_i^2 P(x_i)$
0	0.006	$(0)(0.006) = 0.000$	0	$(0)(0.006) = 0.000$
1	0.092	$(1)(0.092) = 0.092$	1	$(1)(0.092) = 0.092$
2	0.398	$(2)(0.398) = 0.796$	4	$(4)(0.398) = 1.592$
3	0.504	$(3)(0.504) = \underline{1.512}$	9	$(9)(0.504) = \underline{4.536}$
		2.400		6.220

Therefore, in repeated trials of the experiment (each trial is an intrusion into the office), on the average 2.4 sensors will correctly identify the intrusion. The standard deviation of the collection of 0s, 1s, 2s, and 3s is 0.678 units.

In Chapter 2, we introduced the Chebychef theorem as a means of interpreting the standard deviation. For example, if $k = 2$ is used in the theorem, it tells us that *at least* $1 - (\frac{1}{2}^2) = \frac{3}{4}$ of the measurements comprising a set of data lie within $k = 2$ standard deviations of their mean. This theorem can also be applied to values of a random variable, since random variables and their distributions can be thought of as frequency distributions of a set of (arbitrarily large) population data. Using the theorem with $k = 2$, we know that at least ¾ of the values of the random variable should lie within two standard deviations of the mean or expected value of the random variable. Applied to Example 5.7, this means that at least ¾ of the values of $\mathbf{X}$ should lie in the interval

$$2.4 - (2)(0.678) \text{ to } 2.4 + (2)(0.678) \quad \text{or} \quad 1.044 \text{ to } 3.756$$

In fact, 90.2 percent of the values lie in this interval (the interval contains the values 2 and 3 of $\mathbf{X}$). The Chebychef theorem provides a lower bound (¾ in this case) for the number of values of $\mathbf{X}$ in the interval. In this instance, the actual number of values within the interval (90.2 percent) is appreciably higher than this lower bound. But the use of the theorem together with the expected value and standard deviation of a random variable can shed much light on the nature of a discrete probability distribution when the distribution mass function is not completely known.

■ 5.5 CONTINUOUS RANDOM VARIABLES AND THEIR DISTRIBUTIONS

We have thus far been concerned with random variables that can assume a finite or countably infinite number of values. In many applications of statistics, the random variable can assume *any* value on an interval of the real number line. Examples are measurements such as height, weight, length, and speed. In these cases, the random variable is called *continuous* because it can take on an infinite number of values that are not countable.

A continuous random variable must be dealt with in a slightly different way than a discrete random variable for reasons that we shall now make clear. Suppose that a machine is producing ball bearings for gyroscopes in space vehicles and that these ball bearings are intended to be exactly 0.5 inch in diameter. Due to flaws in the composition of the bearings, changes in temperature and humidity in the production cycle, and faults related to the fine-tuning of the machine itself, very few bearings produced will measure exactly 0.5 inch in diameter. As a means of quality control, random samples are selected from production runs, and the bearings in these samples are measured by a caliper to an accuracy of 0.001 inch. Let us suppose that ten bearings are randomly sampled, the diameters are measured, and the histogram shown in Figure 5.6A is formed from the data. Six classes have been

FIGURE 5.6

A. Histogram of the diameters of ten randomly sampled ball bearings

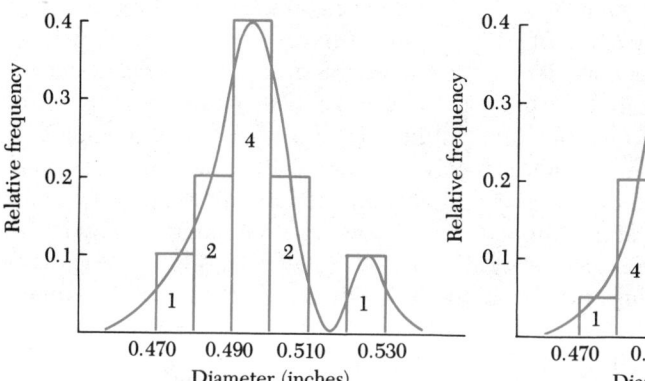

B. Histogram of the diameters of twenty randomly sampled ball bearings

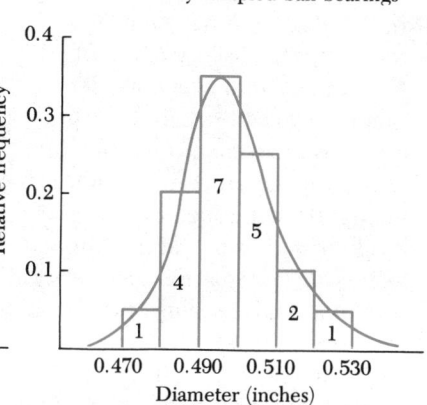

C. Histogram of the diameters of one hundred randomly sampled ball bearings

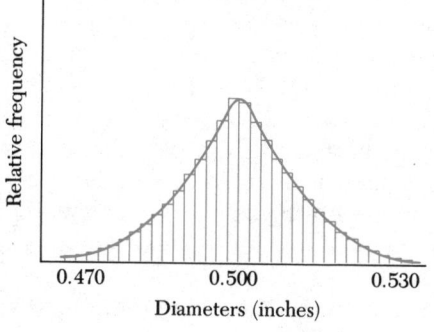

D. The population relative frequency distribution of the diameters of ball bearings

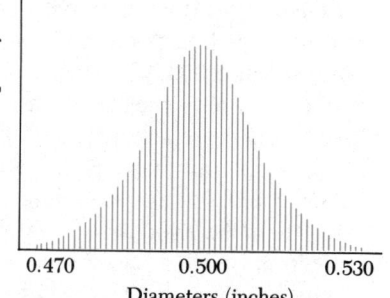

chosen to form the histogram in Figure 5.6A. Now suppose that another ten ball bearings are added to the sample. The histogram formed from using the 20 measurements is shown in Figure 5.6B. A smooth curve is drawn through the midpoints of the tops of the histogram rectangles to accentuate the form of the distribution. Naturally, as more bearings are used to form the random sample, more divisions will appear on the x-axis of the histogram, each representing the diameter of a selected ball bearing. If the sample is increased to include 100 ball bearings, a histogram formed from the resulting diameter measurements and by using more classes may appear as in Figure 5.6C.

There are two important features of the histograms in Figure 5.6A, B, and C that deserve special attention. First, as the sample size increases, many more different values of the diameter will occur. Thus, the set of marks on the histogram axis will more densely cover the interval from 0.470 to 0.530 as the number of sample ball bearings increases. We are assuming here that 0.470 and 0.530 are the minimum and maximum possible diameters that the machine can produce, respectively. Second, as the sample size increases, the values begin to cluster more densely about a central value. If the number of classes used to form the histogram is increased, the width of the histogram rectangles becomes smaller and the midpoints on the tops of the rectangles more closely follow a smooth, continuous curve.

Suppose that we now denote the diameter of the ball bearings by the random variable **X** and continue to take larger and larger samples of ball bearings from the production line. Intuitively, it is reasonable to assume that the larger the sample, the more closely the sample frequency distribution and histogram will resemble the population frequency distribution (the probability distribution of the random variable **X**) and its stick diagram. Figure 5.6D shows the population frequency distribution (stick diagram) for the diameters of all ball bearings produced on the machine. Since the caliper can only measure with an accuracy of 0.001 inch, the population values are 0.470, 0.471, 0.472, . . . , 0.529, and 0.530 inch. This is a set of finite values (61 in all), so Figure 5.6D is the probability distribution of the *discrete* random variable **X,** where its values are given on the x-axis of the graph and the probabilities that **X** assumes these values are given on the y-axis.

The random variable **X** = diameter of ball bearings is, in fact, however, a *continuous* random variable because its values can occur *anywhere* on the real number line from 0.470 to 0.530. The values of **X** are restricted to the finite set above 0.470, 0.471, . . . , 0.529, and 0.530, in this case since our caliper can only measure with an accuracy of 0.001 inch. If the measuring instrument were accurate to 0.00001 inch, then many more values of **X** would be possible and many sticks would occur in the gaps between the sticks in Figure 5.6D. Taken to the limit, if our measuring device is "infinitely" accurate, then the probability distribution of **X** may be represented as a dense collection of sticks as shown in Figure 5.7.

The analogy to the collection of sticks in Figure 5.6D for the discrete

FIGURE 5.7 Probability distribution of the random variable **X** = diameter of ball bearings

Relative frequency

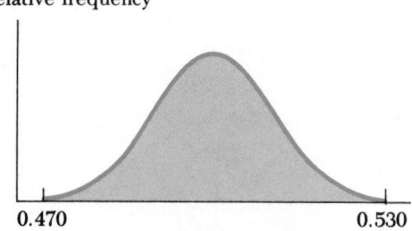

0.470   0.530

FIGURE 5.8 Probability density function $f(x)$ of a continuous random variable

$f(x)$

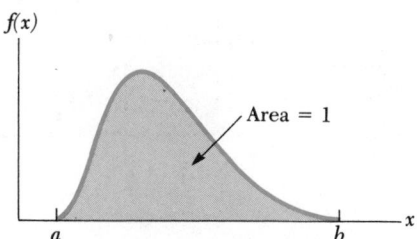

Area = 1

a b

random variable **X** is the shaded area under the curve, denoted by $f(x)$, in Figure 5.7. Since the sum of the heights of the sticks in Figure 5.6D must be 1, by analogy, the area under $f(x)$ must also be 1 in Figure 5.7. This leads us to a formal definition of the probability distribution of a continuous random variable.

Definition 5.8
Probability density function (continuous case)

Let **X** be a continuous random variable defined over an interval of the real number line from a to b as illustrated in Figure 5.8. The *probability density function* of **X**, denoted by $f(\mathbf{X})$, must satisfy two conditions:

1. $f(x) \geq 0$, $a \leq x \leq b$
2. The area under $f(x)$ from $x = a$ to $x = b$ must be 1.

Notice the similarity between the conditions for $P(\mathbf{X})$ to be a probability mass function of a discrete random variable given in Definition 5.4 and the conditions placed on $f(\mathbf{X})$ to be a probability density function for a continuous random variable given in Definition 5.8. In the discrete case, probability is *massed* at the discrete values of the random variable, whereas in the continuous case, the probability is spread *densely* over the range of values of the random variable. In the discrete case, the probability sticks must sum to 1, whereas in the continuous case, the dense set of sticks [the area under the function $f(x)$ from a to b] must have an area of 1.

Recall that if we want to compute $P(c \leq \mathbf{X} \leq d)$ where $d \geq c$ for a discrete random variable, we need only sum the probabilities of **X** taking on the values c through d; that is:

$$P(c \leq \mathbf{X} \leq d) = \sum_{x=c}^{x=d} P(x); \quad d \geq c$$

The analogy to a continuous random variable is straightforward. The

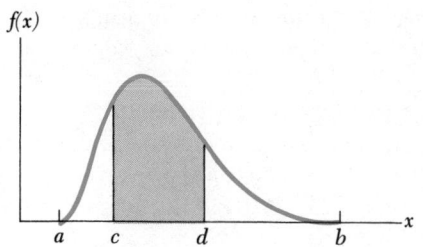

$f(x)$

a c d b x

FIGURE 5.9   Area under the curve corresponding to $P(c \le X \le d)$

probability that a continuous random variable **X** takes on a value between c and d is equal to the *area* under $f(x)$ between c and d, as illustrated in Figure 5.9.

There is *one* significant difference in computing probabilities for a discrete and a continuous random variable. If the discrete random variable **X** can assume the value e, then the probability that **X** does assume this value is $P(e)$, the height of the stick over e on the stick diagram for **X**. However, if e resides within the defined interval for a continuous random variable **X**, the probability that **X** assumes the value of e is 0; that is, $P(X = e) = 0$. This is true regardless of the numerical value of e over the defined interval of real numbers for the random variable **X**. *Thus, $P(X = e)$ is not equal to $f(e)$, the height of the curve at the point $x = e$!*

The reason for this may be argued as follows. Let us assume that the continuous random variable **X** is defined for values on the real number line from $x = a$ to $x = b$. Assume that e is in the interval from a to b and let $P(X = e) = f(e)$. Since we have defined the probability that **X** assumes the value of e in the interval a to b in this manner, this definition must also hold for all other values, say $e_1, e_2, e_3, \ldots$ in the interval from a to b. There are an infinite number of values (not countable) between a and b, so we can see that the sum of the "probabilities" $f(e) + f(e_1) + f(e_2) + f(e_3) + \cdots$ will quickly exceed 1, which means that any particular number, say $f(e_2)$, can no longer be interpreted as a probability. Indeed, any one of the $f(e_i)$ values may exceed 1 by itself if the height of the curve at the point $X = e_i$ is greater than 1.

Another way to look at this is to write $P(X = e)$ as $P(e \le X \le e)$ and use the definition for the probability that **X** assumes a value between two points, say c and d, as illustrated in Figure 5.9. Since there is no *area* between e and e, $P(X = e) = P(e \le X \le e) = 0$.

The student who has had calculus should begin to recognize some familiar notions at this point. First, the area under a continuous curve, say $f(x)$, from c to d can be determined by integration between the limits of c to d. Second, the condition that the area under the curve $f(x)$ from $x = a$ to $x = b$ (where the random variable is defined over the range of values, $a \le X \le b$) must equal 1 can be concisely stated as: the integral of $f(x)$ from $x = a$ to $x = b$ must equal 1.

tively, just as they are in the discrete case. The computation of the expected value and the variance of a continuous random variable is very straightforward, but a derivation of these two quantities requires a knowledge of calculus that is beyond the scope of this text. Hence, we shall state without proof the expected value and the variance of a continuous random variable in this text. For the reader with a calculus background, we present in Appendix A the basic formulas required to determine the mean and variance of a continuous random variable. For a treatment of the derivation of these quantities beyond that in Appendix A, the reader is directed to the references at the end of the chapter.

Example 5.8 Consider the probability density function

$$f(x) = \begin{cases} 1 & 0 \le x \le 1 \\ 0 & x < 0 \text{ or } x > 1 \end{cases}$$

(This density function might arise in a practical situation when the random variable X is representing service times that are assumed to be uniformly distributed over the interval from 0 to 1—measured in time units—for example.) Find:

$$P(0 \le X \le 1) \qquad P(0.25 \le X \le 0.60) \qquad P(X > 0.75)$$

Solution The probability distribution of the random variable X is illustrated in Figure 5.10. Since the probability distribution is a square from

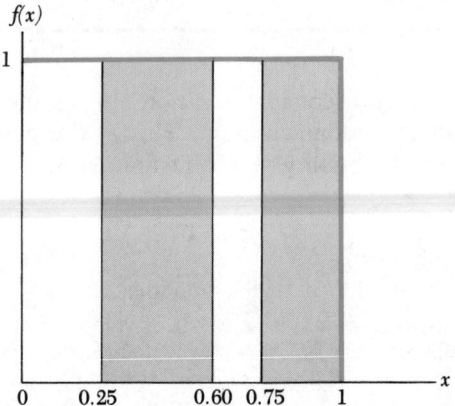

FIGURE 5.10 Probability distribution of the random variable in Example 5.8

$x = 0$ to $x = 1$, we can compute the required probabilities by using the fact that the area of a rectangle is given by the product of its base and its height. Thus,

$$P(0 \le X \le 1) = (1)(1) = 1$$

And since $f(x) \ge 0$ for all x, $f(x)$ is a proper probability density function.

$$P(0.25 \le X \le 0.60) = (0.60 - 0.25)(1) = 0.35$$

$$P(X > 0.75) = (1 - 0.75)(1) = 0.25$$

Notice that this is also $P(X \geq 0.75)$, since $P(X = 0.75) = 0$.

From Example 5.8, it is clear that calculus is *not* needed to answer the required probability questions—the areas corresponding to the probabilities can be determined by using results in plane geometry. Although it will not always be possible to determine the area under a curve for many probability density functions of a continuous random variable such as those discussed in Chapter 7 by using arguments from plane geometry, tables are always provided for the random variables discussed, which enable the student to look up the required probabilities corresponding to the area(s) under a curve.

One last point concerning the distribution of continuous random variables: the student might wonder about representing the distribution of a continuous random variable by the distribution of a discrete random variable through "discretizing" the continuous random variable. If a sufficient number of values are chosen to represent the continuous random variable, its distribution could be approximated arbitrarily by a discrete distribution and then *no* calculus would be required in the calculation of probabilities concerning the random variable. Indeed, this is occasionally done, although, perhaps surprisingly at this point, the reverse representation is more frequently the case—approximating a discrete probability distribution by a continuous probability distribution! There are primarily two reasons for this. First, it is often easier to work with a continuous random variable than with a discrete random variable that has *many* values. Second, there are numerous continuous probability distribution models in statistics that have been used successfully to represent a variety of distributional shapes. We will discuss the properties of a number of discrete and continuous distribution models in the next two chapters.

■ 5.7 MATHEMATICAL EXPECTATION

Computing an expected value is a much more general tool than simply a means of finding the mean or the variance of a random variable. Let us review the process of determining the expected value of a discrete random variable. Suppose the discrete random variable **X** has the probability mass function $P(\mathbf{X})$ given in Table 5.14. The computation of $E(\mathbf{X}) = \Sigma x P(x)$ is also shown in Table 5.14. The probability distribution of **X** tells us that 20 percent

TABLE 5.14 Probability mass function of the discrete random variable X and the computation of its expected value $E(X)$

x	$P(x)$	$xP(x)$
0	0.2	$(0)(0.2) = 0.0$
1	0.3	$(1)(0.3) = 0.3$
2	0.5	$(2)(0.5) = 1.0$
	1.0	$E(X) = 1.3$

of the values of **X** are 0s, 30 percent are 1s, and 50 percent are 2s among the population of all values of **X** generated by repeated trials of the experiment. As noted earlier, by forming the products of the value of the random variable and its relative frequency in this conceptualized population and by adding these products together for each and every value of **X**, we are forming the population mean value—the mean of the random variable **X**. Thus, the values of **X**, weighted by their respective relative frequencies, give us the average of **X**. We can extend this notion to form the expected value of *any* function of the random variable **X**, as the following definition describes.

Definition 5.10
Expected value of a function of a random variable
(discrete case)

Let **X** be a discrete random variable with probability mass function $P(X)$ and consider a function of **X**, denoted by $h(X)$. The *average* or *expected value* of $h(X)$ is given by

$$E[h(X)] = \sum_{\text{all X}} h(x)P(x)$$

where the symbol $\sum_{\text{all X}}$ means that the summation process is to be conducted over all values of the random variable **X**.

First, we note that if $h(X) = X$, then $E[h(X)]$ represents the expected value of **X**. Recall the definition of the variance, $V(X)$, of a discrete random variable (Definition 5.7):

$$V(X) = \sum_{\text{all X}} [x - E(X)]^2 P(x)$$

Thus, if $h(X) = [X - E(X)]^2$, then $E[h(X)]$ represents the variance of the random variable **X**. The variance is, therefore, the *average* of the squared deviations of the values of the random variable from the mean or expected value.

Example 5.9 Consider the distribution of the random variable described in Example 5.4.

x	$P(x)$
0	0.006
1	0.092
2	0.398
3	0.504
	1.000

Determine the expected value of $h(X)$, where $h(X)$ is equal to X, $[X - E(X)]^2$, and $2X$.

Solution   The computations of the required expectations are shown in Table 5.15. Note that the solutions to the first two parts represent the mean and the variance of the random variable X, which we previously found in Table 5.12. Thus, $E(X) = 2.4$, $E[X - E(X)]^2 = V(X) = 0.46$, and $E(2X) = 4.8$.

A similar definition holds for continuous random variables.

TABLE 5.15   Computation of $E(X)$, $E[X - E(X)]^2$, and $E(2X)$

x	$P(x)$	$xP(x)$	$x - E(X)$	$[x - E(X)]^2$	$[x - E(X)]^2 P(x)$	$2xP(x)$
0	0.006	0.000	−2.4	5.76	0.03456	0.000
1	0.092	0.092	−1.4	1.96	0.18032	0.184
2	0.398	0.796	−0.4	0.16	0.06368	1.592
3	0.504	1.512	+0.6	0.36	0.18144	3.024
	1.000	2.400			0.46000	4.800

Occasionally, we may wish to construct a new random variable, say Y, from an old one, X, by forming the linear relationship $Y = a + bX$, where a and b are constants. If we know the mean $E(X)$ and the variance $V(X)$ of X, can we determine the mean $E(Y)$ and the variance $V(Y)$ of Y without knowing the probability distribution of either random variable? The answer is given by the following theorem.

Theorem 5.1

Let X denote a random variable (discrete or continuous) with mean and variance known to be $E(X)$ and $V(X)$, respectively. Define a new random variable $Y = a + bX$, where a and b are known constants. Then,

$$E(Y) = E(a + bX) = a + bE(X) \qquad \text{and} \qquad V(Y) = V(a + bX) = b^2 V(X)$$

Theorem 5.1 tells us that the expected value of a constant $\{E(a)\}$ is the constant $\{a\}$, whereas the expected value of a constant times a random variable $\{E(bX)\}$ is equal to the constant times the expected value of the random variable $\{b \cdot E(X)\}$. Furthermore, it indicates that the variance of a constant $\{V(a)\}$ is 0, whereas the variance of a constant times a random variable $\{V(bX)\}$ is the *square* of the constant times the variance of the random variable $\{b^2 \cdot V(X)\}$. We summarize several important properties of the expectation operator in Table 5.16, and illustrate the use of these properties in the following example.

TABLE 5.16 Properties of the expectation operator

Property number	Property	Interpretation
Property 1	$E(a) = a$	The expected value of a constant is equal to the value of the constant
Property 2	$E(kX) = kE(X)$	The expected value of a constant times a random variable is equal to the constant times the expected value of the random variable
Property 3	$E(X \pm Y) = E(X) \pm E(Y)$	The expected value of the sum (or difference) of two random variables is equal to the sum (or difference) of their expected values
Property 4	$V(k) = 0$	The variance of a constant is equal to 0
Property 5	$V(kX) = k^2 V(X)$	The variance of the product of a constant times a random variable is equal to the square of the constant times the variance of the random variable
Property 6	If **X** and **Y** are independent, then $E(X \cdot Y) = E(X) \cdot E(Y)$	The expected value of the product of two independent random variables is equal to the product of their expected values
Property 7	If **X** and **Y** are independent, then $V(X \pm Y) = V(X) + V(Y)$	The variance of the sum (or difference) of two independent random variables is equal to the *sum* of their variances

Example 5.10 Suppose the discrete random variable **X** has the distribution shown in Table 5.17. From the computations in Table 5.17, $E(X) = 2.0$ and $V(X) = 4.0$. Now suppose that we consider a new random variable **Y**, where $Y = 10 + 2X$. Theorem 5.1 and Properties 1, 2, 4, and 5 in Table 5.16 tell us that

$$E(Y) = E(10 + 2X) = 10 + 2E(X) = 10 + 2(2) = 14$$

and

$$V(Y) = V(10 + 2X) = 0 + (2)^2 V(X) = (4)(4) = 16$$

Verify this by forming the probability distribution of the random variable **Y**.

Solution Since $Y = 10 + 2X$, the values of Y are: $10 + 2(0) = 10$, $10 + 2(1) = 12$, and $10 + 2(5) = 20$. The probabilities corresponding to the values

TABLE 5.17 Computation of $E(X)$ and $V(X)$ for a discrete random variable

x	$P(x)$	$xP(x)$	$x - E(X)$	$[x - E(X)]^2$	$[x - E(X)]^2 P(x)$
0	0.2	0.0	-2	4	0.8
1	0.5	0.5	-1	1	0.5
5	0.3	1.5	3	9	2.7
	1.00	$E(X) = 2.0$			$V(X) = 4.0$

10,12, and 20 are 0.2, 0.5, and 0.3 (from the distribution of **X**), respectively. Table 5.18 gives the probability distribution for **Y** and shows the computation of $E(Y)$ and $V(Y)$, which are 14 and 16, respectively, as Theorem 5.1 told us they would be!

TABLE 5.18 Computation of $E(Y)$ and $V(Y)$ for the discrete random variable $Y = 10 + 2X$

y	$P(y)$	$yP(y)$	$y - E(Y)$	$[y - E(Y)]^2$	$[y - E(Y)]^2 P(y)$
10	0.2	2	-4	16	3.2
12	0.5	6	-2	4	2.0
20	0.3	6	6	36	10.8
	1.00	$E(Y) = 14$			$V(Y) = 16.0$

■ 5.8 BIVARIATE RANDOM VARIABLES

Often in statistics we are interested in studying the relationship between variables. For example, we may be interested in studying the relationship between an individual's height and weight, or in exploring the statistical relationship between dollars expended on advertising effort and sales of a product. When two or more random variables are jointly involved in an experiment, their outcomes may be generated by what is called a *multivariate probability function*. In this chapter, we will discuss only *bivariate* probability functions and only for the discrete case. The development of a multivariate continuous probability function follows directly from the explanation of a bivariate probability function of discrete random variables.

The joint or bivariate probability mass function of two random variables **X** and **Y**, denoted $P(X,Y)$, is delineated in Definition 5.11: $P(x,y) = P(X = x, Y = y)$. That is, $P(x,y)$ denotes the probability that the random variable **X** is equal to x *and* the random variable **Y** is equal to y. Hence the term *bivariate* or *joint* probability function.

To understand the notion of a bivariate probability function, consider the data in Table 5.19. This information represents sales of walnut veneers to home craftsmen in standard widths and lengths. For example, over the interval examined, 100 pieces of veneer were sold, and 18 of these pieces were 48 inches long and 12 inches wide. Fifty of the pieces were 6 inches wide, 40 of

TABLE 5.19 Sales of walnut veneers

Length (inches) Width (inches)	24	36	48	Totals
6	12	16	22	50
12	8	24	18	50
Totals	20	40	40	100

Source: Company records.

the pieces were 36 inches long, and so on. We can convert the data in Table 5.19 to a bivariate probability distribution for the lengths and widths of veneers sold if we divide each quantity given in Table 5.19 by the total number of occurrences, 100. This has been done, and the resulting bivariate probability function, $P(X,Y)$, where the random variable X represents the widths of veneer sold and Y represents the lengths of the veneers sold, is given in Table 5.20. A graph of the resulting bivariate probability mass func-

TABLE 5.20 Probabilities of selling various lengths and widths of walnut veneers based on data in Table 5.19

x \ y	24	36	48	$P(x)$
6	0.12	0.16	0.22	0.50
12	0.08	0.24	0.18	0.50
$P(y)$	0.20	0.40	0.40	1.00

tion is given in Figure 5.11. Note that since two random variables (X and Y) are involved, it becomes necessary to draw the resulting mass function in three dimensions. From Table 5.20, we can determine that the probability that a sheet of veneer 12 inches wide and 24 inches long is sold is equal to $P(12,24) = 0.08$. Other probabilities can also be read directly from the table.

In Section 5.7, we examined the general form for the expectation of a function of a random variable. The counterpart to this when two random variables are involved is given in Definition 5.12. A special case of the expectation of a function of a bivariate random variable occurs when $h(X,Y)$

FIGURE 5.11 Graph of bivariate probability
distribution given in Table 5.20

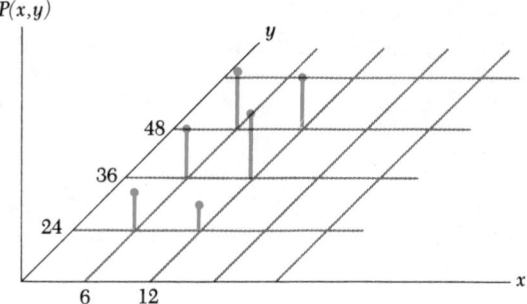

$= [\mathbf{X} - E(\mathbf{X})][\mathbf{Y} - E(\mathbf{Y})]$. In this instance, $E[h(\mathbf{X},\mathbf{Y})]$ is called the *covariance* of the random variables $\mathbf{X}$ and $\mathbf{Y}$ as described in Definition 5.13. The covariance is a statistical measure that indicates how the two random variables vary together, or "co-vary." $C(\mathbf{X},\mathbf{Y})$ will tend to be a large positive number whenever large values of $\mathbf{X}$ [relative to $E(\mathbf{X})$] tend to be associated with large values of $\mathbf{Y}$ [relative to $E(\mathbf{Y})$], and small values of $\mathbf{X}$ tend to be associated with small values of $\mathbf{Y}$. Similarly, $C(\mathbf{X},\mathbf{Y})$ will tend to be a large negative number whenever low values of one random variable tend to be associated with high values of the other random variable, and high values of one random variable tend to be associated with low values of the other random variable. Whenever two random variables are independent (i.e., they are not statistically related), then $C(\mathbf{X},\mathbf{Y}) = 0$.

Definition 5.11

Bivariate probability mass function and cumulative mass function (finite case)

Let the random variable $\mathbf{X}$ assume a finite number of values $x_1, x_2, \ldots, x_r$, where $x_1 < x_2 < \cdots < x_r$, and let the random variable $\mathbf{Y}$ also assume a finite number of values $y_1, y_2, \ldots, y_s$, where $y_1 < y_2 < \cdots < y_s$. Let $P(x,y)$ denote the probability that the random variable $\mathbf{X}$ assumes the value x *and* the random variable $\mathbf{Y}$ assumes the value y.

A *bivariate probability mass function* is a function that assigns probabilities to joint values of the discrete random variables $\mathbf{X}$ and $\mathbf{Y}$ such that the following two conditions on the function $P(\mathbf{X},\mathbf{Y})$ are satisfied:

$$0 \le P(x_i,y_j) \le 1; \qquad 1 \le i \le r, 1 \le j \le s$$

$$\sum_{\text{all Y}} \sum_{\text{all X}} P(x,y) = 1$$

The *bivariate cumulative mass function*, denoted $F(\mathbf{X},\mathbf{Y})$, gives the probability that the random variable $\mathbf{X}$ is less than or equal to x *and* the random variable $\mathbf{Y}$ is less than or equal to y. Mathematically, it is given by:

$$F(x,y) = P(\mathbf{X} \le x, \mathbf{Y} \le y)$$

Experience in computing the covariance of two random variables **X** and **Y** is provided in the following example.

Example 5.11　Using the data in Table 5.20, compute $C(\mathbf{X},\mathbf{Y})$, where the random variable **X** corresponds to the widths of veneer sold and **Y** corresponds to the lengths of veneer sold. Use the first expression in Definition 5.13 for the covariance of two random variables to compute the covariance for this data set, and verify that the second expression for the covariance gives the same result using the data provided. Are the random variables **X** = widths of veneer sold and **Y** = lengths of veneer sold independent?

Solution　Relevant computations are given in Tables 5.21 and 5.22. Note that either of the definitional forms for the covariance of two random variables given in Definition 5.13 produces the result $C(\mathbf{X},\mathbf{Y}) = 0$.

We have to be extremely careful in answering the second part of Example 5.11. Although it is true that if two random variables are independent, their covariance is 0, it is not *in general* true that if the covariance of two random variables is equal to 0, the random variables are independent. To show independence between two random variables, we must demonstrate that $P(x,y) = P(x)P(y)$ for *all* possible combinations of **X** and **Y**. That is, if $P(x,y) \neq P(x)P(y)$ for *any* values of **X** and **Y**, then the random variables are dependent. For example,

$$\left. \begin{array}{l} P(\mathbf{X} = 6, \mathbf{Y} = 24) = 0.12 \\ P(\mathbf{X} = 6) = 0.50 \\ P(\mathbf{Y} = 24) = 0.20 \end{array} \right\} \; P(6, 24) = 0.12 \neq 0.10 = P(\mathbf{X} = 6)P(\mathbf{Y} = 24)$$

TABLE 5.21 Computation of the covariance using the data in Table 5.20 and the first definitional form in Definition 5.13

(x, y)	$P(x, y)$	$x - E(\mathbf{X})$	$y - E(\mathbf{Y})$	$[x - E(\mathbf{X})][y - E(\mathbf{Y})]$	$[x - E(\mathbf{X})][y - E(\mathbf{Y})]P(x, y)$
(6, 24)	0.12	−3	−14.4	+43.2	5.184
(6, 36)	0.16	−3	− 2.4	+ 7.2	1.152
(6, 48)	0.22	−3	9.6	−28.8	−6.336
(12, 24)	0.08	3	−14.4	−43.2	−3.456
(12, 36)	0.24	3	− 2.4	− 7.2	−1.728
(12, 48)	0.18	3	9.6	+28.8	5.184
Totals	1.00				$C(\mathbf{X}, \mathbf{Y}) = \overline{0.000}$

TABLE 5.22 Computation of the covariance using the data in Table 5.20 and the second definitional form in Definition 5.13

x	$P(x)$	y	$P(y)$	$xP(x)$	$yP(y)$	(x, y)	$P(x, y)$	$x \cdot y$	$x \cdot yP(x, y)$
6	0.50	24	0.20	3.0	4.8	(6, 24)	0.12	144	17.28
12	0.50	36	0.40	6.0	14.4	(6, 36)	0.16	216	34.56
		48	0.40		19.2	(6, 48)	0.22	288	63.36
						(12, 24)	0.08	288	23.04
						(12, 36)	0.24	432	103.68
						(12, 48)	0.18	576	103.68
Totals	1.00		1.00	$E(\mathbf{X}) = \overline{9.0}$	$E(\mathbf{Y}) = \overline{38.4}$		1.00		$E(\mathbf{X}, \mathbf{Y}) = \overline{345.60}$

$$C(\mathbf{X}, \mathbf{Y}) = E(\mathbf{X} \cdot \mathbf{Y}) - E(\mathbf{X}) \cdot E(\mathbf{Y}) = 345.60 - 9.0 \cdot 38.4 = 0$$

Hence, we would conclude that the two random variables are dependent, even though their covariance is equal to 0. We claim the random variables are dependent because we discovered at least one value x and y for which $P(x,y) \neq P(x)P(y)$.

Unfortunately, the covariance is not an easy statistical measure to interpret. Quite often, for example, the value of the covariance may change simply because the units used to measure either or both of the quantitative random variables changes. Thus, we may find that the covariance of two random variables differs from a value previously computed because the units used have changed. Still, the covariance of two random variables is an important statistical measure. We will have an opportunity to use this measure again in Chapter 13 when we study correlation, which measures the degree of linear relationship between two random variables. We will learn that the measure of correlation is independent of the units of measurement and is a function of the covariance of the two random variables.

■ 5.9 SUMMARY

In applications of inferential statistics, random variables and their distributions play a vital role. The process may be summarized as follows. A population is described and the numerical characteristics of the population units of interest are identified. A random variable is defined to represent the values of the numerical characteristic. The probability distribution of the random variable is formed and used to answer questions concerning the population characteristic.

There are two kinds of random variables—*discrete* and *continuous*. A discrete random variable takes on a finite or countably infinite number of values, whereas a continuous random variable takes on an infinite number of values that are not countable.

The average value, the standard deviation, and the variance of a characteristic of a random variable can be determined by computing the expected value and standard deviation of the random variable representing the values of the characteristic. The expected value, variance, and standard deviation of a random variable are computed from its probability distribution. The use of integral calculus is often required to compute these numerical measures of a continuous random variable, although tabled values are available for most density functions of a continuous random variable.

The expected value operator may be used to find the average value of any function of a random variable. Particular attention is focused in this chapter on using this notion to compute the mean and variance of a linear function of a random variable. Table 5.16 summarizes several important properties of the expectation operator.

■ REFERENCES

Harnett, D. L. *Introduction to Statistical Methods*. 2d ed. Reading, Mass.: Addison-Wesley Publishing Co., Inc. 1975.

Hoel, P. G. *Introduction to Mathematical Statistics*. 3d ed. New York: John Wiley & Sons, Inc., 1962.

Hogg, R. V., and Craig, A. T. *Introduction to Mathematical Statistics*. 3d ed. New York: Macmillan, Inc., 1970.

Lindgren, B. W. *Statistical Theory*. New York: John Wiley & Sons, Inc., 1959.

Parzen, E. *Modern Probability Theory and Its Applications*. New York: John Wiley & Sons, Inc., 1960.

■ PROBLEMS

5.1. Explain in your own words what is meant by a *random variable*. List several experiences you encountered recently in which reference was made to a random variable without actually calling it by that name. In each instance, try to identify the process that was generating the values of the random variable.

5.2 Distinguish between a probability mass function of a discrete random variable and a probability density function of a continuous random variable. In what ways are these functions similar?

5.3 Explain why, in general, it is not necessary to understand calculus to determine the probability that a continuous random variable assumes a value between two values of the random variable, c and d, with $c < d$.

5.4. Distinguish between a probability mass function and a cumulative mass function of a discrete random variable.

5.5. In a certain population of voters, it is known that 60 percent are Democrats and 40 percent are Republicans. If a sample of three voters is extracted from this population, find the probability distribution of the random variable X = number of Democrats in the sample. Assume that the population is growing sufficiently so that as a unit is extracted from the population, a new one enters such that the percentage of Democrats and Republicans in the population remains constant.

5.6. Assume in Problem 5.5 that the population of voters consists of 100 individuals (i.e., 60 Democrats and 40 Republicans). Find the probability distribution of the random variable X = number of Democrats sampled in a sample of size three. How does this result differ from that obtained in Problem 5.5, and how do you explain this difference?

5.7. A college recruiter wishes to select three of the five job applicants he will interview. Although all three candidates appear to be equally qualified, there is a best candidate, a second best candidate, and so on. Refer to these three unknown best candidates as successes, and let the random variable X = number of successes hired. Determine the probability distribution of the random variable X.

5.8. Sketch in the form of a stick diagram the cumulative density function for the distribution developed in Problem 5.7. Using the cumulative density function developed, derive the mass function for this distribution by taking the differences of appropriate cumulative values. Plot in the form of a stick diagram the mass function for this distribution.

5.9. A multiple-choice exam consists of four questions, each of which has five possible answers (a through e). If a student is forced to guess on all four questions, what is the probability distribution of the random variable X = number of correct guesses? What is the most likely number of correct guesses?

5.10. In an organization consisting of three

Number of children, x	0	1	2	3	4	5	6
Proportion of families, $P(x)$	0.48	0.20	0.15	0.08	0.05	0.03	0.01

women and seven men, a committee of four individuals is to be selected at random from the ten people. Find the probability distribution of the random variable X = number of women on the committee.

5.11. Let X be a random variable with the probability distribution given in the following table. Find the following:

x	$P(x)$
0	0.60
1	0.30
2	0.05
3	0.05

a. The expected value $E(X)$ of X.
b. The variance $V(X)$ of X.
c. The mode of X.
d. The range of X.

5.12. Determine the expected value $E(X)$ and the variance $V(X)$ of the random variable defined in Problem 5.9.

5.13. Plot the cumulative mass function of the random variable described in Problem 5.9.

5.14. Consider the formula $P(x)$ = $(2x - x^2)/2.5$, where x = 0, 0.5, 1.5, and 2.0. Is $P(x)$ a probability mass function? Explain.

5.15. Determine the expected value $E(X)$ and the variance $V(X)$ of the random variable defined in Problem 5.5.

5.16. Determine the expected value $E(X)$ and the variance $V(X)$ of the random

variable defined in Problem 5.6. How do the expected values and variances compare with those determined in Problem 5.15?

5.17. From the most recent national census, it is found that the number of children (X) in American families follows the following probability distribution:

Here it is assumed that the proportion of families with more than six children is negligible.

a. Find the expected value and variance of X.
b. Form a stick diagram of this distribution. Is the distribution skewed?
c. If we let Y represent the number of children in a family, *given* that the family has at least one child, the probability distribution of Y is given by:

y	1	2	3	4	5	6
$P(y)$	0.38	0.29	0.15	0.10	0.06	0.02

How is this distribution formed from the distribution of the random variable X?

d. Find the expected value and variance of Y.
e. Is $E(X)$ or $E(Y)$ more properly the "average family size"? Explain.
f. Define the random variable $Z = Y + 2$. Find $E(Z)$ and $V(Z)$. Interpret $E(Z)$. Might not this be a better number to call the "average family size"? Explain.
g. Use the Chebychef theorem described in Chapter 2 to determine the proportion of values of X that

are contained within two standard deviations of the mean value. Compare this proportion with the actual proportion in this interval determined directly from the probability distribution.

5.18. Given the following probability distribution,

x	$P(x)$
0.........	0.30
1.........	0.40
2.........	0.30

determine:
a. $E(X)$.
b. $V(X)$.
c. $E(10X)$.
d. $E(12 - X)$
e. $E\{[X - E(X)]^2\}$.
f. $E(Y)$, where $Y = 12 + 3X$.
g. $V(Y)$, where $Y = 13 - 2X$.

Are any of the results the same? If so, why? For each of your answers, list those properties from Table 5.16 you used to simplify computations.

5.19. Suppose that the random variable **X** has the following probability distribution. Find:

x	$P(x)$	
0	0.10	a. $E(X)$. 1.00
1	0.10	b. $E(2X)$.
2	0.40	c. $E[(X - 1)^2]$.
3	0.40	d. $E[X - E(X)]$.

5.20. A salesman receives orders for zero, one, two, three, or four units of a particular product each day with probability 0.45, 0.30, 0.15, 0.05, and 0.05, respectively. His salary is $60 per day plus $15 for each unit ordered. What is the salesman's expected monthly salary (based on five days/week, four

weeks/month)? What properties in Table 5.16 did you use (if any) to determine the salesman's expected salary?

5.21. A quality control expert is paid $50 a day to test complex electronic equipment units produced by a small firm. The firm produces five units per day, and the expert is instructed to check each unit thoroughly for defects. For each defective unit found, he is given a bonus of $30 (high incentive—the production workers *love* this guy; the $30 bonus comes out of the salaries of the production workers!). The distribution of the number of defectives is given in the table below.

x	$P(x)$	
0	0.80	a. Find the expected number of defectives per day.
1	0.16	b. Find the expert's expected salary per day.
2	0.02	
3	0.008	
4	0.006	c. What is the expert's most likely daily salary?
5	0.006	

5.22. What is a fair price (the break-even price) to pay to participate in a game in which one can win $25 with probability 0.3 and $10 with probability 0.25?

5.23. In a particular business venture, a man can make a profit of $1,000 with probability 0.2, a profit of $500 with probability 0.2, and a loss of $100 with probability 0.6. Let **X** be his profit or loss if he decides to participate in this game. (*a*) What is the expected value of **X**? (*b*) What is the most likely value of **X**?

5.24. A manufacturing firm is committed to the production of a new type of camera that produces instant-developing pictures. The production lot can be divided into three groups: group I, nondefective cameras; group II, defective cameras, but salvageable; and group III, cameras that cannot be salvaged, except for parts. The mathematical group predicts that 80 percent of all cameras produced will belong to I, 15 percent to II, and 5 percent to III, if the current production schedule is followed. The marketing group predicts that each camera in group I will bring a profit of $50, each camera in group II $10, and the Group III cameras will each result in a loss of $100.

a. What is the expected profit (or loss) per camera?

b. If the company expects to produce 100,000 cameras in the next planning period, what is the total expected profit (or loss) during this period?

c. The mathematical group reports that if production is slowed down, the percentage of cameras in groups II and III can be cut to 10 percent and 3 percent, respectively. On the slowed-down schedule, they estimate that 88,000 cameras can be produced in the next planning period. Should the current production schedule or the slowed-down schedule be adopted?

5.25. If the probability that a man aged 50 will live another year is 0.995, how much should he pay for a $100,000 life insurance policy for the next year if the insurance company is to make a profit of $20?

5.26. The Ajax Paperboard Company, Inc., supplies sheets of cardboard to several manufacturers, which they in turn fold into boxes for distributing their product. Shown here is the number of sheets of each size sold from the last 1,000 delivered.

Width (feet) / Length (feet)	4	5	6
2	25	25	50
3	50	400	50
4	225	75	100

Source: Company records.

Using the table to estimate the probabilities of various sales combinations, determine the following quantities where X = the width of cardboard sold and Y = the length of cardboard sold. Are the random variables X = width and Y = length independent?

a. The expected square feet (area = width × length) of the boxes sold.

b. $P(X = 4)$.

c. $P(Y = 6)$.

d. $P(X = 3, Y = 5)$.

e. $C(X,Y)$.

f. $F(X,Y)$.

5.27. Given a random variable X with $E(X) = 10$ and $V(X) = 5$, one other random variable Y with $E(Y) = 15$ and $V(Y) = 10$, and independence assumed for the random variables X and Y, determine the following quantities:

a. $E(X \cdot Y)$

b. $E(X + 5Y)$.

c. $V(X - Y)$.

d. $V(15 - 5X)$.

e. $C(X,Y)$.

5.28. Assume that the following values of X and Y occur with equal probability:

x	y
7	3
5	3
6	3
11	3
9	3

Determine the following quantities:

a. $E(X)$, $E(Y)$, $V(X)$, $V(Y)$.

b. $E(X \cdot Y)$.

c. $C(X,Y)$.

Are X and Y independent?

Discrete probability distribution models

6

6.1 INTRODUCTION

Some possible representations of a probability mass function for a discrete random variable—a random variable that takes on only discrete values—are given in Figures 6.1 and 6.2. Notice that in Figure 6.1, the random variable can assume only one of two possible values in any single experiment, whereas one of eight possible values can be realized in a single trial of an experiment associated with the distribution in Figure 6.2.

In applied problems, the exact form of the distribution of a discrete random variable may not be known, since only one or a few values of the random variable may be available for analysis. A clue to determining the form of the distribution of a random variable is given by a frequency distribution of the available values. Frequently, additional insight in determining the form of a probability distribution is made available by studying the nature of the experiment that produces specific values of the random variable.

In this chapter, we shall study the nature of several experiments that lead to discrete random variables with distributions that frequently occur in applied problems and are of specific types. In practice, the experiment producing the values of the random variable will often be similar to one of the experiments presented in this chapter. The probability distribution associated with the selected specific experiment is then used as a "model" for the exact, but unknown, distribution of the random variable considered in the problem.

It is important to check how well the selected probability distribution model "fits" the available data—a poor fit can easily lead to erroneous inferences about the properties of the random variable. In Chapter 19, meth-

147

FIGURE 6.1 Several possible representations of a probability distribution of a discrete random variable where the random variable can assume one of two possible values

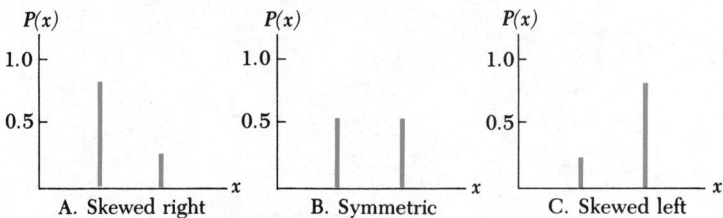

ods will be presented to help assure us that the selected probability distribution model is appropriate to represent the distribution of the random variable of interest. For the present, we shall be concerned with describing the properties of several of the more important discrete probability distributions frequently used in statistics. We will then focus our attention in Chapter 7 on experiments that produce values of a continuous random variable and the resulting probability distribution models.

FIGURE 6.2 Representation of a probability distribution of a discrete random variable that can assume one of eight possible values

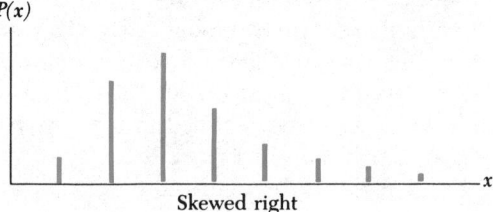

■ 6.2 THE DISCRETE UNIFORM DISTRIBUTION

☐ 6.2.1 Introduction

A distribution that arises frequently in practice—especially in certain gambling situations—is the discrete uniform distribution. The distribution is also used in making decisions (statistical decision theory) when we are uncertain about events that might occur, and it is being used increasingly in computer simulation experiments where "trial environments" are modeled on a computer. Figure 6.3 gives several examples of experiments that produce values of a discrete uniform random variable.

Assume that we assign the values 1, 2, . . . , N to a discrete random variable when it is in state 1, state 2, ... , state N and the random variable is qualitative (e.g., head, tail; heart, spade, diamond, club; etc.), and that a quantitative discrete random variable assumes the values 1 through N. Then

It is not necessary that the reader know calculus to understand the underlying concept of a probability distribution of a continuous random variable. Tables in Appendix B at the end of the text perform the integration required to determine $P(c \leq X \leq d)$ for *all* the continuous random variables treated in the text. However, it is important that the student understand the basic differences between discrete and continuous random variables, and that while it is always possible (theoretically, at least) to determine the probability that a discrete random variable assumes a particular value, the probability that a continuous random variable assumes a particular value is *always* 0, regardless of the value of the random variable or the particular probability density function. *Always!* Table 5.13 summarizes the similarities and the differences between the two types of random variables.

TABLE 5.13 Comparison of properties of discrete and continuous random variables

Property	Discrete random variable	Continuous random variable
Number of values of the random variable	Finite *or* countably infinite	Infinite (not countable)
$P(c \leq X \leq d)$	$\sum_{x=c}^{x=d} P(x)$	Area under the curve $f(x)$ from $x = c$ to $x = d$
$P(X = e)$	$P(e)$	*Always zero*

Before leaving this section, we present in Definition 5.9 the cumulative density function of a continuous random variable. This is the continuous analogue of the cumulative mass function given in Definition 5.5.

> **Definition 5.9**
> Cumulative density function of a continuous random variable
>
> Let the random variable X assume an infinite number of values on the real number line and have a probability density function $f(x)$ as delineated in Definition 5.8. The *cumulative density function* of the random variable X, denoted $F(X)$, is the probability that the continuous random variable assumes a value less than or equal to x. Mathematically, it is given by:
>
> $$F(x) = P(X \leq x)$$

■ 5.6 THE MEAN AND VARIANCE OF A CONTINUOUS RANDOM VARIABLE

Just as a discrete random variable has a mean and a variance, so does a continuous random variable. These are denoted by $E(X)$ and $V(X)$, respec-

FIGURE 6.3   Experiments producing values of a discrete uniform random variable

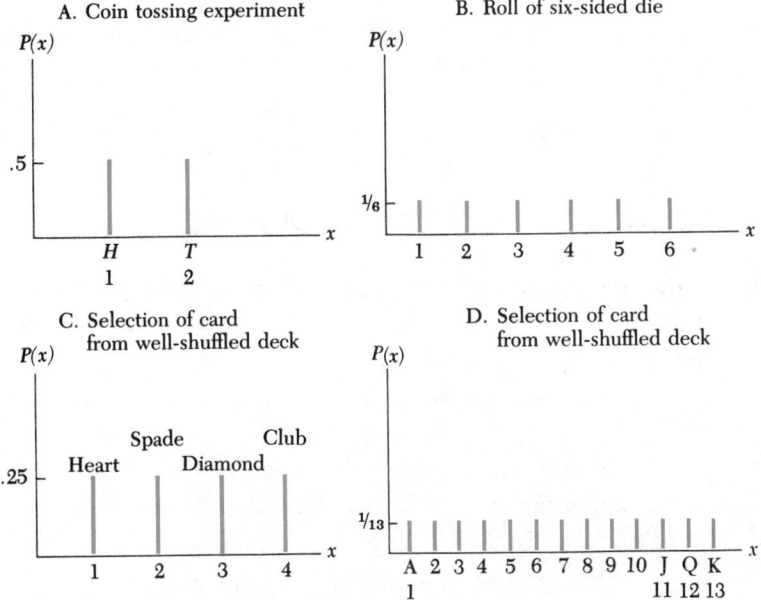

the definition of a discrete uniform random variable is as given in Definition 6.1.

Definition 6.1
Discrete uniform random variable

Let the random variable **X** assume the integer values 1 through N. If the probability that the random variable assumes one of the values 1 through N is $1/N$ for all values i, where $1 \le i \le N$, and if these probabilities do not change on repeated experiments to produce values of the random variable (independence), then **X** is a *discrete uniform random variable*.

The probability mass function of a discrete uniform random variable is given in Definition 6.2.

Definition 6.2
Discrete uniform probability mass function

Let **X** be a discrete uniform random variable. Then the *probability mass function* of **X** is given by

$$P(x) = \frac{1}{N} \qquad x = 1, 2, \ldots, N$$

where N = the number of states or the values the random variable can assume.

6.2.2 The mean and variance of discrete uniform random variables

The mean and variance of a discrete uniform distribution can be computed using the formulas presented in Chapter 5.

Mean:
$$E(\mathbf{X}) = \sum_{x=1}^{N} x \cdot P(x) = \sum_{x=1}^{N} x \cdot \frac{1}{N} = \frac{1}{N} \sum_{x=1}^{N} x$$

Variance:
$$V(\mathbf{X}) = \sum_{x=1}^{N} [x - E(\mathbf{X})]^2 P(x) = \frac{1}{N} \sum_{x=1}^{N} [x - E(\mathbf{X})]^2$$

In general, the mean and variance of a discrete uniform distribution with N states or with N values, 1, 2, . . . , N, are given by the formulas in Theorem 6.1.

Theorem 6.1
Mean and variance of a discrete uniform random variable

If **X** is a discrete uniform random variable with N states or with N values, 1, 2, . . . , N, then its *mean*, $E(\mathbf{X})$, and *variance*, $V(\mathbf{X})$, are given by

$$E(\mathbf{X}) = \frac{N + 1}{2} \quad \text{and} \quad V(\mathbf{X}) = \frac{N^2 - 1}{12}$$

6.2.3 Use of the discrete uniform model

We will illustrate the use of the discrete uniform distribution in the following two examples.

Example 6.1 Let the possible outcomes of the roll of a six-sided die be the *number of dots* that appear on the top side of the die, the integers 1, 2, 3, 4, 5, and 6. Determine the expected value and the variance of this random variable using Theorem 6.1.

Solution

$$E(\mathbf{X}) = \frac{6 + 1}{2} = 3.5 \qquad V(\mathbf{X}) = \frac{6^2 - 1}{12} = \frac{35}{12}$$

Example 6.2 In the die-tossing example, what is the probability of rolling a three *or higher* (i.e., three or more dots appear on the top side) on each roll of the die?

Solution We may use the addition law from Chapter 3 to determine:

$$P(\mathbf{X} \geq 3) = P(3) + P(4) + P(5) + P(6) = \frac{1}{6} + \frac{1}{6} + \frac{1}{6} + \frac{1}{6} = \frac{4}{6} = \frac{2}{3}$$

The process of repeatedly rolling a die and recording certain properties or certain "values" of the roll can also produce values of another discrete random variable, the binomial random variable, as described in the following section.

6.3 THE BINOMIAL DISTRIBUTION

6.3.1 Introduction

Another important distribution of a discrete random variable is the binomial distribution. Perhaps the simplest illustration of a binomial random variable—a random variable with a distribution that is the binomial distribution—arises in the coin-tossing experiment in which a coin is tossed n times and the *number of heads* occurring in the n tosses is recorded.[1] If the n coin tosses are independent events and we define the random variable **X** to be the *number of heads* occurring in the n tosses, then **X** is a binomial random variable.

The key elements of the experiment that produce values of a binomial random variable are:

1. There are n independent trials of the experiment.
2. In each trial, there are only two possible outcomes of the experiment—"success" or "failure."
3. The probability of "success" is the same for each trial.
4. The random variable is defined to be the *number of successes* in the n trials.

Definition 6.3
Binomial random variable

Let the random variable **X** be the number of successes in n independent trials of a simple experiment, where each trial results in one of two possible outcomes—"success" or "failure." Let π be the probability of a success in each and every one of the n trials. Then, **X** is a *binomial random variable*.

The binomial random variable described in Definition 6.3 is often generated by a *Bernoulli process*.

If many trials are conducted, the long-run *fraction* of successes is denoted by π in the Bernoulli process, and if n Bernoulli trials are conducted

---

[1] It is common to refer to a number of "trials" when discussing a binomial random variable. The number of trials is frequently equated to a "sample" of size n. Most introductory statistics texts denote the parameters of the binomial distribution by π and n, therefore, rather than by π and N. We will do the same to retain a consistent notation with other introductory statistics texts. Technically speaking, however, the number of trials is a population parameter, and following the convention we have used in discussing other probability distributions of a random variable, should be denoted by N.

and the number of successes, **X**, is counted, then **X** is a binomial random variable.

Definition 6.4
Bernoulli process

A *Bernoulli process* is a process in which an experiment is repeatedly performed, yielding either a "success" or a "failure" in each trial, and where there is absolutely no pattern in the occurrence of successes and failures. That is, the occurrence of a success or a failure in a particular trial does not affect or is not affected by the outcomes in any previous or subsequent trial—the trials are independent.

Example 6.3 Returning to the die-tossing experiment of Examples 6.1 and 6.2, suppose a net payoff is *received* if four, five, or six dots appear on the top side, and a net payout is *made* if one, two, or three dots appear on the top side. Is this experiment a Bernoulli process, and, if so, how should we define **X** so that it is a binomial random variable?

Solution Define a "success" to be a four, five, or six, and define a "failure" to be a one, two, or three. Repeated trials of the experiment (rolls of the die) are indeed a Bernoulli process because:

1. The die is rolled independently n times (or at least we hope so).
2. In each trial, there are only two possible outcomes: a "success" (four, five, or six), or a "failure" (one, two, or three).
3. The probability of success is the same for each trial [$\pi = P(4) + P(5) + P(6) = 1/6 + 1/6 + 1/6 = 1/2$].
4. The random variable **X** is defined to be the number of successes (i.e., rolls resulting in a four, five, or six) in n trials of the experiment.

The binomial random variable occurs in many other situations. Table 6.1 gives examples of random variables with distributions that are binomial *or with distributions that may be approximated* by the binomial distribution. Assuming the coin is fair and it is flipped fairly, the number of heads in example I is binomially distributed. In examples II, III, and IV, the random variables are *not* binomially distributed since the Bernoulli trials are dependent. To appreciate this, suppose in example III that there are 1,000 units of a product in a warehouse, 100 of which are defective. If we take two units from the warehouse, the probability that the second item withdrawn is defective depends on the outcome in the first trial. If the first trial (selection) produces a defective, then the probability that a second trial produces a defective is 99/999, whereas the probability of a defective in the second trial if the first trial produced a nondefective is 100/999. The appropriate probability distribution for these random variables is the hypergeometric distribution, which is considered in Section 6.4. In many real situations, the binomial distribution is used as a *model* (approximating distribution) for the distribu-

TABLE 6.1 Examples of binomial random variables and random variables that could possibly be modeled by the binomial distribution

Example	Trial	Success	Failure	π	n	X
I.	Tossing a coin	Head	Tail	½ (if the coin is fair)	n tosses	Number of heads
II.	Contacting a woman and recording whether she would purchase a particular consumer product	Decision to purchase	Decision not to purchase	Proportion of women who would purchase the product	Number of women contacted	Number of purchase decisions in the sample selected
III.	Selecting a product from a production line	Nondefective	Defective	Proportion of nondefectives in lot	Size of sample	Number of nondefectives in the sample
IV.	Selecting a registered voter in a poll	Democrat	Republican	Proportion of Democrats in population of registered voters	Size of sample in poll	Number of Democrats in sample
V.	Birth of a child	Boy	Girl	Virtually ½	Number of births during April at a specific hospital	Number of boys born at the hospital during April
VI.	Bidding on a construction project	Winning bid	Losing bid	?	Number of bids during the year	Number of bids won during year

tion of a hypergeometric random variable. We will discuss this more fully in Section 6.4. For the moment, the trials in these experiments would be independent, and the binomial distribution would be the *exact* distribution if each item were replaced in the population before the next selection. For instance, in example III, we could replace each item before the next is drawn, but this would make it possible to examine the same item more than once, an inefficient sampling process.

In examples V and VI, the trials are likely to be dependent as well. In the birth example, the probability of a success (a boy, let us say) is not the same in each and every trial since some families may produce more offspring of one sex than the other (due to genetic considerations). However, it is reasonable to assume that the binomial distribution would fit well to the true distribution of the random variable due to the similarities of the experiments that produce the two random variables (the binomial random variable and the random variable aggregating the number of boy births). In example VI, the probability of winning a bid may drastically change from trial to trial. Thus, the binomial distribution would not be a good model for the random variable *unless* it is reasonable to assume that the probability of winning a bid remains relatively the same for all bids placed.

The labeling of the two outcomes in each trial of the experiment as "success" and "failure" is completely arbitrary as long as the probabilities associated with each outcome are not confused. For instance, suppose in example IV that the proportion of Democrats in the population is 0.60. If the outcome "Democrat" is labeled as a success, then the probability of success in each trial of the experiment, π, is equal to 0.60. If the outcome "Republican" is labeled as a success, then $\pi = 0.40$—the definition of a successful outcome must be consistent with the assignment of the probability of a success in each trial of the simple experiment. The probability mass function of a binomial random variable is delineated in Definition 6.5.

Definition 6.5
Binomial probability mass function

Let **X** be a binomial random variable. Then, the *probability mass function* of **X** is given by

$$P(x) = C_x^n \pi^x (1 - \pi)^{n-x}, \qquad x = 0, 1, 2, \ldots, n$$

where

n = number of trials
π = probability of a "success" in each trial

$$C_x^n = \frac{n!}{x!(n - x)!}$$

The specific form of the binomial distribution depends on the values of the two *parameters*—π and n—of the distribution. After values have been assigned to the parameters π and n, probabilities of the values of the binomial random variable **X** occurring can be computed for the specific distribution. Thus, the probability mass function, $P(x) = C_x^n \pi^x (1 - \pi)^{n-x}$, where $x = 0, 1, 2, \ldots, n$, represents a *family* of probability distributions, each member of which arises when numerical values are specified for the parameters π and n. Figure 6.4 illustrates the form of specific members of the binomial family for selected values of the parameters π and n. Note that the distribution is symmetric when $\pi = 0.5$, skewed to the left when $\pi > 0.5$, and skewed to the right when $\pi < 0.5$.

FIGURE 6.4 Specific members of the binomial family of probability distributions

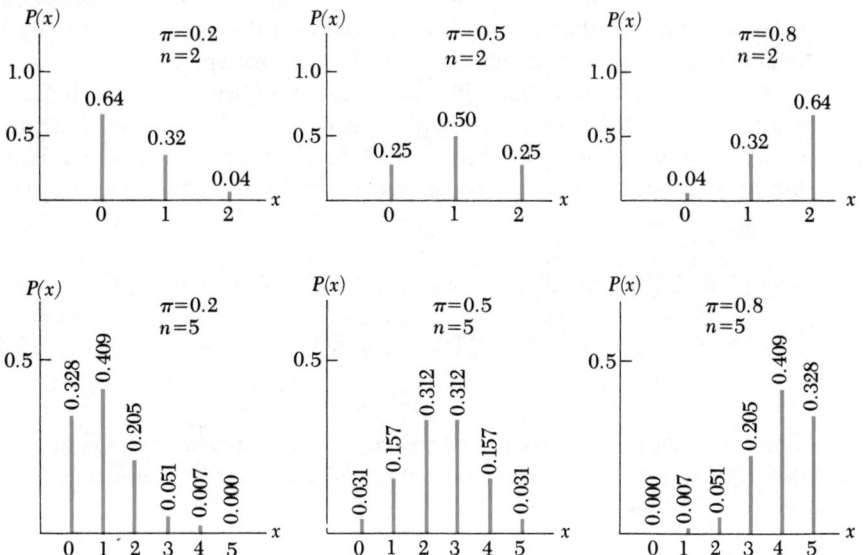

6.3.2 Computing binomial probabilities

Given the values π and n, binomial probabilities may be computed in a number of ways. If n is not too large (less than ten), the probability mass function formula can be used to determine the probability that the binomial random variable **X** assumes a particular value x. If n is very large, the use of the definition of the probability mass function to determine $P(X = x)$ becomes, at best, tedious. By using computers, statisticians have compiled extensive tables of binomial probabilities for a large number of values of π and n. In Table B.1 in Appendix B, cumulative binomial probabilities are given for several values of n and π. Specifically, the tables give

$$F(a) = \sum_{x=0}^{a} P(x) = P(0) + P(1) + \cdots + P(a)$$

or the probability that a random variable **X** assumes a value less than *or equal to x* = *a* for fixed values of *n* and π. Notationally, this probability is given by

$$P(X \leq a/n, \pi).$$

Far more extensive tables have been compiled by statisticians and are available at most libraries. However, the binomial probability mass function is very easy to program on the computer; hence, binomial probabilities are often determined with the aid of a computer in practice.

A method for approximating binomial probabilities for large *n* will be given in Chapter 7. For the moment, we will consider several examples illustrating the use of the binomial distribution.

Example 6.4   In a family of three children, what is the probability that there will be exactly two boys, assuming that the sexes are equally likely to occur in each birth and that the "trials" are independent?

Solution   Let the event "boy" represent a "success" and define **X** to be the number of successes (boys) produced in the *n* = 3 trials. The probability of a success in a single trial, π, is 0.5. Thus, the specific member of the binomial family of distributions, expressed by its mass function, is:

$$P(x) = C_x^3 (0.5)^x (1 - 0.5)^{3-x}, \qquad x = 0, 1, 2, 3$$

The probability that **X** assumes the value of 2 is given by:

$$P(X = 2) = P(2) = C_2^3 (0.5)^2 (1 - 0.5)^{3-2} = \frac{3!}{2!(3 - 2)!} (0.5)^2 (0.5)^1$$

$$= (3)(0.25)(0.5) = 0.375$$

The binomial probability distribution for *n* = 3 and π = 0.5 is given in three different ways (table, stick diagram, and formula) in Table 6.2.

TABLE 6.2   Binomial distribution for π = 0.5 and *n* = 3; tabular, stick diagram, and functional forms

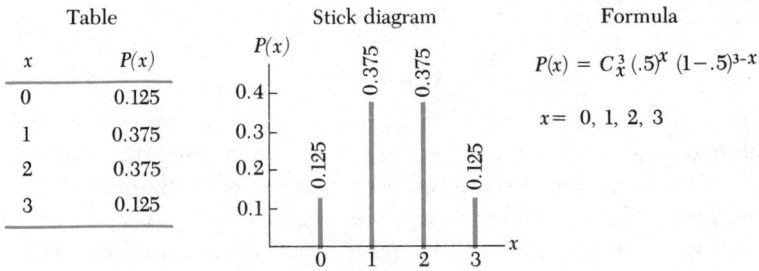

Example 6.5   A large lot of items (10,000) has been produced by a certain production process. It is known by the wholesaler that 20 percent of the items are defective. Herman, the retailer, randomly samples six items from this lot. He will purchase the lot if he finds one or fewer defective items

among the sampled six. Determine the probability that the wholesaler sells the lot to Herman.

Solution[2]   Let a "success" be a defective item and a "failure" be a nondefective item. The probability of success, π, is 0.20, and the probability of a failure—a nondefective item—is 0.80. The approximating probability mass function is:

$$P(x) = C_x^6(0.20)^x(1 - 0.20)^{6-x}, \qquad x = 0, 1, 2, 3, 4, 5, 6$$

Herman will accept the lot if he finds one or fewer defectives in the six trials of the experiment. Therefore, the probability that he will accept the lot is given by:

$$P(\text{accept}) = P(0) + P(1) = C_0^6(0.20)^0(0.80)^6 + C_1^6(0.20)^1(0.80)^5$$
$$= 0.2621 + 0.3932 = 0.6553$$

Similarly, the probability that Herman will reject the lot is given by:

$$P(\text{reject}) = P(2) + P(3) + P(4) + P(5) + P(6) = 1 - [P(\text{accept})]$$
$$= 1 - [P(0) + P(1)] = 0.3447$$

A "success" in this example is defined to be a defective item so that we can work with a direct statement of the problem. If instead we define a success to be a *nondefective* item, then the appropriate probability mass function is:

$$P(x) = C_x^6(0.80)^x(1 - 0.80)^{6-x}, \qquad x = 0, 1, 2, 3, 4, 5, 6$$

(where $\pi = 0.80$), and the probability that Herman will accept the lot is given by:

$$P(\text{accept}) = P(5) + P(6) = C_5^6(0.80)^5(0.20)^1 + C_6^6(0.80)^6(0.20)^0$$
$$= 0.3932 + 0.2621 = 0.6553$$

This stresses the arbitrariness of the labeling of the two outcomes in each trial of the experiment as to which outcome constitutes a "success" and which constitutes a "failure."

Example 6.6   An exam containing ten multiple-choice questions is designed so that the probability of a correct choice on any question by guessing alone is 0.20. What is the probability that a student will get no more than six questions right by guessing? What is the probability that he will get exactly

---

[2] Actually, we are using the binomial distribution to *approximate* the true distribution in this example, for this is similar to example III in Table 6.2. The probability of a success in the first trial (selection of a defective) is 0.20. If a defective is found in the first trial, then the probability of a defective on the second trial is 1,999/9,999 = 0.19992 ≈ 0.20. If the first trial results in a nondefective, then the probability of a defective in the second trial is 2,000/9,999 = 0.20002. Hence, the probability of a success on each trial is *not independent* of what preceded it. However, because the number selected (6) is small in relation to the population size (10,000), the binomial distribution appears to be a reasonable approximation of the true distribution of defectives in this experiment.

five questions right? What is the probability that he will get more than two right?

Solution We wish to find $P(X \le 6)$, $P(X = 5)$, and $P(X > 2)$, where X is defined to be the number of questions answered correctly. These probabilities could be computed by using the probability mass function with $n = 10$ and $\pi = 0.20$, where "success" is correctly answering a question and "failure" is incorrectly answering a question. However, let us use Table B.1 in Appendix B to calculate these probabilities. Since the entries in the tables are cumulative probabilities of the form

$$P(X \le a/n, \pi) = \sum_{x=0}^{a} P(x) = P(0) + P(1) + P(2) + \cdots + P(a)$$

each question must be formulated so that it can be answered from the knowledge of *cumulative probabilities*. The table we wish to use is the one corresponding to $n = 10$ trials; the appropriate column is the one marked $\pi = 0.20$. Thus,

$$P(X \le 6/n = 10, \pi = 0.20) = 0.999$$

This probability can be read directly from Table B.1 and is represented by the stick diagram in Figure 6.5.

$$P(X \le 6) = P(0) + P(1) + P(2) + P(3) + P(4) + P(5) + P(6)$$

FIGURE 6.5 $P(X \le 6/n = 10, \pi = 0.20)$

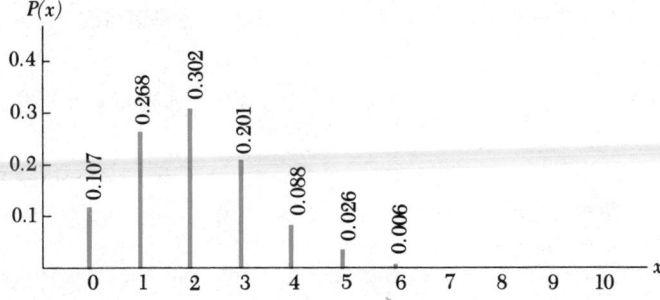

The probability that the random variable X assumes the value $x = 5$ is equal to the probability $X \le 5$ *minus* the probability $X \le 4$. That is,

$$\begin{aligned} P(X = 5/n = 10, \pi = 0.20) &= P(X \le 5/n = 10, \pi = 0.20) \\ &\quad - P(X \le 4/n = 10, \pi = 0.20) \\ &= 0.994 - 0.967 = 0.027 \end{aligned}$$

The calculation of $P(X = 5)$ using Table B.1 is shown in Figure 6.6 using a stick diagram to represent the individual probabilities.

$$\begin{aligned} P(X > 2) &= 1 - P(X \le 2) = 1 - P(X \le 2/n = 10, \pi = 0.20) \\ &= 1 - 0.678 = 0.322 \end{aligned}$$

FIGURE 6.6 Calculation of $P(X = 5/n = 10, \pi = 0.20)$

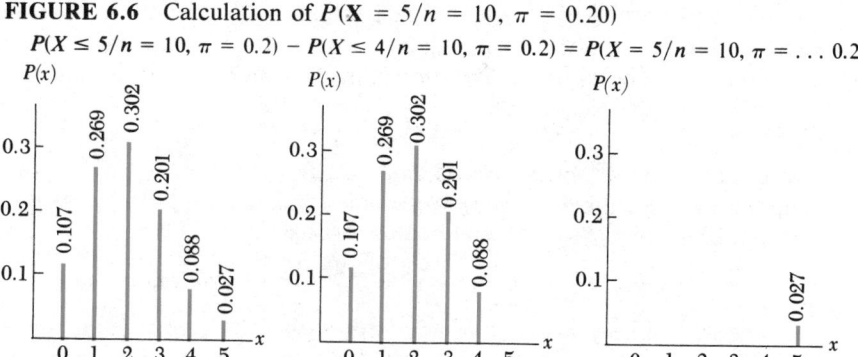

$P(X \le 5/n = 10, \pi = 0.2) - P(X \le 4/n = 10, \pi = 0.2) = P(X = 5/n = 10, \pi = \ldots 0.2)$

Note that $P(X > 2)$ is also equal to $P(X \ge 3) = \Sigma_{x=3}^{10}P(x)$, since the binomial distribution is a discrete distribution in which the values of the random variable are defined only as integers.

☐ 6.3.3 The mean and variance of binomial random variables

The mean and variance of a binomial distribution may be computed by using the formulas presented in Chapter 5:

Mean: $\quad E(\mathbf{X}) = \sum_{x=0}^{n} x \cdot P(x) = \sum_{x=0}^{n} x \cdot C_x^n \pi^x (1 - \pi)^{n-x}$

Variance: $\quad V(\mathbf{X}) = \sum_{x=0}^{n} [x - E(\mathbf{X})]^2 P(x) = \sum_{x=0}^{n} [x - E(\mathbf{X})]^2 C_x^n \pi^x (1 - \pi)^{n-x}$

In general, however, the mean and variance of a binomial distribution with parameters n and π can be simplified to the formulas presented in Theorem 6.2. These formulas are much simpler to work with than are the definitions given in Chapter 5, but they are equivalent.

Theorem 6.2
Mean and variance of the binomial random variable

If $\mathbf{X}$ is a binomial random variable, then its *mean,* $E(\mathbf{X})$ and *variance,* $V(\mathbf{X})$, are given by

$$E(\mathbf{X}) = n\pi \quad \text{and} \quad V(\mathbf{X}) = n\pi(1 - \pi)$$

Example 6.7 Suppose that 70 percent of the members of a particular fraternity smoke cigarettes, and that 25 members are chosen from the fraternity at random. What is the expected number of smokers among the 25 members selected? What is the most likely number of smokers in the selected group of 25? What is the standard deviation expected of the number of smokers in a group of 25 selected, assuming that the binomial distribution is a reasonable approximation in this experiment?

   Solution   The appropriate approximating binomial mass function is:

$$P(x) = C_x^{25}(0.70)^x(1 - 0.70)^{25-x}, \qquad x = 0, 1, \ldots, 25$$

where the random variable **X** is defined to be the number of smokers among the 25 chosen fraternity members. From Theorem 6.2, the average number of smokers, $E(\mathbf{X})$, is:

$$E(\mathbf{X}) = n\pi = (25)(0.70) = 17.5 \text{ smokers}$$

   The most likely number of smokers is the mode of the specific binomial distribution with $\pi = 0.70$ and $n = 25$. From Table B.1 and doing the necessary subtractions, the value of **X** that has the greatest probability of occurring is $x = 18$—it occurs with probability 0.171:

$$P(\mathbf{X} = 18) = \sum_{x=0}^{18} P(x) - \sum_{x=0}^{17} P(x) = 0.659 - 0.488 = 0.171$$

Thus, the mode of the distribution is 18.

   From Theorem 6.2, the variance of the number of smokers is:

$$V(\mathbf{X}) = n\pi(1 - \pi) = 25(0.70)(0.30) = 5.25$$

Since the standard deviation is the square root of the variance, the standard deviation is equal to $\sqrt{5.25} = 2.29$ smokers.

   In the preceding examples, we assumed that the values of the binomial parameters, π and n, are known. Given this information, it is possible to determine the probability that the binomial random variable exceeds some value, is less than or equal to some value, etc. It is also possible to compute its mean, variance, and standard deviation.

   In many problems, the values of the parameters are not known and must be *estimated* from sample information. In Chapter 9, we shall see how π is estimated from sample information (n will be the sample size). After the estimated value of π is computed, the knowledge that the random variable is binomially distributed can be used to say how "good" the estimate of π is.

   In this chapter, we will continue to assume that the parameters of our probability distribution models are known. We now turn to another distribution model for a discrete random variable.

6.4   THE HYPERGEOMETRIC DISTRIBUTION

   Frequently, when sampling from a finite population, the probability of a success or a failure changes as sample units are withdrawn (examined) and are not returned to the population (i.e., sampling without replacement). From Example 6.5, the probability of a success or a defective item for the sixth item sampled is *conditional upon* the number of defectives and nondefectives encountered in the first five units examined, since units are not returned to the population after being inspected so that they can be inspected more than once. If the number of units examined without replacement is small in relation to the population size, then the binomial distribution is a reasonable approximation of the true distribution of sample results. If, on the other hand, the ratio of units examined to units in the population is large, then the binomial distribution may not be a satisfactory approximation to the true distribution. Statisticians use the rule of thumb that if the ratio of units examined without replacement (n) to units in the population (N) is less than 5 percent (i.e., $n/N < 0.05$), then the binomial distribution is a reasonable approximation of the true distribution of selected successes.

It may also be possible to divide the population of N items into k mutually exclusive categories, each containing $N_1, N_2, \ldots, N_k$ elements. However, in this chapter, we will focus our attention on the special case of $k = 2$, where N_1 units in the finite population are labeled "successes," N_2 units are labeled "failures," and $N_1 + N_2 = N$.

The experiment producing values of a random variable X that has a hypergeometric distribution consists of selecting a set of n elements from a finite population of N elements without replacement. The hypergeometric random variable is described in Definition 6.6.

Definition 6.6
Hypergeometric random variable

Let the random variable X be the number of successes observed when n items are selected at random from a finite population of N elements without replacement, and the number of successes in the finite population is equal to $N_1 \leq N$. Then X is a *hypergeometric random variable*.

The description of the hypergeometric probability mass function is given in Definition 6.7, and the mean and variance of the hypergeometric distribution where each element in the population is labeled either a "success" or a "failure" are given in Theorem 6.3.

Definition 6.7
Hypergeometric probability mass function

Let X be a hypergeometric random variable. Then the *probability mass function* of X is given by

$$P(x) = \frac{C_x^{N_1} \cdot C_{n-x}^{N_2}}{C_n^N}, \qquad x = 0, 1, 2, \ldots, min\{n, N_1\}$$

where

N_1 = number of successes in the population
N_2 = number of failures in the population
$N_1 + N_2 = N$
n = number of items (trials) examined (without replacement)
x = number of successes in the n trials
$min\{n, N_1\}$ = minimum of n and N_1

> **Theorem 6.3**
>
> If **X** is a hypergeometric random variable where the population of N items consists of two mutually exclusive and collectively exhaustive groups with N_1 and N_2 items in each group, then its *mean*, $E(\mathbf{X})$, and *variance*, $V(\mathbf{X})$, are given by
>
> $$E(\mathbf{X}) = n \cdot \frac{N_1}{N} \quad \text{and} \quad V(\mathbf{X}) = \frac{N-n}{N-1}\left[n \cdot \left(\frac{N_1}{N}\right) \cdot \left(1 - \frac{N_1}{N}\right)\right]$$

The use of the hypergeometric distribution is illustrated in the following example.

Example 6.8 Assume that a population of ten items contains four defective units. If a sample of three items is selected at random without replacement, what is the probability that zero, one, two, or three defectives will appear in the items selected? What is the expected number (mean) and the variance in the sample? How do these results compare with those that would have been obtained if the binomial distribution had been used to approximate the hypergeometric distribution?

Solution The determination of the various probabilities of occurrence can be accomplished by the application of Definition 6.7. For example,

$$P(0) = \frac{C_0^4 C_3^6}{C_3^{10}} = \frac{\left(\dfrac{4!}{0!4!}\right)\left(\dfrac{6!}{3!3!}\right)}{\dfrac{10!}{3!7!}} = \frac{1 \cdot 20}{120} = 0.167$$

where $P(0)$ is the probability of selecting zero "successes" (defectives) in a sample of three without replacement. Similarly,

$$P(1) = \frac{C_1^4 C_2^6}{C_3^{10}} = 0.500 \qquad P(2) = \frac{C_2^4 C_1^6}{C_3^{10}} = 0.300 \qquad P(3) = \frac{C_3^4 C_0^6}{C_3^{10}} = 0.033$$

From Theorem 6.3,

$$E(\mathbf{X}) = 3 \cdot \frac{4}{10} = 1.2$$

$$V(\mathbf{X}) = \left(\frac{7}{9}\right)(3)\left(\frac{4}{10}\right)\left(1 - \frac{4}{10}\right) = 0.56$$

Using Table B.1 in Appendix B with $n = 3$ and $\pi = N_1/N = 0.4$,

$$P(0) = 0.216 \qquad P(2) = 0.288$$
$$P(1) = 0.432 \qquad P(3) = 0.064$$

From Theorem 6.2,

$$E(\mathbf{X}) = n\pi = 1.2 \quad \text{and} \quad V(\mathbf{X}) = n\pi(1 - \pi) = 0.72$$

These results are summarized in Table 6.3. Note that the means for the two

TABLE 6.3 Summary computations for Example 6.8: hypergeometric distribution and binomial approximation

	Hypergeometric distribution	Binomial approximation
$P(0)$	0.167	0.216
$P(1)$	0.500	0.432
$P(2)$	0.300	0.288
$P(3)$	0.033	0.064
	1.000	1.000
$E(X)$	1.2	1.2
$V(X)$	0.56	0.72

distributions are equal, and that the variances differ by a factor of $(N - n)/(N - 1)$, or by 7/9. If we assume that $\pi = N_1/N$ in the hypergeometric distribution, then the formulas for the mean and variance from Theorem 6.3 are

$$E(X) = n\pi \quad \text{and} \quad V(X) = \frac{N - n}{N - 1}[n \cdot \pi(1 - \pi)]$$

Now the relationship between the binomial and the hypergeometric distributions becomes apparent. The quantity $(N - n)/(N - 1)$, which gives the difference between the variances of the two distributions, is called the *finite population correction factor,* and is further discussed in Chapter 11.

Because of the similarity between the hypergeometric and the binomial probability distributions, the binomial distribution is often used in practice even when the hypergeometric distribution applies. This is because the differences in the two distributions are practically important only when the sample size n, or the number of units examined, is large in relation to the number of units in the population N. Recall that statisticians use the rule of thumb that if $n/N < 0.05$, the binomial distribution is a reasonable approximation of the hypergeometric distribution.

■ 6.5 THE POISSON DISTRIBUTION

□ 6.5.1 Introduction

The final discrete distribution we will discuss in this chapter is the *Poisson distribution*. To describe the Poisson distribution and, in particular, to develop the notion of a Poisson random variable, consider a production process in which insulated wire is produced. In this process, it is known that defects (bare spots) occur at random but are known to average one defect per foot of wire produced. We shall initially view the inspection of the wire as a Bernoulli process in which the inspections of lengths of wire correspond to the trials of an experiment. We will concentrate on determining the probability of two defects in each segment of the wire we examine, and label the discovery of a defect a "success."

In the 1 foot of wire we select at random, we know that the *expected number* of defects is one. If we cut the wire in half—that is, into two equal 6-inch segments—and make the assumption that no more than one defect can occur in each 6-inch segment, then the examination of the wire can be treated as a Bernoulli process. Each segment either has a defect (success) or does not have a defect (failure). Since the expected number of defects per foot of wire is one, the probability of a defect is ½ on each of the 6-inch segments, and the probability of two defectives in a foot of wire (we can have only zero, one, or two defectives since we are assuming that each segment contains at most one defect) is

$$P(2) = \frac{2!}{2!0!} \left(\frac{1}{2}\right)^2 \left(\frac{1}{2}\right)^{2-2} = 0.25$$

We know, of course, that there is some probability that two *or more* defects could be found in each 6-inch segment. Therefore, assume we now cut our wire into six 2-inch segments for a closer approximation, and again assume that no more than one defect can occur in each 2-inch segment. Since we now have six segments and each segment is assumed to contain at most one defect, we can calculate the probabilities of zero, one, two, three, four, five, or six defects in our 1-foot segment of wire. The probability of two defects is given by:

$$P(2) = \frac{6!}{2!(6 - 2!)} \left(\frac{1}{6}\right)^2 \left(\frac{5}{6}\right)^{6-2} = 0.201$$

If the examination of six 2-inch segments seems more valid than the examination of two 6-inch segments because the probability of more than one defect on each segment is less in the former instance, then we might envision cutting our wire into smaller and smaller segments such that the number of defects per piece approaches zero. For example, if we cut our wire into 50 pieces and examine each for a defect (here again we are assuming that each segment can contain no more than one defect), the probability that a 1-foot length of wire contains two defects is given by:

$$P(2) = \frac{50!}{2!(50 - 2)!} \left(\frac{1}{50}\right)^2 \left(\frac{49}{50}\right)^{50-2} = 0.1858$$

Since 50 pieces is a better basis for approximation, we might divide our wire into still smaller segments such that the probability of a segment containing more than one defect does indeed approach 0. Specifically, we might ask the limit of the following expression as n approaches infinity:

$$P(2) = \frac{n!}{2!(n - 2)!} \left(\frac{1}{n}\right)^2 \left(\frac{n - 1}{n}\right)^{n-2}$$

From calculus,

$$\lim_{n \to \infty} P(2) = \frac{e^{-1}1^2}{2!} = 0.184$$

where e is the base of the natural logarithms and is approximately 2.718.

Thus, our approximation of $P(2)$ becomes more accurate as the number of trials of the experiment increases:

n	$P(2)$
2	0.25
6	0.201
50	0.1858
∞	0.184

If we let λ represent the average number of defects per unit of length ($\lambda = 1$ in our example) and let x represent the number of defects encountered, then the general expression for the probability of x defects in the wire example becomes

$$P(x) = \frac{e^{-\lambda}\lambda^x}{x!}$$

which, we will see shortly, is the probability mass function of a Poisson random variable.

In a large number of applied problems, the random variable of interest can assume a countably infinite number of possible integer values, $0, 1, 2, \ldots$, in a continuous interval. The interval might consist of a minute, an hour, a day, etc., and hence need not be specifically tied to a physical interval, such as the length of a wire. Other examples might be the number of customers arriving at a supermarket checkout counter in a minute and the number of aircraft arriving at an airport per minute. More specifically, the experiment generating a Poisson random variable is described in the following definition.

Definition 6.8
Poisson random variable

Let the random variable X be the number of occurrences ($x = 0, 1, 2, \ldots$) of a specific event in a given continuous interval. The random variable X is called a *Poisson random variable* if the following conditions on the experiment generating values of X are satisfied:

1. The numbers of occurrences of the specific event in two nonoverlapping intervals are independent.
2. The probability of an occurrence of the specific event in a small interval is small and is proportional to the length of the interval.
3. The probability of two or more occurrences of the specific event in a small interval is negligible or is 0.

Another classic example of an experiment that generates values of a Poisson random variable is recording the number of telephone calls received at a switchboard in some *time* interval. For example, define the random variable **X** to be the number of calls received during a 5-minute interval at the switchboard. The possible values that **X** can assume are 0, 1, 2, . . .—countably infinite in number. If we consider the 5-minute interval to be composed of many small subintervals, say 1 second long, then the conditions for **X** to be a Poisson random variable described in Definition 6.8 are satisfied. This follows, since in the small time interval of, say, 1 second,

1. The number of calls received in a second time interval is independent of all other intervals of 1 second.
2. The probability of a phone call in a 1-second time interval is small and is proportional to the length of the interval.
3. The probability of two or more phone calls in 1 second is extremely small.

Several examples of other Poisson random variables are listed in Table 6.4.

TABLE 6.4 Examples of Poisson random variables

Example	*Description of* **X**
I	Number of deaths per year due to a rare disease in a large city
II	Number of machines that break down in a large plant during any one day
III	Number of ships arriving per hour at a large dock
IV	Number of typographic errors per page in a large volume of printed material

Definition 6.9
Poisson probability mass function

Let **X** be a Poisson random variable. Then the *probability mass function* of **X** is given by

$$P(x) = \frac{e^{-\lambda}\lambda^x}{x!}, \qquad x = 0, 1, 2, \ldots ; \quad \lambda \geq 0$$

where $e = 2.71828$ and λ is a parameter of the distribution. Numerically, λ represents the average number of occurrences of the specific event in the given interval of measurement.

As with the binomial distribution, we can think of the Poisson probability mass function as a representation of a family of distributions, each member of which is specified by selecting a particular value of the parameter λ.

FIGURE 6.7 Some specific members of the Poisson family of probability distributions (λ = 0.1, 0.5, and 1.0)

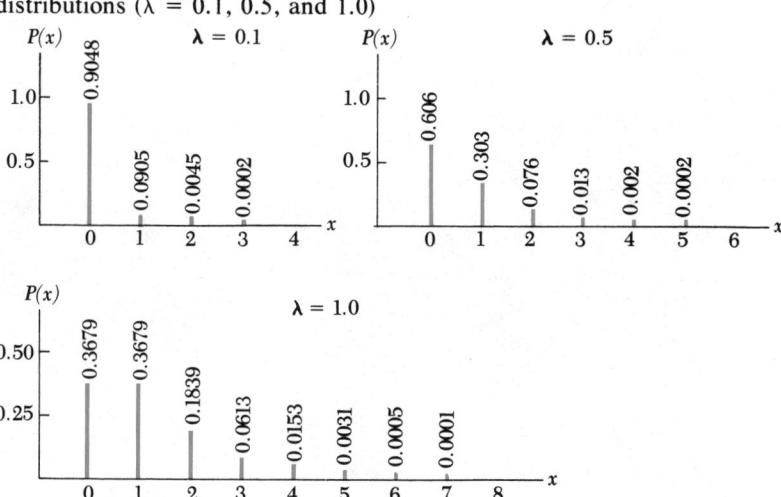

Figure 6.7 illustrates the form of the Poisson distribution for several selected values of the parameter λ.

From Figure 6.7, we see that when λ = 0.1, virtually all the probability is absorbed by the first four values of **X**. Indeed, if we compute $P(0) + P(1) + P(2) + P(3)$ using the values in Figure 6.7, we find that the sum is 1. We must recognize, however, that $P(\mathbf{X} \leq 3)$ is not exactly 1; rather, this is a *rounded* solution to four decimal places. The probability distribution when λ = 0.1 assigns probabilities to *all* the positive integers, but the probabilities assigned to the values of **X** that are greater than 3 are *very* small.

We can see from Figure 6.7 that as the value of λ increases, the probability becomes more spread out over the values of the random variable **X**. In Table B.2 of Appendix B, Poisson probability distributions are presented in tabular form for several values of λ. In most problems, this table can be used to compute Poisson probabilities for specified values of λ rather than using the mass function given in Definition 6.9.

□ **6.5.2 The mean and variance of Poisson random variables**

The mean and variance of the Poisson random variable may be computed from the formulas:

Mean: $$E(\mathbf{X}) = \sum_{x=0}^{\infty} xP(x) = \sum_{x=0}^{\infty} x \cdot \frac{e^{-\lambda}\lambda^x}{x!}$$

Variance: $$V(\mathbf{X}) = \sum_{x=0}^{\infty} [x - E(\mathbf{X})]^2 P(x) = \sum_{x=0}^{\infty} [x - E(\mathbf{X})]^2 \frac{e^{-\lambda}\lambda^x}{x!}$$

By the use of algebraic manipulation and the fact that

$$\sum_{x=0}^{\infty} P(x) = \sum_{x=0}^{\infty} \frac{e^{-\lambda}\lambda^x}{x!} = 1$$

it is possible to show that $E(X) = \lambda$ and $V(X) = \lambda$.

Theorem 6.4
Mean and variance of the Poisson random variable

If **X** is a Poisson random variable, then its *mean*, $E(X)$, and *variance*, $V(X)$, are given by:

$$E(X) = \lambda \qquad \text{and} \qquad V(X) = \lambda$$

☐ **6.5.3   Use of the Poisson model**

We will consider two examples that illustrate the use of the Poisson distribution.

Example 6.9   If the number of telephone calls that an operator receives in a 10-minute interval follows a Poisson distribution with $\lambda = 1$ (an average of one call each 10 minutes), what is the probability that she will receive no calls in a 10-minute interval? What is the probability of less than four calls? What is the most likely number of calls that she will receive?

Solution   With $\lambda = 1$, the appropriate Poisson mass function is:

$$P(x) = \frac{e^{-1}1^x}{x!} = \frac{1}{e(x!)}, \qquad x = 0, 1, 2, \ldots$$

Therefore,

$$P(X = 0) = P(0) = \frac{1}{e(0!)} = \frac{1}{e} = \frac{1}{2.72} = 0.3679$$

Notice that this answer can be read directly from Table B.2.

$P(X < 4) = P(X \leq 3) = P(0) + P(1) + P(2) + P(3)$
$\qquad\qquad = 0.3679 + 0.3679 + 0.1839 + 0.0613$ (from Table B.2) $= 0.9810$

The most likely value of **X** is the mode. From Figure 6.7 with $\lambda = 1.0$, we can see that the distribution is *bimodal:* the two most frequently occurring values of **X** are $x = 0$ and $x = 1$, each of which occurs with probability 0.3679.

Example 6.10   The number of admissions to the emergency ward of a hospital during the time beginning at 12:00 midnight and ending at 2:00 A.M. Saturday is found to be Poisson distributed with an average of 3.5 admissions. During this period on a particular Saturday morning, what is the probability that no one will be admitted? What is the probability that between two and five persons (inclusive) will be admitted? What is the most likely number of admissions?

Solution   The appropriate Poisson mass function is:

$$P(x) = \frac{e^{-3.5}(3.5)^x}{x!}, \qquad x = 0, 1, 2, \ldots$$

Therefore,

$$P(X = 0) = P(0) = 0.0302 \quad \text{(from Table B.2)}$$

$$\begin{aligned} P(2 \leq X \leq 5) &= P(2) + P(3) + P(4) + P(5) \\ &= 0.1850 + 0.2158 + 0.1888 + 0.1322 = 0.7218 \end{aligned}$$

The mode is 3, since $x = 3$ occurs with the greatest probability (0.2158).

The Poisson distribution is used extensively in applied problems to model the distribution of the number of persons joining a *queue* (a line) who wish to receive a service of some kind or purchase a product. With very little thought, it is easy to list numerous instances of this type: a grocery store checkout line, the line at your *friendly* college bookstore, dentist appointments, and so on. One must be careful to ensure that the Poisson distribution is appropriate in these applications. The experiment leading to the values of the random variable in a given applied problem should be critically compared with the Poisson random variable experiment to assess whether or not the conditions given in Definition 6.8 are satisfied. Once it has been determined that the Poisson distribution is an acceptable model for the random variable, the specific member of the Poisson family must be selected; that is, a value of λ must be chosen. Ordinarily, λ must be *estimated* from sample data (the estimation process will be discussed in Chapter 9). However, in many problems, it must be recognized that the value of λ is time-dependent. For example, in Example 6.10, the average number of emergency admissions during the 12:00–2:00 A.M. period on Saturday morning is most likely quite different from the average number of admissions during the period from 2:00 to 4:00 A.M. Tuesday. In this instance, we would have to use two different members of the Poisson family for the two time periods. Care must be taken, therefore, to specify the time period and its length to ensure that the average number of events (λ) occurring in the period is reasonably homogeneous throughout the entire period.

■ 6.6   SUMMARY

In this chapter, several examples of experiments that produce values of a discrete random variable were described. In particular, the specific discrete random variables and their probability mass functions developed were the discrete uniform, binomial, hypergeometric, and Poisson random variables and associated distributions. Table 6.5 summarizes several characteristics of these important discrete random variables.

TABLE 6.5 Summary characteristics of the discrete uniform, binomial, hypergeometric, and Poisson distributions

Distribution	Parameters	Mean $E(X)$	Variance $V(X)$	Calculated probabilities (Appendix B)
Discrete uniform	N	$\dfrac{N+1}{2}$	$\dfrac{N^2-1}{12}$	—
Binomial.............	n, π	$n\pi$	$n\pi(1-\pi)$	Table B.1: $P(X \leq a) = \displaystyle\sum_{x=0}^{a} P(x)$
Hypergeometric	n, N_1, N	$n\left(\dfrac{N_1}{N}\right)$	$\dfrac{N-n}{N-1}\left[n\left(\dfrac{N_1}{N}\right)\left(1-\dfrac{N_1}{N}\right)\right]$	—
Poisson	λ	λ	λ	Table B.2: $P(X = a) = P(a)$

In Chapter 7, we will describe several experiments that produce values of a continuous random variable. We will develop the associated probability distributions of the random variables described.

■ REFERENCES

Harnett, D. L. *Introduction to Statistical Methods*. 2d ed. Reading, Mass: Addison-Wesley Publishing Co., Inc., 1975.

Hoel, P. G. *Introduction to Mathematical Statistics*. 3d ed. New York: John Wiley & Sons, Inc., 1962.

Hogg, R. V., and Craig, A. T. *Introduction to Mathematical Statistics*. 3d ed. New York: Macmillan, Inc., 1970.

Lindgren, B. W. *Statistical Theory*. New York: John Wiley & Sons, Inc., 1959.

Parzen, E. *Modern Probability Theory and Its Applications*. New York: John Wiley & Sons, Inc., 1960.

■ PROBLEMS

6.1 The number of times that a radio station gives the time each hour is known to be uniformly distributed between one and five, inclusive.

   a. What is the average number of times per hour that the time is given?

   b. What is the standard deviation?

   c. What is the probability that the time is given four or more times in a particular hour of broadcasting?

6.2. In Problem 6.1, use the Chebychef theorem to determine what interval contains at least 75 percent of the numbers of times the time is given per hour? What is the actual proportion of hourly time announcements in this interval?

6.3. The number of orders per day for a product is uniformly distributed over the values 0, 1, 2, 3, and 4. What is the average daily order and what is the standard deviation of the daily orders? [*Hint:* Consider the uniform random variable **X** defined over the integers 1, 2, 3, 4, and 5, and the random variable **Y** = **X** − 1. Use Theorem 5.1 to determine $E(Y)$ and $V(Y)$.]

6.4. Using the definition of the expected value and the variance of a random variable given in Chapter 5, verify that the mean and the variance are as given in Theorem 6.1 for the six-sided die experiment in Example 6.1.

6.5. During a day, the number of times a machine fails is known to be uniformly distributed over the values 0, 1, 2, . . . , 10.

   a. What is the average number of times the machine fails?

   b. What is the standard deviation of the number of times the machine fails?

   c. From the Chebychef theorem, what interval contains at least 84 percent of the numbers of times the machine fails daily?

   d. What proportion of numbers of daily failures are contained in the interval determined in part *c?*

6.6. In 800 trials of tossing a fair six-sided die, how many times would you expect three or fewer dots to turn up on the top face of the die?

6.7. In 12 independent tosses of a fair coin, what is the probability of getting the following?

   a. Three heads.

   b. At most three heads.

   c. At least three heads.

d. Two to four heads.

e. What is the expected value of the number of heads in 12 tosses of the fair coin? What is the variance?

6.8. Describe the nature of a random experiment that will produce the following.

a. A binomial random variable.

b. A hypergeometric random variable.

c. What are the differences in these two experiments?

6.9. In each of the following experiments, determine whether the random variable is binomial, hypergeometric, or neither. Explain your answers.

a. An auditor randomly selects 10 of 1,000 tax returns; X = number of returns in which taxes are underpaid.

b. By extensive testing, it has been determined that an electronic sensing device will properly operate when exposed to a specific stimulus 90 percent of the time. Five devices are randomly selected from 1,000 and are exposed to the stimulus; X = number of devices that properly sense the stimulus.

c. A five-person committee is to be randomly formed from a group of four men and four women; X = number of women on the committee.

d. The probability of tossing a 7 in craps (the number of dots on the two top faces of the dice sum to seven) is 1/6. The dice are tossed ten times; X = number of 7s in the ten tosses.

e. A major league baseball player enters a game with a 0.330 batting average. He comes to bat four times during the game; X = number of hits in the four at-bats.

6.10. A population consists of one dozen employment files, six of which need updating. Four files are randomly selected for inspection.

a. What is the probability that three of the four sampled files will be out of date?

b. What is the expected number of files that are out of date in the sample?

6.11. A lot of ten calculators contains two defects. A customer purchases three of these calculators. Let X = number of defectives she has purchased.

a. Using the hypergeometric distribution, find the probability that $X = 2$.

b. Using the binomial distribution, find the probability that $X = 2$.

c. Which answer is "correct"? Why?

6.12. A lot of 15 transistor radios contains four defectives. If a customer buys three transistor radios for gifts and they are randomly drawn from the lot of 15, what is the probability that all three are defective?

6.13. Sketch the binomial probability mass function for $n = 10$ and $\pi = 0.60$. What is the probability that the binomial random variable X will be equal to seven? What is the probability that the binomial random variable X will be less than or equal to seven? Find the mean and the variance of this distribution. Find the mode of this distribution.

6.14. In a certain location in the summer months, rain falls on two out of five days on the average. If three days are selected at random during the summer, what is the probability that rain will fall on exactly one of these three days?

6.15. The probability that a person selected to participate in a survey will respond to a mail questionnaire is 0.20. Assume the binomial distribution is applicable.

a. What is the probability that *less than* five will respond in a sample of 20 persons?

b. What is the most likely number of responses in the sample of 20 persons?

c. What is the expected number of responses in the sample of 20 persons?

6.16. Assuming that the probability of a male birth is 1/2, find the following in a family of five.

a. The probability of at least one boy.

b. The probability of at least one boy and at least one girl.

6.17. The probability that an entering college student will graduate from a certain college is 0.4. Determine the probability of the following out of six students.

a. None will graduate.

b. One will graduate.

c. At least one will graduate.

6.18. Twenty percent of the bolts produced by a machine are defective. If 50 bolts are chosen at random, determine the following probabilities.

a. Five or fewer are defective.

b. Exactly ten are defective.

c. More than 15 are defective.

6.19. A library receives 95 percent of its loaned books back on time. A sample of 50 books now out on loan is taken. Assume that the binomial distribution is applicable.

a. What is the expected number of books among these 50 which will be returned on time?

b. What is the probability that *all* 50 books will be returned on time?

6.20. The probability that a certain part is defective is 0.10. Assume that the binomial distribution is applicable and that 25 parts are randomly selected from a very large lot.

a. What is the probability that five or more are defective?

b. What is the probability that exactly two are defective?

c. What is the expected number of defectives in the sample of 25?

6.21. The probability that a life insurance agent sells a policy to a prospective client is 0.20. On a particular day, the agent contacts 15 prospective clients. Let X = number of policies sold.

a. Is X a binomial random variable? Explain.

b. What assumptions must you make, if any, to use the binomial distribution for X in this case?

c. Find $E(X)$ and $V(X)$ assuming that X is binomially distributed.

d. Using Chebychef's theorem, determine an interval that contains at least 75 percent of the values of X.

e. Assuming the binomial distribution is applicable, what proportion of the values of X actually lie in the interval constructed in part *d?*

6.22. A wholesaler will accept a shipment lot of n items from a producer if d or fewer of the items in the lot are defective. Determine, using the binomial distribution, the probability of accepting the lot, when the lot has the following proportion of defective items, given n and d.

a. 0.00; $n = 5, d = 0$

b. 0.05; $n = 5, d = 0$

c. 0.50; $n = 6, d = 1$

d. 0.10; $n = 10, d = 0$

e. 0.10; $n = 10, d = 1$

f. 0.50; $n = 10, d = 0$

g. 0.50; $n = 12, d = 2$

h. 0.20; $n = 25, d = 2$

i. 0.10; $n = 20, d = 2$

j. 0.50; $n = 20, d = 3$

6.23. A large department store gives a 5 percent discount to customers who pay cash. From past experience, the store determines that 30 percent of its sales are cash sales. This experience is prior to the store's new policy on cash sale discounts. In a sample of 22 transactions taken since the new policy was enacted, it is found that there are 10 cash sales. Assume that the binomial distribution is applicable.

a. What is the probability that the number of cash sales, **X**, is 10 or more in a sample of size 22 if the proportion of cash sales, π, is 0.30?

b. What is the probability that **X** is 10 or more in a sample of 22 if $\pi = 0.40$?

c. What is the probability that **X** is 10 or more in a sample of 22 if $\pi = 0.50$?

d. Is it likely that π is greater than 0.30 since the enactment of the new cash sales policy? Why or why not?

6.24. If three fourths of the students on a certain college campus are freshmen and sophomores, what is the probability that six students selected at random will contain exactly 50 percent freshmen and sophomore members?

6.25. A multiple-choice exam contains 25 questions. Each question contains five possible answers—a, b, c, d, e—only one of which is correct. Let **X** = number of correctly answered questions and π = probability that a question is answered correctly.

a. If a student randomly selects an answer to every question so that $\pi = 0.20$, what is the probability that he or she will get ten or more questions right?

b. In part a, what is the expected number of correct responses reached by guessing?

c. A student receives a score of 15. Is it likely that he or she guessed each answer? Explain.

d. If a student must answer 15 or more questions correctly to pass the examination, what is the probability that a person will pass the examination solely by guessing?

6.26. In a certain community, 60 percent of those who are going to vote in an upcoming election favor candidate A. If a pollster samples 20 people planning to vote, prior to the election, what is the probability that he or she will predict that candidate A will receive *less than* 50 percent of the votes cast? Assume that the binomial distribution is applicable.

6.27. The H. R. Clock Company claims that 70 percent of all persons who seek assistance in completing the federal income tax form in Arlington, Virginia, do so with their offices. In a random sample of 20 persons who are seeking assistance, 10 intend to see H. R. Clock. Does this sample provide sufficient evidence to refute their claim of 70 percent of the market? Explain.

6.28. There are 20 stamping machines in a large plant. The machines operate independently of one another, and breakdowns are random occurrences. The probability of a breakdown for each machine in one day's operation is 0.10.

a. Using the binomial distribution, determine the probability of two or more breakdowns on any given day.

b. Using the Poisson distribution with $\lambda = (20)(0.10) = 2$ breakdowns each day on the average, determine the probability that two or more breakdowns occur on any given day.

c. Which distribution—binomial or Poisson—is more applicable in this problem? Explain.

6.29. Ten percent of the units made in a production process are defective. Suppose that ten units are chosen at random.

a. Using the binomial distribution,

determine the probability that exactly two units are defective.

b. Using the Poisson distribution approximation to the binomial distribution, determine the probability that exactly two units are defective.

6.30. The probability that an individual suffers a bad reaction from taking a new drug orally is 0.001. Using the Poisson distribution, determine the following probabilities out of 2,000 individuals.

a. Three will suffer a bad reaction.
b. At least three will suffer a bad reaction.
c. Could these probabilities be determined by using the binomial distribution? Explain.

6.31. A retailer determines that the average number of orders per day for a certain product is five. Assume the Poisson distribution is applicable.

a. What is the probability that no orders are received on a given day?
b. What is the probability that between two and eight orders, inclusive, will be received on a given day?
c. What is the probability that more than ten orders will be received on any given day?
d. What assumptions are necessary to use the Poisson distribution in parts a–c?

6.32. In preparing mathematical manuscripts, it is found that the average number of typographic errors made per page is 0.4. If ten pages are randomly selected from a large manuscript, what is the probability that two or more errors are found? What assumptions did you make in calculating this probability?

6.33. The emergency squad at a metropolitan fire station receives an average of 0.20 calls per hour. Assume that the Poisson distribution is applicable.

a. What is the probability that the squad will receive more than one call in one hour?
b. Over the 24-hour day, what is the average number of calls the squad receives?
c. In studying the recent experience of squad calls, it is found that an average of four calls are received between the hours of 6:00 A.M. and 10:00 P.M., and an average of one call is received between the hours of 10:00 P.M. and 6:00 A.M. Is this experience consistent with part b and the assumptions required to determine the probability in part a? Explain.

6.34. Suppose that in a carpet manufacturing process, an average of two flaws occur per ten running yards of material. What is the probability that a given ten-yard segment will have one or fewer flaws, if the number of flaws is known to be Poisson distributed?

6.35. A customer service desk receives an average of five customers per hour. Assume that the number of customers arriving per hour is Poisson distributed.

a. What is the probability that more than ten people will request service in a particular hour?
b. What is the mean number of arrivals per hour?
c. What is the standard deviation of the number of arrivals per hour?
d. Using the Chebychef theorem, determine what interval contains at least three fourths of the numbers of arrivals per hour.
e. What proportion of values of the Poisson random variable are actually contained in the interval determined in part d?

Continuous probability distribution models

7

■ 7.1 INTRODUCTION

The distribution of a continuous random variable can assume many forms. Examples of representations of a probability *density function* for a continuous random variable are illustrated in Figure 7.1. Distributions A and B are skewed, while distributions C, D, and E share the property of symmetry but differ with respect to other characteristics (number of modes, dispersion, etc.).

The random variables and their associated distributions to be discussed in this chapter differ primarily from those described in Chapter 6 in that the random variable can assume any one of an infinite number of values. Examples of such random variables are the lengths of steel bars produced by a certain production process, the lifetime in hours of light bulbs, and the time between the arrivals of customers at a service counter.

We will begin our discussion of continuous distributions with the uniform distribution. The uniform distribution is the continuous analogue of the discrete uniform distribution described in Section 6.2.

■ 7.2 THE UNIFORM (CONTINUOUS) DISTRIBUTION

A distribution that is being used increasingly—especially in simulation experiments—is the uniform distribution. Random variates from other continuous distributions can be generated by considering special properties of the uniform distribution. For example, the *sum* of 12 uniform random variables with only slight modification can be made to approximate a normal

FIGURE 7.1 Representations of a probability distribution of a continuous random variable

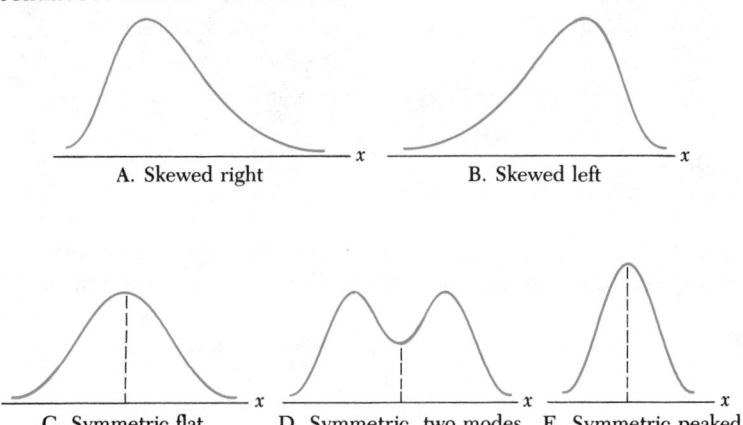

A. Skewed right   B. Skewed left

C. Symmetric flat   D. Symmetric, two modes   E. Symmetric peaked

distribution (described in Section 7.3), and random draws from an exponential distribution (described in Section 7.4) can be made by simply taking the logarithm of a uniform random variable. Many simulated distributions discussed in Chapter 8 on sampling were derived by "drawing" random variables on a computer from a uniform distribution.

The uniform distribution is also used frequently to characterize an "almost informationless" state. That is, we *assume* that all possible values of the continuous random variable are equally likely, and thus the density function is constant over the range of values of the random variable. In this instance, the random variable is said to be *uniformly distributed*. A uniform random variable and its associated probability density function are defined in Definition 7.1.

Definition 7.1
Uniform random variable and its probability density function

Let the random variable **X** be defined over the interval a to b ($a < b$) with *probability density function* given by:

$$f(x) = \begin{cases} \dfrac{1}{b - a} & \text{if } a \leq x \leq b \\ 0 & \text{all other } x \end{cases}$$

Then **X** is a *uniform random variable.*

The mean and variance of a uniform random variable are given in Theorem 7.1.

> **Theorem 7.1**
> **Mean and variance of the uniform random variable**
>
> If **X** is a uniform random variable defined over the interval a to b ($a < b$), then its *mean*, $E(X)$, and *variance*, $V(X)$, are given by
>
> $$E(X) = \frac{b + a}{2} \quad \text{and} \quad V(X) = \frac{(b - a)^2}{12}$$

Two examples of uniform random variables are given in Figure 7.2. Figure 7.2B is a special case of the uniform random variable called the *standardized*

FIGURE 7.2   Two examples of uniform random variables

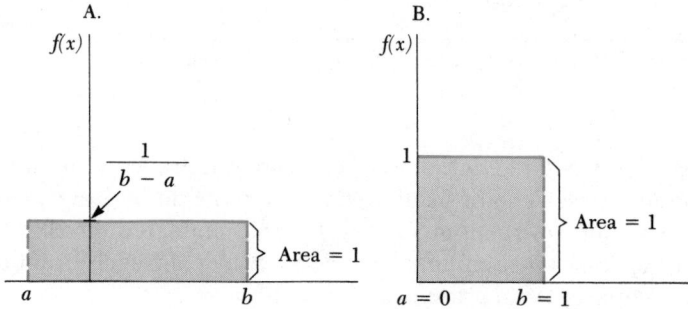

uniform random variable. The mean and the variance of a standardized uniform random variable are developed in Example 7.1.

Example 7.1   A special case of the continuous uniform distribution occurs whenever $a = 0$ and $b = 1$. This distribution is called the *standardized uniform distribution.* Determine the density function, $f(x)$, the mean, $E(X)$, and the variance, $V(X)$, of a standardized uniform random variable.

Solution   Definition 7.1 can be appropriately modified by substituting 0 for a and 1 for b. The density function then becomes

$$f(x) = \begin{cases} \dfrac{1}{1 - 0} = 1 & \text{if } 0 \le x \le 1 \\ 0 & \text{all other } x \end{cases}$$

The mean and the variance of a standardized uniform random variable are

$$E(X) = \frac{1 + 0}{2} = \frac{1}{2}$$

$$V(X) = \frac{(1 - 0)^2}{12} = \frac{1}{12}$$

Note that the mean of this distribution is ½, which is intuitively reasonable since this is the midpoint of the interval from 0 to 1.

Recall from Chapter 5 that because a uniform random variable is continuous, we cannot determine the probability that the random variable is equal to a *particular value*, say $X = x$, but we can calculate the probability that it assumes some value between c and d ($c < d$) by determining the area under the density function from c to d. The determination of the probability that a uniform random variable assumes a value between c and d is illustrated in the following example.

Example 7.2 A uniform random variable is defined over the interval 0 to 6. What is the probability that the random variable assumes a value between 2 and 4?

Solution The required probability is given in Figure 7.3. In essence, we are interested in the ratio of the shaded area to the total area. Since the total area is equal to 1, however, the required probability becomes

$$\frac{4 - 2}{6 - 0} = \frac{2}{6} = \frac{1}{3}$$

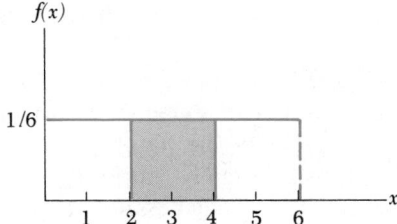

FIGURE 7.3 Determination of $P(2 \leq X \leq 4)$ from Example 7.2

In general, the probability that a uniform random variable defined over the interval a to b ($a < b$) assumes a value between c and d ($c \geq a$, $d \leq b$) is

$$\left(\frac{1}{b - a}\right)(d - c) = \frac{d - c}{b - a}$$

Hence, it is not necessary to have specific tables of the uniform distribution.

7.3 THE NORMAL DISTRIBUTION

7.3.1 Introduction

The normal distribution is *probably* the most important probability distribution in statistics! It is a probability distribution of a continuous random variable, yet it is often used to model the distribution of other continuous random variables and discrete random variables.

The reason for the versatility in using the normal distribution as a probability distribution model is indicated in Figure 7.4. The basic form of the normal distribution is that of a bell—it has a single mode and is symmetric about its central value. The flexibility in using the normal distribution is due

FIGURE 7.4 Three forms of the normal distribution

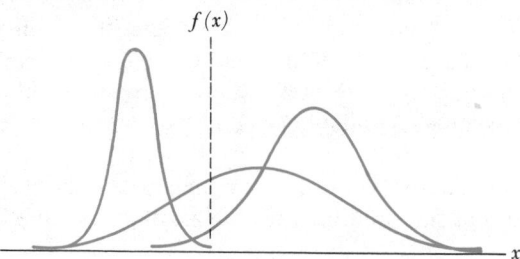

to the fact that the "bell" may be centered over any number on the real line and it may be made flat or peaked to correspond to the amount of dispersion that the values of a random variable may assume. Many quantitative characteristics have distributions similar in form to the normal distribution's bell shape.

Examples of random variables that have been successfully modeled by the normal distribution are the height and weight of persons, the diameter of bolts of a specified size produced on a machine, the IQ of persons, and the lifetime in hours of batteries or light bulbs. Typically, in the type of experiment that produces a random variable that can be successfully approximated by a normal random variable, the values of the random variable are produced by a measuring process, where it is known that the measurements tend to cluster symmetrically about a central value. A random variable that is an average or a sum of values of another random variable is, under very *general* conditions, *almost always* distributed approximately as a normal random variable, *regardless* of the form of the distribution of the random variable with values that are summed or averaged. An example of such a random variable is the average grade point average of a group of students selected at random from the population of students at your university or college. The notion that a random variable that is an average is distributed as a normal random variable is discussed in Chapter 8 when we describe the central limit theorem.

Unfortunately, if it is known that a random variable is symmetrically distributed with a single mode, the random variable may not be a normal random variable. There are other distributions in statistics that are unimodal and symmetric. However, they also can often be successfully modeled by the normal distribution.

For a random variable to be normally distributed, the mathematical expression delineating the form of the bell must be of a specific type as described in Definition 7.2.

Definition 7.2
Normal random variable and its probability density function

A continuous random variable **X** is said to be normally distributed if its *probability density function* is

$$f(x) = \frac{1}{\sigma\sqrt{2\pi}} e^{\left[-\frac{1}{2}\left(\frac{x-\mu}{\sigma}\right)^2\right]}, \qquad -\infty < x < +\infty, \; -\infty < \mu < +\infty, \; \sigma > 0$$

where μ and σ are parameters of the distribution and π and e are mathematical constants equal to 3.14159 and 2.71828, respectively.

Recall that the probability that a continuous random variable assumes a value between two constants a and b ($a < b$) can be determined by finding the area under its density function from a to b. Thus, for a normal random variable, $P(a \leq \mathbf{X} \leq b)$ = area under $f(x)$ above the x-axis between $x = a$ and $x = b$ as illustrated in Figure 7.5.

FIGURE 7.5 Area representing $P(a \leq \mathbf{X} \leq b)$ for a normal random variable

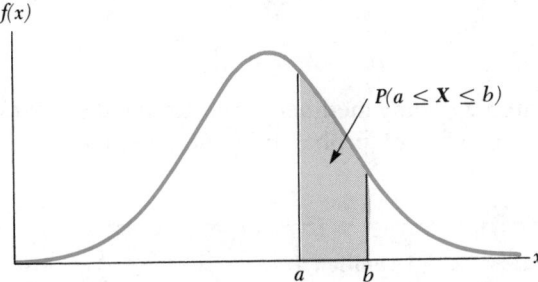

The formula giving $f(x)$ in Definition 7.2 appears to be quite imposing on first glance. But, as we shall see, it is possible to compute probabilities corresponding to the range of values a normally distributed random variable can assume using tables rather than performing mathematical computations involving $f(x)$. Before learning how to do this, we will consider the mean and the variance of a normal random variable.

☐ **7.3.2 The mean and variance of the normal random variable**

The mean and variance of the normal random variable are given in Theorem 7.2.

> **Theorem 7.2**
> **Mean and variance of the normal random variable**
>
> If **X** is a normal random variable, then its *mean*, $E(\mathbf{X})$, and *variance*, $V(\mathbf{X})$, are given by:
>
> $$E(\mathbf{X}) = \mu \qquad \text{and} \qquad V(\mathbf{X}) = \sigma^2$$

Notice that the mean depends *only* on the parameter μ and the variance depends *only* on the parameter σ. Thus, the normal distribution may be located over its central value on the real line independently of the amount of dispersion (σ^2) specified for the distribution! Contrast this with the binomial distribution (and others discussed thus far) in which the mean and the variance jointly depend on the *same* parameters n and π $[E(\mathbf{X}) = n\pi; V(\mathbf{X}) = n\pi(1 - \pi)]$ and hence are *not* independent of one another. This property of the normal distribution adds immeasurably to its flexibility in modeling the distributions of nonnormal random variables, as we shall see.

We now return to the problem of computing probabilities associated with a normal random variable.

□ **7.3.3 The standardized normal distribution**

Probabilities associated with any member of the normal distribution family can be computed from a table of probabilities compiled for the *standard normal distribution*.

> **Definition 7.3**
> **Standard normal distribution**
>
> A normal distribution with $\mu = 0$ and $\sigma = 1$ is called a *standard normal distribution*. When a normal random variable **X** has a mean of 0 and a variance of 1, it is called a *standardized normal random variable* and will be denoted by **Z**. The probability density function of the standardized normal random variable **Z** is:
>
> $$f(z) = \frac{1}{\sqrt{2\pi}} \, e^{[-\frac{1}{2}(z)^2]}, \qquad -\infty < z < +\infty$$

The form of the standard normal distribution is illustrated in Figure 7.6. As indicated in the figure, and for *any* normal distribution, 68.27 percent of the values of **Z** lie within one standard deviation of the mean, 95.45 percent of the values lie within two standard deviations of the mean, and 99.73 percent of the values lie within three standard deviations of the mean.

Probabilities of a standardized normal random variable of the form $P(0 \leq \mathbf{Z} \leq a)$ are provided in Table B.3, Appendix B, and on the inside back cover of your text. By using the fact that the normal distribution is sym-

FIGURE 7.6 Standard normal distribution

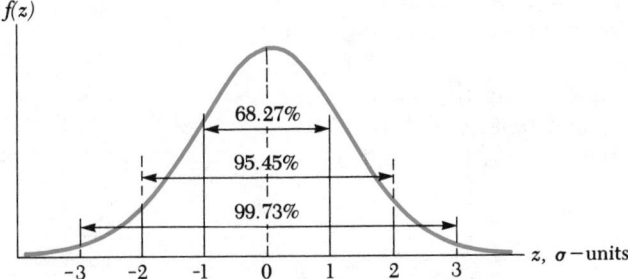

metric about its mean (0 in this case), and that the total area under the curve is 1 (½ to the left of 0, and ½ to the right of 0), the probability that **Z** resides in any interval on the real line may be determined from this table as the following example indicates.

Example 7.3 Find the area under the standard normal distribution curve for each of the following intervals.

a. Between $z = 0$ and $z = 2.0$.
b. Between $z = -1.28$ and $z = 0.0$.
c. Between $z = -0.58$ and $z = 2.54$.
d. Between $z = 1.20$ and $z = 2.44$.
e. Greater than $z = 2.87$.

Solution

a. In Table B.3, Appendix B, proceed downward in the leftmost column to 2.0. Select the first column marked .00 indicating that the second decimal place is 0—the area read from the table is 0.4772.

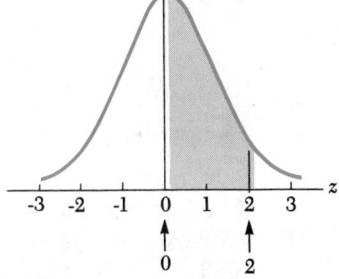

b. Since the normal distribution is symmetric, the area between -1.28 and 0.0 is equal to the area between 0.0 and $+1.28$. Thus, proceed down the leftmost column to 1.2. Select the ninth column marked .08—the resulting number in the table is 0.3997.

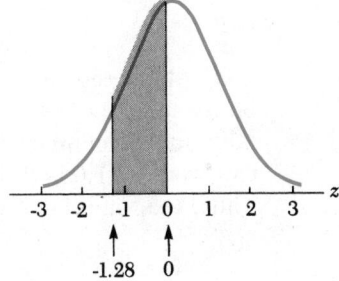

c. We may determine this area in two parts: Total area = (Area between 0.0 and 2.54) + (Area between −0.58 and 0.0). The area between 0.0 and 2.54 is 0.4945. The area between −0.58 and 0.0 is the same as the area between 0.0 and 0.58, which is 0.2190. The answer is 0.4945 + 0.2190 = 0.7135.

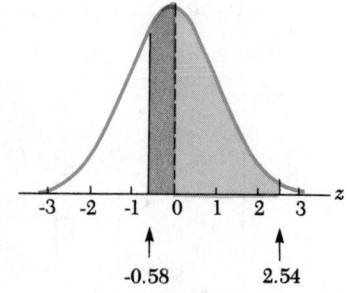

d. We can determine this area by finding the difference between the areas from 0 to 2.44 and from 0 to 1.20. The area between 0 and 2.44 is 0.4927 and the area between 0 and 1.20 is 0.3849. Thus, the area between 1.20 and 2.44 is 0.4927 − 0.3849 = 0.1078.

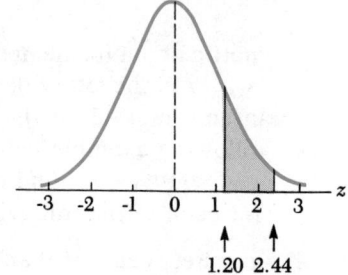

e. Since the area between 0 and $+\infty$ is 0.5, we can determine the area from 2.87 to ∞ by subtracting the area from 0 to 2.87 (0.4979) from 0.5: 0.5000 − 0.4979 = 0.0021.

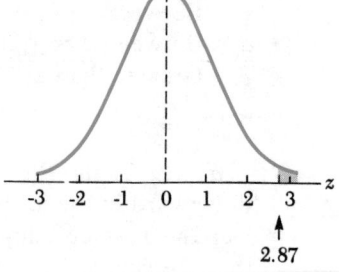

In many problems, we are given the area in a certain interval and then asked to determine the value of **Z** that specifies the interval. This is the "reverse" of the problem solved in Example 7.3. In Example 7.4, the use of the Table B.3, Appendix B, to solve the reverse problem is demonstrated.

Example 7.4 Find the value or values of **Z** on the standard normal distribution axis when (*a*) the area between 0 and z is 0.3413, and (*b*) the area to the right of z is 0.8982.

Solution

a. Now the normal curve area table must be used in reverse. Look in the body of the table for the number 0.3413. It appears in the row marked 1.0 and the column marked 0.0. Thus, the value of **Z** is 1.00.

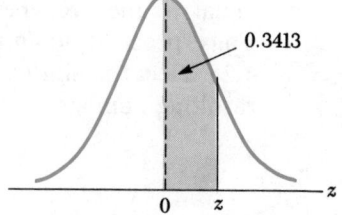

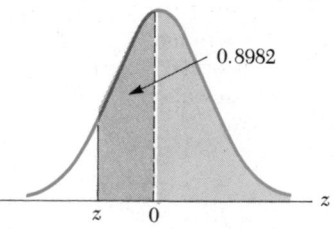

b. Since the area given is greater than 0.5, we know that z must be less than 0. The area between z and 0 is $0.8982 - 0.5000 = 0.3982$. Now assume that z is positive and find z such that the area between 0 and z is 0.3982. The area of 0.3982 does not appear in the table—the closest numbers are 0.3980 and 0.3997. The exact value of z could be determined by interpolation, but we will use $z = 1.27$ since 0.3980 is closer to 0.3982 than 0.3997. Now, we must remember that z must be to the left of 0 (a negative number). Thus, $z = 1.27$.

☐ **7.3.4 Areas under the normal distribution**

Probabilities associated with a normal random variable **X** that is not standardized can be determined by using the results of Theorem 7.3.

Theorem 7.3
Standardization of a normal random variable

If **X** is a normal random variable, the mean of which is μ and the standard deviation of which is σ, then

$$Z = \frac{X - \mu}{\sigma}$$

is a *standardized normal random variable* with a mean of 0 and a standard deviation of 1.

The use of this theorem is illustrated in Figure 7.7. The probability that **X** is between a and b ($b > a$) can be determined by computing the probability that **Z** is between $(a - \mu)/\sigma$ and $(b - \mu)/\sigma$; the area between a and b is preserved by the linear transformation $Z = (X - \mu)/\sigma$!

The following examples illustrate the use of Theorem 7.3 and, more generally, the applicability of the normal distribution model.

Example 7.5 The mean lifetime of 50-watt light bulbs produced by the Stay-Bright Light Bulb Company is 200 hours. It is known that the standard deviation is 20 hours. Assuming that the lifetimes of the light bulbs are normally distributed, what is the probability that a single 50-watt light bulb extracted from the production lot will (*a*) burn out at a time between 180 and 210 hours, and (*b*) burn out at a time greater than 250 hours?

Solution

a. The solution on the **X**-distribution with a mean of 200 and a standard deviation of 20 is the area between $x_1 = 180$ and $x_2 = 210$. This area can be

FIGURE 7.7 Relationship between normal distribution with mean μ and standard deviation σ and the standard normal distribution with mean $\mu = 0$ and standard deviation $\sigma = 1$

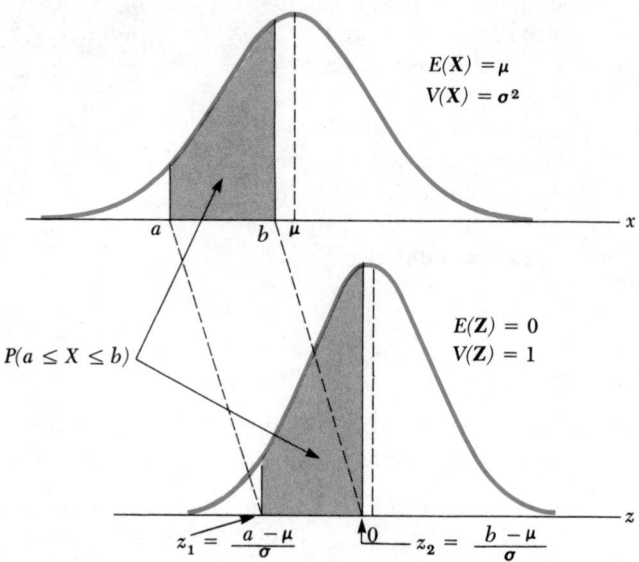

determined by first standardizing x_1 and x_2:

$$z_1 = \frac{x_1 - \mu}{\sigma}$$

$$= \frac{180 - 200}{20}$$

$$= \frac{-20}{20} = -1.00$$

$$z_2 = \frac{x_2 - \mu}{\sigma}$$

$$= \frac{210 - 200}{20}$$

$$= \frac{10}{20} = 0.50$$

The area between -1.0 and 0.50 on the standard normal distribution will equal the area between 180 and 210 on the X-distribution. The area between $z_1 = -1.00$ and $z_2 = 0.50$ is:

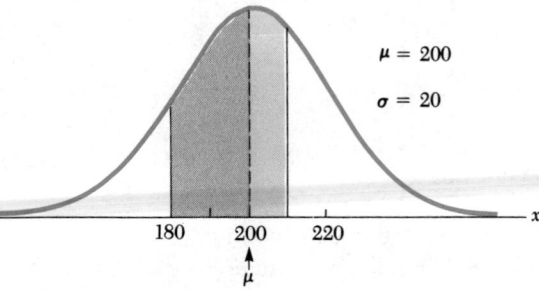

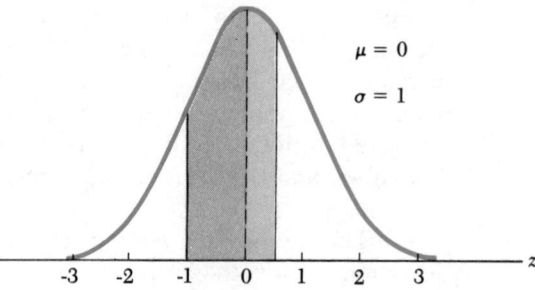

$$\text{Area } (-1.00 \text{ to } 0.00) = 0.3413$$
$$\text{Area } (0.00 \text{ to } 0.50) = \underline{0.1915}$$
$$0.5328$$

Thus, $P(180 \leq X \leq 210) = 0.5328$. This answer tells us that 53.28 percent of the 50-watt light bulbs comprising the population will have times to failure between 180 and 210 hours.

b. The solution on the X-distribution is the area greater than $x_1 = 250$ hours.

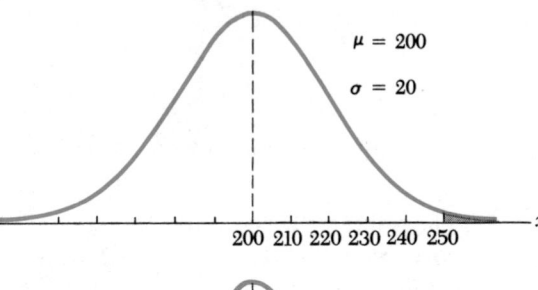

The standardized value is

$$z_1 = \frac{x_1 - \mu}{\sigma}$$

$$= \frac{250 - 200}{20}$$

$$= \frac{50}{20} = 2.5$$

The area from $z_1 = 2.5$ to $+\infty$ is:

$$0.5000 -$$
$$\text{Area } (0.00 \text{ to } 2.5)$$
$$= 0.5000 - 0.4938$$
$$= 0.0062$$

Thus, 0.62 percent of the 50-watt light bulbs will fail at a time exceeding 250 hours.

Example 7.6   A statistics exam is based on 800 points. It is known that the mean of the exam is 500 and the standard deviation is 100. Assuming that the exam scores are normally distributed (i.e., the normal is a good model for this problem), what score must a student make on the exam to be placed in the upper 15 percent of all test scores?

Solution   This is the reverse problem—we are given the area under the normal curve and asked to determine the test score x. If the student is in the upper 15 percent of all students taking the exam, his score x must be such

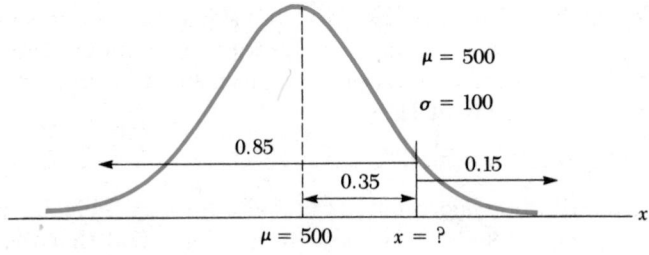

that the area to the right of x is 0.15 and the area between the mean of 500 and x is 0.35.

We first solve the problem on the standard normal distribution:

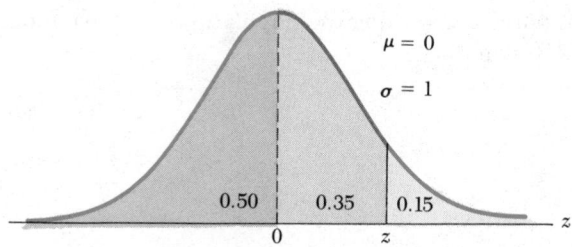

The value of **Z** such that the area between 0 and z is 0.35 is 1.04 ($z = 1.03$ corresponds to an area of 0.3485 and $z = 1.04$ corresponds to an area of 0.3508—0.3508 is closer to 0.3500). Now the standardizing formula

$$z = \frac{x - \mu}{\sigma}$$

is solved for x: $x = \sigma z + \mu$. Thus, the value of **X** that corresponds to $z = 1.04$ is $x = 100(1.04) + 500 = 604$. Thus, a student must score 604 or more points to be placed among the upper 15 percent of the scores.

Whenever computing probabilities associated with normal random variables, the student is advised to draw a picture of the areas to be computed as in Examples 7.4, 7.5, and 7.6—errors are usually greatly reduced if time is taken to make these sketches!

☐ 7.3.5 The normal approximation to the binomial distribution

In Chapter 6, the binomial distribution was introduced; its probability mass function is:

$$P(x) = C_x^n(\pi)^x(1 - \pi)^{n-x}, \ x = 0, 1, 2, \ldots, n$$

In Chapter 6, we mentioned that a method other than using the mass function or the tables to compute binomial probabilities could be applied under certain circumstances. This third method involves fitting a normal curve to the binomial distribution and calculating areas under the curve to *approximate* the sum of binomial probability sticks. To illustrate this method, suppose we flip a fair coin $n = 10$ times and record the number of heads, **X**. The appropriate binomial mass function that describes the probability distribution of **X** in this experiment is:

$$P(x) = C_x^{10}(0.5)^x(1 - 0.5)^{10-x}, \quad x = 0, 1, 2, \ldots, 10$$

This binomial distribution is presented in tabular form in Table 7.1 and as a stick diagram in Figure 7.8. We first note from Figure 7.8 that the sticks

TABLE 7.1 Binomial distribution when $\pi = 0.5$
and $n = 10$

x	$P(x)$	x	$P(x)$
0	0.001	6	0.205
1	0.010	7	0.117
2	0.044	8	0.044
3	0.117	9	0.010
4	0.205	10	0.001
5	0.246		

FIGURE 7.8 Stick diagram of binomial distribution
when $\pi = 0.50$ and $n = 10$

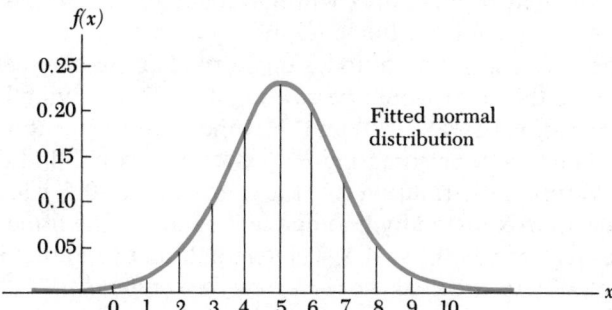

suggest the form of a bell-shaped distribution—this property is strengthened
by drawing a smooth curve along the tops of the sticks. Recall that the mean
and variance of the binomial distribution are given by $E(X) = n\pi$ and $V(X) =$
$n\pi(1 - \pi)$, respectively. When $\pi = 0.50$ and $n = 10$, we have $E(X) = 5$ and
$V(X) = 2.5$. Since, in the normal distribution, the mean is μ and the variance
is σ^2, we will select a normal distribution with $\mu = 5$ and $\sigma^2 = 2.5$
($\sigma = \sqrt{2.5} = 1.58$) to fit over the sticks of this binomial distribution. Then, if
we wish to determine the probability that the binomial random variable **X**
assumes a value between a and b ($a < b$), it can be approximated by the area
under the normal curve from a to b.

There is a problem in this approximation process, however. The normal
distribution assigns probability continuously over the real number line,
whereas binomial probability is massed only at integer values. For example,
the probability of getting more than two heads but less than three heads in
ten tosses of a coin $[P(2 < X < 3)]$ is 0, but the area under the fitted normal
distribution between 2 and 3 would not be 0, of course. A partial solution to
this problem is indicated in Figure 7.9. Suppose we wish to approximate
$P(X \le 4)$ and $P(X \ge 5)$ using the normal distribution. If only the area to the
left of $x = 4$ under the normal curve is used to approximate $P(X \le 4)$, and
only the area to the right of $x = 5$ is used to approximate $P(X \ge 5)$, then the

FIGURE 7.9 Normal approximation to binomial probabilities

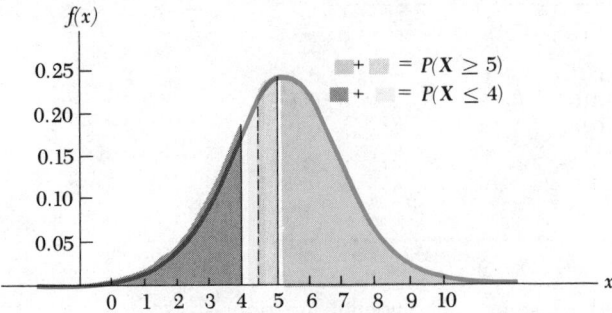

sum of the approximated probabilities will not equal 1. But we know that $P(X \leq 4) + P(X \geq 5) = 1$ for the binomial distribution.

The problem is the "wedge" of area under the normal curve between 4 and 5. A fair allocation of this area would be to roughly split it, giving half this area to the approximation of $P(X \leq 4)$ and the other half to the approximation of $P(X \geq 5)$. Thus, in approximating $P(X \leq 4)$, we would find the area under the fitted normal distribution to the left of $4 + 0.5 = 4.5$. The $P(X \geq 5)$ would be approximated by finding the area under the fitted normal distribution to the right of $5 - 0.5 = 4.5$. The one-half unit (0.5) that is being added and subtracted here is called the *correction for continuity*—it accounts for the gaps between integer values where the binomial mass function is not defined.

   Example 7.7   In the coin-tossing experiment described earlier, approximate the following binomial probabilities using the fitted normal distribution.

a. $P(3 \leq X \leq 8)$, without the correction for continuity.
b. $P(3 \leq X \leq 8)$, with the correction for continuity.
c. $P(X \geq 7)$, with the correction for continuity.

   Solution
   a.   The parameters of the fitted normal distribution are $\mu = 5$ and $\sigma = 1.58$, since the mean and standard deviation of the binomial distribution are 5 and 1.58, respectively. Without the correction for continuity, we wish to find the area under the normal distribution with $\mu = 5$ and $\sigma = 1.58$ between $x_1 = 3$ and $x_2 = 8$. The standardized values are:

$$z_1 = \frac{x_1 - \mu}{\sigma} = \frac{3 - 5}{1.58} = -1.27$$

$$z_2 = \frac{x_2 - \mu}{\sigma} = \frac{8 - 5}{1.58} = 1.90$$

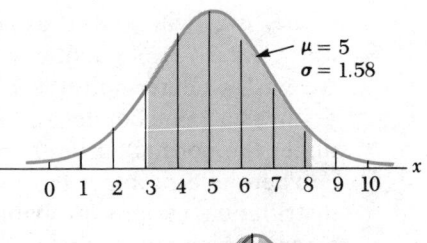

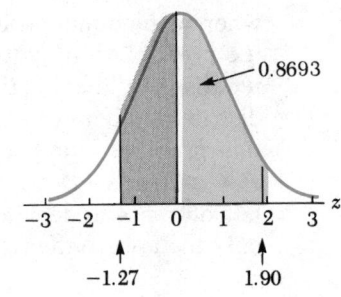

The area between $z_1 = -1.27$ and $z_2 = 1.90$ on the standard normal curve is 0.8693.

b.   Using the correction for continuity, we move down 0.5 unit from 3 ($x_1 = 3 - 0.5 = 2.5$) and up 0.5 unit from 8 ($x_2 = 8 + 0.5 = 8.5$). In this way, the area between 2 and 3 is shared with $P(X \leq 2)$ and the area between 8 and 9 is shared with $P(X \geq 9)$. Now,

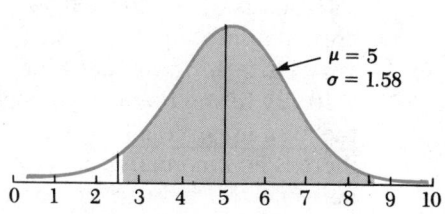

$$z_1 = \frac{x_1 - \mu}{\sigma} = \frac{2.5 - 5}{1.58} = -1.58$$

$$z_2 = \frac{x_2 - \mu}{\sigma} = \frac{8.5 - 5}{1.58} = 2.21$$

The area between $z_1 = -1.58$ and $z_2 = 2.21$ on the standard normal curve is 0.9293.

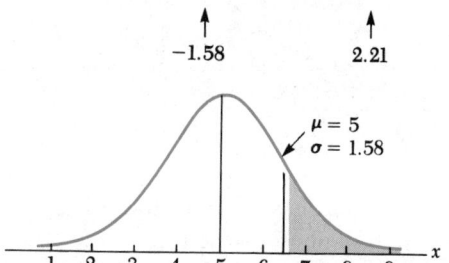

c.   Using the correction for continuity, we move back 0.5 unit to share the area between 6 and 7 with $P(X \leq 6)$. Thus, $x = 7 - 0.5 = 6.5$. Now,

$$z_1 = \frac{x_1 - \mu}{\sigma} = \frac{6.5 - 5}{1.58} = 0.95$$

The area under the standard normal curve from $z_1 = 0.95$ to $+\infty$ is 0.1711.

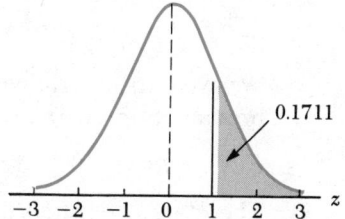

From Table 7.1, we find that $P(3 \leq X \leq 8) = 0.934$ and $P(X \geq 7) = 0.172$. Notice that the correction for continuity gives a much better approximation to $P(3 \leq X \leq 8)$—contrast the solution in part b to the one in part a.

The normal approximation to the binomial distribution should be used only if n is large (the statistician's rule in this case is n must be sufficiently large to satisfy two conditions: 1. $n\pi \geq 5$, and 2. $n(1 - \pi) \geq 5$). The reason it works so well in Example 7.7 is because when $\pi = 0.5$, the binomial distribution is symmetric. If we had chosen $\pi = 0.2$ with $n = 10$ and used the normal approximation, the approximated probabilities would not have been very close to the exact binomial probabilities, as can be seen from Figure 7.10. By fitting the normal curve over the mean value of 2, a large portion of the area must be allocated to the region to the left of $x = 0$ where we know there is no binomial probability. As n becomes very large ($n > 100$), the area between the integers under the fitted normal curve becomes small—it is possible then not to use the correction for continuity because its effect becomes very small.

FIGURE 7.10 Normal curve fitted to the binomial distribution when $\pi = 0.2$ and $n = 10$

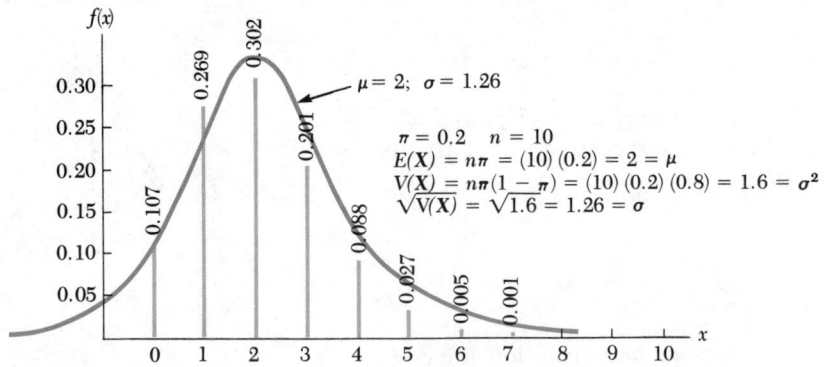

Example 7.8 In a large lot of AM/FM transistor radios produced by the Sunny Corp., it is known that 10 percent are defective. A certain retailer purchases 100 of these radios from Sunny. What is the probability that more than 15 of them are defective?

Solution Let X be the number of defectives in the lot of 100 purchased by the retailer. Assuming that the random variable X is binomially distributed, the appropriate mass function is:

$$P(x) = C_x^{100}(0.1)^x(1 - 0.1)^{100-x}, \qquad x = 0, 1, 2, \ldots, 100$$

We wish to determine $P(X > 15)$ or $P(X \geq 16)$. We will approximate this probability by using the normal distribution. Since $n = 100$ and $\pi = 0.1$,

$$E(X) = n\pi = 100(0.1) = 10 \quad \text{and} \quad V(X) = n\pi(1 - \pi) = 100(0.1)(0.9) = 9$$

Therefore, the fitted normal distribution has parametric values $\mu = 10$ and $\sigma = \sqrt{9} = 3$. Since we wish to approximate the probability that $X \geq 16$, the value on the normal distribution is taken to be $16 - 0.5 = 15.5$; in this way, the area between $x = 15$ and $x = 16$ is shared with $P(X \leq 15)$. Notice that the correction for continuity could be dropped here since n is large

($n = 100$). We include it, however, to show how the 0.50 increment is used. Now the approximated probability on the normal distribution is the area to the right of 15.5 as illustrated in Figure 7.11. The standardized value is:

$$z_1 = \frac{x_1 - \mu}{\sigma} = \frac{15.5 - 10}{3} = \frac{5.5}{3} = 1.83$$

FIGURE 7.11 Normal approximation of the binomial probability $P(X \geq 16/n = 100, \pi = 0.10)$

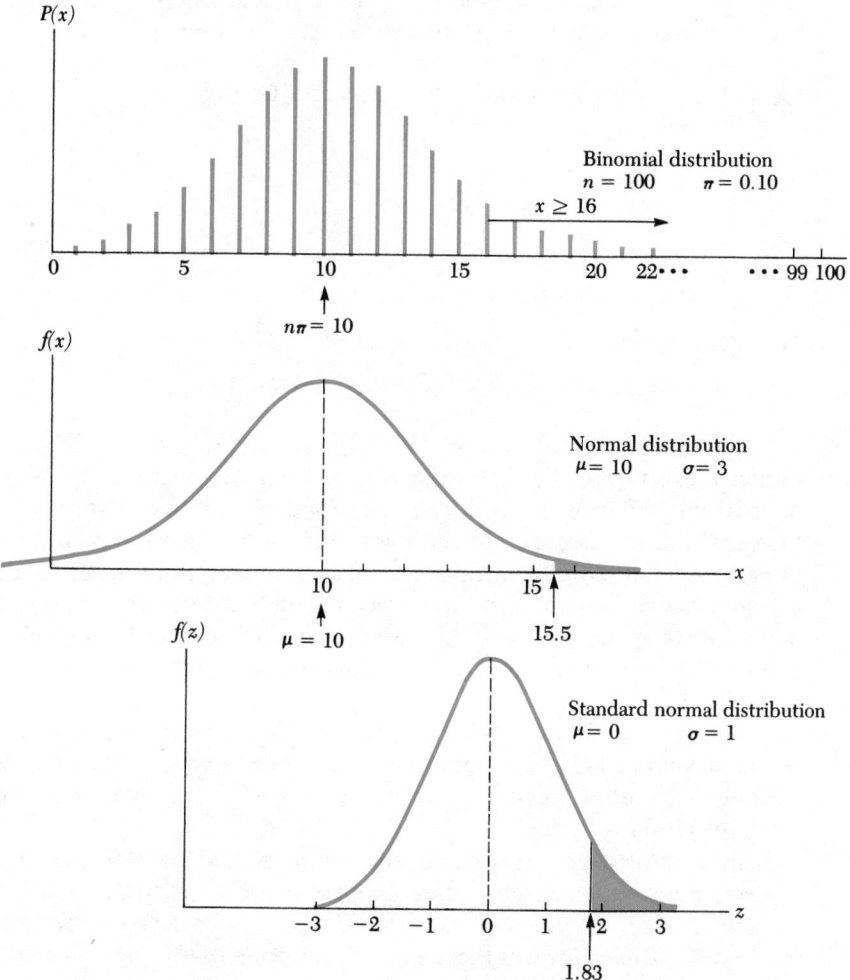

The area to the right of $z = 1.83$ on the standard normal distribution is 0.0336. Thus, the approximate probability that the retailer's lot contains more than 15 defectives is 0.0336.

7.4 THE EXPONENTIAL DISTRIBUTION

The exponential distribution is frequently used to model the distribution of a random variable that is representing service times—the time taken to complete service at a filling station, grocery store checkout stand, automobile repair shop, dentist office, and so on. It is also used in reliability applications to model the lifetimes of components that are subject to wear. Examples are the lifetimes of batteries, transistors, tubes, and bearings. Furthermore, it has been used successfully to model the distribution of the length of time *between* successive random events—the time between the arrival of customers at a service counter, the time between breakdowns of a machine, and the time between admissions to an emergency ward of a hospital, for example. Figure 7.12 suggests why the exponential distribution has been used successfully to

FIGURE 7.12 Two members of the exponential family of distributions

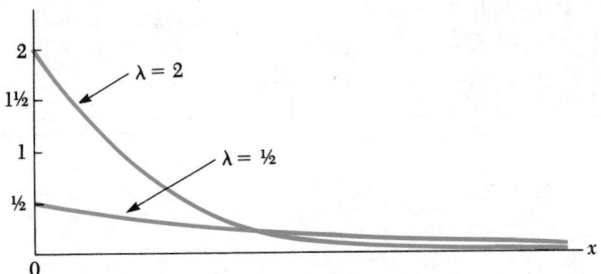

model these types of random variables. First, note that the distribution is defined for only positive values of the random variable **X** representing service times, times between breakdowns, etc. Since time is a positive number, the exponential random variable is often well suited to represent quantitative characteristics measured in time units. (Some textbooks even go so far as to denote the exponential random variable by **T** instead of **X** to emphasize the use of this random variable in modeling events that are *time*-related.) Second, the distribution is rather severely skewed to the right. Since very few service times, or times to failure of components, or times between the occurrence of successive events greatly exceed their expected values, the exponential distribution seems appropriate to model the distribution of these kinds of random variables.

It may come as a surprise, but the exponential distribution is related to a discrete distribution discussed in Chapter 6, the Poisson distribution. It can be shown that if events occur according to a Poisson process, the time *between the occurrences* of these events is exponentially distributed. We show an example of this relationship in Example 7.10.

Two members of the exponential distribution family are shown in Figure 7.12. Note from this figure that the y-intercept of the probability density function is equal to the single parameter of this distribution, λ; that is, $f(0) = \lambda$. The exponential random variable and its probability density function are described in Definition 7.4, and its mean and variance are given in Theorem 7.4. Note that unlike the normal distribution, the exponential distribution is a single-parameter distribution. Because of this, the exponential distribution *cannot* be centered independent of its dispersion. In fact, for this random variable, its variance $V(\mathbf{X})$ is always equal to the square of its expected value or its mean. That is, if $\mathbf{X}$ is an exponential random variable, $V(\mathbf{X}) = E(\mathbf{X})^2$, or the expected value of this random variable is equal to its standard deviation.

Definition 7.4

Exponential random variable and its probability density function

A continuous random variable $\mathbf{X}$ is said to be *exponentially distributed* if its *probability density function* is

$$f(x) = \lambda e^{-\lambda x}, \qquad x \geq 0, \lambda \geq 0$$

where λ is a parameter of the distribution and e is a mathematical constant equal to 2.71828.

Theorem 7.4

Mean and variance of the exponential random variable

If $\mathbf{X}$ is an exponential random variable, then its *mean*, $E(\mathbf{X})$, and *variance*, $V(\mathbf{X})$, are given by

$$E(\mathbf{X}) = \frac{1}{\lambda} \quad \text{and} \quad V(\mathbf{X}) = \frac{1}{\lambda^2}$$

The parameter λ of this distribution is interpreted similarly to its use in the Poisson case—λ is the average or the expected number of occurrences of an event in a given interval of measurement. For example, $\lambda = 5$ could be interpreted as an average of five occurrences of an event in a unit of measurement, say, one minute. In this case, the mean of the distribution is $E(\mathbf{X}) = 1/\lambda = 1/5 = 0.20$, or, if events are generated according to a Poisson process at the rate of five per minute, then the average time between occurrences of events is 0.2 minute and is exponentially distributed.

The probability that an exponential random variable assumes a value between a and b $(a < b)$ is the area under the curve from a to b illustrated in Figure 7.13. Since the computation of this area involves the evaluation of probability integrals between the limits of a and b, tables of the exponential distribution have been provided as they were for the normal distribution. Table B.4 in Appendix B gives the cumulative probability for the quantity λx [i.e., $F(\lambda x)$] for an exponential random variable. Values in this table correspond to the probability that λx is less than or equal to a. The use of this table is illustrated in the following two examples.

FIGURE 7.13 $P(a \leq X \leq b)$ for the exponential
distribution

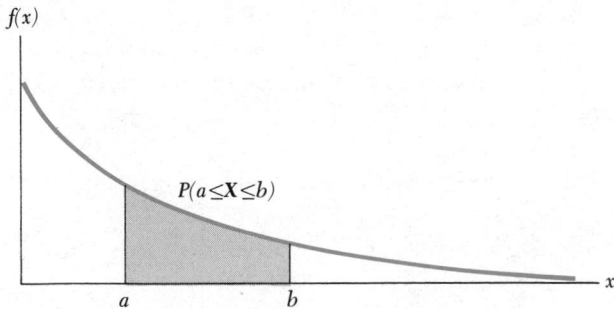

Example 7.9 The length of time required to service customers arriving at
a drive-in bank teller's window is assumed to be exponentially distributed
with an expected service time per customer of 5 minutes. What proportion of
customers are serviced within 1 minute? 5 minutes? Between 4 and 8
minutes?

Solution We must be careful how we define the units of measurement to
be used in computing the required probabilities in this example. We are
interested in the expected number of occurrences of an event in a specified
length of time—in this case, in minutes. We shall use intervals of 1 minute to
compute the required probability. Thus, since customers are served on the
average in 5 minutes (we can think of this as being served in five continuous
1-minute intervals), the expected number of services in 1 minute—our unit of
measurement—is $1/5 = 0.20$. That is, if we expect to service *one* customer in
5 minutes, then we can also expect to service *one-fifth* customer in 1 minute,
or $\lambda = 0.20$. The random variable X is defined to be the time required (in
minutes) to complete a service. Also, λ, the parameter of the exponential
distribution, is equal to the expected number of occurrences in 1 minute,
$\lambda = 0.20$. We want to determine the probability that the random variable $X =$
length of time for service is less than or equal to 1 [i.e., $P(X \leq 1)$]. Since
$\lambda x = (0.2)(1) = 0.2$, we look in Table B.4 for the value of $F(\lambda x)$ to the right of
0.2. This value is 0.181. Thus, we would conclude that the percentage of
customers that get serviced within 1 minute is 18.1 percent.

To determine $P(X \leq 5)$, we first compute $\lambda x = (0.2)(5) = 1.0$. Looking
again in Table B.4 for the value of $F(1.0)$ to the right of 1.0, we find the
required probability to be 0.632. That is, 63.2 percent of the customers will
be serviced in the *average time* or less! (Recall that in a skewed distribution,
the mean will *not* equal the median.)

The probability that the exponential random variable is between 4 and 8
minutes can be determined by subtracting the probability that the random
variable is less than or equal to 4 [$P(X \leq 4)$] from the probability that the
random variable is less than or equal to 8 [$P(X \leq 8)$]. Since $(0.2)(8) = 1.6$
and $(0.2)(4) = 0.8$, the required probability is

$$F(1.6) = 0.798$$
$$F(0.8) = \underline{0.551}$$
$$0.247$$

Thus, 24.7 percent of the customers are serviced between 4 and 8 minutes. Figure 7.14 illustrates the calculated probabilities.

FIGURE 7.14 Calculated probabilities in Example 7.9

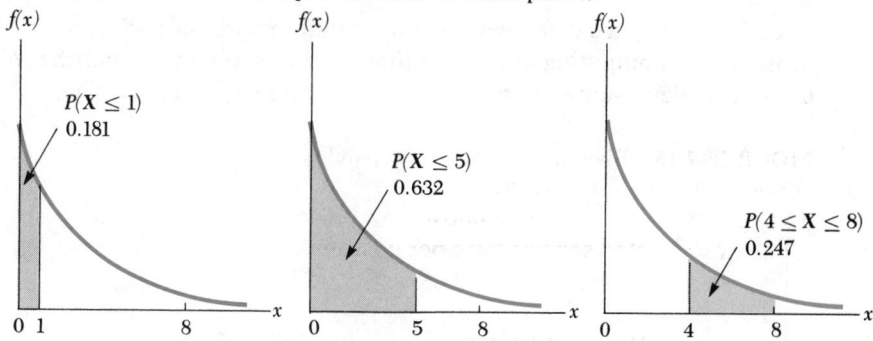

Example 7.10 Example 6.9 required the computation of the probability of receiving no (0) calls by an operator in a 10 minute interval, where the calls the operator receives in a 10 minute interval follow a Poisson distribution with $\lambda = 1$ (an average of one call each 10 minutes), and Example 6.10 required computing the probability that no one would be admitted to the emergency ward of a hospital during a given time interval, where the average number of admissions during this interval are Poisson distributed with an average of 3.5 admissions. Determine each of the required probabilities in these two examples using the exponential probability distribution.

Solution We will do Example 6.9 first. Since "arrivals" of phone calls are Poisson distributed with the mean number of calls being one in each 10-minute interval, the parameter λ of the exponential distribution is also $\lambda = 1$. (They are the same!) The probability of receiving no calls in a 10-minute interval [Poisson process, $P(X = 0)$] is equivalent to the probability that the time between calls is greater than one 10-minute interval [exponential process, $P(X \geq 1)$]. This is equal to $1.0 - F(\lambda x)$ in Table B.4, where $\lambda x = 1.0$. The required probability is:

$$1.0 - F(1.0) = 1.0 - 0.632 = 0.368$$

the same result obtained in Example 6.9.

From Example 6.10, $\lambda = 3.5$ admittances per period, and the probability that no one will be admitted [$P(X = 0)$] is equivalent to the probability that the time between admittances is one unit of measurement, or 2 hours [$P(X \geq 1)$]. Hence, $\lambda x = (1)(3.5) = 3.5$. Since we are assessing the probability in the upper tail of the exponential distribution, the required probability is:

$$1.0 - F(3.5) = 1.0 - 0.970 = 0.030$$

This is the same result obtained in Example 6.10.

■ 7.5 THE BIVARIATE NORMAL RANDOM VARIABLE

In Chapter 5, we defined a bivariate random variable and gave an example of its use in computing such quantities as the probability that the joint random variable assumes particular values (discrete case). In Figure 7.15, we

FIGURE 7.15 Bivariate normal probability density function

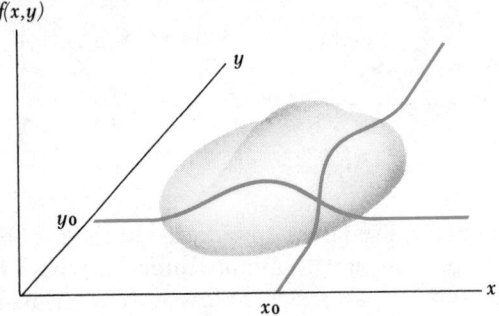

present a bivariate normal random variable with density function $f(X,Y)$. Note that if we were to take a "slice" parallel to either the x-axis, say at x_0, or the y-axis, at y_0, in this figure, the slice, when opened, would have the characteristic bell shape of the normal curve.

The bivariate normal distribution is particularly useful in studying the relationship *between* random variables, such as the heights and weights of individuals and the dollars spent on advertising effort and resulting sales. This distribution receives widespread use in the examination of the *correlation* between random variables, a topic we will discuss in detail in Chapter 13.

We shall not develop the bivariate normal density function in this chapter or give its expected value, $E(X,Y)$, or variance, $V(X,Y)$. However, in Chapter 13, we will use properties of this distribution to assess the degree of linear relationship between two random variables. The reader is referred to the Hogg and Craig reference at the end of the chapter for a detailed development of this distribution and its properties. For the present, the reader should recognize that it is possible to define a multivariate or a bivariate normal probability density function that is a function of more than one normally distributed random variable.

■ 7.6 SUMMARY

Three important continuous probability distributions have been presented in this chapter—the uniform, normal, and exponential distributions. The situations where these distributions can be used as models for the distributions of random variables with unknown exact distribution form were mentioned. In Table 7.2, the more important characteristics of these distributions are summarized.

TABLE 7.2 Summary characteristics of the uniform, normal, and exponential distributions

Distribution	Param-eters	Mean $E(X)$	Variance $V(X)$	Calculated probabilities (Appendix B)
Uniform	a, b	$\dfrac{b + a}{2}$	$\dfrac{(b - a)^2}{12}$	—
Normal	μ, σ	μ	σ^2	Table B.3
Exponential	λ	$\dfrac{1}{\lambda}$	$\dfrac{1}{\lambda^2}$	Table B.4

If the values of the parameters of the distribution are known, we have seen how to use the probability distribution to answer questions regarding properties of the random variable, such as the probability that it exceeds a specified value and the determination of its mean value.

In most cases, the values of the parameters of the probability distribution are unknown and must be estimated from sample data. The process of estimating parametric values is the primary topic of Chapters 8 and 9.

■ REFERENCES

Harnett, D. L. *Introduction to Statistical Methods.* Rev. ed. Reading, Mass.: Addison-Wesley Publishing Co., Inc., 1975.

Hoel, P. G. *Introduction to Mathematical Statistics.* 3d ed. New York: John Wiley & Sons, Inc., 1962.

Hogg, R. V., and Craig, A. T. *Introduction to Mathematical Statistics,* 3d ed. New York: Macmillan, Inc., 1970.

Lindgren, B. W. *Statistical Theory.* New York: John Wiley & Sons, Inc., 1959.

Parzen, E. *Modern Probability Theory and Its Applications.* New York: John Wiley & Sons, Inc., 1960.

■ PROBLEMS

7.1. A random variable **X** is uniformly distributed over the interval between 10 and 26, inclusive.

    a. Find the mean and standard deviation of **X**.

    b. Using the Chebychef theorem,

determine an interval that contains at least eight ninths of the values of **X**.

c. What proportion of the values of **X** are contained in the interval determined in part b?

7.2. Using the normal probability tables, calculate the areas under the standard normal curve for the following z values.

a. Between $z = 0.0$ and $z = 1.2$.
b. Between $z = 0.0$ and $z = -0.9$.
c. Between $z = 0.0$ and $z = 1.45$.
d. Between $z = 0.0$ and $z = -1.44$.
e. Between $z = 0.3$ and $z = 1.56$.
f. Between $z = -1.71$ and $z = -2.03$.
g. Between $z = -1.72$ and $z = 2.53$.
h. Between $z = -0.02$ and $z = 3.09$.
i. Greater than $z = 2.50$.
j. Greater than $z = -0.60$.
k. Less than $z = -1.22$.
l. Less than $z = 1.66$.

7.3. Given the following areas under the standard normal distribution, find the appropriate z-value.

a. Area between 0 and z is 0.4808.
b. Area to the left of z is 0.8621.
c. Area to the right of z is 0.9959.
d. Area to the right of z is 0.0020.

7.4. Determine the following probabilities for the given normal distributions:

a. $\mu = 500$, $\sigma = 100$; $P(X \geq 700)$.
b. $\mu = 40$, $\sigma = 10$; $P(X \leq 25)$.
c. $\mu = 100$, $\sigma = 20$; $P(X \geq 75)$.

7.5. Given the areas under the normal distribution, find the value of the random variable **X**.

a. $\mu = 500$, $\sigma = 100$; $P(X \leq x_0) = 0.4286$.
b. $\mu = 40$, $\sigma = 10$; $P(X \geq x_0) = 0.0110$.
c. $\mu = 100$, $\sigma = 20$; $P(X \leq x_0) = 0.9986$.

7.6. The mean and standard deviation for the lifetimes of a population of light bulbs are 1,200 and 150 hours, respectively. Assuming these lifetimes are normally distributed, what is the probability that a light bulb will last over 1,500 hours?

7.7. The average length of time required to complete the civil service PACE exam is 100 minutes with a standard deviation of 15 minutes. When should the examination be terminated if the examiner wishes to allow sufficient time for 90 percent of the applicants to complete the exam, assuming that the test times are normally distributed?

7.8. Heights of individuals in a certain population are assumed to be normally distributed with a mean of 66 inches and a standard deviation of 4 inches. At what height must door frames be constructed so that no more than 0.1 percent of the individuals in this population will be too tall to pass through them without stooping?

7.9. The heights of students in a certain population are known to be normally distributed with mean $\mu = 68$ inches and standard deviation $\sigma = 3$ inches. If two students are randomly selected from this population, what is the probability that they both will be 76 inches or taller? What assumption is necessary to solve this problem?

7.10. A machine is producing ball bearings with diameters that are normally distributed with mean and standard deviation equal to 0.498 and 0.002, respectively. If specifications require that the bearing diameters equal 0.500 inch plus or minus 0.004 inch, what fraction of the production will be acceptable?

7.11. A soft drink machine can be regulated so that it discharges an average of μ ounces per cup. If the ounces of fill are normally distributed with standard deviation $\sigma = 0.3$ ounce, give the setting for μ so that 8-ounce cups will overflow only 1 percent of the time.

7.12. The Midstate Paper Company employs two salesmen, A and B. Salesman A averages \$20,000 in sales per month with a standard deviation of \$8,000, while salesman B averages \$35,000 in sales per month with a standard deviation of $5,000. During a given month, B's sales amount to $43,000. If it is assumed that the sales volumes for both men are normally distributed, how great would A's sales volume have to be for it to be proportionally as great as B's sales volume?

7.13. All prospective employees at the Naval Supply Depot in Mechanicsburg, Pennsylvania, are required to pass a special civil service examination before they are hired. Over a period of time, the examination scores have been normally distributed with mean $\mu = 85$ and standard deviation $\sigma = 4$. Suppose $n = 50$ applicants for a job sit for the test.

a. What proportion of the 50 can be expected to obtain a score of 95 or greater?

b. If below 80 is considered failing, what is the probability that an individual will fail the exam?

c. Of the 50 sitting for the examination, what is the expected number who will pass?

7.14. The income in a certain community is normally distributed with a mean annual income of $35,000 and a standard deviation of $3,000.

a. What minimum income level does a member of this community have to have to be in the top 10 percent of the citizenry?

b. What is the maximum income one can have and still be in the middle 50 percent of all incomes?

c. What is the minimum income one can have and still be in the middle 50 percent of all incomes?

7.15. Boxes of a specific cereal are packaged with a mean weight of 13.1 ounces and a standard deviation of 0.1 ounce. The box states that it contains 13 ounces of cereal. Regulations allow a box to be no more than 0.2 ounce lighter than the stated weight. If an inspector randomly selects one box from the production line, what is the probability that he will find the box is too light? Assume the package weights are normally distributed.

7.16. The number of customers entering a certain store on any given day is approximately normally distributed with mean $\mu = 50$ and standard deviation $\sigma = 9$. Find the following probabilities during a given day.

a. At least 50 customers arrive.

b. At least 45 customers arrive.

c. Between 45 and 55 customers arrive.

7.17. The average time required to finish a civil service exam is normally distributed with a mean of 60 minutes and a standard deviation of 12 minutes. How many minutes should be allowed for the examination if the supervisor wishes to allow sufficient time for 90 percent of the applicants to complete the test?

7.18. At the Gold Muffler Shop, it takes a servicewoman an average of 20 minutes with a standard deviation of 5 minutes to put on a new muffler. A customer leaves her car and is told to come back in 30 minutes and the car will be ready. Assuming that the muffler-mounting times are normally distributed, what is the probability that her car will not be finished when she returns in 30 minutes?

7.19. A worker submits two jobs at a service window. The average time required to complete a job is 100 minutes with a standard deviation of 20 minutes. Processing is begun on both jobs immediately and the worker decides to return in 2 hours to check whether either or both jobs have been completed. Assuming the processing times are normally distributed, what is the probability that both jobs will be ready when he returns? What assumption is required to answer this question?

7.20. The diameter of ball bearings produced in a manufacturing process is known to have a mean of 0.032 inch and a standard deviation of 0.001 inch.

a. Using the Chebychef theorem, construct an interval that contains at least 75 percent of the diameters of all ball bearings produced.

b. If it is assumed that the diameters are normally distributed, what proportion of ball bearings is contained in the interval developed in part *a*?

c. Compare parts *a* and *b*. Discuss why the proportion of diameters contained in the interval differs in parts *a* and *b*.

7.21. Consider the binomial distribution with $n = 50$ and $\pi = 0.40$. For each part, find the *exact* binomial probability and the normal approximated probability, using the correction for continuity.

a. $P(X \geq 15)$. b. $P(7 \leq X \leq 13)$.
c. $P(X = 9)$.

7.22. In a certain city, it is known that 40 percent of all color TV sets bought will be made by the Xenith Co. In a given week, $n = 25$ color TVs are sold. Find the probability that at least 8 but no more than 11 of the buyers purchased a color TV produced by Xenith.

a. Use the binomial tables.
b. Use the normal approximation to the binomial distribution.

7.23. How many questions *n* would a statistics professor have to put on a multiple-choice examination (where there are five responses—a, b, c, d, and e—to each question) for her to be 95 percent certain that a student making random guesses on each question will miss at least 50 percent of the questions? Use the normal approximation to the binomial distribution without the correction for continuity.

7.24. A newly designed portable radio was styled on the assumption that 50 percent of all purchasers are female. If a random sample of $n = 400$ purchasers

is selected, what is the probability that the number of female purchasers in the sample will be greater than 175? Use the normal approximation to the binomial distribution without the correction for continuity.

7.25. The manufacturer of an electronic part has found that 98 percent of the parts are acceptable. A random sample of 500 parts is tested. What is the probability that 12 or more defective parts will be found? Use the normal approximation to the binomial distribution without the correction for continuity.

7.26. A new serum is 90 percent effective in preventing colds. If 100 people take the serum, what is the probability that between 7 and 12 (inclusive) will catch a cold? Use the normal approximation to the binomial distribution without the correction for continuity.

7.27. Of the people entering a large department store, it is found that 50 percent will make at least one purchase. For a sample of 100 persons, what is the probability that at least 60 will make at least one purchase? Use the normal approximation to the binomial distribution without the correction for continuity.

7.28. A production process is known to produce 10 percent defectives. A buyer purchases 225 of these items. What is the probability that 20 or less are defective? Use the normal approximation to the binomial distribution without the correction for continuity.

7.29. Indicate the type of probability distribution (normal, exponential, Poison) that you would expect to be most applicable in each of the following instances.

a. The pattern of variation in the number of typesetting errors per printed page when the printer averages 0.05 error per page.

b. The pattern of variation in the temperature readings of thermometers produced under the same conditions and experimentally exposed to a temperature of 50°C.

c. The pattern of variation in the lengths of time to receive a hair style at the local barber shop.

7.30. An electronic component is known to have a mean time to failure of 1,000 hours and a standard deviation of 500 hours.

a. Why is the normal distribution an inappropriate model for the distribution of this variable?

b. Is the exponential distribution appropriate in this case? Discuss.

7.31. Given the exponential distribution with parameter λ equal to 0.80.

a. Graph the probability density function for this distribution.

b. Determine the mean and the variance of this distribution.

c. What percentage of the area in this distribution lies within one standard deviation of the mean? Within two standard deviations of the mean?

7.32. The length of time required to service a car at a gas station is exponentially distributed with a mean service time of 4.0 minutes.

a. What proportion of the cars are serviced within 1 minute?

b. What proportion of the cars are serviced within 6 minutes?

c. What is the probability that a car will be serviced within 2 to 8 minutes?

7.33. The life of an electronic component (number of hours before failure) is exponentially distributed with an average time to failure of 550 hours.

a. What is the probability that a component fails before 375 hours?

b. What is the probability that a component fails before 1,000 hours?

c. What is the probability that a component fails between 500 and 1,200 hours?

7.34. Suppose a fast-food hamburger chain can serve its customers on the average at the rate of two customers in 3 minutes. Assume the service rate to be Poisson distributed.

a. What is the probability that a restaurant in this chain will be able to serve four or more customers in a 3-minute period?

b. What is the probability it will take longer than 3 minutes to serve a customer?

c. What is the probability it will take between 3 and 6 minutes to serve a customer?

7.35. Ships arrive at a dock on an average of one ship every two days. What is the probability that after the departure of a ship, four days will elapse before another arrives?

7.36. Each 500-foot roll of fabric produced in a mill contains two flaws on the average. What is the probability that as the fabric is unrolled, the first flaw occurs within the first 50 feet?

7.37. A type C transistor has a mean lifetime of 10,000 hours. Assume that the lifetimes are exponentially distributed.

a. What is the probability that a particular type C transistor will last more than 30,000 hours?

b. Is the exponential distribution an appropriate model for the lifetimes in this case? Why or why not?

Samples and sampling distributions

8

8.1 INTRODUCTION

In Chapters 3–7, we developed the ideas of a random variable and the distribution of a random variable. We have seen that by designating a random variable to represent a quantitative characteristic in a population, properties of the quantitative characteristic including its mean and standard deviation may be determined from the distribution of the random variable.

In this chapter, we will draw on these ideas to lay the foundation for methods of drawing inferences about the properties of population characteristics from a sample.

8.2 THE NATURE OF SAMPLES

A sample is a subset of a population chosen to draw inferences about characteristics in that population. The quality of the inferences drawn about the population characteristics from the sample data is directly related to how well the sample *represents* the population. In sampling, therefore, a major objective is to select a sample plan and implement it in a way that ensures, within reasonable cost and time constraints, that the sample is a *good* representation of the population. There is virtually no way, short of taking a complete census of the population, that we can guarantee that the sample will *exactly* represent the population. By the very act of sampling, we may select for the sample a disproportionate number of atypical units from the population.

What can cause a sample not to be representative of a population? One cause is *sampling error*.

Definition 8.1
Sampling error

Sampling error is the difference between the result of studying a sample and inferring a result about the population, and the result of a census of the whole population.

Sampling error can occur in basically two ways: by chance and by the introduction of sampling bias. Sampling error due to chance underscores the elemental risk inherent in sampling. For example, we may wish to sample the number of full-time students to determine the percentage participating in academic clubs and societies on campus at a specified university. By chance alone (assuming no other sampling or nonsampling errors are made), we may produce a sample that contains proportionately more students who are active in academic clubs and societies on campus than is true for the entire population of full-time students. The only protection against sampling error due to chance is to increase the sample size. The larger the sample, the greater proportion it has of the population and, concomitantly, the more representative of the population it is (provided, of course, that other sampling or nonsampling errors are not made).

Sampling bias is not so easily controlled as is sampling error due to chance.

Definition 8.2
Sampling bias

Sampling bias is exhibited when units are selected by the sampling process that tend to favor certain population characteristics.

Sampling bias can usually be attributed to the selection of a poor or inappropriate sampling plan or to the incompetent execution of an appropriate sampling plan.

An example of a poor sampling plan is the Dewey-Truman election survey of voters discussed in Chapter 1. In the survey, the population was defined to be the registered voters who said they would vote on election day. But telephone directories were used to draw the sample; that is, names were randomly chosen from randomly selected directories, and these persons were called and asked for whom they would vote on election day. The problem that led to sampling bias was that not all registered voters had telephones in 1948, and most who did were not Democrats. The resulting sample was, therefore, not representative of the population. Each element in the target population (the collection of registered voters who voted) did not have an equal chance of being selected on each sample draw because a sizable proportion of the elements did not have telephones. Accordingly, the sample tended to favor the selection of Republicans, which led to the incorrect inference that Dewey would win the 1948 presidential election.

Sometimes the mechanics of implementing the sampling plan can be faulty. An example of this problem is the first draft lottery used to draft men for the armed services in the late 1960s. The idea was to draw days of the year randomly in succession from the 365 days—those having their birthdate on the first date drawn would be the first drafted and so on. The dates of the year were placed on 365 chips (1/1, 1/2, . . . , 12/31), which in turn were placed in a box with separate compartments for each month of the year. As the chips were being dumped into the cylindrical tumbler to be used for the selection of the dates, those in the compartments corresponding to the latter months of the year were first to fall into the cylinder—the box was dumped from the December end rather than the January end. Thus, the chips tended to form layers in the tumbler by the months of the year. The tumbler was turned so fast that centrifugal force tended to maintain the layering. Chips were drawn from the tumbler one after another—each draw presumably being a random sample of one unit from the population. But, on the first few draws, chips with December, November, and October dates were favored over chips with dates corresponding to the other months. The fact that sampling bias led to sampling error in randomly selecting the draft dates led to considerable controversy and, ultimately, to a change in the way the lottery was conducted in subsequent years.

Another form of sampling bias is the *bias of nonresponse*. This form of bias is particularly troublesome in sampling by mail questionnaires—often the characteristics of the nonrespondents are considerably different from those of the respondents. The nonresponse bias in this case can be minimized by second, third, and subsequent mailings to nonrespondents or by conducting telephone or personal interviews with nonrespondents to the first mailing. These procedures are very costly and time-consuming, however.

Minimizing sampling error due to bias is a major objective in any statistical sampling experiment. Realistically, some error due to bias is inevitable in most sample surveys or experiments. In practice, we should attempt to minimize this source of error as best we can subject to cost and time considerations. Often it is necessary to alter the sampling plan by reducing the scope of the study to ensure that the data collected acceptably represent the population.

Errors in the sample may be attributed to causes other than sampling error. In processing sample data, mistakes may be made in acquiring, recording, or tabulating (especially for computer analysis) the data, with the consequence that the sample data set is not representative of the population set of values. For example, data may be mispunched on computer cards or decimal places may be inadvertently changed. If a survey requires responses from people on sensitive issues, such as salary, they may respond erroneously due to an honest mistake or outright lying. A statistician must be aware of this source of error. She should carefully evaluate and check the processing of the sample data from its collection through its analysis to minimize the occurrence of nonsampling error.

Definition 8.3
Nonsampling error

Nonsampling errors are errors that occur in acquiring, recording, or tabulating statistical data that cannot be ascribed to sampling error.

Anyone planning sample surveys or sample experiments must be aware of the sampling and nonsampling errors that can adversely affect the quality of inferences drawn from the sample about the population. By carefully evaluating the potential sources of error in a survey or experiment, a sampling plan can be developed that minimizes the occurrence of the errors.

■ 8.3 SIMPLE RANDOM SAMPLING

Once we have decided to sample, how do we select a portion of the population for our study? There are basically three ways, or types of samples: (1) *probability samples,* (2) *judgment samples* and (3) *convenience samples.*

In this section, we will develop the notion of one kind of probability sample called the *simple random sample.* Other types of samples will be considered later in the chapter. Two definitions of simple random sampling are given; one that applies when the population is *finite* in size and the other when the population is *infinite.*

Definition 8.4
Simple random sample (finite population)

A *simple random sample* from a *finite* population gives each possible sample set of a specified size an equal probability of being selected.

This definition applies only when we are sampling without replacement—that is, when each unit drawn from the population is not returned prior to drawing the next unit. We will be concerned only with sampling without replacement in this text. Sampling with replacement has very limited and special uses in statistics.

To illustrate Definition 8.4, suppose we have a population that contains three elements, A, B, and C, from which we wish to draw a simple random sample of two elements. According to the definition, we must be certain that each of the $C_2^3 = 3$ samples of size two (AB, AC, and BC) has an equal chance of being selected. To draw the simple random sample, we could place the letters AB, AC, and BC on three chips and select one of the chips after mixing them thoroughly. But if the population is sizable, say 1,000 elements, using chips to draw a simple random sample would be impractical. Suppose, for example, we wished to draw a simple random sample of 100 elements from a population of 1,000. To list all possible samples of size 100 would not be feasible, because $C_{100}^{1,000}$ is a huge number. An alternative way of producing a simple random sample from a finite population is to "tag" each unit in the population and draw units one at a time, so that on each draw every unit still in the population has an equal chance of being selected.

When the population is infinite, we must modify our previous definition of a simple random sample, for there is an infinite number of samples of a specified size that can be drawn from the population in this case. Definition 8.5 requires that the act of selecting each sample observation is not affected by, nor does it affect, the act of selecting any other sample observation.

Definition 8.5
Simple random sample (infinite population)

A *simple random sample* from an *infinite* population requires that all sample observations be statistically independent.

An example of selecting a simple random sample from an infinite population is taking a sample of units produced on an assembly line. Conceptually, the line could run forever producing an infinite number of units. To take a simple random sample from this population, we must be certain that the production process is *stable* so the units sampled all come from the same infinite population. If the process changes through time as a result of, say, machine wear, then the sample observations will be statistically dependent if they are drawn while the process is changing.

In practice, almost all populations we may wish to sample from are finite, particularly in business and economic uses of sampling. Thus, Definition 8.4 is applicable in most instances.

In this text, we will be concerned only with simple random samples. Other forms of sampling will be discussed in Section 8.9, but their use requires statistical methods that are beyond the scope of this book. Since we will be using only simple random samples in inference making, we will most often refer to them as "random samples," although simple random samples are usually thought of as one type of random samples.

In the next section, we will develop the role of (simple) random sampling in the inference process.

8.4 POPULATION PARAMETERS AND SAMPLE STATISTICS

In Chapter 2, we presented and discussed numerical measures of data sets, such as the mean and standard deviation. At the time, we considered the data sets to represent *populations*. It is now necessary to distinguish between numerical measures computed in *population* data sets and in *sample* data sets. If a numerical measure is calculated for a population data set, it is called the *value of the population parameter*. If a numerical measure is calculated for a sample data set, it is called the *value of the sample statistic*. In Definition 8.6, the definitions of the *population mean* parameter and the *population standard deviation* parameter are stated for a finite population.

Definition 8.6
Population mean and standard deviation parameters

In a finite population of N values, $x_1, x_2, \ldots, x_N$, of a population characteristic **X**, the value of the *population mean parameter*, denoted by μ, is given by:

$$\mu = \frac{\sum\limits_{i=1}^{N} x_i}{N}$$

and the value of the *population standard deviation parameter*, denoted by σ, is given by:

$$\sigma = \sqrt{\frac{\sum\limits_{i=1}^{N} (x_i - \mu)^2}{N}}$$

If the population in infinite, we will use μ and σ to represent the population mean and standard deviation, respectively, but these quantities cannot be determined from the finite sums as in Definition 8.6.

The analogous numerical measures in the sample, the values of the sample statistics for the mean and standard deviation, are given in Definition 8.7.

Definition 8.7
Sample mean and standard deviation statistics

If a sample of n values, $x_1, x_2, \ldots, x_n$, is taken from a population, the value of the *sample mean statistic*, denoted by $\bar{x}$, is given by:

$$\bar{x} = \frac{\sum\limits_{i=1}^{n} x_i}{n}$$

and the value of the *sample standard deviation statistic*, denoted by s, is given by:

$$s = \sqrt{\frac{\sum\limits_{i=1}^{n} (x_i - \bar{x})^2}{n - 1}}$$

Notice the difference between the definitions of σ and s; the latter has a divisor of $(n - 1)$ within the radical, whereas the former has a divisor of N. The numerical measure s is, therefore, not quite the square root of the average of the squared deviations as is the analogous quantity in σ. The reason for the divisor of $(n - 1)$ in s will be explained in Chapter 9.

There are other population parameters and sample statistics in addition to the mean and standard deviation, of course. We will discuss some of these as additional ideas are introduced and developed later in this chapter through examples.

Suppose we are interested in drawing inferences about the average salary, μ, of 10,000 blue-collar workers in, let us say, the steel industry, based on a simple random sample of $n = 25$ workers. Intuitively, once the sample is drawn from the population, the sample mean $\bar{x}$ would seem to be an obvious numerical measure in the sample to use as an estimate of the value of μ. After all, both numerical measures are means: μ is the mean of all population values, while $\bar{x}$ is the mean of the values in a sample when the numerical values of the sample $(x_1, x_2, \ldots, x_n)$ are determined. Indeed, as we shall see in Chapter 9, when $\bar{x}$ is computed in a specific sample of size n, it is generally used as the "best guess" or estimate of the population mean μ.

In the process of using the sample mean, computed from a sample of n values, to estimate μ, the population mean, what are the benefits and what are the costs? An obvious benefit is computational efficiency. By computing the sample mean based on a sample of $n = 25$ workers from the population of $N = 10,000$ workers in the steel industry, considerable time and expense are saved. Calculating the mean of all 10,000 workers in the population would involve pulling all employee records from the files, checking all figures, and calculating the mean. Although much of this process could be computerized, the time and cost would certainly exceed that required to calculate the mean salary of the 25 sampled salaries. In many cases, it is not even possible to evaluate the mean for the population. An example is destructive sampling. If the lifetimes of a certain 25-watt light bulb were being studied, the testing process (burning the light bulbs until they fail and recording the lifetimes) destroys each light bulb.

An additional benefit of sampling may be a greater amount of accuracy in producing the numerical value of the sample mean due to having more time to check and double check all figures than may be afforded if a complete census of the population is taken and the value of population mean μ is calculated.

What are the costs? The primary cost is the potential for sampling error to occur in the sampling process. If we take a specific sample of $n = 25$ workers and calculate $\bar{x}$, the sample mean, we would have to be extremely fortunate to find that $\bar{x}$ is equal to the value of μ, the mean of all 10,000 workers. More likely, $\bar{x}$, calculated from one sample of $n = 25$ workers, would be either somewhat less or somewhat more than the value of μ. The difference between $\bar{x}$ and μ is due to sampling error, in the absence of nonsampling error.

Thus, the cost we pay by sampling is the risk that the sampling process will not produce a representative sample by design or chance. More specifically, in the one sample taken, the calculated sample mean $\bar{x}$ may not be close to the actual value of the population mean μ. That *different* samples of size n will produce *different* sample means is illustrated in Figure 8.1.

FIGURE 8.1   Distribution of the sample means calculated in samples of size $n = 25$

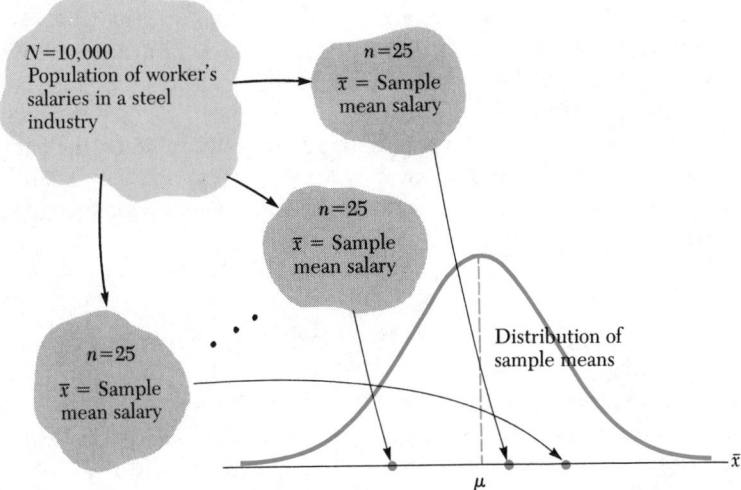

As indicated in Figure 8.1, if we take repeated samples of size $n = 25$, some means will be close to the population mean value μ, while others may quite possibly be far from the value of μ. By taking many repeated samples of size $n = 25$, a *distribution* of sample means is created.

In assessing the risk of incurring sampling error by sampling from the population and to reflect that the various samples of size n will produce *different* sample means $\bar{x}$, we need to develop notation to distinguish between a random variable and the value of a random variable. In Chapters 5, 6, and 7, a *random variable,* say **X,** was denoted by a capital **boldfaced** symbol. A *value* of this random variable is denoted by the corresponding lowercase symbol x. The distinction between random variables and their values arises first in the sampling context when we speak of a random sample drawn from a population. If we are referring to a random sample of size n *before* the sample is actually drawn, the random sample of size n is designated by $\mathbf{X}_1$, $\mathbf{X}_2, \ldots , \mathbf{X}_n$, a set of *random variables.* After a specific, single sample is drawn from the population, the values of these random variables are designated by $x_1, x_2, \ldots , x_n$. The set $x_1, x_2 \ldots , x_n$ represents the collection of n *numbers* that comprise a single sample of size n drawn from the population. For example, suppose we have a finite population of N values $x_1, x_2, \ldots , x_N$

and consider the random sample $X_1, X_2, \ldots, X_n$. That $X_1, X_2, \ldots, X_n$ are random variables may best be understood by considering the following fact: the first sample random variable, X_1, may take on *any one* of the population values $x_1, x_2, \ldots, x_N$ *before* the sample is actually drawn. In a similar fashion, the second sample random variable, X_2, may take on one of the remaining population values before the sample is drawn, and so forth for the rest of the sample random variables. Thus, *before the sample is taken*, $X_1, X_2, \ldots, X_n$ are random variables with values that are determined once a single sample is taken from the population. *After the sample is taken*, the values of the random variables $X_1, X_2, \ldots, X_n$ are known and are denoted by $x_1, x_2, \ldots, x_n$ (note that these values are not necessarily the first n values in the population set of values $x_1, x_2, \ldots, x_N$).

In a similar fashion, we must distinguish between when the sample mean is a random variable and when it is a value of a random variable. *Prior to* taking the sample, the sample mean is a *random variable* and is called a *statistic:*

$$\overline{X} = \frac{\sum\limits_{i=1}^{n} X_i}{n} = \frac{X_1 + X_2 + \cdots + X_n}{n}$$

After the sample is actually drawn and the values of $x_1, x_2, \ldots, x_n$ are known numerically, the sample mean is a *value* of the statistic (random variable):

$$\bar{x} = \frac{\sum\limits_{i=1}^{n} x_i}{n} = \frac{x_1 + x_2 + \cdots + x_n}{n}$$

At first, it may seem unnecessary to be so careful in distinguishing between a statistic, say $\overline{X}$, and a value of this statistic, $\bar{x}$. But a notation is necessary to appreciate that the statistic $\overline{X}$ will vary in value from sample to sample for a fixed sample size n and, as such, is a *random variable* that has a *distribution*. The fact that the statistic $\overline{X}$ is a random variable captures the risk involved in using a value of $\overline{X}$ in a particular sample to estimate the population mean.

Definition 8.8 provides the definition of a sample statistic and gives the formulas for the statistics $\overline{X}$ and S, the sample mean and the sample standard deviation, respectively, and their values in a particular sample.

Definition 8.8
Sample statistic

A *sample statistic* is a random variable that is a *function* of the sample random variables $X_1, X_2, \ldots, X_n$. The random variables $X_1, X_2, \ldots, X_n$ assume the values $x_1, x_2, \ldots, x_n$ when a specific random sample is drawn from the population.

Definition 8.8 (continued)

The sample mean statistic is given by:

$$\overline{X} = \frac{\sum\limits_{i=1}^{n} X_i}{n}$$

and its value, calculated when one sample is actually drawn from the population and the values of $X_1, X_2, \ldots, X_n$ are known, is given by:

$$\bar{x} = \frac{\sum\limits_{i=1}^{n} x_i}{n}$$

The sample standard deviation statistic is given by:

$$S = \sqrt{\frac{\sum\limits_{i=1}^{n} (X_i - \overline{X})^2}{n - 1}}$$

and its value, calculated when one sample is actually drawn from the population and the values of $X_1, X_2, \ldots, X_n$ (and, hence $\overline{X}$) are known, is given by:

$$s = \sqrt{\frac{\sum\limits_{i=1}^{n} (x_i - \bar{x})^2}{n - 1}}$$

In the inference-making process about the population mean μ, knowledge of the distribution of the sample mean statistic $\overline{X}$ will help determine how good the sample mean is in drawing inferences about the population mean μ. This follows because, from the knowledge of the distribution of the sample mean statistic $\overline{X}$, it is possible to determine the probability that $\overline{X}$ will, say, exceed the population mean by a specified amount of sampling error, as shown in Figure 8.2. That is, we can compute the proportion of samples of

FIGURE 8.2 Probability that $\overline{X}$ exceeds the mean by a units

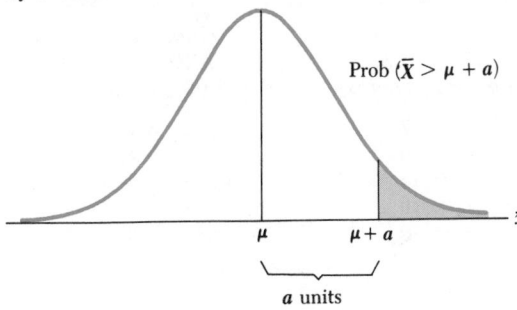

size $n = 25$ in which the sample means $\bar{x}$ exceed the population mean μ by a specified amount, say a units, if the distribution of $\overline{X}$ is known. This will give us some idea of our chance of being stuck with a bad sample; in this context, a sample that gives a value of the sample mean statistic $\overline{X}$ that exceeds μ by a units.

In Figure 8.2, the distribution of the statistic $\overline{X}$ is shown centered over μ. That the average of all sample means $\bar{x}$ is the value of μ is intuitively appealing, and this, as well as other properties of the statistic $\overline{X}$, will be summarized shortly. The distribution of the statistic $\overline{X}$ is called the *sampling distribution* of $\overline{X}$.

Definition 8.9
Sampling distribution of a statistic

A statistic is a random variable with a value that is determined by taking a single random sample of size n from the population. Since a statistic is a random variable, it has a distribution, called the *sampling distribution of the statistic*. Conceptually, the sampling distribution of a statistic can be formed by taking all possible samples of a fixed sample size n, evaluating the statistic in each sample, and constructing a distribution of these values.

If, for example, the statistic is the sample mean $\overline{X}$, the sampling distribution of $\overline{X}$ can be constructed by taking all possible samples of size n, computing the value of $\overline{X}$ in each sample, denoted by $\bar{x}$, and forming the distribution of the $\bar{x}$ values. Example 8.1 illustrates the formation of the sampling distribution of $\overline{X}$ and of another sample statistic when sampling from a small, finite population.

Example 8.1   Suppose we select five students from England who are studying at a large university in the United States and consider them to constitute a population. Two quantitative characteristics are measured for each student: (1) the number of credits (**X**), and (2) a binary variable (**Y**) set equal to 1 if the student's grade point average is above a 3.00 (on a 4.00 scale) and set to 0 otherwise. Table 8.1 lists these characteristics for the five students.

TABLE 8.1   Set of population values of the random variables **X** and **Y**

Student	Values of **X** (number of credits)	Values of **Y** (1 = GPA over 3.00, 0 = 3.00 GPA or less)
A	$x_1 = 12$	$y_1 = 1$
B	$x_2 = 10$	$y_2 = 1$
C	$x_3 = 14$	$y_3 = 1$
D	$x_4 = 9$	$y_4 = 0$
E	$x_5 = 10$	$y_5 = 0$

The values of selected population parameters for each population random variable (**X** and **Y**) will first be calculated:

Mean:

$$\mu_X = \frac{\sum_{i=1}^{5} x_i}{5} = \frac{12 + 10 + 14 + 9 + 10}{5} = 11$$

Standard deviation:

$$\sigma_X = \sqrt{\frac{\sum_{i=1}^{5} (x_i - \mu_X)^2}{5}}$$

$$= \sqrt{\frac{(12 - 11)^2 + (10 - 11)^2 + (14 - 11)^2 + (9 - 11)^2 + (10 - 11)^2}{5}}$$

$$= \sqrt{\frac{16}{5}} = \sqrt{3.2} = 1.789$$

Median:

$$M_X = 10$$

Mean:

$$\mu_Y = \frac{\sum_{i=1}^{5} y_i}{5} = \frac{1 + 1 + 1 + 0 + 0}{5} = \frac{3}{5} = 0.60$$

Standard deviation:

$$\sigma_Y = \sqrt{\frac{\sum_{i=1}^{5} (y_i - \mu_Y)^2}{5}}$$

$$= \sqrt{\frac{(1 - 0.6)^2 + (1 - 0.6)^2 + (1 - 0.6)^2 + (0 - 0.6)^2 + (0 - 0.6)^2}{5}}$$

$$= \sqrt{\frac{1.2}{5}} = \sqrt{0.24} = 0.490$$

Suppose we wish to take a simple random sample of three students from the population. Table 8.2 shows the $C_3^5 = 10$ possible samples of size three.

TABLE 8.2 Possible samples (of size 3 from $N = 5$) and statistic values

Sample composition	Values of X	$\bar{x}$ Mean	M_X Median	Total of Y values	Value of P
ABC	12,10,14	12.00	12	3	1
ABD	12,10,9	10.33	10	2	⅔
ABE	12,10,10	10.67	10	2	⅔
ACD	12,14,9	11.67	12	2	⅔
ACE	12,14,10	12.00	12	2	⅔
ADE	12,9,10	10.33	10	1	⅓
BCD	10,14,9	11.00	10	2	⅔
BCE	10,14,10	11.33	10	2	⅔
BDE	10,9,10	9.67	10	1	⅓
CDE	14,9,10	11.00	10	1	⅓

For each sample, the values of the statistics $\overline{X}$ and P (the proportion of the students with grade point averages greater than 3.00) are calculated. The median for the X values is also computed for each sample.

Notice in particular in Table 8.2 the values of the statistic $\overline{X}$. The population mean is 11, but only two of the ten samples produce a sample mean of 11; the other samples produce sample averages $\bar{x}$ that are closely clustered about 11. The sampling distribution of the statistic $\overline{X}$ is shown in Figure 8.3.

FIGURE 8.3 Sampling distribution of the statistic $\overline{X}$

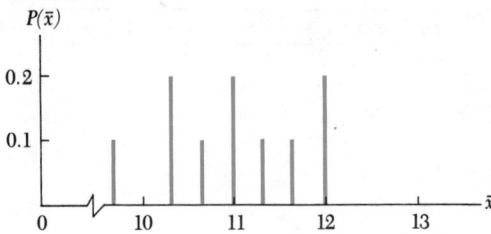

We will now compute the mean and standard deviation of the statistic $\overline{X}$. Denote by $\mu_{\bar{x}}$ the mean of $\overline{X}$:

$$\mu_{\bar{x}} = \frac{\sum_{j=1}^{10} \bar{x}_j}{10} = \frac{12 + 10.33 + \cdots + 11}{10} = 11$$

Denote by $\sigma_{\bar{x}}$ the standard deviation of $\overline{X}$:

$$\sigma_{\bar{x}} = \sqrt{\frac{\sum_{j=1}^{10} (\bar{x}_j - \mu_{\bar{x}})^2}{10}}$$

$$= \sqrt{\frac{(12 - 11)^2 + (10.33 - 11)^2 + \cdots + (11 - 11)^2}{10}} = 0.73$$

Thus, the mean and standard deviation of the statistic $\overline{X}$ and of its sampling distribution are 11 and 0.73, respectively.

Notice that the divisor in $\sigma_{\bar{x}}$ is 10—we are calculating the standard deviation of the entire population of values of $\overline{X}$. It is not too surprising that $\mu_{\bar{x}} = 11$, the population mean. Intuitively, we would expect the sample average to be distributed about the population mean. In fact, its average is equal to the population mean.

The average of the ten sample medians is $\mu_{M_X} = 10.6$. The sample median values have an average that is neither equal to the population mean μ_X nor the population median M_X. In this experiment, the sample median tends to underestimate the population mean and overestimate the population median.

The average of the ten sample proportions is $\mu_P = {}^6/_{10} = {}^3/_5$, which is the proportion of students in the population who have a grade point average greater than 3.00.

We might surmise from this example that the statistic $\overline{X}$ has the property that if its value is determined in each of all possible samples of size n drawn from the population, the average of the $\bar{x}$ values will be μ, the population average. In fact, this is true. Furthermore, the same property holds for the sample proportion statistic **P**. But this property does not hold for the sample median—the average of the sample medians calculated in all possible samples of size n is not generally equal to the population median, nor is it generally equal to the population mean.

We will now summarize the properties of the statistic $\overline{X}$.

Theorem 8.1
Mean and standard deviation of the statistic $\overline{X}$

If we take a simple random sample of n random variables $X_1, X_2, \ldots, X_n$ drawn from a population with mean μ and standard deviation σ, then the *mean* of $\overline{X}$, denoted by $\mu_{\bar{x}}$, and the *standard deviation of* the statistic $\overline{X}$ denoted by $\sigma_{\bar{x}}$, are given by

$$\mu_{\bar{x}} = \mu$$

$$\sigma_{\bar{x}} = \begin{cases} \dfrac{\sigma}{\sqrt{n}} & \text{if population is infinite} \\[2ex] \sqrt{\dfrac{N-n}{N-1}} \cdot \dfrac{\sigma}{\sqrt{n}} & \text{if population is finite} \end{cases}$$

In Example 8.1, we found that $\mu_{\bar{x}} = \mu = 11$ and that $\sigma_{\bar{x}} = 0.73$. Since the population is finite, by Theorem 8.1, $\sigma_{\bar{x}}$ should be given by

$$\sqrt{\frac{5-3}{5-1}} \cdot \frac{1.789}{\sqrt{3}} = 0.73$$

Figure 8.4 summarizes the results of Theorem 8.1.

An important observation from Theorem 8.1 is that the variability of the statistic $\overline{X}$ and the sample size are inversely related. If the population is finite, for example, then

$$\sigma_{\bar{x}} = \sqrt{\frac{N-n}{N-1}} \cdot \frac{\sigma}{\sqrt{n}}$$

As the sample size increases, this quantity becomes smaller. When $n = N$, $\sigma_{\bar{x}} = 0$; there is no variability in $\overline{X}$. Indeed, when we have sampled the entire population, $\overline{X} = \mu$.

If the population is infinite, then the sample size may be increased without bound. Then,

$$\lim_{n \to \infty} \left(\frac{\sigma}{\sqrt{n}} \right) = 0$$

Therefore, this theorem tells us that as n increases, the sampling distribution of $\overline{X}$ converges on its mean $\mu_{\bar{x}}$, which is equal to the value of population

FIGURE 8.4 Relationship between the mean and standard deviation of the population random variable **X** and the mean and standard deviation of the statistic $\overline{X}$

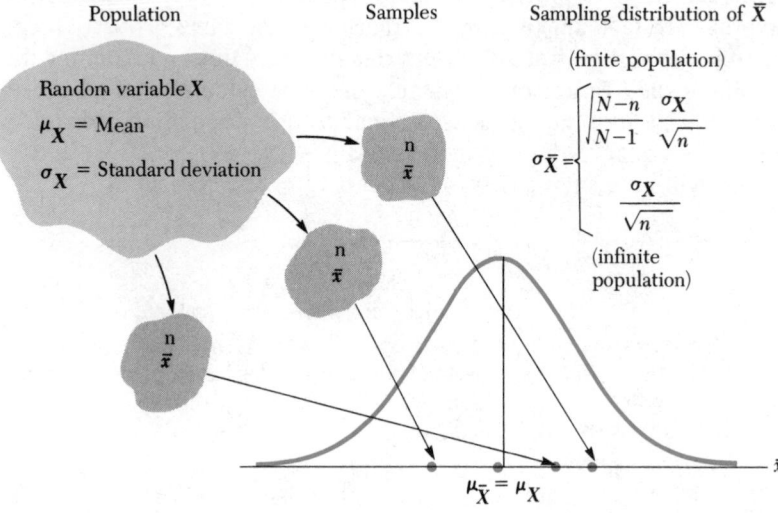

mean μ. This is intuitive as well—the larger the sample size, the closer we should expect the value of $\overline{X}$ to be to the value of μ.

■ 8.5 CENTRAL LIMIT THEOREM

Theorem 8.2
Central limit theorem

Let $X_1, X_2, \ldots, X_n$ be a simple random sample of size n to be drawn from an infinite population with a finite mean μ and standard deviation σ. Then, the random variable $\overline{X}$ has a limiting distribution that is normal with mean μ and standard deviation $\sigma/\sqrt{n}$.

This theorem tells us that the sampling distribution of the statistic $\overline{X}$ will become increasingly close to a normal distribution as the sample size increases, and that when the sample size becomes infinite, the sampling distribution of $\overline{X}$ is the normal distribution. The remarkable aspect of this theorem is that it places only two qualifications on the population distribution of values: (1) the population is infinite and (2) the population has a finite mean and variance. Nothing is specified about the form of the population distribution. It turns out that if the population random variable **X** is normally distributed, then so is $\overline{X}$—exactly, for all sample sizes n. If **X** is not normally distributed and the assumptions of the theorem are met, then for large sample sizes, we may use this theorem to provide an approximate form for the sampling distribution of $\overline{X}$.

To illustrate the results of the central limit theorem, consider the experiment described in the following example.

Example 8.2   Suppose the population characteristic of interest is the time between telephone calls at a switchboard. We could make two assumptions regarding the distribution of the population random variable X = time between calls:

Case 1. X is normally distributed with mean μ = 2.00 minutes and standard deviation σ = 0.25.

Case 2. X is uniformly distributed over the range 0–4 minutes. Its mean is, therefore, μ = 2.00 minutes.

The two distribution models are illustrated in Figure 8.5.

FIGURE 8.5   Two assumed distribution models for the distribution of random variable X = time between calls

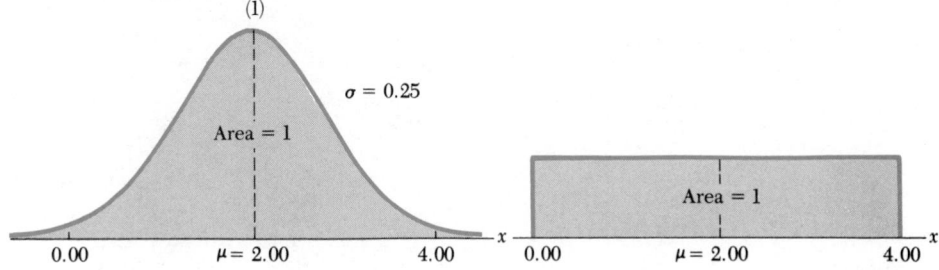

In case 1, the statistic $\overline{X}$ should be normally distributed for all sample sizes, because X is normally distributed. In case 2, the statistic $\overline{X}$ should be approximately normally distributed for large sample sizes. To check these results, random samples of size n = 5, 10, 20, 50, and 100 were drawn from each population for 100 repeated samples of each size. The samples were generated on the computer and the resulting frequency distributions are shown in Figure 8.6. On the left side in the figure are the frequency distributions for case 1. The fact that each distribution is not exactly a normal distribution is due solely to sampling error. By taking more than 100 repeated samples of each size, the frequency distributions would be more closely normal at each sample size. Notice that the convergence to normality of the statistic $\overline{X}$ in case 2 is relatively slow, but by n = 50, the normal distribution form is becoming evident.

We now review the important results concerning the form of the sampling distribution of $\overline{X}$ in Theorems 8.1 and 8.2.

■ **8.6   SAMPLING DISTRIBUTION OF THE SAMPLE MEAN STATISTIC $\overline{X}$**

We will distinguish between the situations where the population random variable X is normally distributed and not normally distributed in summarizing the results about the sampling distribution of the statistic $\overline{X}$.

FIGURE 8.6 Histograms for the sampling distribution of the statistic $\overline{X}$ in 100 repeated samples

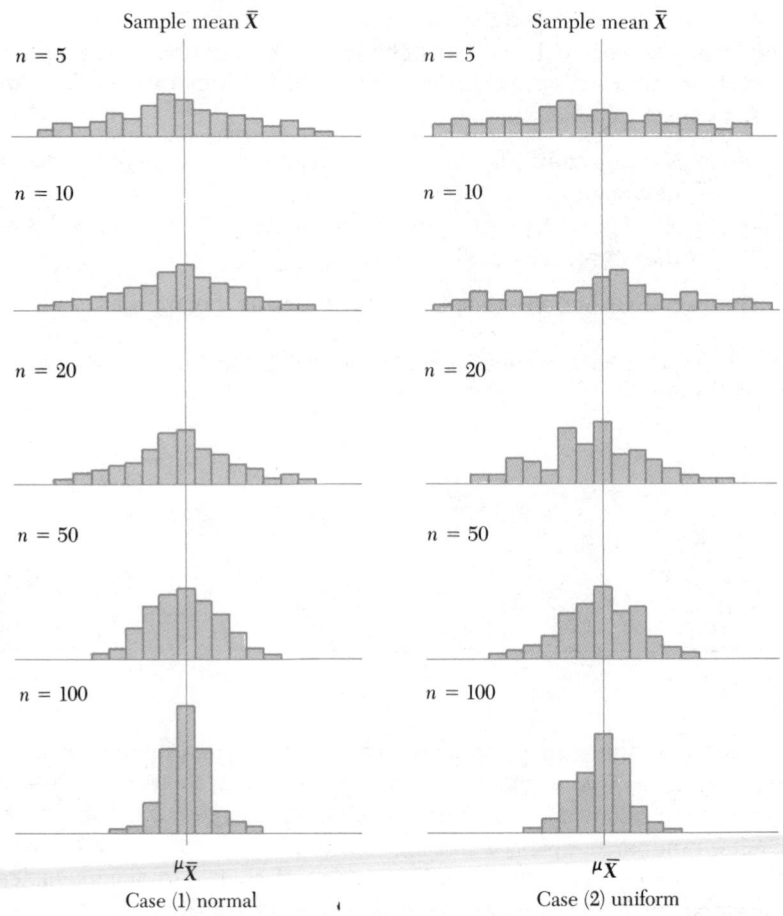

Case (1) normal   Case (2) uniform

Theorem 8.3
Sampling distribution of $\overline{X}$ (normal data)

If the population random variable X is normally distributed with mean μ and standard deviation σ, and a simple random sample of size n is to be drawn, then the sample mean statistic $\overline{X}$ is normally distributed—exactly for all sample sizes n—with a mean $\mu_{\overline{X}} = \mu$ and a standard deviation $\sigma_{\overline{X}} = \sigma/\sqrt{n}$.

This theorem will not be proved. It follows from the fact that the sum of a set of normally distributed random variables will itself be normally distributed.

Example 8.3 Suppose the time between telephone calls at a switchboard is normally distributed with a mean $\mu = 2$ minutes and a standard deviation

$\sigma = 0.25$. If a random sample of size $n = 25$ is drawn, what is the probability that the sample mean statistic $\overline{X}$ will be greater than 2.10 minutes?

Solution From Theorem 8.3, the sampling distribution is formed and is shown in Figure 8.7. To find the probability that $\overline{X}$ will be greater than 2.10 minutes, we must standardize the value of the variable $\overline{X}$ by using

$$z = \frac{\bar{x} - \mu_{\bar{x}}}{\sigma_{\bar{x}}}$$

Then, Table B.3 in Appendix B may be used to determine the desired probability. The answer is 0.0228. Therefore, in only 2.28 percent of all possible samples of size $n = 25$ will the sample mean statistic $\overline{X}$ be 2.10 minutes or greater.

FIGURE 8.7   Calculation of the probability in Example 8.3

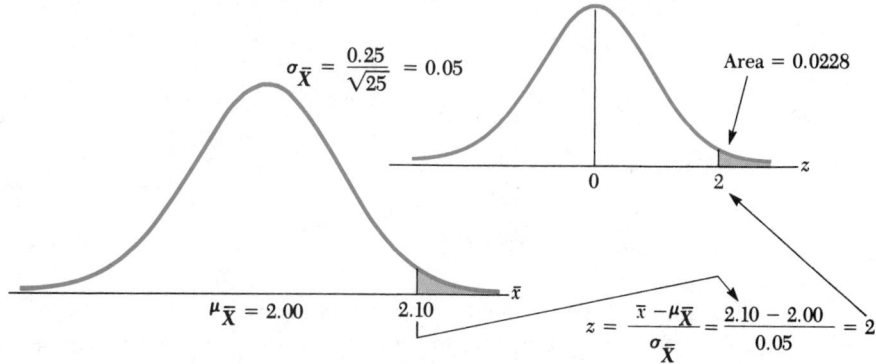

Theorem 8.4
Sampling distribution of the statistic $\overline{X}$ (nonnormal data)

If the population random variable is not normally distributed, but has a finite mean and variance given by μ and σ^2, and a simple random sample of size n is to be drawn, then the sample mean statistic $\overline{X}$ is approximately normally distributed for sufficiently large n with mean $\mu_{\bar{x}}$ and standard deviation $\sigma_{\bar{x}}$, where $\mu_{\bar{x}} = \mu$ and $\sigma_{\bar{x}} = \sigma/\sqrt{n}$.

In Theorem 8.4, the population size must be infinite, although the theorem is applicable to large populations if the sample size is large.

Example 8.4   The number of persons entering a service facility in a plant during 15-minute intervals during the working day is known to have a distribution with a mean rate of two arrivals and a standard deviation of 1.42 arrivals in the 15-minute intervals. If 100 of these 15-minute intervals are randomly selected and the number of arrivals in each interval is recorded, what is the probability that the sample mean number of arrivals will be less than 1.85?

Solution Since the random variable **X** = number of arrivals in a 15-minute interval has an unspecified distribution (the distribution is most likely Poisson; see Chapter 6), Theorem 8.4 is applicable.

Theorem 8.4 tells us that the statistic $\overline{X}$ will be approximately normally distributed with $\mu_{\bar{x}} = \mu = 2$ and $\sigma_{\bar{x}} = \sigma/\sqrt{n} = 0.142$. Thus, the probability that $\overline{X}$ is less than 1.85 can be computed *approximately* by finding the area under the normal curve as indicated in Figure 8.8. The probability is approximately 0.1446 that the average number of arrivals in the sample will be less than 1.85.

FIGURE 8.8 Calculation of the probability in Example 8.4

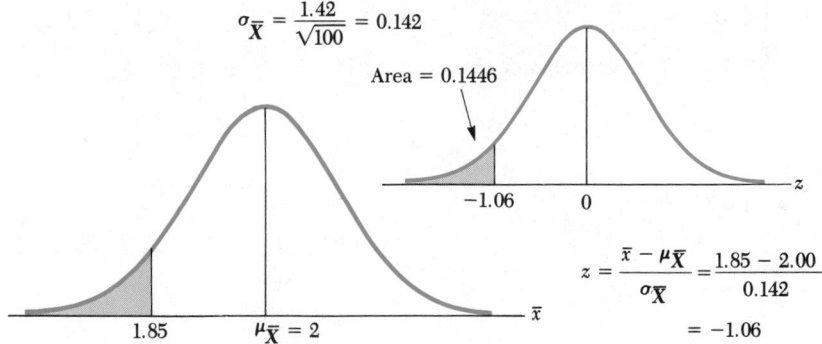

$$\sigma_{\overline{X}} = \frac{1.42}{\sqrt{100}} = 0.142$$

Area = 0.1446

$$z = \frac{\bar{x} - \mu_{\overline{X}}}{\sigma_{\overline{X}}} = \frac{1.85 - 2.00}{0.142}$$

$$= -1.06$$

The importance of Theorems 8.3 and 8.4 is that they allow us to deduce results concerning the outcome in the sample, based on knowledge of the population mean and standard deviation under the stated conditions. Specifically, they allow us to determine the likelihood that the sample mean statistic $\overline{X}$ is larger or smaller than a stated value, or will fall into a stated interval. This result produces a framework with which the direction of inference is "turned around"—instead of *deducing* properties of **X**, we will in subsequent chapters use values of $\overline{X}$ and these theorems to *induce* results about the population characteristic.

■ 8.7 SAMPLING DISTRIBUTION OF THE SAMPLE PROPORTION STATISTIC P

In Chapter 6, we considered the instance when a discrete random variable may be of a very special type—the binomial random variable. If **X** is a binomial random variable, it represents the number of successes in n independent and identical trials where each trial can result in a "success" or a "failure." If a sample of n trials of a binomial experiment is taken, the sample proportion of successes, the statistic **P**, may be computed by dividing the number of successes in the n trials by n; that is,

$$P = \frac{X}{n}$$

Since **X** is the "total" number of successes in the n trials, **P** represents the "average" number of successes in these trials. Since **P** is in the form of an average, it is possible to apply the central limit theorem. The central limit theorem tells us that the form of the distribution of the statistic **P** is approximately normal for large sample sizes (numbers of binomial trials). To use this fact in computing probabilities concerning **P**, it is necessary to first determine which member of the normal family of distributions is the appropriate one for modeling the distribution of **P**. That is, the mean of **P**, μ_P, and the variance of **P**, σ_P^2 must be determined.

Recall that the binomial random variable **X** has a mean of $n\pi$ and a variance of $n\pi(1 - \pi)$, where n is the number of binomial trials and π is the probability of success on each and every trial. Since **P** = **X**/n,

$$\mu_P = E\left(\frac{X}{n}\right) = \frac{1}{n} E(X) = \frac{1}{n} (n\pi) = \pi$$

$$\sigma_P^2 = V\left(\frac{X}{n}\right) = V\left[\left(\frac{1}{n}\right) \cdot X\right] = \left(\frac{1}{n}\right)^2 V(X) = \frac{n\pi(1 - \pi)}{n^2} = \frac{\pi(1 - \pi)}{n}$$

Therefore, the mean of the statistic **P**, μ_P, is given by π; the variance of the statistic **P**, σ_P^2, is given by $[\pi(1 - \pi)]/n$; and the standard deviation of the statistic **P**, σ_P, is given by $\sqrt{[\pi(1 - \pi)]/n}$. The results for the sampling distribution of the statistic **P** are summarized here.

Theorem 8.5
Sampling distribution of the statistic P

If **X** is a binomial random variable representing the number of successes in n independent and identical trials where the probability of success in each and every trial is π, then the sample proportion statistic **P** = **X**/n is, for sufficiently large n, approximately normally distributed with mean $\mu_P = \pi$ and standard deviation

$$\sigma_P = \sqrt{\frac{\pi(1 - \pi)}{n}} ,$$

if the population is infinite, and with standard deviation

$$\sigma_P = \sqrt{\frac{N - n}{N - 1}} \cdot \sqrt{\frac{\pi(1 - \pi)}{n}} ,$$

if the population is finite.

The fact that the sample proportions computed in repeated samples of size n trials average out to π should not be surprising—the statistic **P** is the sample proportion of successes and π is the population proportion of successes. We observed this result in Example 8.1, our sampling experiment—the sample proportions averaged out to be the population proportion.

The standard deviation of the statistic **P**, like the standard deviation of the statistic $\overline{X}$, is a function of $1/n$; as the sample size increases, the statistic **P** tends toward the population proportion. Notice the two forms for the standard deviation σ_P. If the population is finite in size, we must correct the standard deviation by the factor $\sqrt{(N - n)/(N - 1)}$.

We have already used the results of Theorem 8.5 in the text. Recall in Chapter 7 that we approximated binomial probabilities concerning the random variable **X**, the number of successes in n trials, by using the normal distribution. In these cases, we were using the central limit theorem and the results of Theorem 8.5, which ensure that the random variables **X** and **P** will be approximately normally distributed for large samples.

Example 8.5 In a 1980 primary election, a candidate received 60,000 of 100,000 votes, a proportion of "successes" equal to $\pi = 0.6$. If we had taken a simple random sample of 100 voters on the eve of election day, what would have been the probability that the sample proportion would have exceeded 50 percent successes?

Solution From Theorem 8.5, the mean of the statistic **P** is $\mu_P = 0.6$ and the standard deviation of the statistic **P** is

$$\sigma_P = \sqrt{\frac{N - n}{N - 1}} \cdot \sqrt{\frac{\pi(1 - \pi)}{n}} = \sqrt{\frac{100,000 - 100}{100,000 - 1}} \cdot \sqrt{\frac{0.6(1 - 0.6)}{100}}$$

$$= (0.9995)(0.0489) = 0.0489$$

Notice that the effect of the radical $\sqrt{(N - n)/(N - 1)}$ is very small, because the population is large and the sample is only a small portion of the population. The sampling distribution of **P** is illustrated in Figure 8.9.

FIGURE 8.9 Sampling distribution of the statistic **P** in Example 8.5

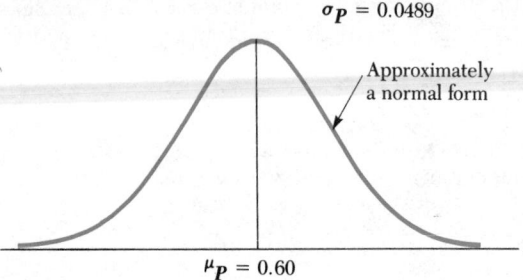

$\sigma_P = 0.0489$

Approximately a normal form

$\mu_P = 0.60$

To find the probability that the statistic **P** is greater than 0.50, we first standardize the value of **P**:

$$z = \frac{p - \mu_P}{\sigma_P} = \frac{0.50 - 0.60}{0.0489} = -2.04$$

From the standard normal tables (Table B.3 in Appendix B), the probability that the random variable **Z** is greater than -2.04 is 0.9793. Therefore, the probability that the value of the statistic **P** is greater than 0.50 in a sample is 0.9793. Another way of interpreting this answer is that if we took all possible samples of size $n = 100$ from the population of $N = 100,000$ voters, in approximately 97.93 percent of these samples, the sample proportion of sampled voters who say they will vote for the candidate is greater than 0.50.

■ 8.8 DISCUSSION

The central limit theorem applies to other sample statistics as well. For example, the sample variance statistic S^2 is an average of sorts; it "averages" the squared deviations of the sample values about their mean. As the sample size increases, the sampling distribution of S^2 will approach a normal distribution. But there are better approximating distributions for S^2 than the normal distribution, so we will not apply the central limit theorem to S^2. The exact and approximate sampling distributions of the statistic S^2 will be given in Chapter 9.

We should also comment about the correction factor for the standard deviation of the sample statistic $\overline{X}$ or **P** when the population is finite. The term,

$$FPC = \sqrt{\frac{N - n}{N - 1}}$$

is called the *finite population correction factor* (FPC). The reason that it is necessary in finite population sampling is intuitive; as each unit is withdrawn from the finite population to be placed in the sample, the variability in the population is reduced before the next draw—there is one less unit in the population. In contrast to this, if the population is infinite, as each unit is withdrawn to be placed in the sample, the remaining population is still infinite—the variability has not been reduced by withdrawing a unit. The FPC factor corrects for this difference by adjusting the standard deviation formula for sample statistics drawn from a finite population.

Notice that if n is relatively small in relation to N, the value of the FPC factor will be close to 1. This situation occurred in Example 8.5. As a general rule, if the sample size n is large ($n \geq 30$) and is less than 5 percent of the population size N, the FPC factor may be ignored—its effect on the standard deviation of the sample statistic will usually be negligible.

■ 8.9 OTHER TYPES OF SAMPLES

A simple random sample is one kind of probability sample. While simple random samples are being extensively used, increasingly more sophisticated probability sampling techniques are being employed. We shall consider three: *stratified sampling, cluster sampling,* and *systematic sampling.*

> **Definition 8.10**
> Stratified sample
>
> A *stratified sample* is obtained by forming strata in the population, and from each stratum, selecting a simple random sample.

> **Definition 8.11**
> Cluster sample
>
> A *cluster sample* is obtained by identifying a set of clusters that comprise the population, randomly selecting a subset of these clusters, and taking a census of each cluster selected.

> **Definition 8.12**
> Systematic sample
>
> A *systematic sample* is formed by selecting one unit at random and then selecting additional units at evenly spaced intervals (every kth population unit, $k > 1$) until the sample has been formed.

The basic idea in stratified sampling is illustrated in Figure 8.10. The population is "stratified," and from each stratum, a simple random sample is drawn. As an example, we may be planning a survey to determine the profitability of offering a lawn care service in a town. Recognizing that wealthy homeowners may be more receptive to the offer than others, we might stratify the population of homes on the basis of value, forming three strata: homes valued at $50,000 and under, over $50,000 but under $75,000, and homes valued at $75,000 and over. From each stratum, we then take a random sample of selected homes. Stratified sampling adds control to the sampling process by decreasing the amount of sampling error. For instance, a simple random sample may produce, by chance, proportionally too many

FIGURE 8.10 Stratified sampling

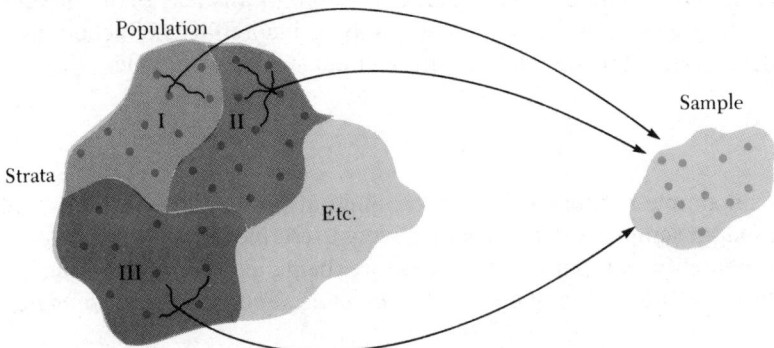

homes in the $75,000 range, which may, in turn, bias our predictions of the number of homes in the town that would accept a lawn care offer.

In general, population units may be stratified on any number of characteristics. Common variables used for stratifying in "people" surveys are age, income, and location of residence.

The notion of cluster sampling is suggested in Figure 8.11. Clusters are identified in the population, a set of clusters is randomly sampled, and a

FIGURE 8.11   Cluster sampling

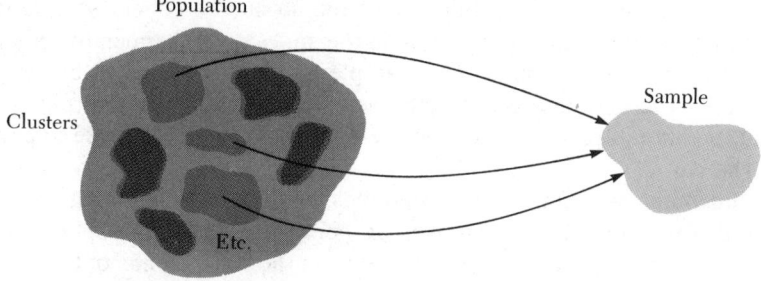

complete census is taken of each to form the sample. As an example, we may be interested in surveying the attitudes of airline passengers concerning service. Of all flights in a given week, we may select ten and take a census of everyone aboard each flight. Each flight is a cluster in the population of flights.

Systematic samples are commonly used in reliability testing—every kth item ($k > 1$) is selected from the assembly line and tested.

The kinds of probability samples all have one thing in common: the sampled population units are chosen according to a probability plan. Two other types of sampling, *judgment* and *convenience,* choose the sampled population units in the ways suggested by their names.

Judgment sampling is common when the sample is to be very small (usually due to the expense of sampling), the population is very heterogeneous, or special skills are required to form a representative subset of the population.

Auditors frequently use judgment samples to select items for study to determine whether a complete audit of the items may be necessary. However, auditors recently have begun to use probability samples increasingly for this purpose.

Definition 8.13
Judgment sample

A *judgment sample* is obtained by having an expert who is familiar with the population characteristics select units from the population.

> **Definition 8.14**
> Convenience sample
> A *convenience sample* is obtained by selecting "convenient" population units.

Suppose we are interested in conducting a study of the attitudes of shoppers at a new shopping mall on the kinds of stores in the mall, the attractiveness of the mall, parking difficulties, and so on. To collect sample information, we ask persons to participate in the survey who happen to walk past the central area of the mall. The sample in this instance is a convenience sample—the people are not being selected according to a probability plan and, presumably, judgment is not being used in selecting those to participate in the survey.

Judgment and convenience samples have the disadvantage when compared with probability samples of being subject to sampling bias. The quality of the judgment sample depends on the competence of the expert who selects the population units to be sampled. Convenience samples are prone to bias by their very nature—selecting population elements that are convenient to choose almost always makes them special or different from the rest of the elements in the population in some way. For example, convenience sampling in attitudinal studies generally results in samples that contain too high a proportion of complainers—persons who are only too happy to let their grievances be known. In almost all cases, it is desirable to take probability samples—they best guard against bias.

■ 8.10 SUMMARY

There are various ways a sample can be withdrawn from a population: *probability sampling* (simple random samples, stratified samples, cluster samples, and systematic samples), *judgment sampling,* and *convenience sampling.* The last two types must be used with care, for they are subject to sampling bias, particularly convenience samples. In this text, we will use simple random samples.

The sample statistics are random variables—their values vary from sample to sample of a fixed size drawn from a population. For sufficiently large sample size, the distribution of the sample average statistic $\overline{X}$ and the sample proportion statistic P, as well as other sample statistics, are approximately normally distributed. By knowing the form of the sampling distribution of a statistic, we are able to answer probability questions about it, such as the probability that the statistic is greater than a stated value. In this way, we can measure the extent of sampling error in the sampling experiment. The amount of sampling error can be reduced by increasing the sample size.

■ **REFERENCES** ─────────────────────────

Cochran, W. *Sampling Techniques,* 3d ed. New York: John Wiley & Sons., Inc., 1977.

Hamburg, M. *Statistical Analysis for Decision Making,* 2d ed. New York: Harcourt Brace & World, Inc., 1977, chaps. 5 and 6.

Lapin, L. *Statistics for Modern Business Decision.* New York: Harcourt Brace Jovanovich, Inc., 1973, chaps. 6 and 7.

Neter, J.; Wasserman, W.; and Whitmore, G. A. *Applied Statistics,* Boston: Allyn & Bacon, Inc., 1973, chaps. 9, 10, 11, and 13.

Winkler, R., and Hays, W. *Statistics: Probability, Inference and Decision.* 2d ed. New York: Holt, Rinehart and Winston, Inc., 1971, chap. 5.

■ **PROBLEMS** ─────────────────────────

8.1. Explain in what sense the statistic $\overline{X}$, the sample average, is a random variable.

8.2. Describe the sampling distribution of the statistic $\overline{X}$ when
  a. The population random variable is normally distributed.
  b. The population random variable is not normally distributed, but the sample size is "large."

8.3. Distinguish among the *population* distribution, the *sample* distribution, and the *sampling* distribution of sample statistics.

8.4. Distinguish among the population mean, μ, the sample mean, $\overline{x}$, and the mean of the sampling distribution of $\overline{X}$, $\mu_{\overline{x}}$. How do these measures compare in size?

8.5. Distinguish among the population variance, σ^2, the sample variance, s^2, and the variance of the sampling distribution of $\overline{X}$, $\sigma_{\overline{x}}$. How do these measures compare in size?

8.6. What is the exact sampling distribution of the sample proportion statistic **P** when the population is infinite in size? Under what circumstances can the normal distribution be used as an approximating sampling distribution for **P**?

8.7. Describe in your own words the central limit theorem. Why is it such an important theorem in statistics?

8.8. What is the finite population correction factor? When should it be used?

8.9. A population consists of the elements listed in the table.

Item	Value
A	0
B	2
C	7
D	4
E	3
F	2

  a. Compute the population mean, median, standard deviation, and proportion of values greater than or equal to 2.
  b. Form a table showing all possible samples of size $n = 2$. For each sample of size 2, compute the mean, the median, and the proportion of values greater than or equal to 2.
  c. Compute the mean and standard deviation of the sampling distributions of the sample mean, median, and proportion. How do the means of these sampling distributions compare with the corresponding population parameters?
  d. Which distribution has less variability—the population distribution or the sampling distribution of the sample mean? Why?

8.10. A population random variable is normally distributed with a mean of 50 and a variance of 16. If a sample of size $n = 25$ is drawn from this population, what are the following probabilities?

a. $\overline{X}$ will exceed 51.

b. $\overline{X}$ will be between 48.5 and 51.5.

8.11. Repeat Problem 8.10, but with $n = 36$. What is the effect on the probabilities in parts a and b of increasing the sample size from 25 to 36?

8.12. Repeat Problem 8.10, but with the variance of the population random variable equal to 25. What is the effect on the probabilities in parts a and b of increasing the population variance from 16 to 25?

8.13. The Sanso Petroleum Company fills steel drums with a lubricating oil in such a way that the weight of the oil in each drum is normally distributed with a mean of 250 pounds and a standard deviation of 2 pounds. If a random sample of nine drums is taken, what is the probability that the average weight per drum in the sample is more than 252 pounds?

8.14. A certain species of apricot tree produces a yield that is distributed with a mean of 2.0 bushels per tree and a standard deviation of 0.5 bushel.

a. If the yields per tree are normally distributed, what is the probability that the average yield per tree in a random sample of $n = 4$ trees will be less than 1.5 bushels?

b. If the yields per tree are *not* normally distributed, is your answer in part a a useful and meaningful one? Explain.

8.15. The mean annual income of a population of union workers in a certain local is known to be $15,000. The population standard deviation is $2,500. A sample of $n = 100$ of these workers is drawn.

a. Describe the sampling distribution of the sample mean statistic $\overline{X}$.

b. What is the probability that $\overline{X}$ will be within $400 of the population mean annual income?

8.16. In a population of 1,000 college students in a particular major at a large university, the mean grade point average is 2.80 and the population standard deviation is 0.20.

a. If a random sample of $n = 100$ students is taken from this population, what is the probability that the statistic $\overline{X}$, the sample mean grade point average, will differ from the population mean grade point average by more than 0.01 unit?

b. Repeat part a with $n = 500$.

c. Repeat part a with $n = 1,000$.

8.17. In a very large production lot, the proportion of defectives is 0.10. If a sample of 100 items in this lot is selected, what is the probability that two or fewer of the items are defective? Is this an approximate or an exact answer? Explain.

8.18. In a set of 10,000 invoices, it is known that 500 contain at least one error. If 100 of the 10,000 invoices are randomly sampled, what is the probability that the sample proportion of invoices with at least one error will exceed 0.08? Is this an approximate or an exact answer? Explain.

8.19. Assume 60 percent of the voters in a large city are going to vote for the Democratic candidate. A random sample of $n = 100$ voters is chosen.

a. Determine the probability that the sample proportion of voters who will vote Democratic is within 10 percentage points of the population proportion (i.e., between 0.50 and 0.70).

b. What assumptions did you make in answering part a?

8.20. Describe each sampling plan and give one example where each is appropriate.
  a. Simple random sample.
  b. Cluster sample.
  c. Stratified sample.
  d. Systematic sample.
  e. Judgment sample.
  f. Convenience sample.

8.21. Why do convenience samples tend to be biased? Is this also true about judgment samples? Explain.

8.22. If you were asked to sample 50 households in a section of a city that contains 1,000 households, explain how you would go about determining which households to include if your sample is to be each of the following.
  a. Simple random sample.
  b. Stratified sample on household annual incomes with three strata: $0 to $5,000, $5,001 to $15,000, and above $15,000.
  c. Convenience sample.
  d. Judgment sample.

8.23. In the following situations, which type of sample—random, judgment, convenience, or a combination of sampling plans—would you recommend and why?
  a. A calculator manufacturer wishes to test the reliability of a certain microcircuit used in its leading model.
  b. A magazine editor selects "Letters to the Editor" to publish monthly.
  c. A university motor pool must purchase a large fleet of cars from one of the leading manufacturers. The primary consideration will be economy of operation.
  d. An accountant must select a sample of invoices in a large department store to determine whether the proportion of invoices with errors is within a set tolerance.
  e. A reporter wishes to receive suggestions about how his daily column in the local newspaper can be improved.

8.24. Consider the population of the first seven integers: 1, 2, 3, 4, 5, 6, and 7; $N = 7$. For this population, $\mu = 4$ and $\sigma = 2$.
  a. How many samples of size three can be extracted from this population (sampling without replacement)?
  b. Form the complete set of samples of size three and for each sample, compute the sample mean and median.
  c. Form the sampling distributions for the sample mean statistic and for the sample median statistic.
  d. Find the mean and standard deviation of each sampling distribution in part *c*.
  e. If you had to select *at random* one of the samples of size three generated in part *b* and had to use either the sample's mean or median to estimate μ, which would you choose and why?

8.25. An automobile dealer has been chastised by the manufacturer of the cars sold by the dealership because of the high percentage of complaints received from customers about service for their cars at the dealership. To assess the seriousness of the problem, the dealer decides to sample individuals randomly who have brought their cars in for service during the past six months. Each sampled individual will receive a mail questionnaire requesting comments about the quality of service received when they brought their cars in. Discuss why the sample set of questionnaires most likely will not be representative of the attitudes toward the service received by all customers who brought their cars in for service during the past six months.

8.26. A production process is known to produce 5 percent defective items. A random sample of 100 items is taken from the process.

    a. What is the probability that the sample proportion of defectives is greater than 0.10?

    b. If it is known that 5 percent of the items produced by a production process in a lot of 1,000 are defective, what is the probability, in a sample of 100 drawn from the lot, that the sample proportion of defectives exceeds 0.10?

    c. Compare the answers in parts *a* and *b*. Why do they differ?

Statistical inference: Estimation

9

In Chapter 2, we observed that a set of numbers can be described by numerical measures, such as representative measures and measures of dispersion. As noted in Chapter 8, if the set of numbers being examined comprises a population, then the numerical measures such as the population mean μ and the population standard deviation σ are called *parameters*. If, on the other hand, we intend to take a sample of size n from the population, the numerical measures such as $\overline{X}$, the mean, and S, the standard deviation, are functions of the random variables that constitute the sample, $X_1, X_2, \ldots, X_n$. The functions $\overline{X}$ and S are called *statistics*. The statistics $\overline{X}$ and S are random variables with values that are determined when a specific sample is taken and the numerical values $x_1, x_2, \ldots, x_n$ of the random variables $X_1, X_2, \ldots, X_n$ are known.

Since a population set of numbers can often be characterized by the values of the population parameters, attention is focused on these parameters in most statistical problems. For example, the population may consist of the lifetimes of a specific electronic component and we may be interested in the mean lifetime, the population parameter μ. The value of the parameter μ could be computed by using each component in the population, recording the lifetime of each and every component, and averaging these lifetimes. It clearly would not be feasible to compute the value of μ in this fashion, because this process would destroy all components—there would be no components left to sell. Alternatively, a sample of components could be selected and their lifetimes used in some manner to "guess," or estimate, the value of μ. In this instance, the statistic $\overline{X}$ would seem like a reasonable

233

function of the sample random variables $X_1, X_2, \ldots, X_n$ to use in estimating the value of the population parameter μ. Indeed, as we shall see later in this chapter, the sample mean statistic $\overline{X}$ is the "best" statistic to use in estimating μ.

In many cases, it is not possible to determine the value of a population parameter by analyzing the entire set of population values. The process of determining the value of the parameter may destroy the population units, or it may simply be too expensive in money and/or time to analyze each unit. In these instances, there is little choice but to use statistical inference to gain information about the values of the population parameters.

The objective of statistical inference is to make inferences about a population based on a subset of it—the sample drawn from the population. More specifically, statistical inference is the process of selecting and using a sample statistic to draw inferences about a population parameter.

In Figure 9.1, the basic framework of the inference process is portrayed. We will identify a population parameter of interest denoted in Figure 9.1 by

FIGURE 9.1 Population parameter θ and the sampling process

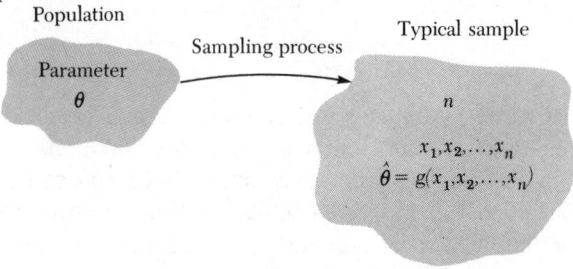

the Greek symbol θ (theta)—θ may be the population mean μ, standard deviation σ, or proportion π, for example. In the sample, we will select a function of the sample random variables $X_1, X_2, \ldots, X_n$, denoted by $\hat{\theta} = g(X_1, X_2, \ldots, X_n)$, to use in drawing inferences about the value of θ. The function $\hat{\theta}$ is a *statistic* with a value that depends on the values of the sample random variables $X_1, X_2, \ldots, X_n$. It is, therefore, a random variable. The symbol over $\theta(\hat{\ })$ will be referred to as a "hat," so that $\hat{\theta}$ reads "theta hat." The hat indicates we are drawing inferences about the symbol beneath it.

For example, suppose we are interested in the specific parameter μ, the population mean. The sample statistic we will use to draw inferences about μ is $\hat{\mu} = \overline{X}$. If a typical sample is taken so that the values of the sample random variables $X_1, X_2, \ldots, X_n$ are known $(x_1, x_2, \ldots, x_n)$, then the estimate of μ, denoted by $\hat{\mu}$, is given by the sample mean $\bar{x}$, as illustrated in Figure 9.2.

There are two basic ways to draw inferences about the value of a population parameter θ. The first way is to *estimate* the value of the parameter. To this point, we have been principally discussing this method of statistical

FIGURE 9.2   Estimating the population parameter μ

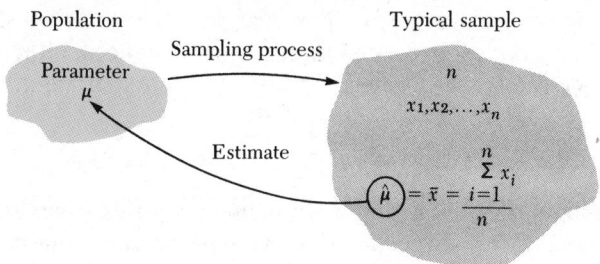

inference. In statistical estimation, the idea is to select a sample statistic with a *value* that will be used as an estimate of the population parameter value.

The second way to draw inferences about the value of θ is called *hypothesis testing*. In this process, we hypothesize a value of θ and use the sample information to make a decision as to whether or not the hypothesis is tenable. For example, suppose we are interested in the population parameter π—the proportion of the population who will vote for Herman in the next presidential election—where the population consists of all registered voters who will vote in the presidential election. Prior to the election, we may hypothesize that Herman will receive 60 percent or more of the votes cast; in this case, we are conjecturing that the value of π will be 0.60 or greater. A sample of voters could be drawn and p, the sample proportion of those voting for Herman, used to decide whether the hypothesis is tenable (reasonable).

Both methods of statistical inference use the same theoretical relationships between sample outcomes and population values. A sample is drawn from the population and a sample statistic is used to drawn inferences about the population parameter. In statistical estimation, the sample information is used to estimate the value of θ. In statistical hypothesis testing, a value or range of values for θ is first hypothesized and the sample information is used to decide whether or not the hypothesis should be rejected. As we shall see, these two types of inference processes are very closely related.

Finally, when we apply the inference process to estimate a population parameter, it is important to say how good the inferences are. In statistical estimation, we will want to know how much difference there *may* be between the estimate of the population parameter, calculated from a specific sample of size n, and the value of the population parameter. In hypothesis testing, the "goodness" of the inference is measured by the *probability* that the decision to reject or not to reject the hypothesized value of the population parameter is correct.

Thus, there are two important elements of statistical inference: the inference and the measure of its goodness.

Statistical estimation will be developed in this chapter and hypothesis testing in the next chapter. This is not to suggest that estimation is more or less important than hypothesis testing. In most statistical problems, it is quite often possible to use either method to resolve an inference question, and the one chosen depends largely on personal preference. As we shall see, the two methods are *so* closely related that in most instances, either may be used to resolve the inference problems in a statistical experiment.

9.2 TYPES OF STATISTICAL ESTIMATION PROCEDURES

Statistical estimation is divided into two main categories: *point estimation* and *interval estimation*. In point estimation, a single number computed from the sample information is used as an estimate of the value of θ. It is called a "point" estimate because one point on the real number line is used to estimate θ. In interval estimation, two points are used to define an interval on the real number line, which hopefully will contain the value of θ. For example, if the parameter is the population mean lifetimes of an electronic component, based on sample information we might arrive at a point estimate of μ, denoted by $\hat{\mu}$, of 550 hours, and an interval estimate of μ of from 530 to 570 hours.

We will consider each type of estimation in more detail, beginning first with point estimation.

9.2.1 Point estimation

The point estimation process is illustrated in Figure 9.3. The random sample is composed of n random variables $X_1, X_2, \ldots, X_n$. The *estimator*

FIGURE 9.3 Point estimator of θ

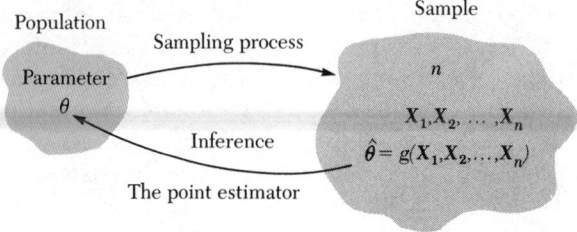

of the population parameter θ is a *function* of the sample random variables X_1, $X_2, \ldots, X_n$ and is denoted by $\hat{\theta} = g(X_1, X_2, \ldots, X_n)$, where g is the selected function of the random variables $X_1, X_2, \ldots, X_n$. Once a particular sample is drawn from the population and the values of the sample random variables $X_1, X_2, \ldots, X_n$ are known numerically $(x_1, x_2, \ldots, x_n)$, then the value of the estimator is known—it is called the *estimate* of the population parameter and is denoted by $\hat{\theta}$.

It is important to differentiate between an *estimator* and an *estimate* of a population parameter. When we use the term *estimator*, we are referring to a function of the sample random variables $X_1, X_2, \ldots, X_n$ *before* a particular sample has been drawn from the population. As such, an estimator is a sample *statistic* and is, therefore, a *random variable*. The value of this random variable is the *estimate* of the population parameter. It is determined numerically after a specific sample has been drawn and the values of the random variables X_1, X_2, . . . , X_n are known numerically (x_1, x_2, . . . , x_n). Since an *estimator* is a *statistic*, we shall denote an estimator by a boldfaced symbol. For example, a possible estimator of the population mean μ is $\hat{\boldsymbol{\mu}} = \overline{X}$, the sample mean statistic. The value of the estimator is $\hat{\mu} = \bar{x}$, the estimate of μ.

Let us illustrate these concepts by considering the population parameter μ, the population mean. We first must settle upon the form of the estimator of μ. Since μ is a measure of central tendency of the population values, three sample functions should come to mind as possible estimators of μ—the sample mean statistic $\overline{X}$, the sample median statistic **M**, and the sample mode statistic $\mathbf{M}_0$. All three statistics measure the central tendency of the sample random variables X_1, X_2, . . . , X_n and could, therefore, be considered for selection as the estimator of μ. Suppose a single sample of size n is taken from the population and the values of the sample random variables X_1, X_2, . . . , X_n are known numerically (x_1, x_2, . . . , x_n). Notice that the values of all three statistics are functions of the sample values x_1, x_2, . . . , x_n. The value of the statistic $\overline{X}$, the mean, is the arithmetic mean or average of x_1, x_2, . . . , x_n; the value of the statistic **M**, the median, is the "middle" value when x_1, x_2, . . . , x_n have been ordered from the smallest to the largest; and the value of the statistic $\mathbf{M}_0$, the mode, is the most frequently occurring sample value in the set x_1, x_2, . . . , x_n.

Which sample statistic should be used as the estimator of μ? It is apparent that we need to establish criteria to tell us which statistic does the "best" job of estimating μ.

We shall use one criterion for selecting a sample statistic as the point estimator of a population parameter θ: *unbiased minimum variance*. An estimator that satisfies this criterion is called an *unbiased minimum variance* (UMV) *estimator*. The description of the criterion is based on the fact that an estimator is a random variable. In Figure 9.4, suppose we select a sample

FIGURE 9.4 The sampling distribution of the estimator $\hat{\boldsymbol{\theta}}$

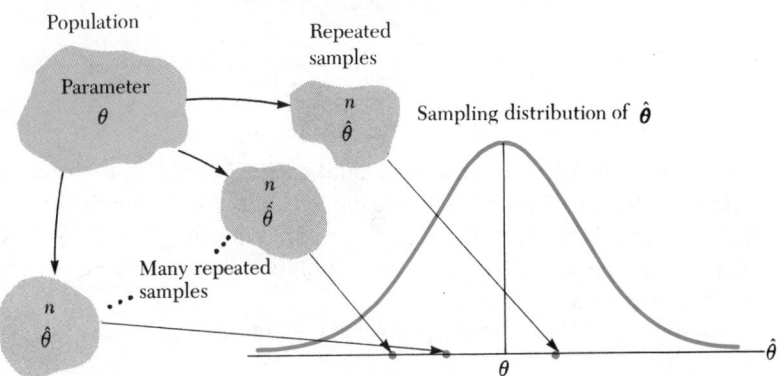

statistic as the estimator of θ and denote it by $\hat{\theta}$. The value that $\hat{\theta}$ assumes depends on the particular set of n sample values $x_1, x_2, \ldots, x_n$ in *one* sample drawn from the population. If we take repeated samples of size n, the value of $\hat{\theta}$ will undoubtedly change from sample to sample. Thus, its value varies over the set of all possible samples of size n.

The selection of the sample statistic to use an an estimator of a population parameter θ usually depends only on the properties of the sampling distribution of the sample statistic. An exception to this is when we select an estimator that is easy to calculate, although it may have inferior properties compared with another estimator. We will usually want, therefore, to select an estimator with values that in repeated samples are clustered closely about θ.

For instance, suppose we could select one of two sample statistics, denoted by $\hat{\theta}_1$ and $\hat{\theta}_2$, with sampling distributions that are illustrated in Figure 9.5, as an estimator of the population parameter θ.

FIGURE 9.5 Sampling distribution of two competing estimators of θ

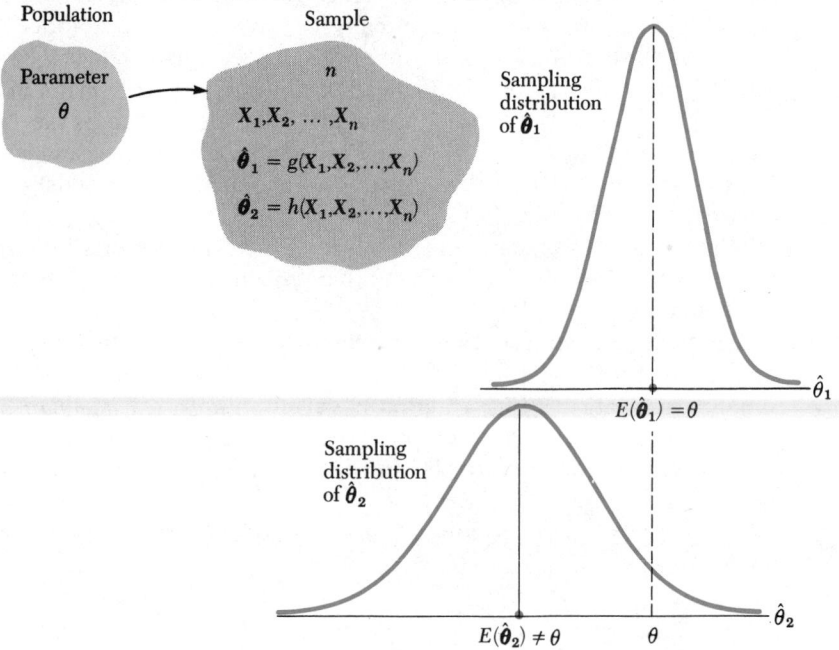

It is intuitively obvious that we would select $\hat{\theta}_1$ over $\hat{\theta}_2$ as the estimator of θ. Most of the $\hat{\theta}_1$ values are closely clustered about θ, which is not the case with the values of $\hat{\theta}_2$. Our chances are much better in getting a good estimate of θ in the sample we select from the population if we compute the value of $\hat{\theta}_1$ rather than the value of $\hat{\theta}_2$.

This argument for selecting the best estimator of θ may be somewhat

FIGURE 9.6   Two estimators, $\hat{\theta}_1$ and $\hat{\theta}_2$, with corresponding estimates of the population parameter

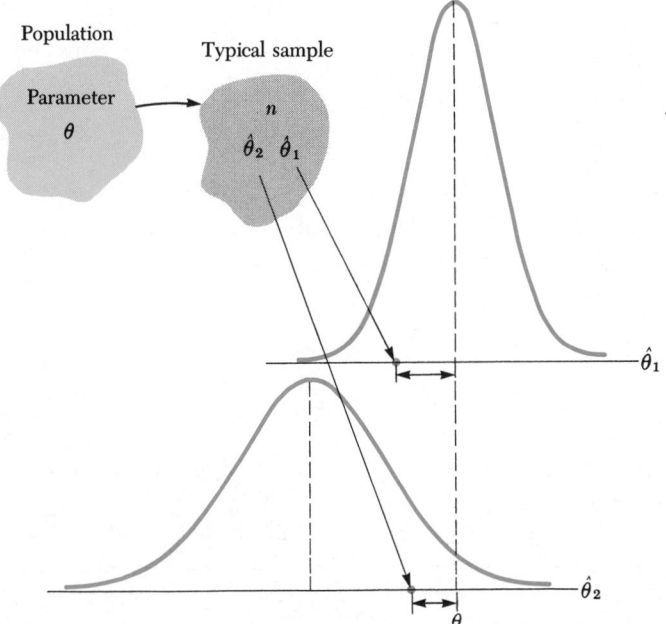

troublesome, for it is certainly possible in a *particular* sample that the value of $\hat{\theta}_2$ may be closer to θ than is the value $\hat{\theta}_1$, as illustrated in Figure 9.6. However, from observing the sampling distributions of $\hat{\theta}_1$ and $\hat{\theta}_2$, it is certainly *more likely* that we will draw a sample in which the value of $\hat{\theta}_1$ is closer to θ than is the value of $\hat{\theta}_2$. We may get "burned" in the sense that we get a bad sample—a sample in which $\hat{\theta}_2$ is closer in value to θ than is the value of $\hat{\theta}_1$—but it is very unlikely.

The UMV criterion for selecting the best estimator of a population parameter θ is based on two factors: *unbiasedness* and *minimum variance*. An estimator $\hat{\theta}$ of a population parameter θ is *unbiased* if $E(\hat{\theta}) = \theta$; that is, the average of the values of $\hat{\theta}$ computed in each of all possible samples of size n is θ. In Figure 9.5, $\hat{\theta}_1$ is an *unbiased* estimator—the mean of its sampling distribution is equal to the value of θ $[E(\hat{\theta}_1) = \theta]$, whereas $\hat{\theta}_2$ is a *biased* estimator—the mean of its sampling distribution is *not* equal to the value of θ $[E(\hat{\theta}_2) \neq \theta]$.

It is not sufficient that an estimator be unbiased for it to qualify as a "good" estimator, as Figure 9.7 illustrates. In this case, both estimators of θ are unbiased, but $\hat{\theta}_1$ is certainly preferable to $\hat{\theta}_2$—its values are more closely clustered about θ. This introduces the concept of minimum variance. It is usually our goal to select among all unbiased estimators of θ the one that has *minimum variance*—the least amount of sampling distribution variability about the value of θ.

FIGURE 9.7 The sampling distributions of two unbiased estimators of θ

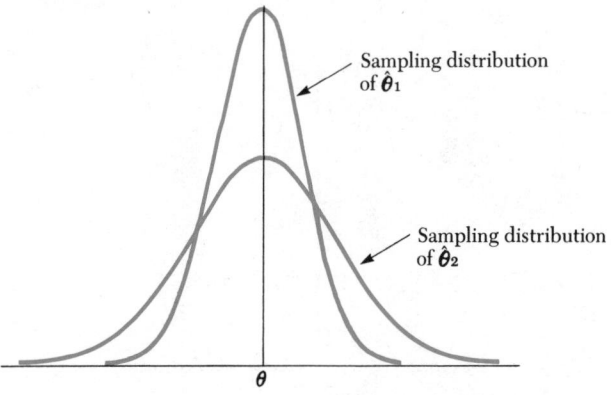

Sampling distribution of $\hat{\theta}_1$	
Sampling distribution of $\hat{\theta}_2$	

Definition 9.1
Unbiased minimum variance estimator

An estimator $\hat{\theta}$ of a population parameter θ is called an *unbiased minimum variance* (UMV) *estimator* if the expected value of $\hat{\theta}$ is θ [$E(\hat{\theta}) = \theta$], and among all unbiased estimators of θ, $\hat{\theta}$ has the least amount of variability for a specific sample size.

Unfortunately, it is not always possible to find a UMV estimator of a population parameter. In these cases, other criteria for selecting a best estimator may be used, or the requirement of unbiasedness or minimum variance may be relaxed. In this chapter, it will be noted when an estimator is UMV. If it is not, the estimator given will generally satisfy other goodness criteria and will be the one most frequently used in statistics.

☐ 9.2.2 Interval estimation

The interval estimation process is illustrated in Figure 9.8. The idea is to use two functions of the sample random variables $X_1, X_2, \ldots, X_n$, denoted by $g_1(X_1, X_2, \ldots, X_n)$ and $g_2(X_1, X_2, \ldots, X_n)$, which will produce two points on the real line, $\hat{\theta}_L$ and $\hat{\theta}_U$, respectively, that define an interval that will contain the value of θ with a specified degree of confidence. For example, we may find $\hat{\theta}_L$ and $\hat{\theta}_U$ so that the value of θ is between their values, $\hat{\theta}_L$ and $\hat{\theta}_U$, with 95 percent confidence. The interval estimator is often called a *confidence interval*—the use of the word *confidence* in this context will be thoroughly explained in Section 9.4.

As with point estimators, we can consider "good" interval estimators. A good interval estimator is one in which the functions $g_1(X_1, X_2, \ldots, X_n)$

FIGURE 9.8   Interval estimator of θ

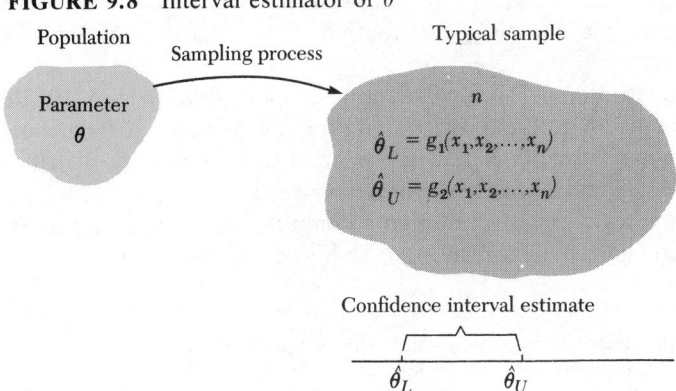

and $g_2(\mathbf{X}_1, \mathbf{X}_2, \ldots, \mathbf{X}_n)$ produce the shortest interval $(\hat{\boldsymbol{\theta}}_L, \hat{\boldsymbol{\theta}}_U)$ for a specified degree of confidence and a specified sample size.

In the next section, we will consider point estimators of the following population parameters: mean (μ), standard deviation (σ), and proportion (π). In Section 9.4, interval estimators for these parameters will be developed.

■ 9.3   POINT ESTIMATION

We will concentrate our attention on the population parameters μ, the population mean, and π, the population proportion, in developing the concepts of a point estimator and its measure of reliability.

□ 9.3.1   Population mean μ

We have already suggested that the best estimator of the population mean μ is the statistic $\overline{\mathbf{X}}$. Its principal competitors are the sample mode and the sample median. We may dispense with the mode as a viable competitor quickly. Intuitively, the mode is not appealing because either a sample set of values may not have a unique mode, or all values may be modal (no value in the sample occurs at least twice). Even if the mode is unique, the sampling distribution of the mode usually will not have a mean of μ, and the distribution will exhibit more variability than the sampling distribution of the statistic $\overline{\mathbf{X}}$ for a specified sample size.

The sample median statistic is a more worthy "opponent" of the statistic $\overline{\mathbf{X}}$ than is the sample mode. The median statistic, denoted by $\mathbf{M}$, is an unbiased estimator of μ if the population random variable is normally distributed. That is, in this case, the mean of the sampling distribution of $\mathbf{M}$ is μ. However, the distribution exhibits more variability than does the sampling distribution of the statistic $\overline{\mathbf{X}}$ for a specified sample size n. Thus, on the basis of our criterion for selecting an estimator of μ—unbiased minimum variance

(UMV)—the statistic $\overline{X}$ is preferred because its sampling distribution is less variable than the sampling distribution of the statistic M for a specified sample size.

As an illustration of this fact, random samples of size n = 5, 10, 20, 50, and 100 have been drawn for 100 repeated samples of each size from a population with a random variable X that is assumed to be normally distributed with mean μ = 2.75 and standard deviation σ = 0.25 (see Example 8.2 in Chapter 8). The samples were generated on a computer, and the resulting frequency distribution histograms for the statistic $\overline{X}$ and the statistic M are illustrated in Figure 9.9.

FIGURE 9.9 Histograms for the statistics $\overline{X}$ and M drawn from a normal distribution with μ = 2.75 and σ = 0.25

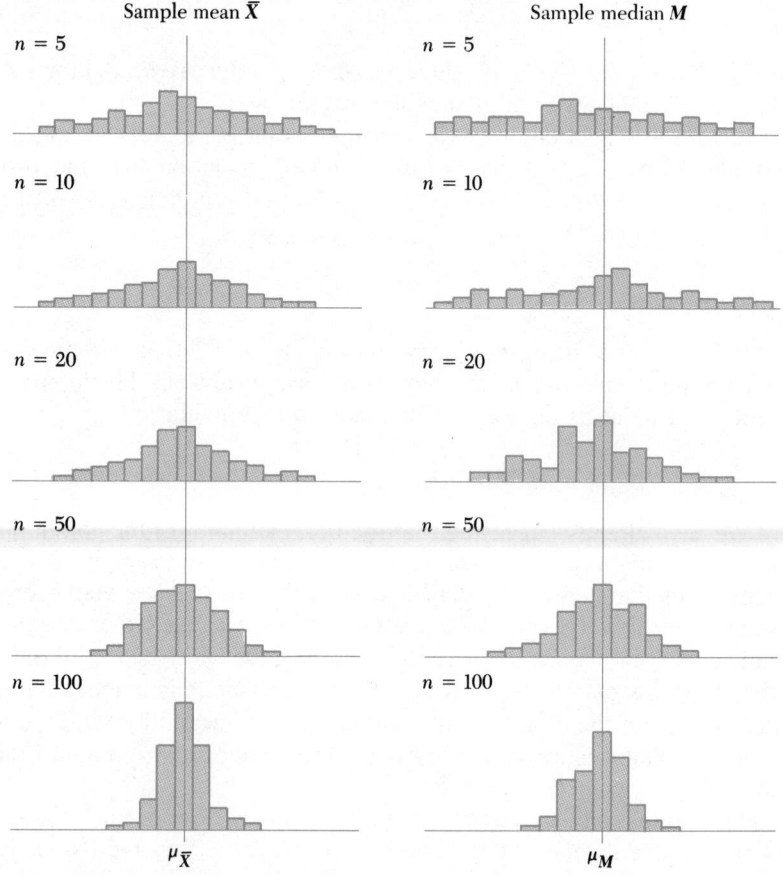

Notice that both sampling distributions have a normal form that becomes more pronounced as the sample size increases. The important observation is, of course, that the sampling distribution of the median statistic exhibits more variability than does the sampling distribution of the mean statistic $\overline{X}$. Thus, there is a better chance that a value of $\overline{X}$ will be closer to μ than will a value of **M** when these values are computed from one set of n sample values, x_1, $x_2, \ldots, x_n$.

The statistic $\overline{X}$ is used as the point estimator of μ, because it is the "best" estimator on the basis of our criterion—unbiased minimum variance. We must now develop a way to determine how good an estimator of μ the sample average $\overline{X}$ is. Recall in Chapter 8 (Theorem 8.1) that the standard deviation of $\overline{X}$ is given by $\sigma_{\overline{X}} = \sigma/\sqrt{n}$ and that the standard deviation of a random variable can be used to determine how dispersed the values of the random variable are. If the population random variable **X** is normally distributed, we know from Theorem 8.3 in Chapter 8 that $\overline{X}$ is normally distributed. If **X** is not normally distributed but the sample size is "large" ($n \geq 30$), the central limit theorem tells us that $\overline{X}$ is approximately normally distributed. In either case, by knowing the form of the distribution of $\overline{X}$, we can determine the proportion of the values of $\overline{X}$ that will be within one, two, or three standard deviations of the value of μ. This information gives us an idea of how close our value of $\overline{X}$ may be to μ, as the following example illustrates.

Example 9.1   A machine is producing ball bearings with diameters of 0.5 inch. Based on lengthy experience with the machine, it is known that the standard deviation of the bearings is 0.005 inch. A sample of 25 ball bearings is selected, and the sample mean diameter is found to be $\bar{x} = 0.498$ inch. Determine a point estimate of the population mean diameter of all bearings produced by this machine.

Solution   The population random variable is **X** = diameter of the ball bearings. The point estimate of the population mean μ is $\hat{\mu} = \bar{x} = 0.498$ inch. But how good is this estimate of the population mean bearing diameter? The standard deviation of $\overline{X}$ is $\sigma/\sqrt{n}$, which is, numerically, $0.005/\sqrt{25} = 0.001$ inch, because $\sigma = 0.005$ inch. The sampling distribution of $\overline{X}$ in this problem is illustrated in Figure 9.10. It is reasonable to assume that the statistic $\overline{X}$ is approximately normally distributed in this case, because **X** is a measurement quantitative characteristic with a distribution that is likely to be approximately normal. Since $\overline{X}$ is approximately normally distributed, we know that approximately 68 percent of the values of $\overline{X}$ lie within one standard deviation

FIGURE 9.10   Sampling distribution of $\overline{X}$ in Example 9.1

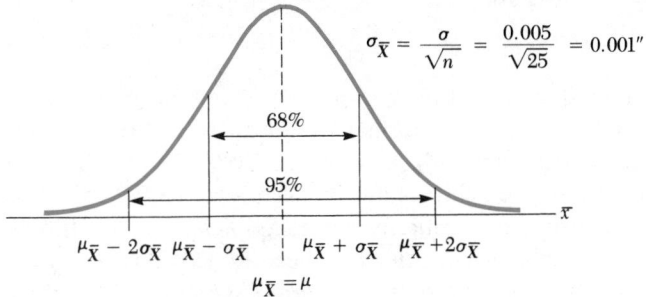

$$\sigma_{\overline{X}} = \frac{\sigma}{\sqrt{n}} = \frac{0.005}{\sqrt{25}} = 0.001''$$

of the mean μ (± 0.001 inch) and 95 percent of the values lie within two standard deviations of μ (± 0.002). Thus, approximately 95 percent of the sample means computed from repeated samples of size $n = 25$ will lie within 0.002 inch of the population mean μ. We can feel reasonably certain, therefore, that our sample mean $\bar{x} = 0.498$ inch lies no more than 0.002 inch from μ. We would answer the question in this example by stating that the point estimate of μ is $\bar{x} = 0.498$ inch, and it is very likely that this estimate lies no more than 0.002 inch from the true value of μ.

The standard deviation $\sigma_{\bar{x}} = \sigma / \sqrt{n}$ is called the *standard error* of the estimator $\overline{X}$. Since 95 percent of the values of $\overline{X}$ lie within two standard deviations of the value of μ, a value of $2\sigma_{\bar{x}}$ is often given as a measure of the "goodness" of the estimate of μ.

Notice in Example 9.1 that the population standard deviation σ is known. In most practical problems, if μ is not known, the value of σ will also be unknown. In these instances, the sample standard deviation s is used as a point estimate of σ so that the standard deviation of $\overline{X}$ is estimated by $s / \sqrt{n}$.

Definition 9.2
Point estimator of the population mean μ

The *point estimator* of the population mean μ is the statistic $\overline{X} = \sum_{i=1}^{n} X_i / n$, where the random variables $X_1, X_2, \ldots, X_n$ represent the random sample. The standard error of the estimator $\overline{X}$ is given by

$$\sigma_{\bar{x}} = \frac{\sigma}{\sqrt{n}}$$

if σ, the population standard deviation, is known, or is estimated by

$$\hat{\sigma}_{\bar{x}} = \frac{s}{\sqrt{n}}$$

if σ is unknown, where s is the sample standard deviation given by

$$s = \sqrt{\frac{\sum_{i=1}^{n} (x_i - \bar{x})^2}{n - 1}}$$

One indication of the "goodness" of the estimator $\overline{X}$ of μ is provided by twice the value of the standard error of the estimate ($2\sigma / \sqrt{n}$ or $2s / \sqrt{n}$). This quantity is called the *error bound* of the estimate.

The error bound on the estimate of the population mean is exact only if the population random variable is normally distributed and the population standard deviation σ is known. If X is not normally distributed but σ is known, then by appealing to the central limit theorem, we may use the error bound $2\sigma / \sqrt{n}$ as an approximate value for large samples ($n \geq 30$). If X is not normally distributed and σ is unknown, then the error bound $2s / \sqrt{n}$ is an approximation that may require larger sample sizes than 30, but we will use the $n \geq 30$ rule in this case as well. In any event, the error bound supplies us with at least an approximate measure of the quality of our estimate.

☐ 9.3.2 Population proportion π

The best estimator of the population proportion π is the statistic $\mathbf{P} = \mathbf{X}/n$, where $\mathbf{X}$ is the number of successes in n trials of an experiment that produces either a success or a failure on each trial.

Since $E(\mathbf{P}) = \pi$ from Theorem 8.5 in Chapter 8, the estimator $\mathbf{P}$ is unbiased. It is also minimum variance; the variability of the sampling distribution of $\mathbf{P}$ is less than the variability of any competing unbiased estimator of π. The standard error of the estimator $\mathbf{P}$ is $\sqrt{\pi(1 - \pi)/n}$, because $\sigma_{\mathbf{P}} = \sqrt{\pi(1 - \pi)/n}$ from Theorem 8.5 (for an infinite population).

If the sample size is large ($n \geq 100$), then the sampling distribution of $\mathbf{P}$ may be approximated by the normal distribution so that the interval $\pi \pm 2\sqrt{\pi(1 - \pi)/n}$ will contain *approximately* 95 percent of the values of $\mathbf{P}$, as illustrated in Figure 9.11.

FIGURE 9.11 Sampling distribution of $\mathbf{P}$

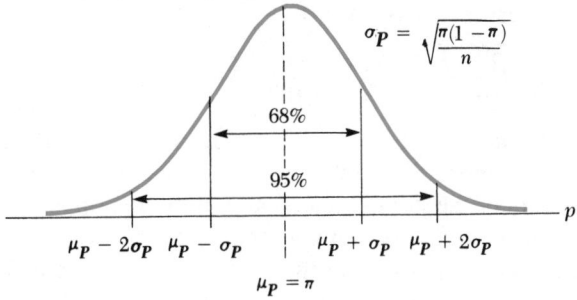

$$\sigma_{\mathbf{P}} = \sqrt{\frac{\pi(1 - \pi)}{n}}$$

68%

95%

$\mu_{\mathbf{P}} - 2\sigma_{\mathbf{P}} \quad \mu_{\mathbf{P}} - \sigma_{\mathbf{P}} \qquad \mu_{\mathbf{P}} + \sigma_{\mathbf{P}} \quad \mu_{\mathbf{P}} + 2\sigma_{\mathbf{P}}$

$\mu_{\mathbf{P}} = \pi$

Since the standard error $\sqrt{\pi(1 - \pi)/n}$ depends on π, it must be estimated before it can be used to give us an idea of how good the estimate p is of π. The estimated standard error is given by $\sqrt{p(1 - p)/n}$; π is estimated by the sample proportion p within the square root.

Thus, the quantity $2\sqrt{p(1 - p)/n}$ will be used as a measure of how well p estimates π in a particular sample.

Definition 9.3

Point estimator of the population proportion π

The point estimator of the population proportion π is the statistic $\mathbf{P} = \mathbf{X}/n$, where $\mathbf{X}$ is the number of successes in a sample containing n sampled population units.

The standard error of the estimator $\mathbf{P}$ is given by

Definition 9.3 (continued)

$$\sigma_P = \sqrt{\frac{\pi(1 - \pi)}{n}} \quad \text{estimated by} \quad \acute{\sigma}_P = \sqrt{\frac{p(1 - p)}{n}}$$

An indication of the "goodness" of the estimator P of π is provided by giving twice the value of the estimated standard error of the estimator $[2\sqrt{p(1 - p)/n}]$, where $n \geq 100$.

Example 9.2   A manufacturing process producing an electronic component is known to produce some defective components, the exact proportion of which is unknown. A random sample of 100 components is selected from the production lot, and it is found that eight are defective. Estimate the population proportion of defectives and provide a measure of its "goodness."

Solution   There are $n = 100$ trials of a binomial experiment, and $x = 8$ defectives ("successes") have been encountered. The population proportion of defectives ("successes") π is point estimated by the sample proportion of defectives ("successes"), $p = x/n = 8/100 = 0.08$. The estimated standard error of the estimator is given by

$$\sqrt{\frac{p(1 - p)}{n}} = \sqrt{\frac{(0.08)(0.92)}{100}} = 0.027$$

Twice the standard error is 0.054. Thus, we are reasonably confident that our estimate $p = 0.08$ is no more than 0.054 unit from the true value of π.

The point estimators of the population mean μ and the population proportion π, and their error bounds are summarized in Table 9.1.

We must use caution in interpreting the error bounds given in Table 9.1. These bounds provide an idea of how close the estimate is to the value of the

---

TABLE 9.1   Point estimates of μ and π

Population parameter	Point estimate*	Error bound on the estimate	Required sample size n
μ (mean)	$\bar{x} = \dfrac{\sum\limits_{i=1}^{n} x_i}{n}$ (sample mean)	$\dfrac{2\sigma}{\sqrt{n}}$ (if σ known) $\dfrac{2s}{\sqrt{n}}$ (if σ unknown)	If **X** is normally distributed, any sample size. If **X** is not normally distributed, $n \geq 30$
π (proportion)	$p = \dfrac{x}{n} = \dfrac{\text{No. of successes}}{\text{No. of trials}}$ (sample proportion)	$2\sqrt{\dfrac{p(1 - p)}{n}}$	$n \geq 100$

* $\bar{X}$ and **P** are UMV estimators.

population parameter, but they are based on the normal distribution (95.44 percent of the values will be within plus or minus two standard deviations of their mean). The only case in Table 9.1 where this strictly applies is in estimating the population mean μ when the population random variable X is normally distributed and the population standard deviation σ is known. For all other cases, the error bounds should be used only for "large" samples, and the error bound probabilities are then only approximate.

A better way to qualify the "goodness" of an estimator is to use a confidence interval, an inference method related to the error bounds. We will develop the confidence interval estimators of the population mean and the population proportion later in this chapter.

□ 9.3.3 Population variance σ^2 and standard deviation σ

The best estimator of the population variance σ^2 is the sample variance statistic S^2, with a value in a specific sample of n values $x_1, x_2, \ldots, x_n$ given by:

$$s^2 = \frac{\sum_{i=1}^{n} (x_i - \bar{x})^2}{n - 1}$$

The sample variance S^2 is an unbiased estimator of σ^2 $[E(S^2) = \sigma^2]$, and the sampling distribution of S^2 exhibits less variability than does any other unbiased estimator of σ^2. Thus, S^2 is a UMV estimator of σ^2.

The fact that S^2 is unbiased means that if we took many repeated samples of size n from the population, the values of S^2 computed in each sample will average out to σ^2. This is the reason the sum of the squared deviations in the formula for s^2 is divided by $(n - 1)$ instead of n. If the divisor were n instead, the expected value of the statistic,

$$\frac{\sum_{i=1}^{n} (X_i - \bar{X})^2}{n}$$

would not be σ^2—this estimator would be biased.

The estimator of the population standard deviation σ most frequently used is S, the sample standard deviation statistic that is the square root of the variance statistic S^2. The statistic S is a slightly biased estimator of σ $[E(S) \neq \sigma]$. The amount of the bias decreases, however, as the sample size increases, which is comforting to know whenever we use a biased estimator. The reason the biased estimator S is used to estimate the population standard deviation σ is its convenient form; it is simply the square root of the statistic S^2.

We will not give error bounds for estimating the population variance and standard deviation due to their complex forms. Table 9.2 gives the point estimates of these two population parameters.

TABLE 9.2 Point estimates of σ^2 and σ

Population parameter	Point estimate*
σ^2 (variance)	$s^2 = \dfrac{\sum\limits_{i=1}^{n}(x_i - \bar{x})^2}{n-1}$ (sample variance)
σ (standard deviation)	$s = \sqrt{\dfrac{\sum\limits_{i=1}^{n}(x_i - \bar{x})^2}{n-1}}$ (sample standard deviation)

* S^2 is a UMV estimator.

Example 9.3 The lifetime of an electronic component used in a computer is known to be approximately normally distributed. The population mean lifetime and standard deviation of the lifetimes are unknown, however. A random sample of $n = 100$ components is selected, and the lifetime of each is measured by using each component until it fails. The average lifetime of the 100 components was found to be 492 hours with a standard deviation of 50 hours.

Find point estimates of the population standard deviation of the lifetimes and the population mean lifetime, and set an error bound on the estimate of the mean.

Solution Let the lifetime of the components be denoted by the random variable **X**. The point estimate of the population standard deviation σ is given by $s = 50$ hours. The point estimate of μ is given by the sample mean $\bar{x} = 492$ hours. The standard error of the estimator is given by $\sigma_{\bar{x}} = \sigma/\sqrt{n}$, which must be estimated because σ is unknown. The estimate of $\sigma_{\bar{x}}$ is $s/\sqrt{n} = 50/\sqrt{100} = 5$ hours, where s is the sample standard deviation. We can be reasonably certain that our point estimate of μ is no more than $2s/\sqrt{n} = 2(5) = 10$ hours from the true value of μ.

■ 9.4 INTERVAL ESTIMATION

☐ 9.4.1 Population mean μ

An interval estimator of μ may be developed from the sampling distribution of the statistic $\overline{X}$. If we assume that $\overline{X}$ is normally distributed, then 95 percent of the values of $\overline{X}$ will be contained within two standard deviations—more correctly, 1.96 standard deviations—of the mean of the sampling distribution of the statistic $\overline{X}$, as illustrated in Figure 9.12. In

FIGURE 9.12 Sampling distribution of $\overline{X}$

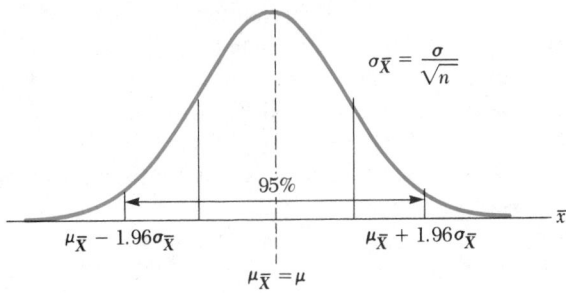

terms of a probability statement, the probability that a value of $\overline{X}$ will fall in this interval may be expressed as

$$P(\mu - 1.96\sigma_{\bar{x}} \leq \overline{X} \leq \mu + 1.96\sigma_{\bar{x}}) = 0.95$$

This probability statement may be "solved" for μ, producing the probability statement:

$$P(\overline{X} - 1.96\sigma_{\bar{x}} \leq \mu \leq \overline{X} + 1.96\sigma_{\bar{x}}) = 0.95$$

That is, the interval $\overline{X} \pm 1.96\sigma_{\bar{x}}$ will contain the value of μ with a probability of 0.95. We used this argument in developing the "goodness" of point estimators in the previous section.

If we compute the lower bound $\overline{X} - 1.96\sigma_{\bar{x}}$ and the upper bound $\overline{X} + 1.96\sigma_{\bar{x}}$ on μ in repeated samples of size n as illustrated in Figure 9.13, then in 95 percent of the samples the computed interval would contain the value of μ, whereas the intervals computed in 5 percent of the samples would not contain μ.

FIGURE 9.13 Interval estimates computed in repeated samples of size n

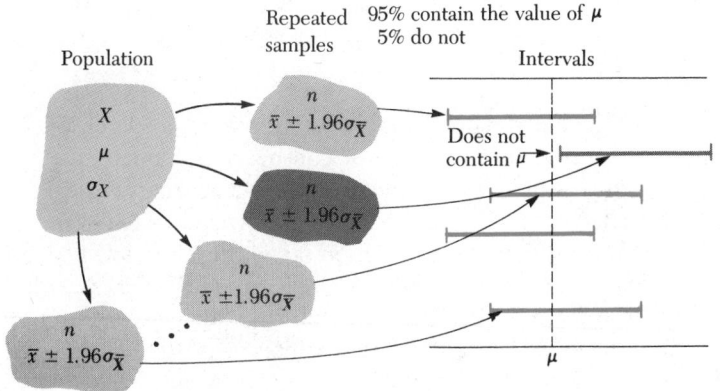

The interval $\bar{x} \pm 1.96\sigma_{\bar{x}}$ is called a *confidence interval* with "confidence" 95 percent. The confidence of 95 percent means that 95 percent of all samples of size n drawn from the population will produce intervals $\bar{x} \pm 1.96\sigma_{\bar{x}}$ which contain the value of μ, while 5 percent of the samples will not. We refer to it as "confidence" because in practice we will be drawing only one sample, and our computed interval $\bar{x} \pm 1.96\sigma_{\bar{x}}$ either will contain the value of μ or will not contain the value of μ with probability one (with certainty). The 95 percent confidence measures the risk involved in the sampling process—we have a chance of 0.95 of securing one of the samples that will produce an interval containing the value of μ.

It should be apparent that we can specify whatever degree of confidence is desired for the confidence interval of μ. If we wish the confidence to be 90 percent, then we must move 1.64 standard deviations above and below the mean μ in the sampling distribution of $\overline{X}$. The 90 percent confidence interval is, therefore, $\bar{x} \pm 1.64\sigma_{\bar{x}}$.

We will now describe a method for constructing a confidence interval for μ with any specified degree of confidence. Define the *confidence coefficient* $(1 - \alpha)$ to be the desired confidence of the interval, measured as a proportion (0.95, 0.90, etc.), and define $z_{\alpha/2}$ to be the value in the standard normal table (Table B.3 in Appendix B) such that the area to the right of $z_{\alpha/2}$ is $\alpha/2$, as shown in Figure 9.14. Then the constructed confidence interval with confidence coefficient $(1 - \alpha)$ is given by

$$\bar{x} \pm z_{\alpha/2}\sigma_{\bar{x}}$$

FIGURE 9.14 Calculation of $z_{\alpha/2}$

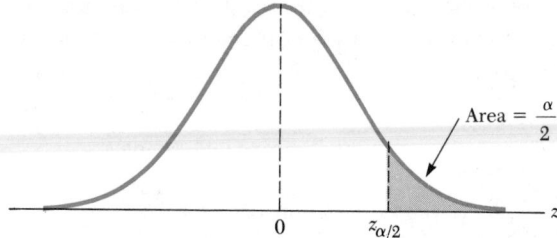

Example 9.4 Referring to Example 9.1, construct a 99 percent confidence interval for the population average of ball bearing diameters.

Solution In Example 9.1, a sample of $n = 25$ bearings was selected and it was found that the sample average diameter was $\bar{x} = 0.498$ inch. Furthermore, we were told that the population standard deviation σ is 0.005 inch.

The confidence interval formula for μ is given by $\bar{x} \pm z_{\alpha/2}\sigma_{\bar{x}}$, where $\sigma_{\bar{x}} = \sigma/\sqrt{n}$. We must first find the value of $z_{\alpha/2}$. Since we want a 99 percent confidence interval, the confidence coefficient is $0.99 = (1 - \alpha)$, so that $\alpha = 0.01$ and $\alpha/2 = 0.01/2 = 0.005$. Thus, we wish to find the value of Z on the standard normal distribution such that the area to the right of it is

FIGURE 9.15 Computation of $z_{\alpha/2}$ for a 99 percent confidence interval

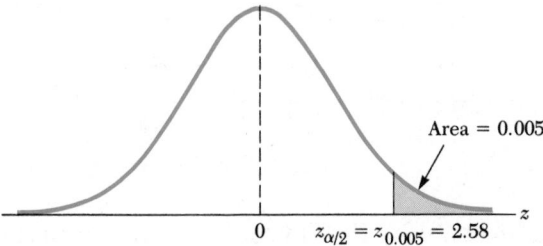

$$z_{\alpha/2} = z_{0.005} = 2.58$$

0.005 as shown in Figure 9.15. From Table B.3 in Appendix B, the value of $z_{\alpha/2} = z_{0.005}$ is 2.58. Thus, the interval is given by

$$\bar{x} \pm z_{0.005} \frac{\sigma}{\sqrt{n}} = 0.498 \pm 2.58 \frac{0.005}{\sqrt{25}}$$

$$= 0.498 \pm (2.58)(0.001) = 0.498 \pm 0.00258$$

The 99 percent confidence interval is $(0.498 - 0.00258)$ to $(0.498 + 0.00258)$. With rounding, this interval becomes 0.4954 to 0.5006.

The value of μ either is in this interval or is not (with certainty). We can only hope that our sample was one of the 99 percent of all possible samples of size $n = 25$ that produce intervals containing the true value of μ.

The most frequently chosen levels of confidence for interval estimates are 0.90, 0.95, and 0.99. In Table 9.3, $z_{\alpha/2}$ and intervals for these levels of confidence are given.

TABLE 9.3 Confidence intervals for μ

Confidence coefficient	$z_{\alpha/2}$	Lower bound	Upper bound
0.90	1.64	$\bar{x} - 1.64\sigma/\sqrt{n}$	$\bar{x} + 1.64\sigma/\sqrt{n}$
0.95	1.96	$\bar{x} - 1.96\sigma/\sqrt{n}$	$\bar{x} + 1.96\sigma/\sqrt{n}$
0.99	2.58	$\bar{x} - 2.58\sigma/\sqrt{n}$	$\bar{x} + 2.58\sigma/\sqrt{n}$

The width of the confidence interval for estimating μ is $2z_{\alpha/2}\sigma/\sqrt{n}$, as shown in Figure 9.16. There are two options for decreasing the width of this interval: (1) by decreasing the confidence coefficient, which decreases $z_{\alpha/2}$—see Table 9.3—or (2) by increasing the sample size.

That the width of the interval is a function of the sample size has intuitive appeal. As the sample size increases, the width of the confidence interval decreases and, taken to the limit as $n \to \infty$, the width of the interval approaches zero. But, the center of the interval is the sample mean $\bar{x}$, which approaches μ as n becomes larger. Therefore, as n becomes large, the confidence interval for μ converges on the value of μ.

FIGURE 9.16   Width of the interval estimate of μ

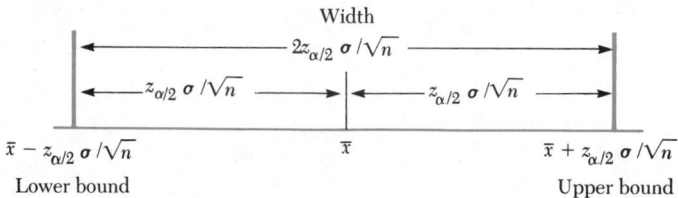

From Table 9.3 it should also be clear that a 100 percent confidence interval is not very meaningful. As the confidence coefficient increases, so does the value of $z_{\alpha/2}$ and, thus, so does the width of the interval. In fact, if we request a 100 percent interval, the answer is the entire real line from $-\infty$ to $+\infty$, which will contain the value of μ with certainty. In practice, of course, the interval would only have to extend over the entire range of values of the population variable.

The formula for the confidence interval for the population mean μ is now summarized.

Formula 9.1
Interval estimate of the population mean μ

If a sample of n values $x_1, x_2, \ldots, x_n$ has been randomly drawn from a population whose random variable **X** is normally distributed, a confidence interval with confidence coefficient $(1 - \alpha)$ for the population mean μ is given by

$$\bar{x} \pm z_{\alpha/2} \frac{\sigma}{\sqrt{n}}$$

where $\bar{x}$ is the sample mean, σ is the population standard deviation, and $z_{\alpha/2}$ is the value on the standard normal distribution such that the area under the normal curve to the right of it is $\alpha/2$.

If the population random variable **X** is not normally distributed, then this formula will provide an approximate confidence interval when n is large ($n \geq 30$ in most cases).

If the population standard deviation σ is not known, then a different formula for determining the confidence interval for μ must be used. It is described in Section 9.5.

□ **9.4.2   Population proportion π**

An approximate confidence interval estimate of the population proportion can be constructed by using the fact that, according to the central limit theorem, the statistic **P** will be approximately normally distributed for large samples. The form of the interval estimator will be:

$$\mathbf{P} \pm z_{\alpha/2}\sigma_P$$

From Section 9.3.2, the point estimator of π is the statistic **P**, and the standard deviation of **P** is given by

$$\sigma_P = \sqrt{\frac{\pi(1 - \pi)}{n}}$$

Since the standard deviation of **P** depends on the unknown value of π, it must be estimated by using p in place of π:

$$\hat{\sigma}_P = \sqrt{\frac{p(1 - p)}{n}}$$

The approximate interval estimate is given by

$$p \pm z_{\alpha/2}\sqrt{\frac{p(1 - p)}{n}}$$

Formula 9.2

Interval estimate of the population proportion π for large samples ($n \geq 100$)

An approximate interval estimate of the population proportion π is given by

$$p \pm z_{\alpha/2}\sqrt{\frac{p(1 - p)}{n}}$$

where p is the sample proportion of successes in n trials of a binomial experiment, and $z_{\alpha/2}$ is the value on the standard normal distribution such that $P(Z \geq z_{\alpha/2}) = \alpha/2$.

Example 9.5 Prior to the last presidential election, Herman was asked to construct a 96 percent confidence interval estimate of the proportion of registered voters who would vote for Slick Sam. Based on a random sample of 2,500 voters, he found that 60 percent said they would vote for Sam. What was the interval estimate of the population proportion π of registered voters voting for Sam that Herman constructed?

Solution The interval estimate formula is

$$p \pm z_{\alpha/2}\sqrt{\frac{p(1 - p)}{n}}$$

Since we want 96 percent confidence, $(1 - \alpha) = 0.96$, $\alpha = 0.04$, and $\alpha/2 = 0.02$. From Table B.3 of Appendix B, the value of z such that $P(Z \geq z) = 0.02$ is 2.05. Thus, $z_{\alpha/2} = z_{0.02} = 2.05$. Furthermore, from the sample information, $n = 2,500$ and $p = 0.60$. Thus, the interval is

$$0.60 \pm 2.05 \sqrt{\frac{0.60(1 - 0.60)}{2500}} \qquad \text{or} \qquad 0.60 \pm 0.02$$

Therefore, the 96 percent confidence interval for π quoted by Herman is 0.58 to 0.62.

9.5 INTERVAL ESTIMATION OF μ WITH AN UNKNOWN POPULATION VARIANCE

In the preceding section, the error bound and the confidence interval for the population mean μ have been based on the assumption that the standardized random variable, $(\overline{X} - \mu)/\sigma_{\overline{x}}$, is approximately distributed as the standardized normal variable Z. If the standard deviation, $\sigma_{\overline{x}}$, is unknown or if the sample size is small ($n < 30$), then the standardized variable will *not* be approximately normally distributed. In this section, a method for constructing a confidence interval for μ when $\sigma_{\overline{x}}$ is unknown will be developed. When the sample size is small (usually, $n < 30$), alternative procedures must be used *whether or not* $\sigma_{\overline{x}}$ is known. The one exception to this is when the population random variable X is normally distributed, in which case the method in this section is appropriate regardless of sample size. Since, in practice, it may be difficult to determine if X is normally distributed, different estimation methods must be used in the small sample case. Some of these methods are presented in Chapter 20.

If the population random variable X is normally distributed, then we know that the statistic $\overline{X}$ will also be normally distributed. Therefore, the standardized random variable

$$Z = \frac{\overline{X} - \mu_{\overline{x}}}{\sigma_{\overline{x}}}$$

is a standard normal variable. If the standard deviation of $\overline{X}$ ($\sigma_{\overline{x}} = \sigma/\sqrt{n}$) must be estimated by $s/\sqrt{n}$, then the standardized variable

$$t = \frac{\overline{X} - \mu_{\overline{x}}}{S/\sqrt{n}}$$

will no longer be a standard normal variable. The sampling distribution of t is called the *Student's* t-distribution if the population random variable X and, hence, $\overline{X}$ are normally distributed. The t-distribution was first reported by Gossett in 1908, who published the result under the pseudonym "Student."

The t-distribution is more "spread out" than the normal distribution, because by using s in place of σ, more uncertainty is introduced. The form of the distribution is a function of the sample size n. The fewer the sample values, the more spread out the distribution becomes, as indicated in Figure 9.17.

The distribution of t is tabulated according to its "degrees of freedom" rather than the sample size n. The degrees of freedom of the t-distribution are determined by the divisor of the estimated standard deviation in the t-statistic. In the t-statistic,

FIGURE 9.17 Comparison of the normal distribution with the t-distribution for $n = 2$, 5, and 10

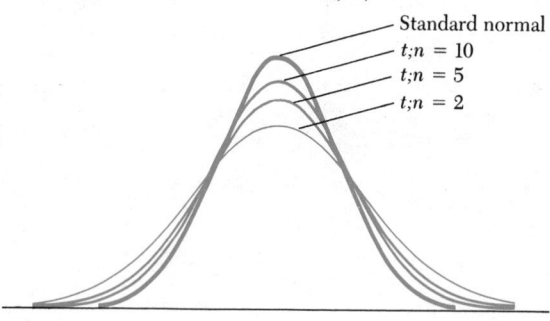

Standard normal
$t;n = 10$
$t;n = 5$
$t;n = 2$

$$t = \frac{\bar{X} - \mu_{\bar{x}}}{S/\sqrt{n}}$$

the sample standard deviation value $s = \sqrt{\sum_{i=1}^{n} (x_i - \bar{x})^2/(n - 1)}$ has $(n - 1)$ in its denominator under the radical. Hence, this t-statistic has $(n - 1)$ degrees of freedom (df).

The t-distribution is tabulated in Table B.5 in Appendix B. It is a symmetric distribution about the value $t = 0$ and Table B.5 gives the values of **t** corresponding to upper-tail areas of 0.10, 0.05, 0.025, 0.01, and 0.005. Notice that as the sample size increases, the t-values converge on the standard normal values corresponding to the same tail areas. Table 9.4 shows the convergence of the **t**-values for an upper-tail probability of 0.025. When the sample size reaches $n = 31$ $(df = n - 1 = 30)$, the **t**-value is quite close to $z = 1.96$.

If the population standard deviation σ is unknown and must be estimated, the interval estimator of μ is changed by using a **t**-value instead of the standard normal value $z_{\alpha/2}$. The interval estimate is:

$$\bar{x} \pm t_{\alpha/2;n-1} \frac{s}{\sqrt{n}}$$

where $t_{\alpha/2;n-1}$ is defined below.

TABLE 9.4 Value of **t** corresponding to an upper-tail probability of 0.025

df	Value of t
1	12.706
10	2.228
20	2.086
30	2.042
120	1.980
$+\infty$	1.960

> **Formula 9.3**
> Interval estimate of the population mean μ for small samples with unknown standard deviation σ
>
> The confidence interval estimate of the population mean μ with a confidence coefficient of $(1 - \alpha)$, based on a random sample of n values $x_1, x_2, \ldots, x_n$ when the population standard deviation σ is unknown, is given by:
>
> $$\bar{x} \pm t_{\alpha/2;n-1} \frac{s}{\sqrt{n}}$$
>
> where $\bar{x}$ is the sample mean, s is the sample standard deviation, n is the sample size, and $t_{\alpha/2;n-1}$ is the value on the t distribution with $n - 1$ degrees of freedom (df) such that the area to the right of it is $\alpha/2$.
>
> An assumption for this interval estimate is that the population random variable **X** is normally distributed, or very nearly so, if n is small (less than 30).

Example 9.6 The length of time required for persons taking the civil service PACE test is assumed to be normally distributed. A random sample of 16 persons taking the test is conducted and their test times are recorded, yielding an average test time of 60 minutes with a standard deviation of 12 minutes. Find a 95 percent confidence interval for the population mean test time μ.

Solution The population standard deviation is unknown. Therefore, the confidence interval formula using the t-distribution should be used:

$$\bar{x} \pm t_{\alpha/2;n-1} \frac{s}{\sqrt{n}}$$

For a 95 percent confidence interval, the confidence coefficient is $(1 - \alpha) = 0.95$, so that $\alpha = 0.05$ and $\alpha/2 = 0.025$. The degrees of freedom are $df = n - 1 = 16 - 1 = 15$. Thus, the value of $t_{0.025;15}$ is 2.13. The required interval is

$$60 \pm 2.13 \frac{12}{\sqrt{16}} \quad \text{or} \quad 60 \pm 6.39$$

Thus, we are 95 percent confident that the population mean test time μ is between 53.61 and 66.39 minutes.

If the standard normal distribution is used in place of the t-distribution, the value of $z_{\alpha/2}$ for a 95 percent interval is 1.96 and the resulting interval (from Section 9.4.1) is

$$60 \pm 1.96 \frac{12}{\sqrt{16}} \quad \text{or} \quad 60 \pm 5.88$$

This interval extends from 54.12 to 65.88 minutes, a shorter interval than the correct one determined above. The fact that the interval based on the t-distribution is wider than the one based on the standard normal distribution reflects that more uncertainty has been introduced due to the estimation of σ by the sample standard deviation s.

In Example 9.6, we must assume that the length of time required to take the exam is normally distributed. If it is not, the t-distribution does not strictly apply because the sample is small. Some procedures for dealing with small samples and nonnormal population random variables are given in Chapter 20.

■ 9.6 INTERVAL ESTIMATION OF π WITH SMALL SAMPLES

In Section 9.4.2, an approximate confidence interval estimator of π based on the standardized variable

$$\frac{\mathbf{P} - \pi}{\sqrt{\dfrac{\mathbf{P}(1 - \mathbf{P})}{n}}}$$

was given. This confidence interval estimator, $\mathbf{P} \pm \mathbf{Z}_{\alpha/2} \sqrt{\mathbf{P}(1 - \mathbf{P})/n}$, should be used only if n is large (100 or more), because the exact sampling distribution of $\mathbf{P}$ is the discrete binomial distribution:

$$\begin{aligned} P(\mathbf{P} = p) &= P(\mathbf{X} = np = x) \\ &= C_x^n \pi^x (1 - \pi)^{n-x}, \qquad x = 0, 1, \ldots, n. \end{aligned}$$

That is, the probability that the statistic $\mathbf{P}$ is equal to a specific value p is equal to the probability that the number of successes $\mathbf{X}$ in the sample is equal to np (because $\mathbf{P} = \mathbf{X}/n$); call this number of successes x. Then the probability of x successes in the sample can be determined from the binomial mass function given earlier.

It is possible to construct confidence intervals for π directly from the binomial distribution, but the computations required are considerable, especially for large samples. Therefore, the normal approximating confidence interval for π is used if the sample size is large. When the sample is less than 100, the binomial distribution has been used by statisticians to construct exact confidence interval bands, which are usually given in tables.

An example of these bands is shown in Table 9.5 for 95 percent confidence intervals. The table is very easy to use and gives exact interval estimates for π for sample sizes $n = 10, 15, 20, 30, 50, 100, 250,$ and $1,000$.

For example, suppose we want a 95 percent interval estimate of π based on a sample of $n = 20$ and a sample proportion $p = 0.60$. The value 0.60 is first located at the bottom of the table. The vertical line corresponding to $p = 0.60$ is followed up through the bands until it intersects with the $n = 20$ bands. At the two points of intersection, the value of $\hat{\pi}$ is read off by moving to the left vertical axis. The lower bound on π is approximately 0.36 and the upper bound is approximately 0.82.

Extensive tables giving the confidence interval bands for many different

TABLE 9.5 95 percent confidence interval bands for the population proportion π

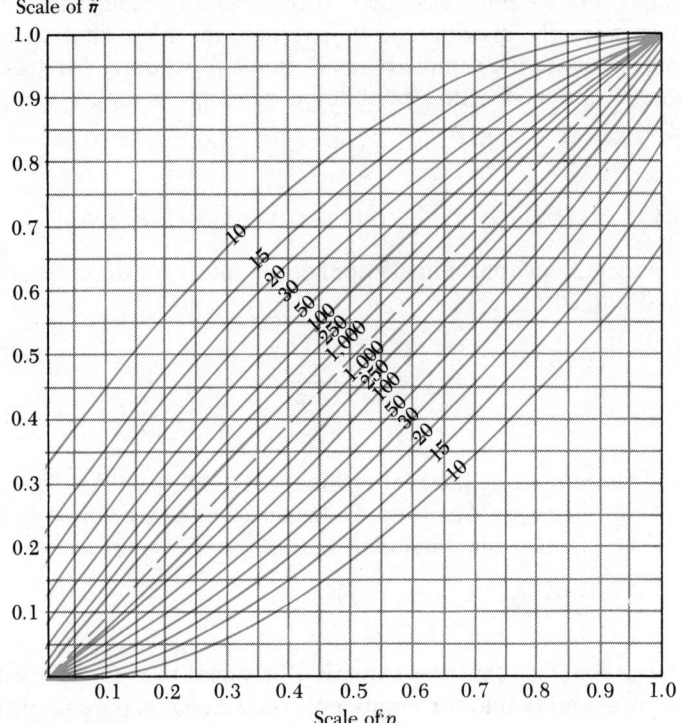

Source: This chart is reproduced with the permission of Professor E. S. Pearson from C. J. Clopper and E. S. Pearson, "The Use of Confidence or Fiducial Limits Illustrated in the Case of the Binomial," *Biometrika*, vol. 26 (1934), p. 404.

levels of confidence and sample sizes have been compiled and are available in most mathematics libraries.

9.7 CONFIDENCE BOUNDS ON μ AND π

In some instances, it is desirable to state either an upper or a lower bound on a population parameter rather than an interval. For example, we may wish to state that we are 95 percent confident that μ *does not exceed* a specified value. This is an example of setting an *upper confidence bound* on μ. The upper and lower confidence bounds are determined in a way similar to how the confidence interval is set. The difference is that we are giving only one end of the confidence interval, with the appropriate change in the determination of the Z or t value in the formula for the bound. Since only one "end" of the confidence interval is given, we do not split α into $\alpha/2$ for the

TABLE 9.6   Confidence bounds for μ and π

Parameter	Lower $100(1 - \alpha)$ percent confidence bound	Upper $100(1 - \alpha)$ percent confidence bound	Conditions
μ.........	$\bar{x} - z_\alpha \dfrac{\sigma}{\sqrt{n}}$	$\bar{x} + z_\alpha \dfrac{\sigma}{\sqrt{n}}$	σ is known, $n \geq 30$ unless X is normally distributed*
	$\bar{x} - t_{\alpha;n-1} \dfrac{s}{\sqrt{n}}$	$\bar{x} + t_{\alpha;n-1} \dfrac{s}{\sqrt{n}}$	σ is unknown, $n \geq 30$ unless X is normally distributed†
π.........	$p - z_\alpha \sqrt{\dfrac{p(1-p)}{n}}$	$p + z_\alpha \sqrt{\dfrac{p(1-p)}{n}}$	n is large (usually 100 or more)

Note: z_α is given by P $(\mathbf{Z} > z_\alpha) = \alpha$; $t_{\alpha;n-1}$ is given by P $(t > t_{\alpha;n-1}) = \alpha$.
* If X is normally distributed, the formula applies to any sample size $(n \geq 1)$.
† If X is normally distributed, the formula applies when $n > 1$.

determination of $\mathbf{Z}$ or t. The error α is placed entirely in one end of the "interval" for a confidence bound. The formulas for the upper and lower confidence bounds on μ and π are given in Table 9.6.

Example 9.7   A small car rental agency is interested in purchasing a fleet of cars of a new diesel model sold by the WV automobile company. One condition for purchasing the fleet of cars of this model is that they produce *no less than* an average of 30 miles per gallon in city driving. To test whether this condition is met, the WV company provides the following data. In a recent test of 30 cars of this model in city driving, the cars produced an average of 32 miles per gallon with a standard deviation of 4 miles per gallon. Should the rental agency purchase the cars of this model?

Solution   It is decided to determine a 99 percent lower confidence bound on μ, the population average miles per gallon for this car model. By doing so, it will be possible to make the statement: "We are 99 percent confident that the population average miles per gallon for this model is *not less than* the lower confidence bound." Since σ is unknown, the t-distribution should be used in setting the bound. The appropriate t-value with $(n - 1) = 29$ degrees of freedom is 2.462 from Table B.5 in Appendix B. The lower confidence bound is:

$$\bar{x} - t \cdot \frac{s}{\sqrt{n}} = 32 - 2.462 \cdot \frac{4}{\sqrt{30}} = 30.2$$

Therefore, we are 99 percent confident that μ, the population average miles per gallon, is not less than 30.2. Accordingly, the rental agency should purchase the fleet of cars of this model.

■ 9.8 CONFIDENCE INTERVAL ESTIMATORS OF THE POPULATION
VARIANCE σ^2 AND THE POPULATION STANDARD DEVIATION σ

The confidence interval estimator of the population variance σ^2 is based on the sampling distribution of the statistic S^2. The sampling distribution of the statistic S^2 has been worked out by standardizing this random variable in a manner similar to the Z standardization. The standardized random variable

$$\chi^2 = \frac{(n-1)S^2}{\sigma^2}$$

is called a *chi-square* random variable if the population random variable X is normally distributed. The sampling distribution of χ^2 is called the chi-square distribution. It is an asymmetric distribution skewed to the right and is defined for positive real numbers. The distribution of χ^2, like the t-distribution, depends on the sample size. Three members of the chi-square family of distributions are illustrated in Figure 9.18. As the sample size increases, the form of the chi-square distribution approaches the normal distribution. The $n = 100$ chi-square distribution is distinctly more like the normal distribution than is the $n = 10$ chi-square distribution.

FIGURE 9.18 Three members of the chi-square family of distributions

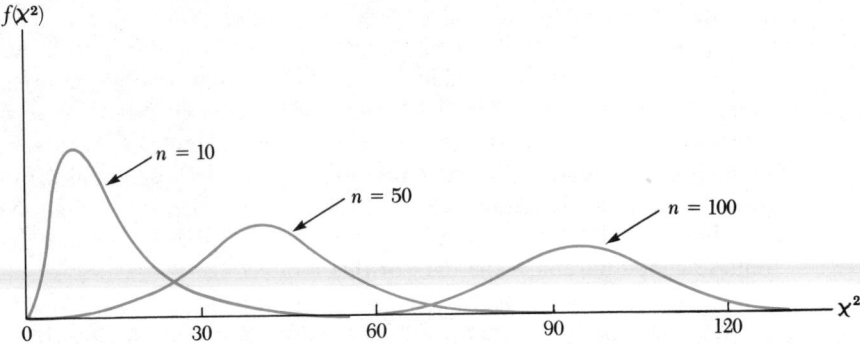

The tabulated values of the chi-square distribution are based on the degrees of freedom, $df = n - 1$, as are the tabulated t-distribution values. Table B.6 in Appendix B gives the values of χ^2. The table is split into two parts—the left page gives the lower-tail values of χ^2 and the right page gives the upper-tail values of χ^2. Thus, Table B.6 gives:

Lower tail: $P(\chi^2 \leq \chi^2) = \alpha$
Upper tail: $P(\chi^2 \geq \chi^2) = \alpha$

The chi-square distribution may be used to construct an interval estimator for σ^2 by solving the probability statement

$$P\left(\chi^2_{L;n-1} \le \frac{(n-1)S^2}{\sigma^2} \le \chi^2_{U;n-1}\right) = 1 - \alpha$$

for σ^2, where $\chi^2_{L;n-1}$ and $\chi^2_{U;n-1}$ are the lower and upper values on the χ^2 distribution with $(n-1)$ degrees of freedom which locate half of α in each tail. The resulting interval estimate is:

$$\frac{(n-1)s^2}{\chi^2_{U;n-1}} \le \sigma^2 \le \frac{(n-1)s^2}{\chi^2_{L;n-1}}$$

Formula 9.4
Interval estimate of the population variance σ^2

An interval estimate of the population variance σ^2 with confidence coefficient $(1 - \alpha)$ is given by

$$\frac{(n-1)s^2}{\chi^2_{U;n-1}} \le \sigma^2 \le \frac{(n-1)s^2}{\chi^2_{L;n-1}}$$

where s^2 is the sample variance computed on the basis of n randomly drawn values from a population that is normally distributed, and where $\chi^2_{L;n-1}$ and $\chi^2_{U;n-1}$ are the lower and upper χ^2 values which locate half of α in each tail of the chi-square distribution with $(n-1)$ degrees of freedom.

Example 9.8 A machine designed to fill soap boxes with 16 ounces of soap is supposed to fill in such a way that the standard deviation of the fills is no more than 0.1 ounce. If the machine operates properly, the variance of the fills must be less than $(0.1)^2$ ounce^2. A sample of 20 soap boxes is selected from a lot produced by this machine; the sample variance s^2 is 0.015. Find a 90 percent confidence interval for the population variance σ^2.

Solution The confidence interval formula for σ^2 is

$$\frac{(n-1)s^2}{\chi^2_{U;n-1}} \le \sigma^2 \le \frac{(n-1)s^2}{\chi^2_{L;n-1}}$$

For a 90 percent confidence interval, $(1 - \alpha) = 0.90$ so that $\alpha = 0.10$ and $\alpha/2 = 0.05$. Thus, we wish to find the values on the χ^2 distribution with $n - 1 = 19$ degrees of freedom so that 5 percent of the area is in each tail. From Table B.6, the values of $\chi^2_{L;19}$ and $\chi^2_{U;19}$ are 10.1 and 30.1, respectively. The resulting interval is, therefore,

$$\frac{(19)(0.015)}{30.1} \le \sigma^2 \le \frac{(19)(0.015)}{10.1} \qquad \text{or} \qquad 0.009 \le \sigma^2 \le 0.028$$

Thus, we are 90 percent confident that the true variance of the fills lies between 0.009 and 0.028 squared ounces.

An exact confidence interval formula for the standard deviation σ for small samples will not be given, because both the development and use of the exact form of the interval are quite complicated. An approximate interval

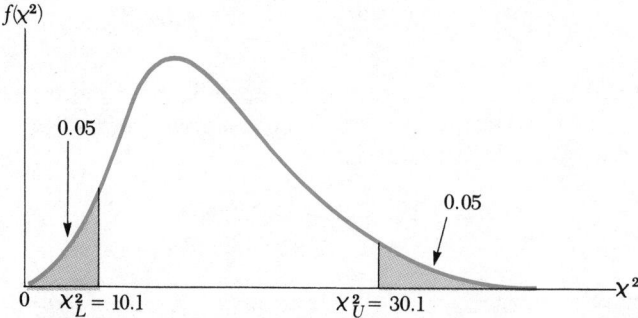

estimate of σ is given by taking the square roots of the upper and lower bounds of the population variance interval estimate. This interval has intuitive appeal, because we are simply taking square roots of the limits given for the interval for σ^2, and it is sufficiently accurate for our purposes. For example, in Example 9.8, an approximate 90 percent confidence interval estimate of σ is given by

$$0.095 = \sqrt{0.009} \le \sigma \le \sqrt{0.028} = 0.167$$

■ 9.9 PLANNING A STATISTICAL EXPERIMENT

As we have seen in this chapter, the sample size is very important in inference making. The sample size plays a major role in determining how "good" our inferences are in a given problem. It certainly makes sense to select the size of the sample *before* the sample is drawn from the population, on the basis of how good we wish the inferences about the population to be.

The determination of the sample size is part of the design of a statistical experiment. Elements of designing an experiment are as follows.

1. *Stating the problem.* The statement should include a careful description of the population, the characteristics of interest, and the parameters about which inferences will be drawn.
2. *Conducting a pilot study.* A pilot study serves two purposes. First, it gives us an idea of whether or not the problem stated in step 1 can be "solved." Often, on the basis of a pilot study, it is necessary to return to step 1 to redefine variables and the like. Second, the pilot study will provide an estimate of the population standard deviation σ, a required quantity in the determination of sample size in step 3. By careful planning of the pilot study, the measurements can be part of the principal study in many cases.
3. *Selecting the sample size and the statistical methods to be used.*
4. *Drawing the sample units.*
5. *Analyzing the results.*
6. *Drawing conclusions on the basis of the statistical analysis.*
7. *Reporting the results.*

Mathematical statisticians have spent a great deal of time developing ways to determine the sample size that achieves a desired level of "goodness" of the inferences. We shall look at two of these results that apply to the estimation of the population mean μ and the population proportion π.

☐ 9.9.1 Determination of sample size; μ estimation

In determining the sample size in a statistical experiment, we must know two things:

1. How close we wish our estimate to be to the true value of the population parameter.
2. How certain we wish to be that our estimate will be within the selected number of units of the value of the parameter.

For example, we may request that our point estimate of the population mean μ not be more than ten units away from the true value of μ with a confidence of 95 percent. By this, we mean that we are willing to take a 5 percent chance that the sample we draw of size n will produce an estimate $\bar{x}$ of μ that will be more than ten units away from the actual value of μ.

We can use the sampling distribution of the statistic $\overline{X}$ to determine what size the sample must be to meet these requirements. We know from the sampling distribution of $\overline{X}$ shown in Figure 9.19 that the interval $\mu \pm 2\sigma_{\bar{x}}$ contains approximately 95 percent of the values of the statistic $\overline{X}$.

FIGURE 9.19 Sampling distribution of the statistic $\overline{X}$

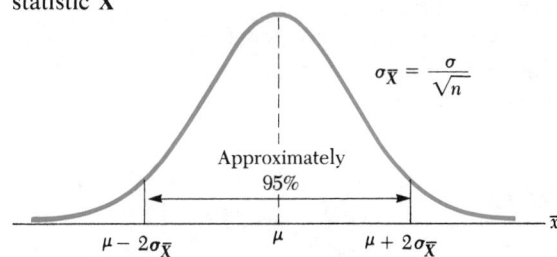

$$\sigma_{\overline{X}} = \frac{\sigma}{\sqrt{n}}$$

Approximately 95%

$\mu - 2\sigma_{\overline{X}}$ μ $\mu + 2\sigma_{\overline{X}}$

If we wish to be no more than A units from μ with our estimate $\bar{x}$, then we must set $2\sigma_{\bar{x}} = A$, or

$$2\,\frac{\sigma}{\sqrt{n}} = A, \qquad \text{or for } n, \quad n = \frac{4\sigma^2}{A^2}$$

This formula presumes a confidence coefficient of $(1 - \alpha) = 0.9544$. If we want a confidence coefficient of $(1 - \alpha)$, then we may set

$$z_{\alpha/2}\sigma_{\overline{X}} = A \qquad \text{or} \qquad z_{\alpha/2}\,\frac{\sigma}{\sqrt{n}} = A$$

which results in the formula

$$n = \frac{z_{\alpha/2}^2 \sigma^2}{A^2} = \left[\frac{z_{\alpha/2}\sigma}{A}\right]^2$$

Since σ is usually not known, an estimate of σ must be used. Notice that the sample standard deviation s cannot be used, because no sample has yet been drawn—we are trying to determine the sample size.

Many times, a rough estimate of σ will be available, usually based on past experience with the population random variable **X**. Alternatively, the fact that the range of the values of many random variables will be approximately 4σ can frequently be used to provide a rough estimate of σ.

In any event, *some* estimate of σ must be provided to use this sample size formula. Most often, a pilot study must be used to provide the estimate.

Formula 9.5
Sample size for estimating the population mean μ

If it is desired to estimate the population mean μ so that the estimate $\bar{x}$ will be no more than A units from the true value of μ with confidence coefficient $(1 - \alpha)$, and $\bar{X}$ is approximately normally distributed, then n should be chosen such that

$$n \geq \left[\frac{z_{\alpha/2}\sigma}{A}\right]^2$$

where $z_{\alpha/2}$ is the value on the standard normal distribution such that $P(Z \geq z_{\alpha/2}) = \alpha/2$ and σ is the population standard deviation.

Example 9.9 Past experience has indicated that the salaries of factory workers in a certain industry are approximately normally distributed with a standard deviation of $500. How large a sample of factory workers would be required if we wish to estimate the population mean salary μ to within $60 with a confidence of 99 percent?

Solution In the sample size formula,

$$n \geq \left[\frac{z_{\alpha/2}\sigma}{A}\right]^2$$

$\sigma = \$500$, $A = \$60$, and $z_{\alpha/2} = 2.58$ (from Table B.3). Thus,

$$n \geq \left[\frac{(2.58)(500)}{60}\right]^2 = 462.25$$

We would require a sample size of at least $n = 463$ factory workers to produce a point estimate of μ with the required degree of precision.

☐ **9.9.2 Determination of sample size; π estimation**

The same argument may be applied to the population proportion π. The point estimator of π is the statistic **P**, which has a standard deviation of

$\sqrt{\pi(1 - \pi)/n}$. If we wish to be within A units of π with our estimate p with confidence coefficient $(1 - \alpha)$, then we must set

$$z_{\alpha/2} \sqrt{\frac{\pi(1 - \pi)}{n}} = A, \quad \text{or for } n, \quad n = \left[\frac{z_{\alpha/2}\sqrt{\pi(1 - \pi)}}{A}\right]^2$$

Since the value of π is unknown, this formula appears to result in a deadend. However, it can be used to provide an approximate bound on n by noting that $\pi(1 - \pi)$ can be at most 0.25—this occurs when $\pi = 0.5$ $(0 \le \pi \le 1)$. Thus, we can replace $\pi(1 - \pi)$ by its largest value, giving

$$n \ge \left[\frac{z_{\alpha/2}(0.50)}{A}\right]^2$$

Formula 9.6

Sample size determination for estimating the population proportion π

If it is desired to estimate the population proportion π so that the estimate p will be no more than A units from the true value of π with confidence coefficient $(1 - \alpha)$, and it is assumed that **P** is approximately normally distributed, then n must be chosen so that

$$n \ge \left[\frac{z_{\alpha/2}(0.50)}{A}\right]^2$$

where $z_{\alpha/2}$ is the value on the standard normal distribution such that $P(Z \ge z_{\alpha/2}) = \alpha/2$.

Example 9.10 Herman wishes to take a random sample of n registered voters who will vote in an upcoming district election. If he wishes the estimate of the proportion of the n voters who will vote for his candidate, Slick Sam, to be no more than 0.005 unit away from the true proportion, what sample size is required if the degree of confidence is set at 0.99?

Solution If we assume that the value of π will be close to 0.5, then

$$n \ge \left[\frac{(2.58)(0.50)}{0.005}\right]^2 = 66{,}564$$

Thus, a sample size of at least $n = 66{,}564$ voters is required.

If it is known that π will be around 0.20, then $\pi(1 - \pi) = (0.20)(0.80) = 0.16$. In this case, a better estimate of n is given by

$$n \ge \left[\frac{(2.58)(0.40)}{0.005}\right]^2 = 42{,}601$$

Notice that *some* knowledge of the value of π prior to the sampling experiment can be very helpful in reducing the size of the sample required to satisfy the stated precision requirements. Notice that Formulas 9.5 and 9.6 require that $\overline{X}$ and **P**, respectively, are approximately normally distributed. This is because the sample size formulas are based on the normal distribution. Thus, if the formulas produce a value of n that is less than 30 for estimating μ

or less than 100 for estimating π, the sample size values *should not be used* that result from these formulas. In those instances, the determined sample size will be too small for the central limit theorem to apply, and other methods of determining the sample size and analyzing the data must be used (some suitable methods are given in Chapter 20).

◼ 9.10 SUMMARY

In this chapter, we have developed the statistical inference method of estimating the value of a population parameter.

There are two types of statistical estimation: point estimation and interval estimation. A point estimator produces one value or point, the estimate, for a particular sample set $x_1, x_2, \ldots, x_n$. As a measure of how good the estimate is, the estimated value of two standard deviations on the sampling distribution of the point estimator is usually given together with the estimate. An interval estimate produces an interval on the real line that will contain the value of the population parameter with a prescribed level of confidence, usually taken to be 90, 95, or 99 percent.

The formulas for both types of estimation procedures arise from the sampling distribution of sample statistics. The point estimator of a given population parameter is the sample statistic with a sampling distribution that has a mean equal to the value of the population parameter and has less variability than the sampling distribution of any other sample statistic under consideration as a point estimator of the population parameter. A point estimator with

TABLE 9.7   Confidence interval formulas for μ and π

Parameter	Formula	Conditions	Nature of interval
μ	$\bar{x} \pm z_{\alpha/2} \dfrac{\sigma}{\sqrt{n}}$	σ is known, X is normally distributed	Exact for any sample size
	$\bar{x} \pm z_{\alpha/2} \dfrac{\sigma}{\sqrt{n}}$	σ is known, X is not normally distributed, but $n > 30$	Approximate
	$\bar{x} \pm t_{\alpha/2;\, n-1} \dfrac{s}{\sqrt{n}}$	σ is unknown, X is normally distributed	Exact for $n > 1$
	$\bar{x} \pm t_{\alpha/2;\, n-1} \dfrac{s}{\sqrt{n}}$	σ is unknown, X is not normally distributed, but $n > 30$	Approximate
π	$p \pm z_{\alpha/2} \sqrt{\dfrac{p(1-p)}{n}}$	σ_P is unknown, $n > 100$	Approximate

Note: For other cases, such as $n < 30$, X not normally distributed and population variance unknown in estimating μ, other methods should be used. Some of these are given in Chapter 20.

a sampling distribution that possesses these two properties is called an *unbiased* and *minimum variance* point estimator.

The interval estimator is almost always based on the sampling distribution of the best point estimator of a given population parameter.

The appropriate confidence intervals for μ and π depend on the size of the sample, the distribution of the population random variable X, and the knowledge of the population variance. Table 9.7 summarizes the confidence interval formulas for μ and π.

An important step in the inference process is determining the sample size—the number of units to be drawn from the population. If it is possible to state the maximum difference between the value of the population parameter and the estimate based on the sample information that will be tolerated and the level of confidence for which this condition will be satisfied, then a rough determination of the sample size is possible. Formulas for determining the sample size when estimating the population mean μ and the population proportion π are given in the chapter.

■ REFERENCES

Ostle, B., and Mensing, R. *Statistics in Research*. 3d ed. Ames: Iowa State University Press, 1975.

Wonnacott, T., and Wonnacott, R. *Introductory Statistics for Business and Economics*. 2d ed. New York: John Wiley & Sons, Inc., 1977.

■ PROBLEMS

9.1. Distinguish between a *parameter* and a *statistic*.

9.2. Distinguish between an *estimator* and an *estimate*.

9.3. What are the two basic ways to draw statistical inferences about a population parameter? How do the two approaches differ?

9.4. What are the two types of statistical estimation? How do the types differ?

9.5. Discuss the criteria for selecting a point estimator of a population parameter.

9.6. Why is a confidence interval estimate usually preferred to a point estimate of a population parameter?

9.7. Why is the sample average statistic preferred to the sample median statistic as a point estimator of the population mean?

9.8. When we form a 95 percent confidence interval on the population mean, what is meant by "95 percent confidence"?

9.9. Under what circumstances is it appropriate to use the *t*-distribution instead of the *z*-distribution in constructing a confidence interval on the population mean?

9.10. Why is the determination of the sample size important in designing statistical experiments?

9.11. A report stated that in a study of a school district's attendance record of its students, a 90 percent confidence interval on the average number of school days missed by each student per year is three to nine days. Two improper interpretations of this statement follow. Explain why each is incorrect.

    a. Ninety percent of the students in

the population miss from three to nine days of school each year.

b. If many random samples are taken, 90 percent of them would produce sample means from three to nine days.

9.12. If, in setting a confidence interval on a population mean, the sample size is doubled, holding the confidence coefficient and the population standard deviation constant, what is the effect on the width of the confidence interval?

9.13. A random sample of 100 starting salaries of MBA graduates employed by financial institutions in New York City produced an average starting salary of $24,000 with a standard deviation of $2,500. Find point estimates of the average starting salary and standard deviation of all MBA graduates working in financial institutions in New York City. Indicate a measure of the "goodness" of your estimate of the mean.

9.14. At a large university, the issue of faculty unionization has become critical. One hundred faculty members are randomly sampled and asked whether or not they favor unionization. Forty say they do. Find a point estimate of the proportion of all faculty members at the university who favor unionization. Indicate a measure of "goodness" of your estimate.

9.15. In a large federal agency, the number of employees with ten years or more of service is 3,000. In a sample of 25 of these employees, the numbers of years of service are:

10, 22, 11, 10, 14, 16, 12, 24, 15, 12, 12, 30, 14, 10, 15, 14, 18, 26, 21, 11, 10, 15, 14, 20, 13

a. Find an estimate of the average number of years of service for these 3,000 employees. Determine an error bound on this esti-

mate. Does the error bound have much credibility in this case? Explain.

b. Find an estimate of the standard deviation of the number of years of service for these 3,000 employees.

c. Estimate the proportion of employees among these 3,000 who have served 20 years or more.

9.16. A sample of 16 observations has been taken from a population in which the random variable is normally distributed. The sample mean is 50 and the sample standard deviation is 10.

a. Determine a 95 percent confidence interval on the population mean.

b. Interpret this interval.

9.17. A machine produces a miniature amplifier capable of producing a maximum of 1.5 watts of power. The amplifier is designed to be used in the space packs of astronauts exploring the moon and other bodies in our solar system. It powers certain sensing devices necessary for the astronauts' survival. Based on prior experience in making this amplifier, it is known that the standard deviation of the maximum power of the amplifiers is 0.1 watt. A random sample of 16 amplifiers is selected from the production line, and it is found that the average maximum power for the 16 is 1.45 watts. Assume that the maximum output power of the amplifier is normally distributed.

a. Construct a 95 percent confidence interval for the average maximum power output among all amplifiers produced by this machine.

b. Is the rated maximum output of 1.5 watts contained in your interval estimate?

c. Based on parts a and b, what would you recommend regarding the production process if the machine is supposed to be producing parts with an average maximum output of 1.5 watts?

9.18. A car rental agency is concerned about the time required to complete a reservation process by telephone by its customers. In a random sample of 36 calls, the average time required to complete the transaction is 4 minutes with a standard deviation of 1 minute.

a. Find a 99 percent confidence interval estimate of the population mean telephone transaction time, using the formula $\bar{x} \pm z_{\alpha/2}s/\sqrt{n}$.

b. Find a 99 percent confidence interval estimate of the population mean, using the formula $\bar{x} \pm t_{\alpha/2;n-1}s/\sqrt{n}$.

c. Compare these two interval estimates. Which is more appropriate? Why?

9.19. A random variable is known to be normally distributed with mean $\mu = 50$, but unknown variance. A random sample of 72 values of the variable is taken, and it is found that the sample variance is 80. Construct a 95 percent confidence interval on the population variance.

9.20. A random sample of 100 teletype operators indicates that their salaries fluctuate quite a bit. The sample standard deviation of their daily salaries is $10. Construct a 90 percent confidence interval on the population standard deviation of the daily salaries.

9.21. The proportion of bad widgets produced in a factory is unknown. A random sample of 100 widgets is selected, and ten bad widgets are found. Determine a 95 percent confidence interval estimate of the proportion of bad widgets by the following:

a. Using normality theory.

b. Using the confidence bands in this chapter.

c. Which confidence interval is more appropriate? Why?

9.22. A statistician is asked to conduct a survey to determine an estimate of the proportion of people who favor the recall of a local politician. He is told that his estimate should not differ from the true proportion by more than 2 percent, with 95 percent confidence. How large should his random sample be to produce an estimate of the proportion satisfying this condition?

9.23. A consumer group is concerned about the truthfulness of a certain producer in stating the net weight of a box of SNUR cereal. It is decided that n boxes of SNUR will be purchased and the sample mean weight used to decide whether or not to press charges against the producer of SNUR. The boxes claim to contain a net weight of 12 ounces. It is decided that the sample size n should be made large enough to ensure that the sample average weight differs from the population average weight by no more than 0.1 ounce with 99 percent confidence. A very conservative guess at the population standard deviation, based on a few boxes of SNUR, is 0.8 ounce. How large should n be?

9.24. Tests on a popular brand of paint gave the following results on the square feet of coverage per gallon: 150, 159, 162, 144, 150, 162, 140, 155, 164, 157. Assuming that the coverages are normally distributed, determine a 95 percent confidence interval on the population average square feet coverage.

9.25. Past experience shows that the standard deviation of the yearly income of garment workers in a certain state is $400. How large a sample of garment workers would one need to take to estimate the population mean to within $50 with a probability of 0.95 of being correct?

9.26. In the next presidential election, it is desired to estimate π, the proportion voting for the Democratic candidate, to within 0.01 unit with a probability of 0.99 of being correct. The estimate is to be made five days prior to the national election. How large a sample size of registered voters who will vote is required?

9.27. A manufacturer claims that the average lifetime of a filter used in a coolant system for nuclear power plants will be at least three months in continuous use before it fails and needs to be replaced. To test this claim, a federal agency places 36 filters in simulated coolant systems and finds that the average time to failure of the filters is 2.8 months with a standard deviation of 0.6 month.

  a. Find a 95 percent upper confidence bound on μ, the average lifetime of the filters.

  b. Based on the upper confidence bound determined in part a, is the manufacturer's claim plausible? Explain.

9.28. Two Democratic candidates, Harry and David, are involved in a runoff election to determine who will run against the Republican candidate in the general election. The campaign manager for Harry takes a random sample of 100 voters who claim they will vote in the runoff election and finds that 42 claim they will vote for Harry.

  a. Find a 90 percent upper confidence bound for π, the proportion of voters who will vote for Harry in the runoff election.

  b. Based on the upper confidence bound determined in part a, does Harry stand a reasonable chance to win the runoff election? Explain.

Statistical inference: Hypothesis testing

■ 10.1 INTRODUCTION

In Chapter 9, we introduced confidence interval estimation as a form of statistical inference. We will now introduce hypothesis testing as a form of statistical inference and show that it is closely related to confidence interval estimation.

The general framework we shall use to develop the method of hypothesis testing is illustrated in Figure 10.1. A value of the population parameter θ is hypothesized, a random sample is collected, and the value of the point estimator $\hat{\theta}$ in the sample is used to determine whether or not the hypothesized value of θ, say θ_0 is reasonable.

FIGURE 10.1 Sampling process in statistical hypothesis testing

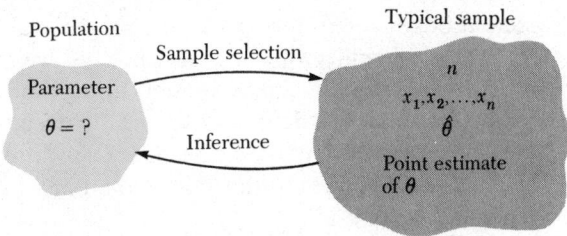

There are three basic forms of the hypothesis statement about θ:

$$
\begin{array}{lll}
\text{Form I:} & H_0: & \theta = \theta_0 \\
& H_A: & \theta \neq \theta_0 \\
\text{Form II:} & H_0: & \theta \geq \theta_0 \\
& H_A: & \theta < \theta_0 \\
\text{Form III:} & H_0: & \theta \leq \theta_0 \\
& H_A: & \theta > \theta_0
\end{array}
$$

In these three forms, the symbol H_0 is called the *null hypothesis*, the symbol H_A is called the *alternate hypothesis*, and θ_0 is the *conjectured* value of θ. It is the null hypothesis, H_0, that we will either reject or not reject on the basis of the sample information in our hypothesis-testing experiment.

We will now consider two examples of constructing a statistical hypothesis.

Example 10.1   The Crunchy Cereals Company has been approached concerning a new packaging machine designed and built by the Ace Packaging Corporation. According to the Ace sales representative, the new machine will package 12-ounce cereal boxes twice as fast as the existing equipment used by the Crunchy people. In addition, the new machine is purportedly less expensive to operate than the old one. The Ace sales representative agrees to allow Crunchy to use the new machine without cost on a one-month trial basis. Of primary concern to the manager of Crunchy is the ability of the new machine to produce an average fill of 12 ounces in the cereal boxes packaged by the machine. During the trial period, he wishes to test the hypothesis that the average fill will indeed be 12 ounces. What is the form of the statistical hypothesis?

Solution   The population parameter of interest is the population mean μ, the average fill for the population of cereal boxes. The manager hypothesizes that μ will assume the value of 12 ounces. Thus, the null and alternate hypotheses are:

$$
\begin{array}{ll}
H_0: & \mu = 12 \quad (\mu_0 = 12) \\
H_A: & \mu \neq 12
\end{array}
$$

In Example 10.1, the population may be conceptualized as all boxes filled by the machine when it is set for the 12-ounce filling weight during the operating cycle of the machine prior to wear out or a required maintenance period. The sample used to test the null hypothesis would be composed of a random selection of cereal boxes filled during the one-month trial period.

In this filling process, the variability of the cereal box fill weight would normally be of equal importance to the mean fill weight. It might be hypothesized, for example, that the standard deviation of the fill weights is 0.4 ounce. This conjecture concerns the population standard deviation σ. The null and alternate hypotheses of interest in this case would be:

$$
\begin{array}{ll}
H_0: & \sigma = 0.4 \quad (\sigma_0 = 0.4) \\
H_A: & \sigma \neq 0.4
\end{array}
$$

The manager for Crunchy would use the sample of selected cereal boxes' fill weights to test both the hypothesis concerning the mean fill weight μ and the hypothesis concerning the standard deviation of the fill weights σ.

Example 10.2   Herman, the entrepreneur, produces plumbing fixtures in a large factory as one of his many sources of income. One of his better-selling items is a device that produces steam (hot air, mostly!) when connected to a standard bathtub/shower unit. This unit converts an enclosed shower to a sauna for home use. For the device to work efficiently, it must increase the water pressure sufficiently to produce steam from the hot water outlet. Herman has determined that, on the average, the pressure must be increased by at least four times the ordinary hot water pressure to produce sufficient steam for the shower unit to serve as a sauna. The device is designed and manufactured accordingly. As a quality control check, Herman wishes to test the hypothesis that the proportion of devices produced in the manufacturing process that will raise the water pressure by at least four times the ordinary hot water pressure is at least 0.96. That is, the process will produce less than 4 percent defective units. What is the form of the statistical hypotheses?

Solution   The population parameter of interest is the population proportion π, where π represents the proportion of devices produced that are nondefective (devices that will raise the water pressure by at least four times the ordinary hot water pressure). In this example, a range of values of π is being hypothesized in the null hypothesis. The null and alternate hypotheses are:

$$H_0: \quad \pi \geq 0.96 \quad (\pi_0 = 0.96)$$
$$H_A: \quad \pi < 0.96$$

This is form II of the hypothesis statement about π. Herman hopes that the null hypothesis cannot be rejected based on sample information (a random sample of devices selected and tested from a production lot).

In those instances in which the null hypothesis involves a range of values of the parameter (forms II and III), it is necessary to decide which way the inequality symbol in the null hypothesis is directed ($\geq$ or $\leq$). We will see later in this chapter how to determine whether form II or form III is appropriate.

Sample information is used to decide whether or not the null hypothesis (H_0) is rejected. The decision is *always* made about the null hypothesis. Before explaining how the decision rule to reject or not to reject the null hypothesis based on sample information is determined, we must understand that there are four possible outcomes resulting from our decision regarding the null hypothesis—two are favorable outcomes and two are not.

These four outcomes are illustrated in Table 10.1. If the null hypothesis is true (column 1), we may either reject H_0 or not reject H_0 based on the sample information. If we reject H_0 when it is true, we have committed an error—it is called the *Type I error*. If we do not reject H_0 when it is true, we have made a correct decision.

TABLE 10.1 States of nature and the decisions in statistical hypothesis testing

		State of nature: Null hypothesis H_0	
		H_0 is true	H_0 is false
D e c i s i o n	Reject H_0	Incorrect decision Type I error	Correct decision
	Do not reject H_0	Correct decision	Incorrect decision Type II error

If the null hypothesis is false (column 2), we may either reject H_0 or not reject H_0 based on the sample information. If we reject H_0 when it is false, we have made a correct decision. If we do not reject H_0 when it is false, we have committed an error—it is called the *Type II error.*

Understandably, we will want to know what our chances are of making either type of error in a given hypothesis-testing experiment. The measurements of the chance of committing these two types of error are denoted by α and β where $\alpha = P(\text{Type I error})$ and $\beta = P(\text{Type II error})$.

Definition 10.1
Type I error

A *Type I error* in a statistical hypothesis-testing experiment is committed by rejecting the null hypothesis when it is true. The probability of committing a Type I error is denoted by α where

$$\alpha = P(\text{Type I error})$$
$$\alpha = P(\text{rejecting } H_0 / H_0 \text{ is true})$$

Definition 10.2
Type II error

A *Type II error* in a statistical hypothesis-testing experiment is committed by not rejecting the null hypothesis when it is false. The probability of committing a Type II error is denoted by β where

$$\beta = P(\text{Type II error})$$
$$\beta = P(\text{not rejecting } H_0 / H_0 \text{ is false})$$

The relationship of α and β to various choices concerning whether to reject or not to reject the null hypothesis is indicated in Table 10.2.

Two points concerning Table 10.2 are important in understanding the

TABLE 10.2 Probabilities of making a correct and an incorrect decision in a statistical test of a hypothesis

		State of nature	
		H_0 true	H_0 false
D e c i s i o n	Reject H_0	α	$1 - \beta$
	Do not reject H_0	$1 - \alpha$	β
	Total probability	1	1

statistical process that will be developed in this chapter to test hypotheses. First, since one of two alternatives must be selected in testing a statistical hypothesis (to reject or not to reject the null hypothesis), by complementary arguments, each *column* in Table 10.2 must sum to 1. That is:

P (reject H_0/H_0 is true) + P (do not reject H_0/H_0 is true) = $\alpha + (1 - \alpha) = 1$
P (reject H_0/H_0 is false) + P (do not reject H_0/H_0 is false) = $(1 - \beta) + \beta = 1$

Second, in examining each *row* in the table, we see that a Type I error can be made *only if* we reject the null hypothesis and that a Type II error can be made *only if* we do not reject the null hypothesis.

■ 10.2 THE STEPS IN THE CLASSICAL HYPOTHESIS-TESTING EXPERIMENT

There are six basic steps in testing a statistical hypothesis:

1. The statement of the null and alternate hypotheses.
2. The selection of a sample statistic on which to base the decision to reject or not to reject the null hypothesis. This statistic is called the *test statistic*.
3. The selection of α, the probability of committing a Type I error. In hypothesis testing, the selected value of α is called the *significance level* of the test.
4. The construction of the decision rule used to decide whether or not the null hypothesis H_0 should be rejected. The decision rule is a function of the test statistic chosen in step 2 and of the significance level selected in step 3.
5. The selection of the sample units and the computation of the value of the test statistic.
6. The decision to reject or not to reject the null hypothesis based on the decision rule and the value of the test statistic.

To illustrate the six steps of hypothesis testing, we will consider an example in which a hypothesis concerning μ is tested.

Example 10.3 Suppose we hypothesize that a population mean μ is equal to 10. Then,

STEP 1

$$H_0: \quad \mu = 10 \quad (\mu_0 = 10)$$
$$H_A: \quad \mu \neq 10$$

Furthermore, let us assume that the population random variable **X** is normally distributed and that its standard deviation σ is equal to 20.

In selecting a sample statistic upon which to base the decision to reject or not to reject the null hypothesis concerning the population mean μ, it seems reasonable to use the sample mean statistic $\overline{X}$ to estimate μ, since it is an unbiased, minimum variance estimator of μ. In general, we will use as the test statistic the unbiased minimum variance point estimator of the population parameter involved in the hypothesis test.

STEP 2   The statistic $\overline{X} = \Sigma_{i=1}^{n} X_i / n$ is selected as the test statistic.

The next step in the hypothesis-testing experiment is the selection of α, the significance level of the test. Since α represents the probability of committing a Type I error (rejecting H_0 given H_0 is true), it would seem reasonable that we would want to select a small value of α. However, α and β (β = the probability of committing a Type II error—not rejecting H_0 given H_0 is false) are inversely related, as we will later demonstrate. That is, as α is made smaller in value, β increases, and vice versa. In classical hypothesis testing, α is usually chosen to be one of the following values: 0.10, 0.05, or 0.01. We will discuss how one of these values of α is chosen later. For the moment, let us suppose that $\alpha = 0.10$ has been selected as the significance level of the test.

STEP 3   We select $\alpha = 0.10$ for the significance level of the test.

FIGURE 10.2 Sampling distribution of $\overline{X}$ centered over the hypothesized value of μ, $\mu_0 = 10$

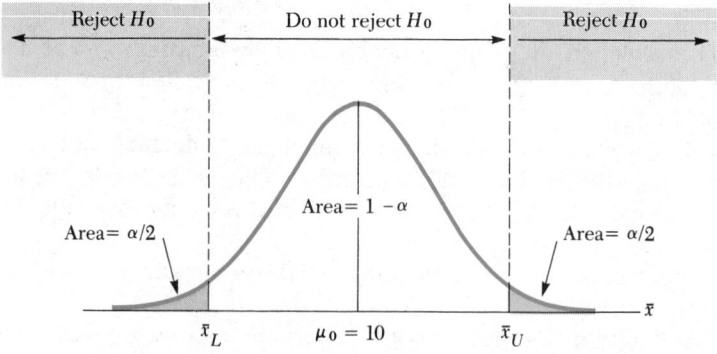

In constructing the decision rule for the form I hypothesis, two steps are required:

1. We assume that the null hypothesis is true. Thus, in our example, we assume that H_0: $\mu = 10$ is true. Since our test statistic $\overline{X}$ is an unbiased estimator of μ, this assumption implies that the sampling distribution of $\overline{X}$ is centered over $\mu_0 = 10$, as illustrated in Figure 10.2.
2. The *action limits* $\bar{x}_L$ and $\bar{x}_U$ are determined on the sampling distribution of $\overline{X}$ so that $P(\overline{X} < \bar{x}_L) = \alpha/2$ and $P(\overline{X} > \bar{x}_U) = \alpha/2$ as illustrated in Figure 10.2. If the computed value of $\overline{X}$ in our sample drawn from the population is less than $\bar{x}_L$ or is greater than $\bar{x}_U$, we will reject the null hypothesis that $\mu = \mu_0 = 10$.

Suppose, in our example, it is decided to randomly sample $n = 25$ units from the population. Following the preceding two steps, the determination of $\bar{x}_L$ and $\bar{x}_U$ is illustrated in Figure 10.3.

Notice that the determination of the action limits $\bar{x}_L$ and $\bar{x}_U$ requires finding values on a normal distribution such that $P(\overline{X} < \bar{x}_L/\mu = \mu_0 = 10, \sigma_{\bar{x}} = 4) = 0.05$ and $P(\overline{X} > \bar{x}_U/\mu = \mu_0 = 10, \sigma_{\bar{x}} = 4) = 0.05$. To find $\bar{x}_L$, for example, we must use the standardizing formula

FIGURE 10.3 Determination of the action limits $\bar{x}_L = 3.42$ and $\bar{x}_U = 16.58$ for Example 10.3

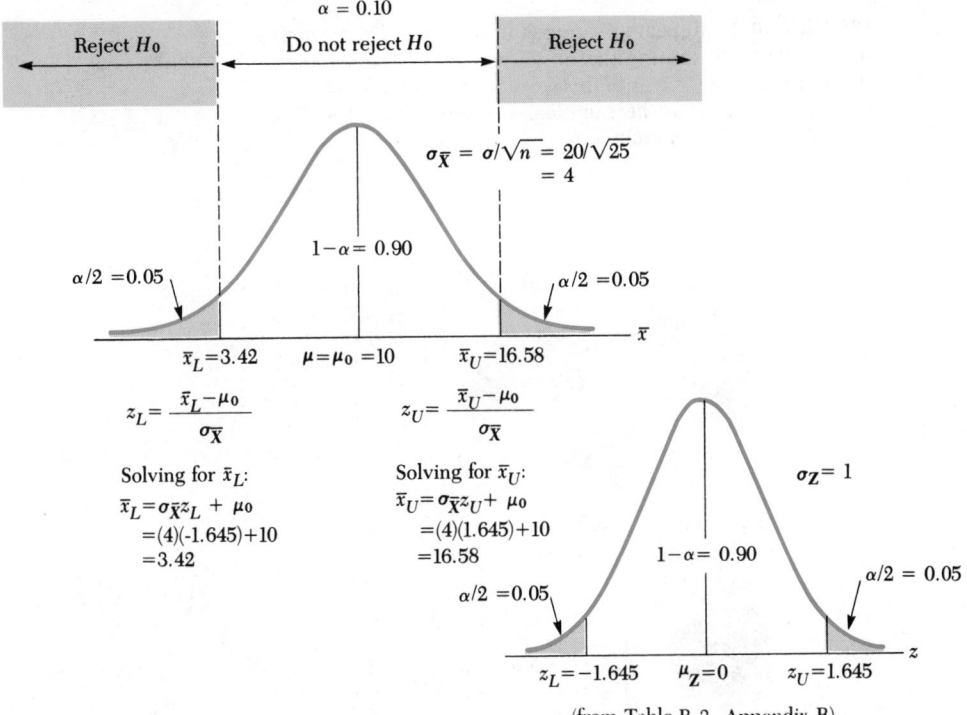

(from Table B.3, Appendix B)

$$z_L = \frac{\bar{x}_L - \mu_0}{\sigma_{\bar{x}}}$$

The solution from the standardizing formula is: $\bar{x}_L = \sigma_{\bar{x}} z_L + \mu_0$, where z_L is found from the standard normal distribution as illustrated in the lower portion of Figure 10.3.

STEP 4 The decision rule: reject H_0: $\mu = 10$ if $\bar{x} < \bar{x}_L = 3.42$ or if $\bar{x} > \bar{x}_U = 16.58$. Otherwise, do not reject H_0.

The reasoning behind the construction of this decision rule is quite simple. If the true value of μ is really $\mu_0 = 10$, the hypothesized value, we should expect that the statistic $\overline{X}$ will assume a sample value close to 10. Indeed, with α set at 0.10, the value of the statistic $\overline{X}$ should be between $\bar{x}_L = 3.42$ and $\bar{x}_U = 16.58$ with probability $(1 - \alpha) = 0.90$. If the value of the statistic $\overline{X}$ is less than $\bar{x}_L = 3.42$, it is very likely that the true value of μ is something less than $\mu_0 = 10$, because $P(\overline{X} < \bar{x}_L / \mu = \mu_0 = 10, \sigma_{\bar{x}} = 4) = 0.05$. Similarly, if the value of the statistic $\overline{X}$ is more than $\bar{x}_U = 16.58$, it is very likely that the true value of μ is something more than $\mu_0 = 10$, since $P(\overline{X} > \bar{x}_U / \mu = \mu_0 = 10, \sigma_{\bar{x}} = 4) = 0.05$. Figure 10.4 illustrates the nature of the sampling distribution of the test statistic $\overline{X}$ if $\mu = \mu_0 = 10$.

Based on the action limits specified in the decision rule (step 4), it may come as a surprise that we will *not* reject the null hypothesis that $\mu = 10$ if

FIGURE 10.4 Repeated samples of size $n = 25$, assuming that $\mu = \mu_0$, illustrating that in 90 percent of the samples, the sample mean $\bar{x}$ will be between $\bar{x}_L = 3.42$ and $\bar{x}_U = 16.58$, and in 10 percent of the samples, $\bar{x}$ will be either less than $\bar{x}_L = 3.42$ or greater than $\bar{x}_U = 16.58$

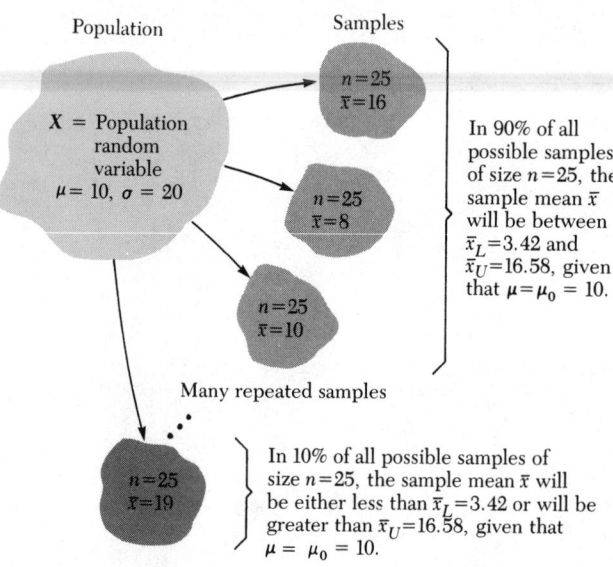

the sample mean $\bar{x}$ is contained *anywhere* in the interval from 3.42 to 16.58. That is, for example, if a sample mean value of $\bar{x} = 3.75$ is obtained in our sample of 25 units taken from the population, we will *fail* to *reject* the null hypothesis that $\mu = 10$. It might appear more reasonable to reject H_0: $\mu = 10$ if the sample mean is this far removed from $\mu = 10$. To see why we *may* not want to set the action limits at some values other than 3.42 and 16.58, say at $\bar{x}_L = 8$ and $\bar{x}_U = 12$, let us determine the probability that the sample mean $\bar{x}$ will be contained in the interval from 8 to 12, *given that* $\mu = 10$. This probability is given by:

$$P(8 \le \bar{X} \le 12/\mu = 10, \sigma_{\bar{x}} = 4) = P\left(\frac{8-10}{4} \le Z \le \frac{12-10}{4}\right)$$

$$= P(-0.50 \le Z \le 0.50) = 0.383$$

That is, if μ *is really* equal to 10, then *in only* 38.3 percent of the samples selected of size $n = 25$ will the sample mean $\bar{x}$ fall in the interval 8 to 12 or, in other words, in 61.7 percent of the samples of size $n = 25$, we will *reject* the null hypothesis when it is true (thereby committing a Type I error). The action limits in statistical hypothesis testing control the amount of error we are willing to assume in rejecting the null hypothesis when it is true. If we are willing to accept only a 10 percent chance ($\alpha = 0.10$) of making a Type I error, we should set the action limits in our statistical hypothesis test at $\bar{x}_L = 3.42$ and $\bar{x}_U = 16.58$ as specified.

Now, let us suppose we have taken our sample and find:

STEP 5 | The $n = 25$ sample units are drawn randomly from the population. Their values are:

0 −12 9 16 −12 45 21 0 5 34 −20 40 70

−10 0 8 20 41 32 50 18 25 28 19 48

For these data, $\bar{x} = 19$. Thus, the value of our test statistic $\bar{X}$ is 19.

Since $\bar{x} = 19 > \bar{x}_U = 16.58$, we would reject H_0: $\mu = 10$ and conclude that $\mu \ne 10$.

STEP 6 | Since $\bar{x} > \bar{x}_U = 16.58$, we reject the null hypothesis that $\mu = 10$ at the $\alpha = 0.10$ significance level and conclude that $\mu \ne 10$.

Our decision to reject H_0: $\mu = 10$ is based on simple logic. We are arguing that it is more likely that μ is different from 10 if $\bar{x} = 19$ than it is for $\mu = 10$, as illustrated in Figure 10.5. Of course, it may well be that $\bar{x} = 19$ came from the sampling distribution of $\bar{X}$ with $\mu = \mu_0 = 10$. But the probability of this occurrence is clearly less than 0.05. Hence, we conclude that it is more likely that $\mu \ne 10$ and reject H_0: $\mu = 10$.

The interpretation of the decision to reject H_0: $\mu = 10$ in our example is as follows: μ is either equal to 10 or is not equal to 10 (the state of nature). Since we have rejected H_0, either μ is not equal to 10 (and we have made a

FIGURE 10.5 More likely sampling distribution of the test statistic $\overline{X}$ if $\bar{x} = 19$

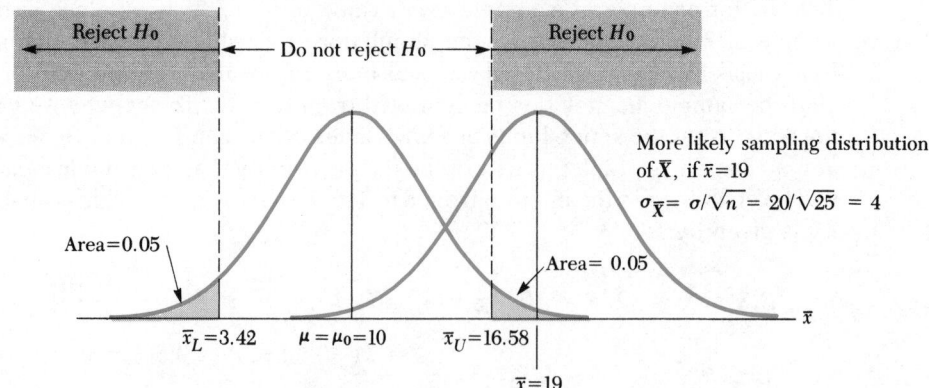

correct decision), or we obtained a sample (with probability less than or equal to $\alpha = 0.10$) in which the sample mean $\bar{x} = 19$ came from the sampling distribution of $\overline{X}$ with $\mu = \mu_0 = 10$ (and we have made an incorrect decision, committing a Type I error).

When the hypothesis test is of form I as in Example 10.3 (H_0: $\mu = 10$, H_A: $\mu \neq 10$), the statistical test is called a *two-sided* or *two-tailed* test, since there are two *rejection regions* that comprise the decision rule on the sampling distribution of the test statistic. One rejection region is in the lower tail of the sampling distribution of $\overline{X}(\bar{x} < \bar{x}_L)$, and the other rejection region is in the upper tail of the sampling distribution of $\overline{X}$ ($\bar{x} > \bar{x}_U$). In the interval between the action limits $\bar{x}_L$ and $\bar{x}_U$, we do not reject the null hypothesis.

We will next illustrate the application of the six steps of classical hypothesis testing for the form II hypothesis test.

Example 10.4 A chemical process is designed to produce at least 100 pounds of high-grade copper per production run. In a sample of 64 production runs, the average yield is $\bar{x} = 98$ pounds. Suppose it is known that the population standard deviation of the runs, σ, is equal to 10, and assume that the population random variable X = yield of copper in pounds per production run is normally distributed. Decide whether or not the null hypothesis, H_0: $\mu \geq 100$, should be rejected at the $\alpha = 0.05$ significance level.

Solution

| STEP 1 | The null and alternate hypotheses are: |

$$H_0: \quad \mu \geq 100 \quad (\mu_0 = 100)$$
$$H_A: \quad \mu < 100$$

Since inferences are being drawn about μ, $\overline{X}$ is the appropriate test statistic.

| STEP 2 | The test statistic is $\overline{X}$. |

STEP 3   The significance level of the test is set at $\alpha = 0.05$.

The decision rule is established by assuming H_0: $\mu \geq 100$ is true. Since there is a range of values of μ that satisfy the condition $\mu \geq 100$, we must select one value within this range to establish the sampling distribution of $\overline{X}$ under the assumption that H_0: $\mu \geq 100$ is true. By convention, we will select $\mu_0 = 100$, the value of μ that *minimally* satisfies the inequality, $\mu \geq 100$. By doing so, the selected value of $\alpha = 0.05$ will represent the *maximum* probability of committing a Type I error, as we shall demonstrate later.

Figure 10.6 illustrates the determination of the decision rule for this hypothesis-testing experiment. Notice that there is only one rejection region—in the lower tail of the sampling distribution of $\overline{X}$. If μ is really equal to or greater than 100, it may be possible to obtain a sample value of $\overline{X}$ that is less than $\mu_0 = 100$, but if the value of $\overline{X}$ is too much less than $\mu_0 = 100$, we will be inclined to reject H_0: $\mu \geq 100$. The action limit turns out to be 97.94, as indicated in Figure 10.6.

FIGURE 10.6   Determination of the decision rule for the hypothesis test in Example 10.4.

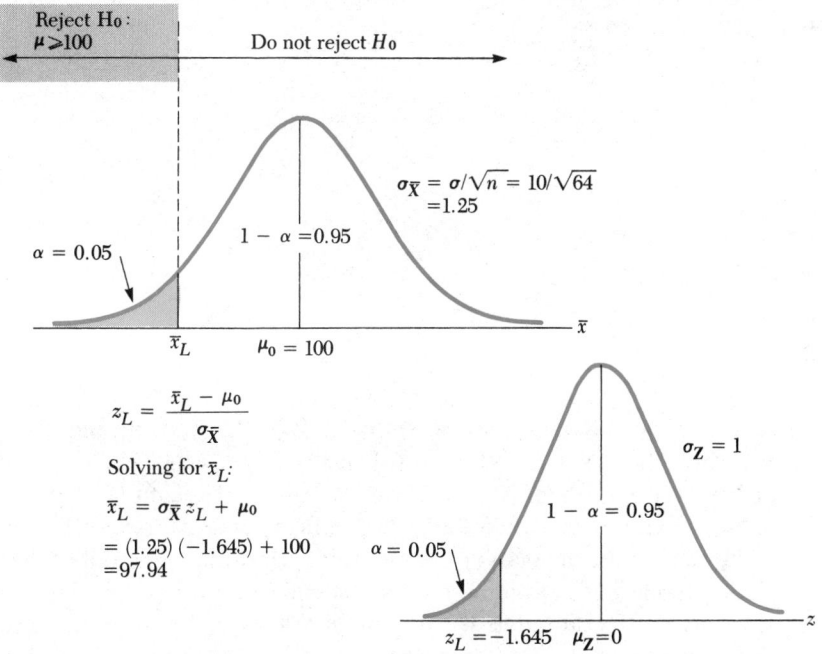

STEP 4   Reject H_0: $\mu \geq 100$ if $\bar{x} < 97.94$. Otherwise, do not reject H_0.

STEP 5   The value of the test statistic is $\bar{x} = 98$, determined from a sample of size $n = 64$.

STEP 6	Since $\bar{x} = 98 \not< 97.94$, do not reject H_0: $\mu \geq 100$.

Our decision to not reject H_0: $\mu \geq 100$ may be interpreted as follows: since $\bar{x} = 98$ did not fall in the rejection region at the $\alpha = 0.05$ significance level, we must conclude that a value of $\overline{X}$ greater than or equal to 98 is reasonably likely if indeed $\mu \geq 100$. Thus, the sampled mean of $\bar{x} = 98$ provides *insufficient evidence* to reject H_0: $\mu \geq 100$.

When the hypothesis test is of form II (as in Example 10.4) or form III, the statistical test is called a *one-sided* or *one-tailed test,* since there is only one rejection region in the decision rule (the left tail for form II and the right tail for form III). The complementary region to the rejection region is the region where the null hypothesis is not rejected.

In the one-sided test, such as in Example 10.4, setting the significance level of the test represents specifying the *maximum* probability of committing a Type I error. To see this, suppose that in Example 10.4, μ is really 101 so that the null hypothesis H_0: $\mu \geq 100$ is true. Now, the sampling distribution of $\overline{X}$ is centered over 101, as illustrated in Figure 10.7.

FIGURE 10.7 Determination of P(Type I error) = P(reject H_0: $\mu \geq 100/\mu = 101$) for Example 10.4

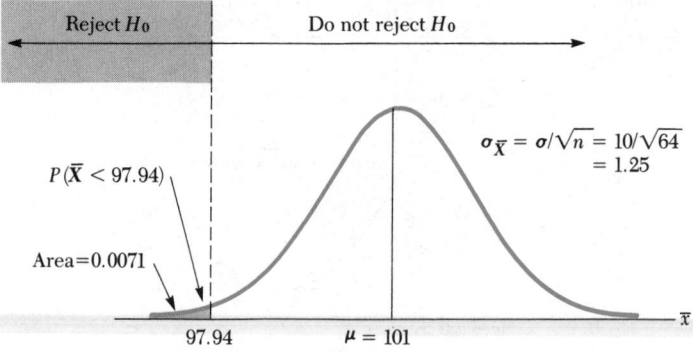

By using the standardizing statistic, $Z = (\overline{X} - \mu)/\sigma_{\bar{x}}$ and Table B.3 in Appendix B, $P(\overline{X} \leq 97.94/\mu = 101, \sigma_{\bar{x}} = 1.25) = 0.0071$; that is, the probability that a value of $\overline{X}$ will fall in the rejection region ($\bar{x} < \bar{x}_L = 97.94$) is 0.0071, which is less than 0.05, the significance level set for the test.

For any value of μ larger than $\mu = 101$, the area in the tail of the distribution to the left of the action limit 97.94 will be *even smaller* than 0.0071. So, if we center the sampling distribution of $\overline{X}$ at $\mu_0 = 100$, we will construct our decision rule such that the probability of a Type I error is no larger than the significance level (α) set for the test.

Therefore, in either form II (H_0: $\mu \geq \mu_0$, H_A: $\mu < \mu_0$) or form III (H_0: $\mu \leq \mu_0$, H_A: $\mu > \mu_0$) of the statistical hypothesis test, by using μ_0 as the mean of the sampling distribution of $\overline{X}$, the selected significance level of the test represents the *maximum* probability of committing a Type I error.

Since the decision in Example 10.4 was to not reject H_0: $\mu \geq 100$, a Type I error could not have been committed (see Table 10.1). By not rejecting the null hypothesis, it is only possible to commit a Type II error, if an error is to be made. What is the probability that a Type II error, β, has been made in this case? As it turns out, we cannot directly answer this question, but we can give some insights into how large β may be in a given hypothesis-testing experiment if we make the decision not to reject the null hypothesis. The method to do this is called the *power curve* of the test. The construction of the power curve of a statistical hypothesis test is developed in the next section.

◼ 10.3 THE POWER OF THE STATISTICAL TEST

In the previous section, we described the six basic steps in hypothesis testing. These steps control the risk of committing a Type I error, since the decision to reject or not to reject the null hypothesis, H_0, is directly based on the selection of the significance level of the test (α).

In Example 10.3, the decision was made to reject H_0, and, accordingly, if an error was made, it could only be a Type I error (rejecting H_0/H_0 is true). In Example 10.4, the decision was made not to reject the null hypothesis. As a consequence of this decision, if an error has been made, it could *only be* a Type II error. The probability of committing the Type II error is denoted by β (refer to Definition 10.2). Example 10.5 illustrates the computation of β for the two-sided test (form I).

Example 10.5 Consider the null and alternate hypotheses in Example 10.3:

$$H_0: \quad \mu = 10$$
$$H_A: \quad \mu \neq 10$$

Assuming that the population random variable **X** is normally distributed and that the population standard deviation σ is equal to 20, a random sample of size $n = 25$ is taken and the sample mean $\bar{x}$ is found to be equal to 19.

Let us assume that the decision maker has set a significance level of 0.05 for the test. Her selection of $\alpha = 0.05$ leads to the decision rule illustrated in Figure 10.8.

The rejection region for the test ($\bar{x}/\bar{x} < 2.16$ or $\bar{x} > 17.84$) and the region not to reject for the test ($\bar{x}/2.16 \leq \bar{x} \leq 17.84$) are also illustrated in Figure 10.8. In Example 10.3, we chose $\alpha = 0.10$ for the significance level of the test. Notice the effects of changing α from 0.10 in Example 10.3 to 0.05 in this example on the rejection region and on the region not to reject for the decision rule. In Example 10.3, the rejection region is ($\bar{x}/\bar{x} < 3.42$ or $\bar{x} > 16.58$) and the region not to reject is ($\bar{x}/3.42 \leq \bar{x} \leq 16.58$)—see Figure 10.3. By *decreasing* α from 0.10 to 0.05, the rejection region has been *decreased* and the region not to reject has been *increased*. Since the region not to reject has been increased, if we took repeated samples of size $n = 25$

FIGURE 10.8 Determination of the action limits $\bar{x}_L$ and $\bar{x}_U$ in Example 10.5

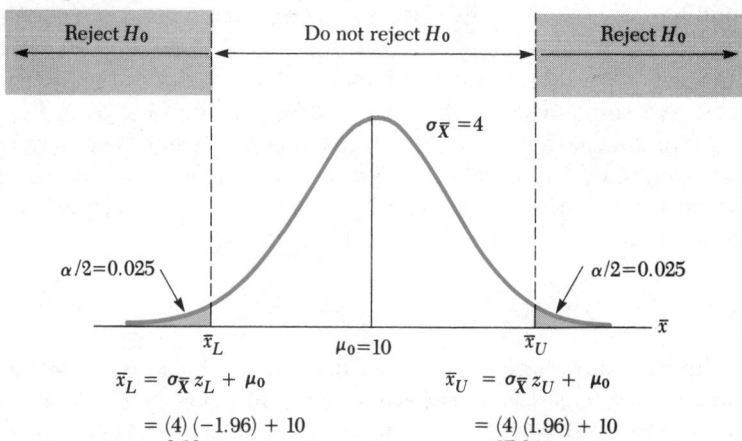

$$\bar{x}_L = \sigma_{\bar{X}} z_L + \mu_0$$

$$= (4)(-1.96) + 10$$
$$= 2.16$$

$$\bar{x}_U = \sigma_{\bar{X}} z_U + \mu_0$$

$$= (4)(1.96) + 10$$
$$= 17.84$$

(the sample size used in Example 10.3), we would not reject the null hypothesis more frequently with the decision rule based on $\alpha = 0.05$. By doing so, we would *increase* the probability of failing to reject a false null hypothesis (β). Therefore, by *decreasing* α from 0.10 to 0.05, we will have, in effect, *increased* β. Concomitantly, by *increasing* α, say from 0.05 to 0.10, we will *decrease* the size of the region not to reject and, thereby, *decrease* the probability of not rejecting a false null hypothesis (β). Thus, for a fixed sample size ($n = 25$ in our examples), α and β are *inversely related;* as one increases, the other decreases.

But, how do we determine the value of β in a particular hypothesis-testing experiment? In Example 10.3, we were given that $\bar{x} = 19$ based on a sample of size $n = 25$. Our decision, based on the new decision rule illustrated in Figure 10.8 using $\alpha = 0.05$, would be to reject H_0: $\mu = 10$ since $\bar{x} = 19 >$ $\bar{x}_U = 17.84$.

Let us now suppose that $\bar{x} = 16$ rather than $\bar{x} = 19$ so that the decision would be to not reject H_0: $\mu = 10$ based on the decision rule illustrated in Figure 10.8. By not rejecting the null hypothesis, it is now possible that we may have committed a Type II error, whose probability is given by β. Can we now compute the probability of committing a Type II error? The answer is no! The reason is apparent by recalling the definition of β: $\beta = P$(do not reject H_0/H_0 is false). The null hypothesis is H_0: $\mu = 10$. In computing β, the "given" event is that H_0 is false. But the complement of the event, $\mu = 10$, is the entire real number line, excluding the point $\mu = 10$! Thus, we do not know where to center the sampling distribution of $\bar{X}$ to compute β. We can, however, find β in a hypothetical sense. Suppose, for example, we *assume* that μ is really equal to 12. In this instance, H_0: $\mu = 10$ is false. Now what is the probability that H_0 is *not* rejected? This probability is illustrated in Figure 10.9. Now,

$$\beta = P(\text{do not reject } H_0/H_0 \text{ is false})$$
$$= P(2.16 \le \bar{X} \le 17.84/\mu = 12, \sigma_{\bar{x}} = 4)$$
$$= P\left(\frac{2.16 - 12}{4} \le Z \le \frac{17.84 - 12}{4}\right)$$
$$= P(-2.46 \le Z \le 1.46) = 0.9210$$

Thus, if the decision maker decides to reject H_0 if $\bar{x} < 2.16$ or if $\bar{x} > 17.84$, based on $\alpha = 0.05$, then if $\mu = 12$, there is a probability of 0.9210 that

FIGURE 10.9 Illustration of β, the probability of committing a Type II error for the hypothesis test in Example 10.5

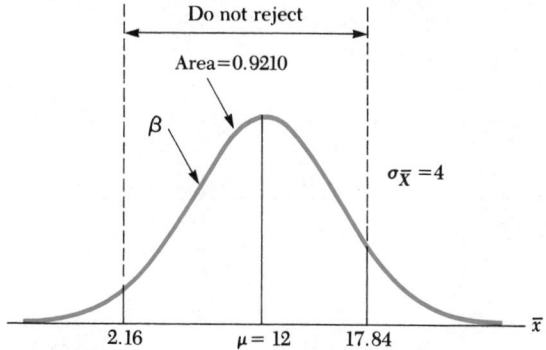

H_0: $\mu = 10$ will be incorrectly not rejected, thereby committing a Type II error.

By selecting a set of values for μ and computing β as illustrated above, the decision maker can get an idea of the likelihood that a Type II error will be made if the null hypothesis is not rejected. Rather than using β for this purpose, however, most statisticians determine the *power function* for the test.

Definition 10.3
The power function

The *power function* for a statistical hypothesis test gives the probability of rejecting the null hypothesis H_0 for the set of assumed values of the population parameter θ:

$$\text{Power} = P(H_0 \text{ is rejected}/\theta)$$

☐ **10.3.1 The determination of the power function in a two-sided test (form I)**

Consider the null and alternate hypotheses in Example 10.3 and continued in Example 10.5:

FIGURE 10.10 Illustration of the probability representing the power of the test for the example 10.5 hypotheses

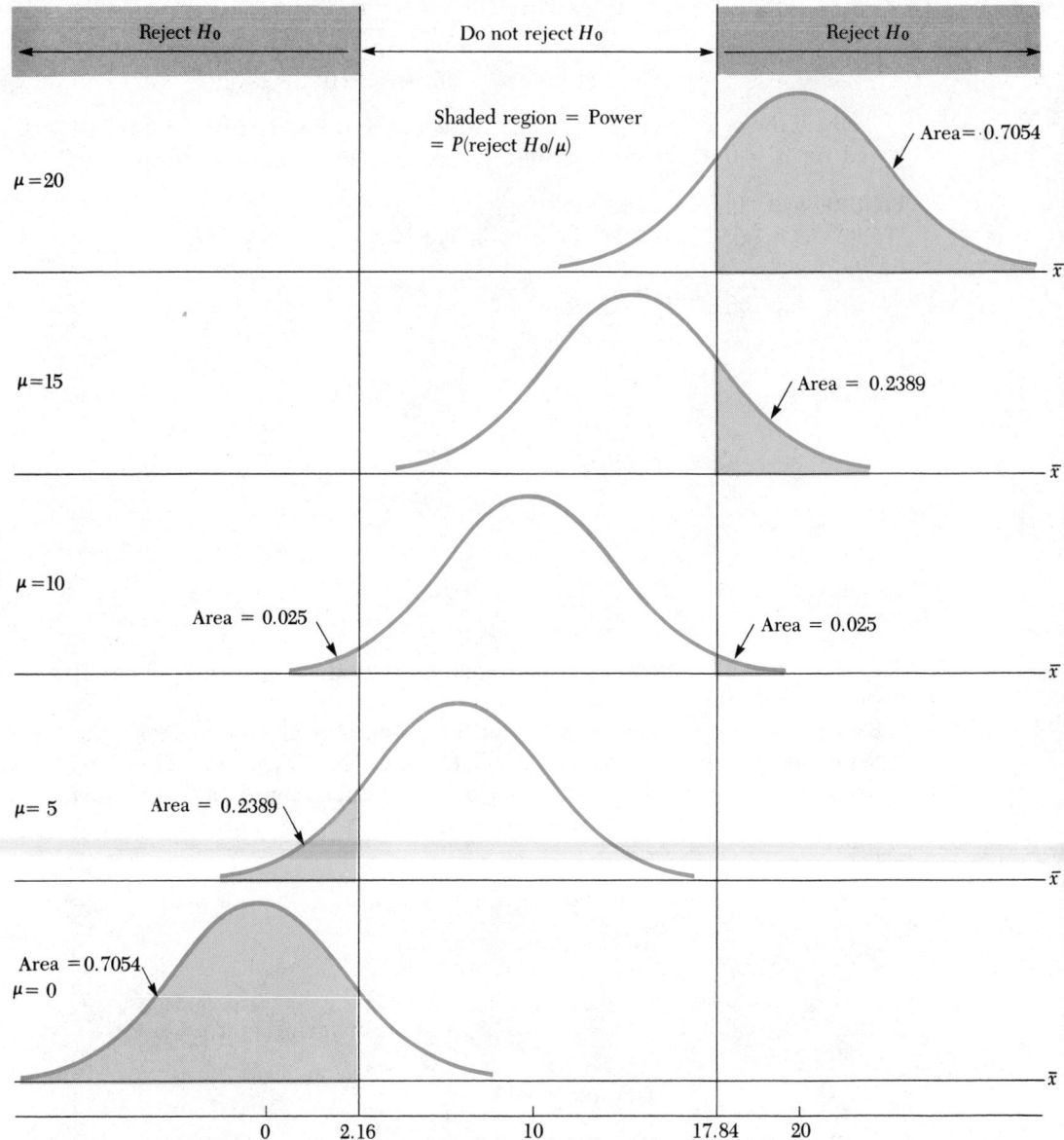

$$H_0: \quad \mu = 10$$
$$H_A: \quad \mu \neq 10$$

Based on the decision rule to reject H_0 with the significance level of the test set at $\alpha = 0.05$, the region in which we would fail to reject H_0 was determined to be $(\bar{x}/2.16 \leq \bar{x} \leq 17.84)$. Figure 10.10 illustrates the computation of values of the power function for this decision rule and the selected values of μ.

By definition, Power = $P(H_0$ is rejected$/\mu)$. For example, in referring to Figure 10.10 the power of the test when $\mu = 20$ is computed as follows:

$$\text{Power} = P(\bar{X} < 2.16 \text{ or } \bar{X} > 17.84)$$

$$= P \left(Z < \frac{2.16 - 20}{4} \text{ or } Z > \frac{17.84 - 20}{4} \right)$$

$$= P(Z < -4.46 \text{ or } Z > -0.54)$$

$$= P(Z < -4.46) + P(Z > -0.54) = 0.0000 + 0.7054 = 0.7054$$

Thus, if $\mu = 20$, the probability that $H_0: \quad \mu = 10$ is rejected is equal to 0.7054 based on the decision rule to reject H_0 if the $\alpha = 0.05$ significance level is used. In a similar fashion, the power can be determined for the other values of μ given in Table 10.3.

TABLE 10.3 Power of the test for selected values of μ and the rejection region

μ	Power = $P(\text{reject } H_0/\mu)$	β
0	0.7054	0.2946
5	0.2389	0.7611
10	0.0500	0
15	0.2389	0.7611
20	0.7054	0.2946

Rejection region: reject H_0 if $\bar{x} > 17.84$ or if $\bar{x} < 2.16$
Sample size: $n = 25$, $\alpha = 0.05$

Using the points in Table 10.3, it is possible to plot the power function by drawing a "smooth" curve through the points so determined, as shown in Figure 10.11. Notice that the power is equal to the significance level of the test ($\alpha = 0.05$) at the hypothesized value of μ, $\mu = 10$. Furthermore, *except when $\mu = 10$, 1 − Power = β*; that is,

$$P(\text{reject } H_0/H_0 \text{ is false}) + P(\text{do not reject } H_0/H_0 \text{ if false}) = 1$$
$$\text{Power} \qquad + \qquad \beta \qquad = 1$$

Several values of β are given in Table 10.3 for the given set of μ values. Notice that the relationship 1 − Power = β does not hold when $H_0: \quad \mu = 10$ is true. If $\mu = 10$, it is not possible to commit a Type II error since H_0 is true.

When H_0 is not true, the power represents the probability that H_0 is rejected given H_0 is false. Thus, the power curve gives us some idea of the

FIGURE 10.11 Power function for the test in Example 10.5

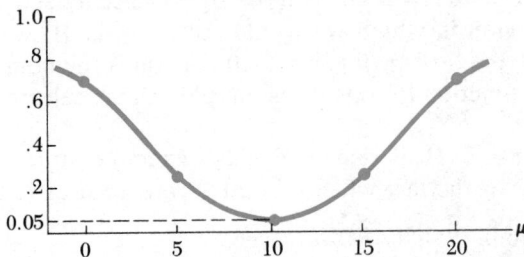

Power = $P(\text{rejecting } H_0/\mu)$

ability of the test to detect when the null hypothesis is not true as a function of μ. It seems intuitive that the power should increase as μ is farther away from μ_0, the hypothesized value in the null hypothesis. This phenomenon is reflected well in Figure 10.11—as the chosen values of μ depart from $\mu = 10$, the power or the probability of rejecting the null hypothesis H_0 increases.

Ideally, we would like the power function to rise very rapidly as the chosen values of μ depart from μ_0. This would imply that the probability of rejecting H_0: $\mu = \mu_0$ is high when μ deviates only slightly from μ_0.

Two factors affect the form of the power function: (1) the sample size and (2) the decision rule (significance level). Let us first consider the effect on the power function of a change in the sample size. Suppose in Example 10.5 the sample size is increased from $n = 25$ to $n = 100$. Now, $\sigma_{\bar{x}} = \sigma/\sqrt{n} = 20/\sqrt{100} = 2$.

The decision rule in terms of the test statistic $\bar{X}$ now changes due to the change in the sample size. The decision rule based on a sample of size $n = 100$ is illustrated in Figure 10.12. Now, the power of the test for selected values of μ is given in Table 10.4.

The power function when $n = 100$ is graphed in Figure 10.13, together with the power function when $n = 25$.

As the sample size increases, the power function will approach the ideal power function illustrated in Figure 10.14. In the ideal power function, the

TABLE 10.4 Power of the test for selected values of μ and the specified rejection region

μ	Power = $P(\text{reject } H_0/\mu)$	β
0	0.9988	0.0012
5	0.7054	0.2946
10	0.0500	0
15	0.7054	0.2946
20	0.9988	0.0012

Rejection region: reject H_0 if $\bar{x} < 6.08$ or if $\bar{x} > 13.92$
Sample size: $n = 100$, $\alpha = 0.05$

FIGURE 10.12 Decision rule in terms of the statistic $\bar{X}$ with a 0.05 significance level and $n = 100$

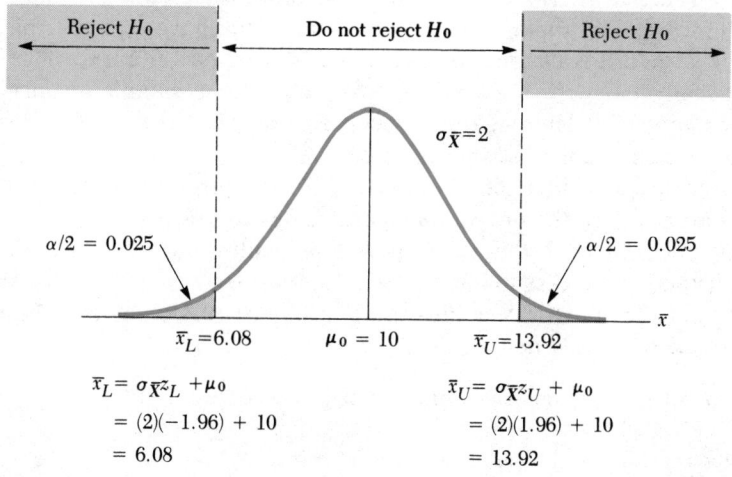

$$\bar{x}_L = \sigma_{\bar{X}} z_L + \mu_0$$
$$= (2)(-1.96) + 10$$
$$= 6.08$$

$$\bar{x}_U = \sigma_{\bar{X}} z_U + \mu_0$$
$$= (2)(1.96) + 10$$
$$= 13.92$$

FIGURE 10.13 Power function for samples of size $n = 25$ and $n = 100$

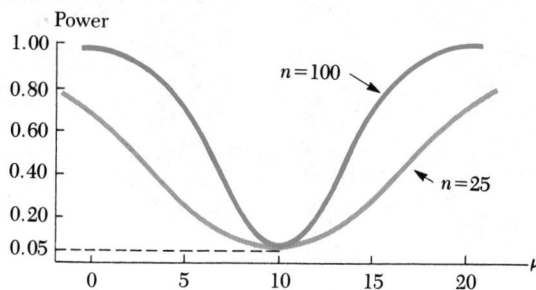

FIGURE 10.14 Ideal power function

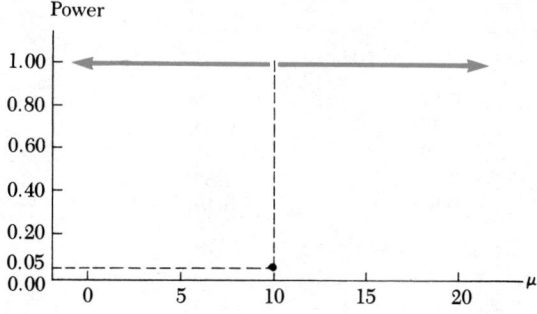

power becomes 1 if μ deviates *at all* from the hypothesized value of $\mu (\mu_0 = 10)$; that is, H_0: $\mu = 10$ will *always* be rejected if μ is not equal to 10. Unfortunately, the number of units that would have to be sampled is *extremely* large if it is desired to have a nearly ideal power curve. The statistician usually balances the cost of obtaining a large sample against the increased power by deciding on an acceptable form for the power curve (and, hence, an acceptable decision rule).

Let us now consider the effect on the power curve of changing the decision rule. Suppose the decision maker decides to reject H_0: $\mu = 10$ with a 0.10 significance level; that is, she is willing to allow a greater risk of committing a Type I error. Assuming we use a sample size $n = 25$, the decision rule in terms of the statistic $\overline{X}$ is illustrated in Figure 10.15. Table 10.5 gives the power and β for this new decision rule.

FIGURE 10.15 Decision rule with 0.10 significance level

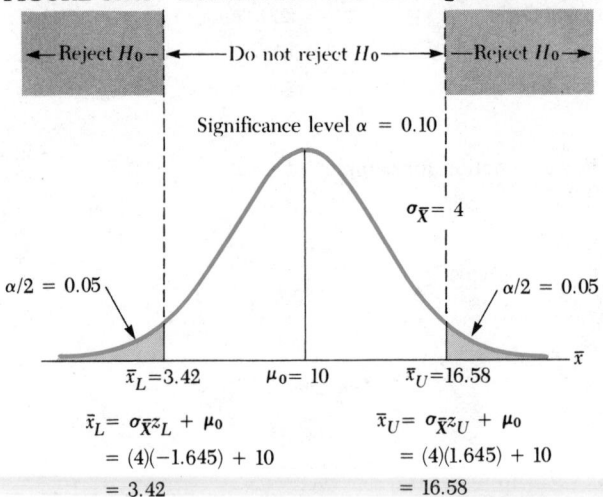

Rejection region: Reject H_0 if $\bar{x} < 3.42$ or if $\bar{x} > 16.58$

Region in which H_0 is not rejected: Do not reject H_0 if $3.42 \leq \bar{x} \leq 16.58$

TABLE 10.5 Power of the test for selected values of μ and for the specified decision rule

μ	Power = $P(reject\ H_0/\mu)$	β
0	0.8051	0.1949
5	0.3465	0.6535
10	0.1000	0
15	0.3465	0.6535
20	0.8051	0.1949

Decision rule: reject H_0 if $\bar{x} < 3.42$ or if $\bar{x} > 16.58$
Sample size: $n = 25$, $\alpha = 0.10$

FIGURE 10.16 Power function graphs for decision rules A and B

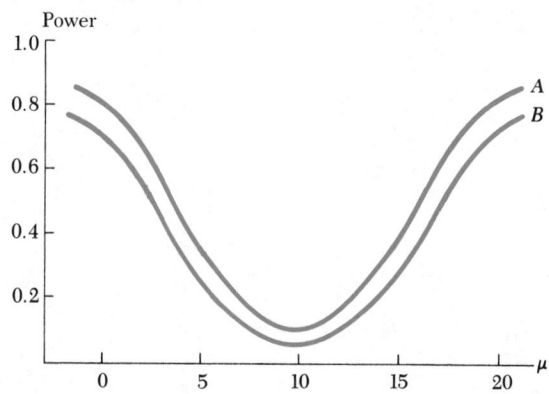

A: Decision rule: Reject H_0 if $\bar{x} < 3.42$ or if $\bar{x} > 16.58$
B: Decision rule: Reject H_0 if $\bar{x} < 2.16$ or if $\bar{x} > 17.84$

This power function is graphed in Figure 10.16 together with the power function for the previous decision rule, where $n = 25$ and $\alpha = 0.05$. Notice in Figure 10.16 that by allowing an increased risk of committing a Type I error (from 0.05 to 0.10), the power of the test is increased and, concomitantly, the probability of committing a Type II error is decreased (compare the columns for β in Tables 10.3 and 10.5).

The power functions in Figure 10.13 and Figure 10.16 illustrate the following facts:

1. For a fixed sample size n, if the probability of committing a Type I error is *increased,* the probability of committing a Type II error is *decreased;* that is, α and β are inversely related.
2. For a fixed sample size n, if the probability of committing a Type I error is *increased,* the power of the test is *increased.*
3. As the sample size n is *increased,* the power of the test is *increased* and the probability of committing a Type II error is *decreased* for a fixed level of the probability of committing a Type I error.

An alternative to the power curve or function is the *operating characteristic (OC) curve.* The OC curve is the "complement" of the power curve and gives the probability of not rejecting the null hypothesis for various values of the population parameter. The OC curve corresponding to the power curve illustrated in Figure 10.11 can be generated by taking the complement of the power curve probabilities given in Table 10.3; that is, $P(\text{do not reject } H_0/\mu) = 1 - P(\text{reject } H_0/\mu)$. The OC curve is shown in Figure 10.17.

FIGURE 10.17 OC curve corresponding to the power curve in Figure 10.11

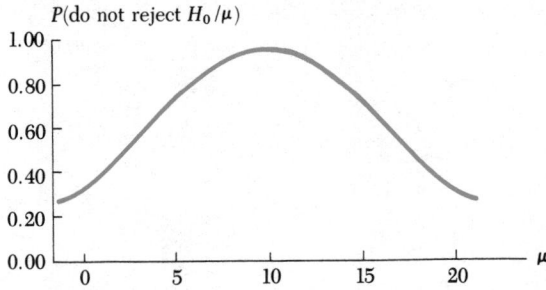

Rejection region: Reject H_0 if $\bar{x} > 17.84$ or if $\bar{x} < 2.16$
Hypotheses: H_0: $\mu = 10$; H_A: $\mu \neq 10$
Sample size: $n = 25$; $\alpha = 0.05$

☐ 10.3.2 The determination of the power function in a one-sided test (forms II and III)

Consider the hypothesis experiment in Example 10.2. Herman wishes to test the null hypothesis that the population proportion π of devices designed to produce steam that are nondefective is at least 0.96. The appropriate null and alternate hypotheses are:

$$H_0: \quad \pi \geq 0.96$$
$$H_A: \quad \pi < 0.96$$

Suppose Herman decides to sample 100 such devices and to use the 0.05 significance level for testing H_0. In terms of the sample proportion p, the rejection region and the region not to reject are illustrated in Figure 10.18.

FIGURE 10.18 Rejection region and the region not to reject for the null hypothesis H_0: $\pi \geq 0.96$ with $n = 100$ and 0.05 significance level

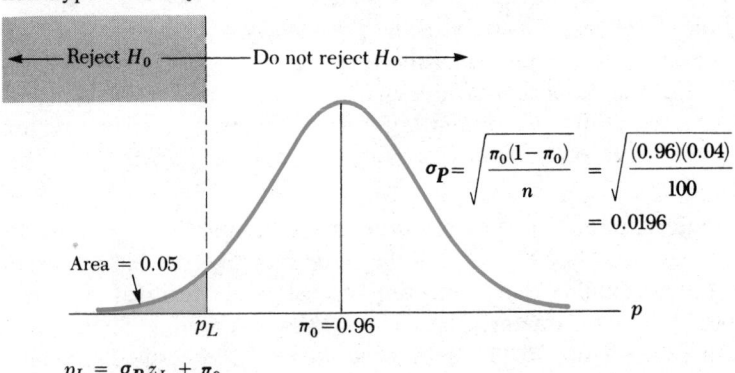

$$p_L = \sigma_P z_L + \pi_0$$
$$= (0.0196)(-1.64) + 0.96$$
$$= 0.928$$

We are assuming that the sampling distribution of the test statistic **P** is approximately normally distributed (via the central limit theorem).

The power of the test for selected values of π is given in Table 10.6. The computation of the power probabilities given in Table 10.6 is illustrated in Figure 10.19.

FIGURE 10.19 Computation of the power probabilities in Table 10.6

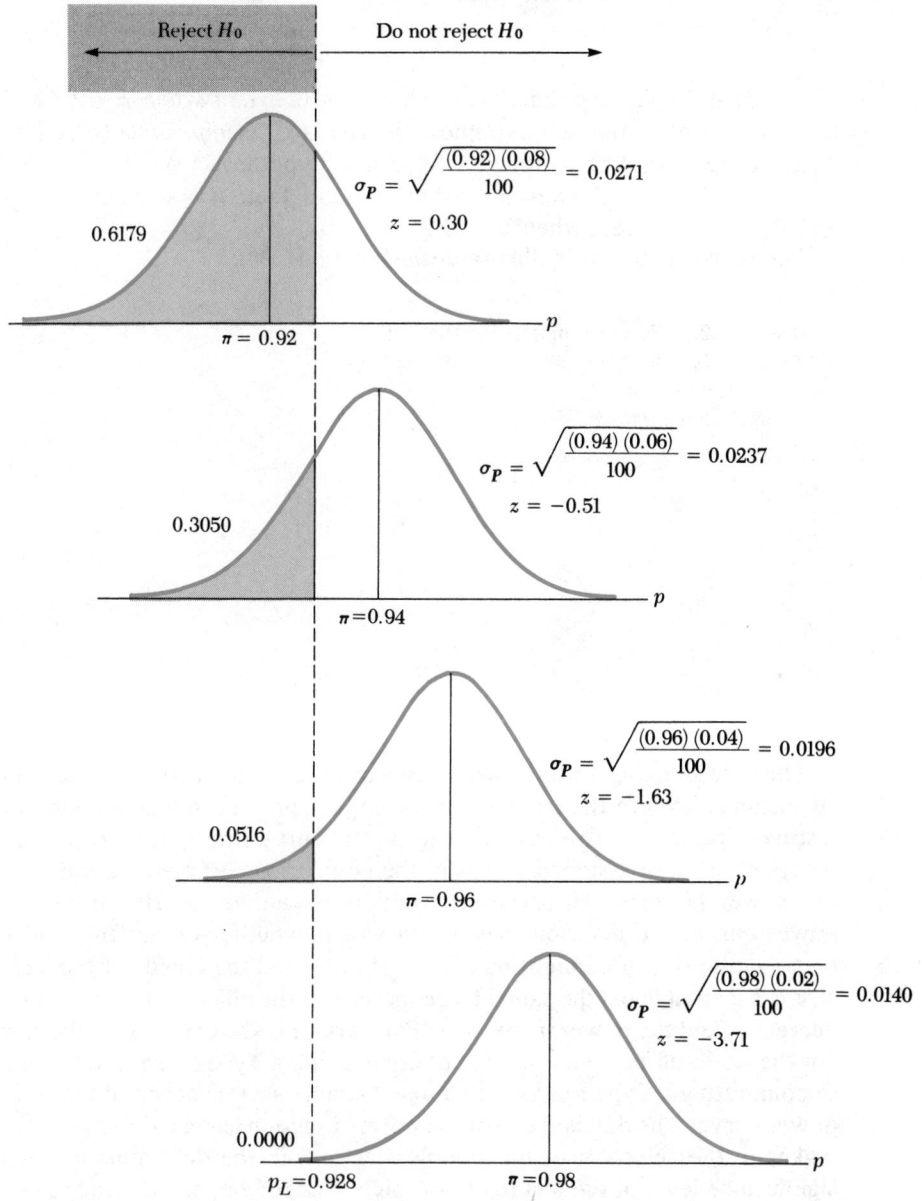

TABLE 10.6 Power probabilities and probability of a Type II error (β) for the rejection region in Figure 10.18

π	$Power = P(reject\ \mathrm{H_0}/\pi)$	β
0.92	0.6179	0.3821
0.94	0.3050	0.6950
0.96	0.0516	0
0.98	0.0000	0

Notice in Table 10.6 that β is 0 when $\pi = 0.96$ *and* when $\pi = 0.98$. For these values of π, the null hypothesis is true, so it is *impossible* to commit a Type II error—failing to reject a false null hypothesis. When $\pi = 0.92$ or $\pi = 0.94$, $\beta = (1 - \text{Power})$ in Table 10.6. In fact, $\beta = 0$ when $\pi \geq 0.96$ and $\beta = (1 - \text{Power})$ when $\pi < 0.96$.

The power function is illustrated in Figure 10.20.

FIGURE 10.20 Power function for the null hypothesis $\mathrm{H_0}$: $\pi \geq 0.96$ with $n = 100$ and 0.05 significance level

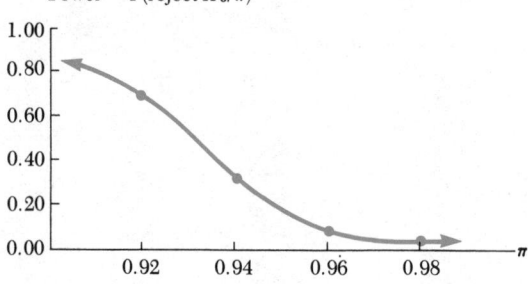

The determination of the power curve or function for various values of the significance level of the test (α) is very important in designing a hypothesis-testing experiment. To construct the power curve (the graph of the power function), a set of assumed values of the population parameter is chosen and the power [P(reject $\mathrm{H_0}$/parameter value)] is calculated. By studying the power curve, the decision maker can assess whether or not the trade-off between the risk of committing a Type II error and the choice of the sample size is acceptable (as the sample size increases, the chance of a Type II error decreases and the power increases). Furthermore, she can assess whether or not the trade-off between the risk of committing a Type II error and the risk of committing a Type I error (for a fixed sample size) is acceptable from the power curve. The detailed use of the power function as an aid to the decision maker in the selection of the sample size and in the determination of the significance level of the test (α) is a topic advanced beyond the scope of our introductory coverage of hypothesis testing. However, it is important for us to understand what the power function represents and how it is constructed at this stage in our development of statistical inference.

■ 10.4 THE *p*-VALUE METHOD OF HYPOTHESIS TESTING

In the past few years, the classical method of hypothesis testing described in Section 10.2 has drawn criticism due to the rather simplistic nature of the decision made in the final step: either to reject H_0 or not to reject H_0. It has been argued that the classical method is a "cookbook" procedure that does not provide insights into the *degree* of the conviction in the decision to reject or not to reject H_0. The following example illustrates the problem associated with the classical method and a suggested solution to the problem, called the *p-value method*.

Example 10.6   A cigarette manufacturer claims that the average tar content of Puff brand cigarettes is less than 5 milligrams per cigarette. To test this conjecture, 100 Puff cigarettes are sampled, and it is found that the average tar content is 4.94 milligrams. From previous experiments, it is known that the population standard deviation of the tar content is $\sigma = 0.2$ milligram. Test this conjecture at the $\alpha = 0.05$ significance level.

Solution   The null and alternate hypotheses are:[1]

$$H_0: \quad \mu \geq 5 \text{ milligrams}$$
$$H_A: \quad \mu < 5 \text{ milligrams}$$

The decision rule for this test is illustrated in Figure 10.21 with the signifi-

FIGURE 10.21   Decision rule for the hypothesis test in Example 10.6

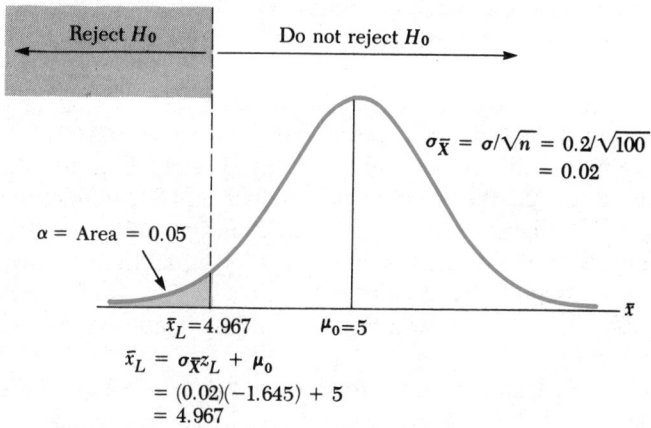

$$\sigma_{\bar{X}} = \sigma/\sqrt{n} = 0.2/\sqrt{100}$$
$$= 0.02$$

$\alpha = $ Area $= 0.05$

$\bar{x}_L = 4.967 \qquad \mu_0 = 5$

$$\bar{x}_L = \sigma_{\bar{X}} z_L + \mu_0$$
$$= (0.02)(-1.645) + 5$$
$$= 4.967$$

---

[1] Since the manufacturer is claiming that μ, the average tar content, is less than 5 milligrams, it would first seem that the null hypothesis should be $H_0: \quad \mu \leq 5$, rather than $H_0: \quad \mu \geq 5$. In one-sided tests (forms II and III), we place the event we wish to demonstrate is true in the alternate hypothesis, H_A. The reason for this will be discussed in Section 10.7.

cance level set at $\alpha = 0.05$. Since $\bar{x} = 4.94 < \bar{x}_L = 4.967$, reject H_0: $\mu \geq 5$ and conclude that $\mu < 5$ (H_A).

In the application of the classical method, we would still decide to reject the null hypothesis, but the method does not give us an idea of the *strength* of the *conviction* that the decision is, in fact, correct. A suggested alternative to this approach is to determine how likely it is to sample a value of the statistic $\overline{X}$ that is less than or equal to $\bar{x} = 4.94$, the sampled average, when, in fact, $\mu = 5$. The probability of this event is called the *p-value* of the test. Its computation is illustrated in Figure 10.22. The probability that $\overline{X}$ is less than

FIGURE 10.22 Computation of the *p*-value in Example 10.6

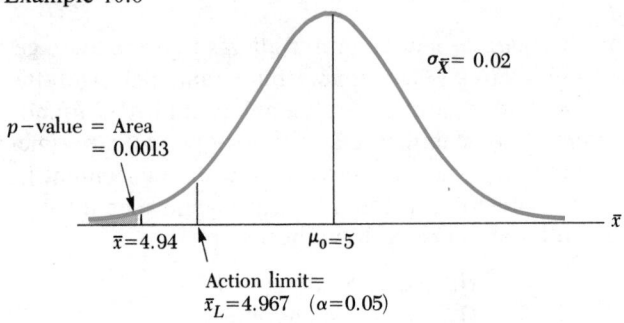

$$P(\overline{X} \leq \bar{x}=4.94/\mu_0=5, \ \sigma_{\overline{X}} = 0.02) = P\left(Z \leq \frac{4.94 - 5.00}{0.02}\right)$$

$$= P(Z \leq -3.00) = 0.0013$$

or equal to the sampled average ($\bar{x} = 4.94$) is 0.0013. Now, this probability conveys important information about the strength of the decision to reject H_0: $\mu \geq 5$. We know from Figure 10.21 that $\bar{x} = 4.94$ falls in the rejection region so that with $\alpha = 0.05$, the probability that we have made a Type I error is no greater than 0.05. But, given that the *p*-value is 0.0013, we know that it is extremely unlikely that we could take a sample of 100 Puff brand cigarettes and find that the average tar content is 4.94 milligrams or less, *given* that $\mu \geq 5$ milligrams. That is, if $\mu \geq 5$, in only 13 of 10,000 random samples would the value of the test statistic $\overline{X}$ be equal to or less than 4.94. Thus, we have strong evidence to reject H_0: $\mu \geq 5$.

Example 10.7 illustrates the computation of the *p*-value in a two-sided test (form I).

Example 10.7   In Example 10.3, the null and alternate hypotheses are:

$$H_0: \quad \mu = 10$$
$$H_A: \quad \mu \neq 10$$

Given that $\sigma = 20$, the decision rule for this test with a sample of 25 units and a significance level of $\alpha = 0.10$ is to reject H_0 if $\bar{x} > 16.58$ or if $\bar{x} < 3.42$, and not to reject H_0 if $3.42 \leq \bar{x} \leq 16.58$. In Example 10.3, $\bar{x} = 19$ and, since $\bar{x} = 19 > 16.58$, the decision was made to reject H_0 at the $\alpha = 0.10$ significance level. What is the p-value for this test?

Solution To find the p-value for a two-sided test, we first determine the probability of sampling a value of the test statistic $\overline{X}$ that is 19 *or greater*, given that $\mu = \mu_0 = 10$. The computation of this probability is illustrated in Figure 10.23. The p-value is determined by *doubling* this probability. Thus,

$$p\text{-value} = 2P(\overline{X} \geq \bar{x} = 19) = 0.0244$$

So, although we decided to reject H_0: $\mu = 10$ at the $\alpha = 0.10$ significance level, we can be more confident that we have made the right decision than the $\alpha = 0.10$ significance level would seem to indicate. This is because the p-value is 0.0244, which gives us the probability of making a Type I error if we decide to reject H_0.

FIGURE 10.23 Computation of $P(\overline{X} \geq 19)$ in Example 10.7

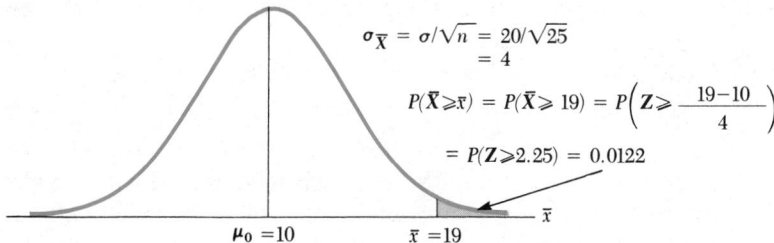

The determination of the p-value in the two-sided test (form I) is similar to the calculation of the p-value in the one-sided test. In Example 10.7, we wish to find the probability that we could sample a value of the test statistic $\overline{X}$ that is $19 - 10 = 9$ or more units removed from the hypothesized value of μ, $\mu_0 = 10$. The one difference is the doubling of this probability. The reason for doubling $P(\overline{X} \geq \bar{x})$ in the two-sided test is the nature of the test itself. The null hypothesis is $\mu = \mu_0$ so that an indication that μ is *either* less than μ_0 or greater than μ_0 leads to rejecting H_0. When setting up the decision rule according to the classical method, we *halve* the significance level α, forming the two rejection regions in the tails of the sampling distribution of the test statistic. The doubling of the tail probability in calculating the p-value provides a quantity that is comparable to the significance level α in the two-sided test. For instance, in Example 10.7, by the classical method, we would reject H_0: $\mu = 10$ as long as the significance level α is chosen to be greater than the p-value of 0.0244. For example, we would *not* reject H_0: $\mu = 10$ in Example 10.7 if we choose $\alpha = 0.01$, since the p-value $= 0.0244 > 0.01$. Table 10.7 summarizes the computation of the p-value in the two-sided (form I) and one-sided (forms II and III) tests.

The five steps in testing a hypothesis using the p-value method are:

TABLE 10.7 The p-value determination for the three hypothesis-testing forms

Hypothesis	Test statistic	p-Value
H_0: $\theta = \theta_0$ H_A: $\theta \neq \theta_0$	$\hat{\theta}$	If $\hat{\theta} > \theta_0$, p-value $= 2P(\hat{\theta} > \hat{\theta})$ If $\hat{\theta} < \theta_0$, p-value $= 2P(\hat{\theta} < \hat{\theta})$
H_0: $\theta \geq \theta_0$ H_A: $\theta < \theta_0$	$\hat{\theta}$	p-value $= P(\hat{\theta} < \hat{\theta})$
H_0: $\theta \leq \theta_0$ H_A: $\theta > \theta_0$	$\hat{\theta}$	p-value $= P(\hat{\theta} > \hat{\theta})$

1. The statement of the null and alternate hypotheses.
2. The selection of the sample statistic on which to base the decision to reject or not to reject the null hypothesis. This statistic is called the *test statistic* ($\hat{\boldsymbol{\theta}}$).
3. The random selection of the sample units, and the computation of the value of the test statistic ($\hat{\theta}$).
4. The computation of the p-value (see Table 10.7).
5. The decision to reject or not to reject the null hypothesis based on the p-value.

Notice the differences in these steps when compared with the steps in the classical method delineated in Section 10.2. In the p-value method, the significance level (α) is not set. In step 5 of the p-value method, the decision to reject or not to reject H_0 can be made by comparing the p-value with a chosen value of the significance level α. If the p-value $< \alpha$, we reject H_0; otherwise, we do not reject H_0.

Most statistical computer software packages now print out the p-value for almost all hypothesis-testing situations. Noteworthy examples are SPSS (Statistical Package for the Social Sciences) and MINITAB, two very popular software packages referenced at the end of this chapter. In Chapter 15, computer printouts for regression inferences give the p-values, rather than deciding to reject or not to reject the null hypotheses based on preset significance levels.

In the remaining sections of this chapter, we will use the p-value method for testing hypotheses concerning the population mean μ, the population proportion π, and the population variance σ^2.

▪ 10.5 HYPOTHESIS TESTS CONCERNING THE POPULATION MEAN μ

The testing of a population mean follows the five hypothesis-testing steps in the p-value method given in Section 10.4. The appropriate test statistic is the sample mean statistic $\overline{X}$. Two cases must be considered when testing a hypothesis concerning the population mean μ, however: (1) case 1: the population standard deviation σ is known, and (2) case 2: the population standard deviation σ is not known. In each case, the appropriate testing procedures will further depend on whether or not the population random variable **X** is normally distributed.

☐ **10.5.1 Case 1: The population standard deviation σ is known**

The computation of the p-value depends on knowing the sampling distribution of the test statistic. Recall that if the population random variable **X** is normally distributed, then the sample mean statistic $\overline{X}$ is normally distributed (see Chapter 8). In this case, the determination of the p-value can be made directly from the normal distribution. Example 10.8 illustrates the application of the five steps in hypothesis testing with the p-value method when the population random variable **X** is normally distributed and the population standard deviation σ is known.

Example 10.8   It is hypothesized that the mean starting salary of accounting graduates with no prior work experience who are hired by one of the Big 8 accounting firms nationally is $19,000. A random sample of $n = 100$ starting salaries of recent accounting graduates hired by these firms is taken, and it is found that the sample average starting salary is $18,800. Test the null hypothesis that μ is $19,000. Assume it is known from comprehensive past studies of these starting salaries that $\sigma = \$800$ and that the population random variable **X** is normally distributed.

Solution   The null and alternate hypotheses are:

$$H_0: \quad \mu = \$19,000 \quad (\mu_0 = 19,000)$$
$$H_A: \quad \mu \neq \$19,000$$

The values of the population random variable **X** represent the starting salaries of all recent accounting graduates hired by the accounting firms. Our sample is a random selection of 100 values of **X**. It is assumed that σ, the population standard deviation, is equal to $800. The population and sample information are summarized in Figure 10.24.

The appropriate test statistic is $\overline{X}$, and the determination of the p-value is illustrated in Figure 10.25. From Figure 10.25, the p-value for the test is

FIGURE 10.24   Population and sample information for Example 10.8

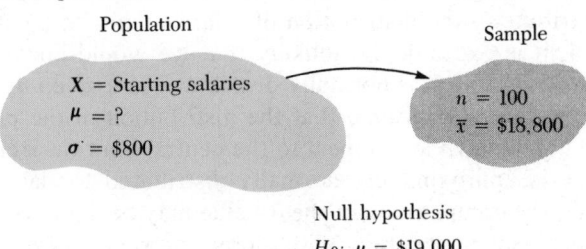

Population   Sample

X = Starting salaries
μ = ? $n = 100$
$\sigma = \$800$ $\overline{x} = \$18,800$

Null hypothesis
$H_0: \mu = \$19,000$

FIGURE 10.25 Determination of the p-value in Example 10.8

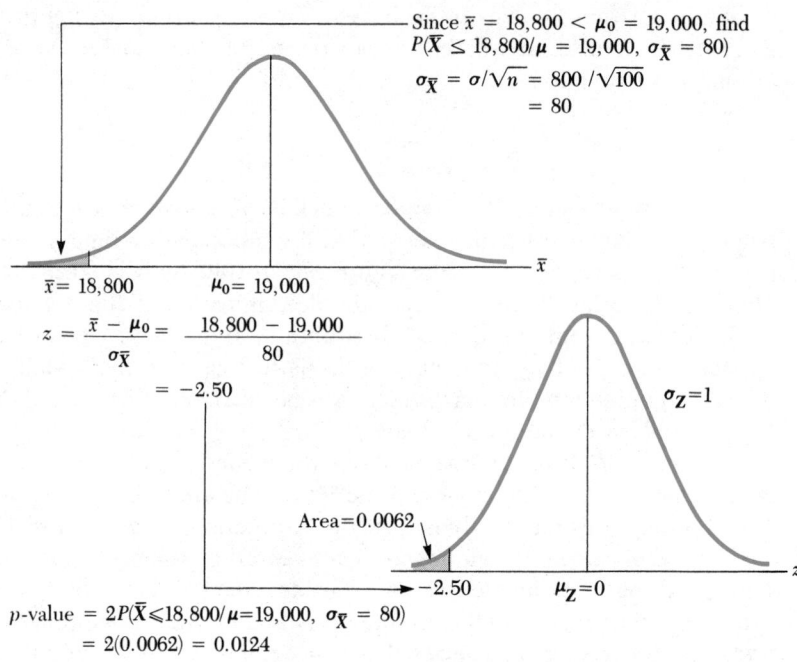

Since $\bar{x} = 18,800 < \mu_0 = 19,000$, find
$P(\bar{X} \le 18,800/\mu = 19,000,\ \sigma_{\bar{X}} = 80)$

$\sigma_{\bar{X}} = \sigma/\sqrt{n} = 800\ /\sqrt{100}$
$= 80$

$\bar{x} = 18,800 \qquad \mu_0 = 19,000$

$z = \dfrac{\bar{x} - \mu_0}{\sigma_{\bar{X}}} = \dfrac{18,800 - 19,000}{80}$

$= -2.50$

$\sigma_Z = 1$

Area$=0.0062$

$-2.50 \qquad \mu_Z = 0$

$p\text{-value} = 2P(\bar{X} \le 18,800/\mu = 19,000,\ \sigma_{\bar{X}} = 80)$
$= 2(0.0062) = 0.0124$

0.0124. Given this small p-value, the decision maker would most likely decide to reject H_0: $\mu = \$19,000$ and conclude that μ is not equal to $\$19,000$. (This decision would be consistent with rejecting H_0 based on the 0.10 and 0.05 significance levels, but not with rejecting H_0 based on the 0.01 significance level.)

Let us review the interpretation of the p-value (0.0124) in the context of this example. If μ really is $\$19,000$, in only 1.24 percent of all possible samples selected would we expect to find that the value of the statistic $\bar{X}$ is more than $\$200$ from $\$19,000$. Since, in our sample, $\bar{x} = \$18,800$, we could have encountered one of these samples. But it is far more likely that the sampling distribution of $\bar{X}$ is centered over some value other than $\$19,000$, as Figure 10.26 illustrates.

A final note concerning this example: since the population random variable X represents starting salaries, it is possible that the distribution of X is not normally distributed—the distribution of salaries tends to be skewed to the right. Indeed, it is exceedingly unlikely that we would know that any population random variable X is normally distributed in a given hypothesis-testing problem. Rather, we assume that the distribution of the population random variable is unknown and appeal to the central limit theorem, which assures us that $\bar{X}$ is approximately normally distributed for large sample sizes. In this case, the computation of the p-value may be approximate. The approximation increases in accuracy with increasing sample size.

FIGURE 10.26 Sampling distribution of $\overline{X}$ under H_0 and a more likely sampling distribution of $\overline{X}$

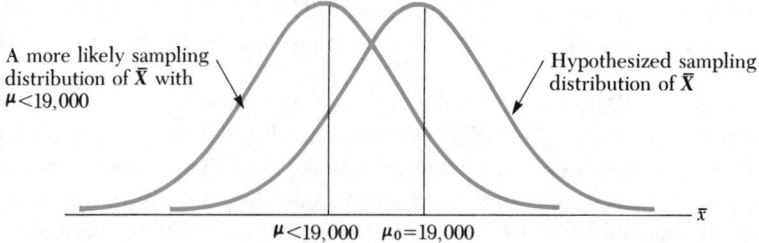

Example 10.9 The mean lifetime of 100 randomly selected fluorescent light bulbs produced by the Stay Bright company is computed to be 1,560 hours. From past studies of the bulbs, it is assumed that the population standard deviation of bulb lifetimes is 120 hours. Let μ be the mean lifetime of all similar bulbs produced by the company. Test the null hypothesis that $\mu \geq 1,600$ hours.

Solution The null and alternate hypotheses are:

$$H_0: \quad \mu \geq 1,600 \quad (\mu_0 = 1,600)$$
$$H_A: \quad \mu < 1,600$$

The test statistic is $\overline{X}$, which we will assume to be approximately normally distributed via the central limit theorem. The determination of the p-value is illustrated in Figure 10.27.

FIGURE 10.27 Determination of the p-value in Example 10.9

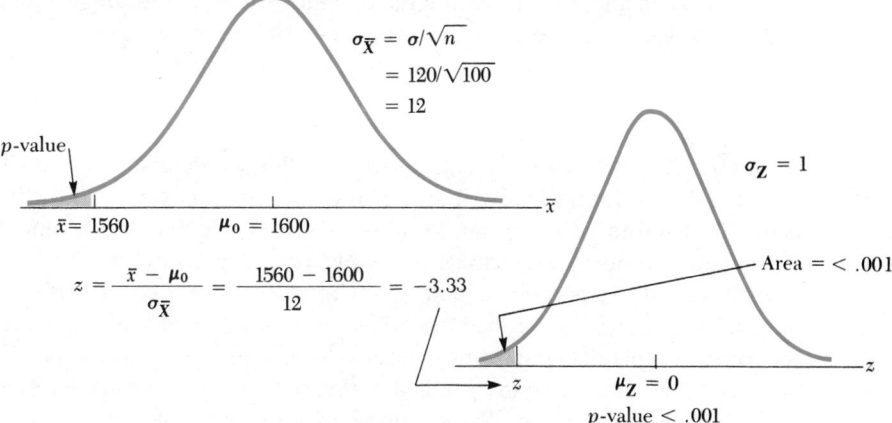

Notice that the p-value cannot be determined in this example because of the limitations on the tabulated values of the normal distribution (Table B.3 in Appendix B). The probability that $\overline{X}$ is less than 1,560 given $\mu = 1,600$ is equivalent to the probability that z is less than -3.33, when $\bar{x}$ is standardized.

From Table B.3 the largest z-value is 3.09 and $P(Z < -3.09) = 0.001$. Thus we can conclude only that the p-value $= P(Z < -3.33) < 0.001$. Based on such a small p-value, we would reject H_0: $\mu \geq 1,600$ and conclude that $\mu < 1,600$—the average lifetime of the bulbs appears to be less than 1,600 hours.

Notice in Example 10.9 that no assumption was made concerning the distribution of the population random variable X. Thus, we must appeal to the central limit theorem for the computation of the p-value. Since $n = 100$, we can be reasonably confident that the sampling distribution of $\overline{X}$ is approximately normal. The p-value is then computed from the assumed normal distribution with a mean of 1,600 and a standard deviation of 12. In this case, the computed p-value is *approximate*, since it has been calculated from a normal distribution when the exact sampling distribution of the statistic $\overline{X}$ is not known.

□ **10.5.2 Case 2: The population standard deviation σ is not known**

When the population random variable X is normally distributed, then $\overline{X}$ is also normally distributed. And, if the population standard deviation σ is known, then the standardized variable

$$Z = \frac{\overline{X} - \mu}{\sigma/\sqrt{n}}$$

is a standard normal (mean $= 0$, standard deviation $= 1$) random variable. If σ is not known and must be estimated by using the sample standard deviation statistic S, then the standardized random variable,

$$t = \frac{\overline{X} - \mu}{S/\sqrt{n}}$$

is t-distributed with $(n - 1)$ degrees of freedom, as we discussed in Chapter 9. Thus, if X is normally distributed and the population standard deviation σ is not known, the t-distribution should be used to calculate the p-value rather than using the normal distribution, as illustrated in Example 10.10.

Example 10.10 A coffee machine is supposed to deliver an average of 7.7 ounces of coffee into 8-ounce cups. The company owning the machine has received complaints that some persons have been receiving cups that are only half full. The company sends a serviceman out to the machine, and he finds that a sample of ten fills produced an average fill of 7.55 ounces with a standard deviation of 0.2 ounce. Should the machine be adjusted or not? Assume that the fills are normally distributed.

Solution Since the company has received reports that the fills are short, a one-sided hypothesis test is indicated. The company establishes the following null and alternate hypotheses:

H_0: $\mu \geq 7.7$ ($\mu_0 = 7.7$)—machine operating properly
H_A: $\mu < 7.7$—machine not operating properly; needs adjustment

The test statistic is $\overline{X}$, with a value in the sample of size $n = 10$ of 7.55 ounces. The computation of the p-value is illustrated in Figure 10.28. Note that the p-value must be found by interpolating from Table B.5 in Appendix B, as illustrated in Table 10.8, noting that $P(t \leq -2.38) = P(t \geq 2.38)$.

FIGURE 10.28 Computation of the p-value in Example 10.10

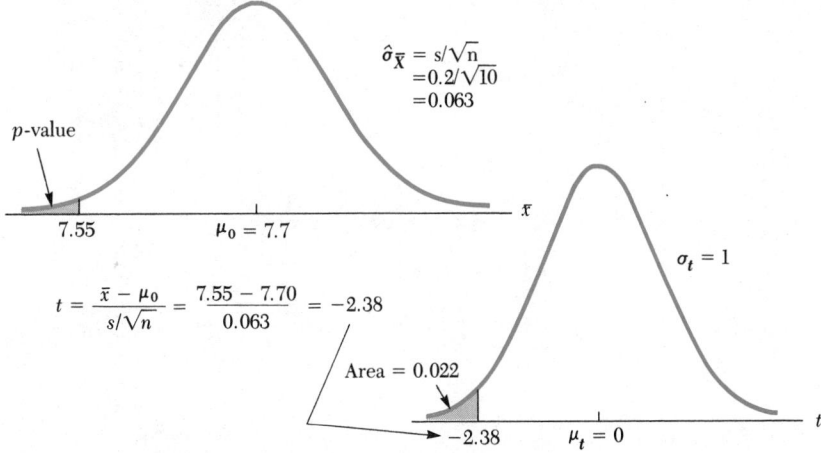

TABLE 10.8 Determining the p-value by interpolation

	t^*	$P(t \geq t)$		
	2.262	0.025		$\dfrac{0.118}{0.559}(0.015) = 0.003$
0.559 [0.118 [2.380	0.022] 0.003] 0.015			
	2.821	0.010		$0.025 - 0.003 = 0.022$

* Values of t with 9 degrees of freedom.

From Table 10.8, $P(t \leq -2.38) = 0.022$. Thus, if H_0: $\mu \geq 7.7$ is rejected, there is only a $100(0.022) = 2.2$ percent chance that a Type I error has been made. Even though this risk is quite low, it would not be a surprise if the company decided *not* to reject H_0: $\mu \geq 7.7$. By doing so, they would not adjust the machine (thereby saving a considerable amount of money), but would very likely incur a high risk of committing a Type II error—not rejecting a false null hypothesis [remember α and β are inversely related; since α is quite small (0.022), β is very likely quite high]. By not rejecting H_0, if a mistake has been made, the *consumer* will suffer the consequences by purchasing short fills.

The computation of the p-value from the t-distribution will almost always require interpolation in the table of t-values. To avoid interpolating, we could

TABLE 10.9 Testing procedures for the population mean

Distribution of the population random variable X	Knowledge of the population standard deviation σ	Sample size	Test statistic value to find the p-value	p-Value
Normal	Known	All	$z = \dfrac{\bar{x} - \mu_0}{\sigma/\sqrt{n}}$	Exact
	Unknown	All	$t = \dfrac{\bar{x} - \mu_0}{s/\sqrt{n}}$	Exact
Unknown	Known	$n \geq 30^*$	$z = \dfrac{\bar{x} - \mu_0}{\sigma/\sqrt{n}}$	Approximate
	Unknown	$n \geq 30^*$	$t = \dfrac{\bar{x} - \mu_0}{s/\sqrt{n}}$	Approximate
	Known or unknown	$n < 30^*$	Special small sample methods must be used (see, for example, Chapter 20 on nonparametric statistics)	

* The sample size of $n \geq 30$ is not a strict rule. If the population random variable X is a continuous random variable with a unimodal distribution, a sample of size 30 will usually be sufficient to ensure that the distribution of $\bar{X}$ is close enough to a normal distribution for the purpose of calculating the p-value. However, it may be necessary in some sampling experiments to require larger sample sizes before the central limit theorem may be depended on with confidence for the purpose of computing the p-value.

simply report that the p-value is less than 0.025 but greater than 0.010 by observing that the computed t-value of 2.380 is between the 2.5 percent value of 2.262 and the 1 percent value of 2.821. Thus, we would reject the null hypothesis at the 0.05 significance level and at the 0.025 significance level, but we would not reject the null hypothesis at the 0.01 significance level.

In Example 10.10, it was assumed that the population random variable **X** is normally distributed. The standardized t-statistic,

$$t = \frac{\overline{X} - \mu}{S/\sqrt{n}}$$

is t-distributed *only if* **X**, the population random variable, is normally distributed. If the distribution of the population random variable **X** is not known, then we must again appeal to the central limit theorem for the approximate sampling distribution of $\overline{X}$. If the sample size is large and **X** is not normally distributed, then the standardized statistic

$$\frac{\overline{X} - \mu}{S/\sqrt{n}}$$

will be *approximately* t-distributed with $(n - 1)$ degrees of freedom. In this case, the calculated p-value will be approximate as well. Table 10.9 summarizes the testing procedures for hypotheses concerning the population mean μ.

When it is known that the population random variable **X** is normally distributed, the computation of the p-value is *exact*. If σ is known, the p-value is computed by using the **Z**-statistic (case 1), and if σ is not known, it is computed using the **t**-statistic, (case 2).

When the population random variable **X** is not normally distributed, the computation of the p-value is *approximate*. Again, notice from Table 10.9 that the **Z**-statistic is used when σ is known and the **t**-statistic is used when σ is not known. If $n < 30$, when **X** is not normally distributed, then the p-value should *not* be computed by the methods described in this chapter. Rather, special methods for small samples should be used, such as those described in Chapter 20.

If there is doubt concerning whether or not $n \geq 30$ is sufficiently large for the central limit theorem to apply, a statistician should be consulted. Usually, if **X** is continuous and its distribution is unimodal and reasonably symmetric, $n \geq 30$ will suffice. But if the distribution of **X** is multimodal or if it is severely skewed, then larger sample sizes will be needed and should be specified by a statistician before any samples are drawn.

10.6 HYPOTHESIS TESTS CONCERNING THE POPULATION PROPORTION π

Recall that the standard deviation of the sample proportion statistic **P** is given by

$$\sigma_P = \sqrt{\frac{\pi(1 - \pi)}{n}}$$

Notice that σ_P will not be known in value *under any circumstances* when testing hypotheses concerning π. This is obvious, because if σ_P was known in value, so also would be the value of π and we would have no need to test hypotheses concerning π. So, unlike testing hypotheses concerning the population mean μ, when testing hypotheses concerning π, whether or not the standard deviation of the sampling distribution of the test statistic is known is not relevant.

The exact sampling distribution of the statistic **P** is either the binomial distribution (if the *population* is infinite in size) or the hypergeometric distribution (if the *population* is finite in size). For "small" samples, one of these two distributions *must be used* to calculate the p-value. If, on the other hand, the sample size is "large," then the central limit theorem may be appealed to for the approximate sampling distribution of **P**. Since the true sampling distribution of **P** is discrete and quite often severely skewed, we will consider $n \geq 100$ to be "large."

We will not describe the small sample ($n < 100$) procedure to calculate the p-value based on either the binomial or hypergeometric distribution in this text. As an alternative, nonparametric methods such as those given in Chapter 20 may be used in this case.

FIGURE 10.29 Approximate sampling distribution of **P** for large $n(n \geq 100)$

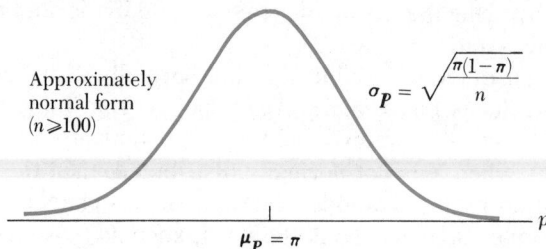

Approximately normal form $(n \geqslant 100)$

$$\sigma_P = \sqrt{\frac{\pi(1 - \pi)}{n}}$$

$\mu_P = \pi$

Figure 10.29 illustrates the *approximate* normal sampling distribution of **P** when $n \geq 100$. Note that the normal distribution is used rather than the t-distribution, though the value of σ_P is not known. The t-distribution is the appropriate sampling distribution to approximate the exact sampling distribution of **P** in this case. But when $n > 100$, the t-distribution is very similar to the standard normal distribution. Recall from Chapter 9 that as the sample size increases, the t-distribution converges to the standard normal distribution. Thus, we will use the normal distribution as the approximating distribution for all computations of p-values concerning hypothesis tests on π when $n \geq 100$.

Example 10.11   The marketing survey department of a large food producer believes that consumers will be indifferent to two different proposed packaging designs for a new cereal. That is, it is believed that 50 percent of the consumers would choose package A and 50 percent would choose package B. In a random sample of 225 potential consumers of the cereal, 130 preferred package A design. Test the hypothesis that the consumers are indifferent to the two package designs.

Solution   Let π = proportion of consumers who prefer package A design. The null and alternate hypotheses are:

$$H_0: \quad \pi = 0.50 \quad (\pi_0 = 0.50)$$
$$H_A: \quad \pi \neq 0.50$$

The appropriate sample statistic is the sample proportion statistic P. In the sample, 130 consumers preferred package A design to package B design. Therefore, $p = 130/225 = 0.578$. The estimate of σ_P is given by replacing π in the formula with the point estimate p:[2]

$$\hat{\sigma}_P = \sqrt{\frac{\hat{\pi}(1 - \hat{\pi})}{n}} = \sqrt{\frac{p(1 - p)}{n}}$$

Thus, the estimate $\hat{\sigma}_P$ is given by:

$$\hat{\sigma}_P = \sqrt{\frac{p(1 - p)}{n}} = \sqrt{\frac{(0.578)(0.422)}{225}} = 0.033$$

The p-value for this test is 0.0182. Its computation is illustrated in Figure 10.30. Since the p-value is quite small (0.0182), it is reasonable to reject $H_0: \quad \pi = 0.500$ and conclude that π is actually not equal to 0.500. The interpretation of this probability is illustrated in Figure 10.31.

Example 10.12   The administrators of the Sunrise Mutual Fund are considering a proposal to buy a significant number of shares of the Hawaii Airways Company stock. Their decision to purchase the stock will be based primarily on the potential for the value of the stock to rise by 25 percent during the next year. Let π be the proportion of stockbrokers who feel that the stock will rise by 25 percent over the next year. The administrators decide that they will purchase the stock if π is at least 0.80 (80 percent of the brokers feel the stock will rise by 25 percent during the next year). In a random sample of 100 brokers, the administrators find that 74 believe that the stock will rise by 25 percent in the next year, while 26 believe that the stock will not rise by 25 percent in value. What should the administrators do?

---

[2] This estimate of σ_P differs from the one used commonly in the past. Rather than using p in place of π in the formula for σ_P, some statisticians argue that π_0, the hypothesized value of π, should be used. In this way, the sampling distribution of P is completely specified (mean and standard deviation) assuming H_0 is true. We will alternatively use the best estimate of σ_P, *based on the sample information*, estimating π by the sample proportion p.

FIGURE 10.30 Computation of the p-value in Example 10.11

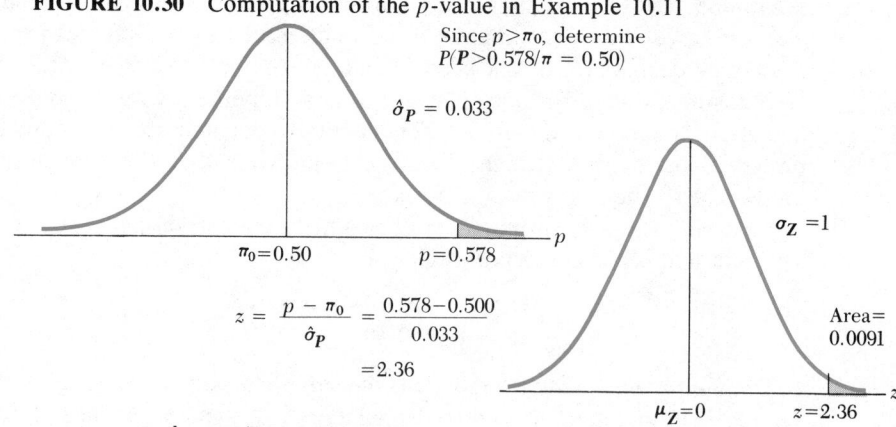

Since $p > \pi_0$, determine
$P(P > 0.578 / \pi = 0.50)$

$\hat{\sigma}_P = 0.033$

$\pi_0 = 0.50$ $p = 0.578$

$\sigma_Z = 1$

Area = 0.0091

$$z = \frac{p - \pi_0}{\hat{\sigma}_P} = \frac{0.578 - 0.500}{0.033}$$

$$= 2.36$$

$\mu_Z = 0$ $z = 2.36$

p-value $= 2P(P > 0.578 / \pi = 0.50)$
$= (2)(0.0091) = 0.0182$

$100(0.0182) = 1.82$ percent of all possible samples of size $n = 225$ will produce values of **P** either **greater** than 0.578 or less than 0.422 if $\pi = 0.50$

$100(1 - 0.0182) = 98.18$ percent of all possible samples of size $n = 225$ will produce values of **P** between 0.422 and 0.578 if $\pi = 0.50$

FIGURE 10.31 Interpretation of the p-value computed in Example 10.11

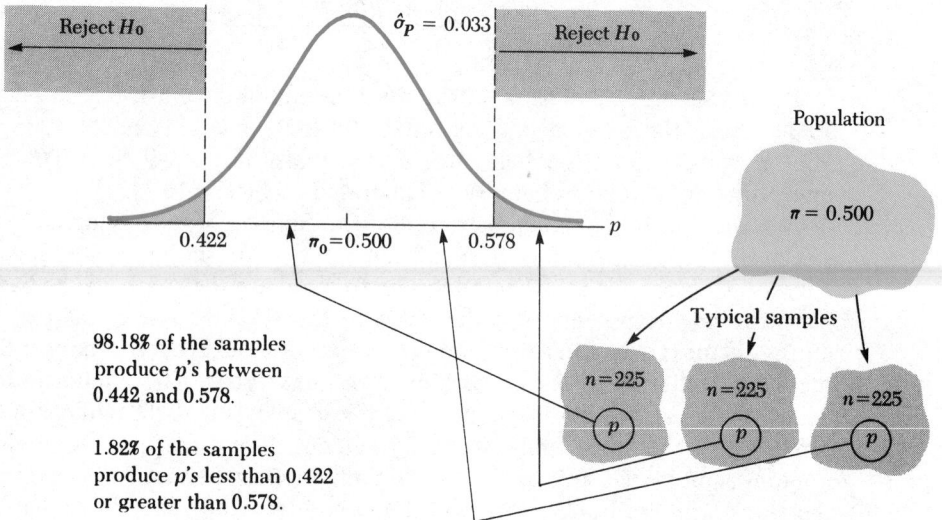

Reject H_0

$\hat{\sigma}_P = 0.033$

Reject H_0

Population

$\pi = 0.500$

0.422 $\pi_0 = 0.500$ 0.578

Typical samples

98.18% of the samples produce p's between 0.442 and 0.578.

$n = 225$ $n = 225$ $n = 225$

1.82% of the samples produce p's less than 0.422 or greater than 0.578.

p p p

Solution The null and alternate hypotheses are:

$$H_0: \quad \pi \geq 0.80 \quad (\pi_0 = 0.80)$$
$$H_A: \quad \pi < 0.80$$

The test statistic is the sample proportion statistic **P**, with a value equal to $p = 74/100 = 0.74$. The p-value computation is illustrated in Figure 10.32.

FIGURE 10.32   Computation of the p-value in Example 10.12

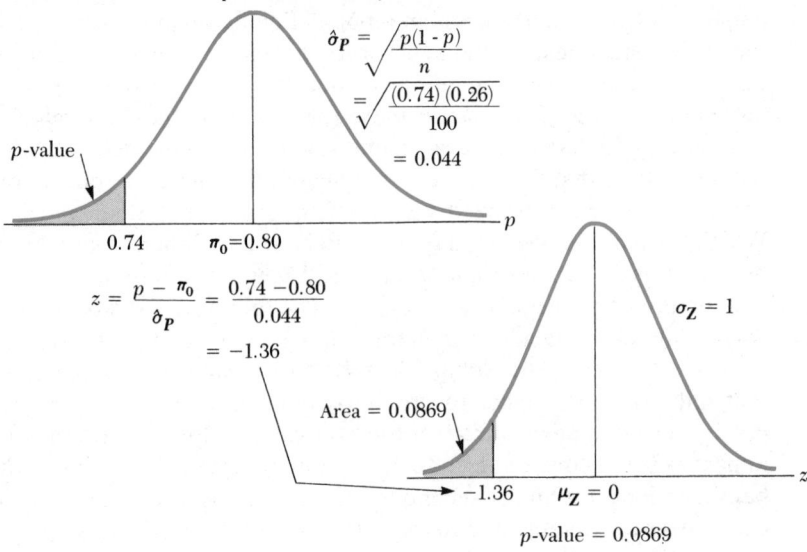

$$\hat{\sigma}_P = \sqrt{\frac{p(1-p)}{n}}$$

$$= \sqrt{\frac{(0.74)(0.26)}{100}}$$

$$= 0.044$$

$$z = \frac{p - \pi_0}{\hat{\sigma}_P} = \frac{0.74 - 0.80}{0.044}$$

$$= -1.36$$

$\sigma_Z = 1$

Area = 0.0869

-1.36 $\mu_Z = 0$

p-value = 0.0869

The p-value is 0.0869. In only 8.69 percent of all possible samples of size $n = 100$ would we expect to find a value of the statistic **P** less than or equal to 0.74 *when in fact* $\pi = 0.80$. On the other hand, if H_0: $\pi \geq 0.80$ is *not* rejected, the probability of committing an error (Type II) is unknown. An assessment could be made of the chance of committing a Type II error for various values of π by constructing the power curve. In this problem, it would be necessary for the administrators to assess the financial consequences of committing a Type I or Type II error. The chance of committing a Type I error in this instance might not be considered to be too high, given the consequences of not rejecting H_0 when in fact it is false (a Type II error). Thus, the administrators would likely reject H_0: $\pi \geq 0.80$. As a consequence of the decision to reject H_0, they would *not* purchase any shares of the stock.

In the next section, we will explore how a decision maker assesses the consequences of committing either a Type I or Type II error. Also, we will study how the decision maker chooses between form II and form III of the null and alternate hypotheses if a one-sided test of hypothesis is to be used.

■ 10.7   THE FORM OF THE STATISTICAL HYPOTHESIS

In the preceding section, we discussed statistical hypothesis-testing procedures based on large samples for the three forms of a statistical hypothesis: a two-sided test (H_0: $\theta = \theta_0$) or a one-sided test (H_0: $\theta \geq \theta_0$ or H_0: $\theta \leq \theta_0$). In a typical hypothesis-testing problem, how do we know which *form* of the hypothesis to use?

The distinction between a two-sided and a one-sided test is reasonably apparent. If we hypothesize that a population parameter assumes one value, then a two-sided test should be used. In this case, we will be led to reject the null hypothesis if the sample information suggests that the true value of the parameter is *either* less than *or* more than the hypothesized value.

In many problems, however, a one-sided test is indicated. We wish to test the hypothesis that the value of a population parameter is equal to or greater (less) than the hypothesized value. The question that arises in this case is: Which form of the one-sided test should be used? Should the null hypothesis be stated as a "greater than or equal to" inequality statement or as a "less than or equal to" inequality statement? The following example illustrates the way a statistician decides which form of the one-sided hypothesis test to use.

Example 10.13 The Savior Seat Belt Company has designed an airbag that inflates upon impact to protect occupants involved in an automobile crash. The company claims that the average time for the bags to inflate upon impact is 0.1 second or less. Tests on similar airbags have shown that if the bags take longer than 0.1 second to inflate, the passengers of the automobile can expect to be subjected to modest to severe "distress" as a result of the impact. The company wishes to test the claim based on 25 trial impacts of autos equipped with airbags. How would you set up the hypothesis statement for this problem?

Solution Let the inflation time of an airbag be denoted by X (measured in seconds). In this problem, we are concerned about the (population) average inflation time μ of the airbags. The Savior Company claims that $\mu \leq 0.1$ second. In Table 10.10, the two forms of the one-sided hypothesis test are given for this problem. For each of these two forms, A and B, the two types of error, Type I and Type II, are described.

TABLE 10.10 Statement of the two forms of the one-sided test and a description of the Type I and Type II errors for each hypothesis statement

	Hypothesis	
	A H_0: $\mu \leq 0.1$ H_A: $\mu > 0.1$	B H_0: $\mu \geq 0.1$ H_A: $\mu < 0.1$
Type I error	We reject the null hypothesis that the airbags will inflate in an average of 0.1 second or less when they really will.	We reject the null hypothesis that the airbags will inflate in an average of 0.1 second or more when they really do.
Type II error	We fail to reject the null hypothesis that the airbags will inflate in an average of 0.1 second or less when they will not.	We fail to reject the null hypothesis that the airbags will inflate in an average of 0.1 second or more when they do not.

In making a decision about this problem, there are potentially two errors that can be made by the testing agency based on the sample information:

1. The testing agency can claim that the average inflation time for the airbags is, indeed, less than or equal to 0.1 second when in fact it is not.
2. The testing agency can claim that the average inflation time for the airbags is greater than or equal to 0.1 second when in fact it is not.

Let us analyze the ramifications of these two errors. If the first error is made, the Savior Company is going to produce airbags they think are safe when, in fact, they are not. Potentially, many automobile occupants (consumers) can be hurt if this error is made. If the second error is made, the Savior Company would probably spend a considerable amount of research money in further testing only to find (eventually) that the airbags had an average inflation time of 0.1 second or less all along. From the consumer's standpoint, the first error is by far the more serious one!

Now refer to Table 10.10. In hypothesis statement A, the first error described is the Type II error, whereas in hypothesis statement B, the first error is the Type I error. Recall that, in constructing the decision rule, the probability of committing a Type I error is computed, but the actual probability of committing a Type II error (β) cannot be computed. Another way of stating this is that the probability of committing a Type I error is "in our control," while the probability of committing a Type II error is not.

On this basis, as consumers, we would select hypothesis statement B since the description of the Type I error is in fact the more serious of the two errors described above. If we wished to be particularly cautious (possibly at considerable expense to the Savior Company), we would require that the p-value be very small, say less than 0.001, to reject the null hypothesis in statement B. We minimize the risk of claiming that the airbags will inflate in 0.1 second or less when they will not, but we increase the probability of the Type II error (β)—claiming that the airbags will inflate in 0.1 second or more when, in fact, they will inflate in less than 0.1 second.

Notice that we selected hypothesis statement B in Table 10.10 on the basis of arguing from the consumer's viewpoint. The Savior Company is quite likely to view this matter from a different perspective. They most probably would consider the second error described above as being the more "serious" error, because if this error were made, it would cost the company money for further research when, in fact, none is needed.

If you were the statistician for the Savior Company, which hypothesis statement would you select, A or B? If you study the vast number of automobile recalls to correct faulty equipment that began in the early 1970s *forced* on auto makers by the government, the answer might well be surmised.

Example 10.14 Consider the problem described in Example 10.10. A coffee machine is supposed to deliver an average fill of $\mu = 7.7$ ounces of coffee in 8-ounce cups. The company has received reports that the fills are less than designed. How should the company set up the test of the hypothesis to determine whether the machine is operating properly?

Solution   Since the company has received reports that the fills are short, a one-sided test is indicated. They are interested in determining whether or not the average fill has fallen significantly below 7.7 ounces. Next, they must decide which form of the one-sided test to use:

A   B

H_0: $\mu \leq 7.7$ or H_0: $\mu \geq 7.7$

H_A: $\mu > 7.7$ H_A: $\mu < 7.7$

From the company's standpoint, it does not want to spend the money adjusting the machine unless it is clear that the machine is out of kilter. Thus, they wish to minimize the probability of rejecting that μ is greater than or equal to 7.7 when it really is. In hypothesis statement B, this is the description of the Type I error, so they would select B. Thus, the null and alternate hypotheses are:

$$H_0:\ \mu \geq 7.7 \quad (\mu_0 = 7.7)$$
$$H_A:\ \mu < 7.7$$

■ 10.8   RELATIONSHIP OF HYPOTHESIS TESTING TO CONFIDENCE INTERVAL ESTIMATION

Hypothesis testing and confidence interval estimation are closely related inference methods. In fact, hypothesis testing can often be conducted by using confidence intervals, as the next example illustrates.

Example 10.15   An air-conditioning/heating system is designed to maintain a 72°F temperature in a room that houses a large mainframe computer system. In 30 randomly sampled temperatures taken over a two-month period, the sample average temperature was 73.0°F with a standard deviation of 2.8°F. Is the heating/cooling system meeting its design specifications? Assume that the temperatures are normally distributed.

Solution   Viewed as a hypothesis-testing problem, we have:

$$H_0:\ \mu = 72 \quad (\mu_0 = 72)$$
$$H_A:\ \mu \neq 72$$

The test statistic is $\overline{X}$, with a value of 73.0°F in the sample of $n = 30$ randomly taken temperatures. The appropriate approximate sampling distribution for the standardized statistic

$$\frac{\overline{X} - \mu_0}{S/\sqrt{n}}$$

is the t-distribution with $(n - 1) = (30 - 1) = 29$ degrees of freedom. The computation of the p-value is shown in Figure 10.33.

FIGURE 10.33   Calculation of the *p*-value in Example 10.15

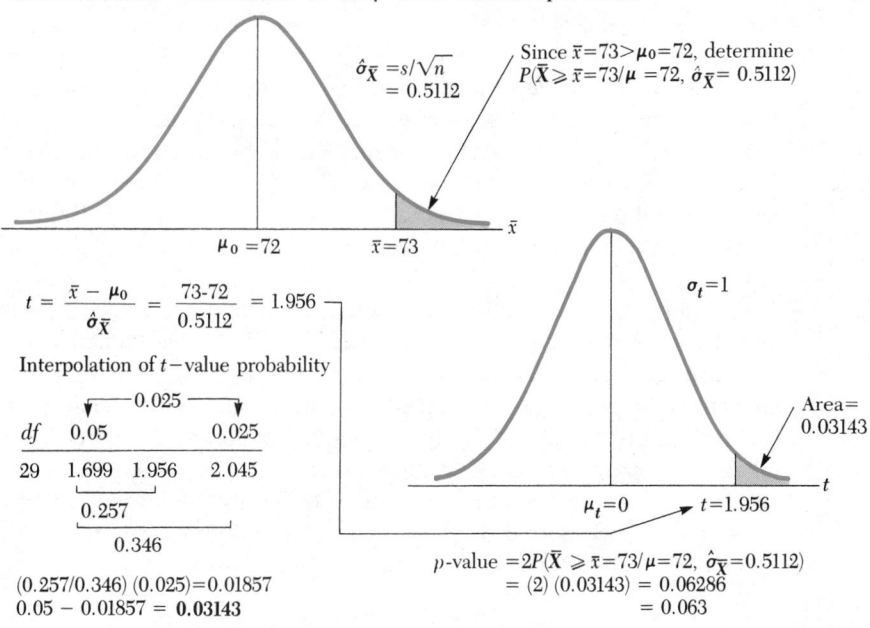

$$t = \frac{\bar{x} - \mu_0}{\hat{\sigma}_{\bar{x}}} = \frac{73\text{-}72}{0.5112} = 1.956$$

Interpolation of *t*–value probability

	0.025	
df	0.05	0.025
29	1.699 1.956	2.045

0.257

0.346

(0.257/0.346) (0.025)=0.01857
0.05 − 0.01857 = **0.03143**

p-value $=2P(\bar{X} \geqslant \bar{x}=73/\mu=72, \hat{\sigma}_{\bar{x}}=0.5112)$
$= (2) (0.03143) = 0.06286$
$= 0.063$

Based on the *p*-value (0.063), we would reject the null hypothesis at the 0.10 significance level, but not at the 0.05 or the 0.01 significance level.

Now, consider 90 percent and 95 percent confidence intervals on μ:

90 percent: $\bar{x} \pm t \left(\dfrac{s}{\sqrt{n}} \right) \Rightarrow 73 \pm 1.699 \left(\dfrac{2.8}{\sqrt{30}} \right) \Rightarrow 72.13$ to 73.87

95 percent: $\bar{x} \pm t \left(\dfrac{s}{\sqrt{n}} \right) \Rightarrow 73 \pm 2.045 \left(\dfrac{2.8}{\sqrt{30}} \right) \Rightarrow 71.96$ to 74.04

Notice that the 90 percent confidence interval does not contain the hypothesized value of μ ($\mu_0 = 72$), whereas the 95 percent confidence interval does contain $\mu_0 = 72$. Since the *p*-value was calculated to be 0.062, we should not find this result surprising. In fact, if the confidence coefficient in the confidence interval is *greater* than $100(1 - 0.062)$ percent $= 93.8$ percent, the confidence interval will contain $\mu_0 = 72$, and if the confidence coefficient is *less* than 93.8 percent, the confidence interval will not contain $\mu_0 = 72$. In the former case, we would not reject H_0 (since $\mu_0 = 72$ is contained in the interval), whereas in the latter case, we would reject H_0 (since $\mu_0 = 72$ is not contained in the interval).

It should be apparent that the significance level of the hypothesis test is related to the "confidence" in a confidence coefficient. In fact, $100[1 - \text{(significance level)}]$ percent $=$ confidence. Thus, if we set the significance level at, say 0.05, so that the null hypothesis is rejected if the *p*-value < 0.05, the associated confidence is $100(1 - 0.05) = 95$ percent. Then, to test the null hypothesis H_0: $\mu = \mu_0$ at the 0.05 significance level, we can construct a 95 percent confidence interval and (1) reject H_0 if the confidence interval does not contain μ_0 or (2) not reject H_0 if the confidence interval does contain μ_0.

A similar relationship exists between one-sided hypotheses and confidence bounds. For example, if we are testing the null hypothesis H_0: $\mu \geq \mu_0$ and decide to reject it at the 0.05 significance level based on the sample information, this decision would be consistent with finding a 95 percent *upper* confidence bound on μ and finding that μ_0 lies *above* this bound. Example 10.16 illustrates the relationship between the one-sided test and the confidence bound.

Example 10.16 A hamburger franchise chain sells a special hamburger called the "bacon and guacomole special." Since this hamburger requires special ingredients, the chain is concerned about the mean number μ of specials sold per week in each of its outlets. It is hypothesized that μ is greater than or equal to 750. In a random sample of 61 outlets in the chain taken during a given week, it is found that an average of 722 specials are sold per week with a standard deviation of 80. By constructing a 95 percent upper confidence bound on μ, should this null hypothesis be rejected?

Solution The null and alternate hypotheses are:

$$H_0: \quad \mu \geq 750 \quad (\mu_0 = 750)$$
$$H_A: \quad \mu < 750$$

The 95 percent upper confidence bound is given by:

$$\bar{x} + t_{df=60;\ \alpha=0.05}\left(\frac{s}{\sqrt{n}}\right) = 722 + 1.671\left(\frac{80}{\sqrt{61}}\right) = 739.12$$

Thus, we are 95 percent confident that μ is not greater than 739.12 units. Since $\mu_0 = 750$ lies above this bound, we would reject H_0 at the 0.05 significance level.

A similar relationship exists between hypothesis testing and confidence

TABLE 10.11 Decision rules for rejecting a null hypothesis by the use of confidence intervals or confidence bounds

Null hypothesis	Estimation	Decision
H_0: $\mu = \mu_0$ or H_0: $\pi = \pi_0$	$100(1 - \alpha)$ percent confidence interval	Reject H_0 if hypothesized value (μ_0 or π_0) *does not* lie in the confidence interval
H_0: $\mu \geq \mu_0$ or H_0: $\pi \geq \pi_0$	$100(1 - \alpha)$ percent *upper* confidence bound	Reject H_0 if hypothesized value (μ_0 or π_0) is *greater than* upper confidence bound
H_0: $\mu \leq \mu_0$ or H_0: $\pi \leq \pi_0$	$100(1 - \alpha)$ percent *lower* confidence bound	Reject H_0 if hypothesized value (μ_0 or π_0) is *less than* lower confidence bound

interval estimation when we are drawing inferences about the population proportion π. Table 10.11 summarizes the use of confidence intervals and confidence bounds to make a decision concerning a null hypothesis about either μ, the population mean, or π, the population proportion.

■ 10.9 SETTING THE SAMPLE SIZE FOR HYPOTHESIS TESTING

When we make the decision, "do not reject H_0," we are saying that based on the sample data set, we have insufficient information to reject H_0. Had we taken a larger sample, however, we may have been able to reject H_0.

If we carefully plan our experiment, we can usually spare ourselves doubt in the case where the decision is not to reject H_0. We can do this by specifying the *maximum* values of α and β we wish to tolerate in a particular hypothesis-testing experiment. Setting the maximum tolerable values of α and β is made possible by selecting the sample size for the experiment, as illustrated in Example 10.17.

Example 10.17 Suppose we wish to test the null hypothesis H_0: $\mu = \mu_0$, such that $\alpha \leq 0.05$, and we wish to be 95 percent confident that if μ is more than two units from μ_0, then our decision rule and sample data will detect this difference ($\beta \leq 0.05$). Notice that the 95 percent confidence sets the *minimum* power for this test [Power = P(reject H_0)]. But if the null hypothesis is false, Power = $1 - \beta$, so that we are in effect setting the *maximum* value of β equal to 0.05. Now suppose that $\mu = \mu_0 + 2$. Figure 10.34 illustrates the

FIGURE 10.34 Illustration of α and β in Example 10.17

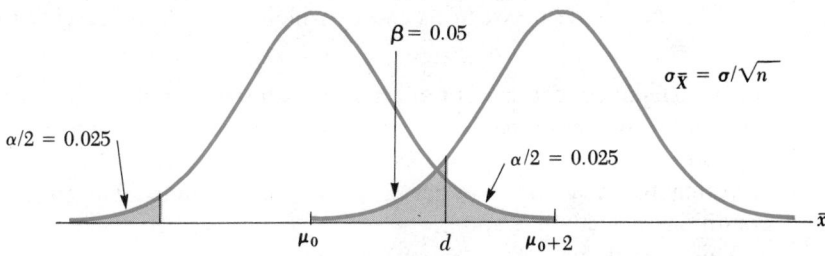

α and β probabilities. In Figure 10.34, the point d on the $\bar{x}$ axis represents the number of standard deviations above μ_0 which leaves $\alpha/2 = 0.025$ area to the right of d and the number of standard deviations below $\mu_0 + 2$ which leaves $\beta = 0.05$ area to the left of d. The quantity d also represents the difference between μ and μ_0 we wish to detect, given the maximum settings of α and β.

Now in terms of the sampling distribution under H_0: $\mu = \mu_0$,

$$d = \mu_0 + 1.96 \left(\frac{\sigma}{\sqrt{n}} \right)$$

and in terms of the sampling distribution when $\mu = \mu_0 + 2$,

$$d = (\mu_0 + 2) - 1.64 \left(\frac{\sigma}{\sqrt{n}}\right)$$

Setting the two equations for d equal to one another and solving for n, we have:

$$n = \frac{\sigma^2(1.96 + 1.64)^2}{2^2}$$

If $\sigma = 10$ from, say, a pilot study, then

$$n = \frac{10^2(1.96 + 1.64)^2}{2^2} = 324$$

Thus, to ensure that $\alpha \leq 0.05$ and $\beta \leq 0.05$, we must sample at least 324 units.

The general formula for a two-sided test is given by

$$n \geq \frac{\sigma^2(z_{\alpha/2} + z_\beta)^2}{d^2}$$

where

σ^2 = population variance

$z_{\alpha/2}$: $P(Z \geq z_{\alpha/2}) = \dfrac{\alpha}{2}$

z_β: $P(Z \geq z_\beta) = \beta$

d = difference between μ_0 and μ, which we wish to detect with Power
$= 1 - \beta$

In a one-sided test, $z_{\alpha/2}$ is replaced by z_α in the formula. When we wish to test a two-sided hypothesis concerning the population proportion π, σ^2 may be replaced by $\pi(1 - \pi) = 0.25$ [the maximum value of $\pi(1 - \pi)$]. The resulting formula will produce a *conservative* value of n to satisfy the conditions placed on α and β.

This formula must be used with caution. If we use it to set a sample size for conducting a hypothesis test on μ, then the formula should produce a sample size of at least 30 to be applicable. This is because the formula assumes that the sampling distribution of the statistic $\overline{X}$ is approximately normal and, via the central limit theorem, a sample of at least 30 is usually required. If the formula produces a sample size smaller than 30, it is not applicable and other methods should be used (methods that depend on the exact sampling distribution of $\overline{X}$, or nonparametric approaches found in Chapter 20). When the formula is used to set a sample size to test a hypothesis concerning π, the sample size produced by the formula should be at least 100. If not, the formula is not applicable in this case.

Example 10.18 Suppose we are interested in testing the null hypothesis H_0: $\pi \geq 0.25$ with $\alpha \leq 0.10$ and we wish to detect a difference of 0.025 unit with a minimum power of 0.99 ($\beta \leq 0.01$). What sample size is required to meet these conditions?

Solution

$$n \geq \frac{[\pi(1 - \pi)][z_\alpha + z_\beta]^2}{d^2}$$

where

$\pi(1 - \pi) = 0.25$

$z_\alpha = z_{0.10}$; $P(\mathbf{Z} \geq z_{0.10}) = 0.10$; $z_{0.10} = 1.28$
$z_\beta = z_{0.01}$; $P(\mathbf{Z} \geq z_{0.01}) = 0.01$; $z_{0.01} = 2.33$
$d = 0.025$

Now,

$$n \geq \frac{(0.25)(1.28 + 2.33)^2}{(0.025)^2} = 5,213$$

Thus, a sample of at least 5,213 units is required to meet these conditions.

◼ 10.10 TESTING HYPOTHESES CONCERNING THE POPULATION VARIANCE σ^2 AND THE POPULATION STANDARD DEVIATION σ

The hypothesis tests concerning the population variance and the population standard deviation are based on the χ^2 distribution introduced in Chapter 9. These tests assume that $\mathbf{X}$, the population random variable, is normally distributed. If $\mathbf{X}$ is not normally distributed, but is approximately so, these tests should be used only for "large" samples ($n = 100$ or more is often required in this case). The degrees of freedom (df) for the χ^2 test is $df = n - 1$. Figure 10.35 illustrates the computation of the p-value for the three forms of the null hypothesis.

Example 10.19   A process that produces ball bearings is considered to be out of control if the standard deviation of the bearing diameters *exceeds* 0.0007 inch. If the production process is not stopped in this case, a considerable cost is incurred to eventually replace the bearings. In a sample of $n = 100$ of these ball bearings, it is found that the sample standard deviation is $s = 0.00083$. Should the production process be stopped and considered to be out of control?

Solution   Since the event, $\sigma > 0.0007$, represents a potentially serious loss if it is not identified, we will place this event in the alternate hypothesis:

H_0: $\sigma \leq 0.0007$ ($\sigma_0 = 0.0007$)
H_A: $\sigma > 0.0007$

The sample χ^2 test statistic value is:

$$\chi^2 = \frac{(n - 1)s^2}{\sigma_0^2} = \frac{(99)(0.00083)^2}{(0,0007)^2} = 139.2$$

FIGURE 10.35 The p-value computation for the three forms of the null hypothesis concerning σ^2 or σ

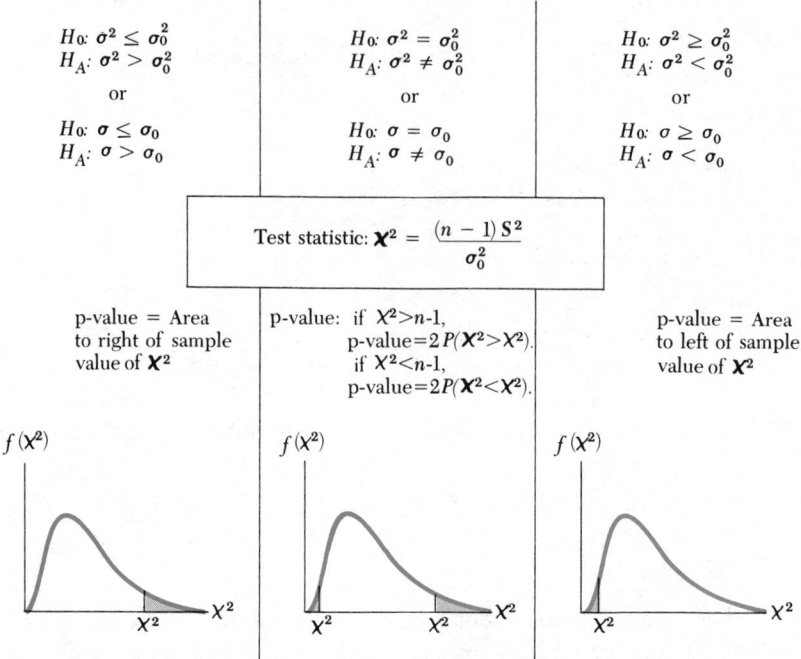

Figure 10.36 illustrates the computation of the p-value, which is the shaded region in the figure. Table B.6 in Appendix B gives the values of the χ^2 test statistic corresponding to the 0.100, 0.05, 0.025, 0.01, 0.005, and 0.001 significance levels. The table entries are given for 95 and 100 degrees of freedom. Without interpolating, it is apparent that the p-value is approximately 0.005. Since the p-value is less than 0.01, we would reject H_0: $\sigma \leq 0.0007$ at the 0.10, 0.05, and 0.01 significance levels. In this prob-

FIGURE 10.36 Computation of the p-value in Example 10.19

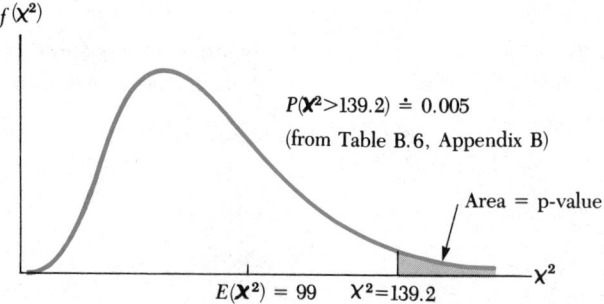

lem, the manager of the production process would no doubt reject the null hypothesis, since by doing so, he would take a risk of only 0.005 that a Type I error has been committed; that is, shutting down the process when in fact $\sigma \leq 0.0007$. Thus, he would decide to shut the process down to make the necessary adjustments to ensure that $\sigma < 0.0007$.

When we are testing a two-sided hypothesis (form I), the following rule should be used:

$$\text{If } \chi^2 > E(\chi^2) = n - 1, \quad \text{find } P(\chi^2 \geq \chi^2) \text{ and double it}$$
$$\text{If } \chi^2 < E(\chi^2) = n - 1, \quad \text{find } P(\chi^2 \leq \chi^2) \text{ and double it}$$

Consider Example 10.20.

Example 10.20 Find the p-value associated with the null and alternate hypotheses

$$H_0: \quad \sigma^2 = 75 \quad (\sigma_0^2 = 75)$$
$$H_A: \quad \sigma^2 \neq 75$$

when, in a sample of size 25, $s^2 = 38.75$. Assume **X**, the population random variable, is normally distributed.

Solution The value of the test statistic is:

$$\chi^2 = \frac{(n-1)s^2}{\sigma_0^2} = \frac{(24)(38.75)}{75} = 12.4$$

The mean of the appropriate χ^2 distribution for the test statistic is $E(\chi^2) = n - 1 = 24$. Since $\chi^2 = 12.4 < 24$, we must find $P(\chi^2 \leq 12.4)$. This computation is illustrated in Figure 10.37.

FIGURE 10.37 Computation of the p-value in Example 10.20

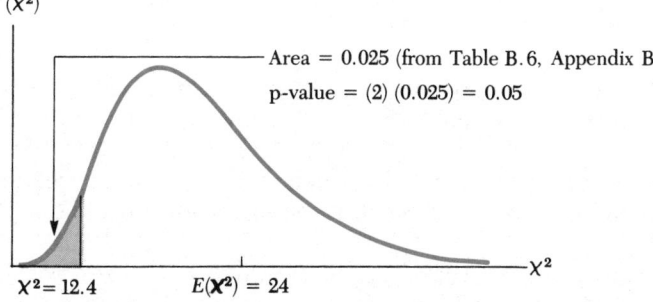

$f(\chi^2)$

Area = 0.025 (from Table B.6, Appendix B)
p-value = (2) (0.025) = 0.05

$\chi^2 = 12.4 \qquad E(\chi^2) = 24 \qquad \chi^2$

Thus, the p-value is 0.05 for this hypothesis test is $[2P(\chi^2 \leq 12.4)]$.

■ 10.11 SUMMARY

In statistical hypothesis testing, we hypothesize a value for a population parameter and use the sample information to determine whether the hypothesis is reasonable. The hypotheses statements may assume one of three forms:

Form I	Form II	Form III
H_0: $\theta = \theta_0$	H_0: $\theta \geq \theta_0$	H_0: $\theta \leq \theta_0$
H_A: $\theta \neq \theta_0$	H_A: $\theta < \theta_0$	H_A: $\theta > \theta_0$

The form chosen for a particular problem frequently is clearly indicated by the nature of the problem. If one value of the parameter ($\theta = \mu$, π, or σ) is being hypothesized, the two-sided test (form I) should be used. If a range of values of the parameter is being hypothesized, one of the one-sided tests (form II or III) should be used. The determination of which one-sided test to use depends on assessing the consequences of committing each of the two types of errors that can be committed in hypothesis testing. The Type I error is made when a true null hypothesis (H_0) is rejected, and a Type II error is made when a false null hypothesis is not rejected. In statistical hypothesis testing, the probability of committing a Type I error is denoted by α, while the probability of committing a Type II error is denoted by β.

In conducting a statistical hypothesis test by the p-value method, five steps are involved:

1. State the hypotheses, H_0 and H_A.
2. Choose the appropriate test statistic ($\overline{X}$ for μ, P for π, and S^2 for σ^2).
3. Select the random sample and compute the value of the test statistic.
4. Compute the p-value for the test.
5. Make the decision based on the calculated p-value. The traditional significance levels are 0.10, 0.05, and 0.01. For example, we would reject H_0 at the 0.05 significance level if the p-value < 0.05.

In this chapter, hypothesis-testing procedures were developed for testing hypotheses concerning μ, π, and σ^2 (or σ) when the population random variable X is normally distributed. Approximate tests based on the central limit theorem were developed for tests concerning μ, π, and σ^2 (or σ) that require "large" sample sizes [$n \geq 30$ for $\overline{X}$, $n \geq 100$ for P and S^2 (or S)]. If the sample sizes do not meet these conditions for the approximate tests, other methods must be used, such as nonparametric testing methods (see Chapter 20).

The selection of the level of significance and the selection of the sample size in a hypothesis-testing experiment are complex but highly interrelated processes. A statistician generally spends considerable time constructing power curves for different significance levels and sample sizes before the final set is selected. An alternative is to control the sizes of the Type I and Type II error probabilities by the selection of the sample size. A formula is developed to determine n in this fashion for hypothesis tests concerning μ and π.

A final note: hypothesis testing is a very intuitive inference-making process. It is perhaps best understood in a two-sided test concerning, say, the population mean μ. If the null hypothesis is H_0: $\mu = \mu_0$, where μ_0 is the hypothesized value of μ, the reasoning develops as follows. We assume that H_0 is true, so that the sampling distribution of $\overline{X}$ is centered at μ_0. Then, we determine how likely it is to experience a difference between the sample value of $\overline{X}$ ($\bar{x}$) and the hypothesized value of μ (μ_0) of $\bar{x} - \mu_0$ units. If the sample value $\bar{x}$ is not close to μ_0 in the center of the sampling distribution, then we are likely to conclude that μ is not, in fact, equal to μ_0. If, on the other hand, the sample value $\bar{x}$ is close to μ_0 in the center of the sampling distribution, then we are likely to conclude that μ is equal to μ_0. The calculation of the p-value simply "mathematizes" this process by determining in the two-sided test, for example, how likely it is to get a value of $\overline{X}$ in the sample that is more than $d = |\bar{x} - \mu_0|$ units from the hypothesized value μ_0, if μ is in fact equal to μ_0.

■ REFERENCES

Nie, N.; Hull, C.; Jenkins, Jean; Steinbrenner, Kavin; Bent, Dale. *SPSS: Statistical Package for the Social Sciences.* New York: McGraw-Hill Book Company, 1975.

Ostle, B., and Mensing, R. *Statistics in Research.* 3d ed. Ames: Iowa State University Press, 1975.

Ryan, T.; Joiner, B.; and Ryan, B. *MINITAB—A Student Handbook.* North Scituate, Mass.: Duxbury Press, 1976.

Winkler, R., and Hays, W. *Statistics: Probability, Inference and Decision.* 2d ed. New York: Holt, Rinehart and Winston, Inc., 1976.

■ PROBLEMS

10.1 How does hypothesis testing differ from statistical estimation?

10.2. What are the three basic forms of statistical hypotheses?

10.3. What is meant by a *one-sided* test? A *two-sided* test?

10.4. Carefully define a Type I error and a Type II error.

10.5. Explain the following terms used in statistical hypothesis testing.
a. Decision rule.
b. Region to not reject.
c. Rejection region.
d. p-Value.

10.6. Discuss the relationship between hypothesis testing and confidence interval estimation.

10.7. How is the sampling distribution used in hypothesis testing?

10.8. What is meant by a *test statistic?*

10.9. What is the relationship between α and β? Explain. How can both be jointly decreased in a hypothesis-testing problem?

10.10. What role does the central limit theorem play in hypothesis testing?

10.11. In testing a hypothesis concerning a population mean, when is it appropriate to use the t-distribution rather than the z-distribution?

10.12. Consider the simple hypothesis:

$$H_0: \quad \mu = 50$$
$$H_A: \quad \mu = 60$$

The decision rule, based on a sample of 100, is reject H_0 if $\bar{x} > 55$; otherwise, do not reject H_0. Assume $\sigma = 25$, and $\overline{X}$ is normally distributed.

a. Determine the probability of a Type I error.

b. Determine the probability of a Type II error.

10.13. Repeat parts *a* and *b* in Problem 10.12 for a sample of $n = 225$. Compare your answers here with those in Problem 10.12.

10.14. A study is to be conducted of MBA achievement scores on a graduate entrance exam. It is hypothesized that the average score of MBAs taking this test is 500:

$$H_0: \quad \mu = 500$$
$$H_A: \quad \mu \neq 500$$

A random sample of $n = 100$ MBA students is selected. The decision rule is: reject H_0 if $\bar{x} > 525$ or if $\bar{x} < 475$. Assume that the standard deviation is 100, and $\bar{X}$ is normally distributed.

a. Find α for this test.

b. Find β for this test if μ is really 520.

10.15. A packing process is designed to fill steel drums with 400 pounds of a chemical. To determine whether the process is working properly, a random sample of 36 drums will be selected for testing. The hypothesis is:

$$H_0: \quad \mu = 400$$
$$H_A: \quad \mu \neq 400$$

It is decided to reject H_0 if $\bar{x}$ is less than 390 or greater than 410. Assume that $\sigma = 42$ pounds and that the drum fills are normally distributed.

a. Sketch the power curve by determining the points on the curve for the following possible values of μ: 380, 390, 400, 410, 420.

b. What is the probability of committing a Type I error for this test?

c. Suppose the decision rule is changed to the following: reject H_0 if $\bar{x}$ is less than 395 or greater than 405. Repeat parts *a* and *b*,

and compare your new results with those in *a* and *b*. Sketch the new power curve. What was the effect on the power curve of changing the decision rule?

10.16. It is known that the daily wages in a particular industry are approximately normally distributed with a mean of $54 and a standard deviation of $6. If a company in this industry employing 36 workers pays these workers an average daily salary of $51, can it be accused of paying inferior wages? Explain.

10.17. The mean lifetime of a sample of 100 fluorescent bulbs produced by a company is computed to be 1,570 hours with a standard deviation of 120 hours. If μ is the mean lifetime of all bulbs produced by the company, test the hypothesis $H_0: \quad \mu \geq 1,600$ hours ($H_A: \quad \mu < 1,600$ hours). Compute the p-value for the test and reject H_0 if the p-value is less than 0.10.

10.18. A manufacturer of CRT (cathode-ray tube) computer terminals purchases the tubes from one of a few large suppliers. The firm will not purchase the tubes from a particular supplier, however, unless it can be demonstrated that the average lifetime of the tubes exceeds 5,000 hours prior to initial failure. A random sample of nine tubes is tested, and the following sample values are obtained: $\bar{x} = 5,060$, $s^2 = 2,500$. Assume that the lifetimes are normally distributed. Test the null hypothesis $H_0: \quad \mu \leq 5,000$ ($H_A: \quad \mu > 5,000$) by calculating the p-value for the test. Explain your decision regarding H_0.

10.19. With routine equipment like light bulbs, which wear out after a time, the standard deviation of the lifetime is an important factor in determining whether or not it is cheaper to replace all pieces at a fixed interval of time or to replace each piece individually when it breaks down. For a certain gadget, an industrial statistician has calculated that it will pay to replace at fixed time intervals if $\sigma < 6$ days. A sample of 71 pieces gives $s = 4.2$ days. Test the hypothesis H_0: $\sigma \geq 6$ (H_A: $\sigma < 6$) by calculating the p-value for the test. If the p-value is less than 0.01, reject H_0.

10.20. From past experience, it has been determined that a qualified operator on a certain machine turning out 400 items per day produces 20 or fewer defective items per day. A new operator is hired to run the same machine and the hypothesis is made that he is a qualified operator.

a. Test the null hypothesis H_0: $\pi \geq 0.05$ (H_A: $\pi < 0.05$) if $p = 32/400 = 0.08$ (that is, in one day's production, the new operator produces 32 defects) by calculating the p-value for the test and rejecting H_0 if the p-value is less than 0.10.

b. We wish to demonstrate that the operator is qualified. Why do we place this statement ($\pi < 0.05$) in the alternate hypothesis?

10.21. In each of the following hypothesis-testing situations, state whether a one-sided or a two-sided statistical test should be conducted, and the reasons for your choice. Give the statements of H_0 and H_A.

a. An auditor wishes to determine whether the proportion of invoices containing at least one error has changed since his audit last year.

b. A government agency is conducting a study to determine whether the proportion of women who smoke has increased since the last study.

c. A business must determine whether its average order size for units of a product has decreased from last year.

d. An executive of a major credit card corporation claims that the average amount owed by all those using the card is $100.

10.22. A cigarette manufacturer claims the average nicotine content of one of its brands of cigarettes is 10 grams. In a sample of 36 cigarettes of this brand, the average nicotine content was 12 grams with a standard deviation of 2 grams. Is the manufacturer's claim tenable? Use the p-value for the test to decide whether the claim is tenable.

10.23. Consider the hypotheses:

$$H_0: \quad \pi = 0.30$$
$$H_A: \quad \pi = 0.40$$

The decision rule, based on a sample of size $n = 100$, is to reject H_0 if p, the sample proportion, is greater than 0.35. Assume that p is approximately normally distributed.

a. What is the probability of a Type I error?

b. What is the probability of a Type II error?

10.24. When a machine is in perfect adjustment, it produces bolts with a mean diameter of 0.0600 inch and a standard deviation of 0.0150 inch. In order to ascertain whether or not the machine is still in adjustment, a sample of 36 bolts is selected. The sample mean is found to be 0.0575.

a. Use a confidence interval with $\alpha = 0.05$ to answer the question.

b. Set up a hypothesis test and calculate the p-value for the test. Use the p-value to decide whether or not the machines need adjustment.

10.25. A company issuing credit cards believes that the holders of its credit cards average at least $20,000 annual salary. A random sample of 145 of its 120,000 cardholders shows an average salary of $19,953 and a standard deviation of $156. The null and alternate hypotheses are stated as:

$$H_0: \quad \mu \geq \$20,000$$
$$H_A: \quad \mu < \$20,000$$

a. Using a 95 percent confidence bound, decide whether or not H_0 should be rejected.

b. Calculate the p-value for the test. Reject H_0 if the p-value is less than 0.05.

10.26. A company claims that at least 20 percent of the public prefers its product to the competing company's product. A sample of 100 persons is taken to check the validity of this claim. With $\alpha = 0.05$, how small would the sample proportion have to be before the claim could be statistically rejected?

10.27. A production process is believed to produce 2 percent defectives. In a random sample of 225 items taken from this process, it is found that 6 items are defective. The null hypothesis is $H_0: \quad \pi = 0.02$.

a. Calculate the p-value for the test. At which significance levels, $\alpha = 0.01, 0.05$, and 0.10, would the null hypothesis be rejected? Explain.

b. Calculate a 99 percent confidence interval for π. Based on the interval, should the null hypothesis be rejected or not rejected? Explain.

10.28. The Ajax Auditing Company has a contract to audit the books of the Star Office Supply Corporation. As part of the audit, Ajax must estimate the average gross profit per sales invoice, μ, by using sample information. In the past year, the average gross profit per sales invoice was $20.52. It is hypothesized that this average did not change in the current year. The risk of concluding that the average differs from that of the past year when in fact it does not is set at $\alpha = 0.05$. Also, it is decided that if the average differs by more than $0.50, the risk of concluding that there is no difference in the average gross profit per sales invoice is set at 0.10. From the data compiled last year, it is determined that the standard deviation of the gross profit per sales invoice is $8.40. What sample size is required to test the hypothesis that μ, the average gross sales profit per invoice, is $20.52 for the current year?

10.29. A production process is considered to be in control if the proportion of defectives, π, produced by the process is 0.02. To determine whether the process is currently in control, it is decided to randomly sample items produced by the process. The risk of rejecting $H_0: \quad \pi = 0.02$ when in fact H_0 is true is set at 0.10. It is further decided that a difference of 0.01 should be detected by the test with a probability of 0.95. What sample size is required to meet these conditions? [Assume that π is approximately 0.02 so that $\pi(1 - \pi)$ in the formula is set at $(0.02)(0.98) = 0.0196.$]

Statistical inference: Two populations

11

▪ 11.1 INTRODUCTION

In many practical problems, we are concerned with comparing two populations with regard to some quantitative characteristic. For example, we may wish to compare the average time required to assemble a unit of a product by one production process with the average time required for assembly by an alternate process. In this instance, the quantitative characteristic is the assembly time per unit, and the two populations are the sets of assembly times corresponding to the two production processes. The comparison of average assembly times for the two processes can be made by extending the methods of statistical estimation and hypothesis testing of Chapters 9 and 10. We may wish to construct a confidence interval on the difference between the two population mean assembly times or test a hypothesis that one mean is different from the other, for example.

The notation for two-population statistical inference is given in Figure 11.1. It is necessary to use subscripts to distinguish between the two population parameters and the values of the sample statistics. For example, in population 1, the population parameter θ and random variable X are subscripted by 1 to indicate that they belong to the first population. In the sample drawn from population 1, n_1 indicates the number of units sampled and $\hat{\theta}_1$ denotes the point estimator of θ_1. The sample values must now be double subscripted, x_{ij}. The first subscript indicates which population the sample value came from (the ith, $i = 1$ or 2), and the second subscript designates the jth sample unit drawn from the ith population.

The methods developed in this chapter will assume that the population random variables X_1 and X_2 are normally distributed. If this is not the case,

325

FIGURE 11.1 Notation for the two-population case

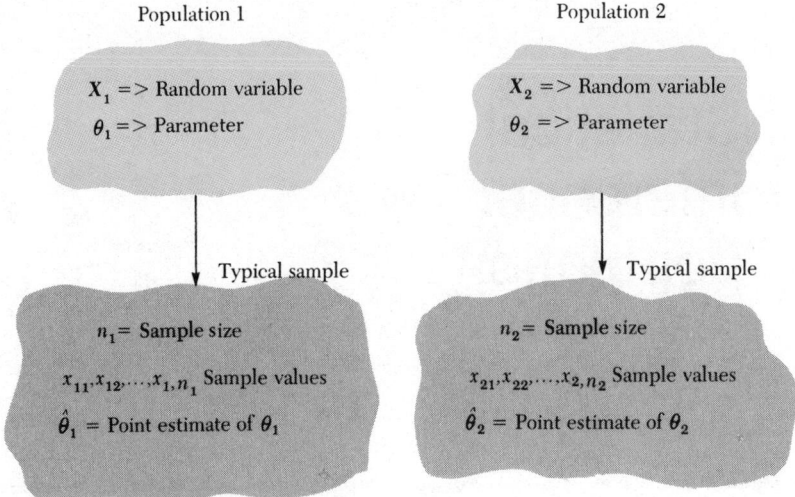

Population 1   Population 2

X_1 => Random variable X_2 => Random variable

θ_1 => Parameter θ_2 => Parameter

Typical sample   Typical sample

n_1 = Sample size n_2 = Sample size

$x_{11}, x_{12}, \ldots, x_{1,n_1}$ Sample values $x_{21}, x_{22}, \ldots, x_{2,n_2}$ Sample values

$\hat{\theta}_1$ = Point estimate of θ_1 $\hat{\theta}_2$ = Point estimate of θ_2

then the procedures will be approximate and will be satisfactorily so only if n_1 and n_2 are "large." How large n_1 and n_2 must be for the methods to provide a good approximation depends on the specific cases considered. But as a general rule, n_1 and n_2 should each be larger than 25.

In this chapter, we shall focus our attention on three cases: Comparing two population means (μ_1, μ_2), comparing two population variances (σ_1^2, σ_2^2), and comparing two population proportions (π_1, π_2). Specifically, we will consider the difference, $\theta_1 - \theta_2$, between parameters from two populations for the population mean and proportion, and we will consider the ratio σ_1^2/σ_2^2 for the variances. The difference, $\theta_1 - \theta_2$, will be point estimated and estimated by confidence intervals. Furthermore, methods for testing hypotheses of the forms

$$H_0: \quad \theta_1 \leq \theta_2 \qquad H_0: \quad \theta_1 = \theta_2 \qquad H_0: \quad \theta_1 \geq \theta_2$$
$$H_A: \quad \theta_1 > \theta_2 \qquad H_A: \quad \theta_1 \neq \theta_2 \qquad H_A: \quad \theta_1 < \theta_2$$

will be developed. In most cases, a more general form of these hypotheses will be used. If we let $\theta_1 - \theta_2 = D$, the above hypotheses may be stated more generally as

$$H_0: \quad D \leq D_0 \qquad H_0: \quad D = D_0 \qquad H_0: \quad D \geq D_0$$
$$H_A: \quad D > D_0 \qquad H_A: \quad D \neq D_0 \qquad H_A: \quad D < D_0$$

where D_0 is the hypothesized value of the difference, $\theta_1 - \theta_2$. If in the second set of hypothesis statements we let $D_0 = 0$, then the second set is equivalent to the first (for example, $\theta_1 - \theta_2 = 0$ is equivalent to $\theta_1 = \theta_2$).

Most methods developed for making inferences concerning differences between parameters from two populations will be based on the following theorem.

Theorem 11.1
Variance of the difference of two random variables

If a random variable X_1 has a variance of $\sigma_{X_1}^2$ and random variable X_2 has a variance of $\sigma_{X_2}^2$ and X_1 and X_2 are independent random variables, then the variance of the difference between X_1 and X_2 is the sum of the variances of X_1 and X_2. That is,

$$V(X_1 - X_2) = V(X_1) + V(X_2) = \sigma_{X_1}^2 + \sigma_{X_2}^2$$

This theorem was first introduced in Table 5.16. The interested reader is directed to the Winkler and Hays reference at the end of this chapter for proof of the theorem.

■ 11.2 INFERENCES CONCERNING TWO POPULATION MEANS, μ_1 AND μ_2

The inference framework for comparing two population means is illustrated in Figure 11.2. The sample random variables $\overline{X}_1$, $\overline{X}_2$, S_1^2, and S_2^2 will be used to make inferences concerning the difference between the two population means, $D = \mu_1 - \mu_2$. We will first consider point estimation, then interval estimation, and finally hypothesis testing for the difference $D = \mu_1 - \mu_2$.

FIGURE 11.2 Two-population framework for comparing μ_1 and μ_2

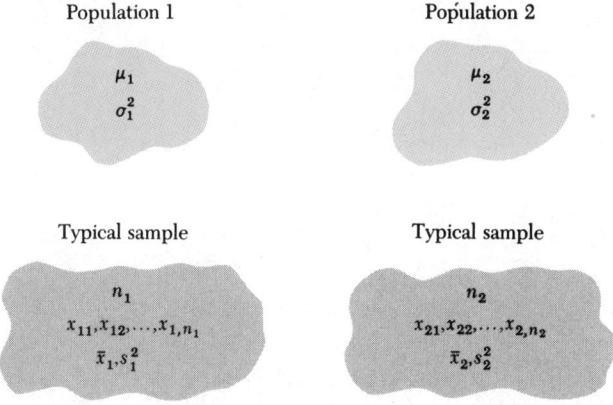

□ 11.2.1 Point estimation of $D = \mu_1 - \mu_2$

Recall from Chapter 9 that our best estimator of a population mean μ is the statistic $\overline{X}$. Thus, it seems intuitive that the best point estimator of $D = \mu_1 - \mu_2$ is the difference between the statistics $\overline{X}_1 - \overline{X}_2$. In fact, the point estimator $\hat{D} = \overline{X}_1 - \overline{X}_2$ is unbiased and minimum variance. That is, there is no "better" function of the two sets of sample values than $\overline{X}_1 - \overline{X}_2$ in estimating $\mu_1 - \mu_2$.

☐ **11.2.2 Confidence interval estimation of $D = \mu_1 - \mu_2$**

In constructing confidence intervals for $D = \mu_1 - \mu_2$, it is necessary to consider three cases:

Case 1: The values of σ_1^2 and σ_2^2 are known.

Case 2: The values of σ_1^2 and σ_2^2 are unknown, but it is known that $\sigma_1^2 = \sigma_2^2$.

Case 3: The values of σ_1^2 and σ_2^2 are unknown, and it is known that $\sigma_1^2 \neq \sigma_2^2$.

The distinction among these cases at least in part can be drawn from the confidence interval formulas concerning one population mean μ. Recall that if σ^2 is known, then the confidence interval for μ developed in Chapter 9 is based on the standard normal variable Z. However, if σ^2 is unknown and is estimated by the sample variance s^2, then the confidence interval for μ is based on the t-distribution.

As might be expected, the case 1 confidence interval formula is based on the Z-statistic, while the interval formulas for cases 2 and 3 are based on the t-statistic. The distinction between cases 2 and 3 is subtle, but important, and will be drawn momentarily.

Case 1: σ_1^2 and σ_2^2 known

When the values of σ_1^2 and σ_2^2 are known, a confidence interval for the difference between μ_1 and μ_2 can be constructed by using the general form of the standard normal statistic

$$Z = \frac{\hat{\theta} - \theta}{\sigma_{\hat{\theta}}}$$

where we let $\theta = \mu_1 - \mu_2$. Since the best point estimator of $\theta = \mu_1 - \mu_2$ is $\bar{X}_1 - \bar{X}_2$, $\hat{\theta}$ in the Z-statistic is given by $\bar{X}_1 - \bar{X}_2$. We need only to find $\sigma_{\hat{\theta}}$, the standard deviation of the point estimator $\bar{X}_1 - \bar{X}_2$. From Theorem 11.1,

$$\sigma^2_{\bar{X}_1 - \bar{X}_2} = V(\bar{X}_1 - \bar{X}_2) = V(\bar{X}_1) + V(\bar{X}_2) = \frac{\sigma_1^2}{n_1} + \frac{\sigma_2^2}{n_2}$$

Therefore,

$$\sigma_{\bar{X}_1 - \bar{X}_2} = \sqrt{\sigma^2_{\bar{X}_1 - \bar{X}_2}} = \sqrt{\frac{\sigma_1^2}{n_1} + \frac{\sigma_2^2}{n_2}}$$

Thus, the form of the Z-statistic for constructing a confidence interval estimator for $\mu_1 - \mu_2$ is

$$Z = \frac{(\bar{X}_1 - \bar{X}_2) - (\mu_1 - \mu_2)}{\sqrt{\dfrac{\sigma_1^2}{n_1} + \dfrac{\sigma_2^2}{n_2}}}$$

By using the fact that $P(-z_{\alpha/2} \leq Z \leq z_{\alpha/2}) = 1 - \alpha$, it can be shown by solving the inequality statements for $(\mu_1 - \mu_2)$ that a $100(1 - \alpha)$ percent confidence interval on $(\mu_1 - \mu_2)$ is given by

$$(\bar{X}_1 - \bar{X}_2) \pm z_{\alpha/2}\sqrt{\frac{\sigma_1^2}{n_1} + \frac{\sigma_2^2}{n_2}}$$

Formula 11.1
Confidence interval estimate $\mu_1 - \mu_2$
Case 1: σ_1^2 and σ_2^2 known

A $100(1 - \alpha)$ percent confidence interval estimate of the difference between two population means, $\mu_1 - \mu_2$, is given by

$$(\bar{x}_1 - \bar{x}_2) \pm z_{\alpha/2}\sqrt{\frac{\sigma_1^2}{n_1} + \frac{\sigma_2^2}{n_2}}$$

where $\bar{x}_i$, σ_i^2, and n_i are the sample mean, population variance, and sample size associated with the ith population and sample (i = 1, 2). Additionally $z_{\alpha/2}$ is the point on the standard normal distribution such that $P(Z \geq z_{\alpha/2}) = \alpha/2$.

The confidence interval is exact if the population random variables X_1 and X_2 are normally distributed. If this is not the case, then n_1 and n_2 should each be greater than 25, in which case the resulting confidence interval is approximate.

Example 11.1 A company is in the process of deciding whether or not to produce a new electronic component. In the plant, there are two machines that could be adapted to produce the component. As a part of the overall decision, the company must select one of the machines to use if they decide to produce the component. In a test conducted on machine 1, the average production time per component was 5.23 minutes for a sample of 100 components. In a sample of 64 components, machine 2 averaged 5.37 minutes per component. From past experience in using the machines to produce similar items, it is known that the standard deviations of the production times on machines 1 and 2 are 0.15 minute and 0.10 minute, respectively.

Determine a point estimate and a 95 percent confidence interval estimate of the difference between the population mean production times for machines 1 and 2.

Solution The first population is composed, conceptually, of the production times associated with the components that could be produced on machine 1. In a sample of 100 production times, the sample average production time is $\bar{x}_1 = 5.23$ minutes, while the population standard deviation of the production times is given as $\sigma_1 = 0.15$ minute. In a sample of 64 production times drawn from the second population, the sample average production time is $\bar{x}_2 = 5.37$ minutes, while the population standard deviation of the production times is given as $\sigma_2 = 0.10$ minute.

The point estimate of the difference $\mu_1 - \mu_2$ is:

$$\widehat{\mu_1 - \mu_2} = \bar{x}_1 - \bar{x}_2 = 5.23 - 5.37 = -0.14 \text{ minute}$$

For a 95 percent confidence interval, $100(1 - \alpha) = 95$, so that $\alpha = 0.05$ and $\alpha/2 = 0.025$. From Table B.3 in Appendix B, $z_{\alpha/2} = z_{0.025} = 1.96$. Thus, the 95 percent confidence interval estimate of $\mu_1 - \mu_2$ is given by

$$(\bar{x}_1 - \bar{x}_2) \pm z_{\alpha/2} \sqrt{\frac{\sigma_1^2}{n_1} + \frac{\sigma_2^2}{n_2}} = (5.23 - 5.37) \pm 1.96 \sqrt{\frac{(0.15)^2}{100} + \frac{(0.10)^2}{64}}$$

$$= -0.14 \pm 1.96(0.0195) = -0.14 \pm 0.038$$

and we are 95 percent confident that the true difference between μ_1 and μ_2 is between -0.102 and -0.178 minute. There is good evidence, therefore, to conclude that μ_1 is less than μ_2, since the difference, $\mu_1 - \mu_2$, appears to be negative.

We should at this point remind ourselves of the meaning of "95 percent confidence." If many repeated pairs of samples of sizes n_1 and n_2 are drawn and a confidence interval for the difference between the two population means is constructed in each sample pair, 95 percent of the confidence intervals will contain the true difference and 5 percent will not.

In most problems where inferences about μ_1 and μ_2 are drawn, the values of the population variances σ_1^2 and σ_2^2 will not be known, of course. This leads us to cases 2 and 3.

Case 2: σ_1^2 and σ_2^2 unknown; $\sigma_1^2 = \sigma_2^2$

In case 2, we assume that the two population variances are unknown, but we know that they are equal in value. Let σ^2 denote the common value of σ_1^2 and σ_2^2. The two population schematic is illustrated in Figure 11.3.

FIGURE 11.3 Two-population framework for case 2

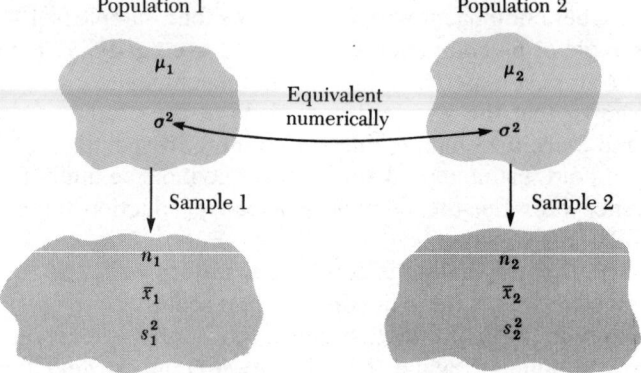

In this case, the standard deviation of the point estimator of $\mu_1 - \mu_2$ can be simplified:

$$\sigma_{\bar{x}_1 - \bar{x}_2} = \sqrt{\frac{\sigma_1^2}{n_1} + \frac{\sigma_2^2}{n_2}} = \sqrt{\frac{\sigma^2}{n_1} + \frac{\sigma^2}{n_2}} = \sigma \sqrt{\frac{1}{n_1} + \frac{1}{n_2}}$$

Since the populations have the same variance σ^2, the sample variances s_1^2 and

s_2^2 both point estimate σ^2. It seems reasonable that s_1^2 and s_2^2 should be "pooled" in some way to form a combined point estimate of σ^2 based on the $n_1 + n_2$ sample units comprising both samples. The pooled variance, denoted by s_p^2, is given by

$$s_p^2 = \frac{(n_1 - 1)s_1^2 + (n_2 - 1)s_2^2}{(n_1 - 1) + (n_2 - 1)}$$

It is the weighted mean of s_1^2 and s_2^2.

The pooled variance S_p^2 is the best point estimator of σ^2. Thus, the estimate of $\sigma_{\bar{X}_1 - \bar{X}_2}$ is

$$\hat{\sigma}_{\bar{X}_1 - \bar{X}_2} = s_p \sqrt{\frac{1}{n_1} + \frac{1}{n_2}}$$

Recall from Chapters 9 and 10 that in the use of the Z-statistic,

$$Z = \frac{\hat{\theta} - \theta}{\sigma_{\hat{\theta}}}$$

when the standard deviation of the point estimator, $\sigma_{\hat{\theta}}$, had to be estimated, the standardized variable with $\sigma_{\hat{\theta}}$ replaced by its estimator $\hat{\sigma}_{\hat{\theta}}$ became the t-statistic,

$$t = \frac{\hat{\theta} - \theta}{\hat{\sigma}_{\hat{\theta}}}$$

In this case with $\theta = \mu_1 - \mu_2$, the form of the t-statistic is

$$t = \frac{(\bar{X}_1 - \bar{X}_2) - (\mu_1 - \mu_2)}{S_p \sqrt{\frac{1}{n_1} + \frac{1}{n_2}}}$$

where S_p is the pooled standard deviation statistic,

$$S_p = \sqrt{\frac{(n_1 - 1)S_1^2 + (n_2 - 1)S_2^2}{(n_1 - 1) + (n_2 - 1)}}$$

The degrees of freedom associated with this t-statistic are $n_1 + n_2 - 2$ since this number represents the divisor in the pooled sample standard deviation formula $[(n_1 - 1) + (n_2 - 1)]$.

Formula 11.2
Confidence interval estimate $\mu_1 - \mu_2$,
Case 2: σ_1^2 and σ_2^2 unknown; $\sigma_1^2 = \sigma_2^2$

A $100(1 - \alpha)$ percent confidence interval estimate of the difference between two population means, $\mu_1 - \mu_2$, is given by

$$(\bar{x}_1 - \bar{x}_2) \pm t_{\alpha/2;n_1+n_2-2} S_p \sqrt{\frac{1}{n_1} + \frac{1}{n_2}}$$

> **Formula 11.2 (continued)**
>
> where $\bar{x}_i$ and n_i are the sample mean and sample size, respectively, of the ith sample ($i = 1, 2$), s_p is the pooled standard deviation given by
>
> $$s_p = \sqrt{\frac{(n_1 - 1)s_1^2 + (n_2 - 1)s_2^2}{(n_1 - 1) + (n_2 - 1)}}$$
>
> and $t_{\alpha/2;n_1+n_2-2}$ is the point on the t-distribution with $n_1 + n_2 - 2$ degrees of freedom such that $P[t > t_{\alpha/2;n_1+n_2-2}] = \alpha/2$.
>
> The confidence interval is exact if the population random variables X_1 and X_2 are normally distributed. If this is not the case, then n_1 and n_2 should each be greater than 25, in which case the resulting confidence interval is approximate.

Example 11.2 A large corporation frequently sends its managers from the corporation home office to Washington, D.C., on business. The managers have customarily traveled on two airlines, Agony and TransLand. It has become apparent that Agony arrives consistently late. While arriving late is better than not arriving at all, some managers have been missing important meetings due to the inability of Agony to arrive on schedule. A random sample of 36 recent flights to Washington on Agony averaged ten minutes late (if the arrival is on time or early, it receives a score of 0) with a standard deviation of three minutes. A random sample of 25 recent flights on TransLand averaged four minutes late with a standard deviation of two minutes. Find a 99 percent confidence interval for the difference between the average minutes late on Agony and TransLand flights.

Solution The sample information is:

Agony	TransLand
$n_1 = 36$	$n_2 = 25$
$\bar{x}_1 = 10$	$\bar{x}_2 = 4$
$s_1 = 3$	$s_2 = 2$

We will assume $\sigma_1^2 = \sigma_2^2$. (Does this seem reasonable? More about this later.) For a 99 percent confidence interval, $100(1 - \alpha)$ percent $= 99$ percent, so that $\alpha = 0.01$ and $\alpha/2 = 0.005$. The degrees of freedom are $n_1 + n_2 - 2 = 36 + 25 - 2 = 59$. From Table B.5 in Appendix B, we wish to find $t_{0.005;59}$; it is approximately 2.66.

The pooled standard deviation is:

$$s_p = \sqrt{\frac{(n_1 - 1)s_1^2 + (n_2 - 1)s_2^2}{(n_1 - 1) + (n_2 - 1)}} = \sqrt{\frac{(35)(3)^2 + (24)(2)^2}{59}} = 2.64$$

The 99 percent confidence interval is:

$$(\bar{x}_1 - \bar{x}_2) \pm t_{0.005;59}S_p \sqrt{\frac{1}{n_1} + \frac{1}{n_2}} = (10 - 4) \pm (2.66)(2.64)(0.2603)$$

$$= 6 \pm 1.83$$

Thus, we are 99 percent confident that the true difference between the average minutes late per flight for Agony and TransLand is between 4.17 and 7.83 minutes.

In Example 11.2, we assumed $\sigma_1^2 = \sigma_2^2$. On what basis can this assumption be made? In case 2, the values of σ_1^2 and σ_2^2 are unknown. Thus, in most instances, the only information available to determine whether $\sigma_1^2 = \sigma_2^2$ is the sample data. In Example 11.2, $s_1 = 3$ and $s_2 = 2$; they are reasonably close, and on this basis, may we assume that $\sigma_1^2 = \sigma_2^2$? Well, what is "reasonably close"? A way to resolve this problem is to test the hypothesis that $\sigma_1^2 = \sigma_2^2$ using the sample data, in particular, s_1^2 and s_2^2. If the null hypothesis H_0: $\sigma_1^2 = \sigma_2^2$ cannot be rejected, then we can feel reasonably confident that the values of σ_1^2 and σ_2^2 are similar and the case 2 method for setting a confidence interval on $\mu_1 - \mu_2$ can be used. In Section 11.3, we will develop the procedure for testing the equivalence of two population variances.

If it is know that σ_1^2 and σ_2^2 are not equal, then we have case 3.

Case 3: σ_1^2 *and* σ_2^2 *unknown;* $\sigma_1^2 \neq \sigma_2^2$

In this case, we must estimate both σ_1^2 and σ_2^2. A logical sequence of steps would seem to be:

1. Since $V(\bar{X}_1 - \bar{X}_2) = \sigma_{\bar{X}_1 - \bar{X}_2}^2 = \sigma_1^2/n_1 + \sigma_2^2/n_2$, if $\bar{X}_1$ and $\bar{X}_2$ are independent, we could estimate $\sigma_{\bar{X}_1 - \bar{X}_2}^2$ by estimating σ_1^2 and σ_2^2 by s_1^2 and s_2^2, respectively. Thus,

$$\hat{\sigma}_{\bar{X}_1 - \bar{X}_2}^2 = \frac{s_1^2}{n_1} + \frac{s_2^2}{n_2}$$

2. Form the standardized statistic,

$$t' = \frac{(\bar{X}_1 - \bar{X}_2) - (\mu_1 - \mu_2)}{\hat{\sigma}_{\bar{X}_1 - \bar{X}_2}} = \frac{(\bar{X}_1 - \bar{X}_2) - (\mu_1 - \mu_2)}{\sqrt{\frac{s_1^2}{n_1} + \frac{s_2^2}{n_2}}}$$

However, due to the unequal population variances, t' will not be t-distributed. There are two approximate approaches to setting a confidence interval on $\mu_1 - \mu_2$ in this case:

1. Take very large samples n_1 and n_2. Eventually, s_1^2 and s_2^2 will converge to σ_1^2 and σ_2^2, respectively, and $\bar{X}_1 - \bar{X}_2$ will approach a normal distribution. Then we have, for all practical purposes, case 1. The resulting confidence interval estimate is:

$$(\bar{x}_1 - \bar{x}_2) \pm z_{\alpha/2} \sqrt{\frac{s_1^2}{n_1} + \frac{s_2^2}{n_2}}$$

But how large should n_1 and n_2 be to use this approach? The answer depends on a number of factors such as the sizes of the two populations, the distributions of X_1 and X_2, and so forth. Generally, n_1 and n_2 should *each* be greater than 100 to use this approach.

2. Correct the t'-statistic so that it is approximately t-distributed for moderate sample sizes (n_1 and n_2 both greater than 25). One such modification is given in the summary table at the end of the chapter. The t'-statistic is used, but the degrees of freedom are modified (Δ). If Δ turns out to be fractional, it is suggested that it be rounded to the next larger integer.

Considerable caution should be exercised in applying the case 3 approaches to setting a confidence interval on $\mu_1 - \mu_2$. Both methods can produce very crude estimates of $\mu_1 - \mu_2$. A statistician must carefully assess whether the approaches are appropriate for a specific experiment. For an excellent and interesting detailed discussion of case 3, see the Snedecor and Cochran reference given at the end of this chapter.

In general, it is better to use a nonparametric procedure in case 3. A nonparametric method for case 3 is given in Chapter 20 (the Mann-Whitney method).

□ **11.2.3 Hypothesis testing concerning $D = \mu_1 - \mu_2$**

Since hypothesis testing is directly related to confidence interval estimation, we must differentiate among the three cases discussed in Section 11.2.2. For cases 1 and 2, we will state the null hypothesis as a difference in one of three forms:

$$H_0: \quad \mu_1 - \mu_2 = D \geq D_0$$
$$H_0: \quad \mu_1 - \mu_2 = D \leq D_0$$
$$H_0: \quad \mu_1 - \mu_2 = D = D_0$$

In these hypothesis statements, D_0 is the hypothesized value of the difference, usually taken to be zero. We will not develop the approximate test for case 3.

Case 1: σ_1^2 and σ_2^2 *known*

The test statistic in this case is

$$Z = \frac{(\bar{X}_1 - \bar{X}_2) - D_0}{\sqrt{\dfrac{\sigma_1^2}{n_1} + \dfrac{\sigma_2^2}{n_2}}}$$

This test statistic follows directly from the Z-statistic

$$Z = \frac{(\bar{X}_1 - \bar{X}_2) - (\mu_1 - \mu_2)}{\sqrt{\dfrac{\sigma_1^2}{n_1} + \dfrac{\sigma_2^2}{n_2}}}$$

used in Section 11.2.1 to develop the confidence interval formula for $\mu_1 - \mu_2$.

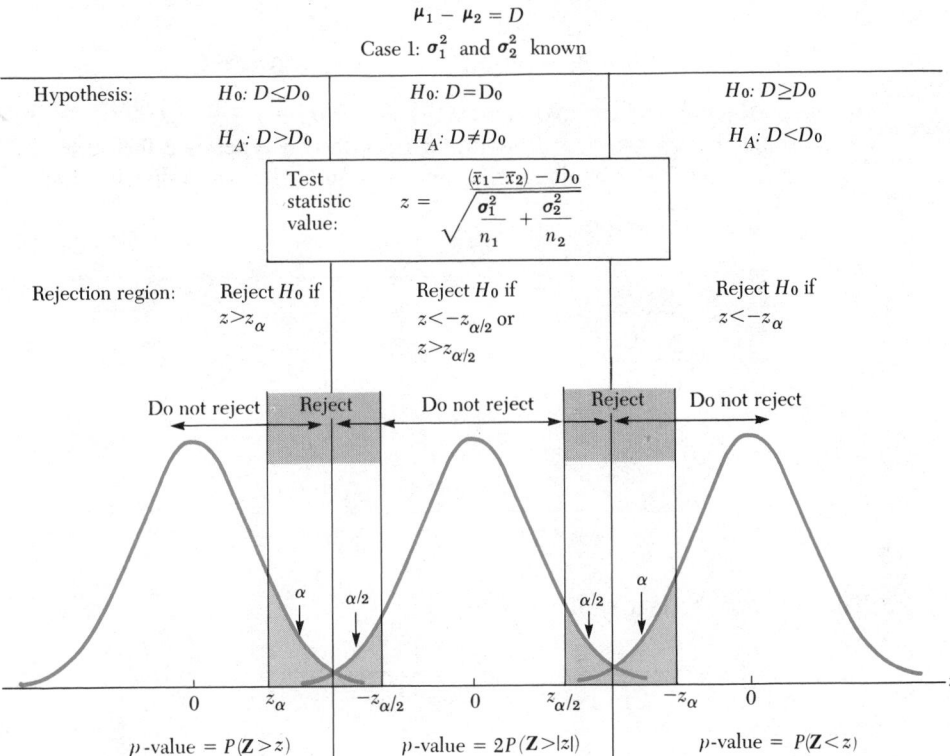

Hypothesis test

$$\mu_1 - \mu_2 = D$$

Case 1: σ_1^2 and σ_2^2 known

Hypothesis:	$H_0: D \leq D_0$	$H_0: D = D_0$	$H_0: D \geq D_0$
	$H_A: D > D_0$	$H_A: D \neq D_0$	$H_A: D < D_0$

Test statistic value:
$$z = \frac{(\bar{x}_1 - \bar{x}_2) - D_0}{\sqrt{\dfrac{\sigma_1^2}{n_1} + \dfrac{\sigma_2^2}{n_2}}}$$

Rejection region:	Reject H_0 if $z > z_\alpha$	Reject H_0 if $z < -z_{\alpha/2}$ or $z > z_{\alpha/2}$	Reject H_0 if $z < -z_\alpha$

Do not reject | Reject | Reject | Do not reject | Reject | Do not reject

α | $\alpha/2$ | $\alpha/2$ | α

0 | z_α | $-z_{\alpha/2}$ | 0 | $z_{\alpha/2}$ | $-z_\alpha$ | 0

p-value $= P(\mathbf{Z} > z)$ | p-value $= 2P(\mathbf{Z} > |z|)$ | p-value $= P(\mathbf{Z} < z)$

Example 11.3 The Flamerock Tire Company has decided to enter the radial tire market. The company decides to copy one of the European radial tire designs. The company has managed to procure the designs of two tires, type A and type B. Both tires are designed to last 40,000 miles. The company decides to produce the one that has the greater average tire mileage. From considerable previous testing with many similar radial tires, the company has found that the standard deviation of each type of radial tire is 3,000 miles. The company places 25 tires of type A on its mileage machine, and finds that the average mileage is 39,780 miles. A sample of 25 type B tires produces an average mileage of 40,650 miles. The company wishes to determine whether there is a significant difference between the average mileage of all tires of types A and B. Test this hypothesis at the $\alpha = 0.05$ significance level.

Solution The hypothesis is:

$$H_0: \quad D = \mu_1 - \mu_2 = 0 \quad (\text{i.e., } D_0 = 0)$$
$$H_A: \quad D = \mu_1 - \mu_2 \neq 0$$

Hypothesizing that the difference is zero is, of course, equivalent to hypothesizing that the two population means are equivalent.

The sample data are:

$$n_1 = 25 \qquad n_2 = 25$$
$$\bar{x}_1 = 39{,}780 \qquad \bar{x}_2 = 40{,}650$$

Based on the information given, we will assume $\sigma_1^2 = \sigma_2^2 = (3{,}000)^2$; that is, it is assumed that the standard deviation of both tire types is 3,000 miles.

With $\alpha = 0.05$, the decision rule on the standard normal distribution is:

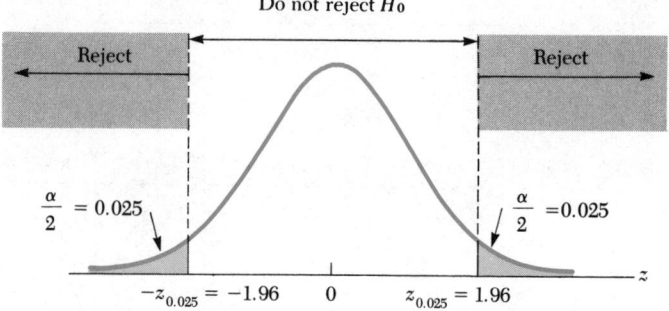

The value of the test statistic is:

$$z = \frac{(\bar{x}_1 - \bar{x}_2) - D_0}{\sqrt{\dfrac{\sigma_1^2}{n_1} + \dfrac{\sigma_2^2}{n_2}}} = \frac{(39{,}780 - 40{,}650) - 0}{\sqrt{\dfrac{(3{,}000)^2}{25} + \dfrac{(3{,}000)^2}{25}}} = -1.03$$

Since the value of the test statistic does not fall into either rejection region, we cannot reject the null hypothesis that $\mu_1 = \mu_2$, or equivalently, that $\mu_1 - \mu_2 = 0$.

Since this is a two-sided test, the p-value may be determined by:

$$p\text{-value} = 2P(\mathbf{Z} > |z|) = 2P(\mathbf{Z} > |-1.03|)$$
$$= 2P(\mathbf{Z} > 1.03) = 2(0.1515) = 0.303$$

Thus, if we had decided to reject H_0: $\mu_1 - \mu_2 = 0$, we would have taken a risk equal to 0.303 of committing a Type I error (rejecting H_0/H_0 is true). (Recall that in the two-sided test, the p-value is determined by doubling the tail probability.)

As we discussed in Chapter 10, the p-value provides useful information in making a decision about the null hypothesis. At the $\alpha = 0.05$ significance level, it is evident that we should not reject H_0. Knowing that the p-value is relatively high (0.303) gives us some comfort in this decision. If we were to reject H_0, the probability of making an error is relatively high (0.303).

Case 2: σ_1^2 and σ_2^2 *unknown;* $\sigma_1^2 = \sigma_2^2$

In case 2, the t-statistic based on the pooled sample standard deviation is used.

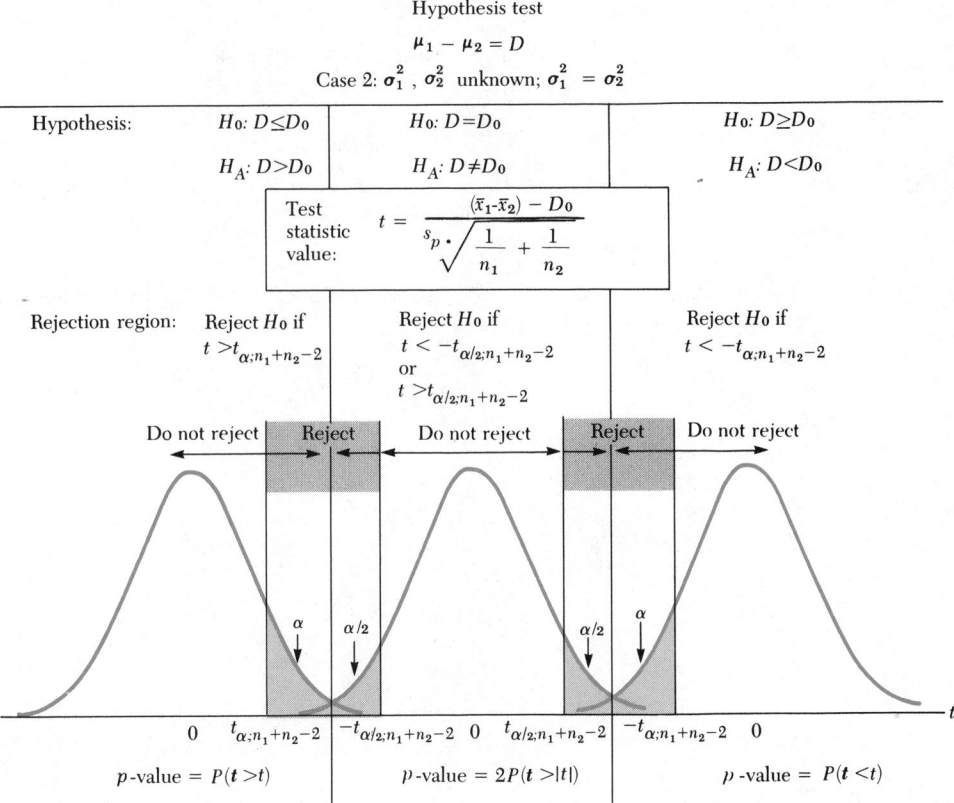

Hypothesis test

$$\mu_1 - \mu_2 = D$$

Case 2: σ_1^2, σ_2^2 unknown; $\sigma_1^2 = \sigma_2^2$

Hypothesis:	H_0: $D \leq D_0$	H_0: $D = D_0$	H_0: $D \geq D_0$
	H_A: $D > D_0$	H_A: $D \neq D_0$	H_A: $D < D_0$

Test statistic value:
$$t = \frac{(\bar{x}_1 - \bar{x}_2) - D_0}{s_p \cdot \sqrt{\dfrac{1}{n_1} + \dfrac{1}{n_2}}}$$

Rejection region: Reject H_0 if $t > t_{\alpha; n_1 + n_2 - 2}$ — Reject H_0 if $t < -t_{\alpha/2; n_1 + n_2 - 2}$ or $t > t_{\alpha/2; n_1 + n_2 - 2}$ — Reject H_0 if $t < -t_{\alpha; n_1 + n_2 - 2}$

p-value $= P(t > t)$ p-value $= 2P(t > |t|)$ p-value $= P(t < t)$

Example 11.4 A manufacturer claims an automobile battery it produces will last at least 100 hours longer, on the average, than competing batteries in the same price range when the batteries are subjected to continuous use in a controlled laboratory experiment. To test this claim, a competitor places samples of its battery and the supposedly superior one in a testing machine in which both batteries are subjected to identical conditions, and measures the lifetimes of the batteries. In a sample of 15 of its own batteries, the competitor finds that the average lifetime is 2,050 hours with a standard deviation of 80 hours, while 15 units of the supposedly superior battery last an average of 2,100 hours with a standard deviation of 100 hours. Test the manufacturer's claim at the $\alpha = 0.05$ significance level. Assume that the lifetimes of both batteries are normally distributed.

Solution Let the supposedly superior batteries produced by the manufacturer making the claim comprise the first population and the competitor's batteries comprise the second population.

We wish to test the hypothesis that the mean lifetime of population 1 is at least 100 hours longer than the mean lifetime of population 2; that is,

$\mu_1 - \mu_2 \geq 100$. Thus, the hypothesis is:

$$H_0: \quad D = \mu_1 - \mu_2 \geq 100 \quad (D_0 = 100)$$
$$H_A: \quad D = \mu_1 - \mu_2 < 100$$

The sample data are:

$$
\begin{array}{ll}
n_1 = 15 & n_2 = 15 \\
\bar{x}_1 = 2{,}100 & \bar{x}_2 = 2{,}050 \\
s_1 = 100 & s_2 = 80
\end{array}
$$

If we assume that $\sigma_1^2 = \sigma_2^2$, then the case 2 test statistic based on the pooled sample standard deviation is appropriate:

$$s_p = \sqrt{\frac{(n_1 - 1)s_1^2 + (n_2 - 1)s_2^2}{(n_1 - 1) + (n_2 - 1)}} = \sqrt{\frac{(14)(100)^2 + (14)(80)^2}{14 + 14}} = 90.55$$

The decision rule is:

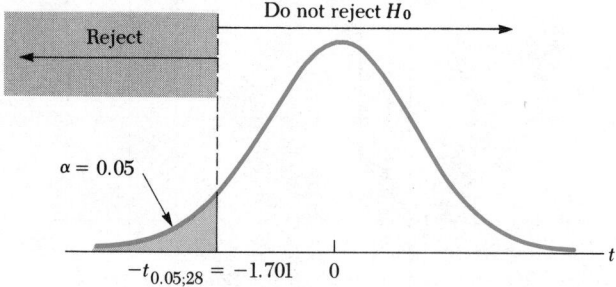

The test statistic value is:

$$t = \frac{(\bar{x}_1 - \bar{x}_2) - D_0}{s_p \sqrt{\dfrac{1}{n_1} + \dfrac{1}{n_2}}} = \frac{(2{,}100 - 2{,}050) - 100}{90.55 \sqrt{\dfrac{1}{15} + \dfrac{1}{15}}} = -1.512$$

Thus, even though the sample mean $\bar{x}_2$ is within 50 hours of $\bar{x}_1$, we cannot reject the hypothesis that the difference between μ_1 and μ_2 is greater than 100 hours.

The p-value for this test is:

$$p\text{-value} = P(t < t) = P(t < -1.512)$$

Since $P(t < -1.313) = 0.10$ and $P(t < -1.701) = 0.05$, from Table B.5 in Appendix B, the p-value for this test is between 0.05 and 0.10. A rough interpolate is 0.075, since -1.512 is about halfway between -1.313 and -1.701. Because the p-value is about 0.075, it is evident that we cannot reject the null hypothesis at the $\alpha = 0.01$ or $\alpha = 0.05$ significance level, but we could reject the null hypothesis at the $\alpha = 0.10$ significance level.

In Examples 11.3 and 11.4, the null hypotheses were not rejected. In Example 11.3, since the p-value was 0.303, the decision to not reject H_0 was reasonably easy to make. In Example 11.4, the p-value was about 0.075. Thus, if we rejected H_0, we would have had a probability of about 0.075 of committing a Type I error (rejecting H_0/H_0 is true). But, if we did not reject H_0, then it is possible that we made a Type II error (not rejecting H_0/H_0 is false) with a probability given by β. To assess the possible magnitude of β, we would have had to construct a power curve as discussed in Chapter 10. Since the p-value in Example 11.4 was relatively small (0.075), we would certainly want to evaluate the power curve prior to making a final decision concerning the null hypothesis.

Although constructing the power curve for cases with two population parameters is a straightforward extension of the power curve construction in the one-parameter case (see Chapter 10), we will not develop power curves in this chapter. But it is important to remember the purpose and value of a power curve for a statistical test of a hypothesis. It is an important factor to consider in conjunction with the p-value (or the setting of the significance level) in the decision to reject or not to reject the null hypothesis.

The tests described in this section for comparing two population means assume that the population random variables X_1 and X_2 are normally distributed. If this is not the case, then n_1 and n_2 should both be 25 or more for the application of these procedures. Furthermore, case 2 assumes that the population variances, σ_1^2 and σ_2^2, are equal. Experimental results have demonstrated that mild departures from the assumptions of normality and equivalent variances do not seriously affect the conclusions drawn from using these tests, provided the sample sizes are equal or nearly equal and the population distributions are not more than moderately skewed. If the departures from the assumptions are mild, it will usually turn out that the probability of committing a Type I error, α, will in actuality be somewhat larger than the value chosen to establish the decision rule for the test.

In Chapter 19, we shall learn how to test the hypothesis that a set of data came from a normal distribution. If the departure from the assumption that the data are normal is not mild, then these procedures for comparing two population means should not be used. Testing procedures in this case will be presented in Chapter 20.

If the data do support the assumption of normality, we must still check the assumption of equal variances, for this dictates whether case 2 or case 3 applies. In the preceding examples, we assumed that $\sigma_1^2 = \sigma_2^2$. In the next section, we will develop a statistical hypothesis-testing procedure for checking this assumption.

■ 11.3 INFERENCES CONCERNING TWO POPULATION VARIANCES, σ_1^2 and σ_2^2

In making comparisons between two population variances, we will focus our attention on the ratio of the variances, σ_1^2/σ_2^2, rather than the difference, $\sigma_1^2 - \sigma_2^2$. For example, if we wish to test the hypothesis H_0: $\sigma_1^2 = \sigma_2^2$, we can test the hypothesis stated as

$$H_0: \quad \frac{\sigma_1^2}{\sigma_2^2} = 1$$

The confidence interval estimator and hypothesis-testing procedures will be based on the ratio of the two sample variances.

☐ 11.3.1 Confidence interval estimator of the ratio of two population variances

In constructing a confidence interval estimator of the ratio σ_1^2/σ_2^2, it seems intuitive that we should use the sample variances s_1^2 and s_2^2. In fact, we shall use the statistic

$$F = \frac{S_1^2/\sigma_1^2}{S_2^2/\sigma_2^2}$$

to construct the confidence interval formula. It can be shown that this random variable has an F-distribution under certain conditions.

The F-distribution is shown in Figure 11.4. It is a skewed (to the right) distribution with values that range from 0 to $+\infty$. There are two sets of degrees of freedom associated with F: the numerator degrees of freedom, denoted by r_1, and the denominator degrees of freedom, denoted by r_2. For the statistic,

$$\frac{S_1^2/\sigma_1^2}{S_2^2/\sigma_2^2}$$

r_1 is equal to $(n_1 - 1)$, the degrees of freedom associated with S_1^2, and r_2 is equal to $(n_2 - 1)$, the degrees of freedom associated with S_2^2.

FIGURE 11.4 F-distribution

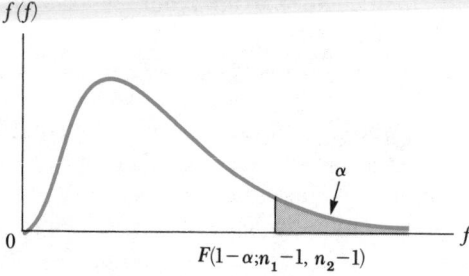

Table B.7 in Appendix B gives the cumulative density function for the F-distribution, denoted by $F(1 - \alpha; r_1, r_2)$, where $F(1 - \alpha; r_1, r_2)$ is given by

$$P[F \le F(1 - \alpha; r_1, r_2)] = 1 - \alpha$$

For example, if $n_1 = 6$ and $n_2 = 5$ with $\alpha = 0.05$, the value of $F(1 - \alpha; r_1, r_2) = F(1 - 0.05; r_1 = n_1 - 1, r_2 = n_2 - 1) = F(0.95; 5, 4) = 6.26$ is illustrated in Figure 11.5.

FIGURE 11.5 $F[1 - \alpha; r_1 = (n_1 - 1),$
$r_2 = (n_2 - 1)]$ on the F-distribution with
$n_1 - 1 = 5$ and $n_2 - 1 = 4; \alpha = 0.05.$
$f(f)$

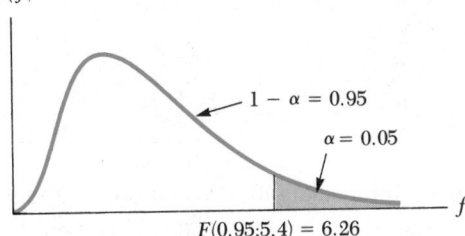

$1 - \alpha = 0.95$

$\alpha = 0.05$

f

$F(0.95; 5, 4) = 6.26$

Table B.7 gives F-values for $r_1, r_2 = 1$ to 10, 12, 15, 20, 24, 30, 60, 120, and ∞, and for $1 - \alpha = 0.5, 0.9, 0.95, 0.975, 0.99, 0.995,$ and 0.999.

For values of α greater than 0.50, the required probability can be determined by using the relationship,

$$F(1 - \alpha; r_1, r_2) = \frac{1}{F(\alpha; r_2, r_1)}.$$

For example, suppose we wish to find $F(1 - 0.95; 5, 4)$—the lower-tail probability such that $P[\mathbf{F} < F(0.05; 5, 4)] = 0.05$. From the above relationship,

$$F(0.05; 5, 4) = \frac{1}{F(0.95; 4, 5)} = \frac{1}{5.19} = 0.1927$$

Notice in this computation that the degrees of freedom are switched when using the relationship to find lower-tail probabilities.

The statistic $(S_1^2/\sigma_1^2)/(S_2^2/\sigma_2^2)$ has an F-distribution if each sample is a random sample drawn from a normal population. From the fact that

$$\mathbf{F} = \frac{S_1^2/\sigma_1^2}{S_2^2/\sigma_2^2}$$

has an F-distribution under these circumstances, the confidence interval estimate of the ratio σ_1^2/σ_2^2, given below, follows.

As with the other methods developed in this chapter, modest departures from normality do not invalidate this procedure for constructing a confidence interval on the ratio of the population variances.

Formula 11.3
Confidence interval estimate of σ_1^2/σ_2^2

A $100(1 - \alpha)$ percent confidence interval estimate of the ratio σ_1^2/σ_2^2 is given by

$$\frac{s_1^2}{s_2^2} \cdot \frac{1}{F(1 - \alpha/2; n_1 - 1, n_2 - 1)} \leq \frac{\sigma_1^2}{\sigma_2^2} \leq \frac{s_1^2}{s_2^2} \cdot \frac{1}{F(\alpha/2; n_1 - 1, n_2 - 1)}$$

> **Formula 11.3 (continued)**
>
> where s_1^2 and s_2^2 are the two sample variances and $F(\alpha/2; n_1 - 1, n_2 - 1)$ and $F(1 - \alpha/2; n_1 - 1, n_2 - 1)$ are values on the F-distribution with $n_1 - 1$ and $n_2 - 1$ degrees of freedom, such that
>
> $$P[F \le F(\alpha/2; n_1 - 1, n_2 - 1)] = \alpha/2$$
> $$P[F \ge F(1 - \alpha/2; n_1 - 1, n_2 - 1)] = \alpha/2$$

Example 11.5 Two methods for packaging "Wheatoos," a new wonder breakfast cereal, produce the same package fill weight. However, the second method is slightly faster and will be used unless there is an indication that the variance of the weights produced by the two machines differ. A random sample of 31 packages produced by the first method and a random sample of 21 packages produced by the second method are analyzed. It is found that the sample standard deviation of weights produced by the first machine is 0.50 ounce while the sample standard deviation of weights produced by the second machine is 0.62 ounce. Find a 90 percent confidence interval estimate for σ_1^2/σ_2^2.

Solution The sample data are:

$$n_1 = 31 \qquad n_2 = 21$$
$$s_1 = 0.50 \qquad s_2 = 0.62$$

The degrees of freedom are $n_1 - 1 = 30$ and $n_2 - 1 = 20$. From Table B.7, $F(0.95; 30, 20) = 2.04$. Also,

$$F(0.05; 30, 20) = \frac{1}{F(0.95; 20, 30)} = \frac{1}{1.93} = 0.518$$

The 90 percent confidence interval is:

$$\frac{(0.50)^2}{(0.62)^2} \cdot \frac{1}{2.04} \le \frac{\sigma_1^2}{\sigma_2^2} \le \frac{(0.50)^2}{(0.62)^2} \cdot \frac{1}{0.518}$$

$$0.318 \le \frac{\sigma_1^2}{\sigma_2^2} \le 1.256$$

Since this interval contains 1, we can conclude that σ_1^2 is not significantly different from σ_2^2.

☐ **11.3.2 Hypothesis testing concerning the ratio σ_1^2/σ_2^2**

If the null hypothesis H_0: $\sigma_1^2 = \sigma_2^2$ (or H_0: $\sigma_1^2/\sigma_2^2 = 1$) is true, then the F-statistic

$$\frac{S_1^2/\sigma_1^2}{S_2^2/\sigma_2^2} = \frac{S_1^2}{S_2^2} \cdot \frac{\sigma_2^2}{\sigma_1^2} \qquad \text{reduces to} \qquad F = \frac{S_1^2}{S_2^2}(1) = \frac{S_1^2}{S_2^2}$$

The hypothesis tests concerning the ratio σ_1^2/σ_2^2 are based on this statistic.

Example 11.6 A study is conducted to determine whether the mean sys-

Hypothesis test

$$\sigma_1^2 / \sigma_2^2$$

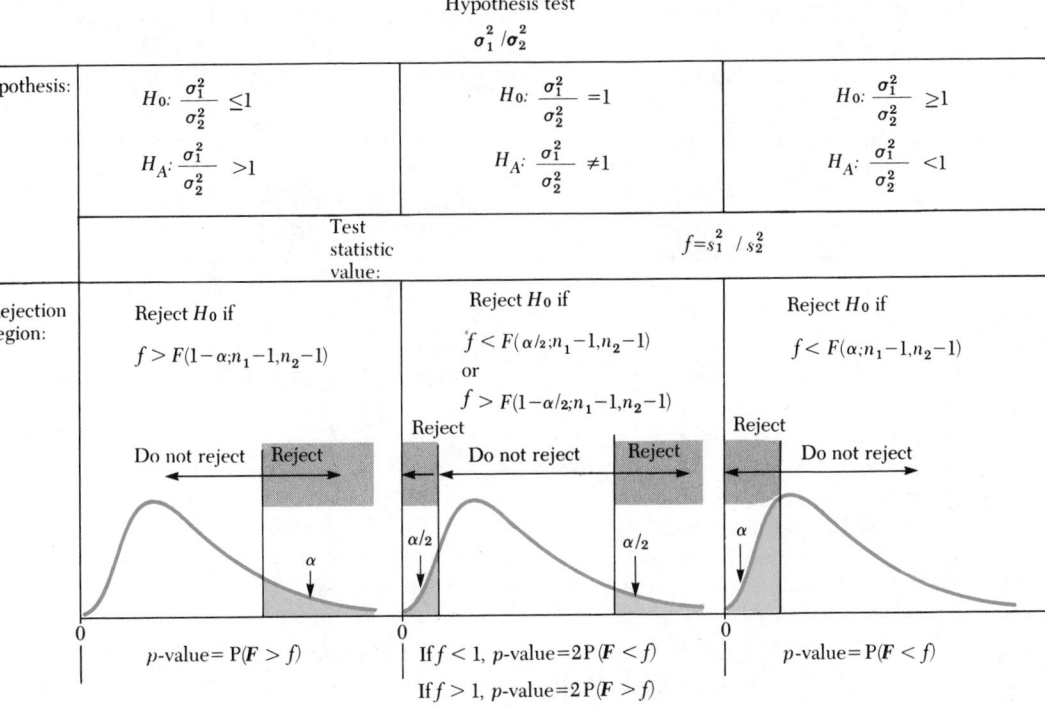

Hypothesis:	$H_0: \dfrac{\sigma_1^2}{\sigma_2^2} \leq 1$ $H_A: \dfrac{\sigma_1^2}{\sigma_2^2} > 1$	$H_0: \dfrac{\sigma_1^2}{\sigma_2^2} = 1$ $H_A: \dfrac{\sigma_1^2}{\sigma_2^2} \neq 1$	$H_0: \dfrac{\sigma_1^2}{\sigma_2^2} \geq 1$ $H_A: \dfrac{\sigma_1^2}{\sigma_2^2} < 1$

Test statistic value: $f = s_1^2 / s_2^2$

Rejection region:

Reject H_0 if
$f > F(1-\alpha; n_1-1, n_2-1)$

Reject H_0 if
$f < F(\alpha/2; n_1-1, n_2-1)$
or
$f > F(1-\alpha/2; n_1-1, n_2-1)$

Reject H_0 if
$f < F(\alpha; n_1-1, n_2-1)$

$p\text{-value} = P(F > f)$

If $f < 1$, $p\text{-value} = 2P(F < f)$
If $f > 1$, $p\text{-value} = 2P(F > f)$

$p\text{-value} = P(F < f)$

tolic blood pressure of football players is different from the mean systolic blood pressure of the general student body at a large university. A normal systolic blood pressure is approximately 120 for this age group. A lower number for an individual in good health usually is an indication of good physical condition. It is decided to sample 11 football players and 21 other students at random to test the hypothesis that the mean systolic blood pressure of football players (μ_1) is equal to the mean systolic blood pressure of the other students (μ_2). The sample results are $\bar{x}_1 = 117$, $s_1 = 4$ (football players) and $\bar{x}_2 = 120$, $s_2 = 5$ (other students). Test the hypothesis that $\mu_1 = \mu_2$ using the $\alpha = 0.10$ significance level. Assume that the measure, systolic blood pressure, is normally distributed.

Solution Since we do not know the population variances, σ_1^2 and σ_2^2, the first step is to determine whether the sample data can support the hypothesis that $\sigma_1^2 = \sigma_2^2$; if so, we can use the case 2 method for testing the equivalence of μ_1 and μ_2.

The null and alternate hypotheses are:

$$H_0: \quad \frac{\sigma_1^2}{\sigma_2^2} = 1$$

$$H_A: \quad \frac{\sigma_1^2}{\sigma_2^2} \neq 1$$

In constructing the decision rule, we must find $F(\alpha/2; n_1 - 1, n_2 - 1)$ and $F(1 - \alpha/2; n_1 - 1, n_2 - 1)$, where $n_1 - 1 = 10$, $n_2 - 1 = 20$, and $\alpha/2 =$

$0.10/2 = 0.05$. From Table B.7 in Appendix B, $F(0.95; 10, 20) = 2.35$, and

$$F(0.05; 10, 20) = \frac{1}{F(0.95; 20, 10)} = \frac{1}{2.77} = 0.36$$

The decision rule is:

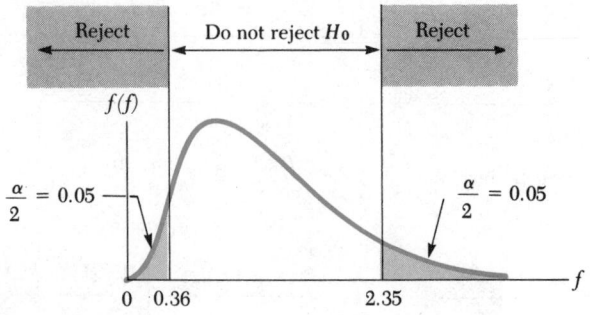

The test statistic value is:

$$f = \frac{s_1^2}{s_2^2} = \frac{(4)^2}{(5)^2} = \frac{16}{25} = 0.64$$

Since $f = 0.64$ falls in the region where H_0 is not rejected, we cannot reject the null hypothesis that $\sigma_1^2/\sigma_2^2 = 1$.

The p-value for this test is:

$$\begin{aligned} p\text{-value} &= 2P(F < f) \quad \text{since } f = 0.64 < 21/(21 - 2) = 1.11 \\ &= 2P(F < 0.64) \end{aligned}$$

Since, from Table B.7 in Appendix B, $F(0.10; 10, 20) = (1)/(2.20) = 0.45$ and $F(0.50; 10, 20) = 0.966$, the p-value is between twice 0.10 and 0.50, or between 0.20 and 1.00. It is certainly sufficiently large not to reject H_0 without too much apprehension that an error (Type I) has been made.

We can now proceed with the test comparing the population means μ_1 and μ_2 by using the case 2 procedure. The hypotheses are:

$$\begin{aligned} H_0: & \quad \mu_1 - \mu_2 = 0 \\ H_A: & \quad \mu_1 - \mu_2 \neq 0 \end{aligned}$$

The decision rule is:

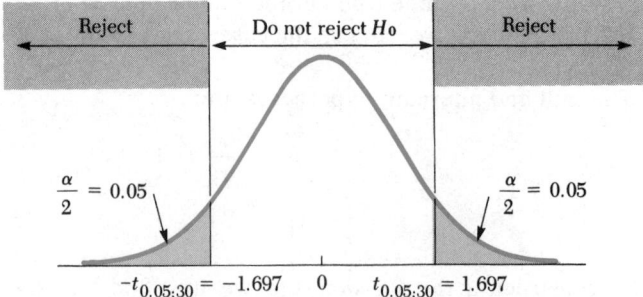

The test statistic value is:

$$t = \frac{(\bar{x}_1 - \bar{x}_2) - D_0}{s_p \sqrt{\dfrac{1}{n_1} + \dfrac{1}{n_2}}} = \frac{(117 - 120) - 0}{\sqrt{\dfrac{(10)(4)^2 + (20)(5)^2}{10 + 20}} \sqrt{\dfrac{1}{11} + \dfrac{1}{21}}} = -1.72$$

Since the value of the test statistic falls in the rejection region, we reject the null hypothesis that $\mu_1 = \mu_2$; it appears that the mean systolic blood pressure of football players is different from that of the rest of the student body.

The p-value for this test is:

$$p\text{-value} = 2P(t > |t|) = 2P(t > |-1.72|) = 2P(t > 1.72).$$

Since $P(t > 1.697/df = 30) = 0.05$ and $P(t > 2.042/df = 30) = 0.025$ from Table B.5 in Appendix B, twice the p-value is between twice 0.025 to 0.05, or between 0.05 and 0.10. Therefore, we would reject H_0: $\mu_1 = \mu_2$ at the $\alpha = 0.10$ significance level (the value set for this test) but not reject H_0 at the $\alpha = 0.01$ and $\alpha = 0.05$ significance levels.

Example 11.6 illustrates an integral part of testing a hypothesis concerning two population means by using the procedures for comparing means developed in Section 11.2. If the population variances are not known, then the F-test to check the assumption of equal variances should be made. If the test is passed (not rejected), then the case 2 method for comparing two means should be used. If this test does not pass, then the case 3 method must be used. A final check that must be made is to ensure that the population random variables, X_1 and X_2, are normally distributed.

■ 11.4 INFERENCES CONCERNING TWO POPULATION PROPORTIONS, π_1 AND π_2

The framework for comparing two population proportions is illustrated in Figure 11.6. Inferences concerning the population proportions π_1 and π_2 are

FIGURE 11.6 Comparing two population proportions

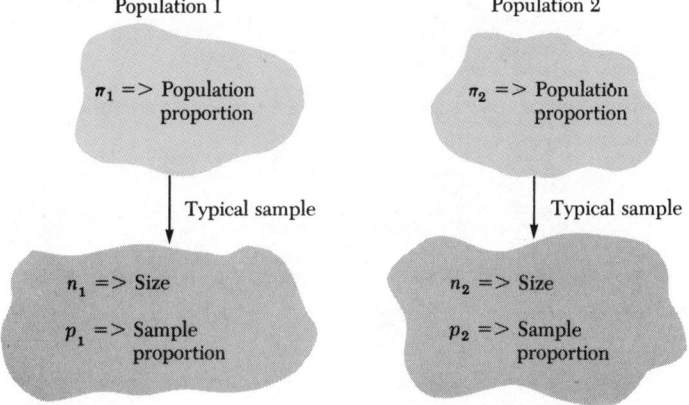

based on the sample proportions p_1 and p_2, since **P** is the best point estimator of π from Chapter 9.

In this section, we shall develop a large sample method for constructing confidence intervals and testing hypotheses on the difference between two population proportions. If the sample sizes are small (n_1 or n_2 or both less than 100), then this method should not be used. A small sample test for comparing population proportions will be given in Chapter 19.

☐ 11.4.1 Confidence interval estimation of $\pi_1 - \pi_2$

The central limit theorem can be used to construct a confidence interval estimator of $\pi_1 - \pi_2$. Recall that if $\hat{\theta}$ is an unbiased estimator of θ and is approximately normally distributed (it will be by the central limit theorem if it is an average or a sum and n is large), then the standardized random variable

$$Z = \frac{\hat{\theta} - \theta}{\sigma_{\hat{\theta}}}$$

is approximately distributed as a standardized normal random variable. In this Z-statistic, let $\theta = \pi_1 - \pi_2$. Then $\hat{\theta} = \widehat{\pi_1 - \pi_2} = P_1 - P_2$; that is, the difference of the sample proportion statistics is the best point estimator of the difference of the population proportions. Now,

$$\sigma_{\hat{\theta}}^2 = \sigma_{P_1-P_2}^2 = V(P_1 - P_2) = V(P_1) + V(P_2) = \sigma_{P_1}^2 + \sigma_{P_2}^2$$

by Theorem 11.1. Thus,

$$\sigma_{\hat{\theta}}^2 = \frac{\pi_1(1 - \pi_1)}{n_1} + \frac{\pi_2(1 - \pi_2)}{n_2}$$

The appropriate Z-statistic is

$$Z = \frac{(P_1 - P_2) - (\pi_1 - \pi_2)}{\sqrt{\dfrac{\pi_1(1 - \pi_1)}{n_1} + \dfrac{\pi_2(1 - \pi_2)}{n_2}}}$$

This statistic can be used to construct the following confidence interval estimate of $\pi_1 - \pi_2$.

Formula 11.4
Confidence interval estimate of $\pi_1 - \pi_2$
A $100(1 - \alpha)$ percent confidence interval estimate of $\pi_1 - \pi_2$ is given by

Formula 11.4 (continued)

$$(p_1 - p_2) \pm z_{\alpha/2} \sqrt{\frac{p_1(1 - p_1)}{n_1} + \frac{p_2(1 - p_2)}{n_2}}$$

where n_i and p_i are the sample size and sample proportion, respectively, associated with the ith sample ($i = 1, 2$) and $z_{\alpha/2}$ is the point on the standard normal curve such that $P(\mathbf{Z} \geq z_{\alpha/2}) = \alpha/2$.

Example 11.7 During a recent flu epidemic, a study was made at a large university to determine whether flu shots help persons resist contracting the flu. A sample of 200 students who had received the series of flu shots at the campus medical center produced 40 victims of the flu. A sample of 100 students who did not receive the shots produced 35 students who got the flu. Find a 90 percent confidence interval for the difference between the proportion of flu victims among those students receiving the shots (π_1) and the proportion of flu victims among those not receiving the shots (π_2).

Solution For a 90 percent confidence interval, $z_{\alpha/2} = z_{0.05} = 1.64$ from Table B.3 in Appendix B. The 90 percent confidence interval for $\pi_1 - \pi_2$ is

$$(p_1 - p_2) \pm z_{\alpha/2} \sqrt{\frac{p_1(1 - p_1)}{n_1} + \frac{p_2(1 - p_2)}{n_2}}$$

$$= \left(\frac{40}{200} - \frac{35}{100}\right) \pm 1.64 \sqrt{\frac{(0.2)(0.8)}{200} + \frac{(0.35)(0.65)}{100}}$$

$$= -0.15 \pm 0.09$$

Thus, we are 90 percent confident that the true difference between π_1 and π_2 is between -0.24 and -0.06. Since 0 is not included in this interval, we would conclude that the flu shots probably were helpful in resisting the flu bug.

☐ **11.4.2 Hypothesis testing concerning $\pi_1 - \pi_2$**

The hypothesis-testing procedure based on large samples for comparing π_1 and π_2 is developed from the Z-statistic used in the preceding section to construct an interval estimator of $\pi_1 - \pi_2$. The test statistic for testing the null hypothesis H_0: $\pi_1 - \pi_2 = D_0$ is

$$\mathbf{Z} = \frac{(\mathbf{P}_1 - \mathbf{P}_2) - D_0}{\sqrt{\frac{\mathbf{P}_1(1 - \mathbf{P}_1)}{n_1} + \frac{\mathbf{P}_2(1 - \mathbf{P}_2)}{n_2}}}$$

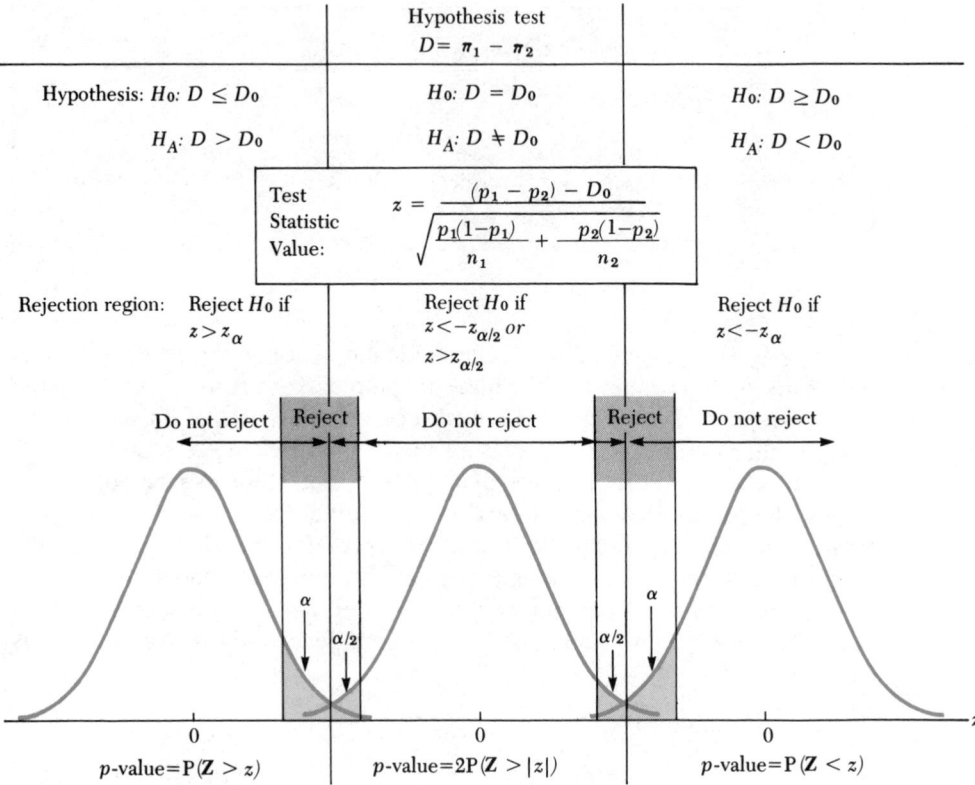

	Hypothesis test $D = \pi_1 - \pi_2$	
Hypothesis: $H_0: D \leq D_0$ $H_A: D > D_0$	$H_0: D = D_0$ $H_A: D \neq D_0$	$H_0: D \geq D_0$ $H_A: D < D_0$

Test Statistic Value:

$$z = \frac{(p_1 - p_2) - D_0}{\sqrt{\dfrac{p_1(1-p_1)}{n_1} + \dfrac{p_2(1-p_2)}{n_2}}}$$

Rejection region: Reject H_0 if $z > z_\alpha$	Reject H_0 if $z < -z_{\alpha/2}$ or $z > z_{\alpha/2}$	Reject H_0 if $z < -z_\alpha$

| Do not reject | Reject | Do not reject | Reject | Do not reject |

α $\alpha/2$ $\alpha/2$ α

| 0 | 0 | 0 |
| p-value $= P(Z > z)$ | p-value $= 2P(Z > |z|)$ | p-value $= P(Z < z)$ |

Example 11.8 A study is conducted to determine whether the same proportion of "hard hats" and college professors voted for Herman, the new sewer commissioner, in a county election. A random sample of 144 hard hats produced a sample proportion of 0.66 who said they voted for Herman, and a random sample of 100 college professors yielded a sample proportion of 0.52 who said they voted for Herman. Test the hypothesis that the two population proportions, π_1 (the proportion of hard hats voting for Herman) and π_2 (the proportion of professors voting for Herman) are equal at the $\alpha = 0.05$ significance level.

Solution The hypothesis is:

$$\begin{array}{ll} H_0: & \pi_1 = \pi_2 \\ H_A: & \pi_1 \neq \pi_2 \end{array} \left\{ \begin{array}{l} \text{or, equivalently} \\ \text{with } D = \pi_1 - \pi_2 \end{array} \right\} \begin{array}{ll} H_0: & D = 0 \\ H_A: & D \neq 0 \end{array}$$

The decision rule is:

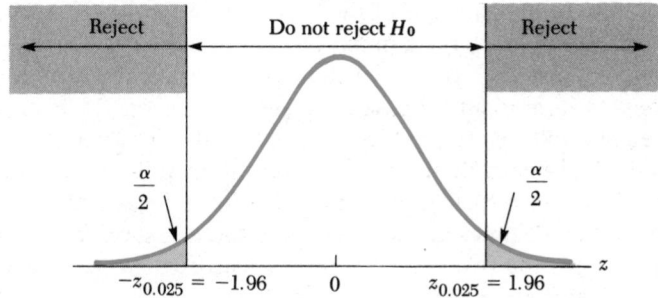

The test statistic value is:

$$z = \frac{(p_1 - p_2) - 0}{\sqrt{\dfrac{p_1(1 - p_1)}{n_1} + \dfrac{p_2(1 - p_2)}{n_2}}}$$

$$= \frac{0.66 - 0.52}{\sqrt{\dfrac{(0.66)(0.34)}{144} + \dfrac{(0.52)(0.48)}{100}}} = 2.20$$

Since $z = 2.20$ falls in the rejection region, we reject the null hypothesis that $\pi_1 = \pi_2$.

The p-value for this test is:

$$p\text{-value} = 2P(\mathbf{Z} > |z|) = 2P(\mathbf{Z} > 2.20) = 2(0.0139) = 0.0278$$

Since the p-value is small, we can be fairly confident that we have made the right decision to reject H_0.

11.5 PAIRED t-TEST FOR COMPARING TWO POPULATION MEANS

In Section 11.2, when comparing two population means, we assumed that the two random samples drawn from the two populations were independent. In many instances, this will not be true. In fact, many times we purposefully design the experiment so that the samples drawn from the two populations will be dependent.

Let us first consider a two-population experiment in which the samples are dependent by the nature of the experiment. We will then discuss an experiment that is *designed* so that the two samples are dependent.

Example 11.9 A doctor wishes to test the efficacy of a new dietary plan. He decides to select ten persons randomly from a large group who are interested in losing weight. To test the efficacy of this plan, he will weigh the persons in the sample at the beginning of the dietary program and again at the end of six weeks. He wishes to demonstrate, of course, that the population mean weight at the end of the six-week program is less than the population mean weight at the beginning of the program.

It is clear that the ten weights comprising the first sample and the ten weights comprising the second sample are not independent. In fact, the weights will be "paired"; a weight in the first sample is taken from one individual who will produce a weight (hopefully less) in the second sample.

Whenever measurements are taken from the same units (usually people) at two different times and the means of the measurements at the two times are compared, the two sets of sample measurements are dependent.

In many experiments, we may design the experiment so that units in the two samples are paired to eliminate effects in which we have no interest.

Example 11.10 A professor in charge of an introductory statistics course at a large university is interested in developing a new teaching method. She decides to form two classes, each containing 25 students, for the purpose of testing the new teaching method. The old teaching method will be used in the first class, and the new teaching method will be used in the second class. In the term during which she will compare the two methods, more than 500 students will take this course. The professor decides to select the 50 students she will need for her study in pairs where the two students comprising each pair are as similar as possible—same sex, roughly the same IQ, about the same set of previous courses, same grade point average, and so on. She will flip a coin to see which member of each pair is assigned to the first class—the "loser" will be assigned to the second class.

The two samples (classes) will be dependent because the assignments of students to the two groups were made in pairs—the formation of the two samples was not completely random. Yet, by pairing the sample observations, many factors such as grade point average and IQ that may cloud the determination of which teaching method is "better" have been eliminated. If the samples had been formed completely at random, we might by chance find that the better students among the 50 selected were placed in the first class (old teaching method) while the less prepared students were placed in the second class. Certainly, this circumstance would not bode well for the new teaching method, even if it was really more effective than the old method.

The analysis of the paired observation experiment is different than if the two samples were chosen at random. Suppose the experiment yields n pairs of values denoted by (x_{11}, x_{21}), (x_{12}, x_{22}), (x_{13}, x_{23}), . . . , (x_{1n}, x_{2n}). That is, x_{ij} is the member of the jth pair drawn from the ith population. For the jth pair, define

$$d_j = x_{1j} - x_{2j}$$

That is, d_j is the difference between the two values forming the jth pair. It seems reasonable that a hypothesis test to determine whether the two population means are equal is based on the differences d_j, $j = 1, 2, . . . , n$, where n is the number of observations in each sample. These differences are used in the following way. We first form the average of the differences, $\bar{d}$, and the variance of the differences, s_d^2, where

$$\bar{d} = \frac{\sum_{j=1}^{n} d_j}{n} \quad \text{and} \quad s_d^2 = \frac{\sum_{j=1}^{n} (d_j - \bar{d})^2}{n - 1}$$

If we assume that the collection of differences, d_j, where $j = 1, 2, \ldots, n$, forms a random sample from a normal population with mean $\mu_d = \mu_1 - \mu_2$, then the statistic

$$t = \frac{\bar{d} - \mu_d}{s_d / \sqrt{n}}$$

is a value from the t-distribution with $(n - 1)$ degrees of freedom.

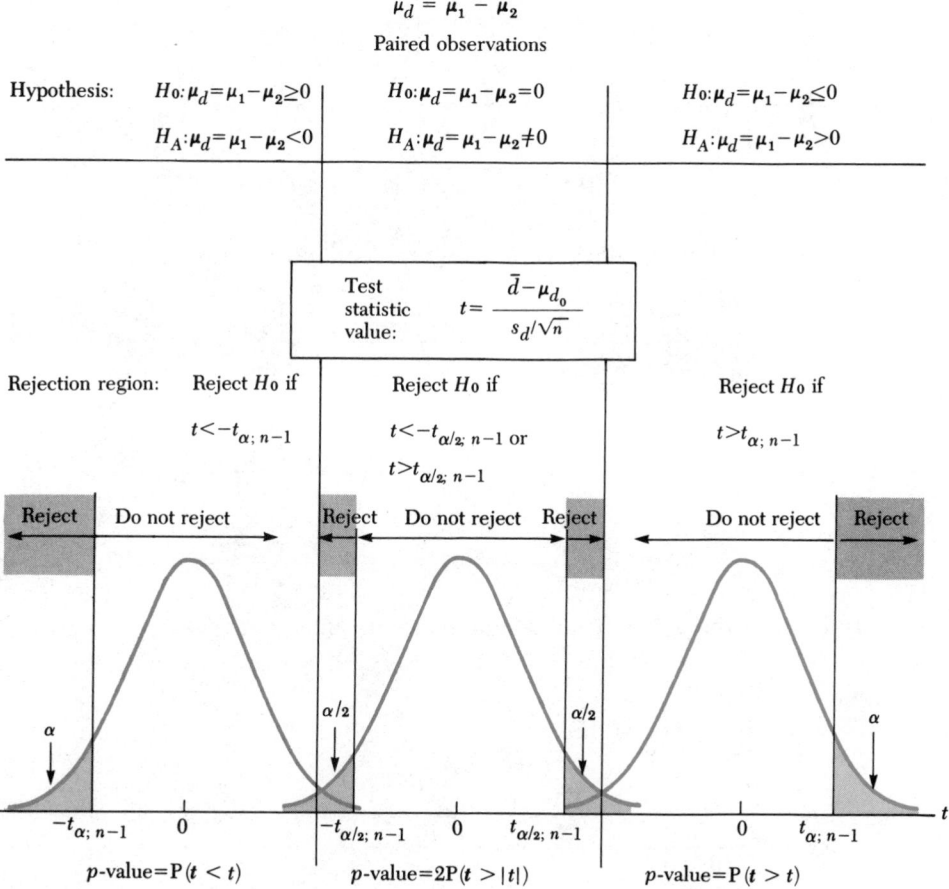

Hypothesis test

$$\mu_d = \mu_1 - \mu_2$$

Paired observations

Example 11.11 Suppose the weights of the ten participating people in the dietary plan described in Example 11.9 are those listed in Table 11.1. Test the hypothesis that $\mu_d = 0$ at the $\alpha = 0.01$ significance level.

TABLE 11.1

	Sample data		
Beginning x_{1j}	*Six weeks* x_{2j}	d_j	d_j^2
155	154	1	1
228	207	21	441
172	165	7	49
141	147	−6	36
162	157	5	25
211	196	15	225
185	180	5	25
122	121	1	1
164	150	14	196
199	204	−5	25
		58	1,024

Solution The hypothesis is:

$$H_0: \quad \mu_1 - \mu_2 = \mu_d = 0$$
$$H_A: \quad \mu_d \neq 0$$

For the $\alpha = 0.01$ significance level, the rejection region is

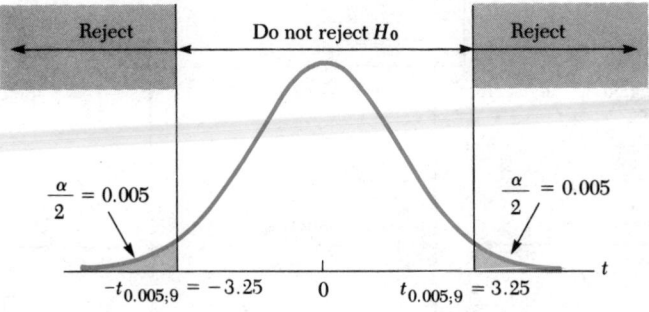

$$\frac{\alpha}{2} = 0.005 \qquad \frac{\alpha}{2} = 0.005$$

$$-t_{0.005;9} = -3.25 \qquad 0 \qquad t_{0.005;9} = 3.25$$

In forming the value of the test statistic, we must compute $\bar{d}$ and s_d:

$$\bar{d} = \frac{\sum\limits_{j=1}^{10} d_j}{10} = \frac{58}{10} = 5.8$$

$$s_d^2 = \frac{\sum\limits_{j=1}^{10} (d_j - \bar{d})^2}{10 - 1} = \frac{\sum\limits_{j=1}^{10} d_j^2 - \frac{\left(\sum\limits_{j=1}^{10} d_j\right)^2}{10}}{9} = \frac{1,024 - \frac{3,364}{10}}{9} = 76.4$$

$$s_d = \sqrt{76.4} = 8.74$$

The test statistic value is:

$$t = \frac{\bar{d} - \mu_{d_0}}{s_d/\sqrt{n}} = \frac{5.8 - 0}{8.74/\sqrt{10}} = 2.10$$

Since $t = 2.10$ falls in the region where H_0 cannot be rejected, we cannot reject the null hypothesis that $\mu_d = 0$. We conclude, therefore, that the doctor's dietary plan was not very effective.

The p-value for this test is given by:

$$p\text{-value} = 2P(\mathbf{t} > |t|) = 2P(\mathbf{t} > 2.10)$$

Since $P(\mathbf{t} > 1.833) = 0.05$ and $P(\mathbf{t} > 2.262) = 0.025$ from Table B.5 in Appendix B, the p-value lies between twice 0.025 and twice 0.05, or between 0.05 and 0.10. Note that the significance level α was set very low for this test ($\alpha = 0.01$). We would not reject the null hypothesis if $\alpha = 0.01$ or 0.05, but we would reject the null hypothesis if $\alpha = 0.10$.

It should be apparent that we can use the t-statistic given above to construct a confidence interval estimator of $\mu_d = \mu_1 - \mu_2$.

Formula 11.5
Confidence interval estimate of $\mu_d = \mu_1 - \mu_2$, paired samples

A $100(1 - \alpha)$ percent confidence interval estimate for $\mu_1 - \mu_2$ when the sample observations $(x_{11}, x_{21}), (x_{12}, x_{22}), (x_{13}, x_{23}), \ldots , (x_{1n}, x_{2n})$ are paired is given by

$$\bar{d} \pm t_{\alpha/2;n-1} \frac{s_d}{\sqrt{n}}$$

where

$$\bar{d} = \frac{\sum_{j=1}^{n} d_j}{n}, \quad s_d = \sqrt{\frac{\sum_{j=1}^{n} (d_j - \bar{d})^2}{n-1}}, \quad d_j = x_{1j} - x_{2j}, j = 1, 2, \ldots , n$$

A few comments regarding the paired experiment are in order. First, the paired experiment is usually more "efficient" than the independent sample experiment. By this, we mean that fewer sample observations are generally required to detect a true difference between μ_1 and μ_2 when the paired sample experiment is used instead of the two independent sample experiment. This is because the pairing usually eliminates effects that may make it difficult for the two independent sample experiments to detect a true difference. Second, the paired experiment does not require that the population variances, σ_1^2 and σ_2^2, be equal. When it is known that σ_1^2 is not equal to σ_2^2, if the sample values can be paired in some meaningful way, then the paired t-test can be used in place of the approximate case 3 method described in Section 11.2.3.

SUMMARY TABLE Hypothesis testing and confidence intervals for two populations

Population parameter	Hypothesis	Confidence interval formula	Test statistic value
μ_1, μ_2 (means)	$H_0: \ \mu_1 - \mu_2 = D_0$		
	Case 1: σ_1^2, σ_2^2 known	$(\bar{x}_1 - \bar{x}_2) \pm z_{\alpha/2} \sqrt{\dfrac{\sigma_1^2}{n_1} + \dfrac{\sigma_2^2}{n_2}}$	$z = \dfrac{(\bar{x}_1 - \bar{x}_2) - D_0}{\sqrt{(\sigma_1^2/n_1) + (\sigma_2^2/n_2)}}$
	Case 2: σ_1^2, σ_2^2 unknown $\sigma_1^2 = \sigma_2^2$	$(\bar{x}_1 - \bar{x}_2) \pm t_{\alpha/2; n_1 + n_2 - 2} \, s_p \sqrt{\dfrac{1}{n_1} + \dfrac{1}{n_2}}$ $s_p = \sqrt{\dfrac{(n_1 - 1)s_1^2 + (n_2 - 1)s_2^2}{(n_1 - 1) + (n_2 - 1)}}$	$t = \dfrac{(\bar{x}_1 - \bar{x}_2) - D_0}{s_p \sqrt{\dfrac{1}{n_1} + \dfrac{1}{n_2}}}$ Degrees of freedom $= n_1 + n_2 - 2$
	Case 3: σ_1^2, σ_2^2 unknown $\sigma_1^2 \neq \sigma_2^2$	(approximate) $(\bar{x}_1 - \bar{x}_2) \pm t_{\alpha/2; \Delta} \sqrt{(s_1^2/n_1) + (s_2^2/n_2)}$ $\Delta = \dfrac{[(s_1^2/n_1) + (s_2^2/n_2)]^2}{\dfrac{(s_1^2/n_1)^2}{(n_1 - 1)} + \dfrac{(s_2^2/n_2)^2}{(n_2 - 1)}}$	$t' = \dfrac{(\bar{x}_1 - \bar{x}_2) - D_0}{\sqrt{(s_1^2/n_1) + (s_2^2/n_2)}}$ Degrees of freedom $= \Delta$
π_1, π_2 (proportions)	$H_0: \ \pi_1 - \pi_2 = D_0$	$(p_1 - p_2) \pm z_{\alpha/2} \sqrt{\dfrac{p_1(1 - p_1)}{n_1} + \dfrac{p_2(1 - p_2)}{n_2}}$	$z = \dfrac{(p_1 - p_2) - D_0}{\sqrt{\dfrac{p_1(1 - p_1)}{n_1} + \dfrac{p_2(1 - p_2)}{n_2}}}$
σ_1^2, σ_2^2 (variances)	$H_0: \ \dfrac{\sigma_1^2}{\sigma_2^2} = 1$	$\dfrac{s_1^2}{s_2^2} \cdot f(1 - \alpha/2; n_1 - 1, n_2 - 1)$ $\leq \dfrac{\sigma_1^2}{\sigma_2^2} \leq \dfrac{s_1^2}{s_2^2} \cdot \dfrac{1}{f(\alpha/2; n_1 - 1, n_2 - 1)}$	$f = \dfrac{s_1^2}{s_2^2}$ Degrees of freedom $= (n_1 - 1), (n_2 - 1)$

n_i = size of ith sample; σ_i^2 = variance of ith population; μ_i = mean of ith population; $\bar{x}_i$ = mean of the ith sample; s_i^2 = variance of the ith sample; π_i = proportion in the ith population; p_i = proportion in the ith sample; z = standard normal statistic value, t = **t** statistic value, f = **F** statistic value.

■ 11.6 SUMMARY

Statistical hypothesis testing procedures are developed for testing the difference between population means (μ_1, μ_2) and population proportions (π_1, π_2), and for testing the equivalence of two population variances. In testing hypotheses concerning the difference between two population means, two different experimental designs may be used. The two samples drawn randomly from the two populations may be independent (the independent samples test) or dependent (the paired samples test). It is important to know the distinction between these two experimental designs and to know which design is appropriate in a given situation.

■ REFERENCES

Conover, W. J. *Practical Nonparametric Methods*. New York: John Wiley & Sons, Inc., 1971.

Snedecor, G. W., and Cochran, W. G. *Statistical Methods*. 7th ed. Ames: Iowa State University Press, 1980.

Winkler, R., and Hays, W. *Statistics: Probability, Inference, and Decision*. 2d ed. New York: Holt, Rinehart and Winston, Inc., 1975.

■ PROBLEMS

11.1. Let $D = \mu_1 - \mu_2$.
  a. Are the hypotheses H_0: $\mu_1 \geq \mu_2$ and H_0: $D \geq 0$ equivalent?
  b. In terms of μ_1 and μ_2, restate the hypothesis H_0: $D \leq 2$.

11.2. If the random variables X and Y are independent, and $\sigma_X^2 = 10$, $\sigma_Y^2 = 25$, what is the variance of the difference $X - Y$? What do we mean in words when we say "X and Y are independent random variables"?

11.3. In constructing confidence intervals on the difference $\mu_1 - \mu_2$, distinguish among the three cases concerning the variances, σ_1^2 and σ_2^2. Why is it necessary to define three cases for setting a confidence interval on $\mu_1 - \mu_2$?

11.4. Two machines, A and B, are used to fill one-pound coffee cans with ground coffee. From lengthy experience with the machines, it is known that the standard deviation of the fills for both machines is 0.20 ounce ($\sigma_A = \sigma_B = 0.20$). But, it is suspected that the mean fills for the two machines differ. A random sample of 25 cans taken from each machine produced the sample average fills, $\bar{x}_A = 15.96$ ounces and $\bar{x}_B = 15.67$ ounces. Assume that the fills are normally distributed.
  a. Determine a 95 percent confidence interval for $\mu_A - \mu_B$.
  b. At the $\alpha = 0.05$ level, test the null hypothesis that $\mu_A = \mu_B$.
  c. What is the p-value for this test?

11.5. A market research analyst believes that items placed on shelves at or near eye level will sell more rapidly than those placed on the bottom shelf in a store. An experiment is set up where ten weeks are randomly selected (during the first half of the year), and the product is placed at eye level during five of these weeks and on the bottom shelf during the remaining five weeks. The sales data in units sold per week are:

Bottom shelf		Eye-level shelf
33	....	41
35	....	40
32	....	45
38	....	42
44	....	48

Assuming that $\sigma_1^2 = \sigma_2^2$,

a. Set a 90 percent confidence interval on the difference between the average number of units sold per week at eye level and on the bottom shelf.

b. At $\alpha = 0.10$, are the means significantly different?

c. What is the p-value for this test?

d. What assumptions must you make in parts a and b to use the methods in this chapter? Are the assumptions valid in this case?

11.6. For each of the following inferences, carefully describe the assumptions required to use the methods developed in this chapter for hypothesis testing.

a. Comparison of two population means; independent samples.

b. Comparison of two population means; paired samples.

c. Comparison of two population proportions.

d. Comparison of two population variances.

11.7. The management wishes to make changes in the assembly process of a particular product. Workers are paid on the basis of their output, so it is necessary to demonstrate that the rate of assembling a unit will be increased under the new system. Ten workers are randomly selected to participate in an experiment in which each worker assembles one unit under the old process and one unit under the new process. The assembly times in minutes are:

Do both ways

Worker		Old process	New process
1	....	23	20
2	....	25	22
3	....	20	20
4	....	18	14
5	....	26	21
6	....	18	21
7	....	15	12
8	....	16	13
9	....	21	21
10	....	24	20

Assume that the assembly times are normally distributed.

a. Are the assembly times for the old and new processes independent samples? Explain.

b. Is the average assembly time for the new process faster than the average assembly time for the old process? (Compute the approximate p-value for this test and reject the null hypothesis if p-value < 0.05.)

11.8. In attempting to sell subscriptions to a local newspaper door to door, two different sales approaches are tried. Each of 20 salesmen use the two approaches alternately for the same period and for the same number of household contacts. The sales data (number of subscriptions sold per period) are given here.

Salesman		Approach 1	Approach 2
1	....	10	8
2	....	15	12
3	....	26	24
4	....	13	15
5	....	18	10
6	....	10	10
7	....	16	20
8	....	21	16
9	....	5	0
10	....	3	7
11	....	12	11
12	....	14	17

Salesman	Approach 1	Approach 2
13...........	24	17
14...........	18	20
15...........	13	13
16...........	12	4
17...........	18	12
18...........	21	19
19...........	6	3
20...........	11	16

a. Is this a paired experiment or an independent sample experiment? Discuss.

b. Do the two sales approaches produce significantly different averages sales? (Compute the approximate p-value for this test and reject the null hypothesis if p-value < 0.05.)

11.9. What are the consequences of analyzing a paired sample experiment as if it were an independent sample experiment?

11.10. If it is possible to use either a paired or independent sample experiment in a particular problem, why is a paired experiment preferred?

11.11. Random samples of two brands of whole milk are checked for the amount of fat content in grams. Twenty-five half-gallon containers of each brand are selected and the fat content in grams is weighed. The data are given here.

Brand A			Brand B		
30	26	36	24	33	17
26	33	35	27	20	21
31	20	32	22	18	18
27	28	27	31	26	25
37	27	29	30	25	27
28	31	33	25	20	24
31	35	27	22	22	24
29	30	30	26	24	20
25			29		

a. Is this a paired experiment or an independent sample experiment? Explain.

b. Assuming $\sigma_A^2 = \sigma_B^2$, do the two brands have different average gram weight fat content? (Calculate the p-value and use it to decide whether H_0 should be rejected or not.)

c. Is the assumption in part b of equal population variances reasonable? (Test the hypothesis $\sigma_A^2 = \sigma_B^2$ at the $\alpha = 0.10$ level. What is the approximate p-value for this test?)

11.12. A car manufacturer is interested in testing the ability of two bumper designs to withstand low-speed crashes. Twenty-five samples of each type of bumper are installed on test cars, which are run into a brick wall at 10 m.p.h. An estimate of the cost to repair the damage is then made based on studying the impact in slow motion and carefully examining the car. Based on the experiment, the following sample statistics are computed:

	Mean	Standard deviation
Bumper type 1	$\bar{x}_1 = \$55$	$s_1 = \$10$
Bumper type 2	$\bar{x}_2 = \$70$	$s_2 = \$ 5$

a. Assuming that $\sigma_1^2 = \sigma_2^2$, find a 95 percent confidence interval on the difference $\mu_1 - \mu_2$. Is one bumper design significantly more effective than the other? Explain.

b. Is the assumption of equal population variances reasonable in this case? Test at the $\alpha = 0.10$ level. What is the approximate p-value for this test?

11.13. It is suspected that two machines that are set up to make ⅛-inch diameter ball bearings do so with different degrees of precision. A random sample of 50 bearings is taken from each machine and the sample standard deviation is calculated. The sample statistics are:

$$n_1 = 50 \qquad n_2 = 50$$
$$s_1 = 0.001 \qquad s_2 = 0.004$$

a. Determine a 90 percent confidence interval on the ratio σ_1^2/σ_2^2.

b. At the $\alpha = 0.10$ level, would you conclude that σ_1^2 is different from σ_2^2? Explain

11.14. Two machines fill bottles with catsup on an assembly line. It is suspected that the fills are more variable with machine A than with machine B. A random sample of 24 fills from A and a random sample of 18 fills from B produces the following data:

$$n_A = 24 \qquad n_B = 18$$
$$s_A = 0.5 \text{ oz.} \qquad s_B = 0.3 \text{ oz.}$$

Assume that the fills are normally distributed. Do these two machines have significantly different fill-variances? (Calculate the p-value for this test and reject the null hypothesis if p-value < 0.10.)

11.15. The braking distances at 60 m.p.h. for two types of automobiles are compared. For the first make (A), 49 randomly selected cars were tested, and for the second make (B), 36 randomly selected cars were tested. The sample statistics are:

Make	Mean	Standard deviation
A	$\bar{x}_1 = 167$ ft	$s_1 = 10$ ft
B	$\bar{x}_2 = 180$ ft	$s_2 = 20$ ft

Is the difference between the mean stopping distances for the two cars significantly different at the $\alpha = 0.05$

level? What is the approximate p-value for this test?

11.16. A random sample of 25 engineers in company A produces a mean salary of $18,000 with a standard deviation of $3,000, and a random sample of 36 engineers in company B produces a mean salary of $22,000 with a standard deviation of $4,000. Can we conclude that company B pays its engineers more than company A? Use the p-value for making the decision. Explain your decision.

11.17. In one area, a random sample of 100 persons at grocery stores produces 60 who prefer Best soap to all others. In another area, 75 out of 225 liked Best best.

a. Determine a 90 percent confidence interval on the difference between the proportions in the two areas who prefer Best.

b. What assumptions are required in part a?

11.18. It is suspected that the proportion of voters who prefer Henry Liberal in the Northeast is greater than the proportion who prefer him in the South. In two pre-election polls, it is found that 100 of 150 prefer Henry to his competitor in the northeast, and 75 of 150 prefer Henry in the South. Do these polls support the conjecture that Henry is more preferred in the Northeast than in the South? Use the p-value for making the decision. Explain your decision.

11.19. In a poll of workers in a large plant, 50 of 100 white-collar workers prefer the adoption of a new retirement plan, while 80 of 120 blue-collar workers prefer the new plan. Are the proportions favoring the new plan among the two groups significantly different at the $\alpha = 0.05$ level? What is the approximate p-value for this test?

11.20. The buying habits in two suburban communities located around a large city were surveyed. Of the 200 housewives in the first suburb who were randomly sampled, it was found that they spend an average of $150 per month on food. In the second suburb, it is found that an average of $180 per month is spent on food among the 225 sampled housewives. The standard deviation in the first suburb is found to be $25 and in the second suburb, $40. Determine a 99 percent confidence interval on the difference between the two population means.

11.21. A student is interested in graduate school education upon her graduation from an undergraduate business administration program. A consideration in her decision is the average salary of the graduates of the particular degree program she has in mind. Among two candidate schools, she finds that the average salary among ten sampled graduates from the first school is $20,000 with a standard deviation of $2,000, and the average salary among ten sampled graduates from the second school is $24,000 with a standard deviation of $3,000. Assuming that the salaries are normally distributed, are the average salaries for graduates of the two programs significantly different at the $\alpha = 0.01$ level of significance? Compute the approximate p-value for this test.

11.22. The administration at a large state university has proposed to the faculty a new policy for promotion. It is believed that the proportion favoring the proposal among faculty who are not tenured is different from the proportion favoring the proposal among tenured faculty. In a sample of 64 faculty without tenure, it is found that only 20 percent favor the proposal, while in a sample of 36 tenured faculty, 55 percent favor the proposal. Determine a 95 percent confidence interval on the difference between the two population proportions, assuming that the sample proportions are normally distributed.

Analysis of variance

12

In this chapter, we shall describe methods that may be used to compare two or more sample means and infer whether or not the corresponding population means are the same. To illustrate the kind of problem we will be addressing in this chapter, consider the following example.

Example 12.1 Farmco Investment Corporation has recently purchased a large chicken farm for the production of broilers intended for overseas markets. An animal breeding consultant engaged by Farmco has advised that three feeds are commonly used to ensure quality broilers and to produce rapid weight gains. He suggests conducting a statistical experiment to determine which feed produces the fastest weight gains, the factor considered most important by Farmco. Taking his advice, Farmco instructs its research staff to apply each feed to a randomly selected group of chicks and measure the weight gains over a specified interval of time. In a pilot experiment, it is decided to apply each feed to five randomly selected chicks from a large brood hatched at approximately the same time. The sample mean weight gains for each feed over the specified time interval will be compared to infer whether or not the corresponding population mean weight gains for the three feeds—Growbig, Max, and Topnotch—are equivalent.

The form of the sample data corresponding to this experiment is outlined in Table 12.1. For the ith feed, five responses (weight gains) x_{ij}, $j = 1, 2, \ldots, 5$, are recorded where x_{ij} = weight gain in grams for the jth chick subjected to the ith feed. The mean weight gain corresponding to the ith feed is denoted by $\bar{x}_{i\cdot}$.

TABLE 12.1 Form of the sample data

		Feed		
$i =$	Growbig (1)	Max (2)	Topnotch (3)	j
	x_{11}	x_{21}	x_{31}	(1)
	x_{12}	x_{22}	x_{32}	(2)
	.	.	.	.
	.	.	.	.
	.	.	.	.
	x_{15}	x_{25}	x_{35}	(5)
Means	$\bar{x}_{1.}$	$\bar{x}_{2.}$	$\bar{x}_{3.}$	

x_{ij} = Weight gain for the jth chick subjected to the ith feed.
$\bar{x}_{i.}$ = Mean weight gain for the ith feed.

The three sample mean weight gains—$\bar{x}_{1.}$, $\bar{x}_{2.}$ and $\bar{x}_{3.}$—may be used to draw inferences about the equivalence of the three population means μ_1, μ_2, and μ_3. The ith population mean μ_i represents the average weight gain in grams for all chicks subjected to the ith feed under the selected experimental conditions. For example, the true relationship among the three population means *may* be as illustrated in Figure 12.1. In the figure, Max produces the greatest average weight gain, while Topnotch produces the least.

If the true relationship among the means is the one depicted in Figure 12.1, then the means are not equivalent and we should reject the null hypothesis

FIGURE 12.1 Comparison of three population means
Density

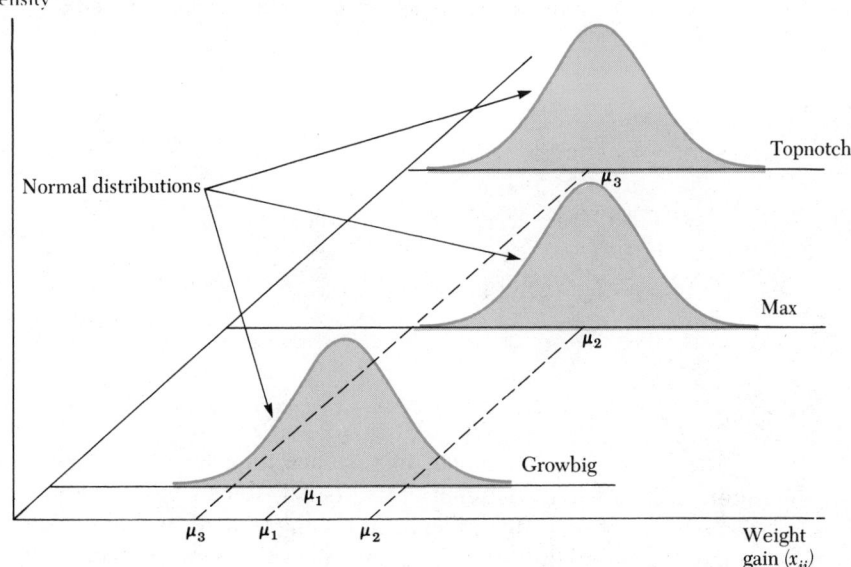

Normal distributions

Topnotch

Max

Growbig

Weight gain (x_{ij})

that $\mu_1 = \mu_2 = \mu_3$. The methods in this chapter provide a statistical procedure for determining whether or not the null hypothesis should be rejected based on the sample information—specifically the values of $\bar{x}_1$., $\bar{x}_2$., and $\bar{x}_3$.

Recall in Chapter 11 that we developed a statistical procedure for testing the equivalence of two population means. If the variances of the two populations are unknown and must be estimated from the sample data, the resulting test statistic in Chapter 11 is based on the t-distribution. The procedures developed in this chapter are called *analyses of variance* and are equivalent to the t-test in Chapter 11 when exactly two means are being compared.

The name of the procedure, *analysis of variance,* may appear to be inappropriate because we wish to compare means, not variances. But, as we shall see, the name does accurately describe the procedure used to test the equivalence of two or more population means.

The analysis of variance procedure may be extended to study more than one factor in the same experiment. For example, Farmco may wish to study the effects of two or more kinds of vitamin supplements on chick weight gain in addition to the feed effects. We will discuss the methods of conducting a two-factor or multiple-factor experiment in this chapter. The application of analysis of variance to a two-factor (or multiple-factor) experiment will allow us to test the equivalence of the means associated with each factor simultaneously. In addition, we can determine whether the two factors significantly *interact.* For example, two vitamin supplements may produce approximately the same weight gain when used in conjunction with the feed Growbig, but they may produce quite dissimilar weight gains in conjunction with another feed, say Topnotch, as illustrated in Figure 12.2. In this instance, we say that the two factors, vitamin supplement and feed, *interact.*

FIGURE 12.2 Two-factor interaction effect between feeds and vitamin supplements

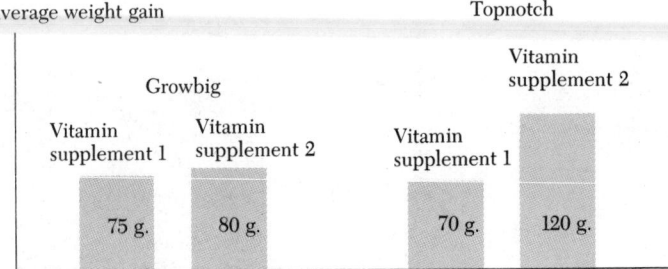

■ 12.2 ASSUMPTIONS OF ANALYSIS OF VARIANCE

The proper use of analysis of variance requires that certain assumptions about the nature of the data are satisfied. Recall from Chapter 11 that the comparison of two population means using the t-test requires that both populations be normally distributed and have the same variance, and that two independent random samples be drawn from the two populations. The assumptions required to use analysis of variance properly for the comparison of two or more population means are extensions of the two-population t-test assumptions.

The first assumption of analysis of variance is that the dependent (or response) variable is normally distributed in each of the populations being compared. To relate this assumption to Example 12.1, it requires that weight gain x_{ij} is a value of a normal distribution for each of the three feeds, $i = 1, 2,$ and 3, as illustrated in Figure 12.1.

The second assumption of analysis of variance is that the distributions of the dependent variable in each population have the same variance. This assumption is called *homogeneity of variance*.

If these two assumptions are satisfied, the distributions of the dependent variable for each feed in Example 12.1 must be normal and have a common variance. In Figure 12.1, notice that the distributions are bell shaped and appear to have the same variability.

We will discuss further assumptions in the context of specific models used to compare, by analysis of variance, a set of population means. In Section 12.7, methods will be presented to check the assumptions of normality and homogeneity of variance. If the assumptions are not satisfied by the data, procedures to transform the dependent variable so that they conform to the assumptions will be discussed.

12.3   ONE-FACTOR, COMPLETELY RANDOMIZED DESIGN

Let us now return to the discussion of Example 12.1. Suppose 15 chicks have been randomly selected from a large brood and have been randomly separated into three groups of five each. Each group is fed one of the three feeds, Growbig, Max, and Topnotch, over a fixed interval of time. At the end of the time, the weight gain of each chick is recorded. The gains are given in Table 12.2.

These data conform to a *one-factor, completely randomized design*. The one factor is the feed given to the chicks. It is a completely randomized design

TABLE 12.2   Weight gain of chicks in grams

	Feed		
	Growbig	*Max*	*Topnotch*
	42	112	70
	96	96	17
	81	88	49
	95	135	24
	76	119	40
Means:	78	110	40

because the experimental units—the 15 chicks—have been randomly assigned to the three kinds of feed.

The statistical hypothesis we wish to test is:

$$H_0: \quad \mu_1 = \mu_2 = \mu_3$$
$$H_A: \quad \text{At least one mean is different from the others}$$

If the null hypothesis is true, then the three population mean weight gains are equivalent. Let μ denote the common population mean. In this case, the three random samples of size five are drawn from one population with mean weight gain μ and the model for the dependent random variable X_{ij} is specified by:

$$X_{ij} = \mu + e_{ij}; \quad i = 1, 2, 3; \quad j = 1, 2, 3, 4, 5$$

where e_{ij} is the random error random variable component associated with the weight gain of the jth chick subjected to the ith feed.

From studying Table 12.2, it is apparent that the dependent random variable X_{ij} values are not randomly fluctuating about a common mean μ—the variation among the responses is too great to be entirely attributed to random sampling error of values drawn from one population. Rather, it *appears* more likely that each set of random samples of five per feed is drawn from a different population. This contention is strongly supported by the values of the three sample means, $\bar{x}_{1.} = 78, \bar{x}_{2.} = 110$, and $\bar{x}_{3.} = 40$; they *appear* to be too dissimilar to have come from one population.

If the null hypothesis is not true, as *appears* to be the case in this example, then we must write the model for the dependent random variable in a way that properly accounts for the fact that the responses are drawn from two or more populations with different means:

$$X_{ij} = \mu_i + e_{ij} = \mu + (\mu_i - \mu) + e_{ij} = \mu + \tau_i + e_{ij}$$

where $\tau_i = \mu_i - \mu$. In this model, τ_i (tau) represents the effect of the ith treatment (feed)—it is the difference between the mean of the ith treatment and the mean of the overall population.

Notice that if the null hypothesis is true ($\mu_1 = \mu_2 = \mu_3 = \mu$), then this model reduces to $X_{ij} = \mu + e_{ij}$, the assumed model under the null hypothesis.

It is now possible to state the null hypothesis in terms of the treatment effects:

$$H_0: \quad \tau_1 = \tau_2 = \tau_3$$
$$H_A: \quad \text{At least one } \tau_i \text{ differs from the rest}$$

Since the τ values represent deviations of the three population means about the overall mean μ, it is apparent that this hypothesis is equivalent to the one expressed in terms of μ_1, μ_2, and μ_3.

The relationship of the sample means to the null hypothesis is illustrated in Figure 12.3. If H_0 is true, then the three sample means have been drawn from the same population with mean weight gain μ. Any differences in

FIGURE 12.3 Relationship of the sample means to the null hypothesis

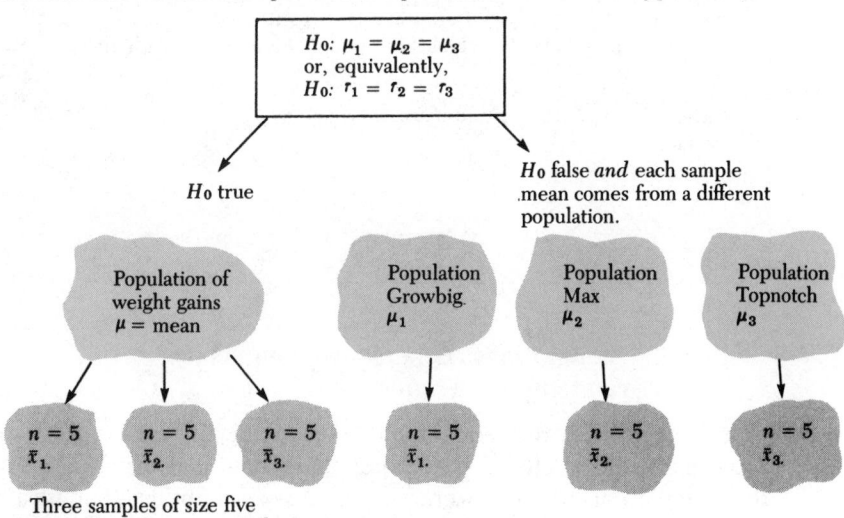

Three samples of size five
drawn from the *same* population

weight gains are attributable to sampling variability. The null hypothesis may be false in a number of ways. In Figure 12.3, the sample means are each drawn from different populations; that is, the three population means μ_1, μ_2, and μ_3 are all different. It is possible, of course, that two of the population means are the same while the third is different. For example $\bar{x}_1$ and $\bar{x}_2$ may be drawn from the same population, while $\bar{x}_3$ is drawn from a second and different population. This would imply that the treatment effects τ_1 and τ_2 corresponding to the feeds Growbig and Max are equal or, equivalently, that μ_1 and μ_2 are equal, while the treatment effect τ_3 corresponding to Topnotch differs from τ_1 and τ_2.

Let us summarize the one-factor, completely randomized design model.

Definition 12.1
One-factor, completely randomized design model

The model is specified by the linear equations:

$$X_{ij} = \mu + (\mu_i - \mu) + e_{ij} = \mu + \tau_i + e_{ij}$$
$$i = 1, 2, \ldots, t; \quad j = 1, 2, \ldots, n$$

where

X_{ij} = dependent response random variable corresponding to the ith treatment and the jth experimental unit subjected to the ith treatment

μ_i = population mean response corresponding to the ith treatment (the ith population mean)

τ_i = effect of the ith treatment; it is equal to the deviation of the ith population mean μ_i from the overall population mean μ

> **Definition 12.1 (continued)**
>
> e_{ij} = random error variable component associated with the response random variable X_{ij}
>
> The (tn) experimental units are *randomly* assigned in groups of size n to the t treatments.
>
> The hypotheses are:
>
> $$H_0: \quad \tau_1 = \tau_2 = \cdots = \tau_t$$
> $$H_A: \quad \text{At least one } \tau_i \text{ differs from the rest}$$

☐ **12.3.1 Assumptions: One-factor, completely randomized design**

In Section 12.2, the requirements that the dependent random variable X_{ij} be normally distributed in the ith population and that the variance in each of the t populations be equal were presented as assumptions that must be met to use the analysis of variance procedure properly. These two assumptions plus the fact that $E(X_{ij}) = \mu_i$, $i = 1, 2, \ldots, t$, can be stated more concisely in terms of the error component e_{ij}. Specifically, we require that:

1. e_{ij} is normally distributed for each i and j.
2. $E(e_{ij}) = 0$ for each i and j.
3. $V(e_{ij}) = \sigma^2$ for each i and j.

These three assumptions concerning the random variable e_{ij} will ensure that X_{ij} is normally distributed and that the homoscedasticity (equal variances) assumption is satisfied, provided that we further assume that μ is a constant and τ_i is a *fixed effect*. If μ and τ_i are fixed (constants), then the normality of e_{ij} will guarantee the normality of X_{ij}, since

$$X_{ij} = \mu + \tau_i + e_{ij} = \text{Constant} + e_{ij}$$

in this case. Therefore, we add as a fourth assumption,

4. The treatment effect τ_i is a fixed effect.

To treat τ_i as a fixed effect implies that we are only concerned with the t treatment effects, τ_i, where $i = 1, 2, \ldots, t$ in the experiment, and inferences drawn from the analysis of variance procedure will be made about these treatment effects only. Alternatively, we could consider the t treatments as representing a sample taken from a larger "population" of treatments, in which case the treatment effects τ_i, where $i = 1, 2, \ldots, t$, are referred to as *random effects*. In our chicken feed example, we could consider the three feeds Growbig, Topnotch, and Max to be a sample of three feeds taken from a larger population of feeds. Then τ_1, τ_2, and τ_3 are random effects. In this case, it would be necessary that each treatment effect τ_i be a normally distributed and independently distributed random variable. In this chapter, we will consider *all treatment effects to be fixed*. Therefore, we will draw inferences only about those treatments run in the experiment. The interested reader is directed to the Ostle and Mensing text referenced at the end of this chapter for a description of the random-effects analysis of variance model and its applications.

Finally, we must assume that the error terms e_{ij} are independently distributed random variables.

5. The error terms e_{ij} are independent random variables.

This assumption requires that the error components in the model are independent from one another and is needed in constructing the hypothesis-testing methods that will follow. The process of randomly allocating the experimental units to the treatments is done to ensure that the errors are independent, so that no relationship exists among the errors which may favor one or more treatment responses.

☐ 12.3.2 Analysis of variance procedure

The statistical test for testing the null hypothesis H_0: $\tau_1 = \tau_2 = \cdots = \tau_t$ is based on the assumption that the populations corresponding to the *t*-treatment effects have the *same variance*. Additionally, the test uses the relationship that exists between the variance of a sample mean and the population variance. Suppose there are $t = 3$ treatments as in Example 12.1 and the null hypothesis H_0: $\tau_1 = \tau_2 = \tau_3$ is true. Then Figure 12.4 illustrates

FIGURE 12.4 Sample statistics assuming H_0 is true

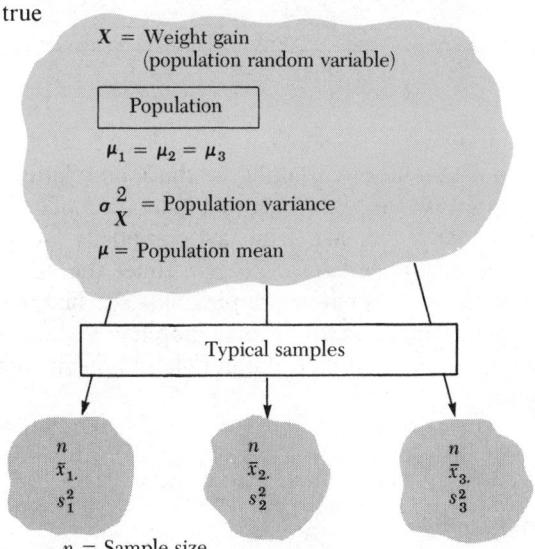

X = Weight gain (population random variable)

Population

$\mu_1 = \mu_2 = \mu_3$

σ_X^2 = Population variance

μ = Population mean

Typical samples

| n $\bar{x}_1.$ s_1^2 | n $\bar{x}_2.$ s_2^2 | n $\bar{x}_3.$ s_3^2 |

n = Sample size

$\bar{x}_i$ = Sample mean corresponding to i^{th} treatment

s_i^2 = Sample variance corresponding to i^{th} treatment

the relationship between the sample data and the population from which they are drawn.

Given that H_0 is true, the three sample means each provide an unbiased estimate of the population mean μ and each of the sample variances provides an unbiased estimate of the population variance σ_X^2. Thus, we are taking, in effect, three repeated random samples, each of size n, from the same population. Recall from Chapter 8 that the variance of the sample mean, denoted by $\sigma_{\bar{X}}^2$, is equal to the population variance σ_X^2 divided by the sample size n:

$$\sigma_{\bar{X}}^2 = \frac{\sigma_X^2}{n}$$

Thus, if the null hypothesis is true, the population variance σ_X^2 should be equal to n times the variance of the sample means, $\sigma_{\bar{X}}^2$. If the null hypothesis is not true, then the equation $\sigma_X^2 = n\,\sigma_{\bar{X}}^2$ will not hold; indeed, $n\sigma_{\bar{X}}^2$ will be greater than σ_X^2 due to the fact that the population means corresponding to the $t = 3$ treatments are different. This relationship may be seen in Figure 12.5. If the three treatment effects are equal, we are drawing the random

FIGURE 12.5 Comparison of the population variability when the treatment effects are equal and unequal

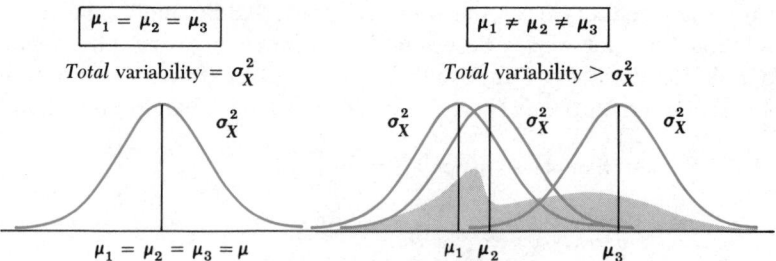

samples from one distribution with variance σ_X^2. If the three treatment effects are not equal, then the total amount of variability in the **X** population must be greater than σ_X^2 as illustrated on the right side of Figure 12.5. The repeated samples of size n would then be drawn from a "composite" population, indicated in Figure 12.5 by the shaded area. Then, n times the variability of the sample average statistic $\bar{X}$ in repeated samples must be larger than σ_X^2, because these samples are being drawn from a population in which the variability is greater than the one in which the population means are equal. Therefore, the hypotheses:

$$H_0: \quad \tau_1 = \tau_2 = \tau_3$$
$$H_A: \quad \text{At least one } \tau_i \text{ differs from the rest}$$

are equivalent to the hypotheses:

$$H_0: \quad \sigma_X^2 = n\,\sigma_{\bar{X}}^2$$
$$H_A: \quad \sigma_X^2 < n\,\sigma_{\bar{X}}^2$$

That is, we can test the equality of population mean effects by comparing estimates of σ_X^2, the population variance, and $\sigma_{\bar{X}}^2$, the variance of the sample mean statistic. The analysis of variance procedure does, in fact, analyze variances to compare means.

The analysis of variance procedure compares an estimate of σ_X^2, denoted by $\hat{\sigma}_X^2$, with an estimate of $\sigma_{\bar{X}}^2$, denoted by $\hat{\sigma}_{\bar{X}}^2$. If $\hat{\sigma}_X^2$ is "much less" than $n\,\hat{\sigma}_{\bar{X}}^2$, then there is reason to suspect that the null hypothesis is not true. To determine whether $\hat{\sigma}_X^2$ is significantly less than $n\,\hat{\sigma}_{\bar{X}}^2$, we compute the probability that the difference $n\,\sigma_{\bar{X}}^2 - \hat{\sigma}_X^2$ could arise by chance (sampling error) if $\sigma_X^2 = \hat{\sigma}_{\bar{X}}^2$.

To illustrate the testing process, consider again Example 12.1 and the specific data in Table 12.2. Since the three sample means are quite dissimilar ($\bar{x}_1 = 78$, $\bar{x}_2 = 110$, and $\bar{x}_3 = 40$), we might expect the analysis of variance procedure to suggest that the null hypothesis should be rejected.

☐ 12.3.3 Estimate of the population variance σ_X^2

To calculate a sample estimate of σ_X^2, we will use the sample variances s_1^2, s_2^2, and s_3^2. The variance for the ith sample is given by:

$$s_i^2 = \frac{\sum_{j=1}^{n} (x_{ij} - \bar{x}_{i.})^2}{n-1} = \frac{\sum_{j=1}^{n} x_{ij}^2 - \dfrac{\left(\sum_{j=1}^{n} x_{ij}\right)^2}{n}}{n-1}$$

Thus,

$$s_1^2 = \frac{(42)^2 + (96)^2 + \cdots + (76)^2 - \dfrac{(42 + 96 + \cdots + 76)^2}{5}}{4}$$

$$= \frac{32{,}342 - \dfrac{(390)^2}{5}}{4} = \frac{32{,}342 - 30{,}420}{4} = \frac{1{,}922}{4} = 480.5$$

Similarly, $s_2^2 = 347.5$ and $s_3^2 = 441.5$. A critical assumption of analysis of variance is that the population variances corresponding to the t treatments are equal, regardless of whether or not the null hypothesis is true (that is, whether or not the t population means are equal). Thus, any differences in the sample variances must *always* be attributable to sampling error. The three sample variances appear to be reasonably similar in this example. In Section 12.7, we will discuss a procedure to test the equivalence of a set of t sample variances.

Since s_1^2, s_2^2, and s_3^2 each estimate the population variance σ_X^2, we can produce an improved estimate of σ_X^2 over each sample variance taken individually by pooling the three estimates (recall the pooled estimate of the population variance in the denominator of the t-statistic used to compare two population means).

The pooled estimate of σ_X^2 is given by:

$$\hat{\sigma}_X^2 = \frac{(n-1)s_1^2 + (n-1)s_2^2 + (n-1)s_3^2}{(n-1) + (n-1) + (n-1)}$$

$$= \frac{(n-1)(s_1^2 + s_2^2 + s_3^2)}{3(n-1)}$$

Substituting in the values of s_1^2, s_2^2, and s_3^2, we obtain

$$\hat{\sigma}_X^2 = \frac{4(480.5 + 347.5 + 441.5)}{3 \cdot 4} = \frac{5078}{12} = 423.17$$

☐ **12.3.4   Estimate of the variance of the sample mean $\sigma_{\bar{X}}^2$**

The three sample means are $\bar{x}_{1.} = 78$, $\bar{x}_{2.} = 110$, and $\bar{x}_{3.} = 40$. To compute the estimate $\hat{\sigma}_{\bar{X}}^2$ of the three sample means, we use the variance formula:

$$\hat{\sigma}_{\bar{X}}^2 = \frac{\sum\limits_{i=1}^{t} (\bar{x}_{i.} - \bar{x}_{..})^2}{t - 1}$$

where t = number of sample means (treatments), $\bar{x}_{i.}$ is the ith sample mean, and $\bar{x}_{..}$ is the overall sample mean computed from:

$$\bar{x}_{..} = \frac{\sum\limits_{i=1}^{t} \sum\limits_{j=1}^{n} x_{ij}}{nt}$$

In our example, $\bar{x}_{..} = (42 + 96 + \cdots + 40)/[(5)(3)] = 76$. Then,

$$\hat{\sigma}_{\bar{X}}^2 = \frac{(78 - 76)^2 + (110 - 76)^2 + (40 - 76)^2}{3 - 1}$$

$$= \frac{4 + 1,156 + 1,296}{2} = 1,228$$

If the null hypothesis is true, $n\hat{\sigma}_{\bar{X}}^2$ should be an unbiased estimate of σ_X^2. Its value is:

$$n\hat{\sigma}_{\bar{X}}^2 = (5)(1,228) = 6,140$$

Since $n\hat{\sigma}_{\bar{X}}^2 > \hat{\sigma}_X^2 (6,140 > 423.17)$, the sample means *appear* to be much too variable to have been drawn from a common population with mean μ. But is the difference between $n\hat{\sigma}_{\bar{X}}^2$ and $\hat{\sigma}_X^2$ statistically significant?

In Chapter 11, we developed a procedure for testing the equivalence of two population variances. To review this procedure, we compute the F-statistic.

$$F = \frac{\hat{\sigma}_A^2}{\hat{\sigma}_B^2}$$

to test the null hypothesis that two variances, σ_A^2 and σ_B^2, are equivalent. If the null hypothesis is true, we should expect the value of **F** to be approximately 1.

Let $\hat{\sigma}_A^2 = n\,\hat{\sigma}_{\bar{X}}^2$ and $\hat{\sigma}_B^2 = \hat{\sigma}_X^2$ and calculate the value of the F-statistic:

$$f = \frac{n\,\hat{\sigma}_{\bar{X}}^2}{\hat{\sigma}_X^2} = \frac{6,140}{423.17} = 14.51$$

To determine whether the calculated value of **F** is significantly different from 1, we look up the critical value of **F** based on the numerator degrees of freedom $r_1 = t - 1$, the denominator degrees of freedom $r_2 = t(n - 1)$, and a selected value of α, say $\alpha = 0.05$. From Table B.7 in Appendix B,

$$F(1 - \alpha; r_1, r_2) = F(0.95; 2, 12) = 3.89$$

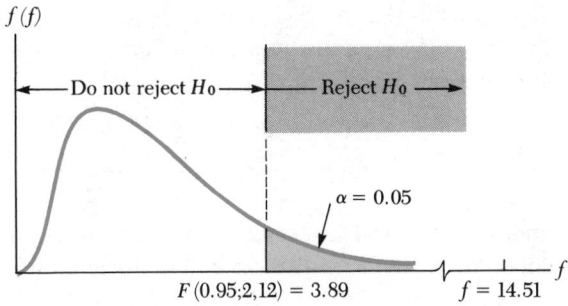

Since the calculated value of the statistic **F** ($f = 14.51$) is greater than the critical value ($14.51 > 3.89$), we reject the null hypothesis H_0: $\tau_1 = \tau_2 = \tau_3$ at the $\alpha = 0.05$ significance level. Thus, we conclude that there appears to be strong evidence that the three feeds do produce different average weight gains among the chicks.

It is possible to calculate the p-value for this test by using the methods developed in Chapter 10. Thus, we desire to determine:

$$p\text{-value} = P(F > 14.51/r_1 = 2, r_2 = 12)$$
$$= 1 - P(F < 14.51/r_1 = 2, r_2 = 12)$$

From Table B.7 in Appendix B, $P(F < 13/r_1 = 2, r_2 = 12) = 0.999$. Thus, we can say that the p-value for this test is less than 0.001 (from a computer-assisted computation, the p-value turns out to be 0.0006). Thus, it is *very* likely that the three feeds do produce different average weight gains.

☐ **12.3.5 Analysis of variance table**

A convenient computational format for calculating the statistics necessary to determine whether the null hypothesis should be rejected is provided by the *analysis of variance table*. Its form is presented in Table 12.3.

The first row of the table, "Among treatments," produces the estimate of the variance $n\,\sigma_{\bar{X}}^2$—it is denoted by MSTR (mean square for treatments) in the table. The second row of the table, "Experimental error," produces the

TABLE 12.3 Analysis of variance table

Source of variation	Degrees of freedom	Sum of squares	Mean square	F-ratio
Among treatments..........	$t - 1$	SSTR	$\text{MSTR} = \dfrac{\text{SSTR}}{t - 1}$	$f = \dfrac{\text{MSTR}}{\text{MSE}}$
Experimental error (within treatments)	$t(n - 1)$	SSE	$\text{MSE} = \dfrac{\text{SSE}}{t(n - 1)}$	
Total.................	$tn - 1$	SST		

pooled estimate of the population variance σ_X^2—it is denoted by MSE (mean square for error) in the table. The F-statistic is the ratio of the mean square for treatments, MSTR, and the mean square for error, MSE.

The formula for MSTR is:

$$\text{MSTR} = n\hat{\sigma}_{\bar{X}}^2 = \frac{n \sum\limits_{i=1}^{t} (\bar{x}_{i.} - \bar{x}_{..})^2}{t - 1}$$

The numerator of MSTR is called the *sum of squares among treatments* and is denoted by SSTR in the table. The formula for MSE is:

$$\text{MSE} = \hat{\sigma}_X^2 = \frac{(n - 1) \sum\limits_{i=1}^{t} s_i^2}{t(n - 1)} = \frac{\sum\limits_{i=1}^{t} \sum\limits_{j=1}^{n} (x_{ij} - \bar{x}_{i.})^2}{t(n - 1)}$$

The numerator of MSE is called the *error sum of squares* and is denoted by SSE in the table. The "Degrees of freedom" column in the table gives the appropriate divisors of the sums of squares to produce the mean squares.

The last row in the table gives the "Total degrees of freedom"— $(tn - 1) = (t - 1) + t(n - 1)$—and the total sum of squares given by:

$$\text{SST} = \sum\limits_{i=1}^{t} \sum\limits_{j=1}^{n} (x_{ij} - \bar{x}_{..})^2$$

Notice that SST gives the sum of the squared deviations of each observation x_{ij} about the grand mean of the data $\bar{x}_{..}$. Thus, this quantity is a measure of the total variability in the dependent variable.

An important relationship given in the analysis of variance table is:

$$\text{SST} = \text{SSTR} + \text{SSE}$$

$$\sum\limits_{i=1}^{t} \sum\limits_{j=1}^{n} (x_{ij} - \bar{x}_{..})^2 = n \sum\limits_{i=1}^{t} (\bar{x}_{i.} - \bar{x}_{..})^2 + \sum\limits_{i=1}^{t} \sum\limits_{j=1}^{n} (x_{ij} - \bar{x}_{i.})^2$$

$$\begin{pmatrix} \text{Total sum} \\ \text{of squares} \end{pmatrix} = \begin{pmatrix} \text{Sum of squares} \\ \text{due to} \\ \text{treatments} \end{pmatrix} + \begin{pmatrix} \text{Error sum} \\ \text{of squares} \end{pmatrix}$$

That is, the total sum of squares can be partitioned into the sum of squares due to treatments plus the error sum of squares.

The deviations of the observations about the grand mean ($\bar{x}_{..} = 76$) and about their respective treatment means ($\bar{x}_{1.} = 78$, $\bar{x}_{2.} = 110$, $\bar{x}_{3.} = 40$) are illustrated in Figure 12.6. The light red lines connecting each observation to the treatment mean give the deviations of the observations about these means—the sum of the squares of these distances produces SSE. Thus, SSE is a measure of the sampling variability within each treatment group of observations. The reddish gray lines give the deviations of the treatment means about the grand mean—n times the sum of the squares of these distances produces SSTR.

In these data, it is clear that the treatment sum of squares accounts for a considerable amount of the total sum of squares (remember the sum of the squares of the reddish gray line deviations is multiplied by $n = 5$ to produce SSTR). After SSTR has been partitioned from SST, what is left is the sum of squared deviations of each observation about its respective group mean— this is the "within-treatment" sum of squares.

FIGURE 12.6 Deviations of observations about the grand mean and group means, and deviations of group means about the grand mean

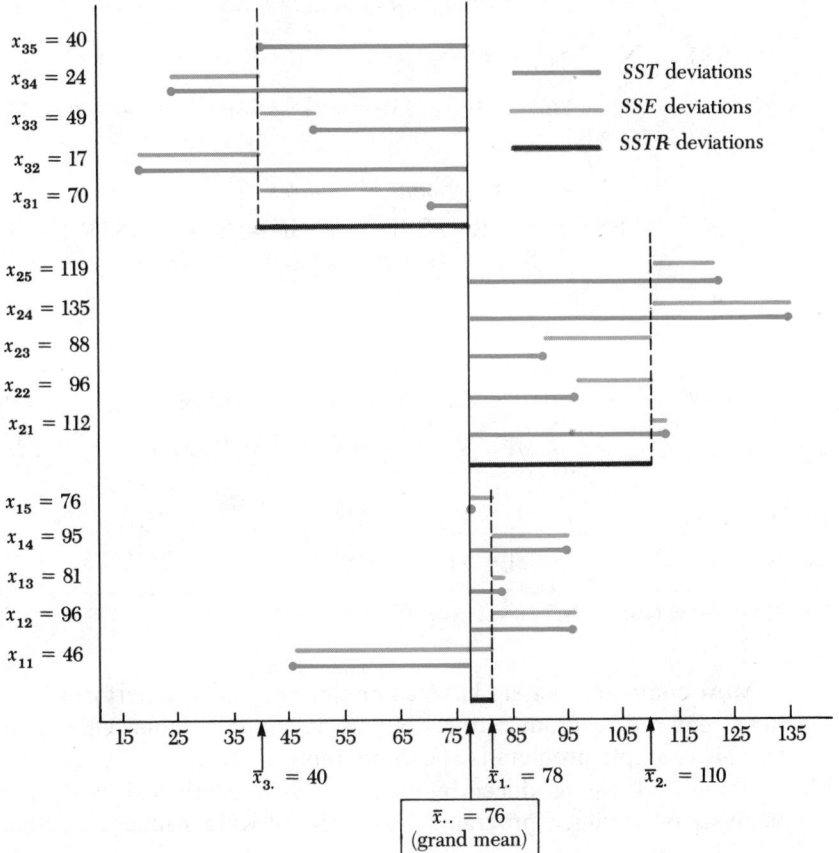

Imagine now that the observations corresponding to treatments 2 and 3 are more closely clustered about the grand mean $\bar{x}_{..} = 76$ so that the values $\bar{x}_{2.}$ and $\bar{x}_{3.}$ are both close to 76. In this case, the treatment deviations (the reddish gray lines) will be small. In fact, when we "average" the treatment sum of squares to produce the treatment mean square MSTR = SSTR/$(t - 1)$ and when we "average" the error sum of squares to produce the error mean square MSE = SSE/$t(n - 1)$, we should find that MSTR and MSE are approximately equal. This will produce an F-ratio of approximately 1 and will not allow us to reject the null hypothesis of equal treatment effects.

Let us now verify that by using the analysis of variance table format we produce the same value of the F-statistic as computed earlier.

Treatment sum of squares, SSTR

$$\text{SSTR} = n \sum_{i=1}^{t} (\bar{x}_{i.} - \bar{x}_{..})^2$$
$$= 5[(78 - 76)^2 + (110 - 76)^2 + (40 - 76)^2]$$
$$= 5(4 + 1{,}156 + 1{,}296) = 12{,}280$$

Total sum of squares, SST

$$\text{SST} = \sum_{i=1}^{t} \sum_{j=1}^{n} (x_{ij} - \bar{x}_{..})^2$$
$$= (42 - 76)^2 + (96 - 76)^2 + (81 - 76)^2 + \cdots + (40 - 76)^2$$
$$= 17{,}358$$

Error sum of squares, SSE

$$\text{SST} = \text{SSTR} + \text{SSE}; \qquad \text{SSE} = \text{SST} - \text{SSTR}$$
$$\text{SSE} = 17{,}358 - 12{,}280 = 5{,}078$$

Analysis of variance table

Source of variation	Degrees of freedom	Sum of squares	Mean square	F-ratio
Among treatments	$t - 1 = 2$	12,280	$\dfrac{12{,}280}{2} = 6{,}140$	$\dfrac{6{,}140}{423.17} = 14.51$
Experimental error	$t(n - 1) = 12$	5,078	$\dfrac{5{,}078}{12} = 423.17$	
Totals	$tn - 1 = 14$	17,358		

$F(0.95; 2, 12) = 3.89$; $14.51 > 3.89$; therefore, reject H_0: $\tau_1 = \tau_2 = \tau_3$.

Most computer centers have computer programs to perform the calculations required to produce the analysis of variance table. One such output for this example problem is shown in Table 12.4.

Table 12.4 was produced by using the SAS (Statistical Analysis System) analysis of variance program. The SAS software package contains many

TABLE 12.4   Computer analysis of Example 12.1 data

ANALYSIS OF VARIANCE PROCEDURE

DEPENDENT VARIABLE: YIELD

SOURCE	DF	SUM OF SQUARES	MEAN SQUARE	F VALUE	PR > F
MODEL	2	12280.000	6140.000	14.51	0.0006
ERROR	12	5078.000	423.167		
TOTAL	14	17358.000			

programs for performing statistical analyses. Notice that the SAS program prints out the p-value (PR > F); that is, $P(F > 14.51) = 0.0006$. Most computer programs written to perform statistical analyses now print the p-value so that interpolation in the statistical tables (e.g., the tables in Appendix B) is not required. We will use the SAS output for selected analysis of variance problems in the remainder of this chapter.

The analysis of variance table, along with the formulas for the sums of squares, is given in Table 12.5.

TABLE 12.5   Analysis of variance table for a one-factor, completely randomized design with equal numbers of observations within each treatment group

Source of variation	Degrees of freedom	Sum of squares	Mean square	F-ratio
Among treatments	$t - 1$	$\text{SSTR} = \sum_{i=1}^{t} (\bar{x}_{i.} - \bar{x}_{..})^2$	$\text{MSTR} = \dfrac{\text{SSTR}}{t - 1}$	$f = \dfrac{\text{MSTR}}{\text{MSE}}$
Experimental error	$t(n - 1)$	$\text{SSE} = \sum_{i=1}^{t} \sum_{j=1}^{n} (x_{ij} - \bar{x}_{i.})^2$	$\text{MSE} = \dfrac{\text{SSE}}{t(n - 1)}$	
Total	$tn - 1$	$\text{SST} = \sum_{i=1}^{t} \sum_{j=1}^{n} (x_{ij} - \bar{x}_{..})^2$		

If $f > F[1 - \alpha; (t - 1), t(n - 1)]$, reject the null hypothesis H_0: $\tau_1 = \tau_2 = \cdots = \tau_t$.

$$p\text{-value} = P[F > f/r_1 = t - 1, r_2 = t(n - 1)]$$

☐ **12.3.6   Computational formulas for the sums of squares**

The computation of SSTR and SST can be somewhat simplified by using the computational formulas:

$$\text{SSTR} = n \sum_{i=1}^{t} (\bar{x}_{i.} - \bar{x}_{..})^2 = \frac{1}{n} \sum_{i=1}^{t} x_{i.}^2 - C$$

$$\text{SST} = \sum_{i=1}^{t} \sum_{j=1}^{n} (x_{ij} - \bar{x}_{..})^2 = \sum_{i=1}^{t} \sum_{j=1}^{n} x_{ij}^2 - C$$

where

$$C = \frac{\left(\sum\limits_{i=1}^{t} \sum\limits_{j=1}^{n} x_{ij}\right)^2}{nt} \quad \text{and} \quad x_{i.} = \text{Total for the } i\text{th treatment}$$

Thus,

$$C = \frac{(42 + 96 + \cdots + 40)^2}{(5)(3)} = \frac{(1,140)^2}{15} = 86,640$$

$$\text{SSTR} = \frac{1}{5}[(390)^2 + (550)^2 + (200)^2] - C$$

$$= \frac{494,600}{5} - 86,640 = 98,920 - 86,640 = 12,280$$

and,

$$\text{SST} = [(42)^2 + (96)^2 + \cdots + (40)^2] - C$$
$$= (1,764 + 9,216 + \cdots + 1,600) - C = 103,998 - 86,640 = 17,358$$

In this problem, the definitional forms of SSTR and SST given in Table 12.5 are easier to use than the computational forms because the sample means $\bar{x}_{1.}, \bar{x}_{2.}, \bar{x}_{3.},$ and $\bar{x}_{..}$ are whole numbers. If the sample means are fractional, the computational formulas for the sums of squares are usually easier to use.

In summary, the analysis of variance procedure tests the hypothesis of equal effects among t-treatments (population means) by analyzing the variation in the dependent (population) variable. If there are no differences among the treatment effects, then the treatment mean square (MSTR) and the error mean square (MSE) should independently provide unbiased estimates of the population variance. If the treatment effects differ, then MSE will still provide an unbiased estimate of the population variance, but MSTR will not. Its expected value will be larger than the population variance due to the differences among the treatment means.

☐ 12.3.7 Unequal sample sizes

In the preceding example, the number of observations per treatment was five. Often we will not be so fortunate as to have a *balanced* experiment—one in which each treatment is applied to the same number of experimental units. In Example 12.1, one or more chicks may die before the time interval has elapsed, producing unequal sample sizes for the three treatments.

In the unequal sample size case, let n_i, where $i = 1, 2, \ldots, t$, denote the number of experimental units subjected to the ith treatment. The appropriate model for a one-factor, completely randomized design now becomes:

$$\mathbf{X}_{ij} = \mu + (\mu_i - \mu) + \mathbf{e}_{ij} = \mu + \tau_i + \mathbf{e}_{ij}$$
$$i = 1, 2, \ldots, t; \quad j = 1, 2, \ldots, n_i$$

TABLE 12.6 Analysis of variance table for a one-factor, completely randomized design with unequal numbers of observations within treatment groups

Source of variation	Degrees of freedom	Sum of squares	Mean square	F-ratio
Among treatments	$t - 1$	$SSTR = \sum_{i=1}^{t} n_i(\bar{x}_{i.} - \bar{x}_{..})^2$	$MSTR = \dfrac{SSTR}{t - 1}$	$f = \dfrac{MSTR}{MSE}$
Experimental error	$\sum_{i=1}^{t}(n_i - 1)$	$SSE = \sum_{i=1}^{t}\sum_{j=1}^{n_i}(x_{ij} - \bar{x}_{i.})^2$	$MSE = \dfrac{SSE}{\sum\limits_{i=1}^{t}(n_i - 1)}$	
Total	$\left(\sum_{i=1}^{t} n_i\right) - 1$	$SST = \sum_{i=1}^{t}\sum_{j=1}^{n_i}(x_{ij} - \bar{x}_{..})^2$		

If $f > F[1 - \alpha; (t - 1), \sum_{i=1}^{t}(n_i - 1)]$, reject the null hypothesis H_0: $\tau_1 = \tau_2 = \cdots = \tau_t$.

This model is identical to the one in the balanced case, except that it allows for unequal sample sizes ($j = 1, 2, \ldots, n_i$).

The formulas in the analysis of variance table do change from those given in Table 12.5 due to the unequal sample sizes. The appropriate analysis of variance table for this case is given in Table 12.6.

Example 12.2   A teacher is experimenting with a new way of teaching a statistical concept. To determine whether the new method produces better comprehension of the concept than does the standard procedure, he randomly selects five students from a large group that have similar abilities and mathematical backgrounds, subjects them to the new teaching method, and tests them for comprehension at the end of the teaching period. In another section, he subjects eight different students drawn from the same population of students to the standard teaching method and tests each for comprehension. The test score range is 0 to 100. The comprehension test scores for the two groups of students are given in the table.

	Treatment	
	(1)	*(2)*
	New method	*Standard method*
	86	75
	91	90
	75	62
	80	65
	88	70
		55
		81
		70

$n_1 = 5$ $n_2 = 8$ Grand mean $\bar{x}_{..} = 76$
$x_{1.} = 420$ $x_{2.} = 568$
$\bar{x}_{1.} = 84$ $\bar{x}_{2.} = 71$

Are the two teaching methods significantly different at the $\alpha = 0.05$ level?

Solution   The null hypothesis is H_0: $\tau_1 = \tau_2$, where τ_1 is the teaching effect of the new method and τ_2 is the teaching effect of the standard method. We now complete the analysis of variance table:

$$\text{SSTR} = \sum_{i=1}^{t} n_i(\bar{x}_{i.} - \bar{x}_{..})^2 = 5(84 - 76)^2 + 8(71 - 76)^2$$
$$= 320 + 200 = 520$$

$$\text{SST} = \sum_{i=1}^{t} \sum_{j=1}^{n_i} (x_{ij} - \bar{x}_{..})^2 = (86 - 76)^2 + (91 - 76)^2 + \cdots + (70 - 76)^2$$
$$= 1{,}538$$

$$\text{SSE} = \text{SST} - \text{SSTR} = 1{,}018$$

Source of variation	Degrees of freedom	Sum of squares	Mean square	F-ratio
Among treatments	$t - 1 = 1$	520	$\dfrac{520}{1} = 520$	$\dfrac{520}{92.54} = 5.62$
Experimental error	$\sum\limits_{i=1}^{t}(n_i - 1) = 4 + 7 = 11$	1,018	$\dfrac{1,018}{11} = 92.54$	
Total	$\left(\sum\limits_{i=1}^{t} n_i\right) - 1 = 5 + 8 - 1 = 12$	1,538	Critical F: $F(0.95; 1, 11)$ $= 4.85$	

Since $f = 5.62$ is greater than 4.85, we reject the null hypothesis H_0: $\tau_1 = \tau_2$; there is evidence to suggest that the new teaching method is different from the standard method.

Note that the critical value $F(0.95; 1, 11)$ must be interpolated from Table B.7 in Appendix B since the table gives only $F(0.95; 1, 10)$ and $F(0.95; 1, 12)$. Also, from Table B.7, it is apparent that the p-value lies between 0.05 and 0.025. Thus, we would *not* reject the null hypothesis at the 0.01 significance level, for example.

Computational forms for the unbalanced case

$$\text{SSTR} = \sum_{i=1}^{t} n_i(\bar{x}_{i.} - \bar{x}_{..})^2 = \sum_{i=1}^{t} \frac{x_{i.}^2}{n_i} - C$$

$$\text{SST} = \sum_{i=1}^{t} \sum_{j=1}^{n_i} (x_{ij} - \bar{x}_{..})^2 = \sum_{i=1}^{t} \sum_{j=1}^{n_i} x_{ij}^2 - C$$

where

$$C = \frac{\left[\sum\limits_{i=1}^{t} \sum\limits_{j=1}^{n_i} x_{ij}\right]^2}{\sum\limits_{i=1}^{t} n_i} \quad \text{and} \quad x_{i.} = \text{Total for } i\text{th treatment}$$

Applied to the data in Example 12.2, we obtain:

$$C = \frac{(86 + 91 + \cdots + 70)^2}{5 + 8} = 75,088$$

$$\text{SSTR} = \left[\frac{(420)^2}{5} + \frac{(568)^2}{8}\right] - C$$
$$= (35,280 + 40,328) - 75,088 = 520$$

$$\text{SST} = [(86)^2 + (91)^2 + \cdots + (70)^2] - C$$
$$= 76,626 - 75,088 = 1,538$$

□ 12.3.8 Relationship of the two-population t-test to analysis of variance

Since there are two treatments in Example 12.2, we should be able to perform the test of the null hypothesis H_0: $\tau_1 = \tau_2$ or, equivalently, H_0: $\mu_1 = \mu_2$, where μ_i is the population average test score for the ith teaching method, by using the two-population independent sample t-test given in Section 11.2.

The t-test for the null hypothesis H_0: $\mu_1 = \mu_2$ is:

1. Compute

$$t = \frac{\bar{x}_1 - \bar{x}_2}{\sqrt{s_p^2 \left(\dfrac{1}{n_1} + \dfrac{1}{n_2}\right)}}$$

where $\bar{x}_i = i$th sample mean, $n_i = i$th sample size ($i = 1, 2$), and s_p^2 is the pooled estimate of the population variance given by:

$$s_p^2 = \frac{(n_1 - 1)s_1^2 + (n_2 - 1)s_2^2}{n_1 + n_2 - 2}$$

where $s_i^2 = i$th sample variance, $i = 1, 2$.

2. Reject H_0: $\mu_1 = \mu_2$ if $t > t_{\alpha/2; n_1+n_2-2}$ or if $t < -t_{\alpha/2; n_1+n_2-2}$. In this problem, $\bar{x}_1 = 84$, $\bar{x}_2 = 71$, $n_1 = 5$, and $n_2 = 8$.

$$s_1^2 = \frac{\sum_{i=1}^{n_1}(x_i - \bar{x}_1)^2}{n_1 - 1} = \frac{(86 - 84)^2 + (91 - 84)^2 + \cdots + (88 - 84)^2}{4}$$

$$= {}^{166}\!/_4 = 41.5$$

$$s_2^2 = \frac{\sum_{i=1}^{n_2}(x_i - \bar{x}_2)^2}{n_2 - 1} = \frac{(75 - 71)^2 + (90 - 71)^2 + \cdots + (70 - 71)^2}{7}$$

$$= {}^{852}\!/_7 = 121.714$$

$$s_p^2 = \frac{(n_1 - 1)s_1^2 + (n_2 - 1)s_2^2}{n_1 + n_2 - 2} = \frac{(4)(41.5) + (7)(121.714)}{5 + 8 - 2}$$

$$= \frac{166 + 852}{11} = \frac{1{,}018}{11} = 92.54$$

Now, we obtain

$$t = \frac{\bar{x}_1 - \bar{x}_2}{\sqrt{s_p^2 \left(\dfrac{1}{n_1} + \dfrac{1}{n_2}\right)}} = \frac{84 - 71}{\sqrt{92.54 \left(\dfrac{1}{5} + \dfrac{1}{8}\right)}} = \frac{13}{\sqrt{30.08}}$$

$$= \frac{13}{5.48} = 2.37$$

From Table B.5 in Appendix B, $t_{0.025; 11} = 2.201$. Since $2.37 > 2.201$, we reject H_0: $\mu_1 = \mu_2$, the same conclusion we reached by analysis of variance. Notice that $t^2 = f$; $(2.37)^2 = 5.62$. This certainly did not occur by chance. If we square the t-statistic with a value given by

$$t = \frac{\bar{x}_1 - \bar{x}_2}{\sqrt{s_p^2 \left(\frac{1}{n_1} + \frac{1}{n_2}\right)}}$$

it can be shown algebraically that $t^2 = MSTR/MSE = f$. Therefore, if we wish to test the equivalence of two population means, we can use either the t-test or analysis of variance. But if more than two population means are to be compared, analysis of variance must be used.

If we are testing the equivalence of t population means ($t \geq 2$) and, based on the analysis of variance, we reject the null hypothesis, we should be interested in the reason for this outcome. One or more population means must be different from the rest, but which one(s)? In Section 12.6, we shall discuss one method for analyzing the nature of the difference among a set of means. We now turn to a design that attempts to control sources of variation that may adversely affect our ability to determine by analysis of variance whether a set of treatment effects do differ or not.

■ 12.4   THE RANDOMIZED BLOCK DESIGN

A randomized block design allocates treatments randomly to experimental units that have first been sorted into homogeneous groups called *blocks*. As an illustration of using this design, reconsider Example 12.2 in which two teaching methods are compared. The two teaching methods are randomly assigned to students selected from a large group with *similar abilities*. Ordinarily, students in a basic statistics course vary considerably in ability, and these differences, unless accounted for, may result in effects that are *confounded* with the treatment effects. For instance, by chance we may select students with less ability for the standard teaching method than those selected for the new teaching method. This may result in the new method "looking better" than the standard method when, in fact, it is not. If this were the case, we would say that student differences are *confounded* with the effect of the teaching method—on the basis of the students' performances on the comprehension test at the end of the term, it is not possible to determine whether any observed differences between the two groups are attributable to teaching methods alone.

To correct this situation, we could *block* on the abilities of the students. That is, we could sort the students into, say, three homogeneous ability groups (blocks) determined by IQ tests, grade point averages, a mathematical abilities test given the first day of class, or a combination of these ability measures. To each block, we would randomly assign the two teaching methods to the students comprising the block.

Thus, an important reason for using a randomized block design is to remove a source of variation (by blocking) from the experimental error. In this design, the sum of squares due to the blocks (differences among student abilities) will be partitioned from the total sum of squares for the dependent variable:

$$SST = SSTR + SSB + SSE$$

If a completely randomized model were used in which the two teaching methods were randomly assigned to the students when differences in student abilities existed, then the sum of squares attributable to these differences (SSB) would be improperly ascribed to the experimental error (SSE).

The need for blocking in experimental designs occurs frequently; the following examples illustrate instances when blocking is required.

Example 12.3 To study the effectiveness of four types of automobile oil filters, an experiment is performed using a specific brand of oil. Twelve quarts of the oil are selected and the same amount of impurities is added to each. Due to the nature of the experiment, only four tests can be performed per day. Since humidity and temperature differences may possibly affect the ability of a filter to trap impurities, it is decided to *block* on days. Designating the four filters by A, B, C, and D, the selected experimental design is:

Blocks (*days*)	*Treatments* (*types of filters*)			
1...............	A	B	D	C
2...............	C	A	B	D
3...............	D	B	A	C

The order in which the four filters are tested each day is randomly selected. For instance, on the first day, the filters are tested in the order A, B, D, and C.

In this design, any effects due to differences among days will be accounted for by blocking on days. The four treatments are randomly assigned to the experimental units (four cans of oil per day) within each block (day). Notice that each block represents a completely randomized design—each treatment is applied randomly to the experimental units in the block. Thus, the blocks play the role of *replicates* of a completely randomized design.

Example 12.4 Frequently, subjects are treated as blocks in social science research. For example, a social scientist may be interested in the effects of five treatments on subjects. Let us suppose that four subjects are selected for the experiment. A possible experimental design is:

Blocks (*subjects*)	*Treatments*				
1.............	A	C	B	D	E
2.............	E	A	C	D	B
3.............	B	D	C	E	A
4.............	E	B	D	A	C

Designating the treatments by A, B, C, D, and E, the five treatments are assigned in random order to the four subjects. In this design, the subjects serve as blocks and the experimental units within each block are considered to be the different times treatments may be applied to the subject.

Although this is a randomized block design, social scientists commonly refer to it as a *repeated measures design*—the treatment measures are repeated for each subject.

We now give the description of the randomized block design.

Definition 12.2
One-factor, randomized block design

The model is specified by the linear equation:

$$X_{ij} = \mu + \tau_i + B_j + e_{ij}$$
$$i = 1, 2, \ldots, t; \quad j = 1, 2, \ldots, b$$

where

X_{ij} = response random variable of the dependent variable corresponding to the ith treatment in the jth block

μ = overall population mean

τ_i = effect of the ith treatment

B_j = effect of the jth block

e_{ij} = random error component associated with the response random variable X_{ij}

The t treatments are *randomly* assigned to the b experimental units in each block. The null and alternate hypotheses are:

$$H_0: \quad \tau_1 = \tau_2 = \cdots = \tau_t$$
$$H_A: \quad \text{Not } \{\tau_1 = \tau_2 = \cdots = \tau_t\}$$

☐ 12.4.1 Assumptions: Randomized block design

1. The treatment and block effects are *fixed.*
2. The error components e_{ij}, where $i = 1, 2, \ldots, t; j = 1, 2, \ldots, b$, are normally and independently distributed random variables with 0 means and a common variance σ^2.
3. There is no treatment by block interaction effect.
4. The design is balanced—each block contains the same number of experimental units subjected to the t treatments in equal numbers.

☐ 12.4.2 The analysis of variance table

The analysis of variance table for the one-factor, randomized block design is given in Table 12.7.

TABLE 12.7. Analysis of variance table for one-factor, randomized block design with one experimental unit per treatment within a block

Source of variation	Degrees of freedom	Sum of squares	Mean square	F-ratio
Among treatments	$t - 1$	$\text{SSTR} = b \sum_{i=1}^{t} (\bar{x}_{i.} - \bar{x}_{..})^2$	$\text{MSTR} = \dfrac{\text{SSTR}}{t - 1}$	$f = \dfrac{\text{MSTR}}{\text{MSE}}$
Among blocks	$b - 1$	$\text{SSB} = t \sum_{j=1}^{b} (\bar{x}_{.j} - \bar{x}_{..})^2$		
Experimental error	$(t - 1)(b - 1)$	$\text{SSE} = \text{SST} - \text{SSTR} - \text{SSB}$	$\text{MSE} = \dfrac{\text{SSE}}{(t - 1)(b - 1)}$	
Total	$bt - 1$	$\text{SST} = \sum_{i=1}^{t} \sum_{j=1}^{b} (x_{ij} - \bar{x}_{..})^2$		

Critical F-value: $F[1 - \alpha; (t - 1), (t - 1)(b - 1)]$ Null hypothesis H_0: $\tau_1 = \tau_2 = \cdots = \tau_t$.

If $f > F[1 - \alpha; (t - 1), (t - 1)(b - 1)]$, reject the null hypothesis H_0: $\tau_1 = \tau_2 = \cdots = \tau_t$.

Example 12.5 Let us suppose that in the Example 12.3 problem, the following responses have been recorded:

Blocks (days)	Amounts of impurities trapped by filter			
1	17(A)	19(B)	14(D)	17(C)
2	18(C)	16(A)	21(B)	14(D)
3	17(D)	23(B)	18(A)	19(C)

Are the amounts of impurities trapped by the four filters significantly different at the $\alpha = 0.05$ level?

Solution The computation of the sums of squares for the analysis of variance table can be simplified somewhat by using the equivalent calculating forms to those given in Table 12.7.

Total sum of squares

$$\text{SST} = \sum_{i=1}^{t} \sum_{j=1}^{b} (x_{ij} - \bar{x}_{..})^2 = \sum_{i=1}^{t} \sum_{j=1}^{b} x_{ij}^2 - \frac{\left(\sum_{i=1}^{t} \sum_{j=1}^{b} x_{ij}\right)^2}{bt}$$

$$= (17)^2 + (19)^2 + \cdot \cdot \cdot + (19)^2 - \frac{[(17) + (19) + \cdot \cdot \cdot + (19)]^2}{(3)(4)}$$

$$= 3,855 - \frac{(213)^2}{12} = 3,855 - 3,780.75 = 74.25$$

Treatment sum of squares

$$\text{SSTR} = b \sum_{i=1}^{t} (\bar{x}_{i.} - \bar{x}_{..})^2 = \frac{\sum_{i=1}^{t} x_{i.}^2}{b} - \frac{\left(\sum_{i=1}^{t} \sum_{j=1}^{b} x_{ij}\right)^2}{bt}$$

Note that $x_{i.} = \sum_{j=1}^{b} x_{ij}$; it is the *sum* of the ith treatment responses over the b blocks.

$$x_{1.} = 17 + 16 + 18 = 51 \qquad x_{2.} = 19 + 21 + 23 = 63$$
$$x_{3.} = 17 + 18 + 19 = 54 \qquad x_{4.} = 14 + 14 + 17 = 45$$

$$\text{SSTR} = \frac{(51)^2 + (63)^2 + (54)^2 + (45)^2}{3} - 3,780.75$$

$$= \frac{11,511}{3} - 3,780.75 = 3,837 - 3,780.75 = 56.25$$

Block sum of squares

$$\text{SSB} = t \sum_{j=1}^{b} (\bar{x}_{.j} - \bar{x}_{..})^2 = \frac{\sum_{j=1}^{b} x_{.j}^2}{t} - \frac{\left(\sum_{i=1}^{t} \sum_{j=1}^{b} x_{ij}\right)^2}{bt}$$

Note that $\bar{x}_{.j} = \sum\limits_{i=1}^{t} x_{ij}$; it is the *sum* of the *j*th block responses over the *t* treatments.

$$x_{.1} = 17 + 19 + 14 + 17 = 67 \qquad x_{.2} = 18 + 16 + 21 + 14 = 69$$
$$x_{.3} = 17 + 23 + 18 + 19 = 77$$

$$\text{SSB} = \frac{(67)^2 + (69)^2 + (77)^2}{4} - 3{,}780.75$$

$$= \frac{15{,}179}{4} - 3{,}780.75 = 3{,}794.75 - 3{,}780.75 = 14$$

Error sum of squares

$$\text{SSE} = \text{SST} - \text{SSTR} - \text{SSB} = 74.25 - 56.25 - 14 = 4$$

The analysis of variance results are summarized in Table 12.8.

TABLE 12.8 Analysis of variance table for Example 12.5

Source of variation	Degrees of freedom	Sum of squares	Mean square	F-ratio
Among treatments	$t - 1 = 3$	56.25	$\dfrac{56.25}{3} = 18.75$	$f = \dfrac{18.75}{0.67} = 28.125$
Among blocks	$b - 1 = 2$	14.00	$\dfrac{14.00}{2} = 7$	
Experimental error	$(b - 1)(t - 1) = 6$	4.00	$\dfrac{4.00}{6} = 0.67$	
Total	$bt - 1 = 11$	74.25		Critical F: $F(0.95; 3, 6) = 4.76$

Since $f = 28.125 > 4.76$, we reject the null hypothesis H_0: $\tau_1 = \tau_2 = \tau_3$. Therefore, it appears from these data that the three oil filters do trap significantly different average amounts of impurities.

Table 12.9 presents the analysis of variance output from the SAS randomized block program. Notice that the sources of variation for the treat-

TABLE 12.9 Computer analysis of the Example 12.5 data

REMOVING OF IMPURITIES DATA
ANALYSIS OF VARIANCE PROCEDURE

DEPENDENT VARIABLE: IMPURE

SOURCE	DF	SUM OF SQUARES	MEAN SQUARE	F VALUE	PR > F
MODEL	5	70.25	14.05000	21.075	0.0010
ERROR	6	4.00	0.66666		
TOTAL	11	74.25			

SOURCE	DF	ANOVA SS	F VALUE	PR > F
DAYS	2	14.00	10.50	0.0110
FILTER	3	56.25	28.13	0.0006

ment and block effects are combined in the upper portion of the table and are denoted by the source of variation "MODEL." The breakdown of the "MODEL" sum of squares into the treatment sum of squares ("FILTER") and the block sum of squares ("DAYS") is given below the main portion of the analysis of variance table. From this table, the p-value for testing the differences among the treatment effects is given by 0.0006; that is, Prob($F > f = 28.13$) = 0.0006.

Notice in Example 12.5 that there appears to be a significant blocking effect since SSB = 14; that is, the filters do appear to behave differently on the three days. We could test to see whether this effect is significant at the $\alpha = 0.05$ significance level by forming the value of the F-statistic,

$$f = \frac{\text{MSB}}{\text{MSE}} = \frac{7.00}{0.67} = 10.5$$

and comparing the calculated f value to the critical F given by $F[1 - \alpha; (b - 1), (b - 1)(t - 1)] = F(0.95; 2, 6) = 5.14$. Since $10.5 > 5.14$, we can conclude that the blocking effect (days) does appear to be statistically significant.

Since the blocks were not chosen in a random fashion, it is really not appropriate to conduct the F-test for the blocking effect. However, by conducting the test, we can get *some* indication of whether or not the use of blocks has absorbed a significant amount of variability in the response random variable **X**.

If the blocking effect had been ignored, then the design would be a completely randomized design with only two sources of variation: among treatments and experimental error. In this case, the sum of squares for blocks, SSB, would appear in the error sum of squares, SSE. The error mean square would then become

$$\text{MSE} = \frac{14.00 + 4.00}{2 + 6} = \frac{18.00}{8} = 2.25$$

instead of the proper value of 0.67. Although we still would reject the treatment effects null hypothesis, the decision to reject H_0 would be much closer. By blocking on days, we have properly partitioned out an important source of variation that could have confounded the treatment effects.

It is possible to use a randomized design in which more than one experimental unit is subjected to each treatment in each block. In Table 12.10, the nature of such a design is indicated. Within each block, each treatment is randomly assigned to n experimental units. This design is called a *randomized block design with subsampling* and requires that the tbn experimental units have been sorted into b homogeneous groups of tn units each. The reader is directed to the Ostle and Mensing reference at the end of the chapter for this and further extensions of the basic one-factor, randomized block design model.

TABLE 12.10 Randomized block design with subsampling

Treatments / Blocks	1	2	$\cdots$	t
B_1	x_{111} x_{112} $\cdot$ $\cdot$ $\cdot$ x_{11n}	x_{211} x_{212} $\cdot$ $\cdot$ $\cdot$ x_{21n}	$\cdots$	x_{t11} x_{t12} $\cdot$ $\cdot$ $\cdot$ x_{t1n}
B_2	x_{121} x_{122} $\cdot$ $\cdot$ $\cdot$ x_{12n}	x_{221} x_{222} $\cdot$ $\cdot$ $\cdot$ x_{22n}	$\cdots$	x_{t21} x_{t22} $\cdot$ $\cdot$ $\cdot$ x_{t2n}
$\cdot$ $\cdot$ $\cdot$	$\cdot$ $\cdot$ $\cdot$	$\cdot$ $\cdot$ $\cdot$	$\cdots$	$\cdot$ $\cdot$ $\cdot$
B_b	x_{1b1} x_{1b2} $\cdot$ $\cdot$ $\cdot$ x_{1bn}	x_{2b1} x_{2b2} $\cdot$ $\cdot$ $\cdot$ x_{2bn}	$\cdots$	x_{tb1} x_{tb2} $\cdot$ $\cdot$ $\cdot$ x_{tbn}

12.5 FACTORIAL EXPERIMENTS

In most studies, we are interested in the changes in the dependent variable as a function of more than one independent variable (factor). To this point, we have been concerned with one-factor designs—the one-factor completely randomized design and the one-factor randomized block design in which the blocking variable is not thought of as a treatment factor, but rather as a variable introduced to control an unwanted source of variation in the dependent variable. We will now generalize the analysis of variance procedure for the analysis of experiments involving two or more factors.

Recall Example 12.1 where Farmco is concerned with the weight gain of chicks raised for overseas shipment. An experiment was conducted to determine whether three feeds—Growbig, Max, and Topnotch—produced different average weight gains among the chicks. As suggested in Section 12.1, Farmco may be interested in the effect on weight gain of a second factor—vitamin supplements. Let us suppose that two vitamin supplements, 1 and 2,

TABLE 12.11 Form of the sample data for a two-factor experiment

Factor B (vitamin supplement), B_j / Factor A (feed), A_i	Growbig $i = 1$	Max $i = 2$	Topnotch $i = 3$	Totals	Means
Vitamin supplement 1, $j = 1$	x_{111} x_{112} . . . x_{11n}	x_{211} x_{212} . . . x_{21n}	x_{311} x_{312} . . . x_{31n}	$x_{.1.}$	$\bar{x}_{.1.}$
Vitamin supplement 2, $j = 2$	x_{121} x_{122} . . . x_{12n}	x_{221} x_{222} . . . x_{22n}	x_{321} x_{322} . . . x_{32n}	$x_{.2.}$	$\bar{x}_{.2.}$
Totals	$x_{1..}$	$x_{2..}$	$x_{3..}$	$x_{...}$	
Means	$\bar{x}_{1..}$	$\bar{x}_{2..}$	$\bar{x}_{3..}$		$\bar{x}_{...}$

are introduced into the experiment. The form of the sample data for this experiment is indicated in Table 12.11.

From Table 12.11, it is clear that the experiment is balanced; that is, n chicks are randomly assigned to each treatment—a combination of one level of each factor. The word *treatment* used in analysis of variance arose from experiments of this kind. The chick associated with the weight gain x_{111} has been "treated" with the Growbig feed and vitamin supplement 1. Treatments are, therefore, combinations of one level of each factor in the experiment. In this experiment, we have six ($i = 3$ times $j = 2$) treatments.

To introduce the appropriate model and analysis of variance table for the two-factor, completely randomized design, we will generalize the preceding experiment to allow for any number of feeds, say a, and any number of vitamin supplements, say b.

Definition 12.3
Two-factor, completely randomized design

The model is

$$X_{ijk} = \mu + A_i + B_j + (AB)_{ij} + e_{ijk}$$
$$i = 1, 2, \ldots, a; \quad j = 1, 2, \ldots, b; \quad k = 1, 2, \ldots, n$$

Definition 12.3 (continued)

where:

X_{ijk} = dependent response random variable associated with the kth experimental unit subjected to the ith level of the first factor (A_i) and the jth level of the second factor (B_j).

μ = overall population mean

A_i = effect of the ith level of the first factor

B_j = effect of the jth level of the second factor

$(AB)_{ij}$ = interaction effect of ith level of the first factor and jth level of the second factor

e_{ijk} = random error component associated with the response random variable X_{ijk}

The abn experimental units are randomly assigned to the ab treatments. Hypotheses of interest are

1. H_0: $A_1 = A_2 = \cdots = A_a$
   H_A: Not $\{A_1 = A_2 = \cdots = A_a\}$

2. H_0: $B_1 = B_2 = \cdots = B_b$
   H_A: Not $\{B_1 = B_2 = \cdots = B_b\}$

3. H_0: $(AB)_{ij} = 0$; $i = 1, 2, \ldots, a$; $j = 1, 2, \ldots, b$
   H_A: Not $\{(AB)_{ij} = 0$; $i = 1, 2, \ldots, a$; $j = 1, 2, \ldots, b\}$

☐ **12.5.1   Assumptions: Two-factor, completely randomized design**

1. The factors, A and B, are fixed.
2. The error components, e_{ijk} where $i = 1, 2, \ldots, a; j = 1, 2, \ldots, b;$ $k = 1, 2, \ldots, n$, are normally and independently distributed random variables with 0 means and a common variance of σ^2.
3. The design is balanced; that is, the same number of experimental units n ($n \geq 2$) are subjected to each of the ab treatments.

☐ **12.5.2   Analysis of variance table**

The analysis of variance table for the two-factor, completely randomized design is given in Table 12.12.

Example 12.6   Let us suppose that 18 chicks are randomly selected from a large brood and are randomly separated into six treatment groups of three each. The resulting weight gains are given in Table 12.13.

The notation in Table 12.13 is as follows:

$$x_{ijk} = \text{weight gain for the } k\text{th chick subjected to the } i\text{th feed}$$
$$\text{and the } j\text{th vitamin supplement}$$

$$x_{i..} = \sum_{j=1}^{b} \sum_{k=1}^{n} x_{ijk} = \text{total weight gain of the chicks subjected to the } i\text{th feed}$$

$$\bar{x}_{i..} = x_{i..}/bn = \text{average weight gain of the chicks subjected to the } i\text{th feed}$$

TABLE 12.12 Analysis of variance table for two-factor, completely randomized design

Source of variation	Degrees of freedom	Sum of squares	Mean square	F-ratio
Factor A	$a - 1$	$SSA = bn \sum\limits_{i=1}^{a} (\bar{x}_{i..} - \bar{x}_{...})^2$	$MSA = \dfrac{SSA}{a-1}$	$f = \dfrac{MSA}{MSE}$
Factor B	$b - 1$	$SSB = an \sum\limits_{j=1}^{b} (\bar{x}_{.j.} - \bar{x}_{...})^2$	$MSB = \dfrac{SSB}{b-1}$	$f = \dfrac{MSB}{MSE}$
AB interaction	$(a-1)(b-1)$	$SSAB = n \sum\limits_{i=1}^{a} \sum\limits_{j=1}^{b} (\bar{x}_{ij.} - \bar{x}_{i..} - \bar{x}_{.j.} + \bar{x}_{...})^2$	$MSAB = \dfrac{SSAB}{(a-1)(b-1)}$	$f = \dfrac{MSAB}{MSE}$
Experimental error	$ab(n-1)$	$SSE = \sum\limits_{i=1}^{a} \sum\limits_{j=1}^{b} \sum\limits_{k=1}^{n} (x_{ijk} - \bar{x}_{ij.})^2$	$MSE = \dfrac{SSE}{ab(n-1)}$	
Total	$abn - 1$	$SST = \sum\limits_{i=1}^{a} \sum\limits_{j=1}^{b} \sum\limits_{k=1}^{n} (x_{ijk} - \bar{x}_{...})^2$		

TABLE 12.13 Weight gain (in grams) in the two-factor experiment

Factor A / Factor B	Growbig	Max	Topnotch	Totals	Means
Vitamin supplement 1	46 75 59 $x_{11.} = 180$ $\bar{x}_{11.} = 60$	120 94 122 $x_{21.} = 336$ $\bar{x}_{21.} = 112$	36 49 50 $x_{31.} = 135$ $\bar{x}_{31.} = 45$	$x_{.1.} = 651$	$\bar{x}_{.1.} = 72.33$
Vitamin supplement 2	92 87 85 $x_{12.} = 264$ $\bar{x}_{12.} = 88$	116 96 88 $x_{22.} = 300$ $\bar{x}_{22.} = 100$	33 47 46 $x_{32.} = 126$ $\bar{x}_{32.} = 42$	$x_{.2.} = 690$	$\bar{x}_{.2.} = 76.67$
Totals	$x_{1..} = 444$	$x_{2..} = 636$	$x_{3..} = 261$	$x_{...} = 1{,}341$	
Means	$\bar{x}_{1..} = 74$	$\bar{x}_{2..} = 106$	$\bar{x}_{3..} = 43.5$		$\bar{x}_{...} = 74.5$

$$x_{.j.} = \sum_{i=1}^{a} \sum_{k=1}^{n} x_{ijk} = \text{total weight gain of the chicks subjected to the } j\text{th vitamin supplement}$$

$$\bar{x}_{.j.} = x_{.j.}/an = \text{average weight gain of the chicks subjected to the } j\text{th vitamin supplement}$$

$$x_{...} = \sum_{i=1}^{a} \sum_{j=1}^{b} \sum_{k=1}^{n} x_{ijk} = \text{total weight gain of all chicks } (abn) \text{ in the experiment}$$

$$\bar{x}_{...} = x_{...}/(a)(b)(n) = \text{average weight gain of all chicks in the experiment}$$

$$x_{ij.} = \sum_{k=1}^{n} x_{ijk} = \text{total weight gain of the } n \text{ chicks subjected to the } i\text{th feed and } j\text{th vitamin supplement}$$

$$\bar{x}_{ij.} = x_{ij.}/n = \text{average weight gain of the } n \text{ chicks subjected to the } i\text{th feed and } j\text{th vitamin supplement}$$

$$a = 3, \ b = 2, \ n = 3$$

The calculations may be somewhat eased by using the calculating forms of the sums of squares:

$$C = \frac{\left(\sum_{i=1}^{a=3} \sum_{j=1}^{b=2} \sum_{k=1}^{n=3} x_{ijk} \right)^2}{nab = (3)(3)(2)} = \frac{(1{,}341)^2}{18} = 99{,}904.5$$

$$SSA = \frac{\sum_{i=1}^{a=3} x_{i..}^2}{nb = (3)(2)} - C = \frac{(444)^2 + (636)^2 + (261)^2}{6} - C$$
$$= 111{,}625.5 - 99{,}904.5 = 11{,}721$$

$$SSB = \frac{\sum_{j=1}^{b=2} x^2_{.j.}}{na = (3)(3)} - C = \frac{(651)^2 + (690)^2}{9} - C$$

$$= 99{,}989 - 99{,}904.5 = 84.5$$

$$SST = \sum_{i=1}^{a=3} \sum_{j=1}^{b=2} \sum_{k=1}^{n=3} x^2_{ijk} - C = [(46)^2 + (75)^2 + \cdots + (46)^2] - C$$

$$= 114{,}627 - 99{,}904.5 = 14{,}722.5$$

$$SSAB = \frac{\sum_{i=1}^{a=3} \sum_{j=1}^{b=2} x^2_{ij.}}{n = 3} - SSA - SSB - C$$

$$= \frac{(180)^2 + (336)^2 + \cdots + (126)^2}{3} - SSA - SSB - C$$

$$= 113{,}031 - 11{,}721 - 84.5 - 99{,}904.5 = 1{,}321$$

$$SSE = SST - SSA - SSB - SSAB$$

$$= 14{,}722.5 - 11{,}721 - 84.5 - 1{,}321 = 1{,}596$$

The analysis of variance table is summarized in Table 12.14, using the output from the SAS two-factor analysis of variance program.

TABLE 12.14   Computer analysis of the Example 12.6 data

FACTORIAL EXPERIMENT ON FEED DATA
ANALYSIS OF VARIANCE PROCEDURE

DEPENDENT VARIABLE: YIELD

SOURCE	DF	SUM OF SQUARES	MEAN SQUARE	F VALUE	PR > F
MODEL	5	13126.50	2625.30	19.74	0.0001
ERROR	12	1596.00	133.00		
TOTAL	17	14722.50			

SOURCE	DF	ANOVA SS	F VALUE	PR > F
FEED	2	11721.00	44.06	0.0001
SUPP	1	84.50	0.64	0.4409
FEED*SUPP	2	1321.00	4.97	0.0268

To test the hypotheses

$$H_0: \quad (AB)_{ij} = 0; \quad i = 1, 2, 3; \quad j = 1, 2$$
$$H_A: \quad (AB)_{ij} \neq 0 \quad \text{for at least one } i, j \text{ pair}$$

at the $\alpha = 0.05$ level, we compare the computed value of F in the table for the interaction effect, $f = 4.97$, with the appropriate tabular value of F, $F(0.95; 2, 12) = 3.89$. Since $4.97 > F(0.95; 2, 12) = 3.89$, we conclude that there is a significant interaction effect. From Table 12.14 the p-value for this test is equal to 0.0268.

The significant interaction effect is readily apparent from studying Table 12.13. The first vitamin supplement appears to produce *less* average weight gain in conjunction with Growbig than does the second vitamin supplement. Yet the first vitamin supplement appears to produce *more* average weight gain in conjunction with Max and Topnotch than does the second vitamin supplement.

If there is no interaction effect between the two factors, then the feeds should produce relatively the same weight gain increase (or decrease) as we move from the first vitamin supplement to the second. An example of sample responses that would suggest a "no-interaction" effect is illustrated in Figure 12.7; the gains indicated by the darkened lines should be nearly equal as we move from the first supplement to the second.

FIGURE 12.7 No-interaction effect

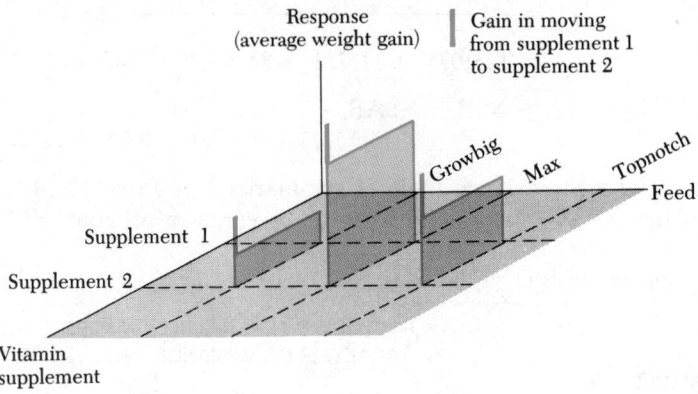

In this figure, the increase in weight gains is the same for each feed when the second vitamin supplement is used instead of the first. The responses for our experiment are illustrated in Figure 12.8. It is apparent that Growbig interacts differently with the vitamin supplements than do Max and Topnotch.

FIGURE 12.8 Mean responses from Table 12.13

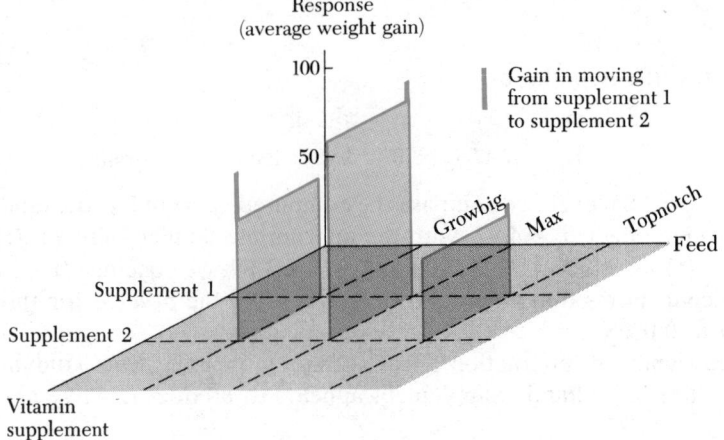

To test the hypotheses concerning feeds:

$$H_0: \quad A_1 = A_2 = A_3$$
$$H_A: \quad \text{Not } \{A_1 = A_2 = A_3\}$$

the appropriate tabular value of **F** at the $\alpha = 0.05$ level is $F(0.95; 2, 12) = 3.89$ and since $f = 44.06 > 3.89$, we conclude that the three feeds do produce different average weight gains among the chicks. From Table 12.14, the p-value is equal to 0.0001.

To test the hypotheses concerning vitamin supplements:

$$H_0: \quad B_1 = B_2$$
$$H_A: \quad B_1 \neq B_2$$

the appropriate tabular value of **F** at the $\alpha = 0.05$ level is $F(0.95; 1, 12) = 4.75$. Since $f = 0.64 < 4.75$, we conclude that there are no significant differences in average weight gains among the chicks for the two vitamin supplements. From Table 12.14, the p-value for this test is equal to 0.4409.

Caution must be exercised when drawing inferences about main effects (the primary factors) in the presence of significant interaction effects. For example, consider the means in Table 12.15 for a 3×3 experiment (two factors, each at three levels). In this experiment, there is a strong interaction effect—note the differences in responses for levels of factor A between the first level of factor B and the other two levels of B. From the margin means, it is clear that the analysis of variance procedure would not find significant differences among the levels of A and would most likely not find significant differences among the levels of B. But suppose that the first level of B were dropped from the experiment. Then the first level of A would appear to produce a significantly different response than would the other two levels of A. That is, excluding the first level of B, the first level of A appears to produce the greatest response in the dependent variable—one that is significantly different from the responses produced by levels 2 and 3 of factor A.

If the interaction is significant, it ordinarily does not make sense to test the hypothesis of significant main effects. As the preceding discussion suggests, the interaction effect may cover up (or uncover) significant main effects. But that is consistent with the nature and intent of multifactor experiments. In the Farmco example, it is understood that the intent is to provide *both* a feed

TABLE 12.15 Means of a 3×3 factorial experiment

Factor B \ Factor A	1	2	3	
1	$\bar{x}_{11.} = 10$	$\bar{x}_{21.} = 90$	$\bar{x}_{31.} = 80$	$\bar{x}_{.1.} = 60$
2	$\bar{x}_{12.} = 90$	$\bar{x}_{22.} = 45$	$\bar{x}_{32.} = 55$	$\bar{x}_{.2.} = 63.33$
3	$\bar{x}_{13.} = 80$	$\bar{x}_{23.} = 45$	$\bar{x}_{33.} = 45$	$\bar{x}_{.3.} = 56.67$
	$\bar{x}_{1..} = 60$	$\bar{x}_{2..} = 60$	$\bar{x}_{3..} = 60$	$\bar{x}_{...} = 60$

and a vitamin supplement for the chicks. While it may be possible to demonstrate that vitamin supplements 1 and 2 produce significantly different average weight gains *alone* by running a one-factor experiment excluding the three types of feed, it would not produce a useful inference since the supplements will be used in *conjunction* with the feed types. As it turned out, the two factors did interact with the result that Max in conjunction with the first vitamin supplement produced the greatest sample average weight gain among the chicks.

The analysis of variance procedure to analyze more than two factors is applied in a straightforward extension of the two-factor case. Although the computations become more tedious, multiple-factor experiments are desirable because of the efficiency they afford—many factors and their interactions can be studied by conducting one experiment. Additionally, a factorial experiment can be conducted in a randomized block design. Suppose we have two factors, A and B; a_1 and a_2 denote the two levels of factor A, and b_1 and b_2 denote the two levels of factor B. In Table 12.16, a randomized block design for the two-factor experiment is indicated.

TABLE 12.16 Two-factor, randomized block design

Block 1		Block 2		Block 3	
a_1b_2	a_1b_2	a_2b_2	a_1b_1	a_2b_2	a_1b_1
a_2b_2	a_2b_1	a_1b_1	a_1b_2	a_1b_2	a_2b_1
a_1b_1	a_1b_1	a_1b_2	a_2b_1	a_2b_1	a_1b_2
a_2b_1	a_2b_2	a_2b_1	a_2b_2	a_1b_1	a_2b_2

In this design, there are three blocks and four treatments (a_1b_1, a_1b_2, a_2b_1, and a_2b_2). Within each block, there are two replicates—each treatment is repeated twice.

The analysis of variance table for the two-factor, randomized block design as well as for more complicated designs can be found in most texts on analysis of variance or experimental design. A particularly good reference for basic designs is the Ostle and Mensing book listed at the end of the chapter.

■ 12.6 MULTIPLE COMPARISONS

When we determine by analysis of variance that the levels of a factor differ significantly, we ordinarily are interested in which mean(s) differ from the rest. Consider, for example, the one-factor, completely randomized experiment to determine whether the three feeds—Growbig, Max, and Topnotch—produce significantly different weight gains among chicks. We found in Section 12.3 that the three feeds do produce significantly different weight gains. Now can we further determine whether all three feeds are different or whether two are similar but differ from the third? The answer is yes, and the procedures that may be used for comparisons among factor level means are called *multiple comparisons*. We will consider one of these procedures: the Scheffe contrast method.

The three sample average weight gains are given in Table 12.17.

TABLE 12.17   Three sample means in Example 12.1

	Growbig	*Max*	*Topnotch*
Population mean	μ_1	μ_2	μ_3
Sample mean	$\bar{x}_{1.} = 78$	$\bar{x}_{2.} = 110$	$\bar{x}_{3.} = 40$

From the sample means, we may conjecture that $\mu_2 > \mu_1 > \mu_3$. Scheffe's method will allow us to test this conjecture by using *contrasts*. If $c_1, c_2, \ldots, c_t$ are known constants such that

$$\sum_{i=1}^{t} c_i = c_1 + c_2 + \cdots + c_t = 0$$

then

$$L = c_1\mu_1 + c_2\mu_2 + \cdots + c_t\mu_t$$

is called a *contrast*. We have already used implicitly the concept of a contrast in testing the equivalence of two populations means in Chapter 11. The hypothesis H_0: $\mu_1 = \mu_2$ can be written as H_0: $\mu_1 - \mu_2 = 0$, and the difference $\mu_1 - \mu_2$ is clearly a contrast with $c_1 = 1$, $c_2 = -1$. With three means ($t = 3$), other possible contrasts are as shown in the table.

	Contrast coefficients		
Contrast	c_1	c_2	c_3
$\mu_1 - \mu_3$	1	0	−1
$\mu_2 - \mu_3$	0	1	−1
$2\mu_1 - \mu_2 - \mu_3$	2	−1	−1
$5\mu_1 - 8\mu_2 + 3\mu_3$	5	−8	3

In fact, an infinite number of contrasts may be constructed with three means.

The Scheffe method produces a confidence interval estimate for a contrast L. The formula for the interval estimate is:

$$(\hat{L} - s\hat{\sigma}_{\hat{L}}, \hat{L} + s\hat{\sigma}_{\hat{L}})$$

where

$$\hat{L} = c_1 \bar{x}_1 + c_2 \bar{x}_2 + \cdots + c_t \bar{x}_t$$
$$s^2 = (t - 1)F[1 - \alpha; (t - 1), t(n - 1)]$$
$$\hat{\sigma}_{\hat{L}}^2 = \text{MSE} \sum_{i=1}^{t} \frac{c_i^2}{n_i}$$

The lower limit of the confidence interval estimate is given by $\hat{L} - s\hat{\sigma}_{\hat{L}}$ and the upper limit is given by $\hat{L} + s\hat{\sigma}_{\hat{L}}$. The statistic $\hat{L}$ is an unbiased estimator of $L = c_1\mu_1 + c_2\mu_2 + \cdots + c_t\mu_t$. The term s^2 is the product of the *total number* of means (t) minus 1 and the tabular F-value with $(t - 1)$ numerator degrees of freedom and $t(n - 1)$ denominator degrees of freedom. The expression for $\hat{\sigma}_{\hat{L}}^2$ is a function of the mean square error (MSE) and the sum of the ratios of the squares of the contrast coefficients to the sample sizes. Note that if the experiment is balanced so that each mean is based on a sample of size n, then

$$\hat{\sigma}_{\hat{L}}^2 = \frac{\text{MSE}}{n} \sum_{i=1}^{t} c_i^2$$

In the Example 12.1 experiment, $t = 3$, $n = 5$, MSE $= 423.17$, $(t - 1) = 2$, $t(n - 1) = 12$, and with $\alpha = 0.05$, $F(0.95; 2, 12) = 3.89$. Thus, $s^2 = 2F(0.95; 2, 12) = (2)(3.89) = 7.78$ and $s = 2.79$.

For the contrast $L = \mu_1 - \mu_2$,

$$\hat{L} = (1)\bar{x}_1 + (-1)\bar{x}_2 + (0)\bar{x}_3 = 78 - 110 = -32$$

$$\hat{\sigma}_{\hat{L}}^2 = \frac{\text{MSE}}{n} \sum_{i=1}^{t} c_i^2 = \frac{423.17}{5}[(1)^2 + (-1)^2 + (0)^2] = 169.27$$

$$\hat{\sigma}_{\hat{L}} = 13.01$$

$$s\hat{\sigma}_{\hat{L}} = (2.79)(13.01) = 36.30$$

The confidence interval estimate is: $(\hat{L} \pm s\hat{\sigma}_{\hat{L}}) = (-32 \pm 36.30)$, or $(-68.3$ to $4.3)$.

For the contrast $L = \mu_1 - \mu_3$,

$$\hat{L} = (1)\bar{x}_1 + (0)\bar{x}_2 + (-1)\bar{x}_3 = 78 - 40 = 38$$

$$\hat{\sigma}_{\hat{L}}^2 = \frac{\text{MSE}}{n} \sum_{i=1}^{t} c_i^2 = \frac{423.17}{5}[(1)^2 + (0)^2 + (-1)^2] = 169.27$$

$$\hat{\sigma}_{\hat{L}} = 13.01$$

$$s\hat{\sigma}_{\hat{L}} = (2.79)(13.01) = 36.30$$

The confidence interval estimate is: $(\hat{L} \pm s\hat{\sigma}_{\hat{L}}) = (38 \pm 36.30)$, or $(1.7$ to $74.3)$.

For the contrast $L = \mu_2 - \mu_3$,

$$\hat{L} = (0)\bar{x}_{1.} + (1)\bar{x}_{2.} + (-1)\bar{x}_{3.} = 110 - 40 = 70$$

$$\hat{\sigma}_{\hat{L}}^2 = \frac{MSE}{n} \sum_{i=1}^{t} c_i^2 = \frac{423.17}{5} [(0)^2 + (1)^2 + (-1)^2] = 169.27$$

$$\hat{\sigma}_{\hat{L}} = 13.01$$

$$s\hat{\sigma}_{\hat{L}} = (2.79)(13.01) = 36.30$$

The confidence interval estimate is: $(\hat{L} \pm s\hat{\sigma}_{\hat{L}}) = (70 \pm 36.30)$, or $(33.70$ to $106.30)$.

Note that only the first interval contains 0; that is,

$$-68.3 \leq \mu_1 - \mu_2 \leq 4.3$$

and recall that we can test a hypothesis of the form H_0: $\mu_1 = \mu_2$ by constructing a confidence interval on the difference between μ_1 and μ_2. If the interval contains 0, we cannot reject the null hypothesis—if the interval does not contain 0, then the null hypothesis is rejected. Thus, we cannot reject the null hypothesis H_0: $\mu_1 = \mu_2$; the implication is that the experiment did not suggest a *significant* difference between the feeds Growbig and Max, but it was close!

For the second contrast,

$$1.7 \leq \mu_1 - \mu_3 \leq 74.3$$

we can conclude that $\mu_1 > \mu_3$ since the interval does not contain 0. From the third contrast,

$$33.7 \leq \mu_2 - \mu_3 \leq 106.3$$

we can conclude that $\mu_2 > \mu_3$ since this interval also does not contain 0. In conclusion, from the Scheffe test, we would infer that the feeds Growbig and Max produce significantly greater weight gains than does Topnotch, but that Growbig and Max are not significantly different.

A multiple comparison procedure such as Scheffe's allows us to compare treatment means after we have determined that the means are significantly different by performing an analysis of variance test. It must be remembered that the Scheffe procedure can be applied only if significant differences in the means have been demonstrated by analysis of variance. If the null hypothesis of equal means cannot be rejected by analysis of variance and Scheffe's method is applied to contrasts of the means, spurious significant contrasts may occur.

One may ask why the three pairwise tests of μ_1, μ_2, and μ_3 could not have been made by using the simple t-test presented in Chapter 11. The answer is that by using the method of contrasts, the $(1 - \alpha)$ 100 percent level of confidence will be maintained for *each* and *every* contrast in the system of contrasts. Had we used three t-tests to compare the three means in pairs, the effective *system* confidence would be reduced to $[(1 - \alpha)100 \text{ percent}]^3$. For example, if the tests were conducted at the $\alpha = 0.05$ level, then the t-tests would each have an effective confidence of $(0.95)^3 = 0.857$, not 0.95. The method of contrasts ensures that the *family* of confidence intervals will have a confidence of 95 percent.

A final note: the Scheffe procedure can be readily applied to factors in a multifactor experiment. In Example 12.6, if we wished to compare the three feed factor level means, the only modification necessary is to change the degrees of freedom for the tabulated *F*-value in the computation to s^2:

$$s^2 = (a - 1)F[1 - \alpha; (a - 1), ab(n - 1)]$$

That is, the numerator degrees of freedom for the F-statistic will always be given by the degrees of freedom in the analysis of variance table corresponding to the set of means (source of variation) being tested. The denominator degrees of freedom for the F-statistic will be given by the degrees of freedom associated with MSE—experimental error source of variation—in the table. The term, n_i, in the expression for $\hat{\sigma}_{\hat{L}}^2$ is always equal to the number of observations used to compute the ith mean involved in the contrast.

12.7 VIOLATION OF ASSUMPTIONS

There are four basic assumptions in applying the analysis of variance method to experimental data: (1) homogeneity of variances, (2) normality, (3) additivity, and (4) independence.

☐ 12.7.1 Homogeneity of variance

This assumption requires that the population variances associated with each treatment in the experiment are equal. If the sample sizes corresponding to each treatment are equal, a very simple test for the hypothesis

$$H_0: \quad \sigma_1^2 = \sigma_2^2 = \cdots = \sigma_t^2$$
$$H_A: \quad \text{Not } \{\sigma_1^2 = \sigma_2^2 = \cdots = \sigma_t^2\}$$

is called the Hartley test, where there are t treatment populations and it is assumed that each population is normally distributed. Hartley's test statistic is:

$$H = \frac{\text{Max } (s_j^2)}{\text{Min } (s_j^2)}$$

where max (s_j^2) is the largest among the t sample variances and min (s_j^2) is the smallest. Values of **H** near 1 support H_0, while large positive values support H_A. The distribution of **H** depends on the number of populations t and the sample size n. The table of critical values is given in Table B.9 in Appendix B.

Example 12.7. For the one-factor, completely randomized design in Example 12.1, test the hypothesis of equal variances.

Solution. In the Farmco problem, there were three treatments (levels of the factor feed) and each sample size was five. Hence, $t = 3$ and $n = 5$. The three sample variances were $s_1^2 = 480.5$, $s_2^2 = 347.5$, and $s_3^2 = 441.5$. Thus,

$$h = \frac{s_1^2 = 480.5}{s_2^2 = 347.5} = 1.38$$

With $\alpha = 0.05$, the critical value of **H** is $H_{0.05;3,5} = 15.5$. Since $h < H_{0.05;3,5}$, we cannot reject the null hypothesis of equivalent population variances. Based on the experimental data, we would conclude, therefore, that the assumption of homogeneous variances (homoscedasticity) appears to be satisfied.

If the design is not balanced (the sample sizes are not equal), a more complicated test due to Bartlett (see the Neter and Wasserman reference) may be used. This test also assumes that the populations are normally distributed.

☐ 12.7.2   Normality

The normality assumption requires that each treatment population is normally distributed. Two methods will be presented to test the "goodness of fit" of sample data to a normal distribution: the chi-square test (see Chapter 19) and the Kolmogorov-Smirnov test (see Chapter 20).

☐ 12.7.3   Additivity

This assumption requires that the effects in the model behave in an additive fashion. Nonadditivity may be caused by multiplicative true effects existing in the model, by the exclusion of significant interactions in the model, or by "outliers"—observations that are inconsistent in value with the majority of the responses in the experiment.

A test for this assumption that is commonly used is due to Tukey, but we shall not introduce it in this text. For the description of the Tukey test and an example of its application to experimental data, the interested reader is directed to the Guenther reference.

☐ 12.7.4   Independence

The independence assumption requires that the errors are statistically independent or, assuming that the treatment populations are normally distributed, that the errors are uncorrelated. This is a crucial assumption in analysis of variance. The use of randomization in experimental design (e.g., randomly assigning experimental units to treatments) is an "insurance policy" against correlated error terms. A test for randomness of errors is given in Chapter 20.

Regarding the assumptions of the analysis of variance, there are two important questions: (1) Are the assumptions satisfied in a particular experiment, and (2) If not, what then?

The satisfaction of the assumptions can often be assured by properly designing the experiment and carefully using randomization throughout the experiment. For each assumption, a test is given in this chapter or elsewhere in the text, or a test is referenced to determine whether the data conform to the assumption. If the assumptions are not satisfied by the data, then we must be cautious in analyzing the data by analysis of variance.

In general, analysis of variance is very robust to modest departures from the assumptions. By *robust,* we mean that if significant differences exist among the factor means, analysis of variance will typically "sense" that the differences exist even if the assumptions are not entirely satisfied. That is, the power of the F-test used in analysis of variance is not seriously affected by small to moderate departures from the assumptions. This is particularly true relative to the normality assumption and, if the experiment is balanced, to the homogeneity of variance assumption.

If the independence assumption is not satisfied, the consequence is somewhat more serious. If the errors deviate slightly from a random set, the analysis of variance will not be seriously affected. If the departure from randomness is considerable, then the results of the analysis of variance may be meaningless.

If the departures from the assumptions in the experiment are serious, often the situation can be corrected by *transforming* the data. For example, we may take the logarithm of each response in the experiment and perform the analysis of variance on the logarithms if the original responses do not conform to the assumptions, but the log responses do. Various forms of transformations are given to correct for nonnormality, nonadditivity, nonhomogeneity of variance, and, to a lesser extent, nonindependence. If the errors are not independent, most often the corrective measure is a modification of the model—the inclusion of another factor or the use of blocking may be necessary to produce random errors.

For a more detailed discussion of the assumptions of analysis of variance, the interested reader is directed to the Neter and Wasserman reference.

■ 12.8 SUMMARY

The analysis of variance procedure is appropriate for testing the equivalence of a set of two or more population means when certain assumptions are satisfied: (1) homogeneity of variances, (2) normality of the error terms, (3) additivity in the model, and (4) randomness of the error terms. Two basic designs are given: the completely randomized design and the randomized block design. In the first design, experimental units are randomly assigned to treatments. In the second design, an extraneous factor that may be confounded with treatments is isolated by blocking. Within each block, the experimental units are randomly assigned to treatments.

More than one factor can be studied in either experimental design. Factorial experiments have the advantages of efficiently utilizing experimental units and allowing for the study of interaction effects.

If it is found that the factor level means differ significantly, Scheffe's multiple comparison procedure may be used to identify how the means differ from one another.

Most analysis of variance problems are solved on the computer by using "canned" computer codes available at most computer facilities. These codes typically perform tests for the assumptions mentioned above as well as other tests. Understanding how to perform these tests and how to compute the analysis of variance table is necessary for properly interpreting the results of a computer analysis and, more importantly, for checking the validity of the results.

■ REFERENCES

Guenther, W. C. *Analysis of Variance*. Englewood Cliffs, N.J.: Prentice-Hall, Inc., 1964.

Neter, J., and Wasserman, W. *Applied Linear Statistical Models*. Homewood, Ill.: Richard D. Irwin, Inc., 1974.

Ostle, B., and Mensing, R. *Statistics in Research*. 3d ed. Ames: Iowa State University Press, 1975.

■ PROBLEMS

12.1. Why is the method to compare t ($t \geq 2$) population means called *analysis of variance*?

12.2. What are the objectives of analysis of variance? Discuss.

12.3. State the assumptions underlying the following analysis of variance models:
   a. One-factor, completely randomized design.
   b. Randomized block design.

12.4. What is the difference between a one-factor, randomized block design and a two-factor, completely randomized design?

12.5. How does the randomized block design minimize the chance that a factor under study is confounded by other factors that are related to the factor under study?

12.6. If the null hypothesis of the equality of population means is rejected by the analysis of variance method, what is the next step in the study of the data?

12.7. To determine whether three brands of light bulbs have equal mean lifetimes, a random sample of five light bulbs of each type is taken and their lifetimes (in hours) are measured. The results are:

Brand A	Brand B	Brand C
75	85	82
63	80	71
67	82	75
70	90	79
72	78	72

At the $\alpha = 0.05$ level, test the hypothesis that the mean lifetimes of the light bulbs are equal. What, approximately, is the p-value for this test?

12.8. New employees of the Lax Company undergo a training program during their first six weeks on the job. Recently, the personnel department has suggested changing the nature of the program. To determine the effectiveness of the new program, 20 new employees are selected from the set of new employees and are randomly divided into two groups. The first group receives the "old program" and the second group receives the "new program." At the end of the training period, all 20 individuals are given an examination to determine how much information has been assimilated. The results are:

Old program	New program
65	80
60	65
45	92
50	78
48	74
62	58
66	72
42	83
45	90
56	78

a. Using analysis of variance, test the null hypothesis of equal population test scores at the $\alpha = 0.10$ level. What, approximately, is the p-value for this test?

b. Using the t-test of Chapter 11, repeat the test in part a.

c. Did you reach the same conclusion in both parts? Should you have? Discuss.

12.9. A car rental agency is in the process of deciding the brand of tire to purchase as standard equipment for its fleet. As a part of the decision process, they are interested in studying the treadlife of five competing brands. Based on testing, the research department determined that each of five tires of each brand lasted the following number of miles (in 1000s):

		Tire brands		
A	B	C	D	E
40	45	30	35	28
42	40	32	40	32
45	40	31	42	34
38	44	35	36	28
40	42	28	38	32

Test the hypothesis at the $\alpha = 0.05$ level that the five tire brands will have identical average miles of wear. What, approximately, is the p-value for this test?

12.10. A meat producer is interested in the ability of various feeds to quickly fatten livestock. Fifteen heifers similar in age and weight are chosen to participate in an experiment to test the effectiveness of three feeds. The results in pounds gained over the duration of the experiment are:

	Feeds	
A	B	C
250	200	400
400	240	420
420	230	360
360	260	*
300	*	*

* Heifers became ill or had to be withdrawn from the experiment for other reasons.

Test the hypothesis of equal feed effectiveness in producing weight gain at the $\alpha = 0.05$ level. What, approximately, is the p-value for this test?

12.11. It is suspected that four machines in a canning operation fill cans to different levels on the average. Random samples of the cans produced from each machine were taken and the fill in ounces was measured. The results were:

Machine			
1	*2*	*3*	*4*
12.10	12.18	12.20	12.01
12.08	12.10	12.25	12.00
12.22	12.10	12.25	12.03
12.12			12.00

Do the machines appear to be filling the cans at different average levels? (Calculate the approximate p-value and decide, based on its value, whether not to reject or to reject the null hypothesis.)

12.12. A marketing research firm conducted an experiment to compare the effect of five different colors on product recognition time. Thirty people were used in the experiment. Each color was used on the product package and was exposed to each person in one of five groups of six people each, formed from the 30 people. The recognition time (in seconds) was then measured. The experiment produced the following data:

$$\bar{x}_{1.} = 0.62 \qquad \bar{x}_{2.} = 0.90$$
$$\bar{x}_{3.} = 1.04 \qquad \bar{x}_{4.} = 0.48$$
$$\bar{x}_{5.} = 0.65$$
$$\text{SSTR} = 1.20 \qquad \text{SSE} = 0.60$$

(Color code: 1 = yellow, 2 = brown, 3 = blue, 4 = orange, 5 = green)
a. State the null and alternate hypotheses.
b. Write the analysis of variance table.
c. Test the null hypothesis at the $\alpha = 0.05$ level.
d. Calculate the approximate p-value for this test.

12.13. Using Hartley's test, determine whether the two population variances are equal for the experiment in Problem 12.6.

12.14. In Example 12.5, use the Scheffe multiple comparisons method to de-termine which pairs of averages for the three oil filters differ. Use $\alpha = 0.05$ for the confidence intervals.

12.15. In Problem 12.7, use the Scheffe multiple comparisons method to determine which mean lifetimes of the light bulbs of brands A, B, and C differ. Use $\alpha = 0.05$ for the confidence intervals.

12.16. In Problem 12.9, use the Scheffe multiple comparisons method to determine which tire brands produce average miles of wear that differ. Use $\alpha = 0.05$ for the confidence intervals.

12.17. A company is interested in buying a fleet of fuel-efficient cars for use by its salespeople. Four makes and models are under consideration: DATMOON 210, TOYA Silica, Chev X-BOD, and VZ Bunny. It is decided to select the make and model that produces the highest average miles per gallon over a 200-mile course that includes both city and highway driving. Four salespersons are selected for an experiment to compare the average miles per gallon for the four cars over the selected course. Each person will drive each car once over the course. Although the salespersons are instructed to drive the cars in a similar fashion (for example, no "rabbit" starts from a standing stop), it is expected that some "people" differences will arise. Therefore, it is decided to treat the salespersons as blocks. The data are shown in the table below.

	Blocks, salespersons			
Treatments	*1*	*2*	*3*	*4*
DATMOON 210	32	30	34	32
TOYA Silica	27	27	25	26
Chev X-BOD	24	29	28	23
VZ Bunny	36	40	39	36

a. Conduct the randomized block design analysis of variance on these data.

b. What is the approximate p-value for this test?

c. Based on the p-value, do the cars appear to produce different average miles per gallon?

d. Apply the Scheffe method of multiple comparisons to these averages to determine which cars differ in average miles per gallon.

12.18. An advertising firm is studying the effects of four different kinds of displays of a product in a grocery store in three different sales areas in the city. Within each sales area, four stores are selected and each receives one of the four displays. Over the duration of the experiment, the number of units of the product sold is recorded. The data are shown in the table.

Display	Sales areas		
	1	2	3
A	100	56	75
B	94	40	82
C	120	65	102
D	82	60	65

a. Which model is appropriate for analyzing these data? Explain.

b. Do the four displays result in different average sales? (Use the approximate p-value to decide whether or not to reject the null hypothesis.)

c. If you found that the four displays do appear to produce different average sales, use the Scheffe method of multiple comparisons to determine which pairs of averages differ.

12.19. In manufacturing a specific part for an automobile, four machines are used in a large plant. To determine whether the machines produce different average daily outputs, output data are collected on five different days. Since daily temperature, humidity, and so forth can affect machine output, the days are used as blocks. The data are shown in the table below.

Days (blocks)	Machines			
	1	2	3	4
1	273	286	303	313
2	240	323	320	310
3	260	303	330	348
4	278	333	323	343
5	268	338	345	325

a. State the null and alternate hypotheses.

b. Test the null hypothesis at the $\alpha = 0.05$ significance level.

c. Calculate the approximate p-value for this test.

d. If you rejected the null hypothesis in part b, use the Scheffe method of multiple comparisons to investigate the differences among the four mean outputs. Use $\alpha = 0.05$ for the confidence intervals.

12.20. To test the efficacy of four brands of fertilizer, the following experiment is designed. Five homogeneous plots of land are selected as blocks and each is divided into five subplots. One subplot receives no fertilizer, and the remaining four receive each of one of the brands of fertilizer. The experiment produced the following data (yields in bushels/acre) on treatment means.

No fertilizer	A	B	C	D
$\bar{x}_{1.} = 23$	$\bar{x}_{2.} = 28$	$\bar{x}_{3.} = 35$	$\bar{x}_{4.} = 38$	$\bar{x}_{5.} = 42$

$SSE = 160$ $SSTR = 1{,}000$ $SSB = 150$

a. Write the analysis of variance model for this experiment.
b. Test the null hypothesis that the mean effects are equal at the $\alpha = 0.05$ significance level. What is the approximate p-value for this test?
c. Use the Scheffe method of multiple comparisons to investigate the differences among the five means and interpret the results. Use $\alpha = 0.05$ for the confidence intervals.
d. Using the Scheffe method, test that the following contrast is 0

$$4\mu_1 - \mu_2 - \mu_3 - \mu_4 - \mu_5$$

Interpret this contrast. Use $\alpha = 0.05$ for constructing the confidence interval.

12.21. Three varieties of corn and four fertilizers were studied in a factorial experiment. The yields (in bushels per acre) were recorded for two subplots exposed to each of the $3 \times 4 = 12$ treatment combinations. The results were as shown in the table.

Fertilizer \ Corn Variety	1	2	3
1	56 47	26 35	39 42
2	67 70	45 60	57 61
3	95 90	92 88	91 82
4	92 85	96 99	98 93

a. Write the analysis of variance model for this experiment.
b. Conduct the appropriate tests of the interaction effects and the main effects (corn variety and fertilizer), if necessary.

12.22. A one-factor, completely randomized design experiment produced the following results: $t = 5$, $n = 4$, $SST = 17.5$, $SSTR = 10$, $\bar{x}_{1.} = 2.5$, $\bar{x}_{2.} = 3$, $\bar{x}_{3.} = 4.5$, $\bar{x}_{4.} = 4$, $\bar{x}_{5.} = 3.5$.

a. Complete the analysis of variance table for these data.
b. Determine the approximate p-value for testing the hypothesis of equal treatment means.
c. Use the Scheffe method of multiple comparisons to test the following contrasts (use $\alpha = 0.05$):

i. $\mu_1 - \mu_2 = 0$
ii. $\mu_1 + \mu_2 - \mu_3 - \mu_4 = 0$
iii. $2\mu_5 - \mu_1 - \mu_2 = 0$

Simple linear regression and correlation

13

Linear regression analysis is a technique used to estimate the value of one quantitative variable by considering its relationship with one or more other quantitative variables. For example, if we know the relationship between height and weight in American adult males, we can use regression analysis to predict weight given a particular value for height.

The relationship between height and weight is familiar to each of us; generally, the taller a person is, the more he weighs. Another example of a familiar relationship is crop yield and the amount of fertilizer applied to the land; the more fertilizer applied to the land, the greater the yield—to a point. If too much fertilizer is applied, the crop will be killed by the fertilizer chemicals; the land will be "burned." An important relationship in business is between the allocation of dollars to advertising effort and the level of sales of a product; the more money expended in advertising, in general, the higher the level of sales. We will develop in this chapter a model for predicting expected sales levels given a fixed amount expended for advertising, for example.

Regression analysis began in the late 1800s, when Sir Francis Galton, an English expert on heredity, was studying the relationship between the heights of fathers and the heights of their sons. He found that there was a tendency for short fathers to have sons of about average height (sons taller than fathers), and fathers with above average height to have sons with heights toward the average (sons were shorter than their fathers). Galton wrote that the heights of sons "regressed" or reverted to the average, thus originating the term *regression*.

The basic idea behind relating the behavior of one quantitative variable to one or more other quantitative variables should not be new to us. Indeed, it is the rare individual who has not heard or used the term *correlation* in making such statements as: "There is definitely an inverse relationship (negative correlation) between how I study and the grade I receive in that class!" While one would hope that effort is positively correlated with exam and other performances, the relationship is *rarely* a perfect one; that is, a host of other factors undoubtedly influence performance in a class and on an examination. We will study the nature of this relationship and will derive quantitative indices that express the strength of the relationship between variables. A description or study of the *nature* of the relationship between variables will be called *regression analysis,* while an investigation of the *relative strength* of such a relationship will be termed *correlation analysis.*

There are generally seven steps to be followed in constructing a regression model, although some steps may occasionally be omitted or even be expanded on in any regression analysis. For example, we may find that the variables to be included in the regression model and the manner in which they interact have been pre-specified for us, and it is our task to estimate the model parameters and check that the assumptions of the model have been met. The steps followed in the model-building and estimation process are given below.

☐ Step 1   Identify the variables to be included in building the
            regression model

In Step 1, we identify the dependent variable and the independent variable(s) we will use in constructing an estimating equation. In this context, the variable that is used to predict the variable of interest will be called the *independent variable,* and the variable of interest will be called the *dependent variable.* Often the independent variables to be used in predicting the dependent variable will be obvious. For example, when we wish to estimate the sales of a product, we may examine the relationship between sales and advertising level, or between sales and the size of the population served, etc., and select the model that we feel comes closest to estimating values of the dependent variable. At other times, we may find that the regression analysis requires that a great deal of time be spent in identifying independent variables, or variables that can be used to estimate values of the dependent variable. We generally will not know until Step 4 is completed whether it will be necessary to determine other independent variables to use in the estimating equation.

☐ Step 2   Collect sample data

More will be said about this step as we work our way through the development of a regression model. Suffice it at this point to mention that the method used for data collection can have an impact on whether the assumptions underlying the use of the simple linear regression model have been met. Frequently, for example, data that have been collected over time violate one of the assumptions of the simple linear regression model. When data are collected over time, it is especially important to check whether one or more of the assumptions underlying the use of the simple linear regression model have been violated.

☐ Step 3 Specify the relationship that exists between the dependent and the independent variable

In Step 3, we specify the nature of the relationship between the dependent variable and the independent variable. For example, is the relationship linear, or should nonlinear terms be included in the model? Frequently, a plot of the data points collected in Step 2 will provide us with clues to the form of the relationship between the variables of interest. We will be concerned in this chapter with developing models in which the presumed relationship between the variables is linear, and in which *one* independent variable is used to estimate the value of the dependent variable. Such analysis is called *simple linear regression and correlation analysis—simple* because only one predictor or independent variable will be used in the estimating equation, and *linear* because of the linear manner in which the parameters of the regression model enter into the estimating equation.

Certainly it is possible to find that the variable of interest is generally related to more than one predictor or independent variable, or that the relationship is not linear. An example of this is the level of sales of a product and the advertising expenditure. The sales level of a product generally "depends" on more predictor variables than advertising expenditure alone. However, we will often find that quite good predictions are possible based on a single independent variable. Multiple linear regression and correlation, involving an analysis of the relationship among several quantitative variables, will be discussed in Chapter 15.

☐ Step 4 Estimate parameters of the model specified

In this step, we will estimate the parameters of the population regression model using sample data, since the population regression function relating the two variables will rarely be known in practice. In this chapter, we will obtain point estimates of many of the regression parameters, and in Chapter 14 we will do hypothesis tests and construct confidence interval estimates of the parameters point estimated in this chapter. In this context, it will be necessary for us to specify the sampling distribution of the point estimators of the population regression parameters.

☐ Step 5 Determine whether the assumptions of the simple linear regression model have been met

Although the determination of whether the underlying assumptions of the model have been met is a complex task, we will find that the construction of *residual plots* is often quite helpful in an initial determination of whether the assumptions have been met, or whether it will be necessary to identify new variables to be included in the model or to specify a different functional form for the manner in which the variables interact. This may lead us back to Step 1 and through the whole process once more if we discover that the model developed is not appropriate or that the underlying assumptions have been violated.

☐ **Step 6   Statistically test the usefulness of the model developed**

In this step, we determine at a stated level of significance whether the model developed or the terms included in the model are statistically significant. This is a part of the hypothesis-testing and interval estimation process, so we will defer a discussion of this step until Chapter 14.

☐ **Step 7   Use the model developed for prediction and estimation**

Once we have developed the simple linear regression model, we will want to use the model for estimating or predicting values of the dependent variable. This step emphasizes one difference (there are others, which include the underlying assumptions for using the technique) between simple linear correlation and simple linear regression. In simple linear correlation, we are interested in specifying the *strength* of the linear relationship between the variables involved, whereas in simple linear regression analysis, we are interested in developing the *form* of the relationship and *deriving* an estimating equation for predicting values of the dependent variable. Thus, in the sales-advertising example, if we are interested in assessing whether there is a strong linear relationship between these variables, then correlation analysis may be appropriate. If we are interested in specifying or estimating the form of the relationship and developing an equation to predict values of the dependent variable, then simple linear regression analysis may be preferred. We will develop further the distinction between correlation and regression in the pages that follow, but first we turn our attention to the nature of a statistical relationship between variables.

■ **13.2   RELATIONSHIPS BETWEEN VARIABLES**

If a *functional relationship* exists between two variables, then it is possible to represent the relationship by a formula $y = f(x)$, where x is the independent or the predictor variable, and y is the dependent or the predicted variable.

Example 13.1   Suppose that for every unit of a product sold, a company makes a profit of $3. Let x = number of units sold and y = total profit. Then, $y = 3x$. Illustrate this linear relationship.

Solution   This linear functional relationship is illustrated in Figure 13.1A. For example, if $x = 10$, $y = 3(10) = \$30$; if $x = 30$, $y = 3(30) = \$90$; and if

$x = 50$, $y = 3(50) = \$150$. Notice that all three of the pairs (x, y) of points fall exactly on a straight line.

In Example 13.1, the functional relationship is *linear*. An example of a *nonlinear* functional relationship is $y = x^2$. If, for example, $x = 2$, then $y = 4$. A graph of the functional relationship $y = x^2$ for $x > 0$ is illustrated in Figure 13.1B.

FIGURE 13.1

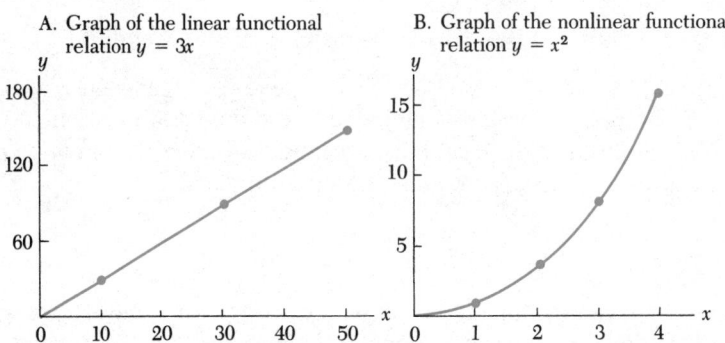

A. Graph of the linear functional relation $y = 3x$

B. Graph of the nonlinear functional relation $y = x^2$

In a *statistical* relationship, the variables are not *perfectly* related as they are in a *functional* relationship. The pairs of points (x, y) will not all lie perfectly on the curve representing the relationship between the variables.

Example 13.2 An example of a *statistical* relationship is the relationship between dollars spent on advertising and sales of a product. Table 13.1 contains sales and advertising figures for a retail sales store in Niwot, Colorado, and several of its branches. Plot these data as a graph similar to Figure 13.1.

Solution These data are plotted in Figure 13.2.[1] Clearly, the greater the number of dollars spent on advertising, the greater the general sales level. But, the relationship is not a perfect one, as is evident in Figure 13.2. A straight line in Figure 13.2 has been drawn to fit reasonably well through the ten points, and the points are scattered about this line. The scattering of points suggests that some of the variation in sales is not accounted for by advertising expenditures alone. The variation in sales not accounted for by advertising dollars may be considered to be random in nature, but may also be due to the failure to include other important independent predictor variables (price, general economic conditions, location, etc.). The randomness of the scattering of points about the fitted line is an important element in

---

[1] This plot was derived by using a "canned" computer program, which will be used throughout the discussion of correlation and regression analysis. Most large-scale regression and correlation analyses require the use of a computer for performing the myriad computations involved, and most universities and colleges have available large-scale computers and "canned" statistical programs for student use. For *simple* linear correlation and regression analysis, a computer is *not* necessary, although it can serve as a worthwhile check of the computations performed.

TABLE 13.1 Sales and dollars expended on advertising for a retail store in Niwot, Colorado, and nine of its branches

Advertising expenditures x ($100)	Sales y ($1000)
18	55
7	17
14	36
31	85
21	62
5	18
11	33
16	41
26	63
29	87

Source: Company records.

FIGURE 13.2 Plot of data in Table 13.1

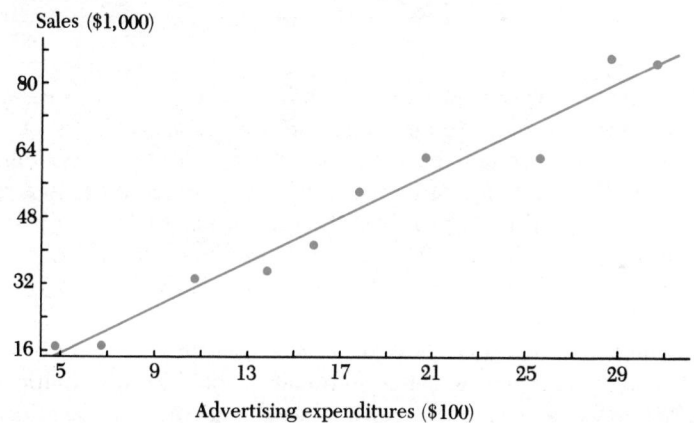

assessing the validity of a regression model. Before explaining how we fit a line such as displayed in Figure 13.2 to the data, we must first describe the regression model and the assumptions necessary for its correct application.

13.3 THE SIMPLE LINEAR REGRESSION MODEL

The simple linear regression model is a mathematical way of stating the statistical relationship between two variables. The two principal elements of this statistical relationship are: (1) the tendency of the dependent random variable **Y** to vary in a systematic way with the independent variable x, and (2) the scattering of points about the "curve" that represents the relationship

414

between x and $\mathbf{Y}$. These two elements of a statistical relationship are represented in the simple linear regression model by assuming that: (1) there is a probability distribution of the random variable $\mathbf{Y}$ for *each value of x,* and (2) the means of these conditional probability distributions fall on a straight line. These two assumptions are illustrated in Figure 13.3 for the Example 13.2

FIGURE 13.3 Graphical form of the simple linear regression model

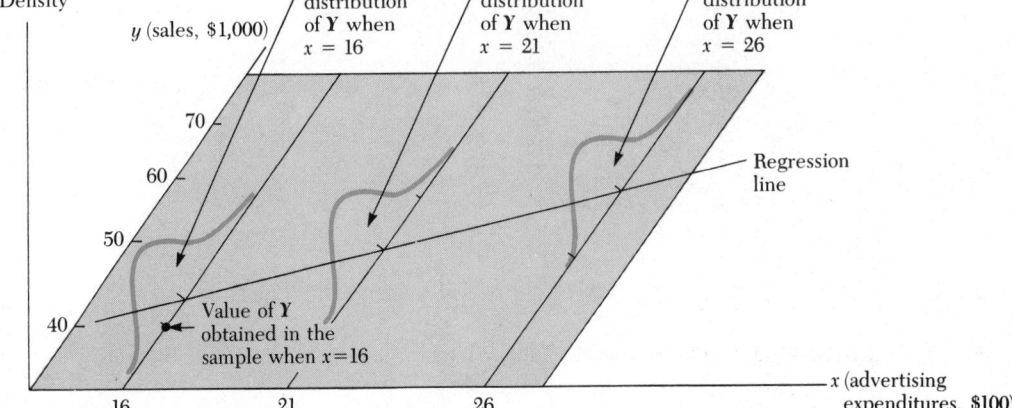

data. The systematic way in which the random variable $\mathbf{Y}$ varies as a function of x is identified as a straight line, the *regression line of* $\mathbf{Y}$ *on x.* The regression line goes through the means of the conditional probability distributions of $\mathbf{Y}$, given a value of x. The data are collected by taking random samples from the conditional probability distributions of $\mathbf{Y}$ for given values of x. For example, from Table 13.1, when $x = 16$, the random variable $\mathbf{Y}$ was observed to be 41. This particular value of the random variable $\mathbf{Y}$ represents a *random sample of size one* drawn from the conditional probability distribution of $\mathbf{Y}$ when $x = 16$. This value of $\mathbf{Y}$ is shown in Figure 13.3 as being somewhat below the mean of the conditional probability distribution.

There are two ways in which we can acquire the needed sample information as given in Table 13.1: by *experimentation* and by *survey.* To generate the sample data experimentally, we would *select* a set of values of x and for each, randomly sample one or more values of $\mathbf{Y}$. Alternatively, we could generate the sample data by taking a survey. For example, we could randomly sample ten branches of the firm in Example 13.2 to determine their advertising expenditures and corresponding sales. The survey method has the disadvantage that we must take whatever values of x occur in the survey; the selection of the set of values, the independent variable, is out of our control. We might be so unfortunate, for instance, to find that *all* ten sales branches in our survey had advertising expenditures of $1,600—we ideally want a spread of x-values over the range of interest to us and over which the regression line will be built.

In regression analysis, it is always better to produce the sample data by experimentation, if possible, for then we can control the independent variable x—the experiment can be designed to suit our needs. When experimentation is not possible, surveys must be used to generate the data.

In discussing regression analysis, we will assume that the values of x, the independent variable, are fixed or are given in advance. This gives rise to the notation Y and x, where the dependent variable Y is a random variable, and the independent variable x is *not* a random variable, but is fixed. This relationship is shown in Figure 13.3—there is a conditional distribution of Y-values *for each* value of x. When discussing correlation in Section 13.9, it will be necessary for us to assume that both Y and X are random variables, and that their joint distribution is bivariate normal as depicted in Figure 7.15. This distinction is more fully developed in Section 13.9. For the present, we will assume that the independent variable x is not a random variable in our discussion of simple linear regression.

The formal statement of the simple linear regression model is given here.

Population regression model

$$Y_i = \beta_0 + \beta_1 x_i + \epsilon_i, \qquad i = 1, 2, \ldots, N$$

where:

$Y_i = i$th dependent random variable
$\beta_0, \beta_1 =$ parameters in the regression model
$x_i = i$th level of the independent variable
$\epsilon_i =$ random error term
$N =$ number of possible values of the independent variable x

---

The assumptions corresponding to the use of the population regression model are given in the following box.

Assumptions underlying the use of the
simple linear regression model

1. For the ith level of x, x_i, the expected value of the error component is 0 $[E(\epsilon_i) = 0]$, and the variance of the error component is σ^2 $[V(\epsilon_i) = \sigma^2]$ and is *constant* for all i, where $i = 1, 2, \ldots, N$.
2. The error components ϵ_i, ϵ_j between *any pair* of values of the dependent variable y_i, y_j are uncorrelated.
3. β_0 and β_1 are typically unknown and must be estimated from the sample data.

The consequences of these assumptions are as follows:

1. The observed value of the random variable Y_i (i.e., y_i) when $x = x_i$ is the sum of two components, a constant and a *value of* a random variable:

$$y_i = \underbrace{\beta_0 \qquad + \qquad \beta_1 x_i}_{\text{constant}} + \underbrace{\epsilon_i}_{\substack{\text{value of random} \\ \text{variable } \epsilon_i}}$$

2. By using the expectation operator rules given in Chapter 5,

$$E(Y_i) = E(\beta_0 + \beta_1 x_i + \epsilon_i) = \beta_0 + \beta_1 x_i + E(\epsilon_i)$$

Since $E(\epsilon_i) = 0$,

$$E(Y_i) = \beta_0 + \beta_1 x_i$$

Thus, the mean of the conditional probability distribution of Y_i given a value of $x = x_i$ and denoted by $\mu_{Y/x}$ (we omit the i subscript here) is equal to $\beta_0 + \beta_1 x$. Therefore, the regression function corresponding to the regression model is:

$$E(Y) = \mu_{Y/x} = \beta_0 + \beta_1 x$$

3.
$$\begin{aligned} V(Y_i) &= V(\beta_0 + \beta_1 x_i + \epsilon_i) \\ &= V(\beta_0 + \beta_1 x_i) + V(\epsilon_i) = 0 + V(\epsilon_i) \end{aligned}$$

Since $V(\epsilon_i) = \sigma^2$,

$$V(Y_i) = \sigma^2$$

Thus, the variance of the conditional probability distribution of Y_i given $x = x_i$ and denoted by $\sigma^2_{Y/x}$ is equal to σ^2, *and each conditional probability distribution has the same variance, σ^2.*

4. The observed value of Y_i, y_i, when $x = x_i$, is larger or smaller than $\mu_{Y/x}$ by the amount ϵ_i, the value of the error component, ϵ_i.

5. By the second assumption, the outcome when $x = x_i$ neither is affected by nor does itself affect the outcome for any other value of x, x_j. Using the covariance operator given in Chapter 5, this assumption implies:

$$C(\epsilon_i, \epsilon_j) = 0 \qquad \text{for all } i \text{ and } j, \ i \neq j$$

Regression coefficients—parameters β_0 and β_1 in the simple linear regression model

Two parameters in the simple linear regression model, β_0 and β_1, are called the *regression coefficients*. The coefficient β_0 is called the *Y-intercept*; it is the value of **Y** (according to the regression line) when $x = 0$. The coefficient β_1 is the *slope* of the regression line; its numerical value gives the change in the dependent random variable **Y** (either positive or negative) when there is a one-unit *increase* in the value of the independent variable, x.

Example 13.3 Suppose that in Example 13.2, the appropriate population regression model is:

$$Y_i = 1.00 + 2.75 x_i + \epsilon_i$$

Then, the regression function is given by $E(Y_i) = 1.00 + 2.75x_i$. Graph this regression function.

Solution This regression function is shown in Figure 13.4. When $x = 11$, $\mu_{Y/x=11} = 1.00 + 2.75(11) = 31.25$; when $x = 21$, $\mu_{Y/x=21} = 58.75$; and when $x = 31$, $\mu_{Y/x=31} = 86.25$. The regression function passes through the condition means $\mu_{Y/x} = 1.00 + 2.75x$. Furthermore, each conditional probability distribution has the same variance σ^2. And, the error component is the deviation between the observed value of Y and $\mu_{Y/x}$. For example, when $x = 31$, from Table 13.1, the observed value of Y is 85; $\mu_{Y/x=31} = 86.25$. The error component is $\epsilon_i = y_i - \mu_{Y/x=31} = 85 - 86.25 = -1.25$ as shown in Figure 13.4.

FIGURE 13.4 Simple linear regression model

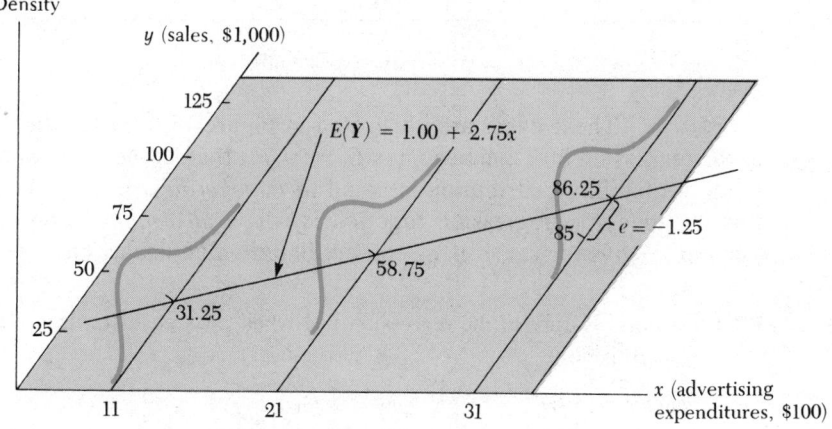

Example 13.4 In Example 13.3, we assumed that the correct regression function relating sales to advertising is $E(Y) = 1.00 + 2.75x$. What are the values of the regression coefficients in this regression function?

Solution This function is graphed in Figure 13.5. The Y-intercept, $\beta_0 = 1.00$, is the value of the regression function when $x = 0$. The slope, β_1, tells us how much sales are *expected* to increase for each dollar increase in advertising expenditure. Since advertising expenditures x are expressed in $100s and sales Y are expressed in $1,000s, the regression coefficient $\beta_1 = 2.75$ indicates that sales are *expected* to increase by $2,750 for each additional $100 spent on advertising.

Example 13.5 The value of β_1, the slope, is particularly important in determining the nature of the regression function. If β_1 is positive, the regression line is *increasing* as the value of x increases. If β_1 is negative, the regression line is *decreasing* as x increases. If $\beta_1 = 0$, the regression line is horizontal (constant) for all values of x. Graph regression lines where $\beta_1 > 0$, where $\beta_1 < 0$, and where $\beta_1 = 0$.

FIGURE 13.5 Graph of the regression function $E(Y) = 1.00 + 2.75x$

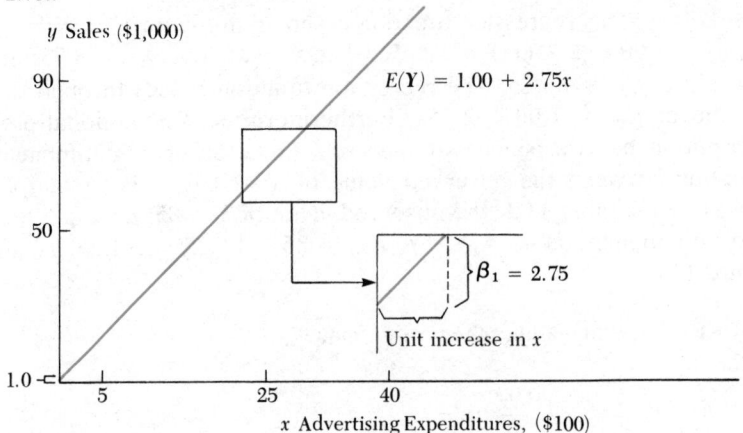

Solution   These cases are illustrated in Figure 13.6. When the slope β_1 is 0, the regression function becomes $E(Y) = \beta_0$; that is, the mean of the conditional probability distribution is equal to β_0 *regardless* of the value of x. In this instance, the regression function is *not at all helpful* in predicting the value of Y given a value of x—the predicted value is *always* β_0.

FIGURE 13.6   Nature of the regression line when β_1 is positive, negative, and 0

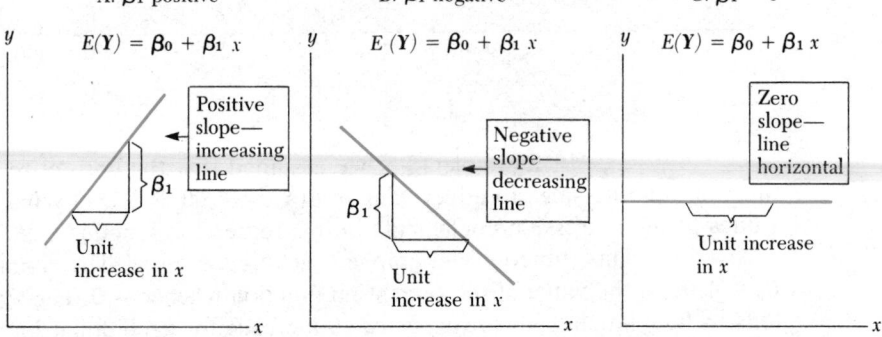

Since β_0 and β_1 will generally not be known in a particular problem, we now turn to a discussion of how a regression line is "fitted" to a set of (x, y) sample data points.

13.4   THE FITTING OF A SIMPLE LINEAR REGRESSION MODEL

Since β_0 and β_1 are generally not known in a regression problem, they must be *estimated* from sample data taken on the dependent random variable Y for a number of values of the independent variable x. These pairs of sample values are obtained either by experimentation or by survey. The data given in Table 13.1 were determined by experimentation—ten branch stores were selected because of the varying amounts they spent on advertising, and their sales amounts and corresponding advertising expenditures were recorded. Each sales branch in this example is called a *unit of association*, because each branch included in the sample provides us with a value of the independent variable x (advertising expenditures) and a corresponding value of the dependent variable Y (sales). It is important to preserve this pairwise identity of (x, y) data pairs as we estimate the population regression line (function) from the sample data.

In Figure 13.7, we have superimposed two "fitted" regression lines through the scatter of points given in Table 13.1 and Figure 13.2,

FIGURE 13.7

A. Scatter diagram and fitted line
   $\hat{y} = 49.7 + 0x$ for data in Table 13.1

B. Scatter diagram and fitted line
   $\hat{y} = 1.02 + 2.73x$ for data in Table 13.1

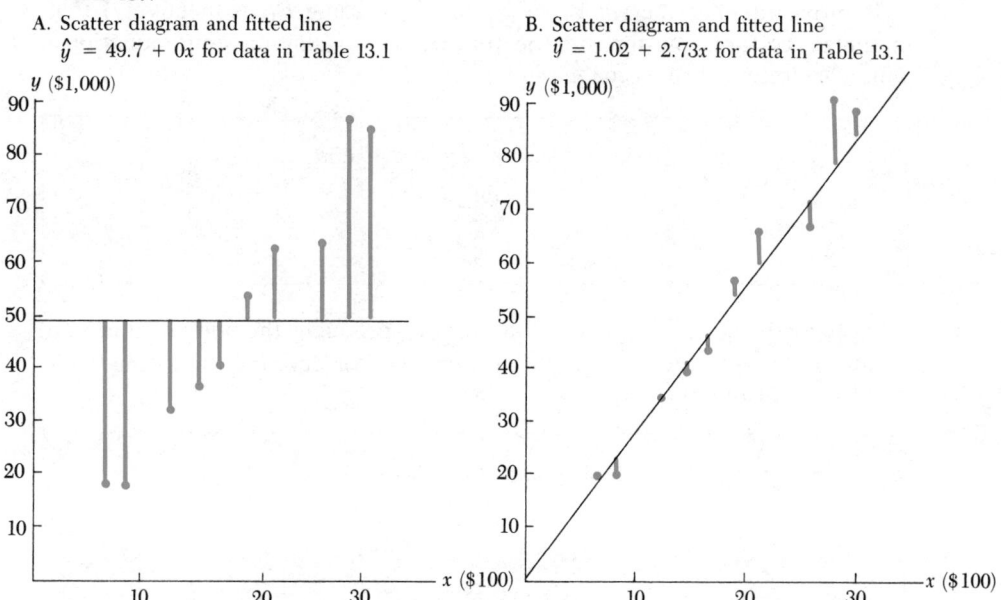

$\hat{y} = 49.7 + 0x$ and $\hat{y} = 1.02 + 2.73x$, respectively. It is apparent in Figure 13.7 that the line $\hat{y} = 1.02 + 2.73x$ fits the given data "better," but we must establish a criterion to evaluate when one line is "better" than another so that we can find the best-fitting line. The criterion we shall use is called *least squares*. For each sample observation pair (x_i, y_i), the least squares (LS) criterion considers the deviation of y_i from its expected value:

$$y_i - E(Y_i) = y_i - (\beta_0 + \beta_1 x_i) = \epsilon_i$$

and requires that *values* of β_0 and β_1 be found that minimize:

$$LS = \sum_{i=1}^{n} (y_i - \beta_0 - \beta_1 x_i)^2 = \sum_{i=1}^{n} \epsilon_i^2$$

This criterion is equivalent to minimizing the sum of the squared vertical "distances," shown by the vertical red lines, in Figure 13.7.

The specific *values* of β_0 and β_1 that minimize LS are the regression coefficient *estimates,* denoted by b_0 and b_1, respectively.[2] Thus, the least squares criterion requires that we find a line, denoted by $\hat{Y} = \mathbf{b}_0 + \mathbf{b}_1 x$, such that the sum of the squared *vertical* deviations between the line and the scatter of points is minimized. In Figure 13.7A, the vertical deviations corresponding to the line $\hat{y} = 49.7 + 0x$, where $b_0 = 49.7$ and $b_1 = 0$, are indicated. Obviously, the line $\hat{y} = 1.02 + 2.73x$ in Figure 13.7B where $b_0 = 1.02$ and $b_1 = 2.73$, does much better in the least squares sense because its vertical deviations from the scatter of points, when squared and summed, will be less than the sum of the squared deviations from the line $\hat{y} = 49.7 + 0x$.

It turns out that the *values* of $\mathbf{b}_0$ and $\mathbf{b}_1$ (b_0 and b_1, respectively) that minimize LS are solutions to the following two simultaneous equations, which are referred to as the *normal equations:*

Least squares normal equations

$$\Sigma y_i = nb_0 + b_1 \Sigma x_i$$

$$\Sigma x_i y_i = b_0 \Sigma x_i + b_1 \Sigma x_i^2$$

Solving the normal equations for b_0 and b_1 produces the point estimates of β_0 and β_1, respectively.[3] Computing formulas for deriving the estimates b_0 and b_1 are as follows:

Computation formulas for least squares estimates b_0 and b_1

$$b_1 = \frac{\Sigma x_i y_i - \dfrac{(\Sigma x_i)(\Sigma y_i)}{n}}{\Sigma x_i^2 - \dfrac{(\Sigma x_i)^2}{n}} \qquad (13.1)$$

$$b_0 = \bar{y} - b_1 \bar{x}, \qquad \text{where } \bar{y} = \frac{\Sigma y_i}{n}, \quad \bar{x} = \frac{\Sigma x_i}{n} \qquad (13.2)$$

---

[2] Following the notation of previous chapters, we will refer to the random variable estimators of the population regression coefficients β_0 and β_1 by $\mathbf{b}_0$ and $\mathbf{b}_1$, respectively. The sample values obtained will be denoted by nonboldface b_0 and b_1. Similarly, the point estimators of Y_i and ϵ_i will be denoted by $\hat{Y}_i$ and $\mathbf{e}_i$, respectively, and the point estimates obtained from a specific sample will be denoted by $\hat{y}_i$ and e_i.

[3] Minimizing the least squares function LS with respect to β_0 and β_1 produces the point estimates b_0 and b_1. Partial differential calculus is required to perform this minimization. See the Draper and Smith reference at the end of the chapter for a derivation of the normal equations.

Example 13.6 Steps 1 through 3 in the regression model process have been performed for us using the sales-advertising data given in Table 13.1. That is, the dependent and the independent variables have been identified, a scatter plot of the data given in Figure 13.2 suggests the reasonableness of the linear model, and the sample data have been collected as shown in Table 13.1. Using the data provided, fit a simple linear regression model using sales as the dependent variable and advertising expenditures as the independent variable.

Solution   The summations required for determining b_0 and b_1 are given in Table 13.2. The format in this table provides a convenient worksheet for finding the necessary components in the formulas for b_0 and b_1.

$$b_1 = \frac{\Sigma x_i y_i - \dfrac{(\Sigma x_i)(\Sigma y_i)}{n}}{\Sigma x_i^2 - \dfrac{(\Sigma x_i)^2}{n}} = \frac{10{,}820 - \dfrac{(178)(497)}{10}}{3{,}890 - \dfrac{(178)^2}{10}} = 2.7347 \doteq 2.73$$

$$b_0 = \bar{y} - b_1 \bar{x} = \frac{497}{10} - (2.7347)\left(\frac{178}{10}\right) = 1.02$$

Thus, the fitted regression line is $\hat{y} = 1.02 + 2.73x$, and this is the best-fitting line based on the least squares criterion.

TABLE 13.2   Computation worksheet for determining b_0 and b_1 in Example 13.6*

Observation	x	y	x^2	y^2	xy	$\hat{y}$	$y - \hat{y}$	$(y - \hat{y})^2$
1	18	55	324.0	3,025.0	990.0	50.247	4.753	22.591
2	7	17	49.0	289.0	119.0	20.165	−3.165	10.015
3	14	36	196.0	1,296.0	504.0	39.308	−3.308	10.942
4	31	85	961.0	7,225.0	2,635.0	85.799	−0.799	0.638
5	21	62	441.0	3,844.0	1,302.0	58.451	3.549	12.594
6	5	18	25.0	324.0	90.0	14.695	3.305	10.922
7	11	33	121.0	1,089.0	363.0	31.104	1.896	3.596
8	16	41	256.0	1,681.0	656.0	44.777	−3.777	14.269
9	26	63	676.0	3,969.0	1,638.0	72.125	−9.125	83.266
10	29	87	841.0	7,569.0	2,523.0	80.329	6.671	44.498
Totals	178	497	3,890.0	30,311.0	10,820.0		0.000	213.332

* The quantities in the column labeled $\hat{y}$ were computer-generated and may differ *slightly* (because of roundoff error) from those determined by use of the equation $\hat{y} = 1.02 + 2.73x$.

The column headed $\hat{y}$ in Table 13.2 gives the *estimated* value of **Y** for each value of x based on the fitted regression line. For example, when $x = 21$, $\hat{y} = 1.02 + (2.73)(21) = 58.45$; that is, for advertising expenditures of 21($100), the estimated sales figure is $58,450. The *actual* value of sales in the sample for an advertising expenditure of $2,100 was $62,000. The error $e_5 = y_5 - \hat{y}_5 = 62 - 58.45 = 3.55$ (in $1000s) represents the estimation error— these errors are given in the next to the last column in Table 13.2. It can be shown mathematically that $\Sigma_{i=1}^{n}(y_i - \hat{y}_i) = \Sigma_{i=1}^{n} e_i = 0$, *always*. Thus, calculating this column in Table 13.2 and showing that it sums to 0 (or nearly 0, allowing for some computational roundoff error) provides an excellent check that b_0 and b_1 have been correctly determined. The quantities headed y^2 and $(y - \hat{y})^2$ are not used in the computation of least squares estimates b_0 and b_1, but are necessary in determining other regression quantities to be described shortly.

The following theorem establishes the properties of the least squares point estimators $\mathbf{b}_0$ and $\mathbf{b}_1$.

Theorem 13.1

Gauss-Markov Theorem (minimum variance, unbiased estimators of population regression coefficients β_0 and β_1)

When the assumptions of the simple linear regression model are satisfied, the least squares estimators $\mathbf{b}_0$ and $\mathbf{b}_1$ are unbiased and have minimum variance among *all* linear unbiased estimators of β_0 and β_1.

Thus, we know from this theorem that $E(\mathbf{b}_0) = \beta_0$ and $E(\mathbf{b}_1) = \beta_1$—the estimators are unbiased. Furthermore, the theorem assures us that the estimators $\mathbf{b}_0$ and $\mathbf{b}_1$ are minimum variance estimators among the class of unbiased and linear estimators. Thus, over this class of estimators, the sam-

FIGURE 13.8 Sampling distributions of $\mathbf{b}_0$ and $\mathbf{b}_1$

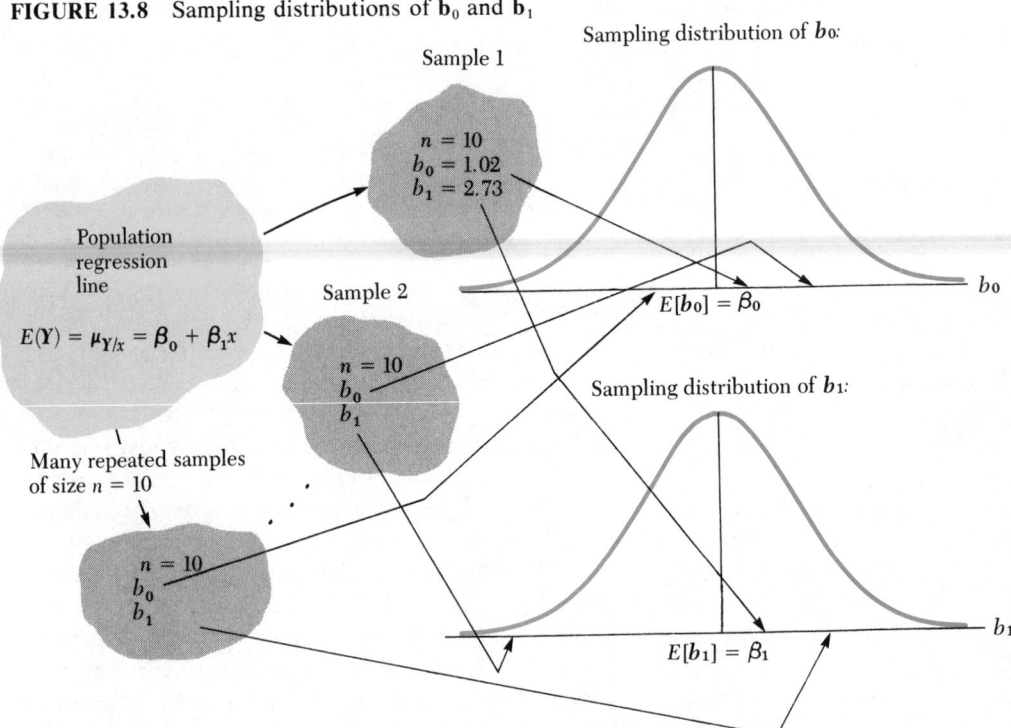

pling distributions of $\mathbf{b}_0$ and $\mathbf{b}_1$ have less variability than do the sampling distributions of other point estimators of β_0 and β_1. The properties of LS estimators $\mathbf{b}_0$ and $\mathbf{b}_1$ are summarized in Figure 13.8. The sampling distribution of $\mathbf{b}_1$, for instance, has a mean of β_1 [since $E(\mathbf{b}_1) = \beta_1$], and its variability is less than that of the sampling distribution of any other linear and unbiased estimator of β_1.

■ 13.5 THE ESTIMATOR OF THE CONDITIONAL PROBABILITY DISTRIBUTION VARIANCE σ^2

We assume in the simple linear regression model that the conditional probability distributions of $\mathbf{Y}$, given a value of x, have the same variance, σ^2. When drawing inferences concerning the fitted regression line, it will be necessary to estimate σ^2 from sample data.

Recall that when we have a single population with an unknown variance σ^2, we use the sample variance

$$s^2 = \sum_{i=1}^{n} \frac{(x_i - \bar{x})^2}{n - 1}$$

as the point estimate of σ^2. The process of forming a point estimate s^2 of σ^2 is reviewed in Figure 13.9.

FIGURE 13.9   Point estimate of σ^2

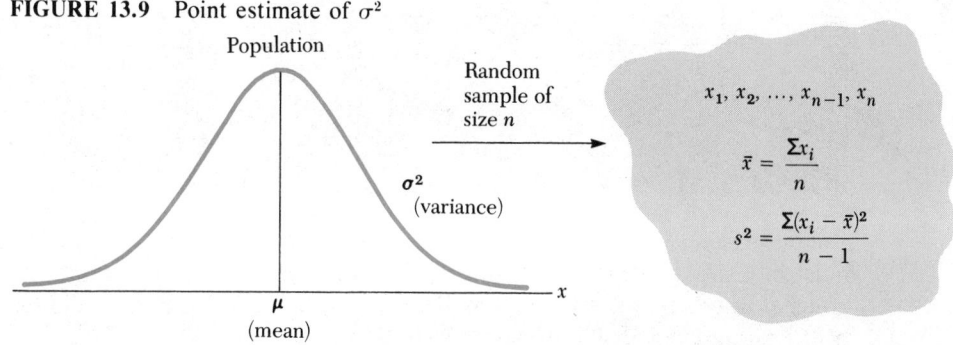

The divisor of $(n - 1)$ reflects that one degree of freedom is lost since the estimate $\bar{x}$ is used in place of the unknown population mean μ in the computation of s^2.

In regression analysis, we have random observations on $\mathbf{Y}$, but they come from *two or more* population distributions as illustrated in Figure 13.10. The number of population distributions from which we sample depends on how many different levels of x we have in the experiment or survey; in Figure 13.10 it is assumed that we have n different levels of x.

To form the point estimator of σ^2, we use the same basic form as the estimate of σ^2 based on a random sample taken from a single population.

FIGURE 13.10 Conditional probability distributions of **Y**

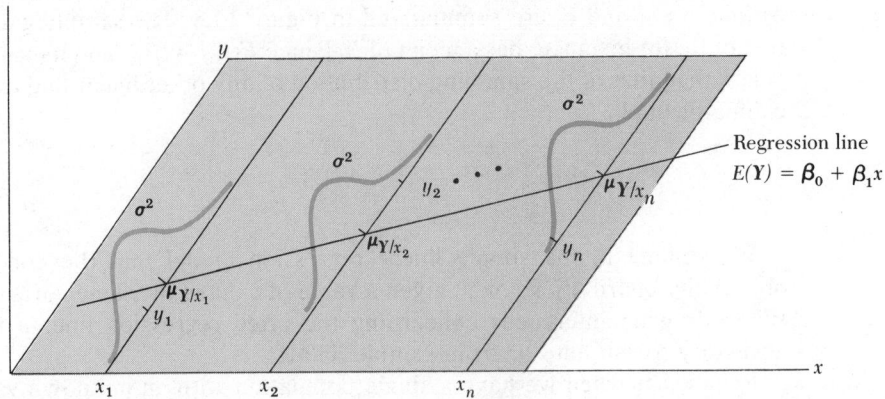

Since the conditional means μ_{Y/x_i} are unknown (as was μ in the single-population case), we will estimate the population variance σ^2 using the predicted value of **Y**, given the value of x. The deviations we will use, therefore, are $y_i - \hat{y}_i = y_i - (b_0 + b_1 x_i)$; for each value of x, $\hat{y}_i = b_0 + b_1 x_i$ estimates the conditional mean μ_{Y/x_i}. Thus, the estimate of σ^2 is given by the following formula (the square root of this quantity is called the *standard error of the estimate* and is given by $s_{Y/x}$).

Estimate of the conditional distribution variance

$$s_{Y/x}^2 = \sum_{i=1}^{n} \frac{(y_i - \hat{y}_i)^2}{n-2} = \sum_{i=1}^{n} \frac{[y_i - (b_0 + b_1 x_i)]^2}{n-2} \qquad (13.3)$$

The reason we divide by $(n - 2)$ in equation (13.3) is that two parameters must be estimated in determining $s_{Y/x}^2$: namely, β_0 and β_1. Hence, two degrees of freedom are lost in estimating σ^2 because b_0 and b_1 are used in place of the unknown population parameters β_0 and β_1.

Two commonly used alternate (computing) formulas for calculating $s_{Y/x}^2$ are given now.

Conditional distribution variance estimate—computing formula no. 1

$$s_{Y/x}^2 = \frac{1}{n-2}\{\Sigma y_i^2 - b_0 \Sigma y_i - b_1 \Sigma x_i y_i\} \qquad (13.4)$$

$S_{Y/x}$ is often called the *standard error of the estimator*. It is a measure of the *absolute* fit of the sample data points to the regression line. This is to be contrasted to the coefficient of determination described in the next section,

Conditional distribution variance estimate—computing formula no. 2

$$s_{Y/x}^2 = \frac{1}{n-2} \left\{ \left[\Sigma y_i^2 - \frac{(\Sigma y_i)^2}{n} \right] - \frac{\left[\Sigma x_i y_i - \frac{(\Sigma x_i)(\Sigma y_i)}{n} \right]^2}{\Sigma x_i^2 - \frac{(\Sigma x_i)^2}{n}} \right\} \quad (13.5)$$

which measures the *relative* goodness of fit of the sample regression line to the sample data points.

Although equation (13.5) looks most imposing for computing $s_{Y/x}^2$, it is usually the most accurate and fastest to use—remember, certain quantities in equation (13.5) have already been calculated in the determination of b_0 and b_1, as the following example illustrates.

Example 13.7   Let us return to the sales-advertising data given in Table 13.2. Using these data and equation (13.3), determine $s_{Y/x}^2$.

Solution   The last column in Table 13.2 gives the squared error deviations, $(y_i - \hat{y}_i)^2$. These deviations may be used to calculate $s_{Y/x}^2$ from equation (13.3). The quantity, $\Sigma_{i=1}^{n}(y_i - \hat{y})^2$, in equation (13.3) is called the *sum of squares due to the errors* and will be abbreviated by SSE. From Table 13.2, SSE = 213.332; therefore,

$$s_{Y/x}^2 = \frac{\sum\limits_{i=1}^{n} (y_i - \hat{y}_i)^2}{n-2} = \frac{SSE}{n-2} = \frac{213.332}{10-2} = 26.67$$

Thus, the point estimate of σ^2 is $s_{Y/x}^2 = 26.67$.

This estimate can also be determined by using equation (13.5) and the data given in Table 13.2:

$$s_{Y/x}^2 = \frac{1}{n-2} \left\{ \left[\Sigma y_i^2 - \frac{(\Sigma y_i)^2}{n} \right] - \frac{\left[\Sigma x_i y_i - \frac{(\Sigma x_i)(\Sigma y_i)}{n} \right]^2}{\Sigma x_i^2 - \frac{(\Sigma x_i)^2}{n}} \right\}$$

$$= \frac{1}{10-2} \left\{ \left[30,311 - \frac{(497)^2}{10} \right] - \frac{\left[10,820 - \frac{(178)(497)}{10} \right]^2}{3,890 - \frac{(178)^2}{10}} \right\}$$

$$= \frac{213.332}{8} = 26.67$$

The terms in braces in equations (13.4) and (13.5) are simply alternate expressions for SSE. Equation (13.4) should be used with care—be certain to use plenty of decimal places for b_0 and b_1 in this formula, or else rather large roundoff errors may occur.

When regression data are generated by experimentation, it is often possible to test statistically the hypothesis that the variances of the n conditional probability distributions are all equal to σ^2 at a stated level of significance using analysis of variance techniques from Chapter 12. Often, plotting the sample data can serve as a "quick" guide in making judgments on the equality of variances. For example, a histogram plot of the data similar to that shown in Figure 13.10 for each level of x would indicate whether there was serious violation of the equality of variance assumption. In Section 13.7, we will describe the use of residual plots in visually testing whether the variances indeed do all appear to be equal.

■ 13.6 THE COEFFICIENT OF DETERMINATION r^2

To this point, we have not dealt with the *relative strength* of the linear relationship between the independent variable x and the dependent variable $\mathbf{Y}$. One measure of the relative strength of the linear relationship between x and $\mathbf{Y}$ is given by the *coefficient of determination*, $\mathbf{r}^2$.

The coefficient of determination $\mathbf{r}^2$ gives the proportion of variability in the sample dependent variable $\mathbf{Y}$ that is explained by the independent variable x through the fitting of the regression line. It is a point estimate of the *population coefficient of determination*, ρ^2. We will now describe how the sample statistic $\mathbf{r}^2$ measures the proportion of variability explained in the sample.

The variation in the observations on the dependent variable $\mathbf{Y}$ is measured in terms of the deviations $y_i - \bar{y}$: we square these deviations, where $i = 1, 2, \ldots, n$, and then sum them. If we divide this sum by $(n - 1)$, we have the sample variance of the random variable $\mathbf{Y}$:

$$s_Y^2 = \frac{\sum_{i=1}^{n} (y_i - \bar{y})^2}{n - 1} = \frac{\sum y_i^2 - \dfrac{(\sum y_i)^2}{n}}{n - 1}$$

Consider the *numerator* of s_Y^2: it is the total sum of squared deviations of the n observations about their mean, $\bar{y}$. We will denote the expression in the numerator of s_Y^2 by SST—*sum of squares total*. Although we typically use s_Y^2 to measure the variation in $\mathbf{Y}$, SST certainly can measure the variation in $\mathbf{Y}$ as well. For the sales-advertising data, the value of SST is 5,610.10. The individual deviations from $\bar{y}$ before they are squared are shown in Figure 13.7A, where $\bar{y} = 49.7$.

$$
\begin{aligned}
\text{SST} = \sum_{i=1}^{n} (y_i - \bar{y})^2 &= (5.3)^2 + (-32.7)^2 + (-13.7)^2 + (35.3)^2 + (12.3)^2 \\
&+ (-31.7)^2 + (-16.7)^2 + (-8.7)^2 + (13.3)^2 + (37.3)^2 \\
&= 5,610.10
\end{aligned}
$$

Figure 13.7B illustrates the regression line fitted to the sales-advertising data, and the computation of SSE, the error sum of squares, is given in Table 13.2 by $\sum_{i=1}^{10}(y_i - \hat{y}_i)^2 = 213.33$. The deviations in Figure 13.7B represent the unexplained deviations in the random variable $\mathbf{Y}$ *after* the sample regression line has been fitted. These deviations, in magnitude, are considerably less than those in Figure 13.7A *before* the regression line was fitted. As we have observed, the regression line does appear to account for considerable variability in **Y**, since there appears to be a rather strong linear relationship between x and **Y**.

The sample coefficient of determination, r^2, is a simple ratio of the amount of variation explained by the regression line to the total variation in **Y**. Its value is given by:

$$\text{Sample coefficient of determination}$$
$$r^2 = \frac{\text{SST} - \text{SSE}}{\text{SST}} \qquad (13.6)$$

For the sales-advertising example,

$$r^2 = \frac{5,610.10 - 213.33}{5,610.10} = 0.962$$

That is, 96.2 percent of the variability in **Y** (sales) has been accounted for by relating it to x (advertising) through the regression line. This means that we have fitted a very strong (relative) linear relationship between sales and advertising expenditures *for these sample data.*

The value of r^2 is bounded between 0 and 1. This can be shown as follows. The total deviation $(\mathbf{Y}_i - \overline{\mathbf{Y}})$ can be partitioned as:

$$(\mathbf{Y}_i - \overline{\mathbf{Y}}) = (\mathbf{Y}_i - \hat{\mathbf{Y}}_i) + (\hat{\mathbf{Y}}_i - \overline{\mathbf{Y}})$$

| total deviation | error deviation | deviation of estimator about the mean |

The last term in this partitioning, $(\hat{\mathbf{Y}}_i - \overline{\mathbf{Y}})$, is required to make the statement an identity; that is, if we add $(\mathbf{Y}_i - \hat{\mathbf{Y}}_i)$ and $(\hat{\mathbf{Y}}_i - \overline{\mathbf{Y}})$, we get $(\mathbf{Y}_i - \overline{\mathbf{Y}})$. This last term represents the deviation of the estimator of the random variable about the mean. These deviations are shown in Figure 13.11 as broken lines. For example, for $y_1 = 55$,

$$(55 - 49.7) = (55 - 50.2) + (50.2 - 49.7)$$

Remarkably, the sums of these squared deviations also bear this same relationship (we will not prove this):

$$\Sigma(\mathbf{Y}_i - \overline{\mathbf{Y}})^2 = \Sigma(\mathbf{Y}_i - \hat{\mathbf{Y}}_i)^2 + \Sigma(\hat{\mathbf{Y}}_i - \overline{\mathbf{Y}})^2$$

| SST | SSE | SSR |

The last term in this expression is called the *sum of squares due to regression,* or **SSR** for short. Since they are sums of squares, SST ≥ 0 and SSE ≥ 0. Also, SST $\geq$ SSE (they could be equal if SSR = 0). These facts

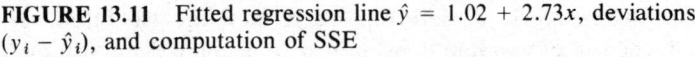

FIGURE 13.11 Fitted regression line $\hat{y} = 1.02 + 2.73x$, deviations $(y_i - \hat{y}_i)$, and computation of SSE

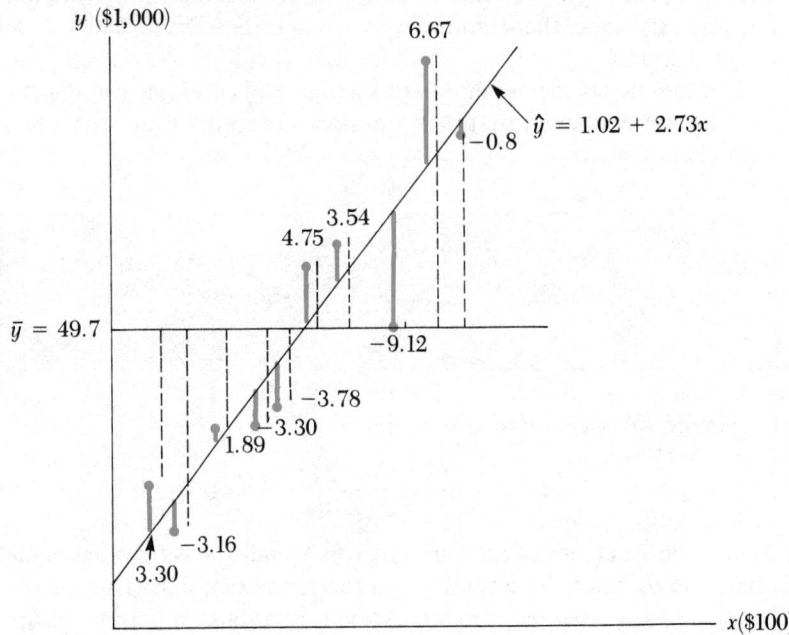

guarantee that the ratio (**SST** − **SSE**)/**SST** can never be less than 0 or greater than 1; that is, $0 \leq \mathbf{r}^2 \leq 1$.

If $r^2 = 0$, then SSE = SST; the error sum of squares is equal to the total sum of squares. In this instance, the regression line has done nothing to reduce the variability in **Y**. If $r^2 = 1$, then SSE = 0; the sample regression line has explained *all* the variability in the sample values. These two cases are illustrated in Figure 13.12. If $r^2 = 1$, the observations fall *perfectly* on the fitted regression line. If $r^2 = 0$, the fitted regression line has a slope of 0.

In our sales-advertising example, we determined that r^2 was 0.96—this indicates that we have fit a very strong linear relationship to the sample data points. It is unusual to experience a value of $\mathbf{r}^2$ that is this close to 1 in practice. In most business problems, for example, we are very fortunate to

FIGURE 13.12 Cases when $r^2 = 0$ and $r^2 = 1$

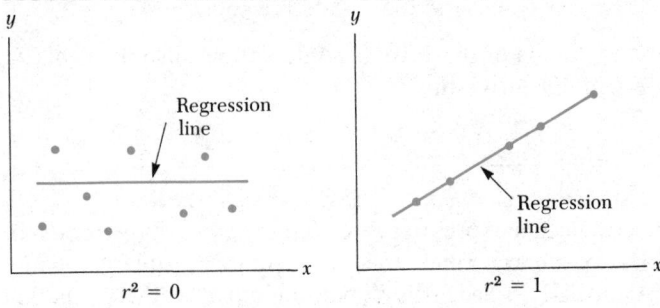

account for as much as 50 percent of the variability in a typical dependent random variable **Y** in our sample. The reason for this is that most often **Y** represents a very complex variable in business applications of regression analysis, and it is not easy to account for its variability through one or even more than one independent variable.

We complete this section by giving an alternate formula for r^2 and verifying that it does produce the same value as determined earlier for the sales-advertising example.

Computation formula for coefficient of determination

$$r^2 = \frac{\left[\Sigma x_i y_i - \frac{(\Sigma x_i)(\Sigma y_i)}{n} \right]^2}{\left[\Sigma x_i^2 - \frac{(\Sigma x_i)^2}{n} \right] \left[\Sigma y_i^2 - \frac{(\Sigma y_i)^2}{n} \right]} \qquad (13.7)$$

For the sales-advertising illustration, the value of r^2 is calculated with this formula by using the data in Table 13.2. Notice that using this formula for computing the value of r^2 justifies the inclusion of the y^2 column in Table 13.2, a column that we have not used up to this point.

$$r^2 = \frac{\left[10,820 - \frac{(178)(497)}{10} \right]^2}{\left[3,890 - \frac{(178)^2}{10} \right] \left[30,311 - \frac{(497)^2}{10} \right]} = 0.96$$

■ **13.7 A DISCUSSION OF THE ASSUMPTIONS IN REGRESSION ANALYSIS**

In this section, we will discuss and review the assumptions underlying the use of the simple linear regression model. Through an examination of *residual plots,* we will complete Step 5 in the development of a regression model—determining whether the underlying assumptions have been met. To summarize, the assumptions required in simple linear regression analysis are:

1. $E(\epsilon_i) = 0$ and $V(\epsilon_i) = \sigma^2$, for all i, where $1 \leq i \leq N$.
2. The error components ϵ_i, ϵ_j are uncorrelated; i.e., $C(\epsilon_i, \epsilon_j) = 0$, $i \neq j$.
3. β_0 and β_1 are parameters. They, and the values of x, are assumed to be *constants.*
4. The error components ϵ_i are normally distributed, where $1 \leq i \leq N$.

The last assumption is *not* required in calculating point estimates of β_0, β_1, σ^2, etc. However, this assumption *is* required if we wish to calculate confidence intervals or to construct hypothesis-testing decision rules for β_0, β_1, etc., as explained in the next chapter.

The key assumptions (1, 2, and 4) focus on the error components or *residuals,* ϵ_i, where $i = 1, 2, \ldots, N$. The quantity ϵ_i is often referred to as a *residual* since it represents the residual deviation after the regression line has been fitted ($\epsilon_i = Y_i - \hat{Y}_i$). A plot of the sample residuals for the sales-advertising data is given in Figure 13.13. This figure was derived by a

FIGURE 13.13 Residual plot of sales-advertising data regression analysis

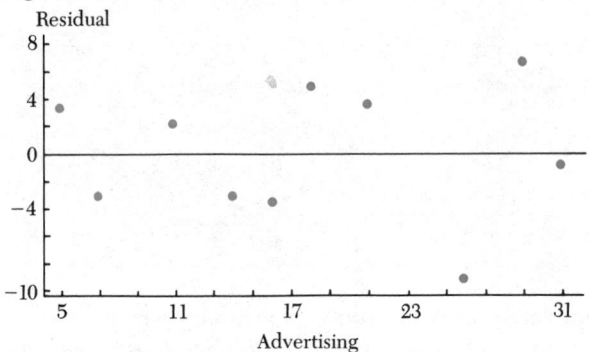

computer using one of the available computer software packages, SAS (Statistical Analysis System).

Can we assess whether or not the three key assumptions made in regression analysis are satisfied from this plot of ten sample residuals? The residuals *always* sum to 0 in the sample and this is consistent with the assumption $E(\mathbf{e}_i) = 0$. The assumption that $V(\mathbf{e}_i) = \sigma^2$ (sample data) requires that each conditional probability distribution of **Y** given x has the same variance, σ^2. To estimate the variance σ^2 in a *particular* conditional distribution, we would require *at least* two residuals since the sample variance formula has a divisor of $(n - 1)$. If we do have repeated observations for given values of x, we can use the set of variance estimates to determine if, indeed, they are estimating the same parameter, σ^2. This procedure will not be described here, but can be found in the Neter and Wasserman reference at the end of this chapter.

The second assumption can be checked using techniques presented in Chapter 14 to test the hypothesis H_0: $\rho = 0$, where ρ represents the population correlation coefficient between pairs of residuals.

The fourth assumption requires that the residuals are sampled from a normal population with a mean of 0 and a variance of σ^2. Techniques to check this assumption are discussed in Chapter 19.

There are, then, specific techniques for checking whether or not the assumptions underlying the use of regression are satisfied based on the set of *sample residuals.* There is, however, a very simple and quick way to determine whether it is *reasonable* to make these assumptions in a particular problem—it is called a *residual plot* like the one shown in Figure 13.13 for the sales-advertising example. In Figure 13.14 there are four prototype residual plots.

FIGURE 13.14   Four prototype residual plots

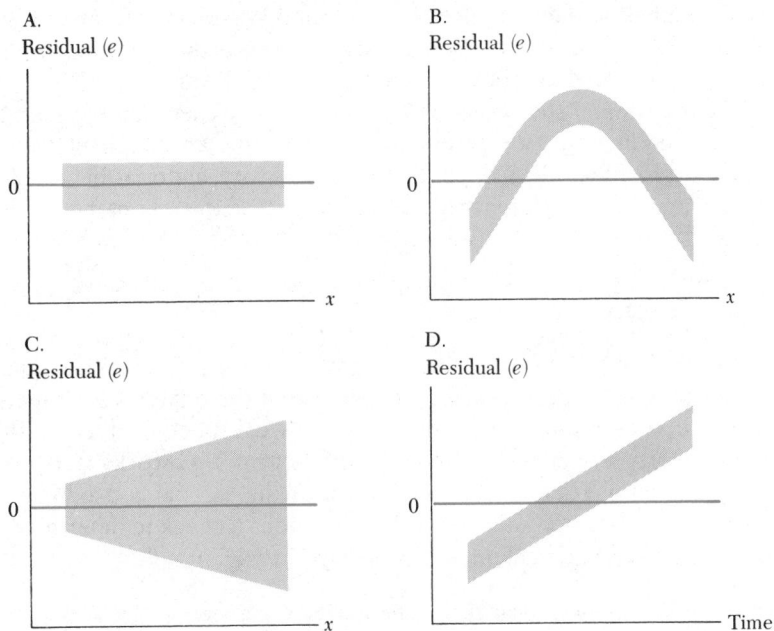

Plot A indicates that the sample data support making the three key assumptions. In a plot of the residuals against *x*, the residuals fall within a rectangle about 0. Plot B is an indication that the regression relationship is probably curvilinear, not linear. Plot C is an indication that the conditional probability distributions of **Y**, given different values of *x*, do not have the same variance σ^2. In this plot, the variability in the conditional distribution increases as *x* increases.

If it is possible to plot the residuals as a function of time, then it may be possible to determine whether or not the error terms are correlated through time. When error terms are correlated, it is often through time, and a plot such as D in Figure 13.14 can identify this situation. In plot D, at the beginning of the experiment, the residuals are negative, whereas at the end of the experiment, the residuals are positive—this systematic relationship of the residuals with time would certainly lead to correlated error terms and the failure of the second regression analysis assumption to hold.

The residual plot in Figure 13.13 appears to be somewhat patterned after plot A in Figure 13.14, an indication that the assumptions underlying the use of the regression model have been met in this example.

What if there is an indication that one or more assumptions are not satisfied? Regression analysis is relatively *robust* to nonnormality of the errors. By this, we mean that modest departures from normality by the residuals will not seriously affect the analysis. Correlated errors or a nonconstant error variance σ^2 are more serious, however. In these cases, corrective measures should be taken and these are usually based on transformations on the dependent variable Y. See the Neter and Wasserman reference at the end of this chapter for an excellent presentation of transformations of Y to correct for correlated errors or a nonconstant error variance σ^2.

The analysis of the residuals in a regression experiment is a fundamental step in conducting the regression analysis. For an excellent treatment of analyzing residual plots beyond the introductory material here, refer to the Draper and Smith reference listed at the end of this chapter.

■ 13.8   A CASE STUDY IN SIMPLE LINEAR REGRESSION MODEL DEVELOPMENT

The analysis of a simple linear regression model is complex, and a great deal of care must be exercised to ensure that the analysis is properly done. From a computational standpoint, using a worksheet as given in Table 13.3 simplifies matters considerably. The totals from this worksheet provide the necessary numbers for calculating most of the estimates used in regression analysis. Furthermore, the worksheet provides a check to determine whether b_0 and b_1 have been calculated properly: $\Sigma(y - \hat{y}) = 0$.

TABLE 13.3   Simple linear regression model worksheet

x	y	x^2	y^2	xy	$\hat{y}$	$y - \hat{y}$	$(y - \hat{y})^2$
x_1	y_1						
x_2	y_2				Filled in *after* fitting the regression line $\hat{y} = b_0 + b_1 x$		
.	.						
.	.						
.	.						
x_n	y_n						
Totals Σx	Σy	Σx^2	Σy^2	Σxy	$\Sigma \hat{y}$	$\Sigma(y - \hat{y})$	$\Sigma(y - \hat{y})^2$

In the following example, we will trace the development of a simple linear regression model through an examination of the underlying assumptions. This example should test your understanding of the steps involved in developing a simple linear regression model and provide you with experience in determining the regression model point estimates (Step 4).

Example 13.8   R. Griffin, a highly-trained accountant, has been given the responsibility by his supervisor in the accounting firm, Ef Lundgren, to estimate federal and state income taxes for individuals for the coming year to serve as a guideline in testing the reasonableness of the computed taxes of its client customers. Ef would like some type of an estimating equation, as well as an indication of the coefficient of determination for the model developed, and an estimate of the conditional distribution standard deviation. He further would like R. to comment on the model developed and whether it appears to fit the underlying regression assumptions.

Solution *Step 1:* *Identify the variables to be included in building the regression model.* Although the income taxes paid by an individual are influenced by a host of factors that include adjusted gross income, unusual medical and other deductions, number of dependents, and other sources of income, R. decides to build a simple regression model as a first step to estimate this tax liability. The obvious choice for the independent variable in this example is the adjusted gross income of the individual, since the model is initially to include only one predictor variable.

Step 2: *Collect sample data.* R. Griffin randomly samples ten of the accounting firm's previous clients and records each client's adjusted gross income and the corresponding federal and state tax liability. The client sampled in this instance is the unit of association, and the data were collected by survey rather than by experimentation. These data are reproduced in Table 13.4.

TABLE 13.4 Survey of client adjusted gross income and tax liability

Client no.	Adjusted gross income	Tax liability
1	$18,945	$ 5,683
2	16,420	4,926
3	13,945	3,765
4	12,554	3,012
5	20,627	7,013
6	28,420	11,223
7	15,626	4,375
8	13,396	3,617
9	17,200	5,504
10	24,319	8,511

Step 3: *Specify the relationship that exists between the dependent and the independent variable.* R. Griffin obtains a scatter diagram of the data sampled and concludes that a linear relationship appears reasonable for these two variables. The scatter diagram constructed is given in Figure 13.15.

Step 4: *Estimate parameters of the model specified.* A worksheet similar to Table 13.3 is given in Table 13.5 for estimating the regression parameters. These estimates are given by:

$$b_1 = \frac{1{,}160{,}943{,}000 - \dfrac{(181{,}452)(57{,}629)}{10}}{3{,}524{,}642{,}000 - \dfrac{(181{,}452)^2}{10}} = 0.496$$

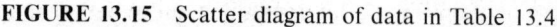

FIGURE 13.15 Scatter diagram of data in Table 13.4

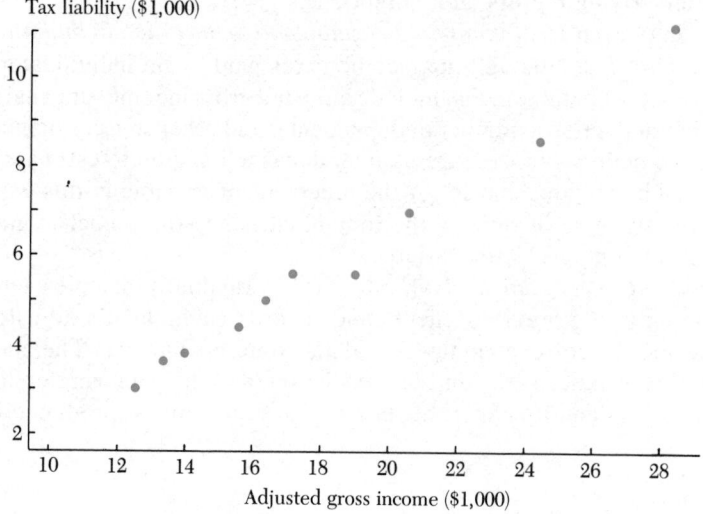

$$b_0 = \frac{57,629}{10} - (0.496)\left(\frac{181,452}{10}\right) = -3,245.15$$

$$\hat{y} = -3,245.15 + 0.496x$$

Furthermore, r^2 and $s_{Y/x}$ are determined to be (verify!):

$$r^2 = 0.990 \qquad s_{Y/x} = 267.82$$

Based on the high value of r^2, R. Griffin is quite pleased with the model developed, since only one independent variable is used to estimate the value of the dependent variable (tax liability).

   Step 5: Determine whether the assumptions of the simple linear regression model have been met.   A residual plot of the data is given in Figure 13.16. This plot is patterned somewhat after plot C in Figure 13.14. As the value of the independent variable x increases, it appears that the variability of the conditional distributions increases also. We might suspect this; it appears reasonable that as adjusted gross income increases, so too would the variability in tax liabilities.

   Step 5 can be one of the most important steps in developing a regression model, since, as the data show, it may be possible to develop a simple linear regression model with a high value of r^2 and a very respectable value of $s_{Y/x}$ for estimation purposes, yet the underlying assumptions of applying the model developed are not satisfied, thus decreasing the applicability of what otherwise would appear to be a very successful model. It may be possible that the model developed is still quite satisfactory for our purposes, but we should not be led into believing the model is any better than it really is. Before applying the model in any meaningful context, we should assure

TABLE 13.5 Simple linear regression model worksheet for data in Table 13.4

Trial	x	y	x^2	y^2	xy	$\hat{y}$	$y - \hat{y}$	$y - \hat{y}^2$
1	18,945	5,683	358,913,000	32,296,480	107,664,400	6,159.95	−476.95	227,484.20
2	16,420	4,926	269,616,300	24,265,470	80,884,910	4,906.43	19.57	382.84
3	13,945	3,765	194,463,000	14,175,220	52,502,910	3,677.74	87.26	7,614.61
4	12,554	3,012	157,602,900	9,072,144	37,812,640	2,987.19	24.81	615.66
5	20,627	7,013	425,473,000	49,182,160	144,657,100	6,994.97	18.03	325.13
6	28,420	11,223	807,696,300	125,955,700	318,957,500	10,863.75	359.25	129,063.30
7	15,626	4,375	244,171,800	19,140,620	68,363,740	4,512.26	−137.26	18,839.71
8	13,396	3,617	179,452,800	13,082,680	48,453,320	3,405.19	211.81	44,862.88
9	17,200	5,504	295,839,900	30,294,010	94,668,800	5,293.66	210.34	44,242.85
10	24,319	8,511	591,413,700	72,437,120	206,979,000	8,827.84	−316.84	100,385.00
Totals	181,452	57,629	3,524,642,000	389,901,300	1,160,943,000		0.03	573,816.00

FIGURE 13.16 Residual plot of regression model developed in Example 13.8

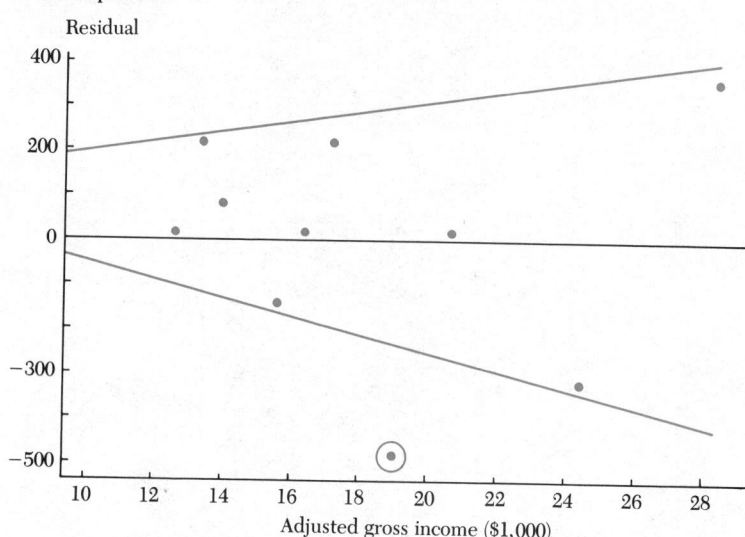

ourselves that the model developed is well suited to the uses to which we intend to put it.

Step 6: Statistically test the usefulness of the model developed. We will defer an analysis of the usefulness of the model developed until Chapter 14, when we learn hypothesis-testing and confidence interval estimation procedures for the simple linear regression parameters.

Step 7: Use the model developed for prediction and estimation. The end result of the simple linear regression analysis is a satisfactory regression model that will allow us to estimate or predict a value of the dependent random variable **Y** given a value of the independent variable x. If we have satisfied ourselves that the underlying assumptions of the simple linear regression model have been met, then we can use the model for estimation and prediction. For example, given a client of the accounting firm with an adjusted gross income of $19,500, we would point estimate her combined state and federal income tax liability to be:

$$\hat{y} = -3,245.15 + 0.496(19,500) = \$6,426.85$$

using the model developed.

Often a fitted regression model is used for extrapolation—estimating a value of **Y** for a value of x that is not within the range considered in the experiment. In general, this is a *very risky* use of regression and should be done with a great deal of caution. Just because a simple linear regression model does a good job of estimating values of the random variable **Y** when the values of x are within the range of values used in model development does not necessarily mean it will do a satisfactory job of estimating values of

Y when values of x are outside the range of those values used for model development. It may be, for example, that the true relationship between x and **Y** is nonlinear, but within the range of x-values examined, a straight line does an excellent job of modeling the behavior of the random variable. As a case in point, we would estimate the tax liability of a person with $0 adjusted gross income to be $-$3,245.15 using the model developed. Although the negative tax liability could be viewed as a transfer payment (the government subsidizes the individual with $0 income), this is well outside the meaning of the model considered. It is more likely that the true relationship between adjusted gross income and tax liability is nonlinear, but is well approximated between the levels of adjusted gross income investigated—$12,554 to $28,420—by a straight line. This situation is illustrated in Figure 13.17.

FIGURE 13.17 Curvilinear relationship and fitted linear regression function

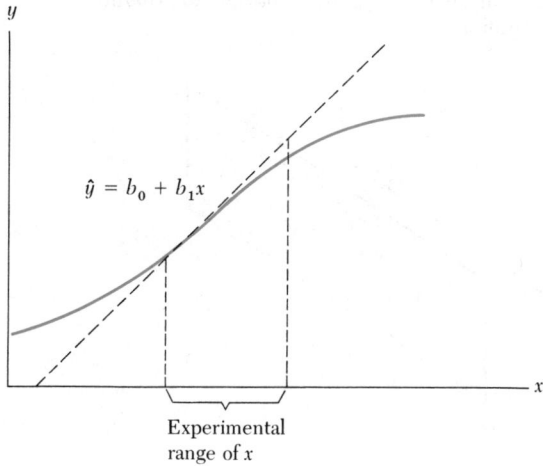

Experimental
range of x

■ 13.9 CORRELATION

We turn now to a study of simple *linear* correlation, a second measure of the relative strength of the relationship between variables. When we speak of using correlation in the analysis of data sets, it suggests to most of us that we wish to study the degree of association between variables in the set. Most correlation methods, therefore, are closely related to regression analysis. In fact, we will show that the most commonly used measure of correlation is directly related to the slope and to the coefficient of determination in simple linear regression. The coefficient of correlation we will develop in this chapter is called *Pearson's product-moment correlation coefficient* and is denoted by ρ for the population and by r (the estimate of ρ) when using a sample data set to estimate ρ. Pearson's product-moment correlation coefficient is only one of several measures of the *relative strength* of the *linear* association between variables. In Chapter 20, we will consider another measure of correlation—*Spearman's rank correlation coefficient*.

The point estimator of ρ—the sample correlation coefficient, **r**—applies regardless of the joint distribution of **X** and **Y** in the population. But, if we wish to test hypotheses or set confidence intervals on ρ, we must assume that **X** and **Y** follow the *bivariate normal distribution* discussed in Section 7.5 and illustrated in Figure 7.15. In this section, we will, therefore, assume that (x, y) pairs are from a bivariate normal distribution. This means, of course, that *both* **X** and **Y** are now random variables. In regression analysis, **X** is often treated as a random variable when data are obtained by survey rather than by experimentation. That is, if units of association are selected randomly yielding (x, y) data pairs, and we exercise no control over fixing the values of x in our study, then **X** is a random variable.

If we slice the bivariate normal distribution along either axis, the "slice" is a univariate normal distribution as illustrated in Figure 13.18. By forming

FIGURE 13.18 "Slice" through the bivariate normal distribution—univariate normal distribution

Density

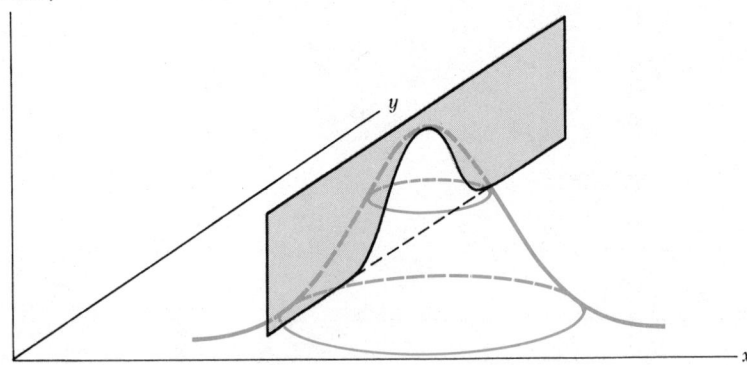

FIGURE 13.19 Contours of the bivariate normal distribution

Density

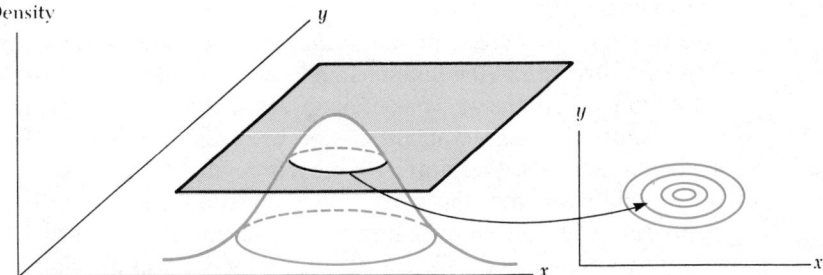

slices that are parallel to the (x, y) plane, we form *contours*—slices as if we were looking down from above the "bell" as illustrated in Figure 13.19. On the left of the figure is shown one possible contour "slice" through the "bell." By repeating this process, we can form a set of contours that gives us in two dimensions a "feeling" for the nature of the three-dimensional sur-

FIGURE 13.20   Four representative contour sets for ρ values

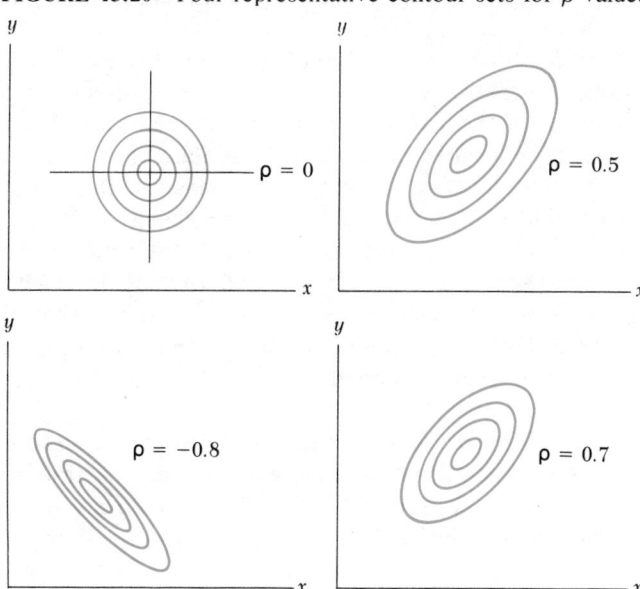

face. In Figure 13.20 are shown four representative contour sets for values of the population correlation coefficient, ρ. As the contours concentrate about their axis, ρ increases in magnitude.

In most practical situations, the value of the population correlation coefficient will not be known and will, therefore, have to be estimated from sample data. The process of forming a point estimate using sample data is illustrated in Figure 13.21. A sample of n (x, y) data pairs are selected from the popula-

FIGURE 13.21   Relationship between ρ and point estimate r

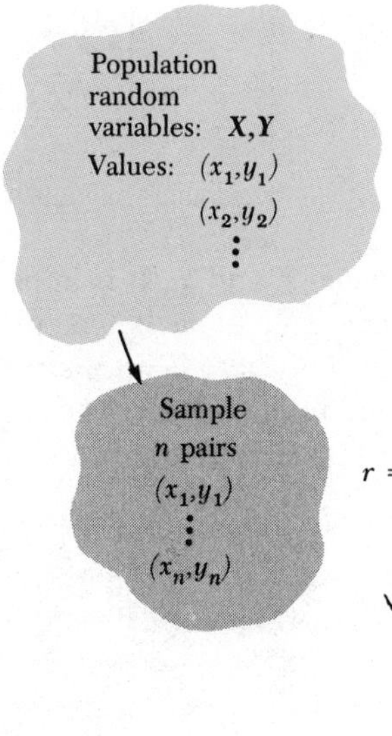

Population random variables: X, Y
Values: (x_1, y_1)
(x_2, y_2)
$\vdots$

$$\rho = \frac{E([X - \mu_X][Y - \mu_Y])}{\sigma_X \, \sigma_Y} = \frac{C[X,Y]}{\sigma_X \, \sigma_Y}$$

Sample
n pairs
(x_1, y_1)
$\vdots$
(x_n, y_n)

$$r = \frac{\left(\displaystyle\sum_{i=1}^{n} (x_i - \bar{x})(y_i - \bar{y}) \right) \left(\dfrac{1}{n-1} \right)}{\sqrt{\dfrac{\displaystyle\sum_{i=1}^{n} (x_i - \bar{x})^2}{n-1}} \sqrt{\dfrac{\displaystyle\sum_{i=1}^{n} (y_i - \bar{y})^2}{n-1}}} = \frac{s_{XY}}{s_X s_Y}$$

tion, and these values are used to form a point estimate, r, of the population correlation coefficient.

One measure of the relationship between variables discussed in Chapter 5 is their covariance, a measure of how the variables "co-vary." Unfortunately, this measure of the linear association between variables is highly influenced by the units used to measure the variables. For this reason, we "standardize" the covariance of two random variables so that the standardized value is not influenced by the units of measurement. This standardization is accomplished by dividing the covariance of the random variables X and Y, $C(X,Y)$, by σ_X and σ_Y in the case of ρ, and dividing the point estimate of $C(X,Y)$, s_{XY}, by point estimates of σ_X and σ_Y, s_X and s_Y, respectively, in the case of a sample. This process yields values of the random variable $\mathbf{r}$, the sample correlation coefficient, bounded by -1 and $+1$, as illustrated in Figure 13.21.

As r approaches $+1$ or -1, the pairs of points must fall closer to a straight line; as r approaches 0, the points show a scatter that demonstrates no linear relationship. Examples of scatter plots with corresponding values of $\mathbf{r}$ are illustrated in Figure 13.22. Notice in particular plot F—$r = 0$, but there is a

FIGURE 13.22 Scatter diagrams with values of r

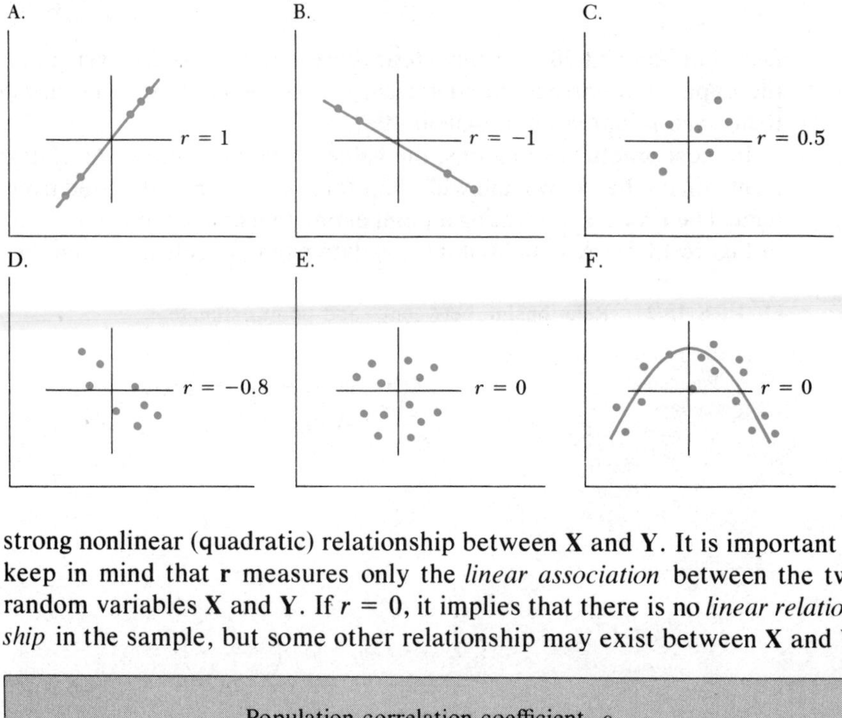

A. $r = 1$

B. $r = -1$

C. $r = 0.5$

D. $r = -0.8$

E. $r = 0$

F. $r = 0$

strong nonlinear (quadratic) relationship between X and Y. It is important to keep in mind that $\mathbf{r}$ measures only the *linear association* between the two random variables X and Y. If $r = 0$, it implies that there is no *linear relationship* in the sample, but some other relationship may exist between X and Y!

Population correlation coefficient, ρ

$$\rho = \frac{E[(X - \mu_X)(Y - \mu_Y)]}{\sigma_X \sigma_Y} = \frac{C(X, Y)}{\sigma_X \sigma_Y}$$

Sample correlation coefficient, r (point estimate of ρ)

$$r = \frac{\left[\sum_{i=1}^{n}(x_i - \bar{x})(y_i - \bar{y})\right]\left(\dfrac{1}{n-1}\right)}{\sqrt{\dfrac{\sum_{i=1}^{n}(x_i - \bar{x})^2}{n-1}}\sqrt{\dfrac{\sum_{i=1}^{n}(y_i - \bar{y})^2}{n-1}}} = \frac{s_{XY}}{s_X s_Y}$$

$$= \frac{\left[\sum_{i=1}^{n}(x_i - \bar{x})(y_i - \bar{y})\right]}{\sqrt{\sum_{i=1}^{n}(x_i - \bar{x})^2}\sqrt{\sum_{i=1}^{n}(y_i - \bar{y})^2}}$$

Computing formula for r, the sample correlation coefficient

$$r = \frac{\sum_{i=1}^{n} x_i y_i - \dfrac{\left(\sum_{i=1}^{n} x_i\right)\left(\sum_{i=1}^{n} y_i\right)}{n}}{\sqrt{\sum_{i=1}^{n} x_i^2 - \dfrac{\left(\sum_{i=1}^{n} x_i\right)^2}{n}}\sqrt{\sum_{i=1}^{n} y_i^2 - \dfrac{\left(\sum_{i=1}^{n} y_i\right)^2}{n}}} \qquad (13.8)$$

Experience in determining the sample correlation coefficient is provided in the following four examples.

Example 13.9 Consider the four points on the line $y = x + 2$ shown here: $(-1, 1), (2, 4), (0, 2),$ and $(1, 3)$. Determine the sample correlation coefficient for these data.

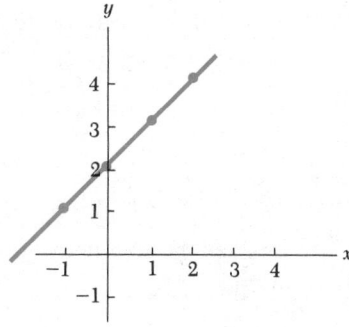

Solution We will work with the direct determination of the sample covariance in this example.

Point	x	$x - \bar{x}$	$(x - \bar{x})^2$	y	$y - \bar{y}$	$(y - \bar{y})^2$	$(x - \bar{x})(y - \bar{y})$
$(-1, 1)$	-1	-1.5	2.25	1	-1.5	2.25	2.25
$(2, 4)$	2	1.5	2.25	4	1.5	2.25	2.25
$(0, 2)$	0	-0.5	0.25	2	-0.5	0.25	0.25
$(1, 3)$	1	0.5	0.25	3	0.5	0.25	0.25
Totals	2	0	5	10	0	5	5.0

$$r = \frac{s_{XY}}{s_X s_Y} = \frac{\dfrac{5}{4}}{\sqrt{\dfrac{5}{4}}\sqrt{\dfrac{5}{4}}} = \frac{5}{\sqrt{25}} = 1$$

Example 13.10 Consider the three points on the line $y = -2x + 1$: $(0, 1)$, $(-1, 3)$, and $(-2, 5)$. These points are illustrated here.

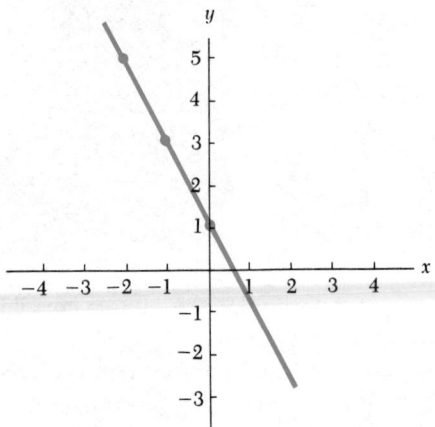

Compute r, the sample correlation coefficient for these data.
Solution

$$r = \frac{-\dfrac{4}{3}}{\sqrt{\dfrac{2}{3}}\sqrt{\dfrac{8}{3}}} = -1 \qquad \text{(Verify!)}$$

Example 13.11 Consider the five data points shown in the graph: $(0, 0)$, $(4, 4)$, $(0, 4)$, $(4, 0)$, and $(2, 2)$.

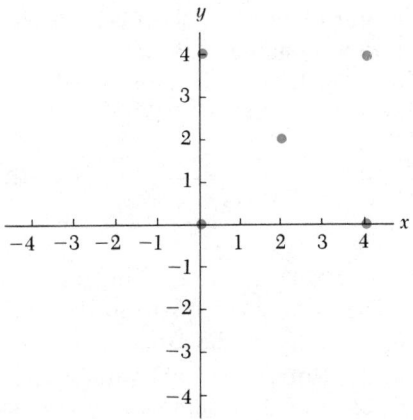

For these data, determine r.

Solution

$$r = \frac{\dfrac{0}{5}}{\sqrt{\dfrac{16}{5}}\ \sqrt{\dfrac{16}{5}}} = 0 \qquad \text{(Verify!)}$$

These three examples suggest what can be proved:[4]

1. If all sample pairs fall on a line with a positive slope, then $r = 1$ (Example 13.9).
2. If all sample pairs fall on a line with a negative slope, then $r = -1$ (Example 13.10).
3. If the sample (x, y) pairs are scattered in the **X, Y** plane with no *linear* association whatever, then $r = 0$ (Example 13.11).

Example 13.12 Using the data in Table 13.2 for the sales-advertising example, compute r, the sample correlation coefficient, using the computational formula (13.8).

Solution From Table 13.2,

$$r = \frac{10{,}820 - \dfrac{(178)(497)}{10}}{\sqrt{3{,}890 - \dfrac{(178)^2}{10}}\ \sqrt{30{,}311 - \dfrac{(497)^2}{10}}} = 0.9808$$

Thus, there is almost a perfect positive linear relationship between **X** and **Y** (note that we *must* now treat the variable x as a random variable **X** in order to use the correlation model) in these sample data.

---

[4] We will not prove these results. The interested reader is referred to the Conover reference at the end of the chapter.

The sample correlation coefficient, $\mathbf{r}$, is directly related to the sample coefficient of determination, $\mathbf{r}^2$, in the simple linear regression model, $\mathbf{Y}_i = \mathbf{b}_0 + \mathbf{b}_1 x_i + \mathbf{e}_i$. If we write r in its *calculating form:*

$$r = \frac{\Sigma(x_i - \bar{x})(y_i - \bar{y})}{\sqrt{\Sigma(x_i - \bar{x})^2}\,\sqrt{\Sigma(y_i - \bar{y})^2}} = \frac{\Sigma xy - \dfrac{(\Sigma x)(\Sigma y)}{n}}{\sqrt{\Sigma x^2 - \dfrac{(\Sigma x)^2}{n}}\,\sqrt{\Sigma y^2 - \dfrac{(\Sigma y)^2}{n}}}$$

we see that $r = \pm\sqrt{r^2}$ from recalling the computing form of r^2 given in Section 13.6. Therefore, the square of the correlation coefficient measures the amount of variability in the random variable $\mathbf{Y}$ when it is related linearly to the variable $\mathbf{X}$. Notice that in Examples 13.9 and 13.10, $r^2 = 1$; that is, the scatter of points in each sample falls *perfectly* on a straight line, so that 100 percent of the variability in $\mathbf{Y}$ is accounted for by the linear relationship with $\mathbf{X}$ ($\mathbf{Y} = \mathbf{X} + 2$ in Example 13.9, and $\mathbf{Y} = -2\mathbf{X} + 1$ in Example 13.10).

Furthermore, the sign of r, the sample correlation coefficient ($+$ or $-$), gives the direction ($+$ or $-$) of the *slope* of the regression line. Since b_1 in the fitted regression line, $\hat{y} = b_0 + b_1 x$, gives the slope, it is not surprising that r is related to b_1:

$$r = b_1\left(\frac{s_X}{s_Y}\right)$$

Thus, there is a *direct relationship* between r and b_1. Since s_X and s_Y are both greater than or equal to 0, the signs ($+$, $-$) of r and b_1 must be the same in a given sample. *In fact, the difference in the values of r and b_1 obtained is simply due to the difference in the measurement scales used!*

Since simple linear regression and correlation are similar kinds of analyses, which one should be used when studying two variables, $\mathbf{X}(x)$ and $\mathbf{Y}$? The difference between the two models depends on the assumptions placed on each. In both models, the dependent variable $\mathbf{Y}$ is assumed to be a random variable. But, in simple regression, the independent variable x is considered to be "fixed"—it is a constant, whereas in correlation, the independent variable $\mathbf{X}$ is considered to be a random variable. Furthermore, for hypothesis testing, correlation analysis requires that the population random variable $(\mathbf{X}, \mathbf{Y})$ be distributed as a bivariate normal random variable, a much more restrictive requirement than the regression inference assumption that only the dependent variable $\mathbf{Y}$ be normally distributed.

Correlation attempts to determine the relative strength of the *linear relationship* between two variables, whereas regression analysis establishes an estimated linear functional relationship between them. The fitted regression line tells us not only *whether* the two variables are linearly related in the sample, but also *how* ($\hat{y} = b_0 + b_1 x$).

In most cases, we would, therefore, prefer regression analysis to correlation analysis; but in fact, there is little difference between them. One difference worth keeping in mind is the sampling process for each method. In regression, we take a random sample of Y-values with the *x*-values being fixed or given, whereas in correlation analysis, we take a random sample of *pairs* of values, (**X, Y**). Thus, we may not receive the "coverage" of the **X** random variable that is possible in correlation (by randomly sampling the *x*-values). This suggests that if we are *only* interested in establishing whether or not two variables are linearly related, correlation may be preferred to regression.

It is also *extremely* important to keep in mind that if two variables are highly correlated, it is not possible to claim an indication of cause and effect *without further study*. An interesting example of a "nonsense" correlation coefficient arose in a study in the Scandinavian countries around the turn of the century. Among the variables studied were **X** = number of births per year and **Y** = stork population per year. It was shown that **X** and **Y** were strongly correlated positively. Thus, as the stork population grew, so did the number of births. It may be passé nowadays, but a possible cause-effect explanation had it that births increased because there were more storks to fly in the babies! Upon further study of the data, it became apparent that there was a third variable that affected both **X** and **Y**; **Z** = the severity of the winter. In severe winters, people had to keep their fireplaces going continuously to keep warm, and since storks build their nests in and about chimneys, this was definitely a plus for the stork population—fewer stork babies perished. And, downstairs, being snowbound with a roaring fire and no TV, the human population flourished as well!

Although this may appear to be a casual example, it brings home the important point that strong correlation *does not imply* causation. There are some very serious examples of this today. What is the relationship between smoking and lung cancer or heart disease? The correlation is positive, but does this imply that smoking "causes," let us say, heart disease? Until we are able to rule out *any* and *all possible causative* variables, such as physiological factors that may make it more likely that a person will become a heavy smoker and will develop heart disease, we cannot make this connection. *Always* be cautious in interpreting results or statements of causality that are based on correlation analysis.

■ 13.10 SUMMARY

This chapter presented the elements of simple linear regression and correlation analysis. Numerous assumptions were described for linear regression and correlation to be appropriate models for statistical analysis; and while statistical tests exist for assessing the reasonableness of these assumptions, residual plots presented in regression analysis are a good "first" step in assessing the reasonableness of the underlying assumptions of the approach. Numerous examples were provided to illustrate the use of the many regression and correlation statistics described.

Computation formulas presented in the chapter often served as shortcut methods for computing the quantities involved. Several of these formulas are summarized in Table 13.6. Table 13.3 gives a convenient format for computing many of the quantities involved in Table 13.6. The check that $\Sigma(y - \hat{y}) = 0$ or nearly equals 0 (allowing for computational roundoff) provides some assurance that the computations have been performed correctly.

The *relative* strength of a linear relationship as described in the chapter is measured by $\mathbf{r}^2$, the coefficient of determination, and $\mathbf{r}$, the coefficient of correlation, while a measure of the *absolute* strength of the linear relationship is given by $\mathbf{S}_{Y/x}$. In the next chapter, we will again address ourselves to measuring the strength of a linear relationship when we describe tests of hypotheses and confidence interval estimation of many of the parameters described in this chapter.

TABLE 13.6 Summary formulas for computing point estimates of various regression and correlation statistics

Point estimate of regression slope:

$$b_1 = \frac{\Sigma x_i y_i - \dfrac{(\Sigma x_i)(\Sigma y_i)}{n}}{\Sigma x_i^2 - \dfrac{(\Sigma x_i)^2}{n}}$$

Point estimate of regression intercept:

$$b_0 = \bar{y} - b_1 \bar{x} = \frac{\Sigma y_i}{n} - b_1 \left(\frac{\Sigma x_i}{n} \right)$$

Point estimate of the variance of the conditional probability distributions:

$$s_{Y/x}^2 = \frac{1}{n-2} \left\{ \left[\Sigma y_i^2 - \frac{(\Sigma y_i)^2}{n} \right] - \frac{\left[\Sigma x_i y_i - \dfrac{(\Sigma x_i)(\Sigma y_i)}{n} \right]^2}{\Sigma x_i^2 - \dfrac{(\Sigma x_i)^2}{n}} \right\}$$

Sample coefficient of determination, r^2:

$$r^2 = \frac{SST - SSE}{SST} = \frac{\left[\Sigma x_i y_i - \dfrac{(\Sigma x_i)(\Sigma y_i)}{n} \right]^2}{\left[\Sigma x_i^2 - \dfrac{(\Sigma x_i)^2}{n} \right] \left[\Sigma y_i^2 - \dfrac{(\Sigma y_i)^2}{n} \right]}$$

Sample coefficient of correlation:

$$r = \frac{\Sigma x_i y_i - \dfrac{(\Sigma x_i)(\Sigma y_i)}{n}}{\sqrt{\Sigma x_i^2 - \dfrac{(\Sigma x_i)^2}{n}} \sqrt{\Sigma y_i^2 - \dfrac{(\Sigma y_i)^2}{n}}}$$

◼ REFERENCES

Conover, W. J. *Practical Nonparametric Statistics.* New York: John Wiley & Sons, Inc., 1971.

Croxton, F. E.; Cowden, D. J.; and Bolch, B. W. *Practical Business Statistics.* 4th ed. Englewood Cliffs, N.J.: Prentice-Hall, Inc., 1969, chaps. 14–16 and 21.

Draper, N. R., and Smith, H. *Applied Regression Analysis.* New York: John Wiley & Sons, Inc., 1966.

Ezekiel, Mordecai, and Fox, Karl A. *Methods of Correlation and Regression Analysis.* 3d ed. New York: John Wiley & Sons, Inc., 1959.

Goldberger, Arthur S. *Econometric Theory.* New York: John Wiley & Sons, Inc., 1964.

Johnston, J. *Econometric Methods.* 2d ed. New York: McGraw-Hill Book Company, 1972.

Neter, John, and Wasserman, William. *Applied Linear Statistical Models.* Homewood, Ill.: Richard D. Irwin, Inc., 1974.

Williams, E. J. *Regression Analysis.* New York: John Wiley & Sons, Inc., 1959.

Yamane, T. *Statistics: An Introductory Analysis.* 3d ed. New York: Harper & Row, Publishers, 1973.

◼ PROBLEMS

13.1. Distinguish between dependent and independent variables in a simple linear regression equation.

13.2. Why is it important to plot a scatter diagram of the relationship between variables in a simple linear regression model?

13.3. What is meant by the term *least squares* in a simple linear regression model?

13.4. Describe the normal equations and how they are derived.

13.5. Discuss the assumptions made in using simple linear regression about the distributions of the conditional mean values.

13.6. What is meant by the coefficient of determination?

13.7. Explain the requirement for the term ϵ_i in the population regression model.

13.8. Describe the properties of the estimators of the regression coefficients, β_0 and β_1.

13.9. In comparing the degree of linear association between two variables, what advantages does correlation provide compared with simple linear regression?

13.10. How are the correlation coefficient and the coefficient of determination, r^2, in a simple linear regression, related?

13.11. What is meant by a "nonsense" correlation coefficient?

13.12. How does correlation differ from simple linear regression?

13.13. Explain and contrast the sampling requirements for simple linear regression and correlation.

13.14. On what scale is the correlation coefficient measured (nominal, ordinal, interval, ratio)? Does a correlation coefficient of 0.6 mean that the degree of linear association is twice as strong as when the coefficient is 0.3? Explain.

13.15. What is the proper interpretation of a correlation coefficient? What is the nature of the sample data when $r = 0$? When $r = +1$? When $r = -1$?

13.16. What is the relationship between the correlation coefficient and the slope in simple linear regression?

13.17. Explain what is meant by the *covariance* of two random variables.

13.18. Given the following four sets of data, determine whether a functional or a statistical relationship exists between the variables, and whether the relationship is linear. Where the relationship is functional, determine the function that relates x and y.

a. x	y	c. x	y
3	109	3	109
4	112	4	116
6	118	5	125
9	127	7	149
12	136	9	181

b. x	y	d. x	y
3	107	3	4
5	117	5	16
7	119	7	36
9	129	8	49
11	131	10	81

13.19. Plot each of the following sets of data as a scatter diagram. Which "curves" appear to best fit the given data?

y, Vehicle registrations*	x, Miles of primary highway*
125	5,022
155	9,984
130	14,738
202	19,921
194	25,021
241	30,550
310	34,729
397	41,001
570	45,143
656	50,002

* In Nittany Valley, 1930–70.

y, Sales of ethyl gasoline (1000s of gallons)*	x, Price of ethyl gasoline (per gallon)
50.5	37¢
57.3	36
60.0	35
60.1	31
67.8	29
70.1	27
68.2	26
74.8	25
75.1	23

* Measurements recorded in Nittany Valley, 1930–70.

y Batting average*	x, Age
.304	21
.299	24
.293	27
.288	28
.280	30
.267	32
.260	33
.252	34
.264	35

* From nine individuals, selected at random, 1980 Happy Valley Softball League.

13.20. The following data show federal individual income tax rates by taxable income (income after exclusions, deductions, and exemptions) for 12 income brackets in 1972.

Income	Tax rate (percent)
$ 0– 2,000	16
2,001– 4,000	19
4,001– 6,000	21
6,001– 8,000	24
8,001–10,000	25
10,001–12,000	27

	Tax rate
Income	(*percent*)
12,001–14,000	29
14,001–16,000	31
16,001–18,000	34
18,001–20,000	36
20,001–22,000	38
22,001–24,000	40

Source: U.S. Bureau of the Census, *Statistical Abstract of the United States: 1972,* Washington, D.C., 1972, pp. 391.

a. Plot the data as a scatter diagram, using the midpoint of each income level as the value of the independent variable.

b. Describe the nature of the relationship that exists from your diagram between taxable income and income tax rate.

c. Derive an estimating equation that predicts individual income tax rates as a function of the midpoint of each income level using the method of least squares.

d. By examining a residual plot of the fitted linear regression equation, state whether the assumptions underlying the use of the simple linear regression model appear to be met in these sample data.

13.21. In a given regression analysis of two random variables, the following quantities were determined: SSE = 129.45, SST = 792.67, and $\hat{y} =$ 1.2 − 4.9x. What is the value of r?

13.22. Given the following values for the two random variables X and Y:

x	y
0	2
1	4
1	3
2	5

a. Determine their covariance.

b. What percentage of the variability in Y can be accounted for or can be explained by relating it (linearly) to X?

13.23. A fuel-oil distribution company has collected data over a series of years to determine the statistical relationship between the average daily temperature and the consumption of fuel oil in single-family dwellings. Given the temperature, they desire to predict the consumption of fuel oil in order to better service customers on their delivery routes. The table shows a sample of the data that have been collected by the firm.

Average consumption of fuel oil in single-family dwellings (gallons)	*Average daily temperature*
7.0	10°
6.2	20°
5.1	30°
4.6	40°
3.5	50°
2.9	60°
1.2	70°
1.0	80°
0.7	90°

Source: Company records, district 3.

a. Plot the data as a scatter diagram and construct a line through the data points that appears to minimize the sum of squares of the vertical deviations from the data points to the line.

b. Now estimate the linear regression equation using the method of least squares and superimpose the line on the graph constructed in part a. How do the two lines compare?

c. What is the interpretation of b_1 in your regression equation?

d. What is the expected level of consumption of fuel oil given an average daily temperature of 55°?

13.24. Joe Super-Jock, big-time star fullback of the local college football team, is deciding whether to accept an offer to play professional football (or else go to work for his father-in-law at $50,000 per year as a busboy in the family restaurant). Joe has collected the following data indicating yards gained in senior year and average annual salary from a sample of the pro-standouts with whom Joe is personally familiar.

Professional player (fullbacks)	Yards gained in senior year in college	Annual salary
1	420	$22,000
2	450	15,000
3	510	27,000
4	620	30,000
5	640	37,000
6	700	25,000
7	740	40,000
8	890	55,000
9	910	57,000
10	955	61,000
11	980	58,000
12	1,210	96,000

a. Plot the data given as a scatter diagram, and estimate the parameters of the simple linear regression equation using the method of least squares.
b. Determine r^2 and $s_{Y/x}$ using the data provided.
c. Joe gained 999 yards his senior year, not counting the Gator Bowl yardage. Estimate Joe's annual salary, given the yardage gained.
d. From a residual plot constructed using the given data and the fitted simple linear regression equation, do the assumptions of the simple linear regression model appear to have been met in this problem?

13.25. A department store, as part of its college recruiting program, regularly supplies college students with its latest survey of wage rates and years of postsecondary education of its current employees. This information is supplied to give *prospective* employees an indication of the wage rate (or equivalent salary) they might *expect* to receive before a formal job offer is made, and to provide them with an indication of the expected "monetary worth" in their firm of obtaining additional education. The results of their latest survey are shown in the table.

Hourly wage rate and years of postsecondary education for ten professional workers selected at random from a department store

Worker	Years of Postsecondary education	Hourly wage rate
1	5.5	$18.20
2	0.0	6.50
3	1.0	3.90
4	4.0	11.70
5	3.5	10.40
6	2.0	7.80
7	6.0	15.60
8	2.5	9.10
9	4.5	13.00
10	5.0	14.30

Source: Sample of company employees.

a. Plot the data provided in the form of a scatter diagram.
b. Obtain an estimating equation of the form $\hat{y} = b_0 + b_1 x$ using the method of least squares.
c. Estimate the expected wage rate in this firm for an individual with four years of postsecondary education.
d. Calculate r^2 using the data provided.
e. Does the simple linear regression model appear to be an appropriate one for modeling the behavior of this random variable?

f. Comment on the "goodness" of the model developed.

13.26. Herman has not been doing very well on statistics examinations. He decides to sample eight of his friends to determine (a) the number of hours spent studying for the last statistics midterm examination, and (b) the points (out of 100) received on that examination. Herman hopes he can surmise that time spent studying is *not* systematically related to exam grades! The results of Herman's sample appear in the table.

Student	Hours of exam preparation	Exam score
Lois	19	58
Don	12	42
Roger	34	70
Ann	42	98
John	9	37
Gail	18	71
Bucky	51	94
Patty	22	85

a. Plot a scatter diagram of the data given.
b. Determine the linear, unbiased, minimum variance estimates b_0 and b_1 of β_0 and β_1 assuming the assumptions of the simple linear regression model are satisfied.
c. Determine $s_{Y/x}$ and r^2.
d. Does a "strong" linear relationship appear to exist between exam scores and hours spent studying? Do the assumptions underlying the use of the simple linear regression model appear to be met?

13.27. A company wishes to determine whether the amount of money spent on advertising is linearly related to the sales of a particular product over a reasonable range of advertising dollars expended. To test this notion, a random sample of ten sales regions produced the data in the table.

Region	Sales ($1000), y	Advertising ($1000), x
1	10	1.0
2	6	0.5
3	12	2.0
4	15	1.8
5	8	1.2
6	10	1.4
7	18	2.8
8	7	0.8
9	20	1.9
10	16	1.8

a. Form a scatter diagram for these data.
b. Determine the fitted linear regression equation.
c. Determine r^2.
d. Estimate sales for the following advertising dollars expended (in $1000):
   (i) $1.8 (ii) $2.8 (iii) $2.5
e. Determine $s_{Y/x}$.

13.28. The data in the following table show the number of weeks of experience in a job involving the production of a circuit board required for TV sets and the number of boards rejected during the past week for 25 workers.

Worker	No. of weeks of experience, x	No. of boards rejected, y
1	8	25
2	11	22
3	1	36
4	2	41
5	10	25
6	18	18
7	5	26
8	10	20
9	4	38
10	22	12
11	15	18
12	21	16
13	3	27
14	20	10
15	6	24
16	12	19
17	40	10

Worker	No. of weeks of experience, x	No. of boards rejected, y
18	20	20
19	13	33
20	5	36
21	10	24
22	25	16
23	32	10
24	11	24
25	16	18

   a. Determine the equation of the fitted linear regression line.

   b. What is the value of r^2?

   c. Estimate the number of rejected boards for each of the following values of x:
(i) 10 (ii) 35 (iii) 1 (iv) 0

   d. What is the value of $s_{Y/x}$?

   e. From an examination of a residual plot, does it appear that the assumptions of the simple linear regression model have been met in this problem?

13.29. An admissions officer in an MBA (Master of Business Administration) program is interested in determining what relationship, if any, exists between the GMAT (Graduate Management Admissions Test) score and the grade point average (GPA) of graduating MBAs. A sample of six recent graduates produced the data in the table.

Student	GMAT score, x	GPA, y
1	610	3.80
2	440	3.33
3	525	3.40
4	555	3.10
5	480	3.65
6	505	3.75

   a. Form a scatter diagram for these data. Do they appear to be linearly related?

   b. Find the fitted regression line.

   c. Determine the value of r^2.

   d. Estimate an MBA applicant's GPA, given that her GMAT score is 600.

   e. Do you have much *faith* in the estimate determined in part *d*? Explain.

13.30. Ten employees were randomly selected in an employee benefits study in a plant. For each, the age and number of sick days used during the past six months were recorded.

Individual	Age, x	Sick days, y
1	32	2
2	34	0
3	27	1
4	25	2
5	21	0
6	44	4
7	48	3
8	22	0
9	36	4
10	52	6

   a. Plot a scatter diagram for the pairs (x, y).

   b. Calculate the coefficient of correlation, r.

13.31. In each of the following cases, state whether you would expect ρ to be positive, negative, or 0.

   a. X = years of education beyond high school, Y = salary.

   b. X = family size, Y = monthly food expenditure.

   c. X = sales in a retail store, Y = inventory on hand.

   d. X = age of a production machine, Y = down time of the machine for repairs.

   e. X = income of family, Y = family size.

   f. X = income of family, Y = monthly food expenditure.

13.32. A firm is interested in determining whether or not a linear relationship exists between hourly pay and production output for a group of its employees. Ten employees in this group

Individual	1	2	3	4	5	6	7	8	9	10
Pay per hour	2.10	3.50	4.00	3.60	4.20	5.00	2.25	3.10	3.50	3.50
Production in units/ hour	20	25	28	21	22	32	27	30	24	30

are randomly selected. The data are shown in the table above.

a. Plot a scatter diagram of these data pairs.
b. Calculate r.
c. Would simple linear regression be preferred to correlation in this problem? Discuss.

13.33. What is the coefficient of correlation between two random variables X and Y under the following conditions?

a. One of the random variables is really a constant.
b. The value of the first random variable is always equal to 7.5 plus the value of the second random variable.
c. The value of the second random variable is always equal to 7.5 plus the value of the first random variable.
d. The random variables are constant multiples of one another.

13.34. A researcher has compiled data on the annual consumption of alcoholic beverages, X, and the number of automobile accidents, Y, in a certain locality. The data are shown in the table.

Year	X (1000 gallons)	Y (100 accidents)
1973	100	10.3
1974	96	9.9
1975	112	11.2
1976	123	11.6
1977	141	12.8
1978	149	13.6
1979	158	13.5
1980	174	14.7

a. Plot a scatter diagram of these pairs.
b. Calculate r.
c. Comment on the meaning (interpretation) of r in this problem.

13.35. The following characteristics: X = years of schooling beyond high school, Y = salary (in $1000), and Z = age, were measured for five randomly selected employees.

	Employee				
	1	2	3	4	5
x	4	4	6	2	4
y	12	20	25	14	30
z	22	30	32	26	35

a. Plot scatter diagrams for the pairs (x, y), (x, z), and (y, z).
b. Determine $r_{X,Y}$ and interpret.
c. Determine $r_{X,Z}$ and $r_{Y,Z}$, and interpret both.
d. Which variable, age (Z) or schooling (X), better explains salary (Y)? Discuss. (*Hint:* Determine the coefficient of determination in each case.)

13.36. A psychologist is interested in determining whether or not two IQ tests produce (linearly) related scores. A random sample of ten subjects is taken, and each subject is administered both tests. A suitable period of time is left between tests to allow subjects to recover; five subjects take test A first, while the remaining five take test B first. The results are shown in the table which follows.

Subject	Test A	Test B
1	120	109
2	144	127
3	100	116
4	124	120
5	132	116
6	108	98
7	114	122
8	132	121
9	110	106
10	128	115

a. Plot a scatter diagram of these ten pairs.
b. Calculate r.
c. Based on the value of **r**, how well do the two tests relate linearly? Explain.

Inferences in simple linear regression and correlation

14

■ 14.1 INTRODUCTION

In Chapter 13 we introduced the topic of simple linear regression and correlation analysis—*simple* because of the inclusion of one independent variable in the estimating equation or in the correlation analysis, and *linear* because of the manner in which the parameters of the regression model enter into the estimating equation. The steps in the regression model building process were defined to be as follows:

Steps in the simple linear regression model-building process

Step 1. Identify the variables to be included in building the regression model.
Step 2. Collect sample data.
Step 3. Specify the relationship that exists between the dependent and the independent variable.
Step 4. Estimate parameters of the model specified.
Step 5. Determine whether the assumptions of the simple linear regression model have been met.
Step 6. Statistically test the usefulness of the model developed.
Step 7. Use the model developed for prediction and estimation.

Chapter 13 covered essentially Steps 1 through 5 and part of Step 7. In this chapter, we will learn how to test statistically for the usefulness of the model developed (Step 6), and we will learn how to construct confidence intervals for the mean and for an individual value of the random variable Y given a value of x (Step 7). We will also learn how to test for the significance of the

455

correlation coefficient ρ based on the sample value obtained. Recall that the coefficient of correlation described—Pearson's product-moment correlation coefficient—measures the relative strength of the *linear* association between variables. Hence, a hypothesis test for ρ will be a test of the *linear* association between **X** and **Y**.

In order to extend our inference techniques in simple linear regression analysis, we will have to add one more assumption to the regression model to those given in Chapter 13—that dealing with the distribution of the error terms or the residuals, ϵ_i. By adding this fourth assumption, we will be able to expand greatly the statements that can be made concerning many regression parameters, and we will be able to complete the sixth step in the regression model-building process.

■ 14.2 INFERENCES IN REGRESSION ANALYSIS

□ 14.2.1 The normality assumption

We have already discussed some aspects of inference making in the regression model in Chapter 13: namely, point estimation of the Y-intercept, β_0, and the slope, β_1, of the regression line, and point estimation of the conditional probability distribution mean value, $\mu_{Y/x}$ and variance, σ^2. We now wish to extend the inference techniques to include confidence intervals and hypothesis tests for these parameters, but to do so, we must add a fourth assumption to our regression model to the three given on page 415.

> Normality assumption
>
> It is assumed that the error components, ϵ_i, that have a mean of 0 $[E(\epsilon_i) = 0]$, a variance of σ^2 $[V(\epsilon_i) = \sigma^2]$, and are uncorrelated are also *normally distributed.*

This assumption should not come as a surprise, since most of the confidence interval and hypothesis-testing techniques discussed in Chapters 9–12 are based on the normal distribution. As a consequence of this assumption, it is possible to say that the conditional probability distributions of **Y**, given x, are normally distributed, since Y_i and ϵ_i are related by an additive constant ($Y_i = \beta_0 + \beta_1 x_i + \epsilon_i$, where $\beta_0 + \beta_1 x_i$ is a constant).

The normality assumption is *not* required to obtain the point estimators of β_0, β_1, $\mu_{Y/x}$, and σ^2 (b_0, b_1, $\hat{Y}$, and $S_{Y/x}^2$, respectively). This assumption is required *only* when constructing confidence intervals and hypothesis-testing decision rules. Throughout the remainder of this chapter *unless otherwise specified,* we will consider the normality assumption to be satisfied.

☐ **14.2.2   Inferences concerning the regression slope, β_1**

The slope of a regression line is of interest to us usually in one of two ways. First, and most important, if we can infer from the sample that β_1 equals 0, then we know that the regression function is of *no use* to us as a predictor [recall that if $\beta_1 = 0$, $E(Y) = \beta_0 + \beta_1 x = \beta_0$, so that for *every* value of x, $E(Y) = \beta_0$]. Second, β_1 gives the amount of increase or decrease in the dependent random variable (Y) per unit increase in the independent variable (x). If x represents advertising expenditures and Y represents sales, for example, then in the model $E(Y) = \beta_0 + \beta_1 x$, β_1 gives the increase (we hope) in sales per unit increase in advertising expenditure. Estimating the value of β_1 becomes quite important to a marketer in this instance!

To establish confidence interval and hypothesis-testing formulas for β_1, we must first describe the sampling distribution of its point estimator, b_1. The sampling distribution of b_1 arises conceptually by taking all possible repeated samples of n (x_i, y_i) pairs from the population regression function when the levels of x *are the same in each sample*, and from each sample, computing the point estimate of β_1 given by:

$$b_1 = \frac{\Sigma x_i y_i - \dfrac{(\Sigma x_i)(\Sigma y_i)}{n}}{\Sigma x_i^2 - \dfrac{(\Sigma x_i)^2}{n}}$$

Refer to Figure 13.8 for an illustration of the process used to generate the sampling distribution of b_1. We now give the characteristics of this sampling distribution without proof.

Theorem 14.1
Sampling distribution of b_1

The *sampling distribution* of b_1:

1. Is normal.
2. Has a mean value of β_1 [i.e., $E(b_1) = \beta_1$].
3. Has a variance, denoted by $\sigma_{b_1}^2$, equal to:

$$\sigma_{b_1}^2 = \frac{\sigma^2}{\Sigma x_i^2 - \dfrac{(\Sigma x_i)^2}{n}}$$

Since σ^2, the variance of the conditional probability distributions of Y, given a value of x, is typically unknown, it must be estimated from sample data. From Section 13.5, the point estimate of σ^2 is:

$$s_{Y/x}^2 = \frac{\displaystyle\sum_{i=1}^{n} (y_i - \hat{y}_i)^2}{n - 2} = \frac{\text{SSE}}{n - 2}$$

By inserting $s_{Y/x}^2$ for σ^2 in the formula for $\sigma_{b_1}^2$, we produce a point estimate of $\sigma_{b_1}^2$, denoted by $s_{b_1}^2$:

Point estimate of $\sigma_{b_1}^2$, the variance of the sampling distribution of b_1

$$s_{b_1}^2 = \frac{s_{Y/x}^2}{\Sigma x_i^2 - \frac{(\Sigma x_i)^2}{n}} \tag{14.1}$$

Since b_1 is normally distributed, the standardized random variable

$$\frac{b_1 - \beta_1}{\sigma_{b_1}}$$

is also normally distributed. If we now substitute $S_{b_1}^2$ for $\sigma_{b_1}^2$ in the above expression, the new standardized random variable

$$\frac{b_1 - \beta_1}{S_{b_1}}$$

is t-distributed with $(n - 2)$ degrees of freedom. This is analogous to the formation of the t-statistic in Chapter 11. This statistic is used to construct confidence intervals and hypothesis-testing decision rules for β_1, and thereby test the usefulness of the model developed.

Confidence interval estimate for β_1, the population slope

From the t-distributed standardized random variable $(b_1 - \beta_1)/S_{b_1}$ may be derived (we will not derive it) the $100(1 - \alpha)$ percent *confidence interval estimate* for β_1:

$$b_1 - t_{\alpha/2:n-2}s_{b_1} \leq \beta_1 \leq b_1 + t_{\alpha/2:n-2}s_{b_1}$$

If this interval contains the number 0, we will conclude that the regression line is *not useful* for prediction purposes.

Example 14.1　Refer to the sales-advertising example first described in Example 13.2. Construct a 95 percent confidence interval estimate for β_1.

Solution　From Example 13.6, the point estimate of β_1 is $b_1 = 2.73$. For a 95 percent confidence interval, $(1 - \alpha) = 0.95$, so that $\alpha = 0.05$ and $\alpha/2 = 0.025$. From Table B.5 in Appendix B, $t_{0.025:10-2} = 2.306$. The value of S_{b_1} is given by:

$$s_{b_1} = \sqrt{\frac{s_{Y/x}^2}{\Sigma x_i^2 - \frac{(\Sigma x_i)^2}{n}}} = \sqrt{\frac{26.67}{3{,}890 - \frac{(178)^2}{10}}} = \sqrt{0.03696} = 0.1922$$

The confidence interval is:

$$2.73 - (2.306)(0.1922) \le \beta_1 \le 2.73 + (2.306)(0.1922)$$
$$2.29 \le \beta_1 \le 3.17$$

We would conclude, therefore, with 95 percent confidence, that the true slope β_1 lies between 2.29 and 3.17 and, since this interval does not contain 0, the sample regression line $\hat{y} = 1.02 + 2.73x$ is a useful predicting equation for estimating sales when advertising expenditures are between \$7 and \$31 (in \$00s).

The construction of a confidence interval for β_1 is equivalent to a test of a hypothesis, as was explained in Chapter 10. The null hypothesis is H_0: $\beta_1 = 0$, and the alternate hypothesis is H_A: $\beta_1 \ne 0$. The decision rule for rejection of the null hypothesis is given in Table 14.1.

TABLE 14.1 Decision rule for testing whether the slope of a regression line is equal to 0

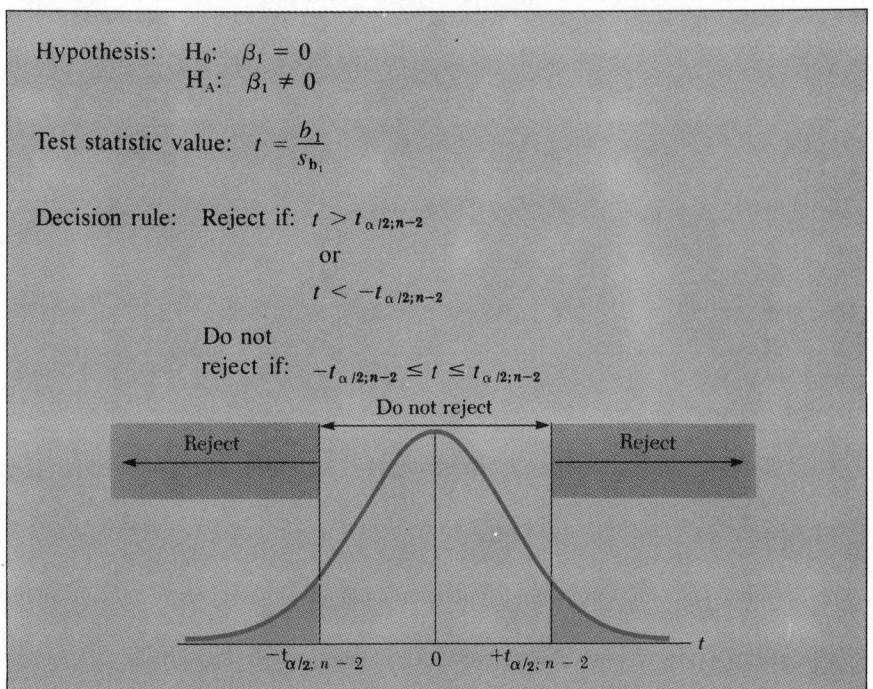

Hypothesis: H_0: $\beta_1 = 0$
H_A: $\beta_1 \ne 0$

Test statistic value: $t = \dfrac{b_1}{s_{b_1}}$

Decision rule: Reject if: $t > t_{\alpha/2;n-2}$

or

$t < -t_{\alpha/2;n-2}$

Do not reject if: $-t_{\alpha/2;n-2} \le t \le t_{\alpha/2;n-2}$

Do not reject

Reject   Reject

$-t_{\alpha/2;\, n-2}$ 0 $+t_{\alpha/2;\, n-2}$ t

The relationships expressed in Table 14.1 concerning the test of a hypothesis can also be shown by reference to the region in which we do not reject and the rejection region of the t-distribution as shown in the accompanying illustration. Thus, if

$$-t_{\alpha/2;n-2} < \frac{b_1}{s_{b_1}} < t_{\alpha/2;n-2}$$

the hypothesis H_0: $\beta_1 = 0$ is *not rejected* at the specified level of significance. Note that the above interval is equivalent to

$$b_1 - s_{b_1} t_{\alpha/2; n-2} < \beta_1 < b_1 + s_{b_1} t_{\alpha/2; n-2}$$

If this interval includes 0, we do not reject H_0: $\beta_1 = 0$; if the interval does not contain 0, we reject H_0: $\beta_1 = 0$.

□ **14.2.3 The analysis of variance test for H_0: $\beta_1 = 0$**

It is also possible to test the null hypothesis H_0: $\beta_1 = 0$ (H_A: $\beta_1 \neq 0$) using the F-distribution discussed in Chapters 11 and 12. The reason for including the test here is that it can be generalized to include more than one independent variable in our estimating equation, a topic we will address in Chapter 15. As might be expected, the F-test for H_0: $\beta_1 = 0$ will produce results equivalent to those from the t-test for H_0: $\beta_1 = 0$. In fact, the *value* of the sample F-statistic is the square of the value of the sample t-statistic, as we will demonstrate in Example 14.2.

TABLE 14.2 Analysis of variance table for simple linear regression

Source of variation	Degrees of freedom	Sum of squares	Mean square	F-ratio
Regression (model)	k	SSR	MSR $=$ SSR$/k$	MSR/MSE
Error (residual)	$n - k - 1$	SSE	MSE $=$ SSE$/(n - k - 1)$	
Total.	$n - 1$	SST		

The general form of the analysis of variance table for simple linear regression is given in Table 14.2. The degrees of freedom for the model (regression) are equal to the number of regression coefficients estimated, *excluding β_0*. Since in *simple* linear regression we are estimating the regression coefficients β_0 and β_1, k, the number of degrees of freedom, will *always* be equal to 1. Similarly, the degrees of freedom error (residual) are equal to one fewer than the difference between the number of sample observations and the number of regression coefficients estimated, *excluding β_0*. In simple linear regression, these degrees of freedom will *always* be equal to $n - (k + 1) = n - 2$.

The F-distribution for testing H_0: $\beta_1 = 0$ is shown in the accompanying figure. If the estimate of MSR/MSE exceeds $F(1 - \alpha; k, n - k - 1)$ given in Table B.7 in Appendix B at the stated significance level (α), we reject H_0: $\beta_1 = 0$; if the estimate MSR/MSE is less than $F(1 - \alpha; k, n - k - 1)$, we fail to reject H_0: $\beta_1 = 0$ and conclude that the independent variable x is of *no value* in predicting values of the independent random variable **Y** (i.e., the model is *not* useful).

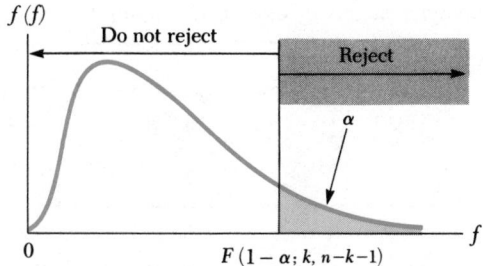

Example 14.2   Using the sales-advertising data given in Chapter 13, test the hypothesis H_0: $\beta_1 = 0$ at the 5 percent significance level using the F-distribution to determine whether to reject or not to reject H_0: $\beta_1 = 0$.

Solution   From Chapter 13,

$$\text{SSE} = \sum_{i=1}^{n} (y_i - \hat{y}_i)^2 = 213.33$$

$$\text{SSR} = \sum_{i=1}^{n} (\hat{y}_i - \bar{y})^2 = 5{,}396.77$$

$$\text{SST} = \sum_{i=1}^{n} (y_i - \bar{y})^2 = 5{,}610.10; \text{ where } n = 10, k = 1$$

The analysis of variance is given in Table 14.3 for the sales-advertising data.

TABLE 14.3   Analysis of variance table for data in Example 14.2

Source of variation	Degrees of freedom	Sum of squares	Mean square	F-ratio
Regression	$k = 1$	5,396.77	5396.77	202.38
Error	$n - 2 = 8$	213.33	26.67	
Total	$n - 1 = 9$	5,610.10		

From Table B.7 in Appendix B, $F(0.95; 1, 8) = 5.32$. Hence, we reject H_0: $\beta_1 = 0$ and conclude that advertising expenditure is a useful predictor of sales for advertising expenditures between $700 and $3,100.

Table 14.4 shows a computer analysis of the data in Example 14.2. Note the correct interpretation of the p-value, $\text{PR} > F = 0.0001$, given in Table 14.4. Since the computer prints only four numbers to the right of the decimal point, either $\beta_1 \neq 0$, or else in sampling, we obtained a sample in which *fewer* than 1 in 10,000 yields an F-value if 202.38 or larger when in fact $\beta_1 = 0$. Hence, we conclude $\beta_1 \neq 0$ because of the odds against (*less than* 1 in 10,000) $\beta_1 = 0$.

TABLE 14.4 Computer analysis of data in Examples 14.1 and 14.2

DEPENDENT VARIABLE: SALES

SOURCE	DF	SUM OF SQUARES	MEAN SQUARE	F VALUE	PR > F
MODEL	1	5396.768	5396.768	202.38	0.0001
ERROR	8	213.332	26.667		
CORRECTED TOTAL	9	5610.100			

Finally, note the relationship between the t-value determined in Example 14.1 and the F-value determined in Example 14.2:

$$t^2 = \left(\frac{b_1}{s_{b_1}}\right)^2 = \left(\frac{2.7348}{0.1922}\right)^2 = 202.38 = f$$

(allowing for some discrepancy because of roundoff error).

☐ **14.2.4 Inferences concerning the population Y-intercept, β_0**

Although confidence intervals and hypothesis tests on β_0 are not frequently used because the Y-intercept does not have important meaning in many practical applications, we include the appropriate formulas here for completeness. The sampling distribution of b_0, the point estimator of β_0, has the following characteristics:

Theorem 14.2
Sampling distribution of b_0

The *sampling distribution* of b_0:

1. Is normal.
2. Has a mean value of β_0 [i.e., $E(b_0) = \beta_0$].
3. Has a variance, denoted by $\sigma_{b_0}^2$, equal to:

$$\sigma_{b_0}^2 = \sigma^2 \left[\frac{1}{n} + \frac{\bar{x}^2}{\Sigma x_i^2 - \frac{(\Sigma x_i)^2}{n}}\right]$$

The point estimate of $\sigma_{b_0}^2$ is given by:

Point estimate of $\sigma_{b_0}^2$, variance of sampling distribution of Y-intercept, β_0

$$s_{b_0}^2 = s_{Y/x}^2 \left[\frac{1}{n} + \frac{\bar{x}^2}{\Sigma x_i^2 - \frac{(\Sigma x_i)^2}{n}}\right] \tag{14.2}$$

The standardized statistic, $(\mathbf{b}_0 - \beta_0)/\mathbf{S}_{\mathbf{b}_0}$ is t-distributed with $(n - 2)$ degrees of freedom. From this standardized t-statistic, the derived $100(1 - \alpha)$ percent confidence interval estimate is as follows:

Confidence interval estimate for β_0, the population Y-intercept

$$b_0 - t_{\alpha/2;n-2}S_{b_0} \le \beta_0 \le b_0 + t_{\alpha/2;n-2}S_{b_0}$$

☐ **14.2.5 Inferences concerning the mean of the conditional probability distribution of Y, given x:** $\mu_{Y/x}$

The mean of the conditional distribution of **Y** given a specific value of x, say x_P, is:

$$\mu_{Y/x = x_P} = \beta_0 + \beta_1 x_P$$

One of the most important applications of regression analysis is the estimation of the conditional mean $\mu_{Y/x}$ for a specified value of x (Step 7). To determine a point estimate of $\mu_{Y/x = x_P}$, we would simply substitute the value of x_P into the fitted regression line $\hat{y} = b_0 + b_1 x$ for $x = x_P$; denote the resulting estimate by $\hat{y}_P = b_0 + b_1 x_P$. Such an estimate is a point estimate and represents our best prediction of the value of the variable of interest. However, point estimates are often in error and often alone are not adequate. We usually desire some indication of the amount of error surrounding this estimate, for example. Such an estimate is provided by a confidence interval constructed for μ_{Y/x_P}.

We can set confidence intervals and establish hypothesis-testing decision rules for μ_{Y/x_P} by using the sampling distribution of $\hat{\mathbf{Y}}_P$.

Theorem 14.3
The sampling distribution of $\hat{\mathbf{Y}}_P$

The *sampling distribution* of $\hat{\mathbf{Y}}_P$:

1. Is normal.
2. Has a mean value of μ_{Y/x_P} [i.e., $E(\hat{\mathbf{Y}}_P) = \mu_{Y/x_P}$].
3. Has a variance, denoted by $\sigma^2_{\hat{Y}_P}$, given by

$$\sigma^2_{\hat{Y}_P} = \sigma^2 \left[\frac{1}{n} + \frac{(x_P - \bar{x})^2}{\Sigma x_i^2 - \dfrac{(\Sigma x_i)^2}{n}} \right]$$

Notice that the variance of $\hat{\mathbf{Y}}_P$ depends on (1) the variance of the conditional probability distribution of **Y** given *any* value of x, σ^2; (2) the reciprocal of the sample size, $1/n$; (3) the distance between x_P, the selected value of x, and the mean, $\bar{x}$; and (4) the reciprocal of the sum of squares due to x,

$\Sigma(x_i - \bar{x})^2 = \Sigma x_i^2 - [(\Sigma x_i)^2/n]$. The dependence of $\sigma^2_{\hat{Y}_P}$ on the first two factors is not surprising. We would expect the variability of $\hat{Y}_P$ to be related to σ^2, the variance of the conditional probability distributions, and to be a function of the reciprocal of the sample size, because $\hat{Y}_P$ is a population mean estimator, itself an expected value. The dependence of $\sigma^2_{\hat{Y}_P}$ on the third factor $(x_P - \bar{x})^2$ is best explained by examining Figure 14.1. In this figure are shown two

FIGURE 14.1   Effect of variability in b_1 on $\hat{Y}_P$

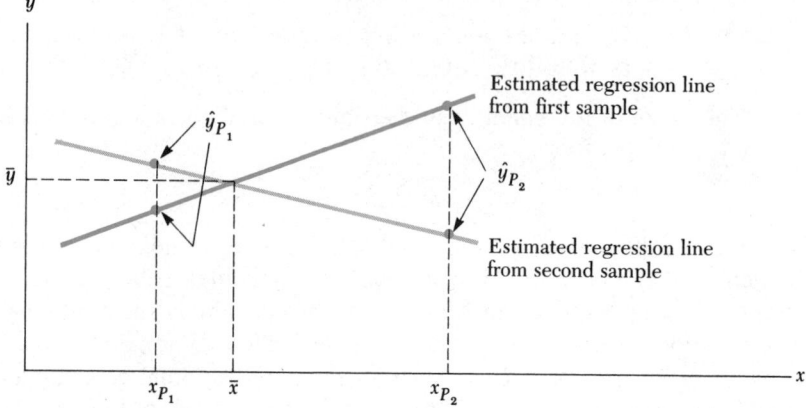

fitted regression lines, both of which pass through the point $(\bar{x}, \bar{y})$. If we are estimating $\mu_{Y/x}$ for a value of x close to $\bar{x}$, say x_{P_1}, then the lines will give similar estimates. But, if we are estimating $\mu_{Y/x}$ for a value of x not close to $\bar{x}$, say x_{P_2}, then the estimates may be quite dissimilar. Thus, "the farther away" x_P is from $\bar{x}$, the greater the variability we should expect in the estimator $\hat{Y}_P$. The effect of the fourth factor, $\Sigma(x_i - \bar{x})^2$, on $\sigma^2_{\hat{Y}_P}$ is to moderate the term $(x_P - \bar{x})^2$. If the x-values are very spread out, then $\Sigma(x_i - \bar{x})^2$ will be large and will provide a small ratio, $(x_P - \bar{x})^2/\Sigma(x_i - \bar{x})^2$. This makes sense because if the x-values are spread out, then it should require a greater deviation $(x_P - \bar{x})$ to significantly increase the value of $\sigma^2_{\hat{Y}_P}$ than if the x-values are closely grouped. If the x-values are closely grouped, then $\Sigma(x_i - \bar{x})^2$ will be small and it will take a relatively small deviation $(x_P - \bar{x})$ to significantly increase the value of $\sigma^2_{\hat{Y}_P}$.

The point estimate of $\sigma^2_{\hat{Y}_P}$ is given by the following:

Point estimate of $\sigma^2_{\hat{Y}_P}$, the variance of the sampling distribution of $\hat{Y}_P$

$$s^2_{\hat{Y}_P} = s^2_{Y/x} \left[\frac{1}{n} + \frac{(x_P - \bar{x})^2}{\Sigma x_i^2 - \dfrac{(\Sigma x_i)^2}{n}} \right]$$

The standardized random variable,

$$\frac{\hat{Y}_P - \mu_{Y|x_P}}{S_{\hat{Y}_P}}$$

is t-distributed with $(n - 2)$ degrees of freedom. This **t**-statistic leads to the derived $100(1 - \alpha)$ percent confidence interval for $\mu_{Y|x_P}$:

$$\hat{y}_P - t_{\alpha/2;n-2}s_{\hat{Y}_P} \leq \mu_{Y|x_P} \leq \hat{y}_P + t_{\alpha/2;n-2}s_{\hat{Y}_P}$$

Example 14.3 In the sales-advertising example from Chapter 13, find a point estimate of mean sales when the amount expended on advertising is $1,100 ($\mu_{Y|x=11}$) and determine a 90 percent confidence interval for $\mu_{Y|x=11}$.

Solution The estimated regression line was determined in Example 13.6—it is $\hat{y} = 1.02 + 2.73x$. Thus, the point estimate of $\mu_{Y|x=11}$ is: $\hat{y}_{11} = 1.02 + 2.73(11) = 31.05$ (in $000). From Table B.5 in Appendix B, $t_{0.05;8} = 1.86$. Thus $s_{\hat{Y}_P}^2$ is:

$$s_{\hat{Y}_P}^2 = s_{Y|x}^2 \left[\frac{1}{n} + \frac{(x_P - \bar{x})^2}{\Sigma x_i^2 - \dfrac{(\Sigma x_i)^2}{n}} \right]$$

$$= (26.67) \left[\frac{1}{10} + \frac{(11 - 17.8)^2}{3,890 - \dfrac{(178)^2}{10}} \right] = 4.376$$

$$s_{\hat{Y}_P} = \sqrt{4.376} = 2.09$$

The 90 percent confidence interval estimate is:

$$31.05 - (1.86)(2.09) \leq \mu_{Y|x=11} \leq 31.05 + (1.86)(2.09)$$
$$27.16 \leq \mu_{Y|x=11} \leq 34.94$$

Thus, we are 90 percent confident that mean sales lie between $27,160 and $34,940 for an advertising expenditure of $1,100.

In Table 14.5, 90 percent confidence interval estimates of $\mu_{Y|x}$ are given for a set of x values. (These estimates may differ slightly from other estimates

TABLE 14.5 Ninety percent confidence intervals for $\mu_{Y|x}$ for a set of x-values in the sales-advertising example

Advertising expenditures, x_P ($00)	Estimated mean sales, $\hat{y}_P$ ($000)	Confidence interval estimate		
		Lower estimate	Upper estimate	Width
5	14.70	9.20	20.19	10.99
11	31.10	27.21	34.99	7.78
16	44.78	41.67	47.88	6.21
$\bar{x} = 17.8$				
21	58.45	55.21	61.70	6.49
26	72.13	67.90	76.35	8.45

FIGURE 14.2 Confidence bands at the 90 percent confidence level for sales given advertising expenditures

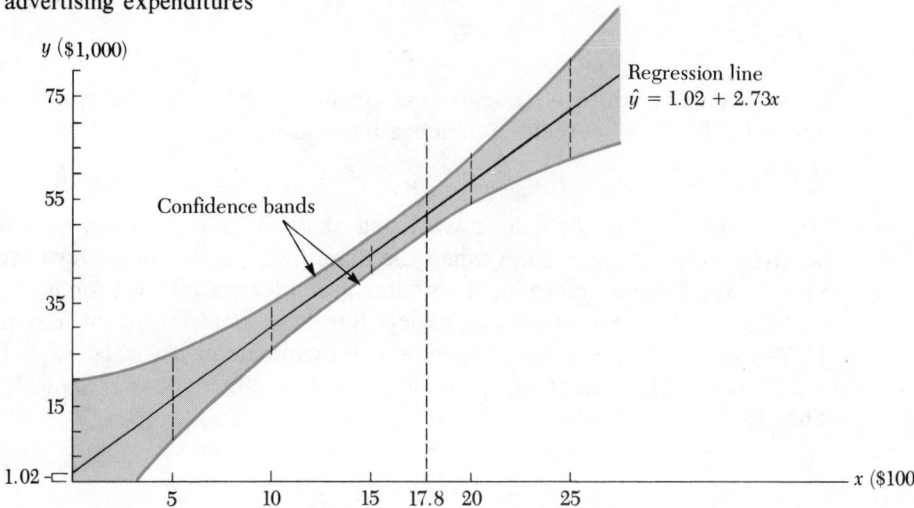

because of the greater number of significant digits retained in calculating them.) These confidence intervals are also shown in Figure 14.2.

First, we note in Table 14.5 that as x_P "moves away" from $\bar{x} = 17.8$, the width of the confidence intervals increases. This is impressively displayed in Figure 14.2—the five confidence intervals are shown as broken lines. If we calculated intervals for *all* values of x between 5 and 31, we would produce the "bands" shown in this figure by connecting all the upper limits and then all the lower limits by two smooth *curves*. The fact that the width of the confidence intervals increases as x_P departs from $\bar{x}$ substantiates our earlier discussion: $\hat{Y}_P$ is more variable as x_P departs from $\bar{x}$.

We will not develop explicitly the hypothesis-testing decision rule for hypotheses concerning μ_{Y/x_P}, since confidence intervals may be used for this purpose. For example, suppose we hypothesize that average sales for this firm with $1,600 in advertising is $50,000, and we wish to test this hypothesis at the $\alpha = 0.10$ significance level. From Table 14.5, we would reject this hypothesis since the 90 percent confidence interval does not contain 50 ($1,000) (the confidence interval estimate is $41,670–$47,880).

□ **14.2.6 Prediction (forecast) interval for the "next" value, Y_{next}**

In the previous section, we were concerned with drawing inferences about the mean $\mu_{Y/x}$ of the conditional probability distribution of **Y** for a given value of x, $x = x_P$. Often, however, we will be interested in inferences about a *single* or *new* observation (value of the random variable). For example, suppose in our sales-advertising illustration, we sample an additional sales branch, the "next" observation past the present ten in our sample, and

$1,900 is spent on advertising. We are interested in predicting sales for *this*, the next branch. This new observation *y* is viewed as the next trial in the experiment that produced the present ten sample observations—this new observation is considered to be independent of the other sample observations. Denote the level of *x*, advertising expenditures, for this next observation as x_{next} and the corresponding Y observation as y_{next}.[1]

The point estimate of Y_{next} is the same as the point estimate of the conditional mean $\mu_{Y/x}$:

$$\hat{y}_{next} = b_0 + b_1 \cdot x_{next}$$

For instance, suppose the amount spent on advertising is $1,900; that is, $x_{next} = 19$, and we are asked to estimate (forecast) corresponding sales. Intuitively, we would give as our answer the estimated mean of the conditional probability distribution of Y, given that $x = 19$. This estimate is: $\hat{y}_{next} = 1.02 + 2.73(19) = 52.89$ (in $000s).

But, the estimation problem here is inherently different from the one presented in the previous section. Here, we are drawing inferences about a *single observation*, Y_{next}, whereas in Section 14.2.5, we were drawing inferences about a *mean* of a set of observations, the conditional mean $\mu_{\hat{Y}/x_P}$. Although the two point estimates are the same, the confidence interval formulas are not—reflecting that it is much more difficult to estimate (predict) a *single* value than a *mean* of a set of values.

Theorem 14.4
The sampling distribution of $\hat{Y}_{next}$

The *sampling distribution* of $\hat{Y}_{next}$

1. Is normal.
2. Has a mean value of $\mu_{Y/x_{next}}$ [i.e., $E(\hat{Y}_{next}) = \mu_{Y/x_{next}}$].
3. Has a variance, denoted by $\sigma^2_{\hat{Y}_{next}}$, given by

$$\sigma^2_{\hat{Y}_{next}} = \sigma^2 \left[1 + \frac{1}{n} + \frac{(x_{next} - \bar{x})^2}{\sum x_i^2 - \frac{(\sum x_i)^2}{n}} \right]$$

Now, consider the variance formula $\sigma^2_{\hat{Y}_{next}}$.

$$\sigma^2_{\hat{Y}_{next}} = \sigma^2 \left[1 + \frac{1}{n} + \frac{(x_{next} - \bar{x})^2}{\sum x_i^2 - \frac{(\sum x_i)^2}{n}} \right] = \sigma^2 + \sigma^2 \left[\frac{1}{n} + \frac{(x_{next} - \bar{x})^2}{\sum x_i^2 - \frac{(\sum x_i)^2}{n}} \right] = \sigma^2 + \sigma^2_{\hat{Y}_P}$$

---

[1] Here, notationally, we are distinguishing the estimation of a mean value and the estimation of an individual observation. When we are estimating a mean value, we denote the level of *x* to be used in the estimation process (and the corresponding Y-statistics and estimates) by subscripting the variable with a *P* (e.g., x_P) and we denote the level of *x* to be used in the estimation process (and the coresponding Y-statistics and estimates) when we are estimating an *individual value* of the random variable by subscripting with the word *next* (e.g., x_{next}).

That is, $\sigma_{\hat{Y}_{next}}^2$ is equal to σ^2, the variance of the conditional probability distribution of Y, given x, *plus* the variance of Y_P, the point estimator of the conditional mean $\mu_{Y/x}$, given a specific value of x. Intuitively, this seems proper. In predicting a single observation y, we are confronted with the variability in estimating the conditional mean $\mu_{Y/x}$ *plus* the *additional variability* of the specific conditional distribution of Y, given a specific value of x. Thus, in estimating a single observation, we must take into account the variability in fixing the *location* of the conditional distribution (by estimating its mean $\mu_{Y/x}$) *and* take into account the variability *within* the conditional distribution (since we are attempting to predict a specific value that belongs to that conditional distribution).

The point estimate of $\sigma_{\hat{Y}_{next}}^2$ is given by the following:

Point estimate of $\sigma_{\hat{Y}_{next}}^2$ the variance of the sampling distribution of $\hat{Y}_{next}$

$$s_{\hat{Y}_{next}}^2 = s_{Y/x}^2 \left[1 + \frac{1}{n} + \frac{(x_{next} - \bar{x})^2}{\Sigma x_i^2 - \frac{(\Sigma x_i)^2}{n}} \right] \qquad (14.3)$$

The standardized random variable,

$$\frac{\hat{Y}_{next} - Y_{next}}{S_{\hat{Y}_{next}}}$$

is t-distributed within $(n - 2)$ degrees of freedom. This t-statistic leads to the derived $100(1 - \alpha)$ percent confidence interval estimate for Y_{next}, called a *prediction or forecast interval* since we are predicting a single value. The quantity $s_{\hat{Y}_{next}}$ is often called the *standard error of the forecast*.

Prediction interval estimate for Y_{next}

$$\hat{y}_{next} - t_{\alpha/2 \, : n-2} s_{\hat{Y}_{next}} \leq Y_{next} \leq \hat{y}_{next} + t_{\alpha/2 \, : n-2} s_{\hat{Y}_{next}}$$

Example 14.4 Determine a 90 percent prediction (*forecast*) interval for sales for an advertising expenditure of 19($100) (the "next" value) using the data provided.

Solution The point estimate of Y_{next} with $x_{next} = 19$ is:

$$\hat{y}_{next} = b_0 + b_1 x_{next} = 1.02 + 2.73(19) = 52.89(\$1,000)$$

The estimate of the variance of $\hat{Y}_{next}$ is given by:

$$s_{\hat{Y}_{next}}^2 = s_{Y/x}^2 \left[1 + \frac{1}{n} + \frac{(x_{next} - \bar{x})^2}{\Sigma x_i^2 - \frac{(\Sigma x_i)^2}{n}} \right] = 26.67 \left[1 + \frac{1}{10} + \frac{(19 - 17.8)^2}{3,890 - \frac{(178)^2}{10}} \right] = 29.39$$

$$s_{\hat{Y}_{next}} = \sqrt{29.39} = 5.42$$

From Table B.5 in Appendix B, $t_{0.05\,;8} = 1.86$. The 90 percent confidence interval is:

$$52.89 - (1.86)(5.42) \le Y_{next} \le 52.89 + (1.86)(5.42)$$
$$42.81 \le Y_{next} \le 62.97$$

Thus, we are 90 percent confident that the next branch's sales will lie between \$42,810 and \$62,970.

The 90 percent confidence interval for the mean value, μ_{Y/x_P}, is given by the interval 49.82 to 55.96 (verify!) when $x_P = x_{next} = 19$. The point estimates of μ_{Y/x_P} and Y_{next} are, of course, the same, since the same value for the independent variable x is substituted into the regression equation. The 90 percent confidence interval for the mean and the 90 percent prediction interval for the next value are given in Figure 14.3. Note the greater width of the prediction interval, reflecting the added variability introduced by estimating a value of the random variable as opposed to estimating a mean value.

FIGURE 14.3 Ninety percent prediction and confidence interval estimates for the data in Example 14.4

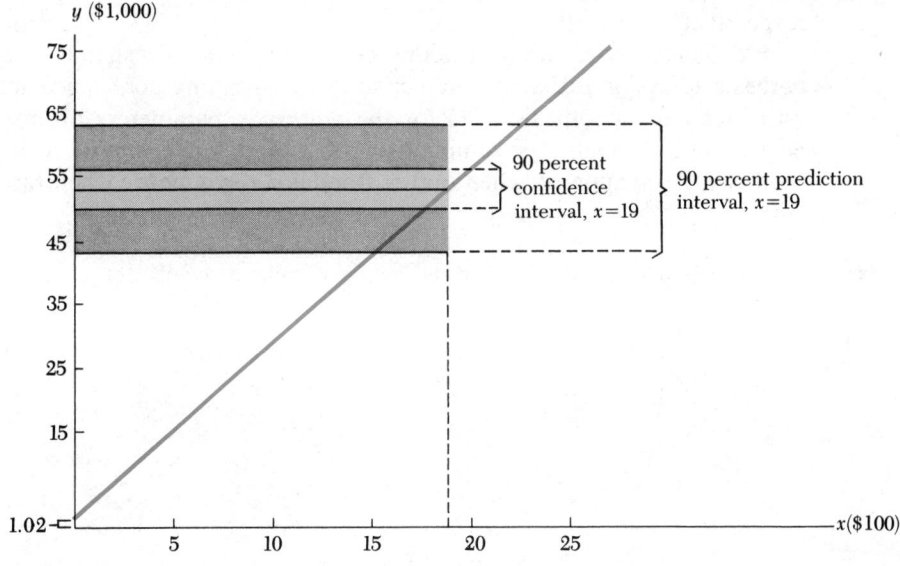

14.3 INFERENCES IN CORRELATION ANALYSIS

14.3.1 Point estimation of the population correlation coefficient ρ

The sample correlation coefficient **r** is the point estimator of the population correlation coefficient, denoted by ρ. Figure 13.21 illustrates the relationship between r (the value of $\mathbf{r}$) and ρ. The population correlation coefficient ρ is given by:

$$\rho = \frac{E[(\mathbf{X} - \mu_\mathbf{X})(\mathbf{Y} - \mu_\mathbf{Y})]}{\sqrt{V(\mathbf{X})} \sqrt{V(\mathbf{Y})}} = \frac{C(\mathbf{X},\mathbf{Y})}{\sigma_\mathbf{X} \cdot \sigma_\mathbf{Y}}$$

The numerator of ρ gives the covariance of the two random variables $\mathbf{X}$ and $\mathbf{Y}$ in the population. The denominator of ρ is the product of the population standard deviation of $\mathbf{X}$ and of $\mathbf{Y}$ ($\sigma_\mathbf{X}$ and $\sigma_\mathbf{Y}$, respectively).

The sample correlation coefficient r is the sample estimate of ρ and is given by:

$$r = \frac{\left(\dfrac{1}{n-1}\right) \sum\limits_{i=1}^{n} (x_i - \bar{x})(y_i - \bar{y})}{\sqrt{\dfrac{\sum\limits_{i=1}^{n} (x_i - \bar{x})^2}{n-1}} \sqrt{\dfrac{\sum\limits_{i=1}^{n} (y_i - \bar{y})}{n-1}}} = \frac{s_{XY}}{s_X s_Y}$$

where $s_{XY}, s_X,$ and s_Y are sample point estimates of $\sigma_{XY} = C(\mathbf{X},\mathbf{Y})$, σ_X, and σ_Y, respectively.

A difficulty arises in constructing confidence interval estimates and hypothesis tests for ρ that did not arise in constructing confidence interval estimates and hypothesis tests for the *regression* parameters discussed in Section 14.2. Namely, the sampling distribution of $\mathbf{r}$, the sample correlation coefficient, is symmetric when $E(\mathbf{r}) = 0$ and is *not* symmetric when $E(\mathbf{r}) \neq 0$.

FIGURE 14.4 Sampling distributions of the sample correlation coefficient, $\mathbf{r}$

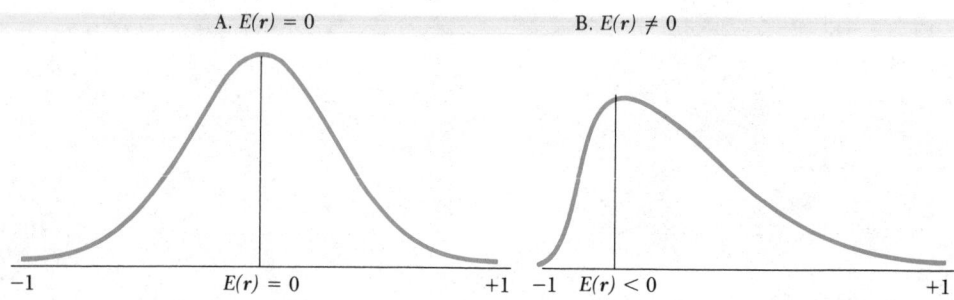

A. $E(r) = 0$ B. $E(r) \neq 0$

This situation is depicted in Figure 14.4. Fortunately, Sir Ronald Fisher, a famous statistician, developed computational formulas for obtaining *approximate confidence interval estimates* for ρ when $E(\mathbf{r}) \neq 0$. We discuss confidence interval estimation of $\rho \neq 0$ in Section 14.3.3 and turn now to a test of the hypothesis H_0: $\rho = 0$.

☐ 14.3.2 Hypothesis tests for the population correlation coefficient, H_0: $\rho = 0$

If we draw a random sample of n pairs, $(x_1, y_1), (x_2, y_2), \ldots, (x_n, y_n)$, from the bivariate normal distribution, then the standardized random variable $t = r/S_r$ is t-distributed with $(n - 2)$ degrees of freedom, where s_r is the point estimate of σ_r, the standard deviation of the sampling distribution of r, and is given by the following:

> Point estimate of σ_r, the standard deviation of the sampling distribution of r
>
> $$s_r = \sqrt{\frac{1 - r^2}{n - 2}} \qquad (14.5)$$

The statistic S_r is used when we wish to test the null hypothesis H_0: $\rho = 0$ when we have drawn a random sample of n pairs (x_i, y_i) from a bivariate normal distribution.

Example 14.5 Assume for the Example 14.4 sales-advertising data that the population joint distribution of (X, Y) is bivariate normal. Test the hypothesis, H_0: $\rho = 0$, using $\alpha = 0.05$.

Solution In Example 13.12, $r = 0.9808$; also,

$$s_r = \sqrt{\frac{1 - (0.9808)^2}{10 - 2}} = 0.069$$

$$t = \frac{r}{s_r} = \frac{0.9808}{0.069} = 14.23$$

Since $t > t_{0.025;8}$, we reject H_0: $\rho = 0$; ρ *appears* to be greater than 0.

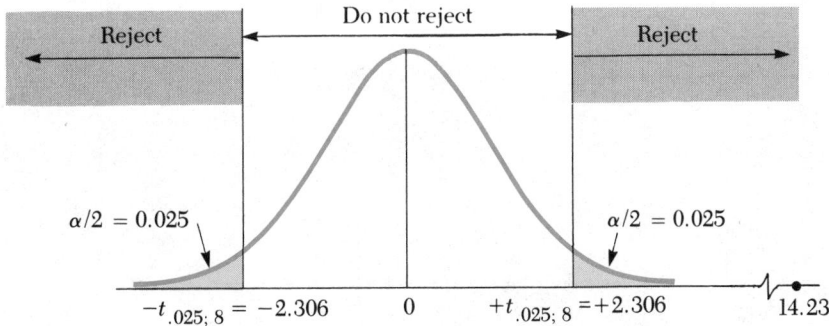

☐ 14.3.3 Confidence interval estimation of ρ; $\rho \neq 0$

If we wish to test the null hypothesis that ρ is a value other than 0 *or* if we wish to set a confidence interval on ρ, the method becomes more complicated. Sir Ronald A. Fisher devised an approximate method for constructing confidence interval estimates for ρ, based on Fisher's Z-transformation:

$$Z = \frac{1}{2} \log_e \left(\frac{1 + r}{1 - r}\right)$$

where $\log_e$ is the logarithm, base e. Fisher showed that Z is *approximately* normally distributed with mean and variance given by, respectively,

$$\mu_Z = \frac{1}{2} \log_e \left(\frac{1 + \rho}{1 - \rho}\right) \quad \text{and} \quad \sigma_Z^2 = \frac{1}{n - 3}$$

The approximation developed by Fisher is reasonably good as long as n is at least 50. By converting the sample correlation coefficient r to Z using this transformation, a confidence interval estimate for μ_Z is given by:

$$l = z - z_{\alpha/2} \frac{1}{\sqrt{n - 3}} \leq \mu_Z \leq u = z + z_{\alpha/2} \frac{1}{\sqrt{n - 3}}$$

The lower and upper limits on μ_Z, l and u, may then be converted to limits on ρ by using Table B.18 in Appendix B.

Example 14.6 Scores on an IQ/achievement test (X) for $n = 100$ randomly selected workers were taken together with experimental measurements of the time (Y) required to put together a complex optical device given

FIGURE 14.5 Confidence interval estimate for ρ

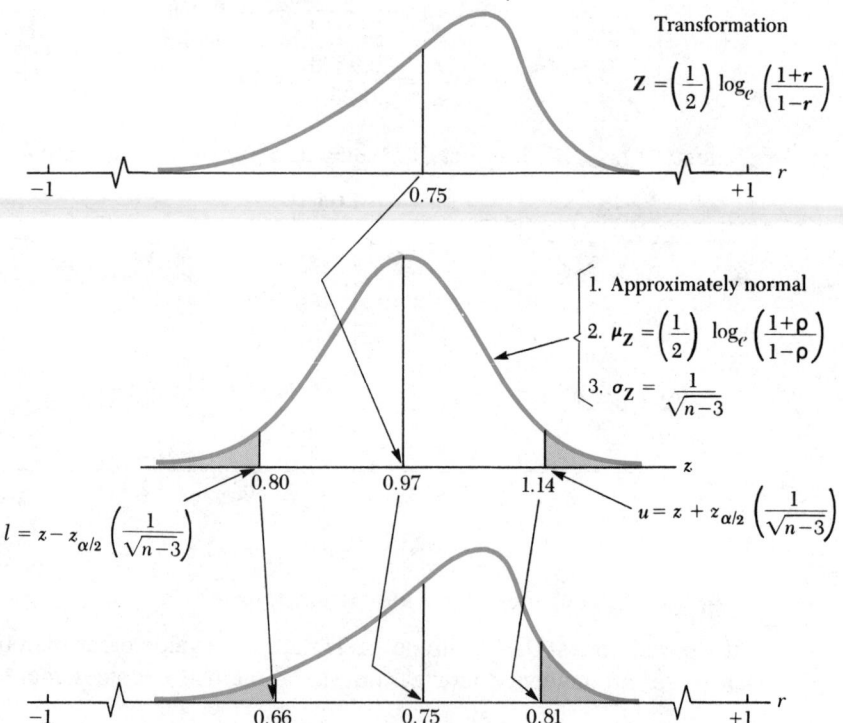

the instructions. The sample correlation coefficient was computed to be $r = 0.75$. Determine an approximate 90 percent confidence interval for ρ.

Solution From Table B.18 in Appendix B, the value of Z that corresponds to $r = 0.75$ is 0.97. Then,

$$l = 0.97 - (1.64)\,\frac{1}{\sqrt{97}} = 0.80 \quad \text{and} \quad u = 0.97 + (1.64)\,\frac{1}{\sqrt{97}} = 1.14$$

Now, referring to Table B.18 and entering the table with $z = 0.80$ and $z = 1.14$, the limits on ρ are:

$$0.66 \le \rho \le 0.81$$

This process is illustrated in Figure 14.5.

Notice in Example 14.6 that Table B.18 does most of the work for us—there is no need to take any natural logarithms! Also, notice that the confidence interval is *not symmetric* about the point estimate, $r = 0.75$. This is because of the upper bound on ρ of $+1$; the sampling distribution of **r** must necessarily become *very skewed* if ρ is close to 1. This is why the **t**-statistic cannot be used to set a confidence interval on ρ if $\rho \ne 0$; in this case, the statistic r/S_r no longer has the symmetric t-distribution as its sampling distribution.

■ 14.4 SUMMARY

In this chapter, we expanded considerably upon the estimation techniques presented in Chapter 13. By adding a fourth assumption to the simple linear regression model, for example—that the error terms or the residuals ϵ_i are normally distributed—we were able to develop tests of hypotheses and confidence interval estimation procedures for many of the regression parameters discussed in Chapter 13. The hypothesis test, whether β_1 equals 0 (i.e., H_0: $\beta_1 = 0$), leads to a direct statement (at a stated level of significance) about the usefulness of the regression model for estimation purposes (Step 6), for example. In Table 14.6 we summarize the variance estimates for the sampling distributions of the regression and correlation statistics presented in the chapter.

In Chapter 15 we shall consider the case in which more than one independent variable is used to predict the value of the dependent variable and where we wish to assess the strength of the linear relationship between a dependent

TABLE 14.6 Summary computational formulas for Chapter 14

Point estimate of the sampling distribution variance of b_1	Point estimate of the sampling distribution variance of b_0
$s_{b_1}^2 = \dfrac{s_{Y/x}^2}{\sum x_i^2 - \dfrac{(\sum x_i)^2}{n}}$	$s_{b_0}^2 = s_{Y/x}^2 \left[\dfrac{1}{n} + \dfrac{\bar{x}^2}{\sum x_i^2 - \dfrac{(\sum x_i)^2}{n}} \right]$

TABLE 14.6 (*continued*)

Point estimate of the sampling distribution variance of $\hat{Y}_P$	Point estimate of the sampling distribution variance of r (for testing H_0: $\rho = 0$)
$$s^2_{\hat{Y}_P} = s^2_{Y/x}\left[\frac{1}{n} + \frac{(x_P - \bar{x})^2}{\Sigma x_i^2 - \frac{(\Sigma x_i)^2}{n}}\right]$$	$$s_r = \sqrt{\frac{1 - r^2}{n - 2}}$$

Point estimate of the sampling distribution variance of $\hat{Y}_{next}$

$$s^2_{\hat{Y}_{next}} = s^2_{Y/x}\left[1 + \frac{1}{n} + \frac{(x_{next} - \bar{x})^2}{\Sigma x_i^2 - \frac{(\Sigma x_i)^2}{n}}\right]$$

variable and several independent variables. These types of analyses are called *multiple linear regression* and *multiple correlation analysis,* respectively.

■ REFERENCES

Croxton, F. E.; Cowden, D. J.; and Bolch, B. W. *Practical Business Statistics.* 4th ed. Englewood Cliffs, N.J.: Prentice-Hall, Inc., 1969, chaps. 14–16 and 21.

Draper, N. R., and Smith, H. *Applied Regression Analysis.* New York: John Wiley & Sons, Inc., 1966.

Ezekiel, Mordecai, and Fox, Karl A. *Methods of Correlation and Regression Analysis.* 3d ed. New York: John Wiley & Sons, Inc., 1959.

Goldberger, Arthur S. *Econometric Theory.* New York: John Wiley & Sons, Inc., 1964.

Johnston, J. *Econometric Methods.* 2d ed. New York: McGraw-Hill Book Company, 1972.

Neter, John, and Wasserman, William. *Applied Linear Statistical Models.* Homewood, Ill.: Richard D. Irwin, Inc., 1974.

Williams, E. J. *Regression Analysis.* New York: John Wiley & Sons, Inc., 1959.

Yamane, T. *Statistics: An Introductory Analysis.* 3rd ed. New York: Harper & Row, Publishers, 1973.

■ PROBLEMS

14.1 What is the proper interpretation of a confidence interval constructed to test whether $\beta_1 = 0$?

14.2. Explain the difference between a prediction or a forecast interval and a confidence interval constructed for $\mu_{Y/x}$.

14.3. Describe the differences in estimation and hypothesis-testing procedures when $E(r) = 0$ as opposed to when $E(r) \neq 0$. What is the nature of the sampling distribution of r in each instance?

14.4. Explain why it is nonsensical to construct a confidence interval for b_1 in a simple linear regression analysis of the linear relationship between variables **Y** and x.

14.5. Given the simple linear regression equation $\hat{y} = 23.0 + 6.21x$ determined by sampling 23 (x, y) data pairs where $s_{b_1} = 1.48$, test the null hypothesis H_0: $\beta_1 = 0$ using a significance level of $\alpha = 0.05$.

14.6. The following data show the dollar value of sales with corresponding advertising levels for a small retail chain in the Midwest.

Advertising	Sales
$584	$6,445
386	4,015
452	4,822
635	7,017
242	2,848
328	3,888
517	5,554

a. Plot these data in the form of a scatter diagram. Does the linear model appear to be appropriate for estimating sales as a function of dollars spent on advertising?
b. Determine the least squares regression line.
c. What would be the level (value) of estimated sales if advertising expenditures were $622?
d. Construct a 95 percent confidence interval for mean sales when advertising expenditures are $555.
e. Construct a 95 percent prediction interval for the value of sales when advertising expenditures are $555.
f. Test the null hypothesis H_0: $\beta_1 = 0$ at the $\alpha = 0.05$ significance level using the t-distribution.
g. Test the null hypothesis H_0: $\beta_1 = 0$ at the $\alpha = 0.05$ significance level using the F-distribution.

h. Would you expect the result determined in part f to be the same as that determined in part g? Why or why not?

14.7. Assume that the data given in Problem 14.6 are distributed according to the bivariate normal distribution.
a. Find the sample correlation coefficient.
b. Test the null hypothesis H_0: $\rho = 0$ at the $\alpha = 0.05$ significance level. At what significance level can H_0 be rejected?

14.8. In an analysis of the linear relationship between two random variables **X** and **Y**, it is found that $r^2 = 0.54$ based on a sample of 25 (x, y) data pairs. Does this indicate a significant (i.e., nonzero) correlation between **Y** and **X** at the $\alpha = 0.05$ significance level?

14.9. Assume that in analyzing the linear relationship between two random variables that are bivariate normally distributed, it is found that SST = 620.25 and SSR = 330.83 based on a sample of 23 observations. Further assume that the slope of a fitted sample regression line is negative.
a. What is the coefficient of determination for this problem?
b. Test the null hypothesis H_0: $\rho = 0$ at the $\alpha = 0.01$ significance level.

14.10. In a simple linear regression analysis involving 19 observations, it is found that SST = 4,910.13 and SSE = 735.87. Use the F-distribution to test the hypothesis H_0: $\beta_1 = 0$ at the $\alpha = 0.01$ significance level.

14.11. An associate dean for undergraduate instruction at a large midwestern university is attempting to determine whether there is a strong linear relationship between undergraduate grade point average (GPA) and the starting salaries received. A sample of ten recent graduates yielded the following results:

GPA	Starting salary
2.21	$10,400
2.23	10,900
2.45	18,450
2.61	14,800
2.63	16,420
2.87	15,990
3.19	17,250
3.26	17,800
3.28	18,400
3.77	19,500

a. Determine the least squares regression equation using GPA as the independent variable.
b. Determine the coefficient of determination, r^2.
c. Find a 95 percent prediction interval for an individual whose grade point average is 3.00.
d. What is the estimated mean starting salary for graduates with a grade point average of 2.80, derived from your estimating equation?
e. What is the 95 percent confidence interval estimate of the mean starting salary for the salary determined in part d?

14.12. Given the regression equation $\hat{y} = 32.96 - 2.65x$ based on 26 observations with $s_{b_0} = 13.45$, test the null hypothesis H_0: $\beta_0 = 0$ at the $\alpha = 0.05$ significance level.

14.13. In a given analysis of two random variables, the following quantities were determined, based on a sample of 100 units:

SSE = 456.78   SST = 3,890.42
$y = 7.705 - 1.43x$

a. Determine a point estimate of ρ.
b. Construct a 95 percent confidence interval for ρ.
c. What is the proper interpretation of the interval constructed in part b?

14.14. In a regression experiment involving 18 observations, it is found that SSE = 45.92 and SSR = 289.61. Use the t-distribution to test the hypothesis H_0: $\beta_1 = 0$ at the $\alpha = 0.05$ significance level.

14.15. A leading discount store sells six major brands of calculators. Unit prices along with corresponding sales amounts are given in the table.

Calculator price	Units sold
$ 19.95	385
29.95	416
34.95	322
59.95	186
79.95	102
139.99	47

a. Plot the data in the form of a scatter diagram. Do units sold appear to be linearly related to the price of the unit?
b. Fit a linear function to the data using the method of least squares.
c. Test the hypothesis H_0: $\beta_1 = 0$ at the $\alpha = 0.05$ significance level. Is price a useful variable to include in a simple linear regression equation for estimating unit sales?
d. What is the p-value (approximate) associated with the test in part c?
e. What other variables might one include in a regression equation for estimating unit sales?

14.16. In a given regression experiment involving 11 observations, it is determined that SSE = 27.45 and SSR = 24.71. Using the F-distribution, determine the approximate p-value (by interpolation) for the hypothesis H_0: $\beta_1 = 0$.

14.17. In a correlation analysis involving 39 subjects, it is determined that $r = 0.36$. Construct a 90 percent confidence interval for ρ. Why do we have to be careful in interpreting the interval constructed?

14.18. As a part of the evaluation process of sales representatives, a consumer goods manufacturer estimates sales in the second year for each representative based on his or her sales during the first year. Data on nine randomly selected sales representatives are given in the table.

First-year sales ($000)	Second-year sales ($000)
85.3	109.6
65.2	88.5
73.9	92.6
109.6	134.3
123.4	143.9
98.7	89.3
50.5	63.1
77.8	95.6
101.1	111.9

a. Construct a linear regression equation to estimate second-year sales as a function of first-year sales.

b. What proportion of the variation in second-year sales is accounted for or explained by first-year sales?

c. Test the null hypothesis H_0: $\beta_1 = 0$ at the $\alpha = 0.10$ significance level.

d. Construct a 95 percent prediction interval for the estimated second-year sales of an individual who reported first-year sales of 98.7($1,000$).

e. Construct a 95 percent confidence interval for mean second-year sales given first-year sales of 73.9($1,000$).

14.19. A firm is interested in determining whether its market share is linearly related to dollars expended on local television advertising in a small geographic area. Data collected on market share and corresponding local television advertising expenditures are shown in the table.

Television advertising expenditure ($000)	Market share (percent)
18	13
21	14
23	16
27	18
31	19
35	19

a. Assuming the data reported in the table are bivariate normally distributed, determine the sample correlation coefficient, r, beginning with the determination of the covariance of these two random variables.

b. Test the null hypothesis H_0: $\rho = 0$ at the $\alpha = 0.05$ significance level. Does there appear to be a high degree of linear relationship between these two random variables?

c. Suppose the firm decided in the next period to increase advertising expenditures to 65($1,000$) to increase its market share. Why would it be dangerous in this instance to use the determined linear regression equation to estimate market share?

14.20. Referring to Problem 13.23, test the hypothesis H_0: $\beta_1 = 0$ at the $\alpha = 0.05$ significance level. What is the p-value for the test, and what is the correct interpretation of this p-value? Is average daily temperature significant in predicting fuel use? Why or why not?

14.21. Refer to Problem 13.24, and assume Joe gained 999 yards in his senior year.

a. Test the null hypothesis H_0: $\beta_1 = 0$ at the $\alpha = 0.01$ significance level. What is the p-value for the test, and how is this number interpreted? At which significance levels, 0.10, 0.05, or 0.01, would we reject the null hypothesis that $\beta_1 = 0$ based on the computed p-value?

b. Construct a 95 percent confidence interval and prediction interval for Joe's expected earnings. Which of these intervals is "wider," and why?

c. Which job should Joe select?

14.22. Based on a random sample of 100 persons, a researcher finds that the correlation between number of pounds overweight and number of days ill during the past year is 0.42.

a. Test the hypothesis at the $\alpha = 0.05$ significance level that $\rho = 0$.

b. Determine a 95 percent confidence interval on ρ.

14.23. An MBA director is interested in determining whether a linear relationship exists between undergraduate grade point average (GPA) and a student's MBA GPA upon graduation. She feels that the correlation between the two grade point averages is positive, and that the correlation coefficient ρ is probably about 0.50. She takes a random sample of 50 MBA students who have recently received their degrees and finds that $r = 0.37$. Using the confidence interval on ρ, test the hypothesis H_0: $\rho = 0.50$ at the $\alpha = 0.05$ significance level.

14.24. Refer to Problem 13.25.

a. Is the number of years of postsecondary education significant in predicting the hourly wage rate in this firm? (Support your premise by interpreting the computed p-value of the test.)

b. Joe College is completing his fourth year at school and wants to determine what wage he might expect to receive if he decides to accept an offer with the firm in question. Joe's situation is complicated somewhat by the fact that the firm will not be able to tell him the specific salary (or equivalent hourly wage rate) he will receive until next week when the personnel manager returns from a college recruiting trip. Another firm has extended an offer to Joe, and he must tell them by the end of the current week whether he intends to accept their offer. If he does not, the other firm wants to make the offer to someone else immediately so that they are assured of getting someone to fill the position. Based on Joe's four years of postsecondary education, obtain a 95 percent prediction interval to help estimate the wage Joe might expect to receive from this firm.

14.25. Consider the test score versus studying time data given in Problem 13.26.

a. Determine a 99 percent confidence interval for β_0. What is the correct interpretation of this interval?

b. Determine a 99 percent confidence interval for β_1. What is the correct interpretation of this interval?

c. Test the null hypothesis H_0: $\rho = 0$ at the $\alpha = 0.01$ significance level.

d. Why would it be inappropriate to construct a confidence interval for ρ in this problem?

e. What other factors might one include in a regression model for estimating exam scores?

14.26. A high school counselor has surveyed 14 members of the senior class of 1970 and recorded (*a*) the individual's yearly income and (*b*) the number of years of higher education completed by the individual. The results of this survey are shown in the table.

Years of higher education	Annual income
0	$ 7,500
0	8,000
1	11,000
1	10,500
2	9,500
3	14,200
4	15,500
4	16,800
4	18,900
5	16,500
6	45,800
7	32,900
7	37,600
7	41,200

a. What is the correlation between years of higher education and annual income in this sample of 14 individuals?

b. Test the null hypothesis that $\rho = 0$ at the $\alpha = 0.01$ significance level.

14.27. Consider the sales versus advertising data given in Problem 13.27.

a. Test the hypothesis H_0: $\beta_1 = 0$ at the $\alpha = 0.05$ significance level.

b. Obtain a 95 percent prediction interval for sales when advertising expenditures are equal to 1.2($1,000).

14.28. An advertising firm believes that for its target market, the relationship between the ages of husbands and wives may be an important factor for consideration if the relationship is "strong": $\rho > 0.6$ or $\rho < -0.6$, where X = age of husband and Y = age of wife. In a random sample of 100 married couples, it is found that $r = 0.70$.

a. Determine an approximate 95 percent confidence interval for ρ.

b. By the firm's definition of a strong relationship, what would you conclude on the basis of part *a*? Explain.

14.29. Consider the data in Problem 13.28.

a. Test the hypothesis H_0: $\beta_1 = 0$ at the $\alpha = 0.05$ significance level. What is the approximate *p*-value for this test?

b. Construct a 95 percent confidence interval for the mean number of boards rejected for an employee with 20 weeks of experience.

14.30. For the data in Problem 13.29, test the null hypothesis H_0: $\rho = 0$ assuming that the paired (x, y) observations are bivariate normally distributed. Use the $\alpha = 0.05$ significance level for reporting your results.

14.31. Assuming the (x, y) pairs of values in Problem 13.30 are bivariate normally distributed, test the null hypothesis H_0: $\rho = 0$ at the $\alpha = 0.05$ significance level.

14.32. An insurance company is interested in determining whether there is a relationship between age and the amount of whole life insurance carried. In a random sample of 250 persons, it is found that $r = 0.12$, where X = age and Y = amount of whole life insurance.

a. Test the hypothesis that $\rho = 0$ at the $\alpha = 0.05$ level.

b. Determine an approximate 95 percent confidence interval for ρ.

c. Interpret your results in parts *a* and *b*. What is your conclusion regarding the relationship between age and amount of whole life insurance carried?

14.33. Assuming the pairs of observations

given in Problem 13.32 are bivariate normally distributed, test the null hypothesis at the $\alpha = 0.10$ level that the random variables **X** and **Y** are *not* correlated.

14.34. A researcher wishes to test the null hypothesis H_0: $\rho = 0$, where ρ is the Pearson product moment correlation coefficient. She finds that $r = 0.20$ and that $s_r = 0.14$, based on a sample of $n = 50$ (x, y) pairs.

What should her decision be, assuming a significance level of $\alpha = 0.05$?

14.35. For the data given in Problem 13.34, test the null hypothesis at the $\alpha = 0.05$ significance level that the annual consumption of alcoholic beverages and the number of automobile accidents in a certain locality are not correlated.

Multiple correlation and regression

15

Multiple linear regression and correlation analyses are extensions of the simple linear regression and correlation models studied in Chapter 13. In multiple linear regression and correlation, however, more than one independent variable is used to account for the variability in the dependent random variable Y.

One obvious advantage in using the multiple linear regression and correlation models over the simple linear regression and correlation models is in prediction and estimation—by having two or more independent variables to predict values of the random variable Y, we should be able to do a better job than by using just one independent variable. Another advantage is that in one statistical analysis, we can study three or more variables together rather than only two. A disadvantage of using multiple linear regression and correlation is the increased complexity of the analysis—in terms of both the computations required and the interpretation of the results. Computers have relieved much of the computational burden, however. The steps in the multiple linear regression model building process are given here.

Steps in the multiple linear regression model-building process

Step 1. Identify the variables to be included in building the multiple regression model.

Step 2. Collect sample data.

Step 3. Specify the relationship that exists between the dependent and the independent variables, and between the independent variables.

481

Step 4.   Estimate the parameters of the model specified.
Step 5.   Determine whether the assumptions of the multiple linear regression model have been met.
Step 6.   Statistically test for the usefulness of the model developed and for the individual terms in the model.
Step 7.   Use the model developed for prediction and estimation.

The majority of these steps are much more involved in the multiple linear regression model-building process than they are in the simple linear regression case. For example, many more variables usually have to be identified to be included in the model, and it is important to specify not only the relationship that exists between the dependent and each of the independent variables, but also the relationship between the independent variables themselves (Step 3). Consider, for example, the estimation of sales where the independent variables to be used are the price of the unit and the dollars spent on advertising. The relationship between advertising and sales may not be independent of the price of the unit. That is, with a higher price, the effect of advertising (in terms of the slope of the regression equation) may not be so great as when the price is low and the influence of advertising is much greater. We show these types of relationships in Figure 15.1. In Figure

FIGURE 15.1   Examples of interaction and no-interaction regression models

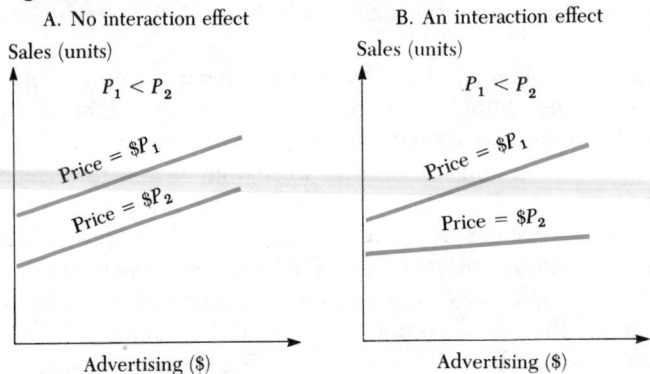

15.1A, the effect of advertising on sales is the same (as shown by the same slope for both regression lines) for each of the prices indicated (here, for explanatory purposes, we have shown only two different prices). In Figure 15.1B, on the other hand, the effect of advertising on sales differs depending on the sales price of the item. This is clearly shown by the different slopes of the regression lines. When the slope of the relationship between the dependent random variable Y and an independent variable x_1 depends on (or is not independent of) the value of one other independent variable, say x_2, the independent variables are said to *interact,* and it may be necessary to introduce a term reflecting the manner in which these variables interact into the regression equation to model the behavior of the dependent variable as a function of both independent variables. For this reason (as well as other reasons to be developed shortly), the development of a multiple linear regression model is much more complex than is the development of a simple linear regression model. Quite often, the statistician will find that he has to *repeat* many of the steps in the development of a multiple regression model to ensure that the model developed is useful for estimation and prediction purposes.

Because multiple linear regression and correlation analyses usually require the use of a computer to calculate estimates of the regression and correlation parameters, and because most colleges and universities have available for student use large-scale computers and appropriate statistical software to perform multiple regression and correlation analyses, our development of these two important techniques will concentrate on the interpretation of the output of one of the software computer routines, the general linear models procedure of the Statistical Analysis System (SAS). Since many other available software packages (computer programs) have output similar to SAS's, it should be relatively easy to interpret the output from these other software packages using the discussion in the text as a guide.

Multiple linear regression and correlation analyses are further best developed and understood by using matrix algebra. Since we do not presuppose a background for many students in matrix algebra, we will demonstrate the principal features of the analysis algebraically (without using matrix algebra) and show that, in fact, they are only extensions of simple linear regression and correlation. The student interested in exploring multiple linear regression and correlation further is advised to take a course in matrix or linear algebra, and *then* a course specifically devoted to multiple linear regression and correlation.

■ 15.2 THE MULTIPLE LINEAR REGRESSION MODEL

The multiple linear regression model is specified by the following:

The population multiple linear regression model

$$\mathbf{Y}_i = \beta_0 + \beta_1 x_{i,1} + \beta_2 x_{i,2} + \cdots + \beta_k x_{i,k} + \epsilon_i, \qquad i = 1, 2, \ldots, N$$

where:

$\mathbf{Y}_i = i$th dependent random variable corresponding to $x_{i,1}, x_{i,2}, \ldots, x_{i,k}$
$\beta_0, \beta_1, \beta_2, \ldots, \beta_k$ are $(k + 1)$ parameters in the model
$x_{i,j} = i$th level of the jth independent variable, $j = 1, 2, \ldots, k$
ϵ_i = random error term

Note that this specification of the population linear regression model does not preclude us from including the interaction effects of different variables, nor does it preclude us from including higher-order terms in the model, such as x^2, x^3, etc. That is, $x_{i,3}$ could just as easily represent the combined effects of $x_{i,1}$ and $x_{i,2}$ in the form $x_{i,1} \cdot x_{i,2}$ as it can represent another independent variable. Similarly, $x_{i,j+1}$ could represent $x_{i,j}^2$, and so on. Some statisticians refer to a model with interaction and higher-order terms as a *complete model*, and refer to a model lacking interaction and higher-order terms as a *reduced effects* or a *main effects model*. The model given allows us to include both of these through a judicious choice of independent variable definitions.

The key assumptions of the multiple regression model are basically the same as those of the simple linear regression model.

Assumptions of the multiple linear regression model

1. The expected value of the error component ϵ_i is 0 $[E(\epsilon_i) = 0]$, and the variance of the error component ϵ_i is σ^2 $[V(\epsilon_i) = \sigma^2]$, for $i = 1, 2, \ldots, N$.
2. The error components are uncorrelated.
3. $\beta_0, \beta_1, \ldots, \beta_k$ are $(k + 1)$ parameters, and $x_{i,1}, x_{i,2}, \ldots, x_{i,k}$ are known constants.

And, for purposes of hypothesis testing and confidence interval estimation,

4. The error component ϵ_i is normally distributed.

We shall illustrate the multiple linear regression model with two independent variables and no significant interaction terms by:

$$Y_i = \beta_0 + \beta_1 x_{i,1} + \beta_2 x_{i,2} + \epsilon_i, \qquad i = 1, 2, \ldots, N$$

By assumptions 1 and 3 given earlier, the regression function is given by $E(Y_i) = \beta_0 + \beta_1 x_{i,1} + \beta_2 x_{i,2}$. The nature of this function is illustrated in Figure 15.2. Since there are now two independent variables, the dependent variable Y becomes the third axis in three-dimensional space; the regression function is a *plane*.

The interpretation of the regression coefficients β_0, β_1, and β_2 in this two independent variable model are analogous to the coefficients in the simple linear regression model: β_0 is the Y-intercept, the point on the y-axis where it is intersected by the *plane*; β_1 gives the change in the value of the dependent variable Y when x_1 is incremented by one unit *and x_2 is held constant*; β_2 is similarly interpreted. The fitted regression plane $\hat{y} = b_0 + b_1 x_1 + b_2 x_2$ is determined by least squares: b_0, b_1, and b_2 are least squares point estimators of β_0, β_1, and β_2, respectively. The least squares function is:

$$\text{LS} = \sum_{i=1}^{n} [y_i - (\beta_0 + \beta_1 x_{i,1} + \beta_2 x_{i,2})]^2 = \sum_{i=1}^{n} \epsilon_i^2$$

FIGURE 15.2 Bivariate regression function

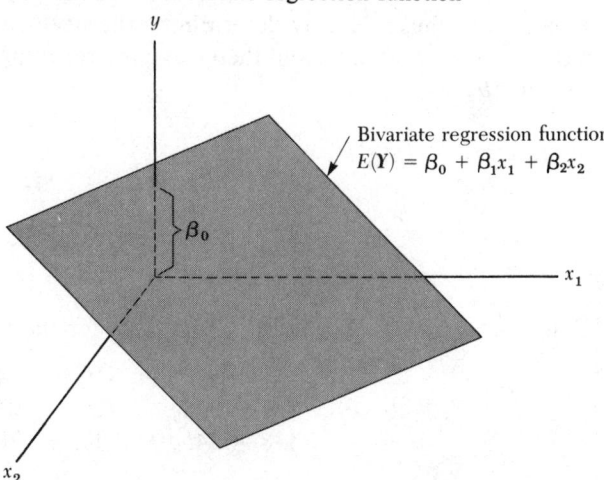

and the values of β_0, β_1, and β_2 that minimize LS are the expected values of the least squares estimators, b_0, b_1, and b_2, respectively.

We will continue our discussion of the bivariate regression model by explaining how one determines algebraically the point estimates of the regression coefficients β_0, β_1, and β_2; the point estimates of the variance of the error terms, σ^2, and the coefficient of multiple determination, ρ^2.

▌ 15.3 POINT ESTIMATION OF THE POPULATION REGRESSION PARAMETERS—BIVARIATE REGRESSION CASE

□ **15.3.1 Point estimation of the regression coefficients, β_0, β_1, and β_2**

The values of b_0, b_1, and b_2 that minimize LS are solutions to the following normal equations:

$$\Sigma y = nb_0 + b_1 \Sigma x_1 + b_2 \Sigma x_2$$
$$\Sigma x_1 y = b_0 \Sigma x_1 + b_1 \Sigma x_1^2 + b_2 \Sigma x_1 x_2$$
$$\Sigma x_2 y = b_0 \Sigma x_2 + b_1 \Sigma x_1 x_2 + b_2 \Sigma x_2^2$$

These equations can be solved algebraically for b_0, b_1, and b_2, but the resulting formulas are quite complicated. Usually, these equations are solved on a computer by using a multiple regression computer program. For illustrative purposes, we will use a "contrived" problem to develop point estimates of the multiple regression parameters. In this problem, the numbers are very "pleasant," since we will be using "hand" solution. In the next section, we will present a multiple regression, computer-analyzed case study to summarize the concepts discussed here.

Example 15.1 Consider the data given in Table 15.1. Fit the regression line $\hat{y} = b_0 + b_1x_1 + b_2x_2$ to these data by determining the necessary summations for use in the normal equations,[1] and then solve the resulting normal equations for b_0, b_1, and b_2.

TABLE 15.1
Bivariate regression data

1	*2*	*3*
x_1	x_2	y
4	3	3
4	4	2
4	3	7
6	4	6
3	2	5
6	4	6
3	2	7
2	2	4

Solution The normal equations for the data are:

$$40 = 8b_0 + 32b_1 + 24b_2$$
$$164 = 32b_0 + 142b_1 + 104b_2$$
$$118 = 24b_0 + 104b_1 + 78b_2$$

Solution of these normal equations yields point estimates $b_0 = 6$, $b_1 = 2$, and $b_2 = -3$ (verify). Thus, the fitted regression line is $\hat{y} = 6 + 2x_1 - 3x_2$.

Scatter plots of x_1 versus y and x_2 versus y are shown in Figure 15.3. The

FIGURE 15.3 Scatter plots of x_1 versus y and x_2 versus y

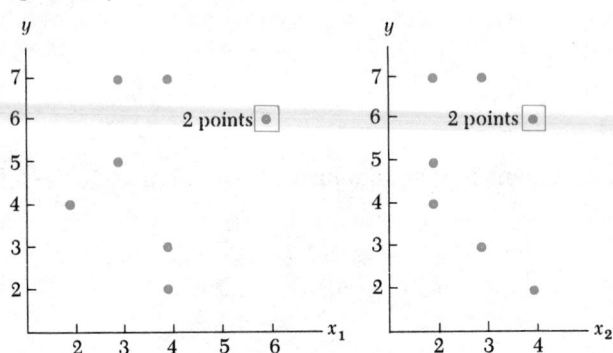

first independent variable, x_1, appears to have little, if any, linear relationship with the random variable **Y**, while the second independent variable, x_2, appears to have a weak negative linear relationship with **Y**.

---

[1] Note that it is possible to drop the i-subscript from the x-values when we are not referring to specific values of x.

□ 15.3.2 Point estimation of the variance, σ^2, of the error terms

The point estimate of the variance of the conditional probability distributions in multiple linear regression analysis is denoted by $s^2_{Y/x_1,x_2,...,x_k}$. Its computation is shown in equation (15.1). Note in equation (15.1) that the numerator is the same as in the simple linear regression model—namely, we square the difference between the observed value of the dependent variable, y, and the estimate derived through the linear regression equation, $\hat{y}$; then sum these squares. The expression shown in the numerator is referred to as the error sum of squares, or SSE, just as in the simple linear regression model. Note in the denominator of equation (15.1), however, that the divisor is $n - (k + 1)$. This is because $(k + 1)$ degrees of freedom are lost in estimating the $(k + 1)$ parameters in the multiple regression model, $\beta_0, \beta_1, \beta_2, \ldots, \beta_k$. Experience in determining this estimate for the bivariate regression case is provided in the following example.

Point estimate of conditional probability distribution
variance—multiple linear regression model

$$s^2_{Y/x_1,x_2,...,x_k} = \frac{\sum_{i=1}^{n} (y_i - \hat{y}_i)^2}{n - (k + 1)} = \frac{SSE}{n - (k + 1)} \quad (15.1)$$

Example 15.2 Using the data given in Example 15.1, determine the point estimate of the conditional probability distribution variance.

Solution Relevant computations are given in Table 15.2. Using the data in this table,

$$s^2_{Y/x_1,x_2} = \frac{SSE}{n - (k + 1)} = \frac{10}{5} = 2$$

TABLE 15.2 Determination of $s_{Y/x_1,x_2}$ using data from Example 15.1

	x_1	x_2	y	$\hat{y}$	$y - \hat{y}$	$(y - \hat{y})^2$
1	4	3	3	5	−2	4
2	4	4	2	2	0	0
3	4	3	7	5	2	4
4	6	4	6	6	0	0
5	3	2	5	6	−1	1
6	6	4	6	6	0	0
7	3	2	7	6	1	1
8	2	2	4	4	0	0
Totals	32	24	40	40	0	10

☐ 15.3.3 Point estimation of the coefficient of multiple determination, ρ^2

The coefficient of multiple determination is defined similarly to the coefficient of determination in the simple linear regression case. That is, the coefficient of multiple determination is the proportion of variability in the random variable **Y** accounted for or explained by the independent variables x_1, x_2, . . . , x_k. This sample statistic measures the *relative* strength of the linear relationship between the dependent random variable and the independent variables. The value of $\mathbf{r}^2$ is computed using equation (15.2).

Sample coefficient of multiple determination, r^2

$$r^2 = \frac{SST - SSE}{SST} \qquad (15.2)$$

Example 15.3 Using the data given in Example 15.1, compute the sample coefficient of multiple determination.

Solution From Table 15.2, SSE = $\Sigma(y - \hat{y})^2$ = 10. And, SST = $\Sigma(y - \bar{y})^2$ = 24 (verify!). Thus,

$$r^2 = \frac{24 - 10}{24} = 0.5833$$

The bivariate regression accounts for 58.33 percent of the variability in the sample Y-values.

☐ 15.3.4 Residual analysis in multiple linear regression

As in simple linear regression, a most important step in the evaluation of the regression model lies in an examination of the sample residuals (Step 5); these are plotted in Figure 15.4. These plots show that the data have been "rigged" in order to present us with "pleasant" numbers—both plots are too

FIGURE 15.4 Residual plots for data in Tables 15.1 and 15.2

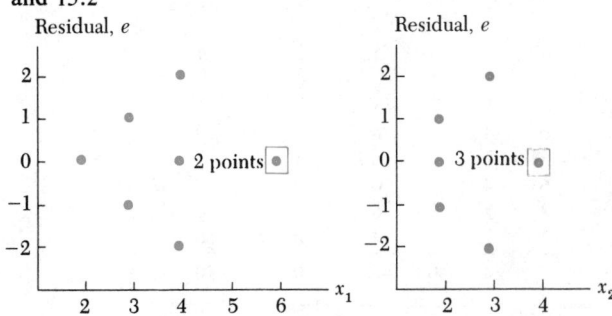

systematic for the Y-observations to have occurred at random! The plot of the residuals versus x_1, for example, shows an increasing variability in the conditional distributions as x_1 increases, and then the variability decreases to 0. This plot is *similar* to plot C in Figure 13.14. The plot of the residuals versus x_2 also shows that the variability is conditional on the level of x_2; as x_2 increases from 2 to 3, the variability increases, and as x_2 increases from 3 to 4, the variability again decreases to 0! The underlying assumptions of the multiple linear regression model simply have not been met with these "rigged" data.

■ 15.4   A COMPUTER-ANALYZED MULTIPLE REGRESSION EXAMPLE

The firm described in Section 13.2 desires to develop a multiple regression model to predict sales of each of its member branches. In addition to the independent variable "advertising expenditure" already included, the firm feels that data on the average income level in each of its sales districts can prove beneficial in estimating sales. Hence, the firm has sampled 15 of its sales branches and recorded the sales amount, the amount spent on advertising, and the average income level in each sampled sales district. This information has been assembled and is presented in Table 15.3. Scatter plots of sales and advertising expenditures and sales and income level are given in Figures 15.5 and 15.6, respectively. Both figures indicate the reasonableness of the straight-line model, although the firm may want to consider the introduction of nonlinear terms x_1^2 and x_2^2 in the model at some future time (we will return to this point later) to assess the impact of these terms on model development.

TABLE 15.3   Survey data for estimating sales

Sales district	Average income level ($000)	Advertising expenditures ($000)	Sales ($000)
1	15.7	1.814	55.420
2	13.1	1.112	45.819
3	6.2	0.718	16.940
4	10.7	1.421	35.818
5	16.3	3.085	85.090
6	17.5	2.119	62.025
7	8.4	0.525	17.918
8	11.1	1.108	32.845
9	13.2	1.621	41.180
10	14.8	2.645	62.910
11	18.9	2.927	87.013
12	7.6	0.847	22.150
13	7.8	0.621	16.660
14	10.0	0.981	27.050
15	12.1	1.394	39.121

FIGURE 15.5 Scatter diagram of sales and advertising expenditures from data in Table 15.3

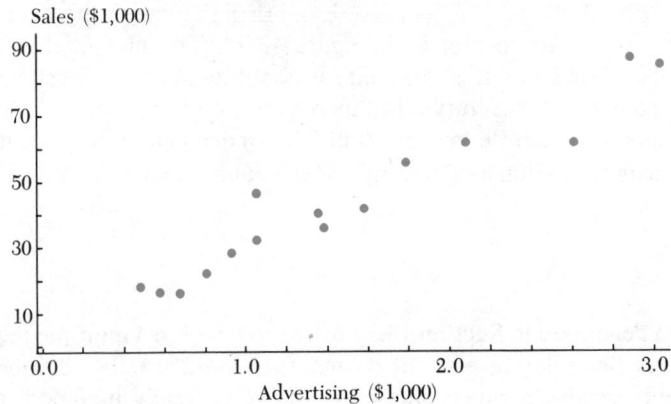

FIGURE 15.6 Scatter diagram of sales and average income level from data in Table 15.3

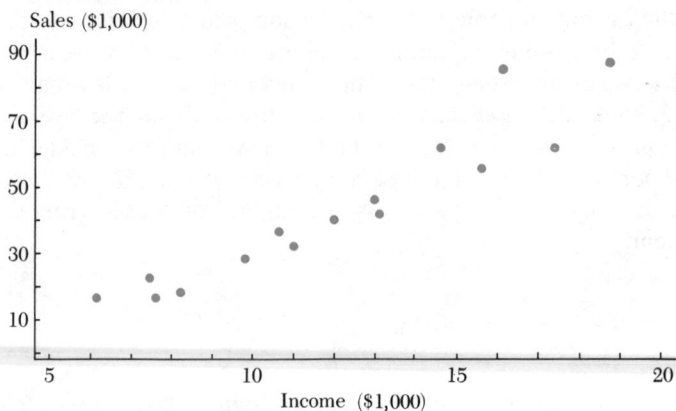

Table 15.4 contains the results of a computer analysis of the data in Table 15.3 using the main effects model. Much of the data shown in the table (printout) have not yet been discussed, but will be discussed later in the chapter. We have shaded portions of Table 15.4 that correspond to the point estimates discussed in Section 15.3; we will describe the other quantities in Table 15.4 in the following section.

The least squares estimates of the β-parameters appear in the column labeled ESTIMATE, and a description of each of these parameters appears in the column labeled PARAMETER. From Table 15.4 one can see that $b_0 = -11.8888$ (INTERCEPT), $b_1 = 17.2336$ (ADDS), and $b_2 = 2.3500$ (INCOME). Hence, the multiple regression estimating equation is:

$$\hat{y} = -11.8888 + 17.2336x_1 + 2.3500x_2$$

TABLE 15.4 Computer analysis of data in Table 15.3

DEPENDENT VARIABLE: SALES

SOURCE	DF	SUM OF SQUARES	MEAN SQUARE	F VALUE	PR > F
MODEL	2	7290.27547531	3645.13773766	215.87	0.0001
ERROR	12	202.63008162	16.88584013		
CORRECTED TOTAL	14	7492.90555693			

R-SQUARE	STD DEV
0.972957	4.10923839

PARAMETER	ESTIMATE	T FOR HO: PARAMETER = 0	PR > ITI	STD ERROR OF ESTIMATE
INTERCEPT	-11.88882579	-2.75	0.0175	4.31808626
ADDS	17.23357834	5.95	0.0001	2.89439663
INCOME	2.34998673	3.79	0.0026	0.62210994

where:

$\hat{y}$ = estimated sales ($000)
x_1 = advertising expenditures ($000)
x_2 = average income level ($000)

To estimate (predict) sales for a sales district with an average income of $16,500 and an advertising expenditure of $1,100, we would substitute the quantity 1.100 for x_1 and 16.5 for x_2 into the estimating equation to obtain:

$$\hat{y} = -11.8888 + 17.2336(1.100) + 2.3500(16.5) = 45.8432,$$

or we would estimate average sales to be $45,843.20 for the given levels of the independent variables.

The estimate of the population coefficient of multiple determination in this regression model is $r^2 = 0.97296$. Thus, we would state that the two independent variables "average income level" and "advertising expenditure" account for approximately 97.30 percent of the variability in the random variable sales. Since r^2 is close to 1 in this example, we would state that there appears to be a strong (relative) linear relationship between sales and advertising expenditures and average income level in these sample data.

The estimate of the conditional distribution variance σ^2 is found in the column labeled MEAN SQUARE in the row labeled ERROR. That is,

$$s^2_{Y/x_1,x_2} = \text{MSE} = \frac{\text{SSE}}{df} = \frac{\text{SSE}}{n - (k + 1)} = \frac{202.6301}{15 - 3} = 16.886$$

where df represents the degrees of freedom used in calculating the estimate of the conditional probability distribution variance. Since we have 15 observations in this example, and are using two independent variables to estimate sales, the degrees of freedom used in the computation of $s^2_{Y/x_1,x_2}$ are 15 − (2 + 1) = 12. The quantity labeled STD DEV in Table 15.4 is the square root of the estimate of the population conditional distribution variance, or:

$$s_{Y/x_1,x_2} = \sqrt{16.886} = 4.1092$$

■ 15.5 HYPOTHESIS TESTING IN MULTIPLE LINEAR REGRESSION

☐ 15.5.1 Individual test of the regression coefficients, β_i

To test the hypothesis that β_1 or β_2 is equal to 0 in the bivariate regression case, we need to estimate the standard deviation of the estimators b_1 and b_2. These formulas are quite complex algebraically but would be printed out in most multiple regression computer programs. For our "contrived" example, they are given in Table 15.5.

A confidence interval placed on either β_1 or β_2 is calculated in the same way as the slope in the simple linear regression model, *except* that the

TABLE 15.5 Point estimates, b_0 and b_1, and their standard errors, s_{b_1} and s_{b_2}

Parameter	Estimate	Standard error
β_1	$b_1 = 2$	$s_{b_1} = \sqrt{(2)(3/10)} = 0.775$
β_2	$b_2 = -3$	$s_{b_2} = \sqrt{(2)(7/10)} = 1.183$

degrees of freedom are now $n - (k + 1) = 8 - 3 = 5$. A 95 percent confidence interval for β_1, for example, is given by

$$b_1 - t_{\alpha/2;n-3}s_{b_1} \le \beta_1 \le b_1 + t_{\alpha/2;n-3}s_{b_1}$$
$$2 - (2.571)(0.775) \le \beta_1 \le 2 + (2.571)(0.775)$$
$$0.007 \le \beta_1 \le 3.993$$

Thus, we are 95 percent confident that β_1 lies between 0.007 and 3.993. Since this interval does not contain 0, we would further conclude at the 0.05 significance level that β_1 is not 0. But, in the multiple regression model, this does not necessarily imply that this factor or this regression line is useful for prediction purposes. To determine whether the fitted regression line is a useful predictor, rather than testing $\beta_1 = 0$ and $\beta_2 = 0$ *separately*, a joint test should be performed. We will describe this joint test after testing whether $\beta_1 = 0$ or $\beta_2 = 0$ in our sales-advertising-income example.

Example 15.4 Using the data in Table 15.4, test the separate hypotheses that $\beta_1 = 0$ and $\beta_2 = 0$ at the 95 percent confidence level.

Solution The form of the hypothesis test is given below. In Table 15.6, we have shaded the necessary quantities for performing these hypothesis tests. The STD ERROR OF ESTIMATE values are the estimates of the standard deviations of the sampling distributions of the point estimators of the regression coefficients, β_i. The quantities T FOR HO: PARAMETER = 0 are the ESTIMATES divided by the STD ERROR OF ESTIMATES. Thus, for example,

$$t = \frac{b_1}{s_{b_1}} = \frac{17.2336}{2.8944} = 5.95;$$

Two-tailed hypothesis test of regression coefficients in multiple regression model

Model: $Y = \beta_0 + \beta_1 x_1 + \beta_2 x_2 + \cdot \ \cdot \ \cdot + \beta_k x_k + \epsilon$

Hypothesis: H_0: $\beta_i = 0$
$\qquad\qquad\quad$ H_A: $\beta_i \ne 0$

Test statistic value: $t = \dfrac{b_i}{s_{b_i}}$

Decision rule: Reject if: $t > t_{\alpha/2;n-(k+1)}$ or $t < -t_{\alpha/2;n-(k+1)}$

> Two-tailed hypothesis test (*continued*)
> Do not reject if: $-t_{\alpha/2 : n-(k+1)} \leq t \leq t_{\alpha/2 : n-(k+1)}$
>
> where:
>
> n = number of observations
> k = number of independent variables

$t_{0.025;12} = 2.179$. Since $t > t_{0.025;12}$, we reject the null hypothesis that $\beta_1 = 0$. For β_2, we have the following:

$$t = \frac{b_2}{s_{b_2}} = \frac{2.3500}{0.6221} = 3.78$$

Since $3.78 > 2.179$, we again reject the null hypothesis that the regression coefficient (β_2 in this case) is equal to 0.

One delightful aspect of using a canned computer program like the one shown (SAS) is that we can request the program to perform the hypothesis test for us. The quantities labeled PR $> |T|$ in Table 15.6 are the p-values for the regression coefficients as they were described in Chapter 10. The interpretation of these p-values is, for example, for β_2, that either β_2 is not equal to 0, or else we obtained one of the 26 samples in 10,000 (0.0026) that yields a t-statistic of 3.78 or larger when in fact $\beta_2 = 0$. Since this is highly unlikely, we reject the null hypothesis that $\beta_2 = 0$. The p-value for β_1 is likely much less than the estimate 0.0001 shown, but the computer program prints only four digits to the right of the decimal point.

We should emphasize that the tests do *not* indicate that both β_1 *and* β_2 are unequal to 0 at the stated significance levels. If, for example, we repeatedly perform hypothesis tests at the $\alpha = 0.05$ level, then inevitably there will be *some* rejections of the null hypothesis when in fact all β_i could be equal to 0. The F-test in the following section is a joint test for all β_i, where $i = 1, 2, \ldots,$ k. The F-test will allow us to conclude either that all β_i are equal to 0 or that some coefficients in the model are significantly different from 0. The F-test solves what is referred to as the "family-wise" problem in determining the significance of the model (i.e., all terms in the model).

We can use the t-test to test for the significance of interaction terms in the model given in Table 15.4. That is, the multiple linear regression model that includes the interaction of advertising expenditures and average income level (denoted by $x_1 x_2$) can be examined, and a test of the hypothesis H_0: $\beta_3 = 0$ performed, where β_3 is the coefficient of the interaction term $x_1 x_2$ in the multiple linear regression model. Table 15.7 gives the multiple regression model with this term included. Note that the hypothesis H_0: $\beta_3 = 0$ would *not* be rejected at the $\alpha = 0.1, 0.05$, and 0.01 significance levels. This result can be read directly from the data given in Table 15.7, since the p-value for this hypothesis test is 0.1258.

TABLE 15.6 Computer analysis of data in Table 15.3

DEPENDENT VARIABLE: SALES

SOURCE	DF	SUM OF SQUARES	MEAN SQUARE	F VALUE	PR > F
MODEL	2	7290.27547531	3645.13773766	215.87	0.0001
ERROR	12	202.63008162	16.88584013	R-SQUARE	STD DEV
CORRECTED TOTAL	14	7492.90555693		0.972957	4.10923839

PARAMETER	ESTIMATE	T FOR H0: PARAMETER = 0	PR > |T|	STD ERROR OF ESTIMATE
INTERCEPT	-11.88882579	-2.75	0.0175	4.31808626
ADDS	17.23357834	5.95	0.0001	2.89439663
INCOME	2.34998673	3.78	0.0026	0.62210994

TABLE 15.7 Multiple linear regression analysis of the model

$Y = \beta_0 + \beta_1 x_1 + \beta_2 x_2 + \beta_3 x_1 x_2 + \epsilon_i$

DEPENDENT VARIABLE: SALES

SOURCE	DF	SUM OF SQUARES	MEAN SQUARE	F VALUE	PR > F
MODEL	3	7330.73788204	2443.57929401	165.75	0.0001
ERROR	11	162.16767489	14.74251590	R-SQUARE	STD DEV
CORRECTED TOTAL	14	7492.90555693		0.978357	3.83959840

PARAMETER	ESTIMATE	T FOR H0: PARAMETER = 0	PR > |T|	STD ERROR OF ESTIMATE
INTERCEPT	-0.66244604	-0.08	0.9346	7.88662644
ADDS(X1)	5.56061422	0.74	0.4767	7.54718047
INCOME(X2)	1.59886111	2.17	0.0529	0.73719684
X1X2	0.73959362	1.66	0.1258	0.44642972

☐ 15.5.2 Tests of the significance of the model—the *F*-test

To test the significance of the model developed, we must perform a *joint* test of the regression coefficients. That is, we must test the following hypothesis:

$$H_0: \quad \beta_1 = \beta_2 = \beta_3 = \cdot \cdot \cdot = \beta_k = 0$$
$$H_A: \quad \text{At } least \text{ one of the coefficients is nonzero}$$

The reason we must perform a joint test as outlined above is that if we, for example, perform a *t*-test at the $\alpha = 0.05$ significance level on both β_1 and β_2 in a bivariate regression model, the joint level of significance will not be $\alpha = 0.05$—it will be slightly greater than 0.05. For more than two independent variables in the regression model, the joint level of significance will be even greater! The correct test for the significance of the regression model is the *F*-test as depicted in Figure 15.7. The reader may recognize the procedure used to test the significance of the model as analysis of variance described in Chapter 12.

The format of the analysis of variance is given in Table 15.8, and Table 15.9 gives the analysis of variance for our "contrived" problem.

FIGURE 15.7 Steps in the analysis of variance for testing the significance of the multiple linear regression model

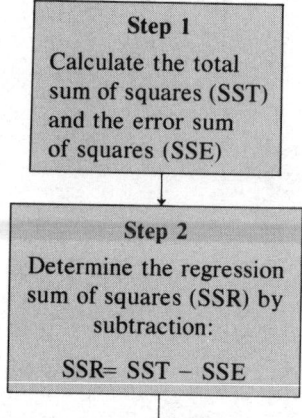

Step 1

Calculate the total sum of squares (SST) and the error sum of squares (SSE)

Step 2

Determine the regression sum of squares (SSR) by subtraction:

SSR= SST − SSE

Step 3

Calculate:

$$f = \frac{\text{SSR}/k}{\text{SSE}/(n - k - 1)} = \frac{r^2/k}{(1 - r^2)[n - (k + 1)]}$$

If $f > F(1 - \alpha; k, n - k - 1)$ then reject the hypothesis that all regression coefficients are equal to 0.

TABLE 15.8 Analysis of variance table for testing the significance of the regression line

Source of variation	Degrees of freedom	Sum of squares	Mean square	F-ratio
Regression (model)	k	SSR	$MSR = SSR/k$	MSR/MSE
Error	$n - k - 1$	SSE	$MSE = SSE/[n - (k + 1)]$	
Total	$n - 1$	SST		

TABLE 15.9 Analysis of variance for example data

Source of variation	Degrees of freedom	Sum of squares	Mean square	F-ratio
Regression (model)	2	14	7	$7/2 = 3.5$
Error	5	10	2	
Total	7	24		

From Table B.7 in Appendix B, the critical F-value at the $\alpha = 0.05$ level is $F(0.95; 2, 5) = 5.79$. Since $f = 3.5 < 5.79$, we *cannot* reject the null hypothesis that *both* β_1 and β_2 are equal to 0. This implies that there is a high probability that the regression function is a *useless* predictor for this dependent variable.

Notice that our t-test (confidence interval) on β_1 leads us *not* to conclude that $\beta_1 = 0$ (although the interval almost contained 0). The problem in using the t-test is that when we perform a t-test, say at the $\alpha = 0.05$ level, on both β_1 and β_2, the *joint* level of significance for the two t-tests will *not* be $\alpha = 0.05$; it will be slightly greater than 0.05. The correct procedure in this case is the F-test as we have outlined it.

Example 15.5 Using the data given in Table 15.4, test the null hypothesis for the sales-advertising-income example that $\beta_1 = \beta_2 = 0$ using the analysis of variance F-test with $\alpha = 0.05$.

Solution In the shaded portion of Table 15.10, we show that portion of Table 15.4 that corresponds to the analysis of variance; $F(0.95; 2, 12) = 3.89$ (see Table B.7 in Appendix B). Since $f = 215.87 > 3.89$, we reject the null hypothesis that *both* $\beta_1 = 0$ *and* $\beta_2 = 0$. Note that the SAS output provides us with the p-value for the test under the column labeled PR > F. Either β_1 and β_2 are not both equal to 0, or else we obtained one sample out of 10,000 (0.0001) that would yield an F-value of 215.87 or larger when in fact β_1 and β_2 are both equal to 0. Since the probability of obtaining one of these samples is extremely small in this case, we reject the null hypothesis that both coefficients are simultaneously equal to 0, and we conclude that the regression equation developed is useful for estimating average sales levels in the firm's branches for the range of advertising levels and the average income levels examined.

TABLE 15.10 Analysis of variance table for data in Table 15.3

DEPENDENT VARIABLE: SALES

SOURCE	DF	SUM OF SQUARES	MEAN SQUARE	F VALUE	PR > F
MODEL	2	7290.27547531	3645.13773766	215.87	0.0001
ERROR	12	202.63008162	16.8858 4013		
CORRECTED TOTAL	14	7492.90555693			

	R-SQUARE		STD DEV
	0.972957		4.10923839

PARAMETER	ESTIMATE	T FOR HO: PARAMETER = 0	PR > ITI	STD ERROR OF ESTIMATE
INTERCEPT	−11.88882579	−2.75	0.0175	4.31808626
ADDS	17.23357834	5.95	0.0001	2.89439663
INCOME	2.34998673	3.78	0.0026	0.62210994

■ 15.6 USING THE MULTIPLE LINEAR REGRESSION MODEL FOR ESTIMATION AND PREDICTION

In Section 14.2.5, we showed how to obtain a point estimate of $\mu_{Y/x}$ as well as a confidence interval estimate of this same quantity. Then, in Section 14.2.6, we showed how to obtain a point estimate of the "next" value of Y, Y_{next}, as well as a prediction interval for this quantity. You will recall that the point estimates in both cases were the same for the simple linear regression model—we simply substituted the appropriate value of x into the derived regression equation to obtain a point estimate of the mean of the Y-values or the next Y-value. Constructing confidence and prediction intervals for these same quantities was quite another matter, however. The prediction interval for the next Y-value was much wider than the confidence interval for the mean, reflecting the added variability of estimating a specific conditional distribution of Y given a specific value of x.

The estimation of a prediction interval for specific values (i.e., the next value) of the random variable Y as well as for the mean of a conditional distribution are quite complicated algebraically in the multiple linear regression case. Fortunately, computer programs sometimes provide these estimates for us by simply indicating that we desire such estimates by inputting a "key" parameter on our data input cards.

In Table 15.11 we show the SAS output of the predicted and observed values of the random variable Y sales for our sales-advertising-income example, as well as the error terms or the residuals used for checking the assumptions of the multiple linear regression model. These estimates were determined by computer by substituting the given levels of the independent

TABLE 15.11 Estimated values of the dependent random variable sales and corresponding residuals

OBSERVATION	OBSERVED VALUE	PREDICTED VALUE	RESIDUAL
1	55.42	56.27	−0.85
2	45.82	38.06	7.76
3	16.94	15.05	1.89
4	35.82	37.74	−1.93
5	85.09	79.58	5.51
6	62.03	65.75	−3.73
7	17.92	16.90	1.02
8	32.85	33.29	−0.45
9	41.18	47.07	−5.89
10	62.91	68.47	−5.56
11	87.01	82.99	4.04
12	22.15	20.57	1.58
13	16.66	17.14	−0.48
14	27.05	28.52	−1.47
15	39.12	40.57	−1.45

Source: Computer analysis of data in Table 15.3.

variables into the regression equation as was done in Section 15.4 to estimate (for example) expected sales when advertising expenditures were $1,100 and the average income was $16,000.

In Table 15.12 we show the SAS-produced 95 percent confidence interval for the mean sales level when the amount spent on advertising is $621 and the average income is $7,800, and the 95 percent prediction interval for the individual or the next sales level for an advertising expenditure of $621 and a corresponding average income level of $7,800. You will note from the estimates given in Table 15.12 that the interval for predicting the "next" Y-value is much wider, reflecting the added uncertainty in predicting an individual observation as opposed to the mean of a distribution.

TABLE 15.12 Ninety-five percent confidence interval and prediction interval for sales when advertising expenditures = $621 and average income level = $7,800

		Confidence Interval	
OBSERVED VALUE	PREDICTED VALUE	LOWER 95% CL FOR MEAN	UPPER 95% CL FOR MEAN
16.66000000	17.14312288	13.54053760	20.74570815

Width of
Confidence Interval

|←——7.205——→|

		Prediction Interval	
OBSERVED VALUE	PREDICTED VALUE	LOWER 95% CL INDIVIDUAL	UPPER 95% CL INDIVIDUAL
16.66000000	17.14312288	7.49223318	26.79401258

Width of
Prediction Interval

|←——————————19.302——————————→|

Source: Computer analysis of data in Table 15.3.

Not all computer software programs have the capacity to produce confidence interval estimates or prediction interval estimates of values of the dependent random variable **Y** in a multiple regression analysis. This is unfortunate, since the culmination of a regression analysis generally consists of estimating a value of the dependent random variable as well as placing a confidence interval or a prediction interval on this estimate. Fortunately, many existing software computer regression programs are being updated to include interval estimation of the dependent random variable. Many of the advanced references listed at the end of the chapter include the formulas necessary for obtaining these interval estimates, although their computation is often quite complex and requires the use of a computer or a *great deal of time* and a calculator to determine their values!

■ 15.7 THE TREATMENT OF QUALITATIVE VARIABLES IN MULTIPLE LINEAR REGRESSION

Our discussion of regression to this point has allowed only for the inclusion of quantitative independent variables such as advertising expenditures and income level in the estimating equation. Frequently, we would like to be able to include qualitative or indicator variables in the regression equation to estimate the value of a dependent random variable. Examples of indicator or qualitative variables (they are also called *dummy* or *binary* variables) are sex, geographic location, and season. For example, if we were to develop an estimating equation for the weight of an individual, we might attempt to relate (in a regression sense) weight to the individual's height—in general, the taller a person is, the more the person tends to weigh on the average. We might also include a variable that indicates the sex of the individual—in general, we would expect a male to weigh more than a female for a given height. The form of the estimating equation might be:

$$Y_i = \beta_0 + \beta_1 x_{i,1} + \beta_2 x_{i,2} + \epsilon_i$$

where

Y_i = random variable weight

$x_{i,1}$ = height of a sampled unit in the population

$x_{i,2}$ = indicator variable for the sex of the sampled unit in the population, where, for example:

$$x_{i,2} = \begin{cases} 1 \text{ if the sampled unit is a male} \\ 0 \text{ if the sampled unit is a female} \end{cases}$$

ϵ_i = random error term

A graph of a population regression equation such as that given above might appear as in Figure 15.8. Note that if the sampled unit is a male, the population regression function becomes:

$$Y = (\beta_0 + \beta_2) + \beta_1 x_1 + \epsilon$$

and if the sampled unit is a female, the population regression function becomes:

$$Y = \beta_0 + \beta_1 x_1 + \epsilon$$

The slope in both estimating equations is β_1. The intercept is $(\beta_0 + \beta_2)$ for a male and β_0 for a female. Thus, β_2 measures the differential effect of sex on average weight. If β_2 is positive, then expected weight is greater for a male equal in height to a female; if β_2 is negative, expected weight is less for a male equal in height to a female. Since β_2 is positive in Figure 15.8, we would expect a male's weight to exceed that of his female counterpart who is equal in height.

We need not restrict ourselves to two classes in the use of the indicator or dummy variables. Indeed, if we desire to model the seasons of the year, for

FIGURE 15.8 Hypothetical population regression lines for weight as a function of height and sex

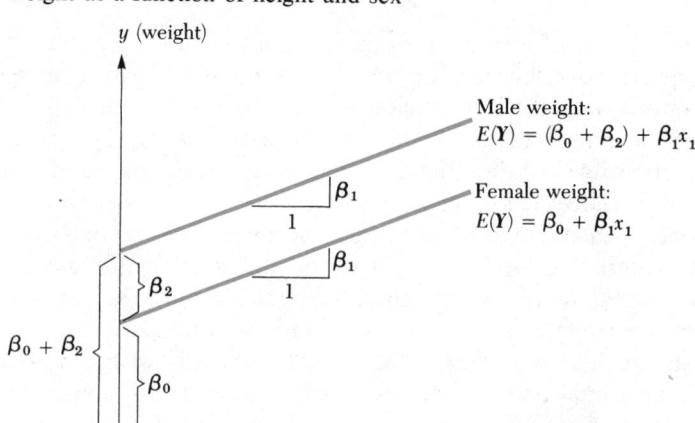

example, (winter, spring, summer, and fall), we would require the use of more than one indicator variable. In general, if there are m classes or distinguishable groups to be used as independent variables in a multiple regression, there will be $j = (m - 1)$ indicator or dummy variables required. To see how to model an indicator or a dummy variable where more than one class is involved, consider the following example.

Example 15.6 A ski and leisure resort in western Colorado has been attempting to predict accidents that require the use of plaster casts in its area as a function of the growth in the number of visitors (mostly skiers) in order to staff its needs for orthopedic medical facilities. Table 15.13 gives data for the past six years (by quarter) on the number of people visiting the facility and the corresponding number of accidents requiring plaster casts at the facility.

As a first step, the firm developed a simple linear regression model to estimate the number of accidents using visitors to the area as the independent variable. The results of this analysis are shown in Table 15.14. As shown in this table, the estimating equation is: $\hat{y} = 9.03 + 0.9589x$, where $\hat{y}$ = predicted number of plaster casts, and x = number of visitors to the facility (in thousands). The firm can fairly accurately predict the number of visitors to its facility, since demand far exceeds supply and reservations must be made well in advance (sometimes as much as 18 months) of the scheduled arrival of the tourists. The firm is disappointed with the proportion of sample variation explained by the simple linear regression equation ($r^2 = 0.7296$), and wonders whether taking seasonal factors into account (more accidents during snow!) might not help in predicting the need for plaster casts.

Construct a table corresponding to the input variables for the multiple

TABLE 15.13 Vistors to a Colorado ski resort and number of accidents requiring the use of a plaster cast

Year and quarter	Vistors (000)	Casts	Year and quarter	Vistors (000)	Casts
1975			1978		
Winter	33.63	53	Winter	74.64	94
Spring	36.46	41	Spring	80.31	86
Summer	41.18	24	Summer	80.97	64
Fall	43.16	57	Fall	87.75	99
1976			1979		
Winter	46.45	70	Winter	88.07	109
Spring	50.63	60	Spring	94.00	101
Summer	54.41	41	Summer	96.16	77
Fall	58.66	77	Fall	96.98	110
1977			1980		
Winter	62.52	81	Winter	103.90	123
Spring	65.55	70	Spring	107.77	120
Summer	69.62	50	Summer	110.42	95
Fall	72.92	87	Fall	114.91	126

Source: Company records.

linear regression equation where the demand for plaster casts is a function of the number of visitors to the facility and the seasons of the year. Be sure to define each variable you use in your estimating equation.

Solution The variables are defined as follows:

$x_{i,1}$ = number (in thousands) of visitors to the facility in period i

$x_{i,2} = \begin{cases} 1 \text{ if period } i \text{ is the first quarter of the year (winter)} \\ 0 \text{ otherwise} \end{cases}$

$x_{i,3} = \begin{cases} 1 \text{ if period } i \text{ is the second quarter of the year (spring)} \\ 0 \text{ otherwise} \end{cases}$

$x_{i,4} = \begin{cases} 1 \text{ if period } i \text{ is the third quarter of the year (summer)} \\ 0 \text{ otherwise} \end{cases}$

$\hat{y}_i$ = estimated demand for plaster casts in period i

Hence, the estimating equation becomes:

$$\hat{y}_i = b_0 + b_1 x_{i,1} + b_2 x_{i,2} + b_3 x_{i,3} + b_4 x_{i,4}$$

Values of the dependent and independent variables are given in Table 15.15. Note that the regression equation for the fall quarter is $\hat{y}_i = b_0 + b_1 x_{i,1}$. In this context, the variables $x_{i,2}$, $x_{i,3}$, and $x_{i,4}$ denote, respectively, the expected increase (or decrease) from the fall quarter in the demand for plaster casts in the winter, spring, and summer.

The reasons we do not use a fourth indicator variable for the fall quarter are that (1) the additional variable is *not necessary* (the fourth season, the fall, is implied when $x_{i,2}$ and $x_{i,3}$ and $x_{i,4} = 0$), and (2) if the variable were included in the model, it would be impossible to solve the resulting normal equations

TABLE 15.14 Simple linear regression of data in Table 15.13

DEPENDENT VARIABLE: CASTS

SOURCE	DF	SUM OF SQUARES	MEAN SQUARE	F VALUE	PR > F
MODEL	1	13004.84571756	13004.84571756	59.37	0.0001
ERROR	22	4819.11261578	219.05057344	R-SQUARE	STD DEV
CORRECTED TOTAL	23	17823.95833333		0.729627	14.80035721

PARAMETER	ESTIMATE	T FOR H0: PARAMETER = 0	PR > |T|	STD ERROR OF ESTIMATE
INTERCEPT	9.02982441	0.93	0.3604	9.66787753
VISITORS	0.95890293	7.71	0.0001	0.12444981

TABLE 15.15 Values of dependent and independent variables in Example 15.6

Period	Casts	Vistors (000)	Quarter 1	2	3	Period	Casts	Vistors (000)	Quarter 1	2	3
1	53	33.63	1	0	0	13	94	74.64	1	0	0
2	41	36.46	0	1	0	14	86	80.31	0	1	0
3	24	41.18	0	0	1	15	64	80.97	0	0	1
4	57	43.16	0	0	0	16	99	87.75	0	0	0
5	70	46.45	1	0	0	17	109	88.07	1	0	0
6	60	50.63	0	1	0	18	101	94.00	0	1	0
7	41	54.41	0	0	1	19	77	96.16	0	0	1
8	77	58.66	0	0	0	20	110	96.98	0	0	0
9	81	62.52	1	0	0	21	123	103.90	1	0	0
10	70	65.55	0	1	0	22	120	107.77	0	1	0
11	50	69.62	0	0	1	23	95	110.42	0	0	1
12	87	72.92	0	0	0	24	126	114.91	0	0	0

Source: Company records.

for least squares estimates b_0, b_1, b_2, etc. This is because, if the additional variable were included in the model, it could be written as a linear combination of the variables already included. This situation is called perfect *multicollinearity* and is briefly discussed in Section 15.9 and in more detail in the advanced references listed at the end of the chapter. For the present, it is sufficient to note that *when there are m categories of a qualitative independent variable, $j = (m - 1)$ dummy or categorical variables* **must** *be used* in deriving an estimating equation for the dependent random variable. The estimating equations for the four seasons of the year become as shown here.

Season	Predicting equation
Winter	$\hat{y}_i = (b_0 + b_2) + b_1 x_{i,1}$
Spring...............	$\hat{y}_i = (b_0 + b_3) + b_1 x_{i,1}$
Summer	$\hat{y}_i = (b_0 + b_4) + b_1 x_{i,1}$
Fall	$\hat{y}_i = (b_0) + b_1 x_{i,1}$

The data in Table 15.15 were input to the SAS multiple linear regression computer program and the results of the analysis are shown in Table 15.16. The derived equation is:

$$\hat{y}_i = 14.225 + 0.99x_{i,1} + 6.443x_{i,2} - 6.442x_{i,3} - 30.592x_{i,4}$$

where the variables are as previously defined. The coefficient of multiple determination in Table 15.16 is 0.99297, indicating that a very strong linear relationship exists in these sample data between the dependent variable casts and the independent variables corresponding to the number of visitors and the seasons of the year. From Table B.7 in Appendix B (and interpolation), $F(0.99; 4, 19) = 4.50$. Since the calculated F-value is 671.16, we conclude that the regression model developed is useful for predicting the demand for

TABLE 15.16 Computer analysis of the data in Table 15.15

DEPENDENT VARIABLE: CASTS

SOURCE	DF	SUM OF SQUARES	MEAN SQUARE	F VALUE	PR > F
MODEL	4	17698.69908874	4424.67477219	671.16	0.0001
ERROR	19	125.25924459	6.59259182	R-SQUARE	STD DEV
CORRECTED TOTAL	23	17823.95833333		0.992972	2.56760430

PARAMETER	ESTIMATE	T FOR H0: PARAMETER = 0	PR > |T|	STD ERROR OF ESTIMATE
INTERCEPT	14.22495659	7.03	0.0001	2.02319422
VISITORS	0.99213765	45.33	0.0001	0.02188722
QUARTER1	6.44293516	4.29	0.0004	1.50134838
QUARTER2	−6.44197011	−4.33	0.0004	1.48945001
QUARTER3	−30.59166399	−20.61	0.0001	1.48450349

plaster casts. Furthermore, each of the regression coefficients is significant as evidenced by the quantities in the column PR > |T|. Hence, we conclude that all regression coefficients are different from 0—all aid in estimating the demand for plaster casts. Note that the p-values given in Table 15.16 would cause us to reject the null hypothesis that $\beta_i = 0$ for all regression coefficients at the 0.10, 0.05, or 0.01 significance levels and would cause us to reject the null hypothesis that all regression coefficients are *simultaneously equal to 0* at the 0.10, 0.05, or 0.01 significance levels (PR > F = 0.0001).

While the value of r^2 increased from 0.73 to 0.99 when the categorical variables corresponding to the seasons of the year were included in the regression model, an interesting question that might be posed at this time is: Can we statistically test for the presence of a *seasonal factor* in our developed model? The answer to this question is yes, and the analysis proceeds as follows. First, rather than testing whether each β_i value is individually significantly different from 0, we will test the hypothesis H_0: $\beta_2 = \beta_3 = \beta_4 = 0$ since these are the regression coefficients (population) corresponding to the seasons of the year.

The test statistic that is useful for testing the above hypothesis is obtained by first taking the difference between the amount of variation (SSE) left unexplained by the regression equation before the indicator variables were introduced and the amount of variation left unexplained (SSE) *after* the categorical variables have been included in the model. Since we are taking the difference between two sum of square error values, it is necessary for us to subscript SSE to distinguish between the models. Let us designate the sum of squares error *before* the categorical variables are introduced into the estimating equation as SSE_B and the sum of squares error *after* the categorical variables have been included in the model as SSE_A. If we divide the above difference by j, the number of additional variables in the revised model, and divide the resulting quotient by $[SSE_A/(n - k - 1)]$, the resulting statistic is F-distributed with j and $(n - k - 1)$ degrees of freedom when the null hypothesis is true. The form of the hypothesis test is shown here.

Hypothesis test for subsets of independent variables in a multiple linear regression model, $j = (m - 1)$ categorical variables

Notation: Let the sum of squares error before the subset of variables is included for which it is desired to test the null hypothesis that they are all simultaneously equal to 0 be denoted by SSE_B, and let the sum of squares error after the subset of variables is included in the model be denoted by SSE_A. Let the number of independent variables in the subset be denoted by j, and let the population regression coefficients for variables in the subset be denoted by $\beta_i, \beta_{i+1}, \ldots, \beta_{i+j-1}$.

Hypothesis: H_0: $\beta_i = \beta_{i+1} = \cdots = \beta_{i+j-1} = 0$

H_A: At least one population regression coefficient β_h is not equal to 0, where $i \leq h \leq i + j - 1$

> Hypothesis test for subsets of independent variables (*continued*)
>
> Test statistic value: $f = \dfrac{(SSE_B - SSE_A)/j}{(SSE_A)/(n - k - 1)}$
>
> Reject if: $f > F(1 - \alpha; j, n - k - 1)$
>
> Do not reject if: $f \leq F(1 - \alpha; j, n - k - 1)$
>
> where:
>
> n = number of observations
> α = significance level of the test
> k = number of independent variables in the model

Example 15.7   Using the data provided in Tables 15.14 and 15.16, test for the presence of a seasonal effect in the demand for plaster casts at the $\alpha = 0.05$ significance level.

Solution   From Table 15.14, $SSE_B = 4819.11$, and from Table 15.16, $SSE_A = 125.26$. Thus, the computed value of the F-statistic is:

$$f = \frac{(4{,}819.11 - 125.26)/3}{125.26/(24 - 4 - 1)} = 237.33$$

$F(0.95; 3, 19) = 3.13$ (by interpolation). Since 237.33 far exceeds 3.13, we reject the null hypothesis and conclude that there is a seasonal effect in the demand for plaster casts.

An advantage of the test described above is that it can be modified for testing hypotheses concerning the significance of *any* subset of independent variables in a multiple linear regression model. For example, if it were felt that the regression coefficients for a certain subset of variables were all equal to 0, a multiple regression could be constructed first with the subset of variables not in the model, and then with the subset of variables in the model. The test just described could then be used to test the significance of the subset of variables. Note that if the subset of variables consists of *one* independent variable, then the test described is equivalent to the *t*-test for the significance of the simple linear regression model when we test H_0: $\beta_1 = 0$. In Chapter 14 (Section 14.2.3) we noted the similarity of the **t**-statistic and the **F**-statistic, indicating that for the test of a single variable, $\mathbf{t}^2 = \mathbf{F}$.

We should also note at this time that it is completely arbitrary which of the m levels of the categorical variable is not assigned a binary (0 or 1) independent variable. In the previous example, the fall quarter was not assigned a categorical variable because fall can be represented as the *absence* of winter, summer, and spring. That is, we represent the fall quarter of each year by assigning the value 0 to $x_{i,2}$, $x_{i,3}$, *and* $x_{i,4}$. We could have just as easily assigned independent indicator variables to the spring, summer, and fall quarters of the year, in which case the winter quarter would have been represented by each of the three binary categorical variables being set equal to

0. From this same line of reasoning, it should be apparent that in the example of relating the weight of an individual to the individual's sex, we could have defined our independent variable to be:

$$x_{i,2} = \begin{cases} 1 \text{ if the sampled unit is a female} \\ 0 \text{ if the sampled unit is a male} \end{cases}$$

Again, the choice is completely arbitrary, and this is why it is extremely important to define carefully each independent variable to aid the user of the developed regression equation.

15.8 MULTIPLE CORRELATION ANALYSIS

15.8.1 The coefficient of multiple correlation

Multiple correlation bears the same relationship to simple linear correlation as multiple linear regression bears to simple linear regression. In multiple correlation, we are attempting to assess the relative strength of the relationship between the dependent random variable Y and the independent random variables $X_1, X_2, \ldots, X_k$. The strength of the linear relationship is measured by the coefficient of multiple correlation, r. Unlike the coefficient of correlation in the single independent variable case, however, the coefficient of multiple correlation is bounded by 0 and 1; that is, $0 \leq r \leq 1$. This is because r does not indicate the slope of the regression equation, since it is not possible to indicate the signs of all the regression coefficients that relate the dependent variable Y to the independent variables X_i. Similar to the simple correlation case, we must treat the independent variables X_i as random variables, since one assumption of the multiple correlation model is that the distribution of Y and the X values is multivariate normal. This assumption is necessary in constructing confidence intervals and performing hypothesis tests on the coefficient of multiple correlation.

As in the case of simple linear correlation, the measure r^2 is much simpler to interpret than is the coefficient of multiple correlation, r. That is, r^2 measures the proportion of variability in the dependent random variable Y explained by or accounted for by the independent random variables. The coefficient of multiple correlation is computed by simply taking the square root of the coefficient of multiple determination. Sometimes the coefficient of multiple correlation is subscripted to indicate the dependent random variable being explained by the independent random variables. Using such notation in the bivariate case, the coefficient of multiple correlation becomes $r_{Y/X_1,X_2}$. Experience in computing the coefficient of multiple correlation is provided in the following example.

Example 15.8 Using the data given in Table 15.6, compute the coefficient of multiple correlation.

Solution The coefficient of multiple correlation is given by:

$$r = \sqrt{r^2} = \sqrt{0.97296} = 0.9864$$

☐ 15.8.2 Partial correlation

The coefficient of multiple correlation measures the strength of the linear relationship between the dependent variable Y and the independent variables $X_1, X_2, \ldots, X_k$. We may also be interested in measuring the strength of the relationship (linear) between the dependent variable Y and *one* of the independent variables, where the effect of all the other variables has been removed or has been held constant. This is to be distinguished from the correlation between pairs of variables, as is the case in simple linear regression. In Table 15.17, for example, we show the correlation matrix giving the simple linear correlation between every pair of variables in the sales-advertising-income example introduced in Section 15.4. Table 15.17 was produced by the

TABLE 15.17 Correlation matrix of data in Table 15.3

CORRELATION COEFFICIENTS AND PROB > |R| UNDER H0:
RHO = 0 FOR N = 15

	SALES	INCOME	ADDS
SALES	1.00000	0.94502	0.96995
	0.0000	0.0001	0.0001
INCOME	0.94502	1.00000	0.89000
	0.0001	0.0000	0.0001
ADDS	0.96995	0.89000	1.00000
	0.0001	0.0001	0.0000

SAS statistical program CORR, which is similar to the regression program we have been discussing thus far. This matrix gives not only the simple correlation between every pair of variables but also the p-value for the significance of the correlation measure under the null hypothesis that the simple coefficient of correlation is equal to 0. As shown in Table 15.17, all pairs of linear correlations are unequal to 0 beyond the 0.10, 0.05, and 0.01 significance levels. From Table 15.17, the correlation of any variable with itself is 1.00, the correlation between income and sales is $r_{YX_2} = r_{X_2Y} = 0.945$, and the simple linear correlation between income and advertising expenditures is $r_{X_2X_1} = r_{X_1X_2} = 0.89$. Thus, these two independent random variables are themselves highly related! Note how we have subscripted the simple coefficient of correlation to indicate the random variables involved.

The partial correlation coefficient measures the strength of the linear relationship between the dependent random variable Y and *one* independent variable X_i where the linear effect of the remaining independent variables is held constant. As was the case with simple linear regression and correlation, it is much easier to interpret the *partial coefficient of determination*, which is the square of the partial correlation coefficient, and so we shall do so. The notation we will use for the partial coefficient of determination is $r^2_{Y/X_i;X_1,X_2,\ldots X_{i-1},X_{i+1},\ldots,X_k}$ where the independent variables after the semicolon indicate the independent variables whose effect is being held constant. The coefficient of partial correlation is, of course, the square root of the above quantity.

In Table 15.18 are given the simple linear regression and correlation analyses of sales and advertising expenditures and of sales and income using the data given in Table 15.3. From Table 15.6, the amount of variation left unexplained by the multiple linear regression when both advertising and income are included as independent variables (SSE) is 202.63. In the bivariate case, the coefficient of partial determination is given by the following:

Coefficient of partial determination—bivariate case

$$r^2_{Y/X_i;X_j} = \frac{\text{Additional variation in } Y \text{ explained by the addition of } X_i}{\text{Variation in } Y \text{ unexplained by } X_j \text{ alone}}$$

From Table 15.18, the total variation in Y to be explained is

$$\Sigma(y - \bar{y})^2 = 7,492.91$$

Based on the simple linear regression when advertising expenditures are included in the estimating equation, the amount of unexplained variation is SSE = 443.58. The amount of variation unexplained when both advertising expenditures *and* average income level are included in the estimating equation is SSE= 202.63. Thus, the *extra* amount of variation explained when average income level is included in the model is 240.95 = 443.58 − 202.63. Thus, the *proportion of previously unexplained variation* that is explained when average income level is included in the model is:

$$r^2_{Y/X_2;X_1} = \frac{240.95}{443.58} = 0.543$$

The partial correlation coefficient for the average income level is thus:

$$r_{Y/X_2;X_1} = \sqrt{0.543} = 0.7369.$$

The partial coefficient of determination for the variable advertising expenditures is shown in Table 15.19. Note that the "area" in Table 15.19 corresponding to the extra variation explained by ads is larger than the "area" for extra variation explained by income, reflecting the larger partial coefficient of determination for the independent variable advertising expenditures. The coefficient of partial determination measures the proportion of variation explained by the variable *of that remaining* after all other independent variables are included in the estimating equation.

TABLE 15.18 Simple linear regression and correlation analyses of data in Table 15.3

$$\hat{y} = b_0 + b_1 x_{ADDS}$$

DEPENDENT VARIABLE: SALES

SOURCE	DF	SUM OF SQUARES	MEAN SQUARE	F VALUE	PR > F
MODEL	1	7049.32975962	7049.32975962	206.60	0.0001
ERROR	13	443.57579731	34.12121518	R-SQUARE	STD DEV
CORRECTED TOTAL	14	7492.90555693		0.940801	5.84133676

PARAMETER	ESTIMATE	T FOR H0: PARAMETER = 0	PR > |T|	STD ERROR OF ESTIMATE
INTERCEPT	1.96334786	0.61	0.5551	3.24106207
ADDS	26.96437275	14.37	0.0001	1.87598194

$$\hat{y} = b_0 + b_1 x_{INCOME}$$

DEPENDENT VARIABLE: SALES

SOURCE	DF	SUM OF SQUARES	MEAN SQUARE	F VALUE	PR > F
MODEL	1	6691.64680572	6691.64680572	108.57	0.0001
ERROR	13	801.25875121	61.63528855	R-SQUARE	STD DEV
CORRECTED TOTAL	14	7492.90555693		0.893064	7.85081452

PARAMETER	ESTIMATE	T FOR H0: PARAMETER = 0	PR > |T|	STD ERROR OF ESTIMATE
INTERCEPT	-25.84262257	-3.73	0.0025	6.92908808
INCOME	5.64666488	10.42	0.0001	0.54192630

TABLE 15.19 Calculation of partial coefficient of determination for sales-advertising-income data

$r^2_{Y/X_2;X_1}$:

Total variation: 7,492.91	94.08% Variation explained by adds alone: 7,049.33	3.22% Extra variation explained by income: 240.95
		2.70% Unexplained variation: 202.63

$$r^2_{Y/X_2;X_1} = \frac{240.95}{443.58} = 0.543$$

Variation unexplained by adds alone: 443.58

$r^2_{Y/X_1;X_2}$:

Total variation: 7,492.91	89.31% Variation explained by income alone: 6,691.65	7.99% Extra variation explained by adds: 598.63
		2.70% Unexplained variation: 202.63

$$r^2_{Y/X_1;X_2} = \frac{598.63}{801.26} = 0.747$$

Variation unexplained by income alone: 801.26

■ 15.9 CAUTIONS IN THE USE OF MULTIPLE LINEAR REGRESSION AND CORRELATION

The t-test for testing the significance of the regression coefficients is often used to eliminate poor predictor variables from the multiple linear regression model—independent variables that do not appear to account for much variability in the dependent variable **Y**. This is not a wise application of the t-test, for if the independent variables are highly interrelated, a seemingly poor predicting independent variable may be eliminated by the t-test when, in fact, it may be a very important factor in explaining the variability in **Y**. This is because when the independent variables are highly interrelated (a concept called *multicollinearity*, to be discussed shortly), the estimated standard error of the coefficient, say s_{b_i}, tends to be inflated. It is inflated because the estimates when the variables are interrelated are sensitive to changes in the sample observations or model specification. When s_{b_i} is larger than it should be and we compute the ratio b_i/s_{b_i}, we will find that it is smaller than it should be, leading us to reject more significant terms than we otherwise would. We noted in the sales-advertising-income example that the independent variables themselves were highly interrelated, with a sample coefficient of correlation between them of 0.89.

This point is really imbedded in a much larger concern in using the multiple regression model—how many independent variables should be used? At first, we might think that the more independent variables we have, the better. In fact, this is true if the model will be used for prediction *only* and no statistical inferences will be drawn from it. That is, it is true in the sense that as more independent variables are added, the larger will become the coefficient of determination r^2. If we take an existing model and add one additional variable, the new r^2 must be at least as large as the one in the model before the variable is added. Therefore, we can make r^2 large (close to 1) by regressing **Y** on a large number of independent variables. But, to say that a regression line is "good" because its r^2 value is close to 1 is misleading for two reasons. First, as the number of independent variables approaches the sample size, the fit of the regression line must improve. This idea is related by analogy to the general problem of fitting a function through a set of points as illustrated in Figure 15.9. A cubic equation (degree 3) will fit perfectly

FIGURE 15.9 Fitting functions through three points

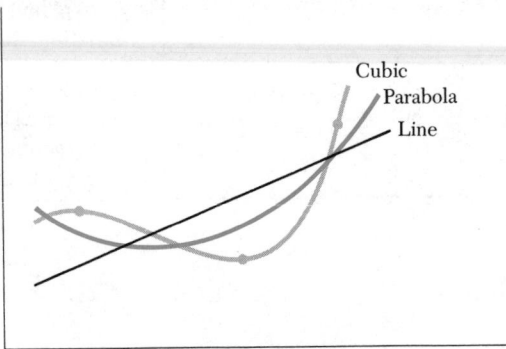

through the three points—the parabola will fit better than the line. The degree of the function is analogous to the number of independent variables in regression. If we have n observations on **Y** and fit a multiple regression function using $(n - 1)$ independent variables as predictors, the function [in $(n - 1)$ space] will fit perfectly—r^2 will be *exactly* 1. But, this only ensures that the function is a perfect predictor of the *observed* values of **Y** in the sample. And if the sample observations are not representative of the population values of Y, x_1, x_2, . . . , x_k, then the fitted regression line would be poor in estimating characteristics in the population—even though $r^2 = 1$! Also, note that we can use at a maximum only one fewer independent variables than the sample size. This fact is also reflected by studying the degrees of freedom in the analysis of variance table—the regression and error degrees of freedom must add to $(n - 1)$.

Another concern in multiple regression analysis when more than one independent variable is used in the estimating equation is that as the number of independent variables increases, the set of variables tends itself to be highly interrelated, and this leads to the problem in multiple regression referred to as *multicollinearity*. If the independent variables are highly interrelated, they are explaining variability in Y in similar, indeed, overlapping ways, and this leads to poorly estimated regression coefficients. The variances of b_0, b_1, . . . , b_k will be much larger than the variances if the set of independent variables is not interrelated. The ideal multiple regression function would have each independent variable explaining variability in Y in a way that is separated from all other independent variables, as illustrated in Figure 15.10.

FIGURE 15.10 Explaining variability in Y by independent variables x_1, x_2, . . . , x_k that are not interrelated

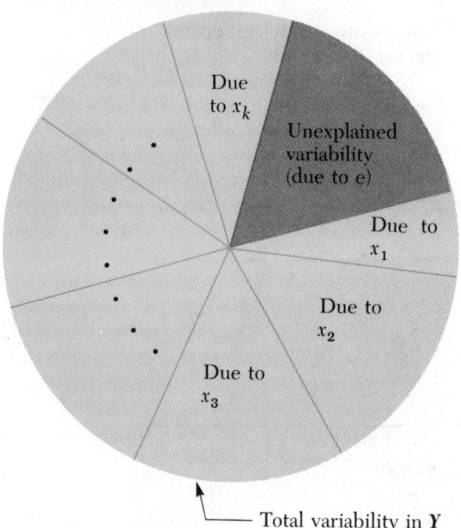

Even when the number of independent variables to be included in the regression model is given, the problem still arises as to how best to combine these variables in the estimating equation. Should interaction terms be included? Should higher-order terms be included? Should variables be eliminated? When the independent variables are themselves highly interrelated, as they are in the sales-advertising-income example, it is possible that one or more of the variables can be removed from the estimating equation and the regression model developed actually improves for prediction and estimation! The examination of residual plots can serve as a guide in eliminating poor predictor variables (Step 5 in the multiple linear regression model-building process).

In Figures 15.11 and 15.12 we give residual plots of the sales-advertising-income model developed in Section 15.4. Figure 15.11 raises the possibility

FIGURE 15.11   Residual plot of regression model given in Table 15.4—residual versus advertising expenditures

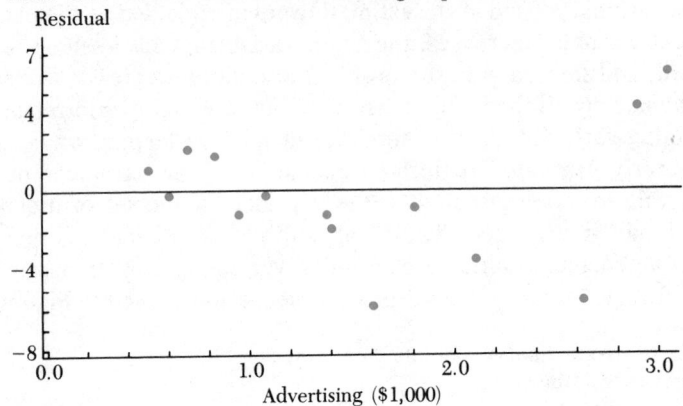

FIGURE 15.12   Residual plot of regression model given in Table 15.4—residual versus average income level

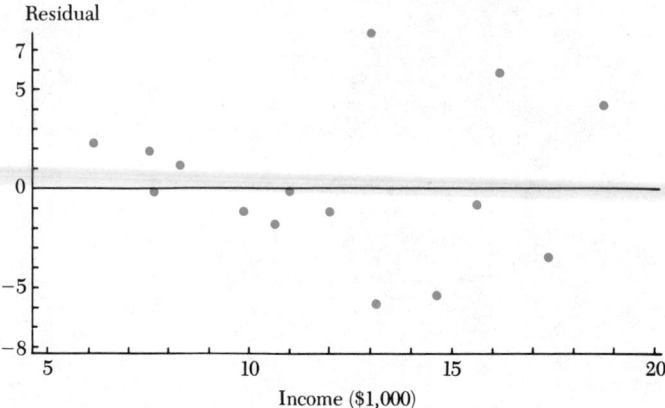

that the relationship between sales and advertising expenditures may be curvilinear. Figure 15.12 raises the possibility that the relationship between sales and income level may also be curvilinear, and that the variances of the conditional distributions may be increasing as the average income level increases. Both plots are inconclusive, however, but do suggest that alternate models should be examined if these two variables are to be retained in any developed estimating equation.

The question is, then: Can the two variables examined be combined in different ways to produce a more acceptable model? To answer this query, we could examine the complete model:

$$Y = \beta_0 + \beta_1 x_1 + \beta_2 x_2 + \beta_3 x_1 x_2 + \beta_4 x_1^2 + \beta_5 x_2^2 + \epsilon$$

as well as subsets of it (i.e., all one-variable models, all two-variable models, etc.) to determine whether a better model is available from the data given. There are in fact some statistical procedures to determine the *best* number and the *best* combination of variables to include in an estimating equation. Using one of these procedures to examine alternate ways of combining the two variables advertising expenditures (x_1) and average income level (x_2) into an estimating equation using the data given and the model specified above, we get the model in Table 15.20. Residual plots from this model are given in Figures 15.13 and 15.14.

FIGURE 15.13 Residual plot of regression model given in Table 15.20—residual versus advertising expenditures squared

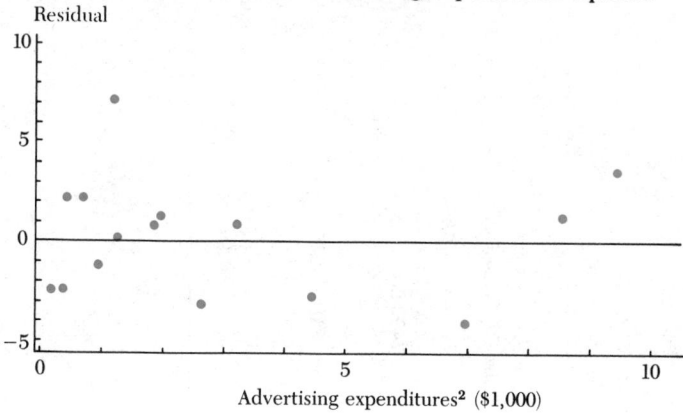

FIGURE 15.14 Residual plot of regression model given in Table 15.20—residual versus average income level

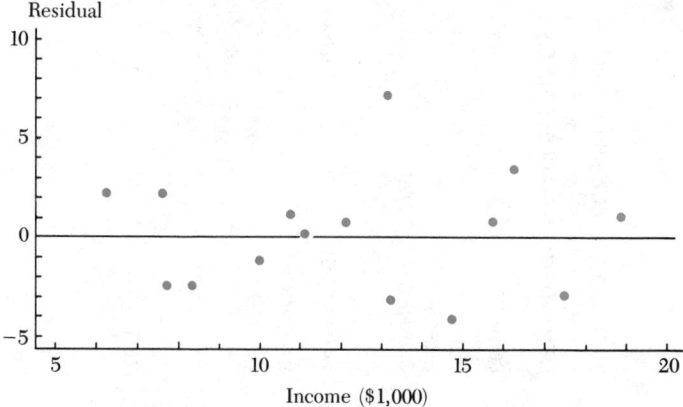

TABLE 15.20 Computer analysis of the Table 15.3 data

DEPENDENT VARIABLE: SALES

SOURCE	DF	SUM OF SQUARES	MEAN SQUARE	F VALUE	PR > F
MODEL	2	7364.89007169	3682.44503584	345.19	0.0001
ERROR	12	128.01548525	10.66795710	R-SQUARE	STD DEV
CORRECTED TOTAL	14	7492.90555693		0.982915	3.26618387

PARAMETER	ESTIMATE	T FOR H0: PARAMETER = 0	PR > |T|	STD ERROR OF ESTIMATE
INTERCEPT	−5.68669644	−1.48	0.1644	3.84024999
INCOME	3.00926970	7.50	0.0001	0.40131189
ADSQ	4.05071215	7.94	0.0001	0.50990116

By examining Tables 15.20 and 15.4, we can see that the model has been improved by including the variables income and the square of advertising expenditures: r^2 has increased from 0.973 to 0.983, the estimate of the conditional probability distribution variance has decreased from 16.89 to 10.67, and based on the p-values for the two independent variables given in both tables, we would conclude that the independent variables included in the model, as well as the model itself, are more significant for the regression equation given in Table 15.20.

The residual plots in Figures 15.13 and 15.14 appear to indicate random residuals more than do the residual plots given in Figures 15.11 and 15.12. The residual plot in Figure 15.14 especially appears to support the notion that the underlying assumptions of the multiple linear regression model have been met, while the evidence in Figure 15.13 is not quite so convincing. The difficulty here in attempting to model sales as a function of advertising expenditures and average income is that the two independent variables are so highly interrelated—the correlation between x_1 and x_2 is equal to 0.89. Thus, in these sample data, there is a tendency for advertising levels to be high in the higher-income communities, and we do not have many observations where just the reverse is true. It may even be that the firm is allocating its advertising dollars in this manner, so that these two variables will always explain the variability in sales in similar ways. Given the data provided, the model given in Table 15.20 is simply the best that can be achieved by using either one or both of these independent variables and the full model given. If this model is not sufficient for some reason, then the statistician has little recourse but to return to Step 1 in the model-building process and identify new variables to be included in the regression model. It is hoped that the newly identified variables will explain the variability in sales in different ways than the two given. It is up to the client firm to decide whether to accept the model developed or to collect new data and repeat the analysis procedure. The statistician has simply done all she can with the information provided!

15.10 SUMMARY

Multiple linear regression analysis is perhaps one of the most widely used statistical procedures today. We cannot do justice to the analysis in this book, but it is hoped that this introduction suggests its wide applicability while discussing the rudiments of the procedure. An excellent reference (and quite readable) is the Neter and Wasserman text cited at the end of this chapter for those desiring more information on multiple linear regression and correlation analysis.

REFERENCES

Croxton, F. E.; Cowden, D. J.; and Bolch, B. W. *Practical Business Statistics*. 4th ed. Englewood Cliffs, N.J.: Prentice-Hall, Inc., 1969, chaps. 14–16 and 21.

Draper, N. R., and Smith, H. *Applied Regression Analysis*. New York: John Wiley & Sons, Inc., 1966.

Ezekiel, Mordecai, and Fox, Karl A. *Methods of Correlation and Regression Analysis*. 3d ed. New York: John Wiley & Sons, Inc., 1959.

Goldberger, Arthur S. *Econometric Theory*. New York: John Wiley & Sons, Inc., 1964.

Johnston, J. *Econometric Methods*. 2d ed. New York: McGraw-Hill Book Company, 1972.

Neter, John, and Wasserman, William. *Applied Linear Statistical Models*. Homewood, Ill.: Richard D. Irwin, Inc. 1974.

Williams, E. J. *Regression Analysis*. New York: John Wiley & Sons, Inc., 1959.

Yamane, T. *Statistics: An Introductory Analysis*. 3d ed. New York: Harper & Row, Publishers, 1973.

■ PROBLEMS*

15.1. Explain the differences and the similarities between simple linear regression analysis and multiple linear regression analysis.

15.2. Distinguish among the coefficient of determination, the coefficient of multiple determination, and the coefficient of partial determination in a bivariate regression model. How are these measures related to the coefficient of determination between the independent variables?

15.3. In a multiple linear regression analysis of income and marital status, it is determined that the number of "levels" of the independent variable "marital status" is five: single, married, separated, divorced, and widowed. Define the binary variables to model the qualitative characteristic marital status. How many binary variables are required to model this qualitative variable?

15.4. Discuss why it might be useful to include more than one independent variable in a regression equation to model the behavior of a dependent variable.

15.5. Explain the meaning of multicollinearity and one of its effects on the model developed.

15.6. In a multiple regression analysis of yield as a function of three different nutrients, it is found that PR> F= 0.0876. Interpret this p-value statistic.

15.7. In a bivariate regression analysis involving 22 observations, it is found that $b_1 = 23.56$ and $b_2 = 0.0052$, with $s_{b_1} = 13.68$ and $s_{b_2} = 0.0013$. Test the null hypotheses H_0: $\beta_1 = 0$ and H_0: $\beta_2 = 0$ at the $\alpha = 0.05$ significance level.

15.8. Which of the variables, x_1 or x_2, is more statistically significant in Problem 15.7?

15.9. Explain why, in a test of H_0: $\beta_1 = 0$ and H_0: $\beta_2 = 0$ where the significance level of each test is 0.05, we cannot claim that the significance of the model with both variables included is 0.05.

15.10. An evaluation has been undertaken of the salary of 12 members of the graduating class of 1970 ten years after graduation. These data are given in the tables, along with each student's undergraduate grade point

---

* Problems 15.21–15.25 require the use of a multiple regression computer program for solution in a reasonable amount of time.

average (GPA) and the number of hours spent per week (average) in extracurricular activities while enrolled as an undergraduate (hours). Also given are the results of three regression analyses to predict salary ten years after graduation. The first model includes only GPA, the second only hours, and the third both GPA and hours.

a. Give the estimating equation associated with each model, and comment on both the significance of the individual terms included in the model and the overall model. Support your comments with appropriate statistical data.

b. Using the third model, estimate an individual's annual salary ten years after graduation if she had an undergraduate grade point average of 3.12 and nine hours per week spent in extracurricular activities. How does this estimate compare with the estimate derived using each of the models with one independent variable? Which model do you have more faith in and why?

c. Compute the correlation be-

tween GPA and hours. What implications (if any) does this correlation have in estimating salary?

Student no.	Annual salary ($000)	GPA	Hours
1	16.5	2.02	0
2	29.4	3.44	7
3	31.9	2.87	17
4	18.4	2.41	4
5	35.6	3.61	15
6	22.8	2.36	5
7	37.1	3.22	2
8	30.5	3.11	10
9	26.3	2.54	9
10	27.9	2.88	12
11	56.9	3.21	22
12	25.4	2.62	6

15.11. In a regression analysis involving 27 observations and three independent variables, it is determined that SSR = 35,023.98 and SSE = 2,219.56. Determine the coefficient of multiple determination and the coefficient of multiple correlation for these data.

15.12. For the data in Problem 15.11, test the null hypothesis H_0: $\beta_1 = \beta_2$

Dependent variable: Salary   Independent variable: GPA

Source	Degrees of freedom	Sum of squares	Mean square	F-value	PR > F
Model	1	581.44675312	581.44675312	9.14	0.0128
Error	10	635.92241355	63.49224135	Standard deviation	
Corrected total	11	1,217.36916667		7.97447436	
			R-square 0.477626		

Parameter	Estimate	t for H_0: Parameter = 0	PR > |t|	Standard error of estimate
Intercept	− 13.41025990	−0.92	0.3770	14.50422069
GPA	15.15378008	3.02	0.0128	5.01150389

Dependent variable: Salary   Independent variable: Hours

Source	Degrees of freedom	Sum of squares	Mean square	F-value	PR > F
Model..............	1	693.03704785	693.03704785	13.22	0.0046
Error...............	10	524.33211881	52.43321188		
Corrected total	11	1,217.36916667	R-square 0.569291		Standard deviation 7.24107809

Parameter	Estimate	t for H_0: Parameter = 0	PR > |t|	Standard error of estimate
Intercept	18.77762376	5.07	0.0005	3.70334359
Hours	1.22356436	3.64	0.0046	0.33655173

Dependent variable: Salary   Independent variables: GPA, hours

Source	Degrees of freedom	Sum of squares	Mean square	F-value	PR > F
Model..............	2	834.31988297	417.15994148	9.80	0.0055
Error	9	383.04928370	42.56103152		
Corrected total	11	1,217.36916667	R-square 0.685347		Standard deviation 6.52388163

Parameter	Estimate	t for H_0: Parameter = 0	PR > |t|	Standard error of estimate
Intercept	−3.30212536	−0.26	0.7987	12.56960850
GPA	8.83697116	1.82	0.1018	4.85025765
Hours	0.87436480	2.44	0.0375	0.35871323

$= \beta_3 = 0$ at the $\alpha = 0.05$ significance level.

15.13. In a regression analysis involving 29 observations and 3 independent variables, the following estimating equation is determined, with standard errors for the coefficients shown in parentheses:

$$\hat{y} = 23 + 7x_1 - 4x_2 + 3x_3$$
$$(2.6) \quad (1.9) \quad (1.8)$$

a. Test the null hypotheses H_0: $\beta_1 = 0$, H_0: $\beta_2 = 0$, and H_0: β_3

$= 0$ at the $\alpha = 0.05$ significance level.

b. Given your results from part a, which variable would you say is the most statistically significant?

c. Assuming x_1 were to increase by 23, x_2 to decrease by 13, and x_3 to increase by 12, what would be the estimated change in y?

15.14. Verify that the correlation between the independent variables advertising expenditures and average income level is 0.89 using the data given in Table 15.3.

15.15. In a simple linear regression analysis of the yield of a chemical process as a function of temperature and pressure, the following quantities are determined: when both independent variables (temperature and pressure) are included in the estimating equation, SST = 3,645 and SSE = 734; when temperature only (x_1) is included in the model, SSE = 1,095; and when pressure only (x_2) is included in the model, SSE = 1,436.

a. Compute the coefficient of multiple determination for these data.

b. Compute the coefficient of multiple correlation for these data.

c. Compute $r^2_{Y/X_2;X_1}$.

d. Compute $r^2_{Y/X_1;X_2}$.

e. Construct a diagram similar to Table 15.19 showing the relationships between the partial coefficients of determination determined in parts c and d.

15.16. The total variation in a random variable Y is determined to be 4,798 units. The model $\hat{y} = b_0 + b_1 x_1 + b_2 x_2$ leaves 854 units unexplained based on 27 observations. When the model is extended to include x_3, 222 more units of the variation in Y are accounted for.

a. Compute $r^2_{Y/X_1;X_2,X_3}$.

b. Compute $r^2_{Y/X_3;X_1,X_2}$.

15.17. Assuming the data in Problem 15.10 are multivariate normally distributed, compute the coefficients of partial determination for each of the independent variables (GPA and hours). What interpretation should be given to these coefficients?

15.18. In a multiple linear regression analysis involving 33 observations and 7 independent variables, it is determined that SST = 5,672.67 and that SSE = 1,945.78. Test for the signifi-

cance of the model developed at the $\alpha = 0.05$ significance level.

15.19. Using 18 observations, a multiple linear regression equation involving four independent variables is estimated using the method of least squares. It is determined that SSR = 700 and SST = 900. Furthermore, the amount of variation explained jointly by x_1, x_2, and x_3 is 550.

a. Compute r^2.

b. Test for the significance of the model developed at the $\alpha = 0.05$ significance level.

c. Compute $r^2_{Y/X_4;X_1,X_2,X_3}$ and interpret its meaning.

15.20. In a multiple linear regression analysis involving 3 independent variables and 21 observations, SSE is determined to be 440.41. The inclusion of a fourth independent variable explains 13.21 more units of variability in the random variable Y.

a. Determine $s_{Y/x_1,x_2,x_3}$.

b. Determine $s_{Y/x_1,x_2,x_3,x_4}$.

15.21. Reproduced below are data on sales and advertising expenditures for a firm marketing a semi-luxury consumer good.

Sales and corresponding advertising levels

Year and quarter	Sales ($000)	Advertising expenditures ($000)
1975		
I	105	10.54
II	96	11.34
III	55	12.70
IV	117	13.25
1976		
I	132	14.18
II	125	15.39
III	82	16.47
IV	145	17.70
1977		
I	151	18.81
II	143	19.66
III	100	20.84
IV	165	21.78

Sales (*continued*)

Year and quarter	Sales ($000)	Advertising expenditures ($000)
1978		
I	177	22.24
II	168	23.89
III	127	24.04
IV	186	26.02
1979		
I	202	26.07
II	194	27.80
III	149	28.40
IV	212	28.59
1980		
I	224	30.62
II	223	31.73
III	176	32.48
IV	234	33.77

a. Fit a simple linear regression line to these data, using advertising expenditures as the independent variable to estimate sales. Do advertising expenditures account for a significant amount of the variability in sales?

b. Using binary variables corresponding to the quarters of the year, fit a multiple linear regression plane to the data, estimating sales as a function of advertising expenditures and the quarter of the year. Has the "predictability" of the fitted regression improved by including the quarter of the year in the estimating equation?

c. Using the methods described in the text, test for the presence of a seasonal (quarter) factor at the $\alpha = 0.05$ significance level.

d. Given the results determined in part c, does the firm appear to be following a wise policy in spending its advertising dollars? Explain.

15.22. Referring to Problems 13.25 and 14.24, Joe College was able to obtain data on the ten employees as to the length of service in the firm. These data, along with their hourly wage rate and years of postsecondary education, are shown in the table.

Worker	Years of postsecondary education	Years of service	Hourly wage rate
1	5.5	12	$18.20
2	0.0	3	6.50
3	1.0	1	3.90
4	4.0	10	11.70
5	3.5	6	10.40
6	2.0	20	7.80
7	6.0	14	15.60
8	2.5	7	9.10
9	4.5	16	13.00
10	5.0	13	14.30

a. Using multiple linear regression, derive an estimating equation for hourly wage rate using years of postsecondary education and years of service in the firm as independent variables.

b. What is the expected wage rate for a person with four years of college and six years of service in the firm?

c. What is the expected wage rate for a person with six years of college and four years of service in the firm?

d. Test the null hypotheses H_0: $\beta_1 = 0$ and H_0: $\beta_2 = 0$ at the $\alpha = 0.01$ significance level.

e. What conclusions can be drawn regarding the efficacy of years of postsecondary education and length of service in the firm in predicting the expected wage rate?

f. Test the null hypothesis H_0: $\beta_1 = \beta_2 = 0$ at the $\alpha = 0.05$ significance level.

g. Do residual plots appear to support the hypothesis that the underlying assumptions of the multiple linear regression model have been met?

15.23. Assuming the data in Problem 15.22 are multivariate normally distributed, determine the following:

    a. The sample coefficient of multiple determination.

    b. The sample coefficient of multiple correlation.

    c. The correlation between the dependent variable and each of the independent variables.

    d. The correlation between the two independent variables.

    e. The coefficient of partial determination for each of the independent variables.

15.24. Sophie Stoltz has recently become concerned about the level of pay of females versus males in the creosoting division of the company in which she works. From "informed sources," she has been told that the average pay of females in her division is less than $12,000 annually,

while the same pay for males is nearly $16,000. She has informed her employer that she is considering suing for sex discrimination in the firm. A sample of 20 individuals in the creosoting division has been taken, producing the data shown in the table.

    a. Develop an estimating equation for salary as a function of length of service in the firm and sex.

    b. Using the *F*-test described in the text, test for the presence of a sex factor (bias) in the data at the $\alpha = 0.10$, 0.05, and 0.01 levels.

    c. Using the results obtained in part *b*, if you were the judge, and using the information provided, would you rule in her favor or not? Explain.

15.25. Assuming the data in Problem 15.24 are multivariate normally distributed, determine the following:

    a. The sample coefficient of multiple determination.

    b. The sample coefficient of multiple correlation.

    c. The coefficient of partial determination for the independent variables sex and length of service.

    d. What interpretation can be placed on the coefficient of partial determination for sex, and how would this influence the ruling on sex discrimination in your opinion?

15.26. In a multiple linear regression analysis involving four independent variables, it is decided to drop one of the independent variables from the model and use three variables for modeling purposes. Explain the effect on the point estimate of the conditional probability distribution variance (σ^2) from dropping this independent variable. Under what conditions will the estimate increase, and under what conditions will the estimate decrease?

Employee number	Annual salary ($000)	Years of service	Sex
1	14.6	5	M
2	7.5	1	F
3	12.1	4	M
4	24.3	11	M
5	13.5	8	F
6	22.1	14	M
7	10.5	1	M
8	21.5	10	M
9	11.2	4	F
10	16.4	6	M
11	11.9	4	F
12	8.7	2	M
13	15.9	5	F
14	17.2	9	M
15	10.5	3	F
16	16.9	8	F
17	9.2	2	F
18	10.8	4	M
19	17.2	7	M
20	9.4	2	F

Time series ⅃16

■ 16.1 INTRODUCTION

Observations of data are frequently made over time. That is, we observe or record the value of a variable as of, or for, a specific instance and then again at some later time. For example, Figure 16.1 shows the growth in the gross national product (the total value of goods and services produced) in the United States recorded for the years 1929–79. As indicated in the figure, there is a general pattern of increase in the level of GNP from 1929 to 1979, but with some variation in specific years or in specific groups of years (e.g., 1940–46). Figure 16.1 is an example of a time series.

> **Definition 16.1**
> Time series
>
> A *time series* consists of statistical data that are collected, recorded, or observed over successive increments of time.

Changes in a time series (such as shown in Figure 16.1) are the result of a variety of forces that make themselves felt in various ways. For example, the long-term growth in GNP can be associated with such factors as a general increase in the population, as well as a host of other factors. Deviations from the general long-term growth pattern of GNP for specific years are usually attributed to a different set of factors, which would include the weather, wars, and political climate. Finally, if measurements of data are made within a year (or other period), seasonal factors often play a somewhat different role in influencing the direction or the magnitude of a given time series.

526

FIGURE 16.1 Gross National Product (GNP) of the United States, 1929–1979

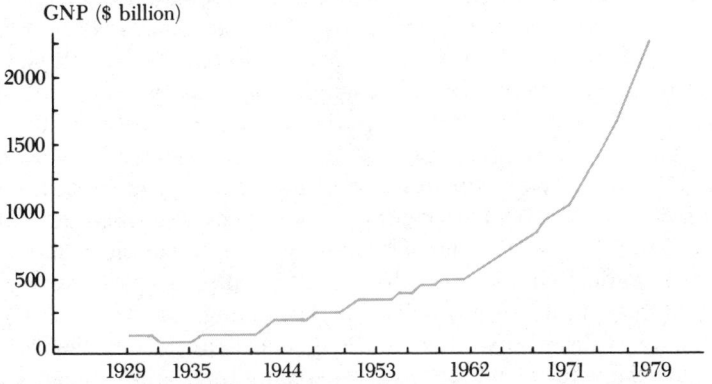

Source: U.S. Department of Commerce, *Survey of Current Business* (various issues).

The objective of studying time series lies in the abilitiy to estimate the value or the level of the series for a specified interval. Sociologists, economists, psychologists, environmentalists, and others ad infinitum often study patterns of growth of many or a few series to gain an understanding of certain phenomena in society and, of course, to plan or to estimate for the future. The estimation or prediction of the total demand for satellites for monitoring the discovery and depletion of specific natural resources, for example, was derived through the use of a logistic or a growth curve time series described in Section 16.3.

Whereas a general historical understanding of a time series enhances our ability to identify and analyze the underlying factors contributing to movements in the series, the greatest potential of a time series does in fact lie in the ability to predict an unknown value of the series. From this information, intelligent choices can be made concerning capital investment decisions or decisions about an appropriate level of production and inventory, etc. If one is willing to assume that there are *regular* and *repeating* components that interact in predictable ways to produce a given time series, he or she can then analyze these specific components to develop a reliable prediction of the series from which a decision can be made.

In this chapter, we will be concerned with describing "models" of time series as well as analyzing in detail the time series' component parts. Section 16.2 describes the factors that are usually attributed to or associated with movements in a time series, and specifies several common models that give the manner in which these factors interact. Section 16.3 then begins a discussion of the long-term growth component (secular trend) of a time series, and Section 16.4 presents statistical techniques necessary for isolating or measuring this component of a series. The effects of seasonal factors are addressed in Section 16.5, and methods are presented for "adjusting" statistical data for the effects of seasonal factors.

The statistical procedures to be employed in the chapter consist of linear regression discussed in the previous three chapters, as well as moving averages to isolate specific components. Because we will attempt to isolate specific components of a time series or to *decompose* it into its specific components, and because we will be employing classical techniques of linear regression and moving averages, the results of our analysis in this chapter will be termed *classical decomposition of time series*. In the following chapter, Chapter 17, we will study other methods of predicting the value of some time series, but will augment the techniques described in this chapter to predict the value of a time series at some future point in time, or with regard to another quantitative variable, etc. For example, we will use the multiple linear regression model for predicting accidents described in Section 15.7 to show how other than a specific "time" variable can be used to estimate the value of a time series at some future point. Because many of the assumptions underlying the use of the multiple linear regression model are violated when studying time series data (especially the assumption that the error terms are not correlated), we will have to content ourselves with obtaining point estimates of the time series only in this chapter and in the next.

■ 16.2 COMPONENTS AND MODELS OF A TIME SERIES

□ 16.2.1 Models of a time series

A model of a time series is a specification of the forces that contribute to movements in the series as well as an analysis of the manner in which these forces interact in influencing the series' direction and magnitude. The most widely used model of a time series is the multiplicative model, in which the series is described as the *product* of four components—trend (T), seasonal (S) (if observations are recorded within a year), cyclical (C), and irregular (I)—although other models may be appropriate in analyzing specific series. This general model is described by the following equation.

Multiplicative model of a time series

$$Y = T \times S \times C \times I$$

In the equation, Y is viewed as the result of the four elements acting in combination to produce the series. This particular model is well suited to situations in which *percentage* changes best represent the movement in the series. It turns out that a wide variety of economic as well as other data are best represented by the multiplicative model—hence its widespread acceptance as the "standard" or the "norm."

Any particular time series *may,* of course, be better represented by some model other than the multiplicative one. For example, if the components of trend, seasonal, cyclical, and irregular forces interact in an additive fashion to produce a given series, then the *additive model* may be appropriate.

Additive model of a time series

$$Y = T + S + C + I$$

If it is observed that certain components interact in an additive fashion in producing a given series, and that other factors interact in a multiplicative manner, then a variation of the two models may be used.

Mixed additive and multiplicative models of a time series

$$Y = T \times C + S \times I$$
$$Y = T \times C \times I + S$$

Because of the widespread use of the multiplicative model and the large variety of different time series it represents, it will be the only model described in our treatment of time series analysis. One should keep in mind, however, that not *all* time series are best represented by this model; if an examination of the data reveals that certain components do not interact in the prescribed fashion, then a different model than the multiplicative one is appropriate.

☐ 16.2.2   **Components of time series**

A given time series consists of the following components:

1. Trend (T).   3. Cyclical (C).
2. Seasonal (S).   4. Irregular (I).

Definition 16.2
Secular trend

The *long-term trend* or *secular trend* of a time series is the smooth component of the series that represents the general long-run growth or decline in the time series over an extended period of time.

For example, Figure 16.1 depicts the growth of the gross national product of the United States from 1929 to 1979. In Figure 16.2, we have superim-

FIGURE 16.2 Gross national product of the United States (1929–1979) and representative long-term growth curve

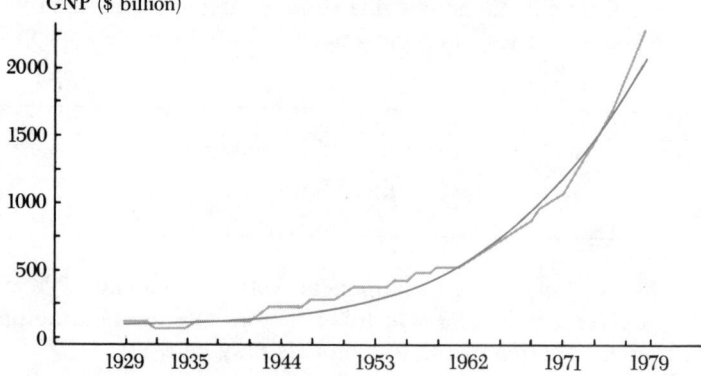

Source: U.S. Department of Commerce, *Survey of Current Business* (various issues), and computer analysis.

posed on this figure a representatitve trend curve, which reflects the general long-term growth in GNP over this interval. Although deviations from the long-term growth of GNP are apparent in Figure 16.2, one can recognize a long-term growth pattern. This long-term growth pattern is called the *trend* or *secular trend* of the series, and it is apparent that a curve as opposed to a straight line might better model the long-term growth in this series. We suggest a time series model for estimating the long-term growth component in GNP in Section 16.4.3.

If observations in data are recorded within a year or shorter period, then *seasonal* variation, which reflects climate, customs (timing of vacations and holidays), and other factors such as the length of calendar months, may be present in the time series.

Definition 16.3
Seasonal variation

Seasonal variations are periodic patterns in a time series that complete themselves within a year and then are repeated according to the same periodic pattern in subsequent years.

Figure 16.3 depicts the demand for plaster casts as a time series using the data given in Table 15.13 in Section 15.7. In addition to the apparent long-term (secular) growth in this series, there is a very definite seasonal pattern within each year as we discovered in Section 15.7. This seasonal pattern is more easily recognized by plotting the time series for selected years by quarter as shown in Figure 16.3B for the years 1978–80. Readily apparent

FIGURE 16.3

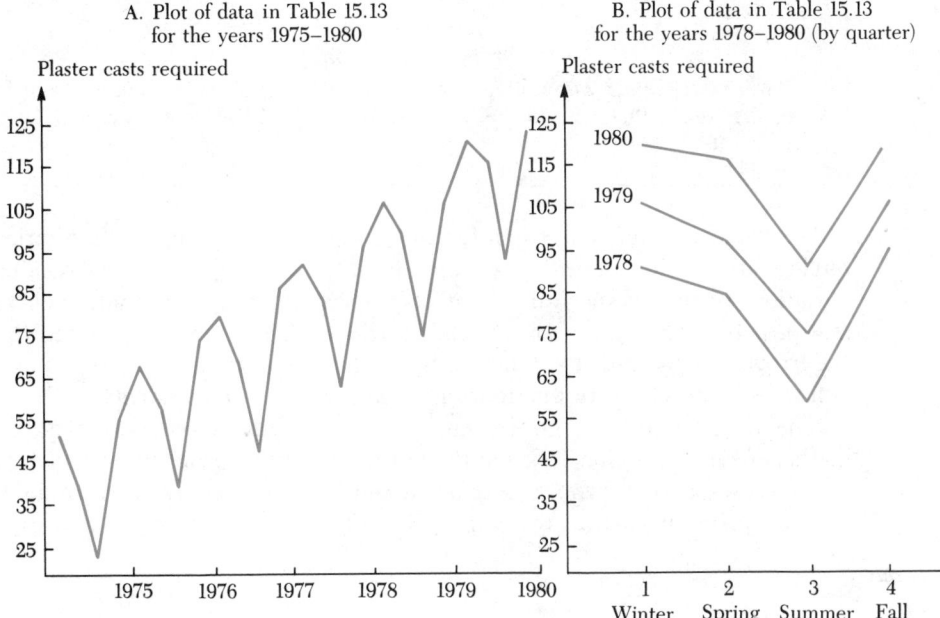

A. Plot of data in Table 15.13
for the years 1975–1980

Plaster casts required

B. Plot of data in Table 15.13
for the years 1978–1980 (by quarter)

Plaster casts required

Source: Company records.

from the data in Figure 16.3B is the tendency for the demand for casts to be higher (relative to other quarters) in the winter and fall quarters, and to be lower (relative to other quarters) in the summer quarter for each year. We can also isolate the seasonal component using residual plots described in Chapters 13 and 15, a topic we will defer until we attempt to assess the effects of seasonal factors in Section 16.5.

Definition 16.4
Cyclical component of a time series

The *cyclical component of a time series* refers to the recurring movements above and below the trend of the time series. These fluctuations last from two to ten years (or even longer) when measured from peak to peak or from trough to trough. The duration of a cyclical component is more than one year.

This wavelike, oscillating pattern is readily apparent in the data shown in Figure 16.2, especially for the years 1940–46. Different factors undoubtedly account for the cyclical movement in this time series and would include wars, uneven growth in the population, and others.

> **Definition 16.5**
> Irregular variation
>
> The *irregular variation* in a time series is composed of nonrecurring, sporadic forces that are not described as or attributed to trend, cyclical, or seasonal factors.

We are rarely able to account for all the variation present in a time series, having broken the series down into its component parts. This residual variation, or the variation that remains after accounting for trend, cyclical, and possibly seasonal factors, is called *irregular variation*. It is hoped that the irregular component will not comprise a large portion of the series and, hence, will be relatively unimportant in analyzing its movements.

The irregular component of a time series is analogous to the error terms ϵ in the linear regression models studied in the previous chapters. In fact, when a residual plot forms a rectangle about 0, we are relatively assured that we have properly isolated the components of the time series under investigation.

16.3 PATTERNS OF SECULAR TREND

In this section we will present several different mathematical functions for measuring or modeling the secular trend component of a time series. The curve used in modeling the trend component depends, of course, on the long-term growth in the series over the period for which trend is to be estimated. Although the selection of a particular mathematical model for assessing the trend component is as much an art as it is a science, two techniques—graphing and using the methods of first and second differences and constant rate increase—are described which are often helpful in selecting an appropriate model to estimate trend. The use of these two approaches for assessing a model of trend is encouraged much in the same way as residual plots were encouraged in Chapter 13—to *suggest* the appropriateness of a certain technique. The student who wishes further information on the analysis of time series is encouraged to take an advanced course either in time series or in econometrics.

16.3.1 Arithmetic straight-line trend

By far the most widely *applied* trend curve is the arithmetic straight-line trend, fitted by the method of least squares introduced in Chapter 13. This particular trend curve is applicable for series in which period-to-period changes are constant in absolute *amount*. (This is to be distinguished from the semilogarithmic trend curve, discussed shortly, which models time series that are changing at a constant *rate*.)

When the given time series is increasing at an approximately constant amount each period, then the straight-line or arithmetic trend represented by the equation

$$E(Y) = \beta_0 + \beta_1 x$$

is used to model the long-term or secular growth of the series. If, for example, we want to estimate the secular trend in the demand for plaster casts shown in Figure 16.3A, then we would most likely use the arithmetic straight-line model since the long-term growth over this period appears from the graph to be a constant amount each period (approximately).

Frequently, a given time series is modeled by a series of arithmetic straight-line trends, especially when a causative factor can be identified that contributes to a definite change in the amount of change per period.

Graphing the data is often the first step in assessing whether the trend component in the series is appropriately modeled by a straight line of the form $E(Y) = \beta_0 + \beta_1 x$. For example, it would appear that a straight line would be a poor "fit" of the growth in GNP in the United States between 1929 and 1979 from a visual inspection of the data plotted in Figure 16.1.

Another "rapid" means for assessing the appropriateness of the straight-line model is the *method of first differences*. If the differences between successive observations of a series are constant (or nearly so), this suggests that an arithmetic straight line may be an appropriate representation of the trend component. The method of first differences is illustrated in Table 16.1 for the given series. Since the differences in successive observations are nearly constant, the arithmetic straight line is an appropriate model for assessing the trend component of the series.

TABLE 16.1 Method of first differences

Year	Sales, y ($1000)	First difference
1971	$160	
		42
1972	202	
		40
1973	242	
		36
1974	278	
		37
1975	315	
		39
1976	354	
		41
1977	395	
		41
1978	436	
		42
1979	478	
		40
1980	518	

Source: Company records.

□ 16.3.2 Second-degree parabola trend

A polynomial of the form

$$E(Y) = \beta_0 + \beta_1 x + \beta_2 x^2$$

is called a *second-degree parabola* and may be an appropriate model for the secular trend component of a time series when the data appear not to fall in a straight line. Figure 16.4 is a graph of a time series represented by the second-degree parabola $E(Y) = \beta_0 + \beta_1 x + \beta_2 x^2$, where $\beta_0 = 12$, $\beta_1 = 4$, and $\beta_2 = 3$. The method used to fit a second-degree parabola to a given time series is multiple linear regression described in Chapter 15, where

$$E(Y) = \beta_0 + \beta_1 x_1 + \beta_2 x_2 \qquad \text{and} \qquad x_1 = x, \, x_2 = x^2$$

A test of the appropriateness of a second-degree parabola in modeling a given time series is based on the *method of second differences*. This method is illustrated in Table 16.2 for the parabola $y = 12 + 4x + 3x^2$. Note that when the differences are taken on the first differences, the result is a constant between successive observations of the given time series. Hence, whenever

FIGURE 16.4 Plot of the second-degree parabola $y = 12 + 4x + 3x^2$ for values of x equal to 1, 2, . . . , 9

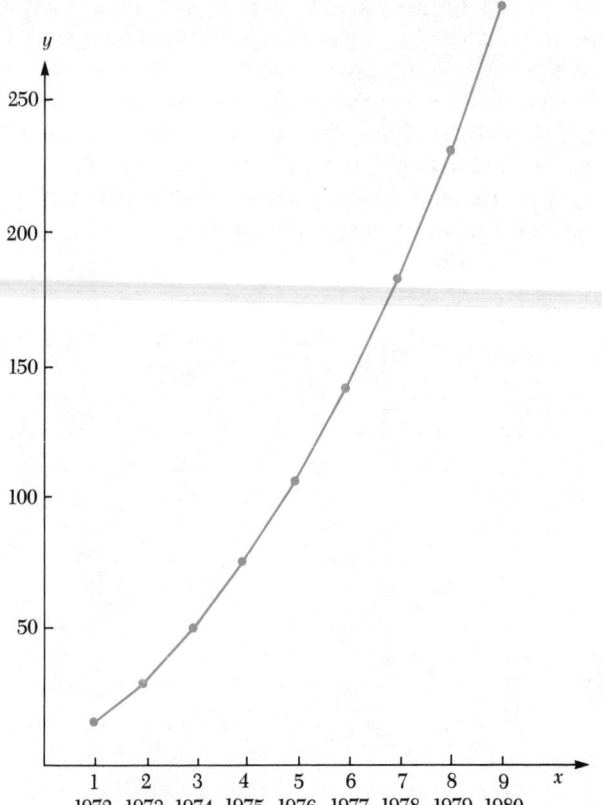

TABLE 16.2 Method of second differences

x	$y = 12 + 4x + 3x^2$	First difference	Second difference
1	19		
		13	
2	32		6
		19	
3	51		6
		25	
4	76		6
		31	
5	107		6
		37	
6	144		6
		43	
7	187		6
		49	
8	236		6
		55	
9	291		

the second differences of a given time series are constant (or are nearly so), a second-degree parabola fitted by the method of multiple linear regression is likely to be an appropriate model for measuring the secular trend of a given time series.

☐ **16.3.3 Semilogarithmic trend**

When a given time series is increasing at a constant *rate* and is *not* approaching some imputed upper limit, it is usually approximated best by an exponential curve, the most widely used form of which is

$$\log E(Y) = \beta_0 + \beta_1 x$$

which is derived from the exponential form

$$E(Y) = \beta_0 \beta_1^x$$

The trend of gross national product in the United States since 1929, for example, is a semilogarithmic trend, increasing at about 3 percent per year since 1929.

Table 16.3 gives information on the sales of electronic calculators by

TABLE 16.3 Sales of electronic calculators—Herman's Wholesale House

Year	Sales ($000)
1973	$30.0
1974	32.1
1975	34.3
1976	36.8
1977	39.4
1978	42.3
1979	45.2
1980	48.5

Source: Company records.

FIGURE 16.5

A. Scatter diagram of sales data in Table 16.3
(arithmetic scale)

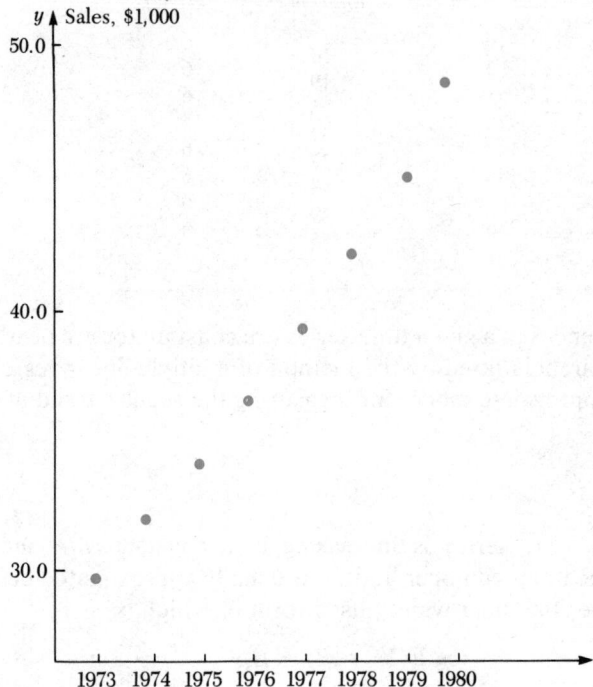

B. Semilogarithmic plot of sales data in Table 16.3

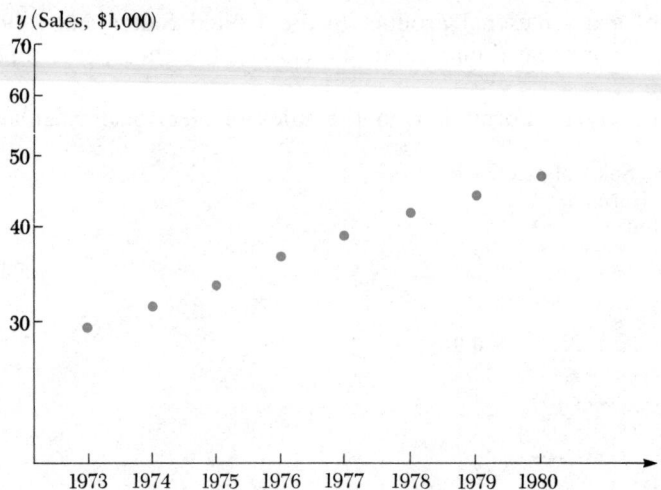

TABLE 16.4 First and second differences for data in Table 16.3

x	y($000)	First difference	Second difference
1973	$30.0		
1974	32.1	2.1	0.1
1975	34.3	2.2	0.3
1976	36.8	2.5	0.1
1977	39.4	2.6	0.3
1978	42.3	2.9	0.0
1979	45.2	2.9	0.4
1980	48.5	3.3	

Herman's wholesale calculator house for the years 1973–80, inclusive. A scatter diagram of the data in Table 16.3 is given in Figure 16.5A. From this figure, it is difficult to determine an appropriate model of the secular trend of this series.

First and second differences of the data in Table 16.3 are given in Table 16.4, and it indeed appears as though a straight-line or a second-degree parabola may not be appropriate models for the secular trend of this series, as neither set of differences appears to be constant.

Table 16.5 gives information on the *rate of increase* of calculator sales using the data in Table 16.3; the rate of increase appears to be fairly constant at around 7 percent each year. The data in Table 16.3 have also been plotted on semilogarithmic graph paper in Figure 16.5B. [With semilogarithmic graph

TABLE 16.5 Assessing the rate of increase in a time series for semilogarithmic model

x	y($000)	Rate of increase (percent)
1973	$30.0	
1974	32.1	7
1975	34.3	6.8
1976	36.8	7.3
1977	39.4	7.1
1978	42.3	7.4
1979	45.2	6.8
1980	48.5	7.3

Source: Company records.

paper, the units on the x-axis are evenly spaced, whereas the vertical scale (y-axis) is logarithmic, hence the term *semilogarithmic*.] Note from Figure 16.5B that when the data in Table 16.3 are plotted on semilogarithmic graph paper, the points appear to fall on a straight line. Hence, whenever the percentage increase in a given time series is constant, or whenever a scatter diagram of the data points of the series on a semilogarithmic scale appears to cluster about a straight line, the semilogarithmic model for measuring the trend component is suggested.

Methods will be presented in the next section for fitting a semilogarithmic trend to a series of data using the method of least squares described in Chapter 13.

☐ 16.3.4 Growth curves

Figure 16.6 depicts two time series that exhibit a pattern of growth in which there is an initial period of slow *absolute* growth, followed by a period of very rapid absolute expansion. Such growth patterns are apparent in the time series of sales in certain industries and in the population growth of various cities and states.

Series such as those in Figure 16.6 are typically modeled by *growth curves;* two of the most widely used are the *Gompertz curve* and the *logistic* or *Pearl-Reed curve*. The formulas for these two families of growth curves are also given in Figure 16.6. Note that both curves depict trend patterns

FIGURE 16.6   Gompertz and Pearl-Reed growth curves

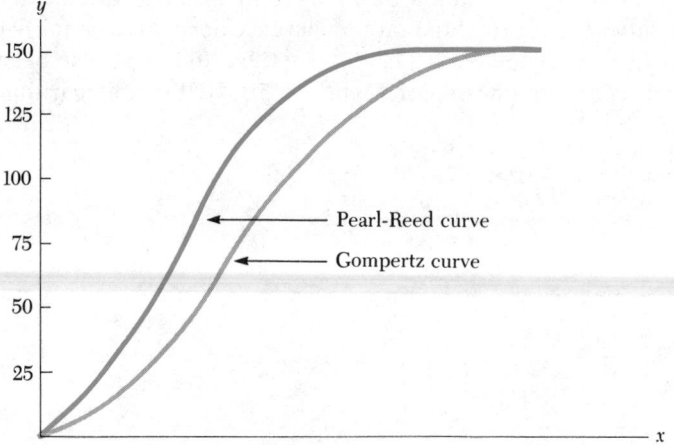

Growth curves

Gompertz curve	$y = ka^{b^x}$ (where k, a, and b are constants)
Pearl-Reed curve	$y = \dfrac{1}{k + ab^x}$ (where k, a, and b are constants)

increasing at a decreasing *rate*, but the approach to maturity is much more rapid with the Pearl-Reed than with the Gompertz curve.

Both growth curves approach a finite limit (as shown), and this must be taken into account when fitting a given time series to one of the curves. Often a key resource is known to exist up to some finite amount, and this can be used to establish a limit on the growth of the time series in question. Increasingly, for example, "new cities" are being planned with an eye toward limiting growth, with the amount of planned land available having an upper limit. Time series analysis using growth curves analyzes or depicts the manner in which (or the rate at which) the series is approaching its limit. Thus, whenever a given time series is increasing at a constant rate, but is understood to be approaching a finite limit in a predictable manner, the use of growth curves may be appropriate.

☐ 16.3.5 Other patterns of secular trend

Other patterns of secular trend than those identified can occur, and special methods exist for fitting trend curves using these advanced techniques. For example, a polynomial of the form

$$E(\mathbf{Y}) = \beta_0 + \beta_1 x + \beta_2 x^2 + \beta_3 x^3 + \cdots$$

may be used to represent a given time series. The various curve-fitting procedures for these advanced models can be found in the references at the end of the chapter. We will now be concerned with fitting a given series using the arithmetic or the semilogarithmic trend curves using the method of least squares.

■ 16.4 CLASSICAL DECOMPOSITION: MEASURING THE TREND COMPONENT OF A TIME SERIES

☐ 16.4.1 Straight-line trend estimation

Whenever a scatter plot of the data or an examination of the first differences of the time series indicates that a straight-line trend of the series is appropriate, the trend component can be estimated using the method of linear least squares presented in Chapter 13. Unfortunately, most assumptions of the simple linear regression model—$E(\epsilon_i) = 0$, constant variance, uncorrelated error terms, normality, etc.—are not met when time series data are examined. Hence, we cannot attach a measure of statistical confidence to the results obtained. We must be content merely to obtain *point estimates* (sometimes biased) of trend components, and we *may not* make statements as to the confidence that can be placed in these point estimates when the linear least squares method is used to assess the trend component of a time series. Example 16.1 introduces the notion of estimating linear trend by the method of least squares.

Example 16.1   The data in Table 15.13 represent the number of plaster casts placed on "patrons" to a ski and leisure resort in western Colorado between 1975 and 1980, by quarter. A graph of these data is given in Figure 16.3. From Figure 16.3, it appears as though a straight-line trend is the appropriate model for the *secular trend component* of this series. Using the method of least squares, compute a trend equation for the data provided. Use a straight-line to measure the trend component, and estimate the trend component for 1981.

Solution   Since the demand for casts is being viewed as a function of time in the trend equation, time (in quarters) becomes the independent variable, x, and the demand for casts becomes the dependent variable, $\mathbf{Y}$, in the regression (trend) equation $E(\mathbf{Y}) = \beta_0 + \beta_1 x$. Because all time series models involve the common element of time, we shall denote the independent variable time by t. Here t is analogous to x in the simple linear regression model. Similarly, we will denote the values of the dependent random variable, $\mathbf{Y}$, by y_t to indicate that $\mathbf{Y}$ is a time series random variable. Thus, our estimating equation will become:

$$\hat{y}_t = b_0 + b_1 t$$

By using the variable t for the independent variable in a trend equation, it will be easy to distinguish between time series variables and variables that are not necessarily time series. The computations, of course, will be the same whether we use x or t to denote the independent variable in our predicting equation.

Since β_0 and β_1 are rarely known in practice, they must be estimated by the least squares estimators $\mathbf{b}_0$ and $\mathbf{b}_1$, respectively. Equations for determining b_0 and b_1 are given as (16.1) and (16.2) to facilitate the computation of the estimates of the trend parameters.

Least squares estimate of the slope in a straight-line trend equation

$$b_1 = \frac{\Sigma ty - \dfrac{(\Sigma t)(\Sigma y)}{n}}{\Sigma t^2 - \dfrac{(\Sigma t)^2}{n}} \tag{16.1}$$

Least squares estimate of the intercept in a straight-line trend equation

$$b_0 = \bar{y} - b_1 \bar{t} = \frac{1}{n}(\Sigma y - b_1 \Sigma t) \tag{16.2}$$

where $\bar{y} = (\Sigma y)/n$ and $\bar{t} = (\Sigma t)/n$.

Before we begin to compute the estimates b_0 and b_1, note that the arithmetic is going to be quite burdensome because of the magnitude of the quantities involved. We can simplify the number of computations performed if we *code the data* and let the periods involved be represented by a sequence of nonnegative, increasing integers, where at the "base" (first) period, $t = 0$. This coding, along with relevant computational data for the trend equation, is given in Table 16.6. The trend equation for the data

TABLE 16.6 Computation worksheet for fitting a straight-line trend to demand for plaster casts, 1975–1980

Year and period	Quarter	t	y	ty	t^2
1975					
1..............	Winter	0	53	0	0
2..............	Spring	1	41	41	1
3..............	Summer	2	24	48	4
4..............	Fall	3	57	171	9
1976					
5..............	Winter	4	70	280	16
6..............	Spring	5	60	300	25
7..............	Summer	6	41	246	36
8..............	Fall	7	77	539	49
1977					
9..............	Winter	8	81	648	64
10..............	Spring	9	70	630	81
11..............	Summer	10	50	500	100
12..............	Fall	11	87	957	121
1978					
13..............	Winter	12	94	1,128	144
14..............	Spring	13	86	1,118	169
15..............	Summer	14	64	896	196
16..............	Fall	15	99	1,485	225
1979					
17..............	Winter	16	109	1,744	256
18..............	Spring	17	101	1,717	289
19..............	Summer	18	77	1,386	324
20..............	Fall	19	110	2,090	361
1980					
21..............	Winter	20	123	2,460	400
22..............	Spring	21	120	2,520	441
23..............	Summer	22	95	2,090	484
24..............	Fall	23	126	2,898	529
		276	1,915	25,892	4,324

$$b_1 = \frac{25,892 - \dfrac{(276)(1,915)}{24}}{4,324 - \dfrac{76,176}{24}} = 3.3648 \qquad b_0 = \frac{1}{24}(1,915 - 3.3648 \cdot 276) = 41.0965$$

provided is:

$$\hat{y} = 41.0965 + 3.3648t, \qquad t = 0 \text{ at quarter I, 1975}$$

where t is in quarter-years and $\hat{y}$ is in single units.

Note that we are obligated (irrespective of the manner used to compute values of t) to state the origin of the trend equation ($t = 0$ at quarter I, 1975) so that trend values can be computed accordingly. Specific units for both t and y are often necessary also.

The trend (arithmetic straight-line) values for each quarter in 1975 to 1980 are given in Table 16.7. These values are derived by substituting the values $t = 0, 1, 2, \ldots$ into the trend equation for the quarters of the years 1975, 1976, 1977, and so on. A projection of the trend component for this time series for each quarter of 1981 is shown in the table.

Quarter	Trend projection
I Winter..............	$41.0965 + 3.3648(24) = 121.852$
II Spring	$41.0965 + 3.3648(25) = 125.217$
III Summer...........	$41.0965 + 3.3648(26) = 128.581$
IV Fall................	$41.0965 + 3.3648(27) = 131.946$

A graph of the derived trend equation and actual values is given in Figure 16.7. The trend line computed appears to represent the long-term growth in this time series. As we learned in Section 15.7, however, seasonal factors play a role in determining the value of the time series *within each year*. This is also readily apparent in the graph in Figure 16.7. In Section 16.5 we will

FIGURE 16.7 Straight-line trend equation for the demand for plaster casts

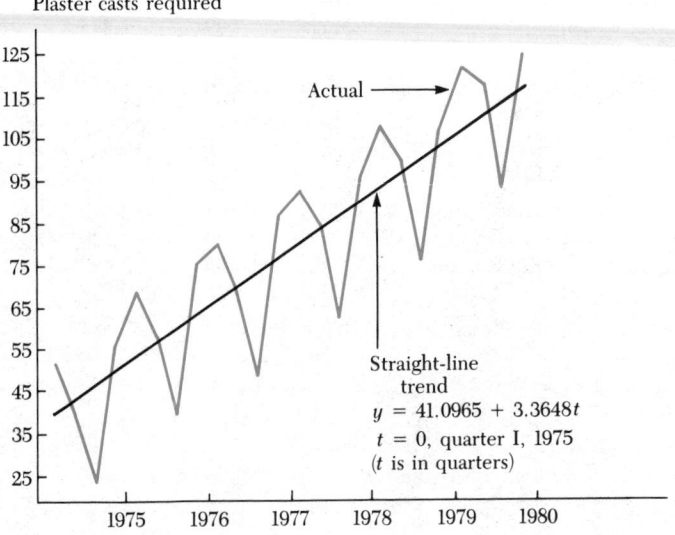

learn another method for determining the seasonal component of this time series, the method of ratio to moving average.

TABLE 16.7 Straight-line trend estimates for data in Table 16.6

Year and quarter	Observed value	Predicted value	Residual
1975			
Winter................	53	41.097	11.903
Spring	41	44.461	−3.461
Summer	24	47.826	−23.826
Fall	57	51.191	5.809
1976			
Winter................	70	54.556	15.444
Spring	60	57.920	2.079
Summer	41	61.285	−20.285
Fall	77	64.650	12.350
1977			
Winter................	81	68.015	12.985
Spring	70	71.380	−1.380
Summer	50	74.744	−24.744
Fall	87	78.109	8.890
1978			
Winter................	94	81.474	12.526
Spring	86	84.839	1.161
Summer	64	88.204	−24.204
Fall	99	91.568	7.431
1979			
Winter................	109	94.933	14.067
Spring	101	98.298	2.702
Summer	77	101.663	−24.663
Fall	110	105.027	4.972
1980			
Winter................	123	108.392	14.608
Spring	120	111.757	8.242
Summer	95	115.122	−20.122
Fall	126	118.487	7.513

Source: Tables 15.13 and 16.6.

☐ **16.4.2 Shifting the origin and the units of measurement in a straight-line trend equation**

We frequently find that we would like to shift the origin or the units of measurement of a trend equation to get a forecast for a specific period in the future. Changing the origin when the units of time (years) do not change poses no special problems. The Y-intercept (b_0-value) simply becomes the trend value for that year, and the value b_1 (slope) remains the same.

Often, we would like a trend value for specific *months* within a year, and this necessitates our changing the units of the trend equation from years to months for an accurate trend extrapolation. The method for accomplishing this is presented in the following example.

Example 16.2 A trend equation for a manufacturer of a consumer durable goods has been assessed by the method of least squares (straight-line trend), yielding the following trend equation:

$$\hat{y} = 19.88 + 2.86t, \qquad t = 0 \text{ at } 1972$$

where t is in years and $\hat{y}$ is in millions of dollars. Obtain an estimate of the trend component of this time series for July and August 1983 using the trend equation provided.

Solution Division of the given trend equation by 12 yields:

$$\hat{y} = 1.66 + 0.2383t$$

Note, however, that t is still expressed in years! That is, with our trend equation (centered at July 1 of each year), $t = 0$ implies July 1972, $t = 1$ implies July 1973, etc. To express the trend equation in consecutive *months*, it is necessary to divide the coefficient of t by 12 once more. The procedure for changing the units of measurement from years to months is

$$\begin{array}{cc} \textit{Years} & \textit{Months} \\[4pt] \hat{y} = b_0 + b_1 t & \hat{y} = \dfrac{b_0}{12} + \left(\dfrac{b_1}{144}\right) t \end{array}$$

Hence, for our example,

$$\hat{y} = 1.66 + \left(\frac{0.2383}{12}\right) t = 1.66 + 0.02t,$$

$t = 0$ at July 1, 1972, t is in months, and $\hat{y}$ is in millions of dollars.

Quite often, one final adjustment is made to the preceding equation. Since the midpoint in each month is more representative of the trend value for that month, the monthly trend equation is centered. For the equation in question,

$$\hat{y} = 1.66 + 0.02 \left(\frac{1}{2}\right) + 0.02t = 1.67 + 0.02t$$

where $t = 0$ at July 15, 1972, t is in months, and $\hat{y}$ is in millions of dollars.

Returning to the original problem, the projection of trend for July and August 1983 becomes:

$$\begin{aligned} \hat{y}_{\text{July 1983}} &= 1.67 + 0.02(132) = \$4.31 \text{ million} \\ \hat{y}_{\text{Aug 1983}} &= 1.67 + 0.02(133) = \$4.33 \text{ million} \end{aligned}$$

☐ **16.4.3 Semilogarithmic trend estimation**

It was noted in Section 16.3.3 that whenever a series is increasing at a constant *rate*, it is usually best represented by a semilogarithmic trend equation. Examples of time series exhibiting a constant rate of growth include U.S. gross national product from 1929 to 1979 (see Figure 16.1) and personal consumption expenditures for automobiles and parts from 1950 to 1978 (see Figure 16.8). The appropriateness of the semilogarithmic model is suggested whenever (1) a plot of the time series data on semilogarithmic graph paper tends to show points clustering about a straight line, and (2) the percentage increment between successive periods is constant or nearly so.

FIGURE 16.8   Personal consumption expenditures for automobiles and parts, 1950–1978 (actual), and semilogarithmic trend

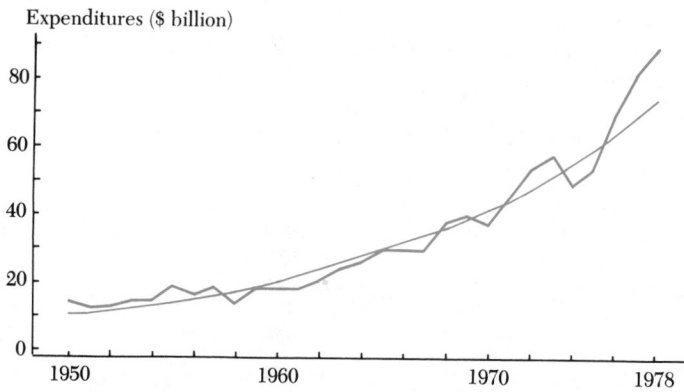

Source: U.S. Department of Commerce, *Survey of Current Business* (various issues), and computer analysis.

Estimates b_0 and b_1 of the parameters of the trend equation modeling a constant rate of growth series are given in equations (16.3) and (16.4). These formulas have been determined by the method of least squares. Note that they differ from the formulas for straight-line trend only in that log y replaces y in the computations.

Least squares estimate of the slope in a semilogarithmic trend equation

$$b_1 = \frac{\Sigma t \log y - \dfrac{(\Sigma t)(\Sigma \log y)}{n}}{\Sigma t^2 - \dfrac{(\Sigma t)^2}{n}} \tag{16.3}$$

Least squares estimate of the intercept in a semilogarithmic trend equation

$$b_0 = \frac{1}{n}\,(\Sigma \log y - b_1 \Sigma t) \tag{16.4}$$

An illustration of the use of these formulas for estimating trend is given in the following example.

Example 16.3 Personal consumption expenditures for automobiles and parts for the years 1950 to 1978 are given in the table below. Estimate the trend component in this series using the semilogarithmic model presented.

Year	Expenditures for automobiles, and parts ($billion)	Year	Expenditures for automobiles, and parts ($billion)
1950	12.9	1965	29.8
1951	11.1	1966	30.3
1952	10.4	1967	30.5
1953	13.2	1968	37.0
1954	12.6	1969	40.2
1955	17.2	1970	37.3
1956	15.8	1971	46.7
1957	17.1	1972	53.1
1958	13.9	1973	57.5
1959	18.1	1974	49.8
1960	18.8	1975	53.4
1961	17.1	1976	69.7
1962	20.4	1977	81.5
1963	24.3	1978	89.8
1964	25.8		

Source: U.S. Department of Commerce, *Survey of Current Business* (various issues).

Solution Relevant computations are shown in Table 16.8. Table B.8 in Appendix B gives the necessary data for computing the logarithms (to base 10) of the observed values. Note that it is again useful to code the data in some fashion. In this case, we have simply let the first year in the series equal 0; the second year equal 1; etc. The relevant trend equation is:

$$\log \hat{y} = 0.98946 + 0.031655t, \qquad t = 0 \text{ at } 1950$$

where t is in years and $\log \hat{y}$ is in billions of dollars.

To determine the trend value from the semilogarithmic model, it is necessary to compute the antilogarithm of $\hat{y}$ for specific years. For example, to compute the trend value for 1955, we must find the antilogarithm of 1.147735 ($0.98946 + 0.031655 \cdot 5$), which is (approximately) $14.05 billion. Trend estimates for the remaining years are given in Table 16.12.

The rate of growth implicit in a semilogarithmic trend is often of interest. It is derived by solving the following equation

$$\log(1 + r) = b_1$$

where b_1 is the slope of the semilogarithmic trend equation, and r is the rate of growth (per year). For the data in question,

$$\log(1 + r) = 0.031655$$
$$1 + r = \text{antilogarithm}(0.031655)$$
$$r = 1.0756 - 1 = 0.0756$$

Hence, we would state that the rate of growth in personal consumption

TABLE 16.8 Computation worksheet for fitting a semilogarithmic trend to data on expenditures for automobiles and parts, 1950–1978

Year	Year in transformed units	Expenditures for automobiles and parts ($ billions)	$\log_{10} y$	$t \log_{10} y$	t^2
1950.......	0	12.9	1.11059	0	0
1951.......	1	11.1	1.04532	1.0453	1
1952.......	2	10.4	1.01703	2.0341	4
1953.......	3	13.2	1.12057	3.3617	9
1954.......	4	12.6	1.10037	4.4015	16
1955.......	5	17.2	1.23553	6.1776	25
1956.......	6	15.8	1.19866	7.1919	36
1957.......	7	17.1	1.23300	8.6310	49
1958.......	8	13.9	1.14301	9.1441	64
1959.......	9	18.1	1.25768	11.3191	81
1960.......	10	18.8	1.27416	12.7416	100
1961.......	11	17.1	1.23300	13.5630	121
1962.......	12	20.4	1.30963	15.7156	144
1963.......	13	24.3	1.38561	18.0129	169
1964.......	14	25.8	1.41162	19.7626	196
1965.......	15	29.8	1.47422	22.1132	225
1966.......	16	30.3	1.48144	23.7031	256
1967.......	17	30.5	1.48430	25.2331	289
1968.......	18	37.0	1.56820	28.2276	324
1969.......	19	40.2	1.60423	30.4803	361
1970.......	20	37.3	1.57171	31.4342	400
1971.......	21	46.7	1.66932	35.0556	441
1972.......	22	53.1	1.72509	37.9521	484
1973.......	23	57.5	1.75967	40.4723	529
1974.......	24	49.8	1.69723	40.7335	576
1975.......	25	53.4	1.72754	43.1885	625
1976.......	26	69.7	1.84323	47.9240	676
1977.......	27	81.5	1.91116	51.6012	729
1978.......	28	89.8	1.95328	54.6917	784
	406		41.54640	645.9111	7,714

$$b_1 = \frac{645.91 - \dfrac{406 \times 41.5464}{29}}{7,714 - \dfrac{(406)^2}{29}} = 0.031655$$

$$\log \hat{y} = 0.98946 + 0.031655t$$
$$t = 0 \text{ at } 1950$$
t is in years; $\log \hat{y}$ is in $billions

$$b_0 = \frac{1}{29}(41.54640 - 0.031655 \times 406) = 0.98946$$

Source: U.S. Department of Commerce, *Survey of Current Business* (various issues).

expenditures for automobiles and parts between 1950 and 1978 (inclusive) has been about 7.6 percent.

There are other coding procedures for the variable "time" than the ones described thus far. For example, if there is an odd number of years (say, 7) and instead of letting the independent variable time assume the values 0, 1, 2, 3, 4, 5, and 6, we let the independent variable time assume the values -3, -2, -1, 0, 1, 2, and 3, the formulas for computing estimates b_0 and b_1 become somewhat simplified. Because computers are performing so many of the calculations in determining time series trend estimates, and because most pocket calculators have preprogrammed functions for measuring trend (straight-line) by the method of least squares, we will not go into detail in describing these other coding schemes for the time variable in a time series model. Several references at the end of the chapter present these other coding schemes for the time variable; the interested reader can find them there.

■ 16.5 CLASSICAL DECOMPOSITION: MEASURING THE SEASONAL COMPONENT OF A TIME SERIES

It is difficult for anyone who regularly reads a newspaper or listens to the news not to have some notion of what are seasonally adjusted data. The government is continually bombarding us with seasonally adjusted data when it reports: "Unemployment for the month of June was up 5.2 percent *on a seasonally adjusted basis*" or "Gross national product rose last month by 3 percent on a seasonally adjusted basis." Naturally, what is meant is not that unemployment rose an *actual* 5.2 percent in June, but that when seasonal factors are taken into account (e.g., recent graduates entering the job market), the rate of unemployment *still* was up 5.2 percent. The actual increase was most likely much larger than 5.2 percent, but because in general we expect it to be higher in June, we make adjustments in the data (deseasonalize) for this.

We first encountered the notion of a seasonal component of a time series in Section 15.7, when, using the analysis of variance procedure, we tested for the presence of a seasonal factor influencing the demand for plaster casts at a Colorado ski resort. A method for *identifying* the presence of a seasonal factor in addition to those already discussed is the use of residual plots described in Chapters 13 and 15. That is, when data are available on a monthly, quarterly, or other seasonal basis, we fit a trend line or curve to the data and examine a plot of the residuals. Such a plot is given in Figure 16.9 for the Table 16.7 data. Here we have "connected" the residuals between successive periods in order to emphasize their recurrent pattern between years. The analysis of the residual plot in Figure 16.9 would lead one to conclude not only that the residuals are not random, but also that there is systematic variability among them that should be able to be modeled. The modeling process consists of the identification and computation of seasonal indices.

FIGURE 16.9   Residual plot of the data in Table 16.7

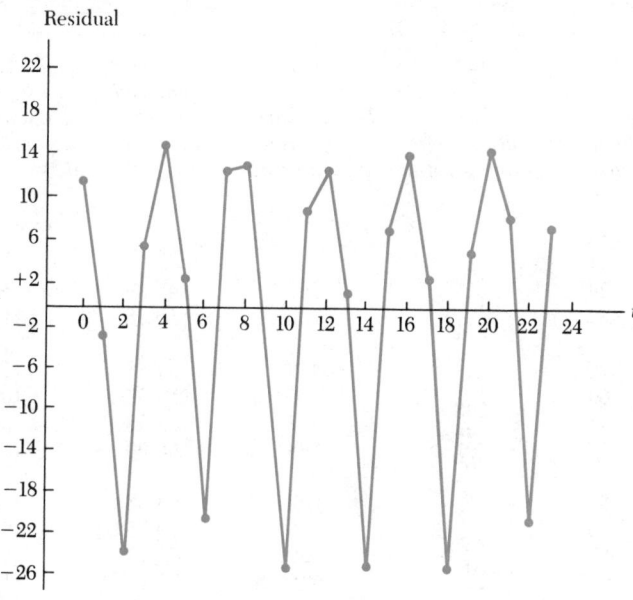

We shall examine the data in Table 15.13 once more and assess the seasonal component by a different method than was discussed in Chapter 15—namely, by the method of *ratio to moving average*.

The first two columns of Table 16.9 give information on the demand for plaster casts repeated as a time series from Table 15.13. These data will be used to describe the ratio to moving average method, as we attempt to determine a quarterly seasonal index for the demand for plaster casts.

In the third column of Table 16.9, the total demand for casts is recorded for consecutive four-quarter periods. These totals are centered between the quarters, because the midpoint in any consecutive four-quarter period lies at the very end of the second quarter in the series. For example, for the four quarters of the year, the midpoint of the period is July 1, or the *start* of the third quarter. Hence, the four-quarter moving total for the four consecutive quarters of a year are placed between the second and the third quarters, or at the midpoint of the year.

To compute the second four-quarter moving total in Table 16.9, we subtract from the current total (175) the value for the first quarter of 1975 (53) and add to this value the time series value for the first quarter of 1976 (70). This moving total = 175 − 53 + 70 = 192 is then centered between the third and fourth quarters of 1975. This procedure for obtaining the four-quarter moving totals is shown in detail in Table 16.10.

In order to center the moving totals so they will correspond to specific quarters, and using the midpoint of each quarter as representative of the value for that quarter, eight-quarter *centered* moving totals are computed by adding consecutive two-period, four-month moving totals, such as indicated

TABLE 16.9 Computation of specific seasonal relatives using method of ratio to moving average

(1) Date	(2) Plaster casts	(3) Four-quarter moving total	(4) Eight-quarter centered moving total	(5) Centered moving average (4) ÷ 8	(6) Specific seasonal relative [(2) ÷ (5)] × 100
1975					
Winter............	53				
Spring............	41	175			
Summer..........	24	192	367	45.88	52.30
Fall	57	211	403	50.38	113.20
1976					
Winter........,....	70	228	439	54.88	127.60
Spring............	60	248	476	59.50	100.80
Summer..........	41	259	507	63.38	64.70
Fall	77	269	528	66.00	116.70
1977					
Winter............	81	278	547	68.38	118.50
Spring............	70	288	566	70.75	98.90
Summer..........	50	301	589	73.63	67.90
Fall	87	317	618	77.25	112.60
1978					
Winter............	94	331	648	81.00	116.00
Spring............	86	343	674	84.25	102.10
Summer..........	64	358	701	87.63	73.00
Fall	99	373	731	91.38	108.30
1979					
Winter............	109	386	759	94.88	114.90
Spring............	101	397	783	97.88	103.20
Summer..........	77	411	808	101.00	76.20
Fall	110	430	841	105.13	104.60
1980					
Winter............	123	448	878	109.75	112.10
Spring............	120	464	912	114.00	105.30
Summer..........	95				
Fall	126				

in column (4) of Table 16.9. Division of this quantity by 8 then gives the centered moving average for the quarter involved. These centered moving averages are indicated in column (5) of Table 16.9. A graph of these centered moving averages as well as the time series values is given in Figure 16.10. Their computation is straightforward; for the third quarter of 1975, the centered moving average is (175 + 192 = 367)/8 = 45.88, for example. Note from Figure 16.10 the "smoothing" effect the moving average has on the time series data. This is a point we will return to in the next chapter, when we discuss moving averages as a method of forecasting the value of a time series.

TABLE 16.10 Computation of four-quarter moving total

Date	Actual data	Four-quarter moving total
1975		
Winter	53	
Spring..........	41	
Summer........	24	175
Fall............	57	192
1976		211
Winter	70	
Spring..........	60	.
Summer........	41	.
Fall............	77	.
	.	
	.	
	.	

FIGURE 16.10 Time series values for the demand for plaster casts and their centered moving averages

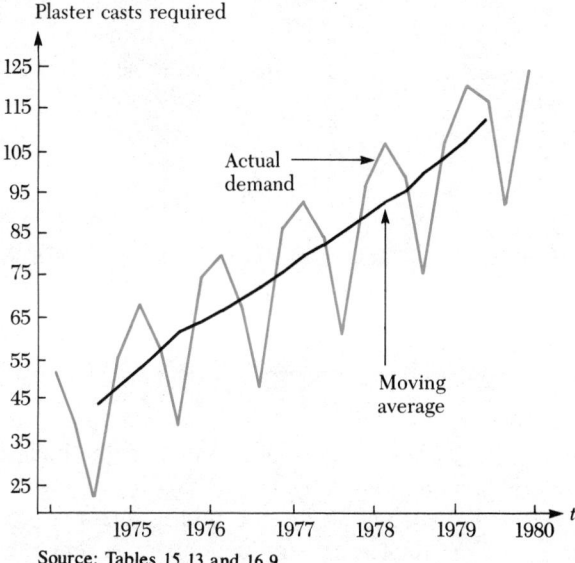

Source: Tables 15.13 and 16.9.

The ratio to moving average method derives its name from the division that takes place between the actual data and the corresponding centered moving average, yielding a *specific seasonal relative*. The term *specific* is attached to these seasonal relatives because each relative refers to a specific quarter and year. Note that by using quarterly data and the method of ratio to moving average, specific seasonal relatives do not exist for the first two quarters of the series and the last two quarters of the series, because it is impossible to construct a four-quarter moving average for these quarters. If instead, we were computing monthly seasonal indices, then specific seasonal relatives would not exist for the first six months and the last six months of the series, etc.

The data in column (6) of Table 16.9 have been rearranged in Table 16.11 listing the relatives by quarter for each of the years involved. Seasonal

TABLE 16.11 Computation of adjusted seasonal indices

| | Specific seasonal relatives | | | | |
Year	Q1 Winter	Q2 Spring	Q3 Summer	Q4 Fall	
1975			52.30	113.20	
1976	127.60	100.80	67.40	116.70	
1977	118.50	98.90	67.90	112.60	
1978	116.00	102.10	73.00	108.30	
1979	114.90	103.20	76.20	104.60	
1980	112.10	105.30			
					Totals
Median of specific seasonal relatives	116.00	102.10	67.90	112.60	398.60
Adjusted seasonal index	116.40	102.46	68.14	113.00	400.00
Average of specific seasonal relatives	117.82	102.06	67.36	111.08	398.32
Adjusted seasonal index	118.31	102.49	67.64	111.56	400.00

(quarterly) indices are then computed by either of two methods from the specific seasonal relatives listed.

1. The *median* of the specific seasonal relatives for each quarter becomes the unadjusted seasonal index for that quarter, and these unadjusted seasonal indices are multiplied by the ratio of 400 (4 quarters, 100 base value per quarter) to their total to normalize their total to 400. The result is an *adjusted seasonal index* based on medians.

2. The *arithmetic mean* of the specific seasonal relatives for each quarter is the unadjusted seasonal index for that quarter, and all unadjusted seasonal indices are normalized (as above) so that they total 400. The result is an *adjusted seasonal index* based on averages.

Quarterly, seasonal indices based on both the median and the arithmetic mean are indicated in Table 16.11. Some economists would argue that a third method should be examined in addition to the two discussed above. Namely, any "outliers" from the data should be omitted, and the arithmetic mean should be taken of the remaining specific seasonal relatives. This method is referred to as computing unadjusted seasonal indices using a *modified mean*. The difficulty in applying this third method lies in the determination of exactly what constitutes an outlier, and what does not. The data in Table 16.11 would appear to have no quarter with a specific seasonal relative that deviates significantly from the rest of the values.

Whichever method is used to compute the unadjusted seasonal indices, part of the irregular variation (I) in the model $Y = T \cdot S \cdot C \cdot I$ is "averaged out" in the process of determining the adjusted seasonal indices. This smoothing effect of the averaging process eliminates many irregular changes from period to period, and hence often dampens much of the irregular variation in the series.

If the data on the demand for plaster casts had been given monthly, then it would have been imperative to compute monthly seasonal indices as opposed to quarterly indices. In the third column of Table 16.9, we would have computed a 12-month moving total, and in the fourth column, we would have computed a 24-month centered moving total. The quantity in column (4) would have then been divided by 24, yielding the centered moving average, and this quantity would have then been divided into the actual demand for the month, yielding the specific seasonal relative for the month. The rest of the analysis would have proceeded similarly.

The use of seasonal indices in decision making is illustrated in the following examples.

Example 16.4   The firm desiring to estimate the demand for plaster casts has obtained an estimate of *yearly* demand for 1981 using a straight-line arithmetic trend of 521 casts. From this annual estimate, derive an estimate for each of the quarters of 1981 using the seasonal indices in Table 16.11 based on the arithmetic mean.

Solution   Since annual demand has been estimated to be 521 units, the demand for each quarter if there were no seasonal influence in these data would be 130.25 units per quarter (521/4). An estimate of the demand for each quarter is obtained by multiplying the average quarterly estimate of 130.25 by the specific seasonal relative, and dividing the result by 100 as shown in the table.

(1) Quarter	(2) Average quarterly demand (estimated)	(3) Seasonal index	(4) Quarterly forecast [(2) · (3)]/100
Winter	130.25	118.31	154.1
Spring	130.25	102.49	133.5
Summer	130.25	67.64	88.1
Fall	130.25	111.56	145.3
	521.0		521.0

We will compare these estimates with those derived from a time series forecasting equation in the next chapter, where indicator or dummy variables will be used in a multiple regression model as described in Chapter 15 to assess the seasonal component in this time series.

Example 16.5 The seasonal index for a particular consumer good for January is 110.24. Actual demand for January 1981 was 1,085 units. Based on only this information, what would you estimate sales to be in 1981?

Solution Since January in general represents a month in which demand is 10.24 percent above an "average" month, average monthly demand based on 1,085 units being demanded in January is

$$\frac{1,085}{110.24} \times 100 = 984 \text{ units}$$

From the information provided, we would estimate total yearly sales to be

$$984 \times 12 = 11,808 \text{ units}$$

Example 16.6 Honest Roger, after four clean years in the Green House, has come under attack for his poor record on unemployment in the greater Happy Valley metropolitan area. "For the last two quarters," his opponent, cigar-smoking Lean Gene, says, "unemployment has risen drastically!" Lean Gene presents the following evidence to support his contention:

Quarter	Number of unemployed in Happy Valley
I	285
II	327
III	340

Honest Roger has been provided with the following quarterly indices of unemployment by the Happy Valley Chamber of Commerce. Not being a very good statistician, however, poor Honest Roger does not know how to use this information to help his cause. How can the statistics quoted by Lean Gene be recast in order to better judge the policies of Honest Roger on unemployment?

Quarter	Index
I	0.90
II	1.10
III	1.15
IV	0.85

Solution In essence, the problem calls for the data presented by Lean Gene to be deseasonalized. That is, we need to assess the net change in unemployment when factors affecting the seasonality of the job market are taken into account. To deseasonalize the data, we must divide the actual unemployment by the seasonal (quarterly) indices provided.

Quarter	Actual unemployment	Quarterly index	Deseasonalized unemployment
I....................	285	0.90	317
II....................	327	1.10	297
III....................	340	1.15	296
IV....................	—	0.85	—
		4.00	

Note that in this case we do not multiply the quotients by 100 because the quarterly indices are expressed on a basis totaling 4 for the four quarters of the year, rather than 400.

When the deseasonalized data are examined, it appears that Honest Roger's policy on unemployment has been effective in actually reducing the seasonal level of unemployment in the Happy Valley area over the last two quarters.

This example should sharpen the distinction between seasonalized and deseasonalized data! Moreover, it should alert the user (voter) to the dangers in interpreting time-phased statistical data *of any type* when all the facts have not been reported. This is especially true with respect to "performance" data when they are measured between successive increments of time.

■ 16.6 CLASSICAL DECOMPOSITION: MEASURING THE CYCLICAL COMPONENT OF A TIME SERIES

The model of a time series when data are collected on an annual basis is given by:

$$Y = TCI$$

(no seasonal factor is involved since measurements are not made within a year). Previous sections have discussed the estimation of the trend component, T, where the trend is given either by a straight line or by a semilogarithmic curve. An estimate of the movement in a time series of the form given above due to cyclical (and irregular) forces can be derived by simply dividing the preceding equation by T, the trend value, because:

$$\frac{TCI}{T} = CI.$$

An estimate of the movements in the time series of personal consumption expenditures for automobiles and parts for 1950–78 examined in Section 16.4.3 is given in Table 16.12. A graph of the fluctuating component (C and I) of this time series is given in Figure 16.11. The CI component of this time series is assessed by dividing the actual value of the time series, y, by the trend estimate, $\hat{y}$. From Figure 16.11 note how the fluctuating component of personal consumption expenditures for automobiles and parts appears to

TABLE 16.12 Fluctuating component in expenditures for automobiles and parts as a percent of trend (semilogarithmic model), 1950–1978

Transformed year (t = 0 at 1950)	Actual ($billions)	Semilogarithmic trend	Actual as a percent of trend
0................	12.900	9.760	132.178
1................	11.100	10.497	105.739
2................	10.400	11.291	92.107
3................	13.200	12.145	108.686
4................	12.600	13.063	96.452
5................	17.200	14.051	122.409
6................	15.800	15.114	104.541
7................	17.100	16.257	105.188
8................	13.900	17.486	79.493
9................	18.100	18.808	96.236
10................	18.800	20.230	92.930
11................	17.100	21.760	78.585
12................	20.400	23.405	87.160
13................	24.300	25.175	96.524
14................	25.800	27.079	95.278
15................	29.800	29.126	102.313
16................	30.300	31.329	96.717
17................	30.500	33.698	90.511
18................	37.000	36.246	102.081
19................	40.200	38.986	103.113
20................	37.300	41.934	88.948
21................	46.700	45.105	103.535
22................	53.100	48.516	109.448
23................	57.500	52.185	110.186
24................	49.800	56.130	88.722
25................	53.400	60.375	88.447
26................	69.700	64.940	107.330
27................	81.500	69.851	116.677
28................	89.800	75.132	119.522

FIGURE 16.11 Personal consumption expenditures for automobiles and parts as a percent of trend (semilogarithmic), 1950–1978

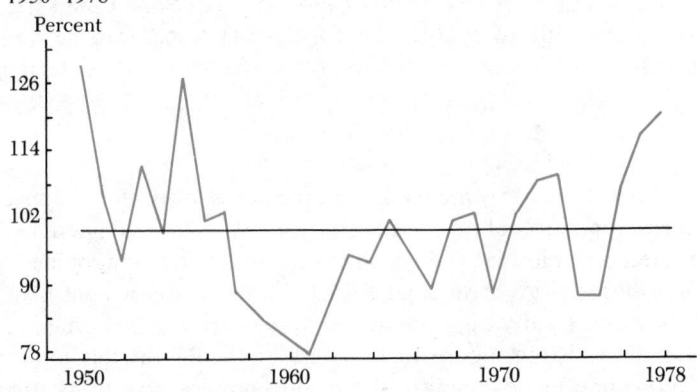

Source: U.S. Department of Commerce, *Survey of Current Business* (various issues), and computer analysis.

follow rather closely the fluctuations in general business activity, being down in periods of near or mild recession (1952, 1958, 1961, 1970) and up in periods of high prosperity (1950, 1955, 1978). Any projection or estimate (forecast) of expenditures for automobiles and parts should thus take into account the general level of business activity.

The process of dividing the actual value of the time series by the trend estimate to assess the *CI* component is valid no matter which of the curve-fitting procedures described in the chapter are used to measure the trend component. Thus, if a straight-line or a power curve would best assess the secular trend component of a given time series, we would still obtain an estimate of the cyclical and irregular component by dividing the actual value of the time series by the trend estimate. Economists often study business and other "cycles" in order to warn of a sharp increase or decrease in general economic activity away from the secular trend component of the time series. Residual plots as described in the text are often an effective means for isolating and measuring the cyclical component after the trend of the time series has been suitably modeled.

■ 16.7 CLASSICAL DECOMPOSITION: MEASURING THE IRREGULAR COMPONENT OF A TIME SERIES

After the quantities $C \cdot I$ have been isolated as described above, an attempt is made to measure the irregular component by computing a measure of the cycle. The following division is then performed to isolate the irregular component:

$$\frac{CI}{C} = I.$$

Several methods exist for "measuring" or assessing the cyclical component, such as averaging across several cycles and using the averages to compute a "cyclical index" for specific years. One such method is called isolating the cyclical component by the *method of residuals.*

After the components of trend, cycle, and season (if measurements are made within a year) are measured, an analysis of the residuals of the difference between the actual and the estimated values should yield a rectangle about 0. This is an indication that all that remains after the modeling process is completed is the irregular or the error component. It is hoped that this will be a small part of the total!

■ 16.8 SUMMARY

This chapter has presented methods for describing data that are collected, recorded, or observed over successive increments of time. Such data are referred to as *time series.*

The components of time series are: (1) trend, which represents the general long-term growth or decline of the data; (2) cyclical, which is the irregular, oscillating pattern of movement in the time series of more than one year's duration; (3) seasonal, which, if data are measured within a year, gives the repetitive, *recurrent* pattern of movement in the time series; and (4) irregular, which is the nonrecurrent, sporadic movement in the series that is not described by the other three factors. Several patterns of trend were reported, and included growth curves, and semilogarithmic and straight-line trends. Least squares regression was used to measure the trend component.

The presentation and computation of several indices alerted the reader to the dangers inherent in interpreting statistical data that are measured over time when these data have not been adjusted for seasonal (and possibly other) factors. This is particularly true regarding measures of performance (both good and bad!)

We hope the chapter stimulated sufficient interest that some of the more advanced methods for analyzing a time series (listed in the references at the end of the chapter) will be investigated at a later time. With the general availability of large-scale computers and software programs, the use of these more advanced techniques is quite simple and straightforward in the majority of computer installations.

One topic that is of special interest regarding the analysis of seasonal indices is analysis of variance presented in Chapter 12. With this technique, we can test the hypothesis that the seasonal indices are all equal, that is, $\mu_{\text{Jan.}} = \mu_{\text{Feb.}} = \cdots = \mu_{\text{Dec.}}$. If the hypothesis is accepted that they are all equal, then the indices we are reporting could be due to chance alone. Rejection of the hypothesis that $\mu_{\text{Jan.}} = \mu_{\text{Feb.}} = \cdots = \mu_{\text{Dec.}}$ adds credibility to the use of the seasonal indices in analyzing seasonal data. This is just one area where one might expand his or her knowledge of the analysis of time series data, and it is similar to the analysis of variance application in Chapter 15 to test for the presence of a seasonal factor in a series.

REFERENCES

Box, G. E. P., and Tiao, G. C. *Time Series Analysis, Forecasting and Control.* San Francisco: Holden-Day, Inc. 1969.

Brown, R. G. *Smoothing, Forecasting, and Prediction.* Englewood Cliffs, N.J.: Prentice-Hall, Inc., 1963.

Morrison, Norman. *Introduction to Sequential Smoothing and Prediction.* Hightstown, N.J.: McGraw-Hill, Inc., 1970.

Summers, G. W., and Peters, W. S. *Basic Statistics in Business and Economics.* Belmont, Calif.: Wadsworth Publishing Co., Inc., 1973, chap. 17.

U.S. Department of Commerce. *Survey of Current Business.* Washington, D.C.: U.S. Government Printing Office.

PROBLEMS

16.1. Describe in your own words what is meant by trend, seasonal, cyclical, and irregular movements in a time series.

16.2. Suppose you were provided with a given time series of data and were asked to analyze its general pattern and fluctuations. Describe in detail the steps you would follow in determining the pattern of trend and whether a seasonal and/or a cyclical component contributed to movements in the series.

16.3. Given a linear trend equation of the form:

$$\hat{y} = 90.5 + 1.027t$$

where $t = 0$ at 1973, and t is in years, estimate the trend value for 1977.

16.4. Given the following trend equation:

$$\log \hat{y} = 1.00554 + 0.029769t$$

where $t = 0$ at 1950, and t is in years, compute the trend value for 1955.

16.5. Given the trend equation of Problem 16.4, estimate the rate of growth implied by this equation.

16.6. Given the following time series of data:

Year	Sales
1977	100.1
1978	112.9
1979	119.6
1980	132.1

estimate the linear trend present using the method of least squares.

16.7. Given the trend equation derived in Problem 16.6, estimate the trend component of this time series for 1981.

16.8. Provided in the table below are sales histories of lumber for two competing companies over a 12-year period.

   a. Determine the pattern of trend for each company, and estimate the trend equation(s) by the method of least squares.

   b. Is there a cyclical component to each of these series, and, if so, are they similar?

   c. Estimate company sales for 1981. On what did you base your projection?

Sales of lumber ($000)

Year	Company 1	Company 2
1969	156.7	100.1
1970	161.2	103.1
1971	162.1	104.2
1972	171.5	109.8
1973	178.9	112.2
1974	181.4	116.3
1975	182.3	118.4
1976	187.2	122.2
1977	190.1	126.8
1978	200.1	129.1
1979	204.1	132.3
1980	212.1	139.8

16.9. Reproduced in the table below are data on total nonagricultural employment in the United States from 1974 to 1979, inclusive.

   a. Estimate the trend component of this time series using a straight-line trend equation.

   b. Does the straight-line trend equation appear to be an appropriate one for modeling the growth component of this time series?

   c. Does there appear to be a cyclical component to this time series, and, if so, what can it be tied to?

   d. Obtain a projection of the value of this time series for 1980.

   e. Obtain the actual employment amount for 1980 from your university library. How does this number compare with the estimate derived in part *d*. Could this difference have been anticipated by considering factors other than the long-term growth in the series? If so, what other factors might have been considered?

Total nonagricultural employment in the United States (000)

Year	Employment
1974...............	78,265
1975...............	76,945
1976...............	79,382
1977...............	82,423
1978...............	86,446
1979...............	89,482

Source: *Economic Indicators*, April, 1980.

16.10. Provided in the table below are sales data for a firm selling a semi-luxury consumer good for the years 1974 to 1980, inclusive. Estimate the trend component of this time series using a second-degree parabola, and comment on the appropriateness of this model for the long-term movements in this time series. Does there appear to be a cyclical component to this time series?

Year	Sales ($000)
1974............	$2,582
1975............	2,613
1976............	2,629
1977............	2,714
1978............	2,775
1979............	2,861
1980............	2,945

16.11. Given the following linear trend equation:

$$\hat{y} = 180.43 + 745.00t$$

where $t = 0$ at 1973, and t is in years, determine the slope of the resulting trend equation assuming it is desired to shift the unit of measurement from years to months.

16.12. What would be the intercept component of the trend equation after the units of measurement were changed to months in Problem 16.11?

16.13. Given your results from Problems 16.11 and 16.12, what is the trend component for this time series for August 1980?

16.14. Given in the table below are corporate profits in the United States from 1969 to 1979.

Corporate profits after taxes ($ billion)

Year	Profits
1969	43.8
1970	37.0
1971	44.3
1972	54.6
1973	67.1
1974	74.5
1975	70.6
1976	92.2
1977	104.5
1978	121.5
1979	144.1

Source: *Economic Indicators*, April 1980.

Estimate the long-term growth component in this time series using a straight-line trend equation. Does a straight-line trend appear to be reasonable for modeling the long-term growth in this time series?

16.15. Using the data given in Problem 16.14, estimate the long-term growth pattern in these data using a semilogarithmic trend equation, and attempt to identify the cyclical component of this time series. How does the trend equation developed compare with the one determined in Problem 16.14?

16.16. Given in the table below are exports of goods and services for the United States for the years 1969–79. Estimate the long-term growth component of this time series using a semilogarithmic trend equation. Obtain a trend estimate for 1980, and compare this estimate with the actual. Does the semilogarithmic model appear to be appropriate for modeling the growth component of this time series?

Exports of goods and
services ($ billion)

Year	Exports
1969	54.7
1970	62.5
1971	65.6
1972	72.7
1973	101.6
1974	137.9
1975	147.3
1976	163.3
1977	175.9
1978	207.2
1979	257.5

Source: U.S. Department of Commerce, Bureau of Economic Analysis, April 1980.

16.17. Figure 16.1 gives data on gross national product in the United States from 1929 to 1979. Obtain the actual estimates of GNP for these years and compute a trend equation using the semilogarithmic model. What is the growth rate implied by your model? How does the rate you computed compare with the one given in the text?

16.18. Unemployment statistics for a large metropolitan area for 1980 are given in the table below. Also shown are seasonal indices for unemployment based on the previous eight years' experience in this area. What is the percentage monthly *change* in unemployment on a seasonably adjusted basis? On a nonadjusted basis?

Month	Seasonal indices	Unemployment
Jan.	97.3	525
Feb.	100.8	536
March	98.9	522
April	100.1	501
May	110.2	599
June	119.5	622
July	116.2	688
Aug.	101.4	616
Sept.	91.5	592
Oct.	94.2	594
Nov.	88.1	590
Dec.	81.5	470

16.19. If a store has sales of 1,043 for October and a seasonal index of 112 for October, what is its adjusted sales level for October?

16.20. In the table below are statistics (by quarter) on housing starts in the United States from 1963 to 1972 (in thousands of units). Compute quarterly seasonal indices using the data provided by the method of ratio to moving average.

Housing starts in the continental United States 1963–1972 (000 units)

Year	Quarter			
	I	II	III	IV
1963	299.0	487.1	447.5	387.2
1964	335.2	476.8	415.9	359.0
1965	298.4	479.3	407.8	357.2
1966	299.2	419.1	307.3	216.5
1967	217.8	381.7	382.1	340.3
1968	298.5	453.2	423.3	372.6
1969	336.2	468.4	386.2	308.5
1970	264.0	399.1	408.4	396.0
1971	388.7	603.9	578.5	513.4
1972	510.3	667.3	692.9	558.0

Source: U.S. Department of Commerce, *Survey of Current Business* (various issues).

16.21. What trend pattern is implied by the data reported in Problem 16.20? Also, isolate the cyclical and irregular components $(C \cdot I)$ of this time series, and comment on the apparent "pattern" in the cyclical movement.

16.22. A trend line for sales of a clothing manufacturer is given in the following equation. Given this information and knowledge that the seasonal index for April is 106.7, estimate sales for April 1981.

$$\hat{y} = 4,560 + 65t$$

where $t = 0$ at 1973, and t is in years.

16.23. What is the estimate of the trend component (i.e., not including an adjustment for seasonal factors) for sales in April 1982 for the Problem 16.22 data?

16.24. If a firm has sales in January of 3,424 units and a seasonal index for January of 89.9, what would be your estimate of sales for the firm for the entire year? What would be your estimate if you did not know the value of the seasonal index for the firm? Which estimate do you have the most faith in, and why?

Forecasting 17

Forecasting is one of the most important functions performed by many managers today. Through forecasts of demand for end items, production schedules can be derived delineating the work force size, the rate of production, and the components to be manufactured. Moreover, each of us uses a forecast daily to decide such routine matters as whether to carry an umbrella (forecast of the weather) and when to leave for work (forecast of driving time).

Previous chapters have touched upon the idea of forecasting or prediction. In the Chapter 13 problems, for example, Herman was interested in forecasting or predicting a future event (exam score) given the number of hours spent studying for the exam. In studying time series in Chapter 16, we were often interested in predicting the value of a series at some future time. In this chapter, we will bring together many of the foundations laid in previous chapters in order to make intelligent choices regarding future events. Because the variable of interest will often be time-oriented in the applications described, we will be concerned primarily with estimating the value of a *time series* at some future instance or span of time. Hence, many of the tools described in Chapter 16 will be relevant to the task at hand. We will not, for example, be concerned with techniques available for forecasting or predicting the weather!

Two basically different approaches exist for estimating the future value of a time series: *prediction* and *forecasting*. Prediction involves the incorporation of subjective factors into the estimate developed. For example, in attempting to predict or estimate a future value of GNP for the United

563

States, it would seem wise to incorporate planned government expenditures, anticipated federal reserve policy, anticipated tax legislation, and other factors into a projection. To estimate or predict future sales of a product, one *may want* to incorporate such qualitative factors as competitive action and the status of the national economy. In many instances, one simply *cannot turn his back* (so to speak) on the qualitative factors that do influence the level of or the value of the time series at some future date.

Alternatively, one can *forecast* the value of a series by *"casting forward"* the past performance or the historical data comprising the series. Forecasting time series then involves the analysis of historical data (usually through mathematical analysis) and the simple extrapolation of the component parts of these data to estimate future values of the series. Forecasting is thus applicable whenever there is no reason *a priori* to suspect that past patterns will not repeat themselves in the forecast period, and when factors not in the model will not appreciably affect the value of the time series in the forecast period.

We will turn now to a discussion of barometric indicators, which, though qualitative, have an appreciable influence on government policy in directing this nation's economy. Subsequent sections will investigate various quantitative forecasting techniques.

Before we begin our discussion of barometric indicators, it will be helpful to introduce some notation that will be used in describing the forecasting models developed. First, we shall use the variable t to denote the time period of interest; t is analogous to the independent variable x in the regression models discussed in Chapters 13–15 and, in fact, many of the forecasting equations developed will be derived by the method of least squares. Since the parameters of the forecasting model $\beta_0, \beta_1, \beta_2, \ldots$ will rarely be known in practice, it will be necessary to obtain point estimates $b_0, b_1, b_2, \ldots$ of them. An observed value of the series at time t will be denoted by y_t, and an estimated or forecasted value of the series at time t will be denoted by $\hat{y}_t$. (Recall that because many assumptions of the least squares linear regression model are not satisfied when time series data are examined, it will not be possible for us to construct confidence intervals for the point estimates obtained.) This notation is similar to that used in Chapter 16.

■ 17.2 BAROMETRIC INDICATORS

□ 17.2.1 Economic indicators

The National Bureau of Economic Research (NBER) has identified groups of economic time series that *in general* lead, are coincident with, or lag behind aggregate economic activity. The usefulness of the leading series is in signaling possible downturns in aggregate economic activity so that remedial action can be taken before the downturn begins or has an opportunity to progress very far. It is these series that are quoted frequently when we hear broadcast, "The government's index of leading economic indicators rose for the third straight month, signaling the healthy recovery of the economy." We will here be concerned with describing the components of the NBER's most frequently quoted composite index of leading indicators, which forewarns of both peaks and troughs in aggregate economic activity. This main aggregate index was revised in 1966 and then again in 1975 to aid in predicting changes in future economic activity. We will describe only the 1975 index.

The components of NBER's composite index of leading economic indicators are listed in Table 17.1. In order for a series to be included in this

TABLE 17.1 New composite index of leading economic indicators

Average workweek of production workers, manufacturing	Net change in inventories on hand and on order, 1967 dollars (smoothed)
Index of net business formation (IV)	Percent change in sensitive prices, WPI of crude materials excluding foods and feeds (smoothed)
Index of stock prices, 500 common stocks	
Index of new building permits, private housing units	Vendor performance, percent of companies reporting slower deliveries
Layoff rate, manufacturing (inverted)	Money balance (M1), 1967 dollars
New orders, consumer goods and materials, 1967 dollars	Percent change in total liquid assets (smoothed)
Contracts and orders for plant and equipment, 1967 dollars	

Source: U.S. Department of Commerce, *Business Conditions Digest,* May 1975, p. x.

composite index, it must have "scored well" on a set of criteria outlined by the NBER. That is, it should signal the upcoming upturn or downturn approximately the same number of months before the upturn or downturn begins, and it should not falsely predict an upturn or a downturn when one is not on the horizon. The performance of this latest composite index for the period 1948–78 is shown in Figure 17.1. Note in general how the series tends to turn down well in advance of a downturn in aggregate economic activity (indicated by the shaded portion of the graph). Note also how the composite series incorrectly signaled a downturn in 1966! Still, barometric indicators such as the NBER's main leading series do provide useful information in alerting public policy officials to potential changes in the *direction* of aggregate economic activity.

☐ **17.2.2   Diffusion indices**

The second barometric indicator we will describe is a *diffusion index,* which is designed to overcome one of the difficulties of leading indicator series—namely, what to do when some leading indicators signal a turn and others do not.

FIGURE 17.1   Composite index of leading economic indicators

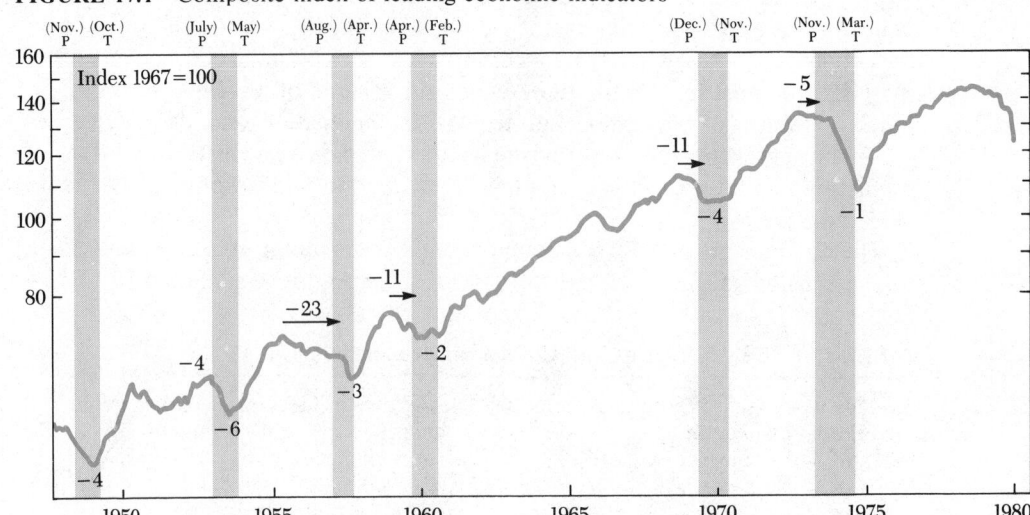

Note: Numbers entered on the chart indicate length of leads (−) and lags (+) in months from reference turning dates.
Source: U.S. Department of Commerce, *Business Conditions Digest*, February 1979, p. 10.

A diffusion index is designed to show the *percentage* of a collection of leading time series that are experiencing rises over some given time interval (e.g., three-months). They are constructed using the following scheme. First, the series that will comprise the diffusion index are selected. Then, each series is differenced over succeeding periods of time. These differences are ignored with the exception of their *sign*. If the difference is positive, the observation is assigned a weight of 1; if the difference is negative, the observation receives a weight of 0; and if there is no difference between succeeding periods, the observation is assigned a weight of 0.5. The weights are then added cross-sectionally, and this total is divided by the total number of series in the index.

The Federal Reserve Board's (diffusion) index of industrial production consists of between 15 and 25 different time series, and the NBER has an all-inclusive index composed of between 600 and 700 different time series. Since the latter-series is dependent upon so many aspects of business, it is considered by many to be one of the best historical indices of the cyclical position of an economy yet devised.

In establishing the efficacy of a given diffusion index, the following criteria are generally assessed:

1.   Average forecasting lead of the indicator.
2.   Number of turning points accurately predicted.
3.   Number of cyclical turns not predicted.
4.   Number of turns that were predicted but never occurred.

Diffusion indices that have generally scored well on these criteria are published monthly in the U.S. Department of Commerce publication *Busi-*

FIGURE 17.2 Diffusion indices

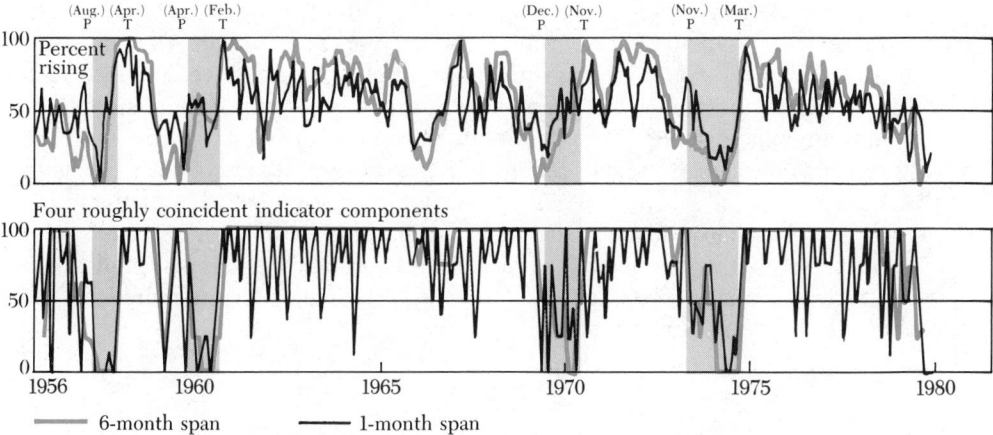

Twelve leading indicator components

Four roughly coincident indicator components

——— 6-month span ——— 1-month span

Source: U.S. Department of Commerce, *Business Conditions Digest,* February 1979, p. 36.

ness Conditions Digest (formerly *Business Cycle Developments*). Figure 17.2 is an example of two diffusion indices that are published monthly by the U.S. Department of Commerce.

It has been said of the diffusion indices that while they reduce irregularities and the diversity of movements in the leading economic indicator series, they do not eliminate these irregularities entirely. (Whenever the number of components in any given index is increased, the number of false leads is likely decreased.) Furthermore, those diffusion indices that use the greater number (3, 6, or 9) of months before peak to identify turning points are smoother than the one-month span indices, but the shorter-spanned indices are particularly important when big changes are taking place—hence, the dilemma that exists when selecting the number of periods (time interval in constructing the index).

The diffusion indices are also said to suffer one of the same, chief disadvantages of the leading indicators—that of predicting change but not the *magnitude* of the change. Hence, many economic forecasters consider the use of barometric indicators a *wise supplement* to any "kit" of forecasting tools. They are by no means self-sufficient, yet they provide a healthy and systematic way of assessing the *direction of movement* in aggregate economic activity.

■ **17.3 MODELS FOR A STATIONARY PROCESS**

In this section we will examine two very similar models for forecasting a future value of a series where the process generating values of the random variable is *stationary*. By a stationary process, we mean that the average of the process is not changing over time. We will describe methods in Section

17.5 for reducing a nonstationary process to a stationary one (called *differencing*), but for the series examined in this section, it will be important to remember that the mean of the process is not changing over time, or else the models used are *incorrect* for the series being examined. An analysis of a residual plot would indicate such a misapplication, and we illustrate such a condition in Section 17.6 when we describe double exponential smoothing and attempt to model a process with a mean changing over time with the methods described in this section. For the present, it is necessary for us to assume that the mean of the process is not changing, and we are attempting to model the movements of the series around this mean value. A scatter plot of the series being modeled is a first indication that the series is stationary, and it is always recommended that a scatter plot be constructed before any forecasting model is used.

☐ 17.3.1 Moving average forecasts

The moving average time series forecasting model is perhaps one of the easiest to use and understand of the time series forecasting techniques. In it, we assume that the pattern followed by a sequence of observations can best be represented by an arithmetic average of past observations. Thus, in terms of a time series model, we assume that the slope component, β_1, is equal to 0 and that the underlying pattern fluctuates randomly around the constant term β_0. A time series exhibiting this type of behavior is depicted in Figure 17.3.

The simple moving average time series model is delineated here. Several properties of this model are worth noting. First, equal weights are given to each of the n most recent observations in determining the forecasted value $\hat{y}_t$, and zero weight is given to observations older than n periods. Second, each "new" estimate of Y_t is simply an updated adjustment of the previous estimate obtained by removing the oldest (nth) and adding the current observation. And third, the rate of response of the model depends greatly on n. Therefore, if slight changes in the underlying pattern are expected, then a large value of n should be selected to reduce the variance property of the forecasting model. If more than slight changes are expected in the underlying pattern, then a smaller value of n should be used to detect turning points and make an appropriate response. The use of the simple moving average time series forecasting model is illustrated in the following example.

Simple moving average forecasting model
Let an observation at time period t be given by

$$Y_t = \beta_0 + \epsilon_t$$

where

β_0 = constant, or level of the series
ϵ_t = random disturbance at time period t with mean 0 and variance σ_ϵ^2

Simple moving average forecasting model (*continued*)

Then the simple moving average forecast of an observation for time period t is given by:

$$\hat{y}_t = \frac{y_{t-1} + y_{t-2} + y_{t-3} + \cdots y_{t-n}}{n}$$

That is, the simple moving average forcast for time period t is the arithmetic mean of the n most recent observations.

Example 17.1 Information on the sales of replacement garbage disposal units from Fort Worth Annie's Plumbing Shop for 1980 is given in Table 17.2. A scatter diagram of these data is given in Figure 17.3. A visual inspection of

TABLE 17.2 Monthly sales of garbage disposal replacement units from Fort Worth Annie's Plumbing Shop

Month (1980)	t	y_t	Month (1980)	t	y_t
Jan.	1	$1,800	July	7	$3,000
Feb.	2	2,000	Aug.	8	2,600
March	3	1,800	Sept.	9	1,700
April	4	3,000	Oct.	10	1,200
May	5	2,700	Nov.	11	2,400
June	6	1,900	Dec.	12	1,500

Source: Annie's records.

FIGURE 17.3 Scatter diagram of data in Table 17.2

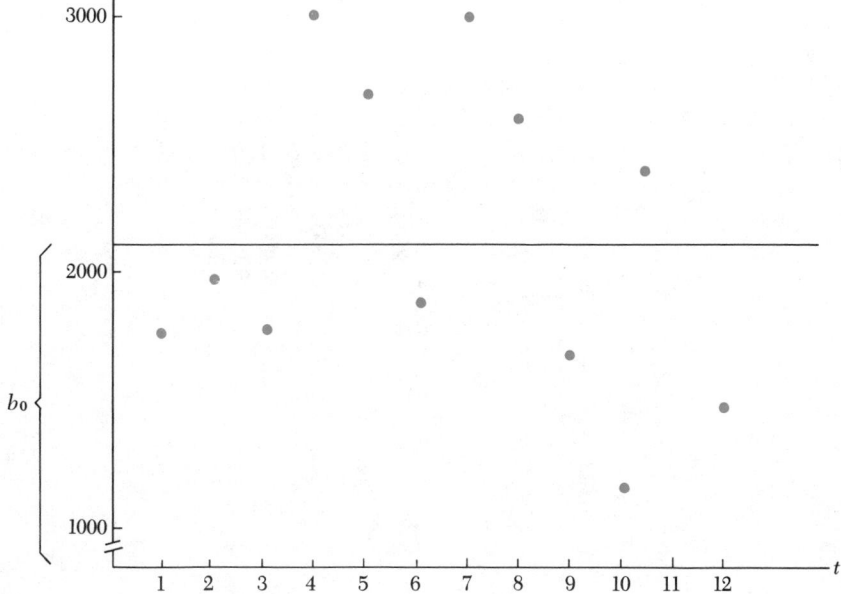

TABLE 17.3 Three-month and six-month simple moving average forecasts for data in Table 17.2

Month	Most recent three-month sales	Forecast	Most recent six-month sales	Forecast
Jan.		—		—
Feb.		—		—
March		—		—
April	1,800 + 2,000 + 1,800	1,867		—
May	2,000 + 1,800 + 3,000	2,267		—
June	1,800 + 3,000 + 2,700	2,500		—
July	3,000 + 2,700 + 1,900	2,533	13,200*	2,200
Aug.	2,700 + 1,900 + 3,000	2,533	13,200 − 1,800 + 3,000 = 14,400	2,400
Sept.	1,900 + 3,000 + 2,600	2,500	14,400 − 2,000 + 2,600 = 15,000	2,500
Oct.	3,000 + 2,600 + 1,700	2,433	15,000 − 1,800 + 1,700 = 14,900	2,483
Nov.	2,600 + 1,700 + 1,200	1,833	14,900 − 3,000 + 1,200 = 13,100	2,183
Dec.	1,700 + 1,200 + 2,400	1,767	13,100 − 2,700 + 2,400 = 12,800	2,133
Jan.	1,200 + 2,400 + 1,500	1,700	12,800 − 1,900 + 1,500 = 12,400	2,067

← Previous six-month moving total.

* 13,200 = 1,800 + 2,000 + 1,800 + 3,000 + 2,700 + 1,900.

these data reveals an apparent absence of growth and seasonal movements in this series.

Using $n = 3$ and then $n = 6$ time periods, compute the three-month and six-month simple moving average forecasts for the remaining months of 1980 and January, 1981. Plot the forecasts obtained on a scatter diagram similar to the one in Figure 17.3. How do these two forecasts compare?

Solution   Relevant computations are given in Table 17.3, and a plot of the forecasted values and a scatter diagram of the actual data are given in Figure 17.4. Note how the moving total is "updated" each month by deleting the oldest observation and adding the most recent observation to the moving total in order to simplify the computations involved in computing the six-month moving average forecast. Note also the "smoothing" effect of the six-month moving average as compared to the three-month moving average as depicted in Figure 17.4. The three-month moving average forecast is much more responsive to movements in the observed values of sales over time.

FIGURE 17.4   Scatter diagram and three-month and six-month moving average forecasts for data in Table 17.2

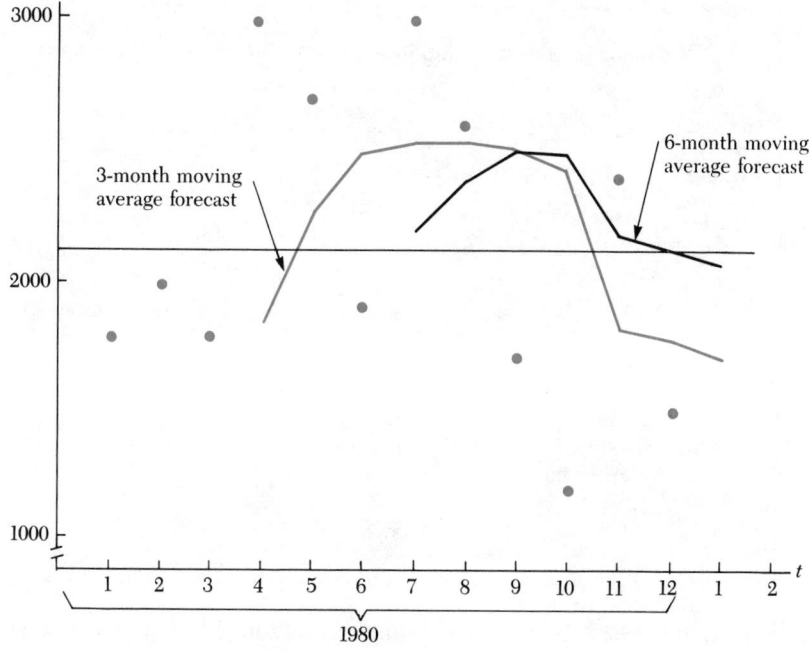

☐ 17.3.2   Single exponential smoothing

Moving average forecasts are a special case of single exponentially smoothed forecasts, and we will develop the correspondence between these two very important forecasting techniques shortly. Single exponential smoothing is perhaps the most widely used time series forecasting technique today. It is *very easily* programmed for computer application and offers several computational advantages over the moving average time series forecasting model. Yet it differs from the simple moving average model only in the method of assigning weights to previous observations. In this section, we will describe the single exponential smoothing forecasting model under the assumption that the series being modeled is stationary. In Section 17.6, we will see how to generalize this technique for a process in which the mean *is* changing over time, called *double exponential smoothing*.

A description of the single exponential smoothing model is given here. The forecasted value of the series for time t, $\hat{y}_t$, is equal to a fraction α of the forecast error of the previous period, $(y_{t-1} - \hat{y}_{t-1})$, plus the forecasted value of the previous period. Thus, to predict the value of the time series at time $(t + 1)$, we would use

$$\hat{y}_{t+1} = \alpha(y_t - \hat{y}_t) + \hat{y}_t$$

Use of the single exponential smoothing model is illustrated in the following example.

Single exponential smoothing—constant process (stationary)

Let an observation at time period t be given by

$$Y_t = \beta_0 + \epsilon_t$$

where

β_0 = constant, or level of the series
ϵ_t = random disturbance at time period t with mean 0 and variance σ_ϵ^2

Then the single exponential forecast of an observation for time t is given by

$$\hat{y}_t = \alpha(y_{t-1} - \hat{y}_{t-1}) + \hat{y}_{t-1}$$

where α is a smoothing constant $(0 \leq \alpha \leq 1)$ and is related approximately to the number of terms in a simple moving average by the relationship

$$\alpha = \frac{2}{n + 1}$$

Example 17.2 Using the data given in Table 17.2 and a forecast of sales for January 1980 of $2,100, forecast sales for February 1980 through January 1981 using α-values of 0.1 and 0.9. Plot the forecasts obtained on a scatter diagram similar to Figure 17.3. How do the forecasts compare?

Solution Relevant computations are given in Table 17.4 and the forecasts obtained are depicted in Figure 17.5. Also shown in Figure 17.5 is a scatter diagram of the observed data. Note that in order to "start" the exponential smoothing process, it was necessary to obtain an initial forecast for the first month. This "seed" forecast can be based on a "best guess" or a moving

TABLE 17.4 Single exponential forecast for data in Table 17.2

Time period t	Actual sales y_t	$\alpha = 0.1$		$\alpha = 0.9$	
		$\alpha(y_t - \hat{y}_t)$	$\hat{y}_{t+1}$	$\alpha(y_t - \hat{y}_t)$	$\hat{y}_{t+1}$
1.............	$1,800	−30*	2,070	−270	1,830
2.............	2,000	−7†	2,063	153	1,983
3.............	1,800	−26	2,037	−165	1,818
4.............	3,000	96	2,133	1,064	2,882
5.............	2,700	57	2,190	−164	2,718
6.............	1,900	−29	2,161	−736	1,982
7.............	3,000	84	2,245	916	2,898
8.............	2,600	36	2,281	−268	2,630
9.............	1,700	−58	2,223	−837	1,793
10............	1,200	−102	2,121	−534	1,259
11............	2,400	28	2,149	1,027	2,286
12............	1,500	−65	2,084	−707	1,578

* $0.1(1,800 - 2,100) = -30.$
† $0.1(2,000 - 2,070) = -7.$

FIGURE 17.5 Smoothed exponential forecasts for data in Table 17.2

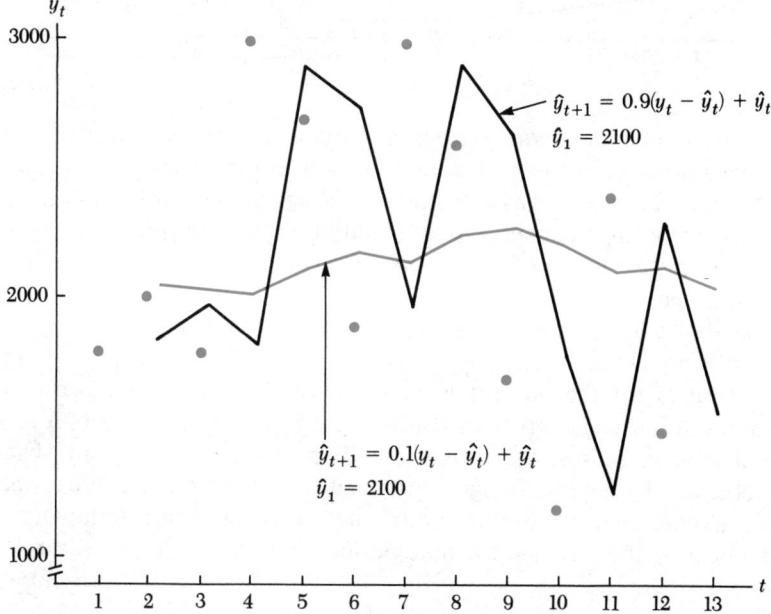

$$\hat{y}_{t+1} = 0.9(y_t - \hat{y}_t) + \hat{y}_t$$
$$\hat{y}_1 = 2100$$

$$\hat{y}_{t+1} = 0.1(y_t - \hat{y}_t) + \hat{y}_t$$
$$\hat{y}_1 = 2100$$

average forecast determined from data from the n months preceding the start of the forecast period.

From a visual inspection of Figure 17.5, it is apparent that when α, the smoothing constant, is close to 0, the forecast behaves like the average of a large number of data. When α is close to 1, the forecast responds rapidly to changes in the pattern underlying the observations.

FIGURE 17.6 Weight attached to previous observations in the exponential smoothing model

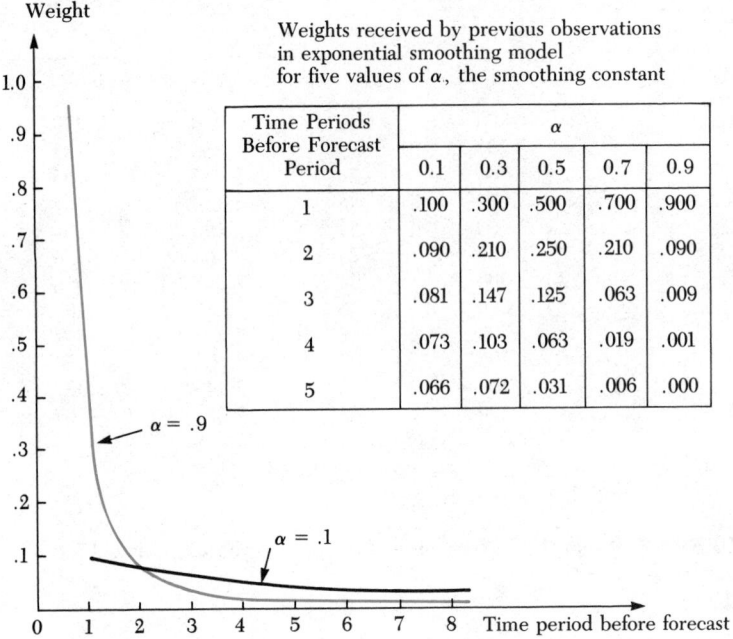

Weight

Weights received by previous observations in exponential smoothing model for five values of α, the smoothing constant

Time Periods Before Forecast Period	α				
	0.1	0.3	0.5	0.7	0.9
1	.100	.300	.500	.700	.900
2	.090	.210	.250	.210	.090
3	.081	.147	.125	.063	.009
4	.073	.103	.063	.019	.001
5	.066	.072	.031	.006	.000

$\alpha = .9$

$\alpha = .1$

Time period before forecast

The term *exponential smoothing* is derived from the weight attached to preceding observations. Figure 17.6 is a graph of the weight attached to previous observations as a function of the age of the observation and the value of the smoothing constant, α. It should be apparent from this graph that *all* previous observations receive some weight in deriving the forecast, and not just the previous n as in the case of the simple moving average. This figure also shows how a plot of the weights yields an exponential curve, hence the term *exponential* smoothing.

Weights for the current period and the four previous periods for five values of α are also given in Figure 17.6. From these numbers it is apparent that as α increases, the weight attached to the more current observations increases. In general, weights attached to each of the previous n periods in the exponential smoothing model are as given in the following equation delineating the forecast for time period $(n + 1)$:

$$\hat{y}_{n+1} = \alpha \left[\sum_{t=0}^{n-1} (1 - \alpha)^t y_{n-t} \right] + (1 - \alpha)^n \hat{y}_0$$

The exponential smoothing model offers several computational advantages over the simple moving average. We need only retain, for example, the most recent forecast, the current observation, and the smoothing constant in order to obtain a forecast for the next period. We need not retain the preceding n observations required in the simple moving average model. Both models, through the averaging process, tend to smooth irregularities in the observations. If the underlying pattern in the data is changing just slightly, then we would choose to use a large value for n in the simple moving average model and a small value of α in the exponential smoothing model. Similarly, if more than slight changes are anticipated in the underlying pattern, we might select a small value of n in the simple moving average model and a large value of α in the exponential smoothing model to detect turning points and make a more rapid adjustment to the change in the underlying pattern.

If we define an exponential smoothing model equivalent to a moving average model where equivalency is used in the sense of a weighted average of the age of past observations, then

$$\alpha = \frac{2}{n+1}$$

Table 17.5 gives the "equivalent" smoothing constant for several values of n in the moving average model. Note that with α set to some large value, little actual smoothing of the data takes place.

TABLE 17.5 Equivalent values of n in a simple moving average forecasting model with α in exponential smoothing

α	No. of observations n in equivalent moving average model
0.1	19
0.3	5.67
0.5	3
0.7	1.86
0.9	1.22

In practice, the value of α to use in the exponential smoothing model is frequently determined in an iterative fashion as follows: α is incremented from 0.1 to 0.2 to 0.3, etc., and the sum of the squares of the observed minus the forecasted values of the series is retained. The value of α yielding the minimum sum of squared error deviations, $\Sigma(y_t - \hat{y}_t)^2$, is then used in the forecasting equation.

■ 17.4 AUTOREGRESSIVE FORECASTING APPROACHES

If observations in a given time series are *highly* correlated over time, then it may be possible to forecast a future value of the series using past observations. If, for example, sales of a product bear a strong statistical relationship to (1) sales of the previous month, (2) sales of six months ago, and (3) sales of one year ago, then a forecasting equation of the form

$$\hat{y}_{t+1} = b_0 + b_1 y_t + b_2 y_{t-5} + b_3 y_{t-11}$$

may yield an accurate forecast of the time series in period $(t + 1)$.

A *correlogram* is a useful statistical tool for assessing the time-lagged correlations, or the *autocorrelation* present in a time series of data. The correlogram is constructed using the techniques described in Chapter 13 for estimating the sample correlation coefficient where the "variables" are simply time-lagged observations of the given series. Figure 17.7 gives two possible correlograms based on an analysis of a series. In Figure 17.7A, the

FIGURE 17.7 Possible correlograms for a given data series

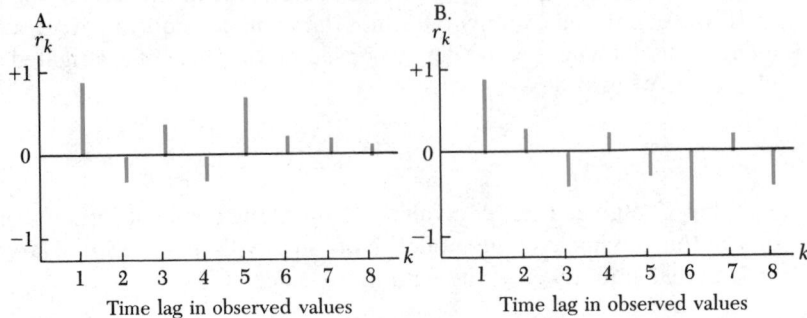

observation for the current month bears a strong statistical relationship (in the sense described in Chapters 13 and 14) to the observation of the previous month and the observation that is five periods old. In Figure 17.7B, the observation of the current period is highly correlated with the observation of the previous period, and it is highly negatively correlated with the observation that is six periods old (possible seasonal factor). These high autocorrelations can be useful in forecasting a future value of the series.

A kth-order autoregressive time series forecasting model is denoted by

$$\hat{y}_t = b_0 + b_1 y_{t-1} + b_2 y_{t-2} + \cdots + b_k y_{t-k}$$

A first-order ($k = 1$) autoregressive forecasting model is delineated here. Its use is illustrated in the following example.

First-order autoregressive forecasting model

Let an observation at time t be given by

$$Y_t = \beta_0 + \beta_1 y_{t-1} + \epsilon_t$$

where

β_0 = constant, or level of the series
β_1 = growth or trend component of the series
ϵ_t = random disturbance at time t with mean 0 and variance σ_ϵ^2

Then the first-order autoregressive time series forecast of an observation for time t is given by

$$\hat{y}_t = b_0 + b_1 y_{t-1}$$

where b_0 and b_1 are point estimates of β_0 and β_1, respectively, determined by the method of least squares.

TABLE 17.6 Sales of stereo equipment in Happy Valley, September 1978–December 1980

Date	Time period	Sales ($000)	Date	Time period	Sales ($000)
1978			**1979**		
Sept.	1	235.89	Nov.	15	292.80
Oct.	2	236.81	Dec	16	301.85
Nov.	3	248.79	**1980**		
Dec.	4	254.16	Jan.	17	297.33
1979			Feb.	18	305.74
Jan.	5	247.47	March	19	307.20
Feb.	6	258.12	April	20	307.34
March	7	253.97	May	21	314.93
April	8	267.84	June	22	319.59
May	9	261.41	July	23	323.16
June	10	271.20	Aug.	24	322.94
July	11	266.47	Sept.	25	335.41
Aug.	12	286.69	Oct.	26	339.46
Sept.	13	299.34	Nov.	27	343.45
Oct.	14	292.34	Dec.	28	345.30

Example 17.3 Table 17.6 gives information on the sales of stereo equipment by a large retail stereo outlet in Happy Valley over a 28-month period. Using the data provided, develop a first-order autoregressive forecasting model and estimate sales for January, 1981. Does a high degree of first-order autocorrelation exist in these sample data?

Solution Relevant computations are given in Table 17.7. Using the data provided,

$$b_1 = \frac{\Sigma x_t y_t - \frac{(\Sigma x_t)(\Sigma y_t)}{n}}{\Sigma x_t^2 - \frac{(\Sigma x_t)^2}{n}} = \frac{2,306,666.45 - \frac{(7,791.69)(7,901.10)}{27}}{2,275,753.95 - \frac{(7,791.69)^2}{27}} = 0.975$$

$$b_0 = \bar{y} - b_1 \bar{x} = \left(\frac{7,901.10}{27}\right) - 0.975\left(\frac{7,791.69}{27}\right) = 11.131$$

$$\hat{y}_{t+1} = 11.131 + 0.975 y_t$$

$$r = \sqrt{\frac{SST - SSE}{SST}} = \sqrt{\frac{25,901.75 - 1,316.96}{25,901.75}} = 0.974$$

The forecast for January, 1981 thus becomes

$$\hat{y}_{\text{Jan., 1981}} = 11.131 + 0.975(345.30) = \$347.799$$

The sample coefficient of correlation is 0.974 ($r^2 = 0.95$). Hence, we would conclude that there is indeed a strong statistical relationship between sales in one period and sales in the following period. The autoregressive model developed appears to be a good predictor of sales as judged by the high autocorrelation between one-period lagged observations.

TABLE 17.7 Computation worksheet for autoregressive forecasting model

$x_t = y_{t-1}$	y_t	x_t^2	y_t^2	$x_t y_t$	$\hat{y}_t$	$y_t - \hat{y}_t$	$(y_t - \hat{y}_t)^2$
235.89	236.81	55,654.04	56,077.56	55,860.88	241.24	−4.43	19.62
236.81	248.79	56,077.56	61,894.47	58,914.27	242.13	6.66	44.36
248.79	254.16	61,894.47	64,599.34	63,232.44	253.81	0.35	0.12
254.16	247.47	64,599.34	61,239.17	62,896.82	259.06	−11.59	134.33
247.47	258.12	61,239.17	66,624.64	63,875.18	252.53	5.59	31.25
258.12	253.97	66,624.64	64,502.28	65,554.88	262.92	−8.94	79.92
253.97	267.84	64,502.28	71,740.41	68,025.14	258.87	−1.03	1.06
267.84	261.41	71,740.41	68,337.54	70,018.31	272.40	−10.99	120.78
261.41	271.20	68,337.54	73,547.81	70,894.83	266.13	5.06	25.60
271.20	266.47	73,547.81	71,008.13	72,266.81	275.68	−9.20	84.64
266.47	286.69	71,008.13	82,189.72	76,394.62	271.07	15.62	243.98
286.69	299.34	82,189.72	89,605.63	85,817.61	290.79	8.56	73.27
299.34	292.34	89,605.63	85,462.97	87,509.79	303.13	−10.79	116.42
292.34	292.80	85,462.97	85,729.20	85,595.98	296.30	−3.51	12.32
292.80	301.85	85,729.20	91,111.91	88,379.59	296.74	5.10	26.01
301.85	297.33	91,111.91	88,403.94	89,747.71	305.57	−8.25	68.06
297.33	305.74	88,403.94	93,475.72	90,904.46	301.16	4.57	20.88
305.74	307.20	93,475.72	94,369.69	93,921.64	309.37	−2.17	4.71
307.20	307.34	94,369.69	94,460.64	94,415.15	310.79	−3.45	11.90
307.34	314.93	94,460.64	99,178.07	96,790.62	310.94	3.99	15.92
314.93	319.59	99,178.07	102,134.57	100,645.47	318.33	1.25	1.56
319.59	323.16	102,134.57	104,432.06	103,276.93	322.88	0.28	0.08
323.16	322.94	104,432.06	104,293.15	104,362.58	326.36	−3.42	11.70
322.94	335.41	104,293.15	112,496.51	108,317.20	326.15	9.25	85.56
335.41	339.46	112,496.51	116,232.07	113,856.08	338.31	1.15	1.32
339.46	343.45	115,232.07	117,961.78	116,588.89	342.26	1.19	1.42
343.45	345.30	117,961.68	119,234.85	118,596.56	346.16	−0.86	0.74
7,791.69	7,901.10	2,275,753.95	2,339,343.77	2,306,660.45		−10.01	1,237.56

Shown in the table below are the input data (Example 17.3) for a second-order autoregressive forecasting model. Note that in this case, $x_t = y_{t-2}$. That is, the dependent observations are lagged by two time periods in computing the autoregressive model. Also note that for every increase in the autoregressive order (k), one additional data item is "lost" in computing the sample autocorrelation coefficient, or we have one less sample value to work with in computing the autoregressive model.

$x_t = y_{t-2}$	y_t	$x_t = y_{t-2}$	y_t
235.89	248.79	292.34	301.85
236.81	254.16	292.80	297.33
248.79	247.47	301.85	305.74
254.16	258.12	297.33	307.20
247.47	253.97	305.74	307.34
258.12	267.84	307.20	314.93
253.97	261.41	307.34	319.59
267.84	271.20	314.93	323.16
261.41	266.47	319.59	322.94
271.20	286.69	323.16	335.41
266.47	299.34	322.94	339.46
286.69	292.34	335.41	343.45
299.34	292.80	339.46	345.30

■ 17.5 AUTOREGRESSIVE INTEGRATED MOVING AVERAGE (ARIMA) MODELS

It is possible to combine the moving average model of Section 17.3.1 with the autoregressive model of the previous section into an integrated model that contains both components. Such models are called *ARIMA models*, which is an acronym for AutoRegressive, Integrated Moving Average. The methodology for combining these approaches was developed by George Box and Gwilym Jenkins and has become known as Box-Jenkins forecasting.

If we denote the order of an autoregressive process by p and the number of terms in a moving average by q, then an ARIMA(p, q) model contains p autoregressive terms and q terms in the moving average. An ARIMA(2, 0) model, for example, would have two autoregressive parameters and no moving average terms. Thus, the ARIMA models are far more general than the autoregressive AR(p) or the moving average MA(q) models, as both are a subset of ARIMA models.

The Box-Jenkins methodology consists of three basic steps, plus the use of the model in actual forecasting.

Steps in the Box-Jenkins forecasting methodology

1. In step 1, the forecasting model is tentatively identified by analyzing sample autocorrelations (SAC) and sample partial autocorrelations (SPAC) to

Steps in the Box-Jenkins forecasting methodology (*continued*)

identify likely candidates (models) for predicting values of the series under investigation.

2. In step 2, the parameters of the ARIMA model are estimated using a nonlinear least squares procedure.

3. In step 3, diagnostic checking of the model takes place, in essence answering the question: Is the model adequate and correct? By analyzing the sample autocorrelations of the *residuals*, the model builder determines whether it is necessary to re-enter the model identification step and repeat the procedure.

In identifying (tentative) the appropriate model, correlograms are constructed similarly to the correlograms of the previous section for both the autocorrelation functions and the *partial autocorrelation functions*. Recall from the discussion of partial correlation in Chapter 15 that partial correlation is a measure of the association (linear) between two variables when the effect of other variables is held constant. Partial *auto*correlations measure the degree of linear association between an observation at time t and at time $(t - p)$ where the effect of other *time lags* is held constant. The purpose of examining the partial autocorrelation functions is to help identify the appropriate ARIMA model.

In Figure 17.8, we indicated possible values of autocorrelation and of partial autocorrelation for three ARIMA modeled processes. In process A, the autocorrelation functions decay exponentially as a function of the

FIGURE 17.8 Autocorrelation and partial autocorrelation functions for three ARIMA processes

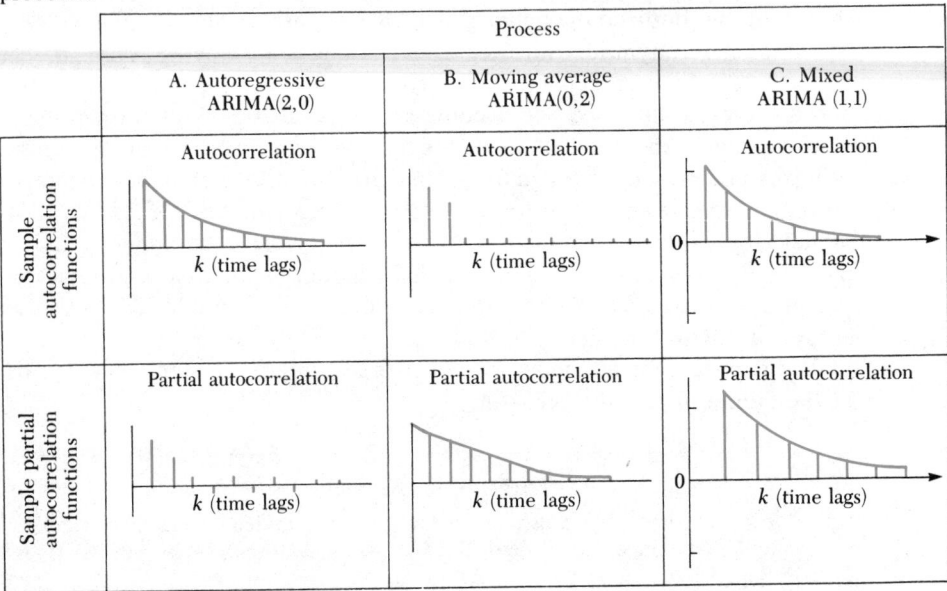

time lags of the observations, and the partial autocorrelation functions are initially large and then become instantly insignificant (it is possible to test statistically for the significance of these terms using the chi-square distribution discussed in Chapter 19). The order of the process is equal to the number of initial significant partial autocorrelations. For process A, the model is thus ARIMA(2, 0). In process B, the autocorrelations become 0 or insignificant after the second time lag, and the partial autocorrelation function decays exponentially as a function of the time lag of the observations. Such processes are modeled by moving averages, and because the autocorrelation function becomes insignificant after two time lags, the order of the process is 2. Hence, the ARIMA model for process B is ARIMA(0, 2). In Figure 17.8C, we have indicated a mixed ARIMA model; since both the partial autocorrelation and the autocorrelation functions decline exponentially, a model of the form ARIMA(1, 1) is suggested. Computers are usually used to provide the analyst with both values and graphs of the sample autocorrelation and sample partial autocorrelation functions, leaving the analyst free to concentrate on model development rather than on arithmetic calculations.

The graphs in Figure 17.8 are not intended to be all-inclusive representing all likley sample autocorrelation functions and sample partial autocorrelation functions. They are intended to provide the reader with examples of the types of these functions that correspond to various processes. In Figure 17.9, for example, we show other sample autocorrelation functions for an autoregressive process of order one and of order two.

FIGURE 17.9   Theoretical autocorrelation functions—autoregressive models

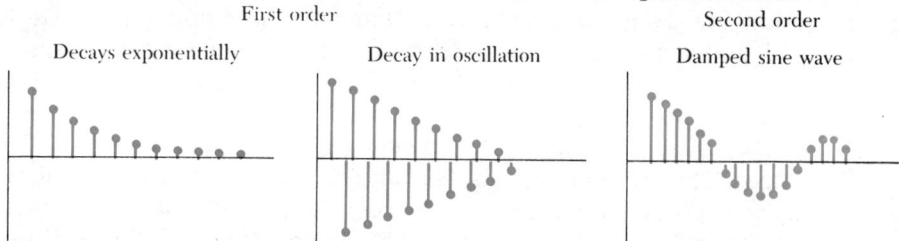

To use the ARIMA models, it is imperative that the series being modeled is *stationary*. By *stationary*, we mean that the average of the process is not changing over time. Such a process is shown in Figure 17.3, as the mean of this time series is not changing over time. The reason that the process must be stationary is that any type of trend that is present tends to introduce spurious autocorrelations into the data that dominate or mask the autocorrelation pattern. Fortunately, methods are available for reducing a nonstationary series to a stationary one for purposes of analysis, and the sample autocorrelation function can aid us in detecting the presence of a nonstationary series so that it can be reduced to a stationary one.

The procedure for removing trend to render a process stationary is called *differencing*. In Chapter 16, a method for removing trend was indicated that divided each actual value of the time series by the trend estimate determined by the method of least squares. The method of differencing is an alternative to this method, which is more computationally efficient and better suited for ARIMA models. We used differencing in Chapter 16 to determine the representativeness of a straight-line trend or a second-degree parabola in measuring the secular trend component in a time series, for example.

In Figure 17.10A, we have indicated a time series that is definitely not stationary—an easily recognized trend is apparent. In Figure 17.10B, we

FIGURE 17.10   Stationary and nonstationary time series processes

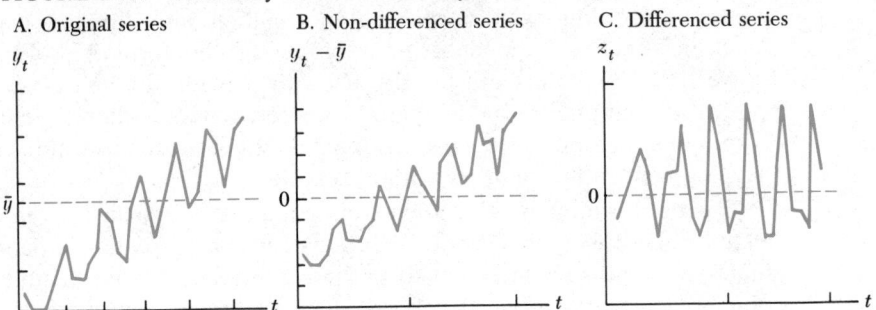

A. Original series

B. Non-differenced series

C. Differenced series

have indicated the same series minus the mean value for each observation in the series—this series too is nonstationary, as there is a definite trend in the data. Figure 17.10C is a scatter plot of the data $z_t = y_t - y_{t-1}$, where the Y-values are given in Figure 17.10A. Note how this figure is similar to Figure 17.3 in that the observations tend to scatter randomly around some non-changing mean value. Such a series is *stationary*, and it is this *latter series* we use in developing our forecasting model. The autocorrelation function of the series z_t in Figure 17.10C would tend to have the first one or two autocorrelation values possibly significantly different from 0, but all remaining autocorrelation values would be very close if not equal to 0. In the event the first differenced series was not stationary, then differences of the first differences can be taken, and so on, until a stationary process is identified. Rarely does one have to proceed beyond second differencing in deriving a stationary series, or one that is approximately so.

Once a series has been identified as being stationary and the form of the ARIMA model has been identified through an examination of the sample autocorrelation and sample partial autocorrelation functions as described, a nonlinear least squares procedure is used to estimate the model coefficients. A computer is almost always required in estimating these quantities.

After the parameters of the model are determined, diagnostic checking of the *residuals* takes place to identify the appropriateness of the model. Diagnostic checking of the residuals is accomplished through an examination of the sample autocorrelations of the residuals. If the model tentatively identified is adequate, the sample residuals will be normally distributed with mean 0 and variance σ^2. They will also be independent. If the residuals are not normally distributed, the pattern of the sample autocorrelation functions should indicate the direction to follow for improvement of the model. For example, a large value (spike) of the autocorrelation function at the jth lag might indicate the need for a moving average parameter of order j. This could occur because there is a seasonal movement in the data that was not initially identified. Hence, we see once more the importance of residual analysis in complete model development.

The Box-Jenkins methodology is quite complex and usually requires the use of a computer to perform the numerous computations required for identifying the model, estimating model parameters, and checking the model developed. Several of the references at the end of the chapter go much further into the development of Box-Jenkins methodology than we can in these few pages. Rather than attempting to make the reader a sophisticated user of these techniques, we have attempted to indicate how two forecasting techniques can be integrated using the tools described in the text to derive a model that is much more sophisticated than the sum of the two individual components. These advanced references also describe the extensions of the Box-Jenkins methodology to cover the incorporation of independent variables other than strictly time-related ones by the use of *transfer functions*.

◼ 17.6 DOUBLE EXPONENTIAL SMOOTHING

If there is a growth pattern or a trend in the series under investigation, the single exponential forecasting model will *always tend* to lag movements in the actual observations. This is because, of course, the single exponential forecasting model is based on a stationary series with only random disturbances. When there is a definite trend in the given time series (either positive or negative), double exponential smoothing reflects and adapts to this changing component. The double exponential smoothing model for a linear trend is described here. Its use is illustrated in the following example.

Double exponential smoothing—linear trend

Let an observation at time period t be given by

$$Y_t = \beta_0 + \beta_1 t + \epsilon_t$$

where

β_0 = constant, or level of the series
β_1 = growth, or trend component of the series
ϵ_t = random disturbance at time period t with mean 0 and variance σ_ϵ^2

Double exponential smoothing—linear trend (*continued*)

Then the double exponential smoothing forecast of an observation for time period t is given by

$$D\widehat{Y}_t = b_{0_t} + b_{1_t}$$

where

$$b_{0_t} = y_{t-1} + (1 - \alpha)^2(D\widehat{Y}_{t-1} - y_{t-1})$$
$$b_{1_t} = b_{1_{t-1}} - \alpha^2(D\widehat{Y}_{t-1} - y_{t-1})$$

and b_{0_1} and b_{1_1} are initial estimates of the intercept and slope, respectively, of the series. (Estimates b_{0_1} and b_{1_1} often are the least squares estimates of β_0 and β_1 in a time series model.)

Example 17.4   Given in the table below are sales levels of a product for nine periods. Using estimates $b_{0_1} = 10$, $b_{1_1} = 5$, and initial forecasts of Y_1 of $\hat{y}_1 = 55.29$ and $D\widehat{Y}_1 = 15$ with an α value of 0.3, predict sales for each of the remaining periods for which actual data are available (one-period-ahead forecasts) using the single and the double exponential forecasting models. Compare the two forecasting models developed on the basis of the average absolute deviation of the actual versus estimated values of the series.

Sales of electronic components

Period	Sales ($000)
1	68.27
2	76.06
3	79.06
4	89.55
5	109.25
6	110.93
7	114.10
8	135.72
9	162.63

Solution   Relevant computations are given in Table 17.8. Because there is a definite trend or growth pattern in these data, the double exponential smoothing model produces a much lower average absolute deviation from actual results than does the single exponential smoothing model. Note also that all the forecast errors using the single exponential smoothing model are negative, indicating the lag present in the forecasted values.

In Figure 17.11, we give the residual plots of the data in Table 17.8. Note in the double exponential smoothing model (Figure 17.11A) how the residuals tend to fall in a random pattern about 0 after the first two periods (reflecting the "poor" initial forecast). Note also how the residuals in Figure

TABLE 17.8 Computation of forecast errors for Example 17.4

Period	Actual value ($000)	Single exponential forecast	Forecast error	Double exponential forecast	Forecast error
1	68.27	55.29	− 12.98	15.00	− 53.27
2	76.06	59.18	− 16.87	51.96	− 24.10
3	79.06	64.25	− 14.81	76.21	− 2.85
4	89.55	68.69	− 20.86	89.88	0.33
5	109.25	74.75	− 34.30	101.90	− 7.35
6	110.93	85.24	− 25.69	118.50	7.57
7	114.10	92.95	− 21.15	126.81	12.71
8	135.72	99.29	− 36.42	131.35	− 4.37
9	162.63	110.22	− 52.41	144.99	− 17.64
		Average absolute error		*Average absolute error*	
		$\dfrac{235.51}{9} = 26.17$		$\dfrac{88.97}{9} = 9.89$	

FIGURE 17.11 Residual plots of data in Table 17.8

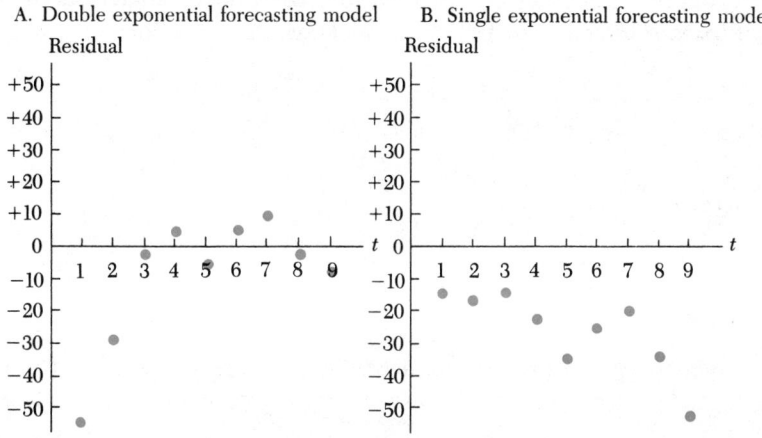

A. Double exponential forecasting model   B. Single exponential forecasting model

17.11B for the single exponential smoothing model all fall below the line separating the positive residuals from the negative ones! An examination of a residual plot such as the one in Figure 17.11B would most certainly alert the user or the developer of the forecasting technique to the fact that the model has been improperly specified in this instance!

It is also possible to define a triple exponential smoothing model, which is applicable when the trend component is nonlinear, and an exponential smoothing model, which is applicable when seasonal and other movements are present in the underlying pattern of the data. A discussion of these more advanced models is given in several of the references listed at the end of the chapter.

■ 17.7 ECONOMETRIC MODEL BUILDING: TIME SERIES AND REGRESSION ANALYSIS REVISITED

The techniques of simple linear regression examined in Chapter 13 and multiple linear regression examined in Chapter 15 can be used to derive estimating equations for the value of a dependent variable we wish to predict as a function (linear) of one or more independent variables whose values are known. For example, in Section 15.7 estimating equations were developed for predicting the number of casts set on broken limbs at a ski and leisure resort in Colorado. The estimating equation developed for assessing the number of casts required using one independent variable was:

$$\hat{y} = 9.03 + 0.9589x$$

where $\hat{y}$ = predicted number of plaster casts required, and x = number of visitors to the facility (000). If we can accurately predict the number of visitors to the facility, then we can estimate the demand for plaster casts as a function of the number of visitors, as the following example illustrates.

Example 17.5 The firm in Example 15.6 is able to estimate fairly accurately the number of visitors to its facility because demand far exceeds supply and reservations must be made well in advance of the scheduled arrival date. Using current information, the firm estimates the number of visitors to the facility in 1981 to be as shown in the table.

Quarter	Expected number of tourists (000)
Winter	118.42
Spring	121.54
Summer	125.84
Fall	128.75

Using the predicting equation given, estimate the expected number of casts that will be required as a function of the number of visitors to the facility.

Solution To estimate the number of casts required, we simply enter the expected number of visitors into the derived equation to estimate demand.

Quarter	Expected number of casts
Winter	$9.03 + 0.9589 \cdot 118.42 = 122.58$
Spring	$9.03 + 0.9589 \cdot 121.54 = 125.57$
Summer	$9.03 + 0.9589 \cdot 125.84 = 129.70$
Fall	$9.03 + 0.9589 \cdot 128.75 = 132.49$

We learned in Section 15.7 that there is a definite seasonal pattern in the necessity for setting broken limbs, and we were able to improve our estimates by taking seasonal factors into account. Using the four quarters of the year to predict the demand, the following estimating equation was developed:

$$\hat{y}_i = 14.225 + 0.99x_{i,1} + 6.443x_{i,2} - 6.442x_{i,3} - 30.592x_{i,4}$$

where

$\hat{y}_i$ = demand for plaster casts (estimated)

$x_{i,1}$ = number (000) of visitors to the facility in period i

$x_{i,2}$ = {1 if period i is the first quarter of the year; 0 otherwise}

$x_{i,3}$ = {1 if period i is the second quarter of the year; 0 otherwise}

$x_{i,4}$ = {1 if period i is the third quarter of the year; 0 otherwise}

The binary or dummy variables in the estimating equation allowed us to "capture" the effect of seasonal factors in influencing demand. In a similar fashion, the use of an independent variable allows us to better estimate the value of a dependent variable whenever there is a linear relationship between the dependent and the independent variable. If we know, for example, that the demand for automobiles and parts closely follows the pattern of overall economic activity as measured by, say, GNP, then it is possible to derive an estimating equation for automobiles and parts as a function of GNP. For such an estimating equation to be useful in predicting a future observation of the dependent variable, we must obtain an estimate of the value of the independent variable for the same time period in the future. As we build simple linear and multiple linear regression models for estimating some future value of a dependent variable, we should be aware that to predict some future value of the dependent variable, we must have the value of the independent variable(s) if the predicting equation is to be useful to us.

In the same vein, an economist might develop an estimating equation for predicting sales of refrigerators, using as explanatory variables or independent variables a forecasted increase in disposable personal income, housing starts from two periods ago, and the price of the refrigerator units. His model specifies that sales of refrigerators in period t are estimated by disposable personal income in period t, housing starts in period $(t - 2)$, and price in period t. In estimating refrigerator sales, the economist must obtain an estimate of disposable personal income in period t, which is usually supplied by government agencies; also, he must know the price charged in period t for the refrigerators, and the housing starts of two periods ago. Note the beauty in using a "lagged" estimate such as housing starts of two periods ago—this quantity does itself not have to be estimated since its value is known. The rationale for including this variable may be that the refrigerator is one of the last items added to the house, so its demand is not affected until two periods after the house begins construction. This is the essence of econometric model building—the identification and specification of causative factors to be used in a predicting equation. Recall, however, from the discussion in Chapter 13 that the use of statistical analysis—notably correlation and partial correlation analysis—does not allow us to claim cause and effect. In econometric model building, the economist is building his or her model from the construct that independent variables included should "influence" the behavior of the dependent variable in ways that can be explained on intuitive grounds.

When the values of the independent variables in a linear regression model are unknown, how can the regression model be used to estimate future values? First, we *may* be able to build an accurate model for predicting the value of the *independent* variables using forecasting model-building procedures described in the chapter. These forecasted values of the independent variables are then substituted into the regression equation to estimate the value of the dependent variable. Second, we can build our forecasting equation with independent variables in the model that are *either already known* or relatively simple to estimate. For example, we may be able to treat the dependent variable strictly in a time series fashion, as the next example illustrates.

Example 17.6 Assume that the firm in Example 17.5 is unable to estimate the demand for its facilities (tourists) in a reliable fashion. Construct a table of values of time series variables to be used to estimate the demand for plaster casts.

Solution A table of independent variable values is given in Table 17.9. The time series, multiple regression equation is:

$$\hat{y}_t = b_0 + b_1 t + b_2 x_{t,1} + b_3 x_{t,2} + b_4 x_{t,3}$$

where

$\hat{y}_t$ = expected demand for plaster casts in time period t

$t = 1$ for first quarter 1975 (t is in quarters)

$x_{t,1} = \{1$ if quarter t is the first quarter of the year; 0 otherwise$\}$

$x_{t,2} = \{1$ if quarter t is the second quarter of the year; 0 otherwise$\}$

$x_{t,3} = \{1$ is quarter t is the third quarter of the year; 0 otherwise$\}$

The results of a computer analysis of the data in Table 17.9 are given in Table 17.10. In comparing this model (strictly time series) with the one given in Table 15.1, which uses visitors as one of its independent variables, the time series model developed appears quite favorable in relation to the multiple linear regression model developed in Chapter 15. This is due in part to the fact that the number of visitors to the facility is itself a time-related variable, increasing fairly steadily over the last six years. Note also that the seasonal estimates are quite similar in both models, and that the time series model has only a slightly larger SSE than the regression model developed in Chapter 15 (132.90 > 125.26).

We might now ask how the seasonal estimates given in Table 17.10

TABLE 17.9 Values of dependent and independent variables in Example 17.6

Casts	Time	Quarter			Casts	Time	Quarter		
		1	*2*	*3*			*1*	*2*	*3*
53	1	1	0	0	94.............	13	1	0	0
41	2	0	1	0	86.............	14	0	1	0
24	3	0	0	1	64.............	15	0	0	1
57	4	0	0	0	99.............	16	0	0	0
70	5	1	0	0	109.............	17	1	0	0
60	6	0	1	0	101.............	18	0	1	0
41	7	0	0	1	77.............	19	0	0	1
77	8	0	0	0	110.............	20	0	0	0
81	9	1	0	0	123.............	21	1	0	0
70	10	0	1	0	120.............	22	0	1	0
50	11	0	0	1	95.............	23	0	0	1
87	12	0	0	0	126.............	24	0	0	0

Source: Company records.

TABLE 17.10 Computer analysis of time series data in Table 17.9

DEPENDENT VARIABLE: CASTS

SOURCE	DF	SUM OF SQUARES	MEAN SQUARE	F VALUE	PR > F
MODEL	4	17691.06190476	4422.76547619	632.32	0.0001
ERROR	19	132.89642857	6.99454887	R-SQUARE	STD DEV
CORRECTED TOTAL	23	17823.95833333		0.992544	2.64472094

PARAMETER	ESTIMATE	T FOR HO: PARAMETER=0	PR > |T|	STD ERROR OF ESTIMATE
INTERCEPT	43.99166667	28.46	0.0001	1.54589916
TIME	3.47678571	44.00	0.0001	0.07902615
QUARTER 1	6.09702381	3.95	0.0009	1.54522571
QUARTER 2	−6.04642857	−3.94	0.0009	1.53508854
QUARTER 3	−30.68988095	−20.07	0.0001	1.52897398

compare with the seasonal estimates computed by the ratio to moving average method in Chapter 16. The estimates given in Table 16.11 correspond to the *multiplicative time series model;* that is, to obtain an estimate by quarter, we would multiply the annual estimate divided by 4 (four quarters in a year) by the seasonal index, and divide by 100. In the time series model developed in this chapter, we would *add* an amount equal to the coefficient of the quarterly independent variable in the regression equation. In the former case, we are multiplying a trend estimate by a fixed *percentage;* greater than 1 if the seasonal effect indicates an increase in the given quarter, and less than 1 if the seasonal effect indicates a decrease over the average quarter. In the latter case, we are *adding* to our time series trend estimate a *constant amount* corresponding to the expected increase in the value of the dependent variable due to seasonal factors. Thus, the former is in the scheme of a *multiplicative model,* and the latter is in the scheme of an *additive model.* Note, however, that if we momentarily consider the fall quarter to be 100, then the estimate for winter (quarter I) from Table 17.3 would be 106.10 (100 + 6.1); for spring it would be 93.95 (100 − 6.05); and for summer it would be 69.31 (100 − 30.69).

In both cases, the ascending sorted seasonal effect would be in the order summer, spring, fall, and winter. Both approaches give the same *relative ranking* of the effects of seasonal factors on the value of the time series. Both are measuring this component in a slightly different fashion, even though the comparative results are similar.

An advantage of the multiple regression methods over the "strict" time series methods frequently cited is that the forecasts obtained need not necessarily be time-dependent. This allows, of course, for the prediction of a horizon that may be considerably longer term than is appropriate for time series models. Furthermore, we can include any number of explanatory variables in our multiple linear regression model. We may find, for example, that sales are related to a host of independent variables, which may include GNP, advertising, prices, competition, and research and development expenditures. We can include several nontime-related variables in a multiple linear regression equation for estimation purposes. Still, if an estimate of the dependent variable is to be made for a definite period in the future, it may be necessary to obtain estimates of the independent variable values, putting us right back where we started: estimating the value of a variable in the future.

Today large-scale econometric models are being used to model the economy, selected industries within the economy, and even specific firms within the industry. Just as simple linear regression is a special case of multiple linear regression, multiple linear regression is a special case of econometrics. Econometric models can (and do) include any number of *simultaneous multiple linear regression equations.* Thus, econometric models are systems of simultaneous equations involving several independent variables. We will not examine econometric modeling in depth in this text, as this is outside the scope of introductory statistics. We will continue in the following sections to augment our kit of forecasting techniques to include other regression and time series approaches for estimating the value of a dependent variable at some future time.

■ 17.8 SINUSOIDAL MODELS

Just as regression models may offer some advantages over strict time series approaches, sinusoidal forecasting models may offer certain advantages over other methods in forecasting the future value of some series. In the sinusoidal models, sine and cosine functions are used to model periodic or *seasonal* movements in the data. A computational advantage of the sinusoidal models over the method of ratio to moving averages for computing seasonal indices is that the amount of data storage and manipulation is greatly reduced with the sinusoidal methods. In this regard (minimal computer storage requirements), the sinusoidal models are similar to exponential smoothing and offer the same comparative advantages over other approaches. And, like the dummy or indicator variable techniques for handling or modeling seasonality, sinusoidal models most frequently use multiple linear regression techniques to estimate the parameters in the forecasting equation. Although the Box-Jenkins forecasting methodology is generalizable to include seasonal movements through differencing, sinusoidal models represent a very different approach to estimating the future value of a series that possesses definite seasonal patterns that repeat themselves from year to year.

Figure 17.12 is a plot of the standard sine and cosine functions. Recall that these transcendental functions repeat themselves every 2π radians or every $360°$ (one full cycle).

In Figure 17.13 is given a plot of the function $y_t = 80 + 10 \sin 30t°$. In terms of a forecasting model, this equation would represent a time series

FIGURE 17.12   Graph of the standard sine and cosine functions

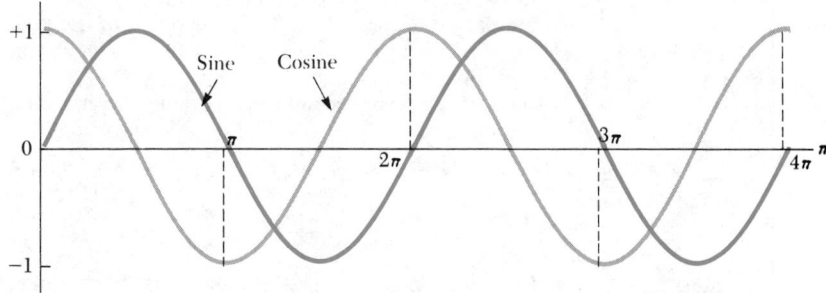

FIGURE 17.13   Graph of $y_t = 80 + 10 \sin 30t°$ and $y_t = 80 + 10 \sin 30t° + 10 \cos 30t°$

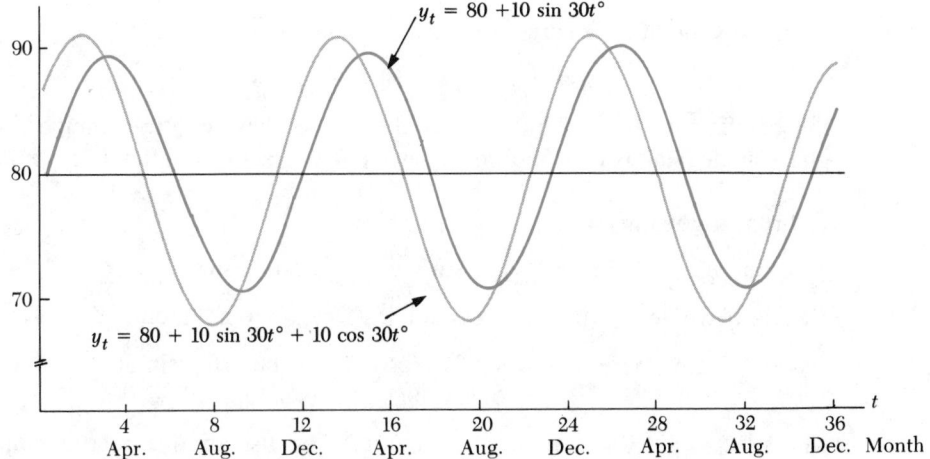

with a level or a constant value of 80, a seasonal movement corresponding to an observation dictated by ten times the standard sine function every month $(360/30 = 12$ months), and a noted absence of trend, cyclical, and irregular factors. We have also superimposed months of the year on this figure to indicate how this sine wave can model movements in a given time series corresponding to months of the year. The *four-term* sinusoidal time series forecasting model corresponding to a time series with level, trend, *fundamental seasonal,* and irregular variation is described here.

In Figure 17.13 is also given a plot of the function: $y_t = 80 + 10 \sin 30t° + 10 \cos 30t°$.

Four-term sinusoidal forecasting model

Let an observation at time period t be given by

$$Y_t = \beta_0 + \beta_1 t + \beta_2 \sin 30(t - t_0)° + \epsilon_t$$

where

$\beta_0 = $ constant, or level of the series
$\beta_1 = $ growth, or trend component of the series
$\beta_2 = $ amplitude of the seasonal peak (trough) in the series
$\sin 30(t - t_0)° = $ standard sine function measured at monthly (30°) intervals, with phase angle given by $(t - t_0)$
$\epsilon_t = $ random disturbance at time period t with mean 0 and variance σ_ϵ^2

Then the *four-term sinusoidal time series forecast* of an observation for time period t is given by

$$\hat{y}_t = b_0 + b_1 t + b_2 \sin 30t° + b_3 \cos 30t°$$

where $b_0, b_1, b_2,$ and b_3 are estimated from past data using the method of least squares (multiple linear regression).

In the general time series forecasting model of the form

$$\hat{y}_t = b_0 + b_2 \sin 30t° + b_3 \cos 30°t$$

the coefficients b_2 *and* b_3 together determine the amplitude (height) and phase angle (start) of a standard sine wave with one observation every month or every 30°.

From trigonometry

$$\sin(x - y) = \sin x \cos y - \cos x \sin y$$

Hence, if we desire to shift the standard sine wave t_0 "months," we have

$$\sin 30(t - t_0)° = \sin 30t° \cos 30t_0° - \cos 30t° \sin 30t_0°$$
$$= b_2 \sin 30t° + b_3 \cos 30t°$$

where $b_2 = \cos 30t_0°$ and $b_3 = -\sin 30t_0°$. The use of this relationship is illustrated in the following example.

Example 17.7 A given time series is known to have an absence of trend, cyclical, and generally irregular movements, a level of 200 units, and a seasonal pattern that follows the general sine function with an amplitude of 50 units, a peak in December of each year, and an absence of seasonal movements in September. Measurements of the time series are made each month. What is the general sinusoidal model for this time series?

Solution In order to shift the peak of the standard sine function from March (90°) to December (360°) and have the seasonal movement equal to 0 in September (270°), t_0 must be set equal to 9 (verify by examining Figure 17.12).

$$\hat{y}_t = 200 + 50 \sin 30(t - 9)°$$
$$= 200 + 50(\cos 30 \cdot 9° \sin 30t° - \sin 30 \cdot 9° \cos 30t°)$$
$$= 200 + 50(\cos 270° \sin 30t° - \sin 270° \cos 30t°)$$
$$= 200 + 50(0 \sin 30t° - \cos 30t°) = 200 - 50 \cos 30t°$$

It is possible, of course, to include more terms in the estimating equation than

$$\hat{y}_t = b_0 + b_1 t + b_2 \sin 30t° + b_3 \cos 30t°$$

FIGURE 17.14 Graph of the equation $y_t = 80 + 10 \sin 30t° + 10 \cos 30t° + 10 \sin 60t° + 10 \cos 60t°$

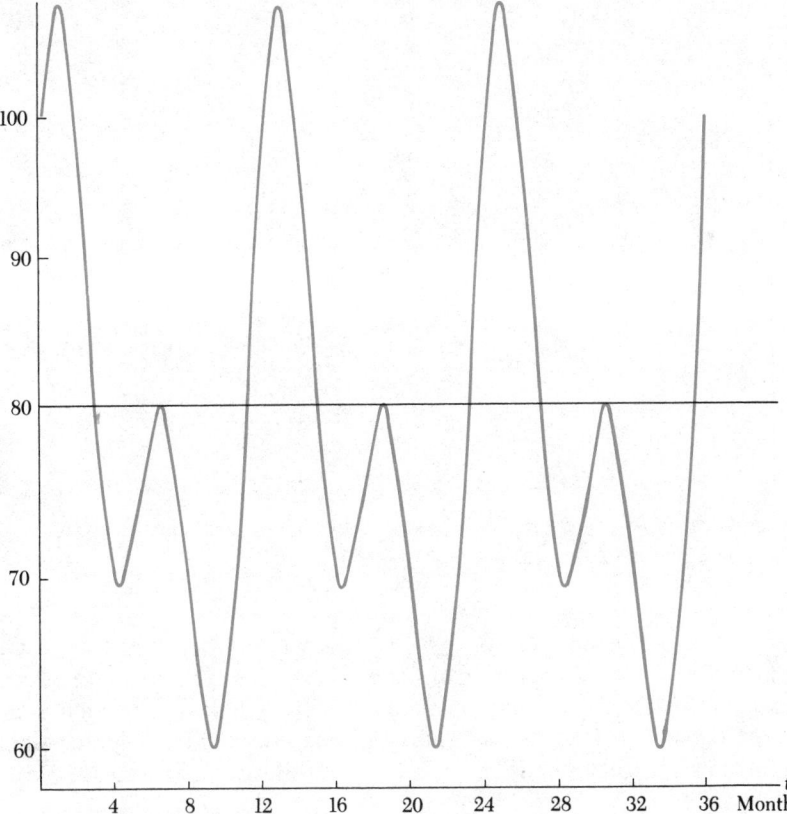

Figure 17.14, for example, represents a time series of the form

$$\hat{y}_t = 80 + 10 \sin 30t° + 10 \cos 30t° + 10 \sin 60t° + 10 \cos 60t°$$

where the latter two terms determine the amplitude and phase angle of a sine wave that completes a cycle in only six months. If trend and irregular variations are also present in the time series, the appropriate model becomes

$$Y_t = \beta_0 + \beta_1 t + \beta_2 \sin 30t° + \beta_3 \cos 30t° + \beta_4 \sin 60t° + \beta_5 \cos 60t° + \epsilon_t$$

The general six-term sinusoidal forecasting model is defined below. The latter two terms in this model are sometimes referred to as *harmonics* and are included to match a variety of shapes and sizes of seasonal or cyclical patterns in a given time series that are not adequately modeled by the *fundamental* component alone.

It is possible, of course, to include more terms than even the six given in the above model, but this is rarely done in practice since the preceding equation will generally model most series satisfactorily. One exception might be the inclusion of a $\beta_5 t^2$ term if the trend component turned out to be a parabola. It can be shown that no more than six sine functions would ever be required to represent a seasonal pattern of the 12 months of the year.

Six-term sinusoidal forecasting model

Let an observation at time period t be given by

$$Y_t = \beta_0 + \beta_1 t + \beta_2 \sin 30(t - t_0)° + \beta_3 \sin 60(t - t_1)° + \epsilon_t$$

where

β_0 = constant, or level of the series
β_1 = growth, or trend component of the series
$\beta_2 \sin 30(t - t_0)°$ = amplitude and phase angle of a standard sine function measured at monthly (30°) intervals
$\beta_3 \sin 60(t - t_1)°$ = amplitude and phase angle of a standard sine function that repeats itself in six months
ϵ_t = random disturbance at time period t with mean 0 and variance σ_ϵ^2

Then the *six-term sinusoidal time series forecast* of an observation for time period t is given by

$$\hat{y}_t = b_0 + b_1 t + b_2 \sin 30t° + b_3 \cos 30t° + b_4 \sin 60t° + b_5 \cos 60t°$$

where b_0, b_1, b_2, b_3, b_4, and b_5 are estimated from past data using the method of least squares (multiple linear regression).

Since the parameters β_0, β_1, β_2, . . . will not be known in practice, it is necessary to obtain point estimates b_0, b_1, b_2, . . . of them. The method used to obtain these point estimates is multiple linear regression as described in Chapter 15. The use of multiple linear regression in determining these forecasting equations is provided in the following examples.

Example 17.8 In Table 17.11 are given sales values of electronic components for small calculators for the 12 months of the years 1978, 1979, and 1980. A linear equation of the form

$$\hat{y}_t = b_0 + b_1 t + b_2 \sin 30t° + b_3 \cos 30t°$$

was fit to the given data (using the method of least squares), yielding the following results:

$$\hat{y}_t = 120.73 + 0.96t + 11.75 \sin 30t° + 6.82 \cos 30t°$$
$$s_{b_1} = 0.08 \qquad s_{b_2} = 1.17 \qquad s_{b_3} = 1.14$$

TABLE 17.11 Sales of electronic components (in $00)

Month	Sales	Month	Sales
1978		1979	
Jan.	130.75	July	131.18
Feb.	136.01	Aug.	123.67
March	139.66	Sept.	132.74
April	127.55	Oct.	137.53
May	118.15	Nov.	142.00
June	120.79	Dec.	152.93
July . . . . , ...	119.50	1980	
Aug.	109.80	Jan.	158.24
Sept.	121.12	Feb.	158.81
Oct.	120.31	March	151.74
Nov.	138.50	April	155.06
Dec.	135.41	May	148.79
1979		June	143.45
Jan.	145.86	July	138.64
Feb.	138.13	Aug.	132.29
March	151.69	Sept.	144.12
April	155.00	Oct.	136.83
May	137.68	Nov.	156.24
June	128.80	Dec.	165.25

Source: Company records.

Plot a scatter diagram of the given data and the equation values. Does the equation determined appear to be a good fit of the data?

Solution Relevant computations are given in Table 17.12, and a scatter plot of the actual data and the estimating equation is given in Figure 17.15. Also shown in this figure is an estimate of a simple linear trend equation, $\hat{y}_t = 123.9 + 0.79t$. The sinusoidal model derived appears to be a good fit of the data given.

Table 17.13 gives the SAS output for the multiple linear regression analysis of the data in Table 17.11 using both the simple linear regression model and the four-term sinusoidal model. Note the significance of the trend component (b_1) in both models. Note also how the sinusoidal model fits the

TABLE 17.12 Sinusoidal trend calculations for Example 17.8

t	$\sin 30t°$	$\cos 30t°$	$\hat{y}_t{}^*$	t	$\sin 30t°$	$\cos 30t°$	$\hat{y}_t{}^*$
1	0.5	0.866	133.47	19	−0.5	−0.866	127.15
2	0.866	0.5	136.24	20	−0.866	−0.5	126.30
3	1.00	0.00	135.37	21	−1.00	0.00	129.09
4	0.866	−0.5	131.34	22	−0.866	0.5	135.03
5	0.5	−0.866	125.50	23	−0.5	0.866	142.79
6	0.00	−1.00	119.66	24	0.00	1.00	150.54
7	−0.5	−0.866	115.65	25	0.5	0.866	156.47
8	−0.866	−0.5	114.80	26	0.866	0.5	159.24
9	−1.00	0.00	117.58	27	1.00	0.00	158.37
10	−0.866	0.5	123.53	28	0.866	−0.5	154.34
11	−0.5	0.866	131.29	29	0.5	−0.866	148.50
12	0.00	1.00	139.04	30	0.00	−1.00	142.67
13	0.5	0.866	144.97	31	−0.5	−0.866	138.65
14	0.866	0.5	147.74	32	−0.866	−0.5	137.80
15	1.00	0.00	146.87	33	−1.00	0.00	140.59
16	0.866	−0.5	142.84	34	−0.866	0.5	146.53
17	0.50	−0.866	137.00	35	−0.5	0.866	154.29
18	0.00	−1.00	131.16	36	0.00	1.00	162.05

$^* \hat{y}_t = 120.73 + 0.96t + 11.75 \sin 30t° + 6.82 \cos 30t°$

FIGURE 17.15 Scatter plot of data in Table 17.11 and sinusoidal model estimate

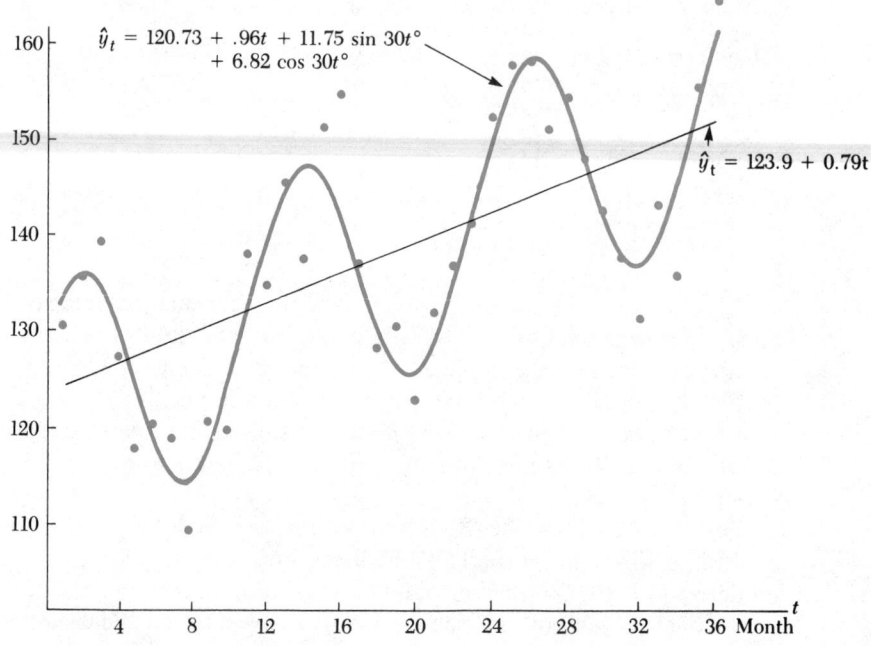

data better as assessed by the lower sum of squares error (SSE) in this model and the higher value of r^2, the sample coefficient of determination.

In Figure 17.16 we give a residual plot for the linear trend model given in Table 17.13, and in Figure 17.17 the residual plot for these same data based on the four-term sinusoidal model given in Table 17.13. Note how the residual plot in Figure 17.16 denotes an incorrect model specification (or in this case an omission—the seasonal movements have not been appropriately accounted for). The pattern of the residuals in Figure 17.16 is too systematic for them to have occurred by chance! The residuals given in Figure 17.17 based on the four-term sinusoidal model, on the other hand, occur randomly within a rectangle about 0, indicating that the model has most likely been correctly specified in this instance.

The use of the six-term sinusoidal forecasting model is illustrated using the data in Table 17.14, which gives sales of selected plumbing fixtures from Fort Worth Annie's Plumbing Shop for 1978, 1979, and 1980. A

TABLE 17.13 Least squares fit of data in Table 17.11

SIMPLE LINEAR TREND

DEPENDENT VARIABLE:
ELECTRIC

SOURCE	DF	SUM OF SQUARES	MEAN SQUARE	F VALUE	PR > F
MODEL	1	2401.30088095	2401.30088095	20.67	0.0001
ERROR	34	3949.77530794	116.16986200	R-SQUARE	STD DEV
CORRECTED TOTAL	35	6351.07618889		0.378094	10.77821237

PARAMETER	ESTIMATE	T FOR HO: PARAMETER = 0	PR > |T|	STD ERROR OF ESTIMATE
INTERCEPT	123.90603175	33.77	0.0001	3.66891704
TIME	0.78619048	4.55	0.0001	0.17292239

FOUR-TERM SINUSOIDAL
      MODEL

DEPENDENT VARIABLE:
ELECTRIC

SOURCE	DF	SUM OF SQUARES	MEAN SQUARE	F VALUE	PR > F
MODEL	3	5608.93062886	1869.64354295	80.62	0.0001
ERROR	32	742.14556003	23.19204875	R-SQUARE	STD DEV
CORRECTED TOTAL	35	6351.07618889		0.883146	4.81581237

PARAMETER	ESTIMATE	T FOR HO: PARAMETER = 0	PR > |T|	STD ERROR OF ESTIMATE
INTERCEPT	120.73095089	71.65	0.0001	1.68497709
T	0.95781926	11.96	0.0001	0.08008258
SIN 30T°	11.75149951	10.01	0.0001	1.17378401
COS 30T°	6.81627444	5.99	0.0001	1.13792099

FIGURE 17.16 Residual plot based on simple linear trend of data in Table 17.11

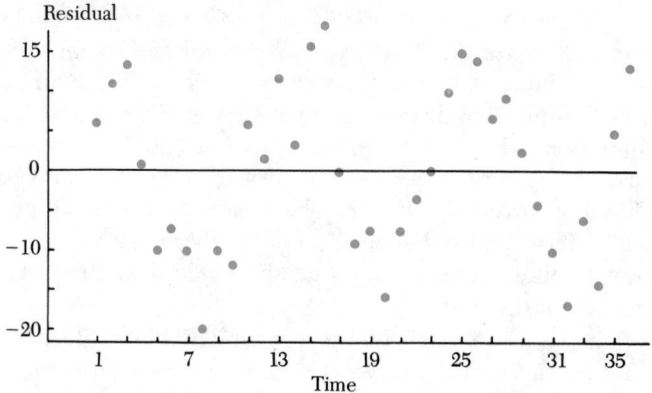

FIGURE 17.17 Residual plot based on four-term sinusoidal model of data in Table 17.11

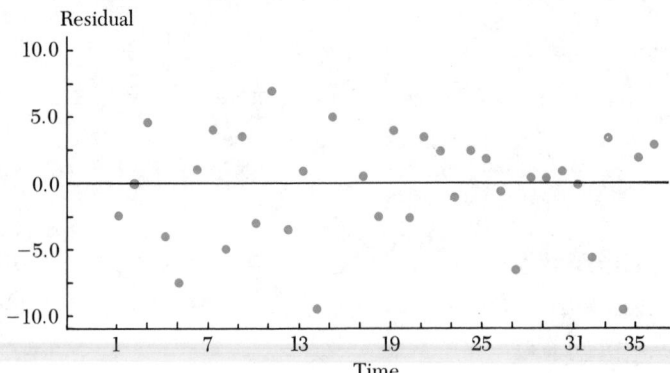

scatter plot of these data is given in Figure 17.18 and Table 17.15 gives a two-factor (level and trend only), four-factor, and six-factor time series forecasting fit, respectively, of the data in Table 17.14. The error sum of squares in these models decreases as more terms are added to the forecasting equation. A visual inspection of the actual versus estimated values of this series given in Figure 17.18 verifies the general superiority of the six-term model for the data given in minimizing the error sum of squares.

The sinusoidal models described offer several computational advantages over the seasonal indices computed in Chapter 16 by the method of ratio to moving average. First, recall that the most recent six-months of data are not used in the computation of the seasonal indices for the latter months. This is not true for the sinusoidal models described. Furthermore, with the four-term or six-term sinusoidal model, many fewer

TABLE 17.14 Sales of selected plumbing fixtures (in $00)

Month	Sales	Month	Sales
1978		**1979**	
Jan.	174.41	July	174.85
Feb.	169.67	Aug.	157.33
March	159.66	Sept.	152.74
April	143.89	Oct.	153.88
May	144.49	Nov.	168.34
June	160.79	Dec.	192.93
July	163.16	**1980**	
Aug.	143.47	Jan.	201.91
Sept.	141.12	Feb.	192.48
Oct.	136.65	March	171.94
Nov.	164.84	April	171.41
Dec.	175.41	May	175.14
1979		June	183.45
Jan.	189.52	July	182.30
Feb.	171.79	Aug.	165.95
March	171.69	Sept.	164.12
April	171.34	Oct.	153.17
May	164.02	Nov.	182.58
June	168.80	Dec.	205.25

Source: Fort Worth Annie's records.

FIGURE 17.18 Scatter diagram, simple linear trend, and four-term and six-term sinusoidal forecasting models for the data in Table 17.14

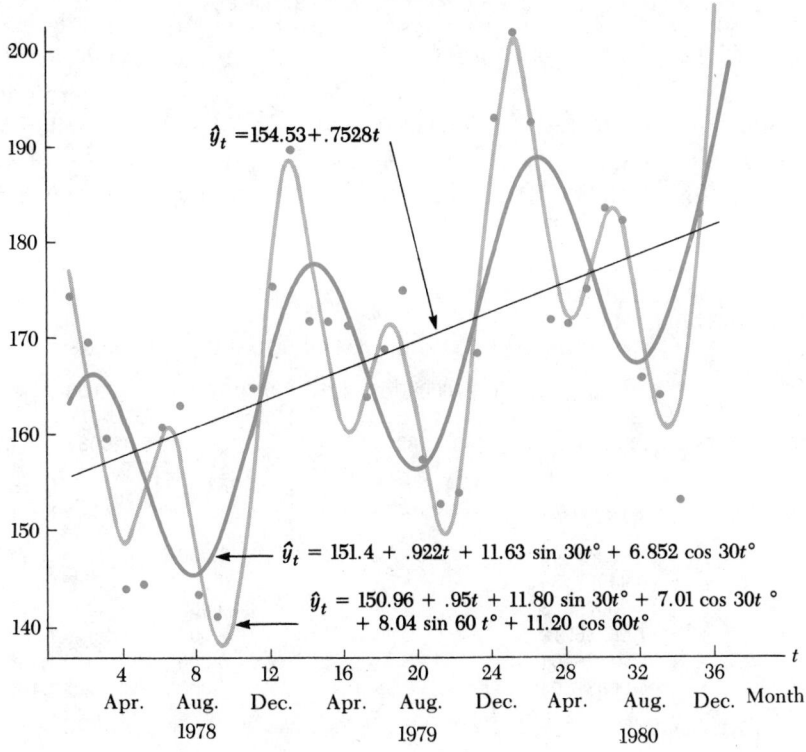

$\hat{y}_t = 154.53 + .7528t$

$\hat{y}_t = 151.4 + .922t + 11.63 \sin 30t° + 6.852 \cos 30t°$

$\hat{y}_t = 150.96 + .95t + 11.80 \sin 30t° + 7.01 \cos 30t°$
$+ 8.04 \sin 60t° + 11.20 \cos 60t°$

coefficients have to be retained in order to develop a forecast by *months*. Where forecasts are made for large quantities of items and the coefficients are stored on a computer-readable medium, the savings in the amount of storage required can be significant.

TABLE 17.15 Least squares fit of data in Table 17.14

SIMPLE LINEAR TREND
DEPENDENT VARIABLE: PLUMB

SOURCE	DF	SUM OF SQUARES	MEAN SQUARE	F VALUE	PR > F
MODEL	1	2201.56510662	2201.56510662	10.21	0.0030
ERROR	34	7328.87505727	215.55514874	R-SQUARE	STD DEV
CORRECTED TOTAL	35	9530.44016389		0.231004	14.68179651

PARAMETER	ESTIMATE	T FOR HO: PARAMETER = 0	PR > |T|	STD ERROR OF ESTIMATE
INTERCEPT	154.53155556	30.92	0.0001	4.99770199
T	0.75278378	3.20	0.0030	0.23555031

FOUR-TERM SINUSOIDAL MODEL
DEPENDENT VARIABLE: PLUMB

SOURCE	DF	SUM OF SQUARES	MEAN SQUARE	F VALUE	PR > F
MODEL	3	5369.63984064	1789.87994688	13.77	0.0001
ERROR	32	4160.80032325	130.02501010	R-SQUARE	STD DEV
CORRECTED TOTAL	35	9530.44016389		0.563420	11.40285096

PARAMETER	ESTIMATE	T FOR HO: PARAMETER = 0	PR > |T|	STD ERROR OF ESTIMATE
INTERCEPT	151.39842983	37.95	0.0001	3.98967841
T	0.92214472	4.86	0.0001	0.18961903
SIN 30T°	11.62978465	4.18	0.0002	2.77927857
COS 30T°	6.85150010	2.54	0.0160	2.69436233

SIX-TERM SINUSOIDAL MODEL
DEPENDENT VARIABLE: PLUMB

SOURCE	DF	SUM OF SQUARES	MEAN SQUARE	F VALUE	PR > F
MODEL	5	8768.79360687	1753.75872137	69.08	0.0001
ERROR	30	761.64655702	25.38821857	R-SQUARE	STD DEV
CORRECTED TOTAL	35	9530.44016389		0.920083	5.03867230

PARAMETER	ESTIMATE	T FOR HO: PARAMETER = 0	PR > |T|	STD ERROR OF ESTIMATE
INTERCEPT	150.96044448	84.94	0.0001	1.77716294
T	0.94978136	11.22	0.0001	0.08466042
SIN 30T°	11.80079073	9.60	0.0001	1.22925263
COS 30T°	7.00656912	5.88	0.0001	1.19077941
SIN 60T°	8.04276101	6.73	0.0001	1.19548617
COS 60T°	11.20037703	9.39	0.0001	1.19230932

Example 17.9 Using the six-term sinusoidal model given in Table 17.15, forecast sales of the selected plumbing fixtures for January, February, and March, 1981.

Solution Relevant computations are given in Table 17.16. The given model forecasts a general decrease in sales over the period corresponding to anticipated seasonal movements in the given time series as indicated.

TABLE 17.16 Estimated sales for data in Example 17.9

Month (1981)	t	$\sin 30t°$	$\cos 30t°$	$\sin 60t°$	$\cos 60t°$	$\hat{y}_t$*
Jan.	37	0.5	0.866	0.866	0.50	210.92
Feb.	38	0.866	0.500	0.866	−0.50	203.36
Mar.	39	1.00	0.00	0.00	−1.0	189.35

* $\hat{y}_t = 150.96 + .95t + 11.80 \sin 30t° + 7.01 \cos 30t° + 8.04 \sin 60t° + 11.20 \cos 60t°$.

It is possible to test for the significance of the fundamental seasonal component in the four-term sinusoidal model (β_2) and to test for the significance of the fundamental component (β_2) and the harmonic component (β_3) in the six-term sinusoidal model, just as we tested for the significance of the presence of a seasonal factor in our dummy variable regression model using analysis of variance in Chapter 15. It turns out that the distribution of the fundamental and the harmonic components in the six-term sinusoidal model are chi-square distributed (Chapter 19) with two degrees of freedom. A description of the test for the significance of the seasonal components in these sinusoidal models is given in the second Brown reference at the end of the chapter.

Just as we saw that multiple linear regression is a special case of econometrics, so too are the sinusoidal models just discussed a subset or a special case of another statistical procedure called *spectral analysis*. Its name is derived from physics, where white light is passed through a prism and broken down into components of its "spectrum." In time series analysis, we may wish to forecast a random variable whose mean is μ_t and whose variance is σ_t^2. If we can reduce σ_t^2, then a more accurate forecast can result. Using spectral analysis, σ_t^2 is broken down into components by "passing" the time series through "prisms, filters, or windows." The variance is broken down in such a way that we are able to associate the most important component with a particular period. An advantage of spectral analysis is that the variance is broken down into spectral components that are *statistically independent*. Thus, one component can be manipulated without affecting the movement of other components.

In spectral analysis, the variance component is broken down into series of sine-cosine functions. Thus, the concept itself is not too unlike the sinusoidal models discussed in this chapter. Both are particularly applicable in the analysis of seasonal or cyclical data. Several references at the end of the chapter include excellent discussions of this fast-emerging forecasting technique.

■ 17.9 A BRIEF COMPARISON OF FORECASTING MODELS

Numerous forecasting techniques have been presented in the chapter. An evaluation or comparison of the various approaches for forecasting a future observation necessarily involves the specification of a criterion upon which the techniques are to be assessed or compared. Unfortunately, there is widespread difference of opinion as to the criterion to use in making a valid comparison.

One criterion that has been put forth for assessing the relative efficacy of alternate approaches is the minimization of the mean square error (MSE),

$$\text{Min MSE} = \frac{\sum_{t=1}^{n} (y_t - \hat{y}_t)^2}{n}$$

Thus, in a least squares sense, we choose the technique that yields the minimum mean square error (MSE).

Another criterion that is used frequently to assess the merits of various time series forecasting techniques is the minimization of the mean absolute deviation (MAD),

$$\text{Min MAD} = \frac{\sum_{t=1}^{n} |y_t - \hat{y}_t|}{n}$$

The relevance of this second criterion is that large deviations from actual results do not "penalize" the forecasting model as much as the squaring of these errors in the computation of MSE.

Finally, a third criterion often used to assess various forecasting models is the minimization of the average absolute forecast error (AFE). That is, only part of the data is used in the computation of model coefficients. Then the model attempts to predict the latter observations that were not used in coefficient determination. Assuming the first m pieces of data are used in model specification, this latter criterion becomes

$$\text{Min AFE} = \frac{\sum_{t=m+1}^{n} |y_t - \hat{y}_t|}{n - m}$$

Many of the models described in the chapter simultaneously produce low values of MSE and MAD, but relatively large values of AFE. And those models that produce relatively low values of AFE simultaneously produce large values of MSE and/or MAD. The following general guidelines can be offered for selecting an appropriate forecasting technique.

1. Time series models in general are poor at detecting turning points and changes in the underlying pattern of data. Where underlying pattern changes

can be anticipated, commonsense adjustments of forecasts are often very effective. The use of leading indicators can prove quite helpful in alerting the forecaster to a probable turning point, for which an examination of the underlying cause for the change can be assessed.

2. If no change in the underlying pattern of the data is expected and strong seasonal and cyclical movements in the data are present, use of the sinusoidal models should be encouraged. For low-value, high-volume items, these models when fit by the method of least squares offer great computational advantages over the other models discussed.

3. Where volatile changes in the series are anticipated, exponential smoothing may produce lower values of forecast errors than other techniques examined. Double exponential smoothing has been found to be among the best forecasting techniques in several simulation studies, for example. Exponential smoothing is even being used to predict general crime centers in large cities at different times of the day, and it is particularly helpful in routing surveillance vehicles in this regard.

4. And finally, where a correlogram indicates a high correlation (either positive or negative) between lagged variables, an autoregressive forecasting model may be a correct choice of a forecasting technique, or, if resources (analyst time and computer time) permit, a mixed moving average autoregressive forecasting model (ARIMA) will likely be the best choice. Quite often the application of ARIMA models involves the greatest expense in terms of computer and analyst resources, and so is reserved for high-value inventory items or for making decisions where particularly large sums of money are involved in planning and the consequences of being right (or wrong!) are substantial.

■ 17.10 SUMMARY

The chapter presented several techniques for estimating the value of a dependent variable, usually at some future period of time, say the next month, the next year, or the next quarter. The techniques described included time series approaches often using multiple regression techniques, and multiple regression approaches using, in addition to time-related variables, other explanatory variables that are felt to contribute to movements in the series under investigation.

Point estimates only were provided for estimating the value of some series, since many of the assumptions underlying the use of the multiple linear regression model for inference purposes are violated when examining time series data. Indeed, when observations are highly correlated over time, autoregressive or mixed autoregressive moving average models may provide the analyst with powerful approaches for estimating the future value of some series.

When the data do not satisfy many assumptions of the multiple linear regression and correlation model, it may be possible to transform the data to conform to the assumptions. For example, when the error terms for given levels of the independent variable do not have the same variance component in the predicted value (conditional distribution), the variances are said to be *heteroscedastic*—unequal between different values of the independent variable(s). Taking logarithms of the data will often correct this situation. When autocorrelation is present in the data and it is desired to compensate for it, we can often do this by specifying a different model. For example, if the true fit of the series should be a second-degree parabola, then the residuals or the error terms will definitely be correlated over time if we fit a straight-line trend or other straight-line function to the data. Simply fitting the correct model will correct for autocorrelation of the error terms in this situation. Transformations of the data can often be used to remove the correlation between successive errors (autocorrelation) if the correct model has been specified. Multicollinearity is present in a model when the independent variables themselves are correlated. We described this situation briefly in Chapter 15. When independent variables are highly correlated, it may be possible to eliminate one of the explanatory variables since the two variables describe the movement in the dependent variable is similar ways. The method of differencing was described in the chapter as one method of data transformation to create a stationary series from a nonstationary one (one with a mean that is changing over time). More transformations of data are possible other than taking differences of the data, taking logarithms of the data, etc. Several references at the end of the chapter describe the other approaches for transforming data values. The result of these transformations is generally a model that meets the assumptions of the multiple linear regression model, so that confidence limits can be set on estimated values of the dependent variable.

■ REFERENCES

Adam, Everett, E. Jr. "Individual Item Forecasting Model Evaluation." *Decision Sciences,* vol. 4, no. 4 (1973), pp. 458–470.

Box, G. E. P., and Jenkins, G. M. *Time Series Analysis: Forecasting and Control.* San Francisco: Holden-Day, Inc. 1970.

Brown, R. G. *Smoothing, Forecasting, and Prediction of Discrete Time Series.* Englewood Cliffs, N.J.: Prentice-Hall, Inc., 1963.

Brown, R. G. *Decision Rules for Inventory Management.* New York: Holt, Rinehart and Winston, 1967.

Chambers, John C.; Mullick, Satinder K.; and Smith, Donald D. "How to Choose the Right Forecasting Techniques." *Harvard Business Review* (July–August 1971), pp. 45–74.

Chow, C. M. "Adaptive Control of the Exponential Smoothing Constant." *Journal of Industrial Engineering,* vol. 16, no. 4 (1965), pp. 314–17.

Ferratt, T. W., and Mabert, V. A. "A Description and Application of the Box-Jenkins Methodology." *Decision Sciences,* vol. 3 (1972).

Groff, Gene K. "Empirical Comparison of Models for Short-Range Forecasting." *Management Science,* vol. 20, no. 1 (1973), pp. 22–31.

Kirbly, R. M. "A Comparison of Short and Medium Range Forecasting Methods." *Management Science,* vol. 13, no. 4 (1966), pp. 202–10.

Lewis, John P., and Turner, R. C. *Business Conditions Analysis.* 2d ed. New York: McGraw-Hill Book Company, 1967.

Mabert, V. A. "An Introduction to Short Term Forecasting Using the Box-Jenkins Methodology." Monograph No. 2, Production Planning and Control Division Series, AIIE, 1975.

Mabert, V. A., and Radcliffe, R. C. "A Forecasting Methodology as Applied to Financial Time Series." *The Accounting Review,* vol. XLIM (1974).

Makridakis, Spyros, and Wheelright, Steven C. *Forecasting Methods and Applications.* Santa Barbara: John Wiley & Sons, Inc., 1978.

Nelson, C. R. *Applied Time Series Analysis for Managerial Forecasting.* San Francisco: Holden-Day, 1973.

Pack, D. J.; Goodman, M. L.; and Miller, R. B. "Computer Programs for the Analysis of Univariate Time Series Using the Methods of Box-Jenkins." Technical Report Number 296, Department of Statistics, University of Wisconsin, Madison, April 1972.

Tiao, G. C., and Thompson, H. E. "Analysis of Telephone Data: A Case Study of Forecasting Seasonal Time Series." *The Bell Journal of Economics and Management Science,* vol. 2 (1971).

Winters, P. R. "Forecasting Sales by Exponentially Weighted Moving Averages." *Management Science,* vol. 6 (1960).

■ PROBLEMS

17.1. Using the data given in the table,

Time period (months)	Actual sales
1	100.4
2	110.9
3	118.6
4	132.9
5	139.9
6	151.1
7	163.2
8	170.5
9	179.4
10	191.7

   a. What would be the forecast for time period 11 using a moving average forecast of six months' duration?

   b. What would be the forecast for time period 11 using a moving average forecast of three months' duration?

17.2. Using the data in Problem 17.1, what would be the forecast for time period 11 using an exponentially smoothed

forecast with an α- value of 0.3 and a forecast for time period 10 of 185.55?

17.3. Using the data in Problem 17.1, what would be the forecast for time period 11 using an exponentially smoothed forecast with an α-value of 0.1 and a forecast for time period 9 of 183.4?

17.4. How many series are in the NBER's composite index of leading economic indicators? When were the components of this series last revised?

17.5. In the computation of a diffusion index, if a given series comprising the index is differenced, and the difference obtained is 0, what weighting does the observation receive?

17.6. An exponential smoothing model with an α-value of 0.4 is related to a moving average forecasting model with how many terms, n?

17.7. Using the data provided in the table below, assess the first-order autocorrelation in the given series. Fit a first-order autoregressive forecasting model to the data using the method of least squares, and graph the results. Does the autoregressive model developed appear to be a good predictor of the series?

Period	t	Sales ($00)
1979 Nov.	1	45.2
Dec.	2	47.1
1980 Jan.	3	45.3
Feb.	4	51.3
Mar.	5	51.9
Apr.	6	55.5
May	7	56.1
June	8	60.1
July	9	58.2
Aug.	10	64.3
Sept.	11	63.6
Oct.	12	66.3
Nov.	13	69.9
Dec.	14	72.1

17.8. Assess the second-order autocorrelation in the data given in Problem 17.7, and fit a second-order autoregressive model to the data. How do the results compare with those determined with the first-order autoregressive model?

17.9. Using the data in Problem 17.7 and a forecasted value of sales of 39.8 for November 1979, obtain forecasts for each of the remaining months using single exponential smoothing and α-values of 0.1, 0.2, 0.3, . . . , 0.9. How do the results obtained compare with those in Problem 17.8?

17.10. Obtain an estimate of trend for the data in Problem 17.7 using the method of least squares. Given your trend estimate and an analysis of the residuals using the single exponential smoothing model, comment on the appropriateness of the double exponential smoothing model for the data given.

17.11. Derive the relationship $\alpha = 2/(n + 1)$ in the exponential smoothing model.

17.12. Using the results from Problem 17.10 to obtain initializing values for $\hat{D}Y_1$, b_{0_1} and b_{1_1}, obtain forecasts of the months involved using the double exponential smoothing model with an α-value of 0.3. Using all the models considered, which one would you use to forecast movements in this time series, and why?

17.13. Assume in Example 17.3 that the given time series peaks in June and attains its average value (level) in September. The appropriate sinusoidal model for predicting a given value of the time series is given by what equation?

17.14. Given here is a computer analysis of a least squares fit of a six-term sinusoidal forecasting model for the data in Table 17.4. Using the results provided, obtain least squares estimates, $\hat{y}_t$, of the actual observations. Compare the six-term model with the four-term model described in Section 17.4 on the basis of mean square error and on the basis of mean absolute deviation. What do the results *suggest* regarding inference on the significance of the harmonic terms in the model?

VAR.	COEFF.	ST. ERROR	T-VALUE
t	0.9502	0.08267	11.49
sin 30t°	11.73506	1.20037	9.78
cos 30t°	6.82321	1.16298	5.87
sin 60t°	−0.79673	1.16885	−0.68
cos 60t°	0.3066	1.16298	0.26

INTERCEPT   120.87896

REGRESSION
  DEGREES OF FREEDOM   5
  SUM OF SQUARES   5630.61509
  MEAN SQUARE   1126.12302

ERROR
  DEGREES OF FREEDOM   30
  SUM OF SQUARES   726.67698
  MEAN SQUARE   24.22257

S.E. OF ESTIMATE   4.92164
F-VALUE   46.49
MULTIPLE R-SQUARED   88.57

17.15. Distinguish between prediction and forecasting. When should each be used in attempting to estimate the future value of a series?

17.16. Describe what is meant by a process being *stationary*, and list one method available for reducing a nonstationary series to a stationary one.

17.17. Describe the steps involved in the Box-Jenkins forecasting methodology. When does one usually conclude that the model developed is the correct one for the series under investigation?

17.18. Given the following four-term sinusoidal forecasting model, determine $\hat{y}_1$ and $\hat{y}_{14}$. If the assumptions underlying the use of the multiple linear regression model are met using these data, which of the terms included are significant?

VAR.	COEFF.	ST. ERROR	T-VALUE
t	4.9586	0.07994	62.03
sin 30t°	11.76639	1.17172	10.04
cos 30t°	6.81481	1.13592	6

INTERCEPT   180.72365

REGRESSION
   DEGREES OF FREEDOM   3
   SUM OF SQUARES   92228.9088
   MEAN SQUARE   30742.9696

ERROR
   DEGREES OF FREEDOM   32
   SUM OF SQUARES   739.54011
   MEAN SQUARE   23.11063

S.E. OF ESTIMATE   4.80735
F-VALUE   1330.25
MULTIPLE R-SQUARED   99.2

17.19. Given the following time series data of sales,

Time period	Sales
1	2,105
2	2,000
3	1,905
4	1,995
5	2,090
6	2,005
7	1,895

obtain forecasts of sales for periods 2 through 7 using the single exponential smoothing model with an α-value of 0.3 and a forecast for the first period of 2,125.

17.20. Develop a first-order autoregressive forecasting model for the data given in Problem 17.19.

17.21. What type of model is suggested by the data given in Problem 17.19? What terms would you include in a model of this time series, and which of the methods described in the text would you use to estimate future values of this series?

Index numbers 18

18.1 INTRODUCTION

An index number is a statistical measure usually involving the simple ratio of two numbers. Index numbers are designed to show changes in one variable or in a group of related variables over time, or with respect to geographic location, or in terms of some other characteristic. For example, Table 18.1 provides information on the cost of living in selected U.S. cities in 1980. These indices were constructed by sampling living costs in the given cities (the cost of housing, food, transportation, medicine, etc.), then dividing the living costs sampled in each of the cities by the living costs in the base location (Dallas), and multiplying the result by 100 (in order to quote them in percentages). The interpretation of the indices given in Table 18.1 might be, for example, that it costs 29 percent more to live in Los Angeles than in Dallas in 1980 because of higher expenditures for the items sampled in determining living costs in each city. Thus, a move from Dallas to Los Angeles with a salary increase of 20 percent (large!) may not seem quite so beneficial as on first glance if one considers the additional amount it is going to cost to live in Los Angeles.

This discussion raises several issues that need to be addressed in constructing an index number. Namely, why was Dallas selected as the base location (its index equals 100), and what effect does this have on the other indices? What goods specifically were used in the construction of the indices, and how were the prices determined for each of the sampled goods? Are there *quality* differences between the goods and services in the different cities, and does the index number reflect this? These and other issues are central to the construction of an index number, and we will attempt to

608

answer many of these questions in the following discussion. For example, we will learn that it is important that the base location or the base period involve relatively normal or representative conditions, although "normal" or "representative" may be extremely difficult to define.

TABLE 18.1   Index of living costs in selected
U.S. cities, 1980

City	Cost of living index
Los Angeles, Calif.	129
Washington, D.C.....................	124
Chicago, Ill.........................	121
Phoenix, Ariz.......................	106
Dallas, Tex.	100
St. Louis, Mo.	97
Atlanta, Ga.........................	97
Denver, Colo.	96
New Orleans, La.....................	92
Louisville, Ky.	90
Jacksonville, Fla.	83

It is difficult to imagine anyone who has not at some time either used or made reference to some type of an index number. For example, the Consumer Price Index, which we hear mentioned so often and which attempts to assess the degree to which purchasing power has been eroded by price increases, looms central as the most talked about and possibly least understood index number in use today. The incomes of more than 45 million people in the United States (including over 5 million laborers, 29 million social security beneficiaries, 13 million food stamp recipients, and untold numbers of retired military personnel, postal workers, and alimony recipients) are affected by this little-understood index.

As we build toward the construction of an index number series, we should keep in mind the following questions and attempt to assess for ourselves whether the index number reported truly measures the changes in the series it is purporting to assess.

☐ **18.1.1   What items should be included in the index?**

The items included in the index should be representative of what the index is attempting to measure. For example, if we are attempting to assess the relative change in the price of rib roast over some period, the construction of the index number to accomplish this goal is fairly straightforward. But if we are attempting to assess the change in meat prices for a consumer over this same interval, then it is important for us to establish which meat products are to be included in the index, and how they should be weighted in determining the value of the index number. If we are attempting to assess the change in prices of a food "basket" of goods for a consumer, then the determination of which food products and which weight each receives in the determination of the index becomes even more complex. The Consumer Price Index discussed in Section 18.5 is based on approximately 400 items out of a potential of about 2,000 potential purchases, for example.

☐ 18.1.2 What is the base period of the index?

The base period or the base location of an index number should reflect a fairly normal period or location in which the goods and their prices were not subject to unusual fluctuations that would have had a significant impact on the value of the index. For example, in the early 1970s, the quantity of tuna consumed worldwide decreased substantially because of a mercury contamination scare. In sampling households for their consumption of tuna, one would likely find that the level of consumption would have been lower than normal because of the reported presence of mercury in these fish.

The base period of an index number should also be in the not too distant past, thus enabling the user of the index number to recall conditions existing in the reference period. Some policymakers regard a base period more than 15 years old as being too dated to allow users to have familiarity with the reference or the base period.

☐ 18.1.3 What are the relative weights of the items to be included in the index?

In constructing a price index with quantity weights, for example, one should carefully assess whether the weights reflect approximate amounts of purchases to be made. Table 18.6 gives the relative weights of major groups of items in the Consumer Price Index over its latest three revisions. Note how the relative weights of the items in the index have changed as consumption patterns have changed. For this index to be a meaningful measure and reflect changes in the "purchasing power" of the dollar, the items purchased should closely parallel those that reflect current buying habits.

☐ 18.1.4 How will the data be compiled, and how often should the index be revised?

The sampling plan used in collecting data is of central importance in obtaining an unbiased statistical measure of the change in the index. Many of the statistical sampling techniques described in Chapter 8 are used in obtaining estimates of changes in prices and quantities. Important questions in data collection reflect where the information is to be collected, when the information is to be collected, and how adjustments for sales are to be made. Whenever an appreciable change in buying patterns is detected, adjustment should be made in the construction of the index number. As a case in point, how should the purchase of a video recorder be reflected in the Consumer Price Index, an item that was not even around for the mass market when the latest weights for the Consumer Price Index were constructed in 1972–73? A well-designed sampling plan will detect changes in buying habits and alert the compositor of the index number to the change so that the index can be revised to reflect substantive changes in buying habits.

□ **18.1.5   How are quality changes to be handled?**

This is undoubtedly one of the most difficult issues in the construction of an index number. How, for example, do we treat the increase in the price of an automobile when the price increases due to improvements such as catalytic converters and airbags? Many would argue that quality has decreased because of a catalytic converter, although one should keep in mind that *air* quality has likely increased by the presence of this item. Is the television set one purchases today of the same quality as the one purchased ten years ago? In the geographic example given in Table 18.1, how should one factor in differences in the "quality of living" among the various cities? The Consumer Price Index, in its latest revision, attempts to account for some of the quality changes that have occurred by pricing quality changes separately, but admittedly, the CPI only approximates measuring some of the more important quality changes that have occurred over the years. As users of index numbers, it is important for us to keep in mind the changes in quality that have accompanied the macro changes in the index series and to interpret these changes accordingly.

We shall adopt the following notation in our discussion of index numbers. The index number itself will be denoted by a capital letter, and the components that determine the *value* of the index number will be denoted by lowercase letters. We will denote the attribute price by p, the attribute quantity by q, and the attribute value by $v = pq$. We will also refer to a base *period* of an index number because the greatest use of index numbers stems from their use in interpreting time series data. Note, though, that we could have just as easily referred to the base *location* of an index number, as we did in Table 18.1. The base period of an index number, or the period against which reference is made (given in Definition 18.1), shall be denoted by a 0 subscript, and the given period shall be denoted by an n subscript. The notation $P_{n/0}$ indicates a price index in period n measured against the base period 0. We alternately indicate the price index for 1980 with 1970 as a base year as $P_{1980/1970=100}$, for example.

Many index numbers require the summation of various prices, quantities, values, etc. We could, then, indicate the sum of M prices in a base year by

$$\sum_{i=1}^{M} p_0^{(i)} = p_0^{(1)} + p_0^{(2)} + \cdots + p_0^{(M)}$$

However, such notation is very cumbersome and confusing. We will thus indicate the sum of prices of a "basket of goods" in a base year, for example, by Σp_0, where it is understood by reference to the problem the items to be included in the summation.

To express index numbers as percentages, we multiply their arithmetically-determined value by 100. This will ensure that the index number value in the base period will be equal to 100 (percent) rather than 1 (proportion). The majority of index numbers we hear reported each day are stated in this manner.

We will begin our discussion of index numbers with simple relative index numbers.

■ 18.2 SIMPLE RELATIVE INDEX NUMBERS

□ 18.2.1 Simple price relatives

The simple relative index number is the simplest to compute of the index numbers. A price relative or a simple price relative, for example, is the ratio of the price of a *single commodity* in a given period to its price in the base period, multiplied by 100. The definition of a price relative is given in Definition 18.2, and some properties of this index number are given in Theorem 18.1. These properties are illustrated in the following example.

Example 18.1 Herman's wife has kept records on the prices paid for various meat products from 1977 to 1980 in order to budget Herman's paycheck from inspecting garbage disposals at the plumbing factory. These records are given in Table 18.2. Using the data provided for rib roast, compute $P_{1980/1977=100}$ and verify that the four properties given in Theorem 18.1 are satisfied for these data where $a = 1977$, $b = 1978$, $c = 1979$, and the index number is expressed as a proportion.

Definition 18.2
Simple price relative index

Let p_0 represent the price of a commodity in a base period o, and p_n represent the price of the commodity during the given period n. Then the *simple price relative index* is defined to be:

$$P_{n/0} = \left(\frac{p_n}{p_0}\right) \times 100$$

Theorem 18.1
Properties of a simple price relative index number

Let the simple price relative index, $P_{n/0}$, be defined as given in Definition 18.2, where the index is expressed as a proportion rather than as a percentage (i.e., before multiplying the index value by 100). Let p_a, p_b, p_c, . . . represent the price of a commodity in periods a, b, c, ... Then $P_{n/0}$ possesses the following properties:

Identity property: $\quad P_{a/a} = 1$

Time reversal property: $\quad (P_{a/b})(P_{b/a}) = 1 \quad$ or

$$P_{a/b} = \frac{1}{P_{b/a}}$$

Circular property: $\quad (P_{a/b})(P_{b/c})(P_{c/a}) = 1$

Modified circular property: $\quad (P_{a/b})(P_{b/c}) = P_{a/c}$

Solution Using the data for rib roast in Table 18.2, and 1977 as the base year and 1980 as the given year,

$$P_{n/0} = \left(\frac{2.29}{1.63}\right) \times 100 = 140.5$$

TABLE 18.2 Price of selected meat products consumed by a family of four for the years 1977–1980*

Meat product	Year			
	1977	1978	1979	1980
Rib roast	$1.63	$1.79	$1.99	$2.29
Ham	1.14	1.62	2.29	2.59
Chicken	0.61	0.64	0.53	0.58
Tuna	0.72	0.89	1.25	1.39

* Price per pound, except for tuna, which is in 7.0 ounce can units.
Source: Family records.

Thus, the simple price relative index for the price Herman's wife pays for rib roast is 140.5, or prices have risen by 40.5 percent in this interval.

1. Identity property:

$$P_{a/a} = \frac{1.63}{1.63} = 1$$

This property simply states that the price relative for a commodity in a specific year a with respect to the same period is 1.

2. Time reversal property:

$$(P_{a/b})(P_{b/a}) = \left(\frac{1.63}{1.79}\right)\left(\frac{1.79}{1.63}\right) = 1$$

This property states that if the two periods are interchanged, the price relatives are reciprocals of one another.

3. Circular property:

$$(P_{a/b})(P_{b/c})(P_{c/a}) = \left(\frac{1.63}{1.79}\right)\left(\frac{1.79}{1.99}\right)\left(\frac{1.99}{1.63}\right) = 1$$

4. Modified circular property:

$$(P_{a/b})(P_{b/c}) = \left(\frac{1.63}{1.79}\right)\left(\frac{1.79}{1.99}\right) = \frac{1.63}{1.99} = P_{a/c}$$

While simple price relatives are "simple" to compute, their usefulness is limited by the fact that they consider only one commodity. That is, Herman's wife would usually be more interested in how the prices of *all meat products* have changed, rather than in how the price of rib roast alone has changed. Certain industry groups nevertheless *are* interested in how the price of a single commodity is changing over time, as were many of the Latin American countries after the "freeze" of the mid-1970s influenced the price of coffee beans. Industry associations often accumulate price and quantity movements for a selected commodity for their member institutions as part of their routine functions. But the majority of us are interested in the change in a "market basket of goods" and in how this change affects us.

☐ 18.2.2 Simple quantity relative

Just as certain groups may be interested in the relative change in prices over an interval of time for a single commodity, they may also be interested in the relative change in the quantity of a commodity produced, consumed, purchased, etc., over an interval of time. For example, the tuna industry was *very interested* in the relative change in the quantities of tuna consumed after the contamination news severely depressed the market for tuna in 1970! The definition of a simple quantity relative is given in Definition 18.3, and

TABLE 18.3 Average weekly quantity of selected meat products consumed by a family of four for the years 1977–1980*

Meat product	Year			
	1977	1978	1979	1980
Rib roast..............	5	5	4	3
Ham	2	3	2	3
Chicken	4	3	4.5	4.5
Tuna	1	0.5	0.5	0.25

* In pounds.
Source: Family records.

properties of this index are provided in Theorem 18.2. In Table 18.3 are provided estimates of the weekly consumption of meat products listed in Table 18.2 for the years indicated. The computation of a simple quantity relative index number is left to the exercises at the end of the chapter.

Definition 18.3
Simple quantity relative index

Let q_0 represent the quantity of a commodity produced, consumed, purchased, etc., in a base period o, and let q_n represent the quantity produced, consumed, purchased, etc., during the given period n. Then the *simple quantity relative index* is defined to be:

$$Q_{n/0} = \frac{q_n}{q_0} \times 100$$

Theorem 18.2
Properties of a simple quantity relative index number

Let the simple quantity relative index, $Q_{n/0}$, be defined as given in Definition 18.3, where the index is expressed as a proportion rather than as a percentage (i.e., before multiplying the index value by 100). Let q_a, q_b, q_c, . . . represent the quantity of a commodity produced, consumed, purchased, etc., in periods $a, b, c, ...$ Then $Q_{n/0}$ possesses the following properties:

Identity property: $\quad Q_{a/a} = 1$

Time reversal property: $\quad (Q_{a/b})(Q_{b/a}) = 1 \quad$ or

$$Q_{a/b} = \frac{1}{Q_{b/a}}$$

Circular property: $\quad (Q_{a/b})(Q_{b/c})(Q_{c/a}) = 1$

Modified circular property: $\quad (Q_{a/b})(Q_{b/c}) = Q_{a/c}$

☐ 18.2.3 Simple value relative

The simple value relative index is the ratio of the price times the quantity of a commodity in a given year, n, divided by the price times the quantity of the commodity in the base year, multiplied by 100. The simple value relative thus measures the relative value of a single commodity between two periods.

A definition of a simple value relative is given in Definition 18.4, and a list of the properties of the value relative are given in Theorem 18.3. In addition to the properties indicated for the simple relative price and simple relative quantity indices, the simple value relative satisfies the factor reversal test in which the product of the price and the quantity simple relatives is equal to the value relative, or,

$$V_{n/0} = \left(\frac{p_n q_n}{p_0 q_0}\right)(100) = \left(\frac{p_n}{p_0}\right)\left(\frac{q_n}{q_0}\right)(100)$$

Definition 18.4
Simple value relative index

Let q_0 represent the quantity of a commodity produced, consumed, purchased, etc., in a base period, and let p_0 represent its price in the same period. Let q_n represent the quantity produced, consumed, purchased, etc., in a given period n, and let p_n represent its price in the same period. Then the *simple value relative index* is defined to be:

$$V_{n/0} = \left(\frac{v_n}{v_0}\right)(100) = \left(\frac{p_n q_n}{p_0 q_0}\right)(100) = \left(\frac{p_n}{p_0}\right)\left(\frac{q_n}{q_0}\right)(100)$$

Theorem 18.3
Properties of a simple value relative index number

Let the simple value relative index, $V_{n/0}$, be defined as given in Definition 18.4, where the index is expressed as a proportion rather than as a percentage (i.e., before multiplying the index value by 100). Let v_a, v_b, v_c represent the values (i.e., $v_a = p_a q_a$, $v_b = p_b q_b$, etc.) of a commodity produced, consumed, purchased, etc., in periods a, b, c, ... Then $V_{n/0}$ possesses the following properties:

Identity property:	$V_{a/a} = 1$
Time reversal property:	$(V_{a/b})(V_{b/a}) = 1$ or
	$V_{a/b} = \dfrac{1}{V_{b/a}}$
Circular property:	$(V_{a/b})(V_{b/c})(V_{c/a}) = 1$
Modified circular property:	$(V_{a/b})(V_{b/c}) = V_{a/c}$
Factor reversal property:	$V_{a/b} = (P_{a/b})(Q_{a/b})$

The simple value relative attempts to measure the relative *change in value* of a commodity between periods. Thus, while the quantity of lobster caught may have declined, the price may have increased sufficiently so that the *value* of the catch may have actually increased!

The use of the simple value relative is illustrated in the following example.

Example 18.2   Using the data provided in Tables 18.2 and 18.3, compute a simple value relative for rib roast for 1980 with 1977 as a base.

Solution

$$V_{n/0} = \left(\frac{p_n q_n}{p_0 q_0}\right)(100) = \left(\frac{2.29 \cdot 3}{1.63 \cdot 5}\right)(100) = \left(\frac{6.87}{8.15}\right)(100) = 84.29$$

Thus, the relative value of rib roast consumed by Herman's family each week has declined from 1977 to 1980. While the price has increased from $1.63 to $2.29 per pound in this period, the average quantity consumed has declined from five to three pounds in this same period; hence, the decline in the relative value of rib roast consumed.

18.3   SIMPLE AGGREGATE AND SIMPLE AVERAGE INDEX NUMBERS

18.3.1   Simple aggregate price index

A simple aggregate price index is the ratio of total commodity prices in a given period to total commodity prices in a base period (multiplied by 100), where the summation of prices is over some well-understood set of commodities. The definition of a simple aggregate price index is given in Definition 18.5. This index possesses the same properties as the simple relative price index described in Section 18.2

Definition 18.5
Simple aggregate price index

Let Σp_0 represent the sum of all commodity prices of a given set in a base period, and let Σp_n represent the sum of the same set of commodity prices in the given period n. Then the *simple aggregate price index* is defined to be:

$$\Sigma P_{n/0} = \left(\frac{\Sigma p_n}{\Sigma p_0}\right)(100)$$

Example 18.3   Compute a simple aggregate price index for meat products using the data in Table 18.2, with 1980 as the given year and 1977 as the base year.

Solution   From Table 18.2, we have

$$\Sigma P_{n/0} = \left(\frac{2.29 + 2.59 + 0.58 + 1.39}{1.63 + 1.14 + 0.61 + 0.72}\right)(100) = \left(\frac{6.85}{4.10}\right)(100) = 167.1$$

☐ **18.3.2 Simple aggregate quantity index**

A simple aggregate quantity index is defined similarly to a simple aggregate price index with the word *quantity* substituted for *price*. This index possesses the same properties as the simple aggregate price index described previously. The determination of a simple aggregate quantity index is provided in the following example.

Example 18.4 Compute a simple aggregate quantity index for the meat products given in Table 18.3, with 1977 as the base year and 1980 as the given year.

Solution From Table 18.3,

$$\Sigma Q_{n/0} = \Sigma Q_{1980/1977=100} = \left(\frac{3 + 3 + 4.5 + 0.25}{5 + 2 + 4 + 1}\right)(100) = \left(\frac{10.75}{12}\right)(100) = 89.58$$

Thus, we would conclude that the average weekly consumption of the four meat products in Herman's house was, in 1980, 89.58 percent of its level in 1977.

There are two basic disadvantages of simple aggregate price and quantity indices:

1. They do not take account of the relative importance of the various commodities.
2. The particular units used in the price or quantity quotations can exert a big influence on the value of the index.

Consider the following example.

Example 18.5 In Example 18.3, a price index was computed based on units of one pound of rib roast, one pound of ham, one pound of chicken, and seven ounces of tuna. Recompute the index based on the price of tuna being expressed in units of a pound, and based on the price of tuna being expressed in units of cases (48 seven-ounce cans).

Solution If seven ounces of tuna cost $0.72 in 1977, then the price of one pound of tuna would have been $p_0 = (16/7)0.72 = \$1.65$. For 1980, the price would be $p_n = (16/7)1.39 = \$3.18$. Hence, $\Sigma P_{n/0}$ becomes

$$\Sigma P_{n/0} = \left(\frac{2.29 + 2.59 + 0.58 + 3.18}{1.63 + 1.14 + 0.61 + 1.65}\right)(100) = \left(\frac{8.64}{5.03}\right)(100) = 171.77$$

which is 4.70 percentage points above the index computed in Example 18.3, simply by changing the units in which prices are assessed.

If one seven-ounce can of tuna sold for $0.72 in 1977, one case sold for $p_0 = 48(0.72) = \$34.56$. For 1980, the price would be $p_n = 48(1.39) = \$66.72$. Hence, $\Sigma P_{n/0}$ becomes

$$\Sigma P_{n/0} = \left(\frac{2.29 + 2.59 + 0.58 + 66.72}{1.63 + 1.14 + 0.61 + 34.56}\right)(100) = \left(\frac{72.18}{37.94}\right)(100) = 190.25$$

Because the "value" of simple aggregate indices can be highly influenced by the units of measurement, simple aggregate indices are not used frequently in practice. One of the few simple aggregate indices that is still published is the *Dun & Bradstreet Wholesale Food Price Index,* which is based on the prices of one pound *each* of 31 basic commodities. This index appears monthly in *Dun's Statistical Review.*

☐ 18.3.3 Simple average of price relatives index

The simple average of price relatives index number is delineated in Definition 18.6. (The simple average of quantity relatives index is similarly defined, substituting the word *quantity* for *price*.) An advantage of the simple average of price relatives index over the simple aggregate price index is that the former is independent of the units of measurement. That is, the same "value" would be obtained for the index no matter what the units of measurement (ounces, pounds, tons, etc.). As just seen, this is not true for the simple aggregate index.

Unfortunately, the simple average of price relatives index also suffers one of the chief disadvantages of the simple aggregate price index—both fail to account for the *relative importance* of the items comprising the index. Chicken, for example, would receive the same weight as tuna in the determination of the index, even though much less tuna is consumed by the average family in a given time period. *Weighted* index numbers attempt to remedy this by weighting commodities by their importance. Weighted price and quantity index numbers are addressed in the next section.

Definition 18.6
Simple average of price relatives index

Let $\Sigma(p_n/p_0)$ represent the sum of the quotients for a given set of N commodities of the price in the given period divided by the price in the base period. Then the *simple average of price relatives index* is defined to be:

$$\bar{P}_{n/0} = \left(\frac{\sum \dfrac{p_n}{p_0}}{N} \right)(100)$$

■ 18.4 WEIGHTED AGGREGATE PRICE INDEX NUMBERS

☐ 18.4.1 Introduction

To overcome one of the difficulties encountered with simple aggregate price index numbers—namely, that the units of measurement can have a big influence on the value of the index—several weighted aggregate price indices have been developed, where the effect of the weights is to reflect (in some sense) the importance of the commodity. Thus, for example, in determining a price index for meat, we would weight the price of rib roast by the quantity of it consumed, the price of tuna by the quantity consumed, etc.

A difficulty encountered in computing a weighted price index concerns the weights or the quantities that will be used. For example, should a weighted price index for food for 1980 with a base year of 1977 reflect consumption patterns in 1980 or in 1977? As we will see shortly, the determination of the weights to be used is chiefly influenced by the economies of data collection. And, while a weighted index presumably would assume more importance for a broader audience than would an unweighted one, the weighted indices possess few of the desirable properties of unweighted indices given in Theorems 18.1–18.3.

We will begin our discussion of weighted aggregate price indices by considering the Laspeyre price index—a price index with base-year weights.

☐ 18.4.2 Laspeyre index—a price index with base-year weights

The Laspeyre price index is a weighted aggregate price index where the weights are determined by the quantities of the base period. A definition of the Laspeyre price index is given in Definition 18.7. Of those properties given in Theorem 18.1, the Laspeyre price index satisfies only the identity property.

Definition 18.7
Laspeyre price index

Let $\Sigma p_n q_0$ represent the sum of the quantities of a given set of commodities produced, consumed, purchased, etc., in the base period multiplied by their respective prices per unit in the given period, and let $\Sigma p_0 q_0$ represent the sum of the same quantities of the set of commodities multiplied by their respective unit prices in the base period. Then the weighted aggregate price index with base-year quantity weights or the *Laspeyre index* is defined to be:

$$L^*_{n/0} = \left(\frac{\Sigma p_n q_0}{\Sigma p_0 q_0}\right)(100)$$

Example 18.6 Using the data in Tables 18.2 and 18.3, compute the Laspeyre price index for 1980 with 1977 as a base.

Solution The computation of the Laspeyre price index for the data provided is given in Table 18.4. Note that the price of tuna has been converted to pounds since the quantities consumed are in pounds.

The Laspeyre price index is frequently criticized because, it is argued, it tends to overstate or have an upward bias. When prices increase, there is a tendency to reduce the consumption of the higher-priced items. Hence, by using base-year weights, too much weight is given to those items that have

TABLE 18.4 Computation of Laspeyre price index for data in Tables 18.2 and 18.3 (1977 = 100)

(1)	(2)	(3)	(4)	(5)	(6)
	Weekly quantity consumed	Price in		Price in	
Product	in base year, 1977	1977	$p_0 q_0$	1980	$p_n q_0$
Rib roast	5	$1.63	8.15	$2.29	11.45
Ham	2	1.14	2.28	2.59	5.18
Chicken	4	0.61	2.44	0.58	2.32
Tuna	1	1.65	1.65	3.18	3.18
			14.52		22.13

$$L^*_{1980/1977=100} = \left(\frac{22.13}{14.52}\right)(100) = 152.41$$

increased the most. Note, for example, how the quantity of rib roast consumed in Table 18.3 has declined as the price has increased.

Similarly, when prices decline, consumers will shift their purchases to those items that have declined the most. By using base-period weights, sufficient weight will not be given to those items that have fallen the most in price, again overstating the index.

☐ 18.4.3 Paasche index—a price index with given-year weights

The Paasche price index is a weighted aggregate price index in which the weights are determined by quantities in the *given* year. A definition of the Paasche price index is given in Definition 18.8. Again, the identity property is the only property exhibited by the Paasche index when the quantities between the base and the given year are not equal, as is generally the case. Use of the Paasche index is illustrated in the following example.

Definition 18.8
Paasche price index

Let $\Sigma p_n q_n$ represent the sum of the quantities of a given set of commodities produced, consumed, purchased, etc., in the given period multiplied by their respective unit prices in the given period, and let $\Sigma p_0 q_n$ represent the sum of the same quantities of the set of commodities multiplied by their respective unit prices in the base period. Then the weighted aggregate price index with given period quantity weights or the *Paasche index* is defined to be:

$$P^*_{n/0} = \left(\frac{\Sigma p_n q_n}{\Sigma p_0 q_n}\right)(100)$$

Example 18.7 Using the data in Tables 18.2 and 18.3, compute the Paasche price index for 1980, with 1977 as a base. How does this compare with the Laspeyre price index determined in Example 18.6?

Solution A computation worksheet for determining the Paasche price index for the data in Tables 18.2 and 18.3 is given in Table 18.5. Using the Paasche price index, we would conclude that prices for this particular group of commodities have increased 57.36 percent over the base year; using the Laspeyre price index, we would conclude that prices have increased 52.41 percent over the base year.

TABLE 18.5 Computation of Paasche price index for data in Tables 18.2 and 18.3

(1)	(2) Weekly quantity consumed in given year, 1980	(3) Price in 1977	(4) $p_0 q_n$	(5) Price in 1980	(6) $p_n q_n$
Product					
Rib roast	3	$1.63	4.89	$2.29	6.87
Ham	3	1.14	3.42	2.59	7.77
Chicken	4.5	0.61	2.745	0.58	2.61
Tuna	0.25	1.65	0.4125	3.18	0.795
			11.4675		18.045

$$P^*_{1980/1977=100} = \left(\frac{18.045}{11.4675} \right) (100) = 157.36$$

The difficulty in computing the Paasche index in practice is that revised weights or quantities have to be computed each year or each period, adding to the data-collection expense in preparation of the index. For this reason, the Paasche index is *not* used frequently when the number of commodities is large.

It is also possible to define a Laspeyre and a Paasche *quantity* index. These two are often referred to as the "symmetric" quantity indices. A discussion of these two symmetric quantity indices is left to the references at the end of the chapter.

☐ 18.4.4 Typical-period price index—a price index with typical-period weights

The typical-period price index is a weighted aggregate price index in which the quantity weights are based on demand, consumption, etc., in a "typical" period. Frequently, the typical period (year) is taken to be the average of some number of consecutive periods. The typical-period price index is applicable in those instances in which the quantity weights for the base or for the current period are known to be unrepresentative. For example, the inclusion of quantity weights for tuna in the base year of 1970 may not be representative because of the mercury contamination scare of 1970, which reduced the consumption of tuna considerably. The Consumer Price Index described in the next section is basically a typical-period price index where the current weights are based on a consumer expenditure survey completed at a cost of $20 million in 1972–73, and the base period is 1967.

The typical-period price index is defined in Definition 18.9.

Definition 18.9
Typical-period price index

Let $\Sigma p_n q_t$ represent the sum of the quantities of a given set of commodities produced, consumed, purchased, etc., in a "typical" period t multiplied by their respective unit prices in the given period n, and let $\Sigma p_0 q_t$ represent the sum of the same quantities of the set of commodities multiplied by their respective unit prices in the base period o. Then the weighted aggregate price index with typical-period weights is defined to be:

$$T^*_{n/0} = \left(\frac{\Sigma p_n q_t}{\Sigma p_0 q_t} \right) (100)$$

☐ **18.4.5 Fisher ideal price index**

The Fisher ideal price index delineated in Definition 18.10 and Theorem 18.4 was designed to overcome several disadvantages of the Laspeyre and the Paasche price indices. Note, for example, from Theorem 18.4 that the Fisher ideal index possesses the time reversal and the factor reversal properties not possessed by the Laspeyre and the Paasche price indices, which gives it certain theoretical advantages over these two price indices. Note also that the Fisher ideal index requires the computation of quantity weights in the given year, giving it the added data-collection costs of the Paasche price index.

Definition 18.10
Fisher ideal price index

Let the quantities $\Sigma p_n q_0$, $\Sigma p_0 q_0$, $\Sigma p_n q_n$, and $\Sigma p_0 q_n$ be as defined in Definitions 18.7 and 18.8. Then the *Fisher ideal price index* is defined to be:

$$F^*_{n/0} = \left[\sqrt{ \left(\frac{\Sigma p_n q_0}{\Sigma p_0 q_0} \right) \left(\frac{\Sigma p_n q_n}{\Sigma p_0 q_n} \right) } \right] (100)$$

Recalling that the geometric mean of two numbers a and b is $(ab)^{1/2}$, the Fisher ideal price index is the geometric mean of the Laspeyre and the Paasche price indices.

> **Theorem 18.4**
> Properties of the Fisher ideal price index
>
> Let the Fisher ideal price index, $F_{n/0}^*$, be defined as given in Definition 18.10, where the index is expressed as a proportion rather than as a percentage (i.e., before multiplying the index value by 100). Let the subscripts $a, b, c, \ldots$ refer to relevant prices or quantities in periods $a, b, c, \ldots$. Then the Fisher ideal price index satisfies the following properties:
>
> Identity property: $\qquad F_{a/a}^* = 1$
>
> Time reversal property: $\qquad (F_{a/b}^*)(F_{b/a}^*) = 1 \quad$ or
>
> $$F_{a/b}^* = \frac{1}{F_{b/a}^*}$$
>
> Factor reversal property: $\quad V(F_{a/b}^*) = P(F_{a/b}^*)Q(F_{a/b}^*)$

Use of the Fisher ideal price index is illustrated in the following example.

Example 18.8 Using the data in Tables 18.2 and 18.3, compute the Fisher ideal price index for 1980, with 1977 as a base.

Solution We can greatly simplify the computation of the Fisher ideal price index after the Laspeyre and the Paasche price indices are computed by realizing that the Fisher index is the geometric mean of the Laspeyre and the Paasche price indices. That is,

$$F_{n/0}^* = \sqrt{L_{n/0}^* P_{n/0}^*}$$

From Tables 18.4 and 18.5,

$$F_{n/0}^* = \sqrt{(152.41)(157.36)} = 154.86$$

The Fisher ideal price index indicates that prices in 1980 for the commodities involved are 54.86 percent above base-year (1977) prices.

In general, we would expect the Fisher ideal price index to lie between the values of the Laspeyre and the Paasche indices, since it is the geometric mean of these two quantities. Recall from the discussion in Chapter 2 that the geometric mean is less affected by extremely large or small values than is the arithmetic mean. This adds to the desirability of the Fisher ideal price index, especially when one of the components of the index experiences an unusually large increase or decrease in the given or the base period.

■ 18.5 THE CONSUMER PRICE INDEX

Perhaps the single index number that touches our lives most is the Consumer Price Index (CPI) compiled by the Bureau of Labor Statistics (BLS). Indeed, provisions for "escalating" wages (escalator clauses) in the union contracts of more than 5 million workers are tied directly to the CPI. Politicians love to quote this index to show how well they are doing (or how badly the other party is doing!), and divorcees frequently receive a change in alimony payments with a change in the CPI! The CPI is further a major yardstick against which government economic policy is judged. A graph showing this index for the years 1960–79 is given in Figure 18.1. Note from this graph that the effect of inflation on the cost of living over the last several years has been quite dramatic! For example, an individual earning $10,000 per year in 1967 would have to earn well in excess of $19,000 per year in 1978 just to have the same buying power!

FIGURE 18.1   Consumer Price Index, 1960–1979

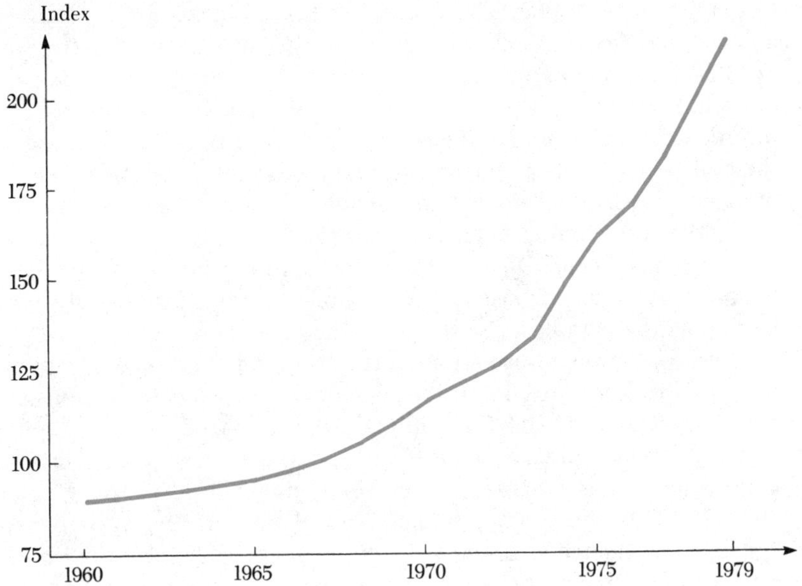

Source: U.S. Department of Commerce, *Survey of Current Business* (various issues).

The CPI has been published continually since 1913, although major revisions in the method of computation, in the data-collection methodology, and in the "market basket of goods" that comprise the index have changed considerably since this time. Many commodities in the current CPI (there are about 400 of them at present) were not even around in 1913! Furthermore, with the most recent revision of the index (there have been five since its inception in 1913), more sophisticated statistical sampling procedures have been used both to measure the prices paid for the goods and services comprising the index and to detect shifts in buying habits so the quantity weights used in compiling the index can be updated more systematically. For example, the selection of retail stores for obtaining prices of selected goods is based on a point-of-purchase survey of households, rather than on secondary sources. Food is now priced throughout the month, rather than during the first week of the month, as before. These and other refinements should lead to a more accurate picture of the change in the purchasing power of the dollar over time.

The index began during the inflationary conditions of World War I, when the Bureau of Labor Statistics investigated the cost of living in a number of shipbuilding centers and other industrial centers. Shortly thereafter, the index was given its official title of "Index of Changes in Prices of Goods and Services Purchased by City Wage Earner and Clerical Worker Families to Maintain Their Level of Living." The type of workers surveyed in the index has changed recently to include more than city wage earners and clerical workers. Today, for example, two consumer price indices are published instead of just the one. The Consumer Price Index for City Wage Earners and Clerical Worker Families attempts to measure changes in the purchasing power of the dollar for about 40 percent of the noninstitutional civilian households in the United States. The more inclusive Consumer Price Index for All Urban Consumers measures the change in the purchasing power of the dollar for *all* urban consumers, and covers in excess of 80 percent of the total noninstitutional civilian households in the United States. To obtain prices for commodities contained in these two indices, the Bureau of Labor Statistics has recently increased the number of urban areas sampled from 56 to 85. These areas are based on a probability sample using the 1970 census as its population data base.

Table 18.6 gives information on the relative importance of major components of the Consumer Price Index over the previous three revisions. Note how the relative weighting of the proportions of income expended for housing for individuals contained in the wage earner and clerical worker index sample has increased from 32.0 percent in 1952 to 39.8 percent in 1972–73. This, of course, reflects the change in buying habits of these consumers detected over the last several years. Note also that for the latest revision of the index, the proportion expended for housing differs depending on whether one is in the wage earner, clerical worker sample or the all urban population sample.

Three base periods for the wage earner, clerical worker index have existed since World War II: 1947–49, 1957–59, and 1967. The base period for the newer all urban population index, which first began publication in January 1978, is also 1967. The current weights for both of these indices, which are given in Table 18.6, are based on a Consumer Expenditure Survey conducted in 1972 and 1973.

Both consumer price indices just described are actually chain index numbers, where the index numbers for successive time periods are combined. This facilitates the updating of the index when there is a detected change in purchasing habits. The CPI very closely approximates a typical-year price index described in the previous section, however, since the base period of the index does not coincide with the dates of the items included in the index based on the 1972–73 Consumer Expenditure Survey. The use of the CPI in decision making is illustrated in the following example.

TABLE 18.6 Relative importance of major components of the Consumer Price Index—1952, 1971, 1972–1973

Major groups	1952 Wage earner, clerical worker population	1971 Wage earner, clerical worker population	1972–1973 All urban population	1972–1973 Wage earner, clerical worker population
Total	100.0	100.0	100.0	100.0
Food and alcoholic beverages	32.1	22.2	18.8	20.4
Housing	32.0	33.8	42.9	39.8
Apparel	9.7	10.5	7.0	7.0
Transportation	11.0	13.3	17.7	19.8
Medical care	4.7	6.5	4.6	4.2
Entertainment	5.4	5.7	4.5	4.3
Other goods and services	5.1	8.0	4.5	4.5

Source: "The Consumer Price Index: A Short Description of the Index as Revised, 1953," Department of Labor, Bureau of Labor Statistics, January 1953; "Relative Importance of Components of the Consumer Price Index, 1970–1971," Department of Labor, Bureau of Labor Statistics, 1972; "The Revision of the Consumer Price Index," W. John Layng, U.S. Department of Commerce, *Statistical Reporter*, no. 78-5, February 1978, p. 145.

Example 18.9 The tables below give the CPI for the years 1974–78 (1967 = 100) and Herman's wages in the plumbing factory for the same period.

CPI

Year	CPI
1974	147.7
1975	161.2
1976	170.5
1977	181.5
1978	195.3

Source: U.S. Bureau of Labor Statistics, *Monthly Labor Review.*

Herman's wages

Year	Wage rate
1974.	$6.99
1975.	7.77
1976.	8.18
1977.	8.70
1978.	9.10

Source: Herman's boss, Fort Worth Annie.

On balance, has Herman's wage rate kept pace with the general increase in prices as measured by the CPI?

Solution Relevant computations are shown in the table.

(1) Year	(2) Wage rate	(3) CPI*	(4) Deflated wage rate [(2) ÷ (3)] (100)
1974	$6.99	147.7	$4.73
1975	7.77	161.2	4.82
1976	8.18	170.5	4.80
1977	8.70	181.5	4.79
1978	9.10	195.3	4.66

* 1967 = 100.

On balance, Herman's wage increases have not kept pace with the cost of living as measured by the CPI. In fact, real income (deflated wage rate) is lower in 1978 than it was in 1974! (Herman is in the same "boat" as most other workers!)

18.6 DEFLATION OF TIME SERIES

Time series data are statistical data that are measured over time; an example would be the level of the gross national product in the United States for the most recent past. Most time series data, such as GNP, include the effect of price changes in their level or their value. To assess the amount of actual *real* growth in the series, it becomes necessary to adjust the data for increases in the price level. This adjusting process is called *deflation of time series.* Mathematically, deflation of time series is applicable only if the price index numbers satisfy the factor reversal test, as does the Fisher ideal price index.

The use of price index numbers for deflating time series data is illustrated in the following example.

Example 18.10   Given in the table below are unadjusted values of GNP for the United States for the years 1975–79.

Year	1975	1976	1977	1978	1979
GNP ($ billion)	1,528.8	1,702.2	1,899.5	2,127.6	2,368.8

Source: *Statistical Abstract of the United States, 1980.*

Table 18.7 gives information on gross national product implicit price deflators (aggregate price index numbers) for these same five years.

TABLE 18.7   Gross national product implicit price deflators (1972 = 100)

Year	1975	1976	1977	1978	1979
Price deflator	127.2	133.8	141.7	152.0	165.5

Source: "Economic Report of the President," January 1980.

Using the information provided, estimate the percentage growth in *actual* GNP and in *real* GNP from 1975 to 1979 inclusive, and determine the level of GNP in "1972 dollars".

Solution   We will do the second part first. In order to express GNP in 1972 dollars, we must divide the actual values of GNP by the implicit price deflators given, and multiply the result by 100 as shown in the table.

Year	1975	1976	1977	1978	1979
Actual GNP ($ billions)	1,528.8	1,702.2	1,899.5	2,127.6	2,368.8
Implicit price deflator (1972 = 100)	127.2	133.8	141.7	152.0	165.5
GNP in 1972 dollars ($ billions)	1,201.9	1,272.2	1,340.5	1,399.7	1,431.3

$$\text{Percentage growth in actual GNP, 1975–79} = \frac{2{,}368.8 - 1{,}528.8}{1{,}528.8} = 0.55$$

$$\begin{array}{c}\text{Real growth in GNP (adjusted}\\ \text{for price level changes),}\\ \text{1975–79}\end{array} = \frac{1{,}431.3 - 1{,}201.9}{1{,}201.9} = 0.19$$

Thus, although actual GNP increased 55 percent in this interval, 65 percent of the increase is due solely to an increase in prices and does not represent *real* growth. Therefore, 19 percent represents the growth in real GNP over this interval.

■ 18.7 INDEX NUMBERS AS RANDOM VARIABLES

Throughout our discussion of index numbers, we have not emphasized that index numbers are random variables, and this may be unfortunate. To gain an appreciation of the inherent variability associated with the computation of index numbers, one has only to consider how compilers of index numbers arrive at the prices and quantities that determine the value of the index. For example, Herman's wife in Example 18.1 bases the prices of the amounts she spends for food items (meats) on trips to the grocery store. The prices she pays for meat products are bound to differ throughout the year, by the store where the purchase was made, whether or not there was a sale on the product the day she shopped, and other factors. Our discussion of the Consumer Price Index in Section 18.5 included a brief discussion of some of the statistical sampling issues involved in determining both the quantity weights used in the construction of the index and in accumulating the prices of the goods priced. Thus, the Consumer Price Index reported each month is in reality one observation from a probability distribution of consumer price index values.

The danger in not treating an index number as a random variable is that policymakers may make important policy decisions based on the value of the index reported, when in fact the change reported may be due to nothing more than errors inherent in any sampling process.

Recently, statisticians have recognized the need not only for accumulating price index information per se, but also for reporting on whether the change in an index number is due to changes in the prices and goods comprising the index, or to inherent errors in statistical sampling. Noteworthy in this regard is the work of Hayya and of Hayya, Saniga, and Schaul (see the references at the end of the chapter). Under some not too restrictive assumptions regarding the distribution of the error terms in index number computation, these authors were able to derive a quadratic expression, which, when solved, yields confidence intervals and confidence bands for price index numbers. Table 18.8, for example, gives confidence intervals for the Bureau of the Census' price index for new one-family houses sold. The new houses sold must be similar to those sold in 1967 with respect to eight characteristics: floor area, number of stories, number of bathrooms, air conditioning, type of parking facility, type of foundation, geographic division within region, and location within a metropolitan area. (The value of ρ denotes the unknown correlation coefficient between the numerator and the denominator of the index.) Note how the width of the interval varies as a function of the unknown correlation coefficient between the numerator and the denominator of the index number expression (as given in the bottom row of Table 18.8).

Unfortunately, much more work needs to be done in this area before meaningful results can be used on a large scale. One may envision a newscaster of the future saying, "The Consumer Price Index rose by 0.3 percent last month, but this increase could be attributable to nothing more than errors inherent in the sampling process in gathering the data." For the time being, we, as users of index number data, should content ourselves with recognizing that index numbers are random variables and are but one realization from the distribution of index number values, and interpret them accordingly.

TABLE 18.8 Ninety percent probability intervals for the price index for new one-family houses sold, third quarter 1976 (1963 = 100)

Estimated index value:			$P_{1976-73/1963=100} = 1.940\ddagger$							
Confidence bounds:			$a < P < b$							
	Correlation, ρ^*									
	−1.0	−0.8	−0.6	−0.4	−0.2	0.0	0.2	0.4	0.6	0.8	1.0
a	1.914	1.915	1.917	1.918	1.920	1.922	1.923	1.926	1.928	1.931	1.937
b	1.967	1.965	1.964	1.962	1.961	1.959	1.957	1.955	1.952	1.949	1.943
$Percent\dagger$	2.7	2.6	2.4	2.3	2.1	1.9	1.8	1.5	1.2	0.9	0.3

* The correlation ρ measures the degree of linear association between the numerator and the denominator of the index number expression.
† The percents give the width of the probability interval as a percentage of the point estimate of the index.
‡ Here, the index number is being quoted as a proportion, rather than as a percentage.
Source: Jack C. Hayya, Erwin M. Saniga, and Ronny A. Schaul, "Uncertainty in Price Indices," Working Paper OM-78-9, The Pennsylvania State University, October 1978.

■ 18.8 SUMMARY

This chapter has presented a discussion of the use, determination, and interpretation of index numbers—statistical measures designed to show relative changes in a variable or in a group of variables with respect to some base point or base period. The types of index numbers covered included simple price, quantity and value relatives, simple aggregate price and quantity indices, and four weighted aggregate price relatives—Laspeyre, Paasche, typical year, and the Fisher ideal index. The Consumer Price Index (CPI), which touches so many of our lives, was also described; it is a special case of a weighted aggregate price index.

Five tests or properties of index numbers were described that contribute to their theoretical development and interpretation. These tests or properties include: the identity test, the time reversal test, the circular and modified circular tests, and the factor reversal tests. A summary table delineating

TABLE 18.9 Summary formulas for computing index numbers

Index Number type	Index number	Computation formula	Tests satisfied				
			Identity test	Time reversal test	Circular test	Modified circular test	Factor reversal test
Simple relative (single commodity)	Simple price relative	$P_{n/0} = \left(\dfrac{p_n}{p_0}\right)(100)$	Yes	Yes	Yes	Yes	—
	Simple quantity relative	$Q_{n/0} = \left(\dfrac{q_n}{q_0}\right)(100)$	Yes	Yes	Yes	Yes	—
	Simple value relative	$V_{n/0} = \left(\dfrac{p_n q_n}{p_0 q_0}\right)(100)$	Yes	Yes	Yes	Yes	Yes
Simple aggregate	Simple aggregate price index	$\sum P_{n/0} = \left(\dfrac{\sum p_n}{\sum p_0}\right)(100)$	Yes	Yes	Yes	Yes	No
	Simple aggregate quantity index	$\sum Q_{n/0} = \left(\dfrac{\sum q_n}{\sum q_0}\right)(100)$	Yes	Yes	Yes	Yes	No
Weighted aggregate price index	Laspeyre index	$L^*_{n/0} = \left(\dfrac{\sum p_n q_0}{\sum p_0 q_0}\right)(100)$	Yes	No	No	No	No
	Paasche index	$P^*_{n/0} = \left(\dfrac{\sum p_n q_n}{\sum p_0 q_n}\right)(100)$	Yes	No	No	No	No
	Typical-year price index (CPI)	$T^*_{n/0} = \left(\dfrac{\sum p_n q_t}{\sum p_0 q_t}\right)(100)$	Yes	No	No	No	No
	Fisher ideal index	$F^*_{n/0} = \left[\sqrt{\left(\dfrac{\sum p_n q_0}{\sum p_0 q_0}\right)\left(\dfrac{\sum p_n q_n}{\sum p_0 q_n}\right)}\,\right](100)$	Yes	Yes	No	No	Yes

computation formulas for the index numbers described and which tests are satisfied by each is given in Table 18.9. Unfortunately, those index numbers that should be the most useful to large groups of people are the ones that fail to pass many of the tests described.

REFERENCES

Fisher, I. *The Making of Index Numbers.* Boston: Houghton Mifflin Company, 1923.

Hayya, Jack C. "Confidence Bands for Price Indices," American Statistical Association 1977 Proceedings of the Business and Economics Statistics Section, part I, pp. 362–67.

Hayya, Jack C.; Saniga, Erwin M.; and Schaul, Ronny A. "Uncertainty in Price Indices," Working Paper OM-78-9, The Pennsylvania State University, October 1978.

Layng, W. John. "The Revision of the Consumer Price Index," U.S. Department of Commerce, *The Statistical Reporter*, no. 78-5 (February 1978), pp. 140–48.

Mudgett, Bruce D. *Index Numbers.* New York: John Wiley & Sons, Inc., 1951.

Several documents are published periodically by the Department of Labor, Bureau of Labor Statistics, Washington, D.C., on

descriptions of or revisions to the Consumer Price Index. These publications include the following:

The Consumer Price Index—January 1953

The Consumer Price Index as Related to Other Consumer Statistics—1959

An Abbreviated Description of the Revised Consumer Price Index—March 1964

Major Changes in the Consumer Price Index—March 1964

Conversion of the Consumer Price Index to 1967 Standard Reference Base Period—1971

Relative Importance of Components in the Consumer Price Index, 1970–1971—1972

Updating and Revising the Consumer Price Index—April 1974

PROBLEMS

18.1. Given the data in the table to the right, compute the following:

    a. $P_{n/0}$ for $n = 1973$ and $0 = 1969$.

    b. $\Sigma P_{n/0}$ for $n = 1973$ and $0 = 1969$, where the summation is over all cereals and bakery products listed.

    c. For $n = 1973$, $0 = 1969$, $a = 1969$, $b = 1970$, and $c = 1971$, show that the properties listed in Theorem 18.1 hold for flour, wheat.

18.2. For the first three products given in Problem 18.1, find the average U.S. per capita consumption of each of the commodities for the years indicated.

Average retail prices of selected foods,* 1969–1973

	1969	1970	1971	1972	1973
Cereals and bakery products					
Flour, wheat .	11.6	11.8	12.0	11.9	15.1
Rice	18.8	19.1	19.6	19.6	26.0
Corn flakes . .	41.7	42.9	44.5	41.6	42.9
Bread, white . .	23.0	24.3	25.0	24.7	27.6

* Cents per pound.
Source: *Statistical Abstract of the United States 1974.*

Year	1967	1968	1969	1970	1971	1972	1973	1974
CPI (1967 = 100)	100.0	104.2	109.8	116.3	121.3	125.3	133.1	144.0
Salary (average per hour)	$5.93	$6.25	$6.48	$6.63	$7.00	$7.35	$7.79	$8.30

Then, for $n = 1973$ and $0 = 1969$, determine the following:

a. The Paasche index, $P^*_{n/0}$.
b. The Laspeyre index, $L^*_{n/0}$.
c. Fisher ideal index, $F^*_{n/0}$.
d. In general, prove that the Fisher ideal price index falls between $P^*_{n/0}$ and $L^*_{n/0}$.

18.3. Prove that $(P_{a/b})(P_{b/c}) = P_{a/c}$ and that $(P_{a/b})(P_{b/a}) = 1$.

18.4. The simple quantity relative for the year 1974 with 1970 as a base is 110, while the simple quantity relative for 1974 with 1972 as a base is 130. What is the simple quantity relative for 1972, with 1970 as a base?

18.5. Using the data provided in Table 18.3, compute a simple quantity relative for rib roast for 1980 using 1977 as a base, and for 1977 with 1980 as a base. Do these two indices satisfy the time reversal test (i.e., exhibit the time reversal property)?

18.6. Which of the properties listed in Table 18.9 are satisfied by the simple average price relative index? Show by counter-example which properties do not hold.

18.7. Develop arguments similar to those developed in Section 18.4.2 to describe why the Paasche index tends to understate or have a downward bias.

18.8. Show that the Paasche price index does not in general satisfy the time reversal and the factor reversal tests using the data in Example 18.7.

18.9. Verify that the time reversal and the factor reversal tests are satisfied by the Fisher ideal price index using the data provided in Example 18.8.

18.10. Using the definition, prove that the Fisher ideal price index satisfies the time reversal test.

18.11. The table above shows the Consumer Price Index for 1967–74. Also shown is the hourly salary received by a statistics instructor at a large midwestern university.

a. Recast the salary amounts into constant 1967 dollars.
b. The *actual* percentage growth in salary over this interval is equal to what amount?
c. The *real* percentage growth in salary over this interval is equal to what amount?

18.12. Reproduced in the table below are data on sales prices of electronic calculators for three models for the years 1976–80, inclusive.

Unit prices of electronic calculators, 1976–1980

Calculator type*	1976	1977	1978	1979	1980
X..	$110	$105	$ 90	$ 83	$ 45
Y..	450	420	260	120	80
Z..	750	650	525	460	320

* Usually relates to number of programmed functions available.

a. Compute the simple relative price index number for calculator type Y for 1980, with 1976 as the base year.

b. Compute the simple aggregate price index for calculator sales, using 1980 as the given year and 1977 as the base year.

c. Compute the simple average of price relatives index number, $\bar{P}_{n/0}$, for 1980, with 1976 as the base year.

18.13. Given in the table below are data on unit sales of electronic calculators, the prices of which are given in Problem 18.12.

Unit sales of electronic calculators, 1976–1980

Calculator type	1976	1977	1980	1979	1980
X...	20	40	60	65	60
Y...	5	10	8	15	60
Z...	1	5	10	20	40

a. Compute the simple quantity relative, $Q_{n/0}$, for calculator type X for 1980, with 1976 as the base year.

b. Compute the simple aggregate quantity index for calculator sales, using 1980 as the given year and 1977 as the base year.

18.14. Use the data from Problems 18.12 and 18.13.

a. Compute the Laspeyre price index for 1980, with 1977 as the base year.

b. Compute the Paasche price index for 1980, with 1977 as the base year.

c. Compute the Fisher ideal price index for 1980, with 1977 as the base year.

18.15. The GNP of a developing nation has increased from $40 billion in 1977 to $44 billion in 1980. The price index used in calculating GNP has increased from 145 to 160 in this same period. What is the *real* growth rate of this country's GNP after accounting for the erosion

of the purchasing power of their currency?

18.16. The Consumer Price Index for City Wage Earner and Clerical Worker Families and the Consumer Price Index for All Urban Consumers survey, respectively, what percentage of the noninstitutional civilian households in the United States at the present time?

18.17. The consumer price index for selected services increased from 156 to 167 over a certain period.

a. Given that Calvin Consumer budgeted $50 per month for these services at the beginning of this period, how much should he budget for these same services at the end of this period?

b. Suppose Calvin's monthly salary increased from $1,200 to $1,340 over this same period. What is the *real* change Calvin has experienced in the purchasing power for these services?

18.18. Consider the information in the table.

Year	Price I	Quantity I	Price II	Quantity II
1977 ...	25¢	10	50¢	40
1978 ...	30	20	45	50
1979 ...	35	30	48	55
1980 ...	40	40	50	50

a. Construct a Laspeyre price index for 1980 using 1977 as the base year.

b. Construct Paasche's price index for each of the years indicated using 1977 as the base year.

c. Construct the Fischer ideal price index for 1980, with 1977 as the base year.

18.19. A stock you have been observing over the last several years has exhibited the year-end closing prices shown in the table.

Year	Closing price
1975...........	$54.30
1976...........	62.35
1977...........	64.86
1978...........	79.94
1979...........	86.45
1980...........	75.32

Using 1975 as the base year, compute the simple price relative index number for this stock between 1975 and 1980.

18.20. Obtain Consumer Price Index estimates for the years 1975 to 1980, inclusive, from your university library. Given this information, compute the deflated year-end closing price for the stock given in Problem 18.19.

18.21. Suppose an executive's income increases from $56,000 to $73,000 over a period of three years while the Consumer Price Index increases from 167 to 197. What is the change in her real income over this period?

18.22. Given the wage rate and consumer price index shown in the table below for the years indicated, Herman's *real* wage rate (adjusted for inflation) has increased by what percentage from 1970 to 1973?

Year	Herman's wages (per hour)	CPI (1967 = 100)
1970..........	$5.40	116.3
1971..........	5.87	121.3
1972..........	6.30	125.3
1973..........	6.87	133.1

Chi-square 19

■ 19.1 INTRODUCTION

In Chapter 10, we used the chi-square distribution to construct decision rules for hypotheses concerning the population variance, σ^2. In this chapter, we introduce several tests of hypotheses that are generally referred to as *chi-square tests*. Most of these testing procedures were first suggested by Karl Pearson in 1900 and were among the earliest methods of statistical inference. While the chi-square tests are frequently misused, they are among the most popular statistical inference procedures today, and for good reason. They are typically easy to implement and can tell us a great deal about the characteristics of a random variable or sets of random variables.

■ 19.2 GOODNESS-OF-FIT TESTS

The chi-square distribution may be used to test the hypothesis that a random variable has a specified theoretical statistical distribution. Perhaps the most important application of the chi-square distribution in this context is testing whether or not a random variable is normally distributed, because virtually all statistical inference procedures presented in the text to this point make the normality assumption about the distribution of the population random variable.

The chi-square goodness-of-fit test is based on the difference between the frequencies in the classes of the observed sample and the class frequencies we would expect if the random variable conforms to the assumed theoretical distribution. The chi-square test statistic is:

$$\chi^2 = \sum_{i=1}^{k} \frac{(O_i - E_i)^2}{E_i}$$

where O_i represents the observed frequency in the ith class E_i represents the expected frequency in the ith class, and k is the number of classes. The distribution of the χ^2 statistic, if the sampled observations do come from the assumed theoretical (population) distribution, is *approximately* the chi-square distribution with $(k - 1 - m)$ degrees of freedom, where m is the number of parameters that must be estimated to determine the expected frequencies.

Since the test statistic is approximately distributed as a chi-square random variable, all hypothesis tests for which the decision rule is based on this test statistic are *approximate statistical tests*. We must keep this fact in mind when using the chi-square goodness-of-fit testing procedures.

If the observed frequencies O_i do not differ much from the expected frequencies E_i, the value of the test statistic χ^2 is small. Indeed, the *minimum* value of χ^2 is 0, and this value of the test statistic occurs when $O_i = E_i$ in each of the k classes. As the observed frequencies O_i begin to differ from the expected frequencies E_i, the value of χ^2 will increase, because the statistic squares these differences, weights them by the reciprocal of the expected frequencies, and adds the resulting ratios. Thus, a small value of χ^2 supports the null hypothesis that the random variable conforms to the specified theoretical statistical distribution, while a large value of χ^2 supports the alternate hypothesis that the random variable does not conform to the specified statistical distribution. Consequently, the decision rule for the chi-square goodness-of-fit test will have an upper-tail rejection region; its location is determined by the form of the chi-square distribution (determined by df) and the selected significance level α of the test. A typical decision rule is illustrated in Figure 19.1. The form of the chi-square distribution is solely a function of its degrees of freedom, df. Several chi-square distributions are illustrated in Figure 19.2 for different values of the degrees of freedom.

FIGURE 19.1 Decision rule for the chi-square goodness-of-fit test

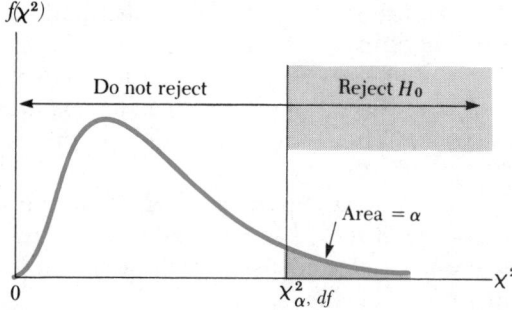

FIGURE 19.2 Four χ^2 distributions

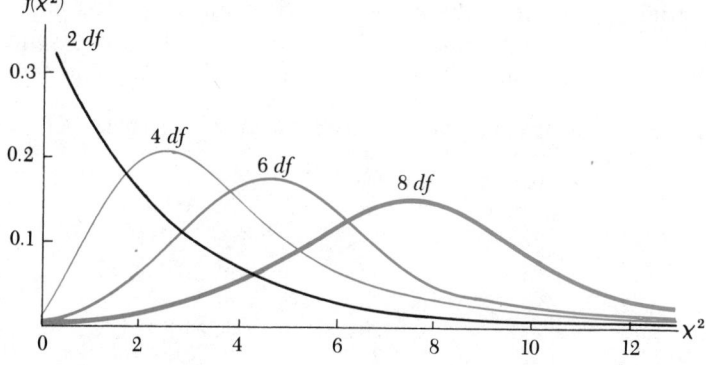

The form of the chi-square statistic,

$$\chi^2 = \sum_{i=1}^{k} \frac{(O_i - E_i)^2}{E_i}$$

should not be unfamiliar to us. It adds weighted squared deviations and is similar in form to the population and sample variance formulas, which add weighted squared deviations of observations about their respective means. It is a very easy statistic to compute, which has no doubt contributed to its widespread use.

Example 19.1   A fair die is one in which each of the six faces has an equal probability of showing on the top when the die is tossed. Suppose we wish to test the "fairness" of a particular die experimentally. We decide to toss the die 120 times and record the number of dots showing on the top face each time. If the die is fair, we should expect to find that each number occurs with an equal frequency in the 120 tosses. Any deviation of the expected frequencies of $(\frac{1}{6})(120) = 20$ for each number from the observed frequencies will give us cause to question the fairness of the die.

Table 19.1 gives the observed frequencies of the numbers in the 120 trials. The expected frequencies are easily computed in this problem. If the die is

TABLE 19.1 Observed frequencies in Example 19.1

Number of dots showing on top face	Observed frequency O_i
1	12
2	14
3	31
4	29
5	20
6	14
Total	120

fair, then the probability that each number of dots will appear on the top face in any trial is $1/6$. Since there are 120 trials, the expected frequency of each number in the experiment is $(1/6)(120) = 20$. The computation of the chi-square statistic value for this experiment is given in Table 19.2.

TABLE 19.2 Computation of the chi-square statistic for Example 19.1

Class	i	Observed frequency, O_i	Expected frequency, E_i	$O_i - E_i$	$(O_i - E_i)^2$	$\dfrac{(O_i - E_i)^2}{E_i}$
1	1	12	20	-8	64	3.20
2	2	14	20	-6	36	1.80
3	3	31	20	11	121	6.05
4	4	29	20	9	81	4.05
5	5	20	20	0	0	0.00
6	6	14	20	-6	36	1.80
Totals		120	120	0		$\chi^2 = 16.90$

The degrees of freedom for this test are $df = k - 1 - m = 6 - 1 - 0 = 5$; that is, $m = 0$ since no parameters had to be estimated to determine the expected frequencies. Under the null hypothesis of a fair die, the expected frequencies are determined [$(1/6)(120) = 20$ in each class] without additional computations.

If we select $\alpha = 0.05$ for the significance level of this test, then from Table B.6 in Appendix B, the critical tabular χ^2-value is $\chi^2_{0.05;5} = 11.1$. The decision rule is:

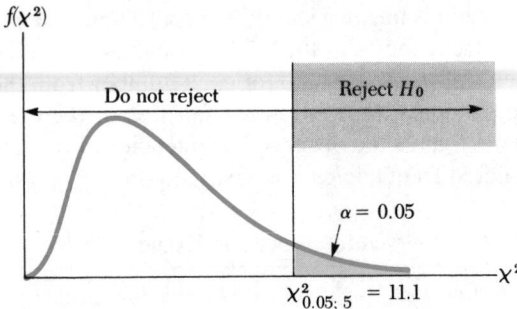

Since $\chi^2 = 16.9 > 11.1$, we reject the null hypothesis that the die is fair and conclude that the die is not fair—it appears to be "loaded" to produce more 3s and 4s than the other values.

The p-value for the chi-square tests in this chapter is given by $P(\chi^2 > \chi^2/df)$, where χ^2 is the calculated χ^2 value for the test and df is the degrees of freedom. Frequently, it will be necessary to interpolate in Table B.6 in Appendix B to calculate the p-value. Alternatively, we can state approximately that the p-value is between, say, 0.05 and 0.025 to avoid interpolation.

For this test, the p-value is given by:

$$p\text{-value} = P(\chi^2 > 16.9/df = 5)$$

Since $P(\chi^2 > 16.7/df = 5) = 0.005$ and $P(\chi^2 > 20.5/df = 5) = 0.001$ from Table B.6, we may conclude that the p-value is between 0.005 and 0.001 (by interpolation, it is 0.0048). Thus, by rejecting the null hypothesis, the probability of committing a Type I error (reject H_0/H_0 is true) is quite small—0.0048. Accordingly, we can be confident that we have made the correct decision by rejecting H_0. The chance that we have made an error is quite remote.

Now, let us make some observations concerning the use of the chi-square goodness-of-fit test in this example. First, the test is equivalent to testing the null hypothesis that the frequencies of the die numbers are uniformly distributed. In Chapter 6, we studied the discrete uniform distribution, the probability mass function of which is given by:

$$P[X = x_i] = \frac{1}{k}, \qquad i = 1, 2, 3, \ldots, k$$

In this example, $k = 6$ and the possible values of the random variable X are $x_1 = 1$, $x_2 = 2$, $\ldots$, $x_6 = 6$. A stick diagram for this discrete uniform distribution is illustrated in Figure 19.3. We can conclude, therefore, that the

FIGURE 19.3 Discrete uniform distribution

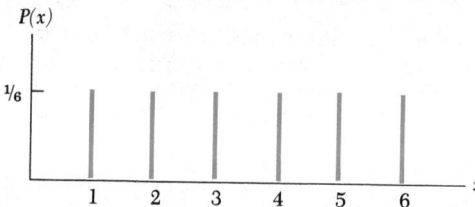

observed frequency distribution does not fit well to the discrete uniform distribution. In this sense, we are testing the goodness of fit of the observed frequency distribution to a specified theoretical probability distribution in Example 19.1.

Our second observation is that this test is related to testing the hypothesis that the population proportions corresponding to the number of dots on the die faces are equal. Define π_i = population proportion of i's, or probability that the number of dots i appears face up. The hypotheses we have tested are equivalent to the hypotheses:

$$H_0: \quad \pi_1 = \pi_2 = \cdots = \pi_6$$
$$H_A: \quad \text{At least one } \pi_i \text{ is different}$$

When used in a situation like Example 19.1, the chi-square test can be considered as an *approximate* test of the equivalence of a set of population proportions. We would conclude in this example that the population proportions do not appear to be equal; the proportions of 3s and 4s appear to be greater than the other proportions.

A final note—the chi-square test must always be based on frequencies, not proportions. In Example 19.1, we could not compute the chi-square statistic as the squared difference between the observed and expected proportions, divided by the expected proportion. If proportions were used, the resulting chi-square statistic would be too small (by a factor of $1/n$).

There is one requirement for the appropriate use of the chi-square tests— each class should have an expected frequency of at least five. This requirement is related to the fact that the chi-square statistic is approximately chi-square distributed. If the expected frequency becomes small (less than five in one or more classes, the quality of the approximation suffers. Although the ''at least five'' minimum expected frequency is not a hard and fast rule (some statisticians use three, and a few use one as the minimum), we shall adopt it in this text. What if a class has less than five as an expected frequency? The solution, as we will see, is to group classes until the minimum of five is achieved.

Example 19.2   The number of service requests per hour in a plumber's shop is thought to be Poisson distributed with an average of three requests per hour. One hundred randomly selected hourly periods during a four-week interval produced the demands for service given in Table 19.3.

Let the random variable X = number of service requests per hour. Since we are assuming that the average number of requests per hour is three, the Poisson parameter λ equals three and the appropriate probability mass

TABLE 19.3   Number of requests for plumbing service per hour

Number of requests per hour	Number of hourly periods having the number of requests per hour
0	3
1	17
2	28
3	16
4	18
5	9
6	2
7	5
8	1
9	1
10	0
More than 10	0
Total	100

function is:

$$P(X = x) = \frac{e^{-\lambda}\lambda^x}{x!} = \frac{e^{-3}(3)^x}{x!}, \quad x = 0, 1, 2, \ldots$$

To determine the expected frequency in the first class, 0 requests per hour, we first compute the probability that $X = 0$, using Table B.2 in Appendix B to find the value of e^{-3}.

$$P(X = 0) = \frac{e^{-3}3^0}{0!} = e^{-3} = 0.0498$$

The expected number of hourly periods out of 100 sampled hours in which 0 service calls will be experienced is $E_1 = (100)(0.0498) = 4.98$ hourly periods.

Table 19.4 gives the computation of the value of the chi-square statistic for these data. Notice that the first two classes and the last six classes must be combined to produce *expected* frequencies of five or more. The degrees of freedom are $k - 1 - m = 6 - 1 - 0 = 5$; there are $k = 6$ classes actually used in the computation of the value of the chi-square statistic, and $m = 0$ since no parameters had to be estimated from the data (the value of the parameter λ needed to calculate the expected frequencies is given—$\lambda = 3$).

The hypotheses are

H_0: The data are Poisson-distributed with a mean of $\lambda = 3$ requests per hour.

H_A: The data are not Poisson-distributed with a mean of $\lambda = 3$ requests per hour.

To test the null hypothesis at the $\alpha = 0.10$ level, the decision rule is:

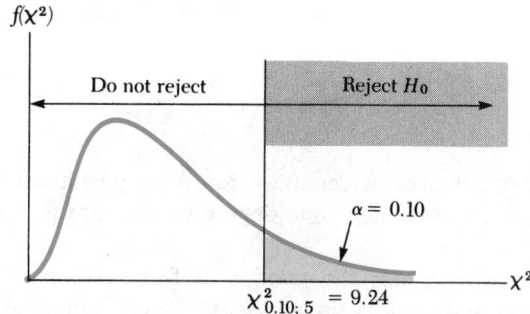

Since $\chi^2 = 3.48 < 9.24$, we cannot reject H_0 and must conclude, therefore, that the data conform reasonably well to a Poisson distribution with a mean of three.

From Table B.6 in Appendix B, since $P(\chi^2 < 1.61/df = 5) = 0.10$ and $P(\chi^2 > 9.24/df = 5) = 0.10$, and the table does not give values of χ^2 between 1.61 and 9.24, we can only say that the p-value is between 0.10

and 0.90. Since its value appears to be closer to 0.10, we can be quite confident that we have made a good decision to not reject H_0.

Note that the number of classes k used to determine the degrees of freedom is equal to the number of classes actually used to compute the chi-square statistic value (6), *not* the original number of classes (12).

In Example 19.2, the Poisson parameter λ is specified ($\lambda = 3$), but frequently this is not the case. We will wish to determine whether a set of data conforms to a theoretical probability distribution in which the parameters have not been specified. What if λ had not been specified in Example 19.2? How, then, could we compute the expected frequencies? One answer is to estimate the parameter from the data. For example, an estimate of the average number of service requests (λ) from the data in Table 19.3 is $^{302}/_{100} = 3.02$. The estimate may be used in the computation of the probabilities as in Table 19.4 in place of the parameter, λ. The consequence of estimating λ is

TABLE 19.4 Computation of the chi-square statistic for Example 19.2

x	O_i	$P(X = x)$	E_i	$O_i - E_i$	$\dfrac{(O_i - E_i)^2}{E_i}$
0.......	3 ⎫ 20	$e^{-3}3^0/0! = 0.0498$	4.98 ⎫ 19.92	0.08	0.000
1.......	17 ⎭	$e^{-3}3^1/1! = 0.1494$	14.94 ⎭		
2.......	28	$e^{-3}3^2/2! = 0.2241$	22.41	5.59	1.394
3.......	16	$e^{-3}3^3/3! = 0.2241$	22.41	-6.41	1.833
4.......	18	$e^{-3}3^4/4! = 0.1681$	16.81	1.19	0.084
5.......	9	$e^{-3}3^5/5! = 0.1009$	10.09	-1.09	0.118
6.......	2 ⎫	$e^{-3}3^6/6! = 0.0504$	5.04 ⎫		
7.......	5 ⎪	$e^{-3}3^7/7! = 0.0216$	2.16 ⎪		
8.......	1 ⎬ 9	$e^{-3}3^8/8! = 0.0081$	0.81 ⎬ 8.36	0.64	0.049
9.......	1 ⎪	$e^{-3}3^9/9! = 0.0027$	0.27 ⎪		
10.......	0 ⎭	$e^{-3}3^{10}/10! = 0.0007$	0.07 ⎪		
More than 10..	0	0.0001*	0.01 ⎭		
Total	100	1.000	100.00		$\chi^2 = 3.480$

* By subtraction, because the probabilities must total 1.

the loss of a degree of freedom—for every probability distribution parameter that must be estimated, one degree of freedom in the chi-square statistic is lost.

We now consider the goodness-of-fit test for normality, an extremely important and useful application of the chi-square statistic.

Example 19.3 A firm's reliability expert believes an electric component used in a switching device has a lifetime that is normally distributed. A sample of 500 component lifetimes provides the data in Table 19.5. The reliability expert determines that the sample average and sample standard deviation of the 500 observations are $\bar{x} = 100$ and $s = 20$, respectively.

The assumed form of the normal distribution is based on the estimates $\bar{x} = 100$ and $s = 20$, since the parameters μ and σ have not been specified. This

TABLE 19.5 Frequency distribution of 500 component lifetimes

Lifetime in hours	Number of components
40 but under 60	25
60 but under 80	50
80 but under 100	225
100 but under 120	125
120 but under 140	75
	500

FIGURE 19.4 Assumed theoretical distribution for component lifetimes

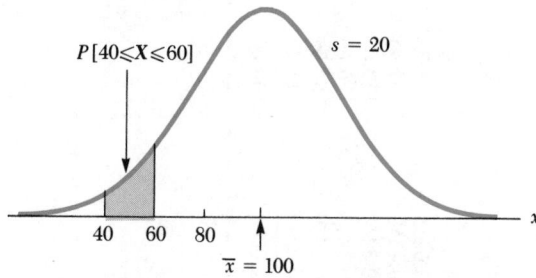

$P[40 \leqslant X \leqslant 60]$ $s = 20$

40   60   80 $\bar{x} = 100$ x

normal distribution is illustrated in Figure 19.4. To find the expected frequency for, say, the class 40 but under 60, we would first find the probability that a component lifetime **X** lies between 40 and 60 hours. This probability is equal to the shaded area in Figure 19.4. By standardizing x, $z = (x - \bar{x})/s$, this area is equivalent to the shaded area in the standard normal distribution:

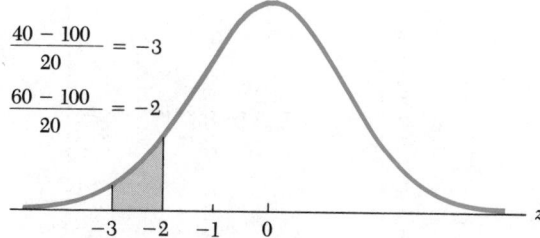

$$\frac{40 - 100}{20} = -3$$

$$\frac{60 - 100}{20} = -2$$

-3 -2 -1 0 z

From Table B.3 in Appendix B, this area is equal to 0.0215. Thus, P(40 $\leq$ X $\leq$ 60) $\doteq$ 0.0215 if the data are normally distributed (this probability is approximate, because we used $\bar{x}$ for μ and s for σ). The expected frequency in this class is given by (500)(0.0215) = 10.75. The chi-square computations for these data are given in Table 19.6. Notice that two open-ended classes must be added to accommodate the tails of the nor-

TABLE 19.6 Chi-square computations for Example 19.3

Lifetime (in hours)	O_i	Probability	E_i	$O_i - E_i$	$\dfrac{(O_i - E_i)^2}{E_i}$
Less than 40	$\left.\begin{array}{r}0\\25\end{array}\right\}25$	$P(X < 40) = 0.0013$	$\left.\begin{array}{r}0.65\\10.75\end{array}\right\}11.40$	13.60	16.225
40 but under 60		$P(40 \leq X < 60) = 0.0215$			
60 but under 80	50	$P(60 \leq X < 80) = 0.1359$	67.95	−17.95	4.742
80 but under 100	225	$P(80 \leq X < 100) = 0.3413$	170.65	54.35	17.310
100 but under 120	125	$P(100 \leq X < 120) = 0.3413$	170.65	−45.65	12.212
120 but under 140	75	$P(120 \leq X < 140) = 0.1359$	67.95	7.05	0.731
Over 140	0	$P(X \geq 140) = 0.0228$	11.40	−11.40	11.400
Total	500	1.0000	500.00		$\chi^2 = 62.620$

mal distribution, though one is lost (the first) because its expected frequency is less than five.

H_0:   The data are normally distributed
H_A: The data are not normally distributed

The degrees of freedom are given by $k - 1 - m = 6 - 1 - 2 = 3$. Six classes are used to calculate the chi-square statistic value and two degrees of freedom are lost due to the estimation of μ and σ by $\bar{x}$ and s, respectively. If we select the $\alpha = 0.01$ significance level, the decision rule is:

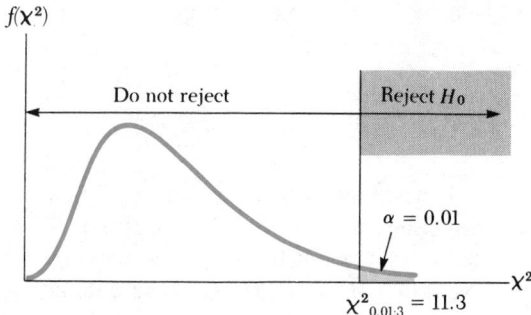

Since $\chi^2 = 62.62 > 11.3$, we would reject H_0 and conclude that the data are not normally distributed. This is evident from studying the frequency distribution of observed responses in Table 19.5—the data are definitely skewed toward the smaller lifetimes. Since $P(\chi^2 > 16.3/df = 3) = 0.001$ from Table B.6, we know that p-value = $P(\chi^2 > 62.62)$ is *much* less than 0.001. Therefore, we can be very confident that we have made the right decision in rejecting H_0.

The use of the chi-square statistic to test the goodness of fit of data to the normal distribution has one serious drawback—the data first must be grouped into a frequency distribution before the test can be conducted. Unfortunately, the chi-square statistic may be affected by how the frequency distribution is constructed (the number of classes, class widths, etc). As a general rule, construct the frequency distribution with equal-width intervals and select a sufficient number of intervals to cover the range of the data (usually five to ten will do), subject to the condition that each class has an expected frequency of five or more.

Chapter 20 describes alternative goodness-of-fit tests to the chi-square test. These tests do not require placing the data into frequency distributions, and typically they are more powerful than the chi-square test.

▮ 19.3   TESTS FOR INDEPENDENCE: CONTINGENCY TABLES

In Chapter 4, we introduced the concept of a *contingency table,* a table used to display the dependence of two or more variables on one another. We repeat the initial Chapter 4 contingency table example here.

Example 19.4 A survey of 100 patients at a local Veterans Administration hospital is conducted to determine whether there is a connection between smoking and lung cancer. Each patient is classified as a smoker or non-smoker and as having lung cancer or not having lung cancer. The resulting frequencies are given in the contingency table.

Smoker	Lung cancer Yes	No	Totals
Yes	15	25	40
No	5	55	60
Totals	20	80	100

In Chapter 4, we showed that the events, S (smoker) and C (lung cancer) are statistically dependent events; that is, $P(S \cap C) = 0.15 \neq P(S) \cdot P(C) = (0.4)(0.2) = 0.08$. This result follows from viewing the 100 VA patients as a *population*. We now wish to consider these data as a *sample* from which we may draw inferences concerning the dependence of the two categories, smoking and cancer. Let π_{ij} be the population proportion in the (i, j)th cell. For example, π_{12} = proportion of persons who are smokers ($i = 1$; first row of the table) and who do not have lung cancer ($j = 2$; second column of the table). Let π_i^r and π_j^c represent the marginal probability distributions for the smoking category and the lung cancer category, respectively. The null hypothesis of independence is:

$$H_0: \quad \pi_{ij} = (\pi_i^r)(\pi_j^c); \quad i = 1, 2, \quad j = 1, 2$$

and the alternate hypothesis is:

$$H_A: \quad \pi_{ij} \neq (\pi_i^r)(\pi_j^c) \text{ for at least one pair } (i, j)$$

That is, if the two categories are independent in the population, then any cell probability (π_{ij}) should be the product of the ith row and jth column marginal probabilities ($\pi_i^r \pi_j^c$).

Let p_{ij} be the sample proportion in the (i, j)th cell, and p_i^r and p_j^c be the row and column sample proportions, respectively. Table 19.7 gives these sample proportions. Also shown in Table 19.7 are the sample expected cell proportions if the two categories are independent in the population. The discrepancies between p_{ij} and $p_i^r p_j^c$ certainly suggest that the two categories are not statistically independent in the population.

The chi-square test for independence is based on the difference between the *observed frequencies* and the *expected frequencies* if the null hypothesis, $H_0: \quad \pi_{ij} = (\pi_i^r)(\pi_j^c)$, is true. The fundamental idea of this test is, therefore, the same as the chi-square goodness-of-fit tests. Let O_{ij} be the observed frequency and E_{ij} be the expected frequency in the (ij)th cell,

TABLE 19.7 Sample proportion and the expected sample cell proportions (in parentheses) assuming independence

Smoker \ Lung cancer	Yes	No	Totals
Yes	$p_{11} = 0.15$ $(p_1^r p_1^c = 0.08)$	$p_{12} = 0.25$ $(p_1^r p_2^c = 0.32)$	$p_1^r = 0.40$
No	$p_{21} = 0.05$ $(p_2^r p_1^c = 0.12)$	$p_{22} = 0.55$ $(p_2^r p_2^c = 0.48)$	$p_2^r = 0.60$
Totals	$p_1^c = 0.20$	$p_2^c = 0.80$	1.00

respectively. The expected cell frequencies are given by

$$E_{ij} = np_i^r p_j^c$$

To find the expected frequency in the (ij)th cell, we multiply the marginal sample proportion in the ith row by the marginal sample proportion in the jth column—this gives the estimated cell proportion, assuming the categories are independent—and multiply this proportion by the sample size to produce a frequency. The observed and expected frequencies for these data are given in Table 19.8.

TABLE 19.8 Observed and expected frequencies

Smoker \ Lung cancer	Yes	No	Totals
Yes	$O_{11} = 15$ $E_{11} = (0.4)(0.2)(100)$ $= 8$	$O_{12} = 25$ $E_{12} = (0.4)(0.8)(100)$ $= 32$	40
No	$O_{21} = 5$ $E_{21} = (0.6)(0.2)(100)$ $= 12$	$O_{22} = 55$ $E_{22} = (0.6)(0.8)(100)$ $= 48$	60
Totals	20	80	100

The value of the chi-square statistic is:

$$\chi^2 = \sum_{i=1}^{a} \sum_{j=1}^{b} \left[\frac{(O_{ij} - E_{ij})^2}{E_{ij}} \right]$$

where a and b are the number of rows and columns in the contingency table, respectively. The degrees of freedom for this chi-square statistic are $df =$

$(a - 1)(b - 1)$. The degrees of freedom for the test are determined as follows:

$$df = \text{(Number of cells)} - 1 - \text{(Number of estimated parameters)}$$

This is similar to the way the degrees of freedom are determined for the chi-square goodness-of-fit test. One degree of freedom is lost in specifying the sample size. The number of estimated parameters is $(a + b - 2)$; we must estimate π_{ij} for each cell and its point estimate is $p_i^r p_j^c$ under the null hypothesis. Since $\Sigma_{i=1}^a p_i^r = 1$ for the sample proportions, only $(a - 1)$ must be determined—this follows because the sum of all row sample proportions must be 1. Similarly, only $(b - 1)$ column sample proportions must be determined. Therefore, the degrees of freedom are:

$$df = (ab) - 1 - [(a - 1) + (b - 1)] = (a - 1)(b - 1)$$

For these data,

$$\chi^2 = \frac{(15 - 8)^2}{8} + \frac{(25 - 32)^2}{32} + \frac{(5 - 12)^2}{12} + \frac{(55 - 48)^2}{48}$$

$$= 6.125 + 1.531 + 4.083 + 1.021 = 12.76$$

From Table B.6 in Appendix B, $\chi^2_{\alpha=0.05; df=1} = 3.84$ with $\alpha = 0.05$, $df = (2 - 1)(2 - 1) = 1$. The decision rule is:

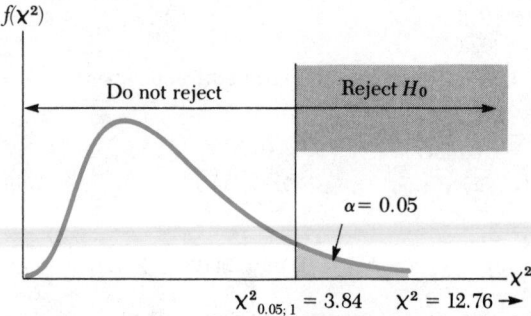

Since $\chi^2 = 12.76 > 3.84$, we reject the null hypothesis and conclude that in the population from which these data are drawn, there appears to be a significant statistical dependence between the two categories in the contingency table, smoking and lung cancer. Since $P(\chi^2 > 10.8/df = 1) = 0.001$ from Table B.6, we know that the p-value $= P(\chi^2 > 12.76/df = 1) < 0.001$. Therefore, we can be confident that we have made the right decision in rejecting H_0.

An important point to remember from our discussions of statistical dependence in Chapters 4 and 13 is that dependence does not imply cause and effect. These data suggest that in this population, there appears to be a greater proportion of lung cancer victims among smokers than among nonsmokers. This relationship may be caused, however, by some other factor, say a physiological one, that increases a person's chance of contracting lung cancer and of becoming a smoker.

An additional point—the expected cell frequencies can be determined directly from the row and column frequencies and the sample size. For example, the expected frequency in the (1, 1) cell may be determined by

$$(\cancel{100})\left(\frac{40}{\cancel{100}}\right)\left(\frac{20}{100}\right) = \frac{(40)(20)}{100} = 8$$

That is, the expected frequency is determined by the product of the first-row and first-column marginal frequencies, divided by the sample size. We now summarize the chi-square test for independence for two categories.

Summary of the chi-square test for independence, given a two-way contingency table

Category B

	1	2	$\cdots$	b	
1	O_{11}	O_{12}	$\cdots$	O_{1b}	R_1
2	O_{21}	O_{22}	$\cdots$	O_{2b}	R_2
$\cdot$	$\cdot$			$\cdot$	$\cdot$
a	O_{a1}	O_{a2}	$\cdots$	O_{ab}	R_a
	C_1	C_2	$\cdots$	C_b	T

Category A labels the rows.

where

T = total sample size
O_{ij} = observed frequency in (i,j)th cell
R_i = total of ith row
C_j = total of jth column
a = number of rows
b = number of columns

The value of the chi-square test statistic for the hypotheses

H_0: Category A is independent of category B
H_A: Categories A and B are statistically dependent

is

$$\chi^2 = \sum_{i=1}^{a} \sum_{j=1}^{b} \left[\frac{(O_{ij} - E_{ij})^2}{E_{ij}}\right]$$

where

$$E_{ij} = \frac{(R_i)(C_j)}{T}$$

$$df = (a - 1)(b - 1)$$

Example 19.5 In a certain population, a psychologist wishes to determine whether salary is affected by level of education attained. Based on a random sample of 150 persons in this population, she forms the contingency table shown here.

Educational level \ Salary (in $000s)	0–4.99	5–9.99	10–14.99	Above 15	Totals
High school education or less	10	10	14	16	50
High school education plus some college training	10	20	28	17	75
College degree or greater (some graduate work)	0	0	18	7	25
Totals	20	30	60	40	150

Are the two categories, salary and education level, statistically independent in this population?

Solution We first determine the expected frequencies.

Education category \ Salary category	1	2	3	4	Totals
1	$\frac{(50)(20)}{150} = 6.67$	$\frac{(50)(30)}{150} = 10$	$\frac{(50)(60)}{150} = 20$	$\frac{(50)(40)}{150} = 13.33$	$R_1 = 50$
2	$\frac{(75)(20)}{150} = 10$	$\frac{(75)(30)}{150} = 15$	$\frac{(75)(60)}{150} = 30$	$\frac{(75)(40)}{150} = 20$	$R_2 = 75$
3	$\frac{(25)(20)}{150} = 3.33$	$\frac{(25)(30)}{150} = 5$	$\frac{(25)(60)}{150} = 10$	$\frac{(25)(40)}{150} = 6.67$	$R_3 = 25$
Totals	$C_1 = 20$	$C_2 = 30$	$C_3 = 60$	$C_4 = 40$	$T = 150$

The value of the chi-square statistic is:

i	j	O_{ij}	E_{ij}	$O_{ij} - E_{ij}$	$(O_{ij} - E_{ij})^2/E_{ij}$
1	1	10	6.67	3.33	1.66
1	2	10	10.00	0.00	0.00
1	3	14	20.00	-6.00	1.80
1	4	16	13.33	2.67	0.53
2	1	10	10.00	0.00	0.00
2	2	20	15.00	5.00	1.67
2	3	28	30.00	-2.00	0.13
2	4	17	20.00	-3.00	0.45
3	1	0	3.33	-3.33	3.33
3	2	0	5.00	-5.00	5.00
3	3	18	10.00	8.00	6.40
3	4	7	6.67	0.33	0.02
					$\chi^2 = 20.99$

If we conduct the test at the $\alpha = 0.05$ level, the decision rule is:

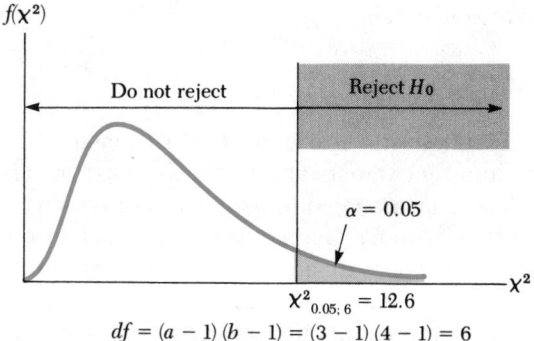

$$df = (a - 1)(b - 1) = (3 - 1)(4 - 1) = 6$$

Since $\chi^2 = 20.99 > 12.6$, we would reject H_0 and conclude that the salary of a person in this population is affected by his or her education level. Since $P(\chi^2 > 18.5/df = 6) = 0.005$ and $P(\chi^2 > 22.5/df = 6) = 0.001$, the p-value lies between 0.001 and 0.005 (by interpolation, p-value $= 0.002$). Thus, by rejecting H_0, we have incurred a small risk that we have made an error.

As with the chi-square goodness-of-fit test, there are two cautions that should be exercised in using the chi-square statistic for tests of independence. First, the expected cell frequencies should be greater than or equal to five. In Example 19.5, $E_{31} = 3.33$. To correct this problem, either we must increase the sample size with the hope that E_{31} will reach five or more, or we must change the classifications for the salary category. By extending the first class beyond 4.99 (thousand) and readjusting the subsequent classes as necessary, we can ensure that all $E_{ij} \geq 5$. The second caution is that this is an *approximate* test for independence. It is best suited when both factors (categories) are *qualitative*. When one or both factors are quantitative, the chi-square statistic does not exploit the numerical character of the factor. For instance, in Example 19.5, salary is quantitative and its specific numerical values in the sample data have been lost by placing these values into four salary categories. Other techniques that directly use the numerical character of salary may be preferable to chi-square in this case. As an example, we could consider the average salary in the population μ_i for each education level $i = 1, 2, 3$, and test the hypothesis H_0: $\mu_1 = \mu_2 = \mu_3$ to determine whether the mean salary is affected by education level. The analysis of variance procedures in Chapter 12 would be well suited for this approach. To repeat the point, if one or both factors are quantitative, methods that exploit the numerical character of the factor(s) should be considered before using chi-square.

Finally, it is possible to develop exact tests for the 2×2 contingency table and to "correct" for the approximation in the general two-way table. The interested reader is directed to the Snedecor and Cochran reference at the end of this chapter for information on this technique and further uses of the chi-square statistic.

■ 19.4 SUMMARY

The chi-square statistic lends itself to constructing *approximate* methods for two important statistical hypothesis tests: the test for goodness of fit of sample data to an assumed probability distribution, and the test for independence of two or more factors. It is important to remember that these chi-square tests are approximate, and before using them it should be ascertained that "better" procedures, if they exist, are not feasible in a given problem.

These tests are commonly referred to as "nonparametric" or "distribution-free" statistical tests. Unlike the tests in Chapters 9–11, these do not assume that the population random variable is distributed according to any specified distribution, such as the normal distribution whose form is controlled by the parameters μ and σ. Rather, the only restriction is that the counts in each cell should be five or greater. Additional nonparametric tests are presented in Chapter 20.

■ REFERENCES

Hamburg, Morris. *Statistical Analysis for Decision Making*. New York: Harcourt, Brace & World, Inc., 1970, pp. 413–37.

Pearson, Karl. "On the criterion that a given system of deviations from the probable in the case of a correlated system of variables is such that it can be reasonably supposed to have arisen from random sampling." *Philosophy Magazine*, series 5, vol. 50 (1900), p. 157.

Rao, C. R. *Linear Statistical Inference and Its Application*. New York: John Wiley & Sons, Inc., 1965, pp. 325–46.

Snedecor, G. W., and Cochran, W. G. *Statistical Methods*. 7th ed. Ames: Iowa State University Press, 1980, pp. 124–127; 194–210.

■ PROBLEMS

19.1. Persons arriving at a service facility are believed to arrive randomly; that is, the number of arrivals in a fixed interval of time, say five minutes, should be uniformly distributed over a set of intervals. Five five-minute intervals are selected, and the following frequencies of the numbers of persons arriving in these intervals are recorded:

Five-minute interval	Number of arrivals
1	20
2	15
3	23
4	14
5	28
Total	100

Are the frequencies of arrivals uniformly distributed over the five intervals? Use $\alpha = 0.01$. What is the p-value for this test?

19.2. A set of four coins is tossed 960 times. The number of times that 0, 1, 2, 3, and 4 heads were obtained is given in the table.

Number of heads	Number of times
0	50
1	230
2	330
3	290
4	60
	960

Assume the random variable X = number of heads when four coins are tossed follows the binomial distribution.

$$P(X = x) = C_x^4(\tfrac{1}{2})^x(1 - \tfrac{1}{2})^{4-x},$$
$$x = 0, 1, 2, 3, 4$$

Do the data suggest that these four coins allow us to use this probability distribution for X? Use $\alpha = 0.05$. What is the p-value for this test?

19.3. The marketing research firm for a large foods manufacturer is interested in determining whether there is any preference for the brand of vanilla ice cream purchased at the grocery store in a certain region. It is believed that among four brands, customers do not express a preference. That is, the four brands have approximately an equal proportion of this vanilla ice cream market. Samples of each vanilla ice cream are given to 100 randomly selected customers and they are asked to identify the one they like best. The data are recorded in the table.

Brand of ice cream	Customers preferring
A	38
B	20
C	20
D	22
Total	100

Does it appear that customers in this vanilla ice cream market have no brand preference? Use $\alpha = 0.05$. What is the p-value for this test?

19.4. It is believed that the average number

of service calls received at a service facility is two per every 15-minute interval. One hundred 15-minute intervals are sampled, and the frequencies of the numbers of calls are recorded in the table.

Number of calls	Number of intervals
0	20
1	50
2	10
3	10
4	5
5	3
6 or more	2
Total	100

Do these data fit to a Poisson distribution with a mean of two calls per each 15-minute interval? Calculate the p-value for this test. Should you reject or not reject H_0? Explain.

19.5. It is believed that the number of persons entering a store per hour is Poisson distributed. Two hundred hourly periods when the store is open are randomly chosen, and the number of persons entering the store in each period is recorded in the table.

Persons per hour	Hourly periods
0	10
1	30
2	60
3	40
4	20
5	18
6	10
7	2
8	5
9	4
10	1
11 or more	0
Total	200

Do these data support the contention that the random variable X = number of persons entering the store per hour is Poisson distributed? (Note: Use the data to estimate the average number of persons entering the store per hour, λ.) Use $\alpha = 0.05$. What is the p-value for this test?

19.6. In the past year, a company had 850 persons employed on an hourly basis. A frequency distribution of their wages is given in the table.

Hourly wage	Frequency
$2.70–$2.89	68
2.90– 3.09	135
3.10– 3.29	366
3.30– 3.49	187
3.50– 3.69	80
3.70– 3.89	10
3.90– 4.09	4
	850

The company statistician intends to use these data to draw inferences regarding the hourly wage scale for the company. The inference-making methods he plans to use assume that the variable is normally distributed. From the 850 values, he finds that the sample mean and standard deviation are $3.25 and $0.25, respectively. Do these data support the normality assumption, using the *estimated* mean and standard deviation? Calculate the p-value for this test. Should you reject or not reject H_0? Explain.

19.7. The lifetimes in hours of 25 60-watt lightbulbs in quality control tests are:

1,310	1,233	1,203	1,240	759
944	872	1,067	984	1,252
1,248	1,105	956	1,233	1,385
1,262	1,303	985	1,122	1,490
1,234	816	1,173	1,028	1,067

a. Group these data into five classes of equal width.

Lot \ Findings	Defectives	Nondefectives	Totals
1	11	64	75
2	14	36	50
Totals	25	100	125

b. The lifetimes are suppsed to have a mean $\mu = 1,100$ hours and a standard deviation $\sigma = 200$ hours. Using $\mu = 1,100$ and $\sigma = 200$, test the null hypothesis at the $\alpha = 0.10$ level that these data are normally distributed.

19.8. An aptitude test for graduate study in business is designed to have a mean test score of $\mu = 500$ with a standard deviation of $\sigma = 100$. It is assumed that test scores are normally distributed. A random sample of 100 recent test scores produced the frequency distribution shown in the table.

Score	Frequency
Below 300	1
300 but under 350	4
350 but under 400	7
400 but under 450	18
450 but under 500	28
500 but under 550	24
550 but under 600	12
Over 600	6
Total	100

a. Test the hypothesis at the $\alpha = 0.05$ level that these data are normally distributed with a mean $\mu = 500$ and a standard deviation of $\sigma = 100$.

b. The mean and standard deviation of these data are $\bar{x} = 485$ and $s = 75$, respectively. Using the estimated mean and variance, may we now conclude that these data

are normally distributed at the $\alpha = 0.05$ level?

19.9. Two shipment lots of manufactured items are randomly sampled to determine whether the proportion of defective items is different in the two lots. The data are shown in the matrix at the top of the page. Using the chi-square contingency table statistic, test the hypothesis that the proportion of defectives in lot 1, denoted by π_1, is equal to π_2, the proportion of defectives in lot 2. Use $\alpha = 0.05$. What is the p-value for this test?

19.10. In Problem 19.9, test the equivalence of π_1 and π_2 by constructing a 95 percent confidence interval on the difference $\pi_1 - \pi_2$, using the formula from Chapter 11:

$$p_1 - p_2 \pm z_{\alpha/2} \sqrt{\frac{p_1(1 - p_1)}{n_1} + \frac{p_2(1 - p_2)}{n_2}}$$

where $p_1 = {}^{11}/_{75}$, $p_2 = {}^{14}/_{50}$, $n_1 = 75$, and $n_2 = 50$. Compare your answer with the one determined in Problem 19.9.

19.11. Students selected at random from public and private high schools were given a standardized achievement test. The results were as shown in the matrix at the top of the next page.

a. Test the hypothesis at the $\alpha = 0.05$ level that an achievement test score is independent of the kind of school a student attends. What is the p-value for this test?

b. Can you suggest a better way to

Combine to get rid of zero
cells 0-400

School \ Test scores	0–300	301–400	401–500	501–600	601–1,000	Totals
Private	0	5	10	20	15	50
Public	5	15	30	35	15	100
Totals	5	20	40	55	30	150

determine whether achievement scores differ between students at private and public schools?

19.12. Four kinds of automobiles (A, B, C, and D) were shown to 200 married couples. Husbands and wives were then asked separately to select their first choice among the four cars. The data are shown in the matrix.

Husband's choice \ Wife's choice	A	B	C	D	Totals
A	10	18	10	20	58
B	16	12	14	2	44
C	0	22	20	7	49
D	8	16	7	18	49
Totals	34	68	51	47	200

Are the two choices independent factors? Calculate the p-value for this test. Should you reject or not reject the null hypothesis? Explain.

19.13. A consulting firm is asked to determine whether the preference for one of three weekly news magazines (A, B, and C) is dependent on the region in the United States (South, West, North, East). One thousand persons randomly sampled from these four geographic regions were asked to state their first choice among the three magazines. The data are shown in the matrix at the bottom of this page. Do these data suggest that the preference is dependent on geographic region? Calculate the p-value for this test. Should you reject or not reject H_0? Explain.

19.14. A company is concerned that the length of its training program has no influence on later job ratings of its production-line workers. One hun-

Region \ Magazine	A	B	C	Totals
North	150	75	25	250
South	100	100	50	250
East	150	50	50	250
West	125	75	50	250
Totals	525	300	175	1,000

Ratings Training	Excel- lent	Aver- age	Poor	Totals
One week	15	20	15	50
Two weeks	15	25	10	50
Three weeks	20	20	10	50
Totals	50	65	35	150

dred fifty new workers are chosen and are randomly divided into three groups of 50 each. The first group receives one week of training, the second group receives two weeks, and the third group receives three weeks. Three months later, all 150 workers are rated on a three-point scale, "Excellent," "Average," and "Poor." The data are shown in the matrix at the top of the page. Do the ratings and amount of training appear to be statistically independent? Use $\alpha = 0.05$.

19.15. A manufacturer produces units of a product in three shifts: day, evening, and night. Quality control teams check the production lots for defects at the end of each shift. The teams have compiled the data in the matrix at the bottom of the page. Do these data suggest that the reliability of the produced units is dependent on which shift produces the unit? Use $\alpha = 0.05$. What is the p-value for this test?

19.16. It is suspected that the number of accidents per year is dependent on the age of drivers. One thousand drivers who are insured by a major insurance company are randomly sampled. The contingency table at the top of the next page is formed. At the $\alpha = 0.05$ level of significance, is age independent of the number of annual accidents? (Although two cells have expected frequencies less than five, do not regroup the categories—use the expected frequencies that are less than five.) What is the p-value for this test?

19.17. In a famous experiment by the geneticist Mendel, when two types of peas are crossed, the following types should occur in the proportion $9:3:3:1$—

Defects Shift	Major	Minor	None	Totals
Day	10	40	450	500
Evening	10	30	260	300
Night	10	30	160	200
Totals	30	100	870	1,000

Number of accidents \ Age	16–21	22–30	31–40	41–50	51–70	Totals
0	170	250	170	120	30	740
1	50	40	20	25	5	140
2	30	25	5	25	5	90
More than 2	10	5	5	10	0	30
Total	260	320	200	180	40	1,000

round and yellow, round and green, wrinkled and yellow, and wrinkled and green. In a cross producing 1,000 peas, the following frequencies were observed:

Type	Frequency
Round and yellow........	575
Round and green.........	193
Wrinkled and yellow......	182
Wrinkled and green.......	50
Total...................	1000

At the $\alpha = 0.05$ significance level, do these data appear to follow Mendel's proportions?

Nonparametric statistics 20

In Chapters 9–15, the statistical inference methods described all have a common thread—a population parameter (μ, σ^2, or ρ, for instance) is identified about which we wish to draw inferences, a random sample is collected, a point estimator of the parameter ($\bar{X}$, S^2, or r) is selected, and its sampling distribution is used to construct hypothesis-testing decision rules or confidence interval formulas. These methods commonly are called *parametric* statistics, for they require the identification of a population parameter, a point estimator, and its sampling distribution. Most of the parametric statistical methods we have described in this text depend on knowing the form of the sampling distribution of the point estimator of the parameter to be estimated. That is, the *parameters* of the sampling distribution must be specified, usually by producing estimates of the parameters from the sample. In fact, we usually assume that the sampling distribution is the normal distribution, or at least approximately so (via the central limit theorem).

In this chapter, we will describe a number of so-called *nonparametric* methods—methods that may be appropriate when one or more of the parametric method (see Chapters 9–15) assumptions are not satisfied by the data. The determination of the p-value for tests of hypotheses using these methods can be extremely difficult due to the nature of the significance tables. Thus, the classical method of testing will be used in which the significance level α is pre-set, thus determining the decision rule for the test. In most instances, we will give only a bound on the p-value (e.g., p-value $>$

661

0.001) because of the difficulties associated with interpolation in the nonparametric test tables.

To properly characterize the circumstances for which nonparametric methods may be appropriate, we first describe the four basic scales of measurement.

■ 20.2 SCALES OF MEASUREMENT

Values of a random variable may belong to any of four measurement scales: *nominal, ordinal, interval,* or *ratio.* We will describe these scales in turn, from the "weakest" (the nominal scale) to the "strongest" (the ratio scale).

The *nominal* scale uses numbers only to name categories to which the observations belong. For example, consider the qualitative variable, sex. We may use a 1 for male and a 0 for female, but clearly the number assignment is arbitrary, for the numbers serve only as category names. We could just as well have used 100 to name the female category and 0 to name the male category.

The *ordinal* scale uses numbers for measurements, where the ordering of the numbers is relevant. For example, we may have designed a survey question that asks the respondent to indicate a preference among three brands of ice cream, where 1 indicates most preferred and 3 indicates least preferred. The ordering of the three numbers (1, 2, and 3) is now relevant, but their magnitude is not. We could have used any three numbers, say 1, 50, and 100, as long as the ordering of the numbers reflects relative preference for the brands of ice cream.

The *interval* scale takes into account the difference between measurements as well as their ordering. An interval scale requires fixing an arbitrary 0 point and a unit distance to measure the difference between measurements. Good examples of interval scales are the Fahrenheit and Celsius temperature scales. These two scales have different 0 points and different unit distances. In general, one interval scale may be transformed to another by changing its scale or the location of 0 or both.

The *ratio* scale applies when the order and distance between measurements are important as in the interval scale, but further requires that the ratio between two measurements is important. For example, if the ratio of one measurement to another is two, then the first is twice as "big" as the second. This requires the ratio scale to have a fixed and *natural* 0 point. Height is a measurement that is properly made on the ratio scale, for example. If we measure height in inches, 0 inches is the natural 0 point, and a person who is 60 inches tall is three times as tall as a person who is 20 inches tall. The unit distance on the ratio scale is arbitrary. For heights, we could use one inch or one foot, for example.

To determine to which scale a set of measurements belongs, we must take into consideration the nature of the quantities being measured and how the measurements are taken. Are we simply naming categories of a qualitative variable with numbers, or do the numbers themselves have meaning, as in their order, distance among measurements, and/or their ratios?

Most parametric statistical methods require that the measurements belong to at least the interval scale. Based on the four scales of measurement, we now define nonparametric statistical methods as in Conover.[1]

Definition 20.1
Nonparametric statistical methods

A statistical method is called *nonparametric* if it satisfies at least one of the following conditions:

1. It is appropriate for nominally measured data.
2. It is appropriate for ordinally measured data.
3. It is appropriate for interval or ratio scale data, but the population distribution function of the random variable from which the data are taken is unspecified.

The set of statistical procedures used for condition 3 situations in Definition 20.1 have frequently been called "distribution-free" statistical methods—methods that do not depend on the specifications of the probability distribution of the population random variable. Consistent with Definition 20.1, we shall call these *nonparametric methods*.

Nonparametric methods have a number of advantages over parametric methods. First, nonparametric statistics are generally easy to compute, although describing the statistics is often somewhat involved. Second, nonparametric statistics can be applied to data in situations where parametric statistics may not be applicable. This is usually the case when the data measurement scale is nominal or ordinal. Third, nonparametric methods do not assume that the population random variable has a specific probability distribution form. These methods are based on sampling distributions, but the form of the sampling distribution is not dictated by assuming a form for the population probability distribution. Fourth, if a nonparametric method applies to a "weak" scale of measurement, it will typically also apply to all "stronger" scales. For example, if a method is applicable for data measured on the ordinal scale, then it also applies to data measured on the interval and ratio scales.

Nonparametric methods are at a disadvantage compared with parametric procedures when a data set satisfies the assumption of a parametric method. In these cases, the *power* of nonparametric test methods is almost always less than the power of the appropriate parametric test. Therefore, as a general rule, if a parametric test is appropriate for the required statistical

---

[1] W. J. Conover, *Practical Nonparametric Statistics* (New York: John Wiley & Sons, Inc., 1971).

analysis, it should be used over any nonparametric method that may also be appropriate.

We will now consider some of the more important and frequently used nonparametric methods.

■ 20.3 ONE SAMPLE: THE MEDIAN TEST

The median test can be used to test the hypothesis that a set of n randomly drawn measurements came from a population with a specified median. If it is known that the distribution of the population random variable X is symmetric, the test on the median is equivalent to a test on the population mean, because the mean is equal to the median in a symmetric distribution.

Definition 20.2
Median test

Assumptions:

The data consist of n measurements $X_1, X_2, \ldots, X_n$. Let D_i denote the difference between X_i and the hypothesized median M. Then,

1. Each D_i must be a continuous random variable.
2. The distribution of each D_i must be symmetric.
3. The measurements $X_1, X_2, \ldots, X_n$ must represent a random sample from the population distribution.
4. The measurement scale for X must be at least interval.

Hypotheses:

H_0: The population median is M.
H_A: The population median is not M.

Test statistic value:

Determine the differences $d_i = x_i - M, i = 1, 2, \ldots, n$. If any $d_i = 0$, drop it from the set and decrease n by one. Rank the absolute values, $|d_i|$. If ties occur among the ranks, average the ranks of the items involved in the tie and use the average as the rank of each tied item. Each rank is suffixed with the sign of the difference corresponding to it. Let r^+ be the total of the positive ranks, and let r^- be the total of the negative ranks. The value of the test statistic is the smaller of r^+ and r^-, let this number be r.

Decision rule:

Reject H_0 at the α level of significance if r exceeds $W_{1-\alpha/2}$ or r is less than $W_{\alpha/2}$, where $W_{\alpha/2}$ is given in Table B.12 in Appendix B and $W_{1-\alpha/2} = [n(n + 1)/2] - W_{\alpha/2}$. Otherwise, do not reject H_0.

Example 20.1 Ten randomly selected cars of a specific year, make, and model and with similar equipment are subjected to an EPA gasoline mileage test. The resulting miles per gallon are: 24.6, 30.0, 28.2, 27.4, 26.8, 23.9, 22.2, 26.4, 32.6, and 28.8. Using the median test, test the hypothesis at the α = 0.10 level that the population median is 30 miles per gallon.

Solution The measurements, d_i, $|d_i|$, and the ranks are given in the table.

Measurement	d_i	$|d_i|$	Rank	r^+	r^-
24.6	-5.4	5.4	7		7
30.0	0.0	0.0			
28.2	-1.8	1.8	2		2
27.4	-2.6	2.6	3.5		3.5
26.8	-3.2	3.2	5		5
23.9	-6.1	6.1	8		8
22.2	-7.8	7.8	9		9
26.4	-3.6	3.6	6		6
32.6	2.6	2.6	3.5	3.5	
28.8	-1.2	1.2	1		1
			Total:	3.5	41.5

Eliminate $d_2 = 0$

$r^+ = 3.5; r^- = 41.5$

Since r^+ is smaller than r^-, $r = 3.5$ is the value of the test statistic. From Table B.12, $W_{\alpha/2} = W_{0.05} = 9$ with $n = 9$ (eliminating observation 2 because $d_2 = 0$) and $W_{1-\alpha/2} = W_{0.95} = 36$. Since $r < 9$, reject the null hypothesis that the population median is 30 miles per gallon. Since the value of the test statistic ($r = 3.5$) lies between $W_{0.005}$ and $W_{0.01}$, we can conclude that the p-value for the test is between $2(0.005) = 0.01$ and $2(0.01) = 0.02$. (Recall that we must double the lower-tail probabilities to get the p-value for a two-sided test.)

Notice the treatment of the tied ranks for $|d_4| = |d_9| = 2.6$. These two absolute differences tie for ranks 3 and 4; thus, each is assigned the average rank of 3.5. Also note that the first assumption of the test does not allow for ties. Since the d_i values must be continuous, ties should not occur. If the d_i vaues are not continuous, such as in Example 20.1, we can use the test as an approximate test by dealing with the ties in the manner described above.

Two observations are important regarding Example 20.1. First, if we can assume that the distribution of miles per gallon is symmetric, then the median test is equivalent to testing the hypothesis that the mean miles per gallon is 30. Second, these data have been measured on the ratio scale, and the t-test given in Chapter 10 is appropriate if we assume that the distribution of miles per gallon is symmetric and normal or, the sample average is approximately normal (via the central limit theorem). The first problem at the end of this chapter will ask you to test the hypothesis in Example 20.1 by using the t-test.

20.4 TWO INDEPENDENT SAMPLES: THE MANN-WHITNEY TEST

The Mann-Whitney test is the nonparametric analogue of the two independent sample t-test presented in Chapter 11.

Definition 20.3
Mann-Whitney test

Assumptions:

1. $X_1, X_2, \ldots, X_n$ and $Y_1, Y_2, \ldots, Y_m$ are two independent random samples of sizes n and m, respectively.
2. Both samples consist of continuous random variables.
3. The measurement scale is at least ordinal.

Hypotheses:

    H_0: The two random samples have been drawn from the same population distribution.

    H_A: The two random samples have been drawn from different population distributions.

Test statistic value:

Assign ranks 1 to $m + n$ to the combined sample. That is, order all $(m + n)$ observations from the smallest to the largest and assign ranks $1, 2, \ldots, (m + n)$ to these $(m + n)$ ordered observations. Let $r(x_i)$ and $r(y_j)$ denote the ranks assigned to the $x_i, i = 1, 2, \ldots, n$, and to the $y_j, j = 1, 2, \ldots, m$. Let $s = \Sigma_{i=1}^{n} r(x_i)$. The test statistic value is:

$$t = s - \frac{n(n + 1)}{2}$$

Decision rule:

Reject H_0 with significance level α if t is less that $W_{\alpha/2}$ or t is greater than $W_{1-\alpha/2}$, where $W_{\alpha/2}$ is given in Table B.11 in Appendix B and $W_{1-\alpha/2} = nm - W_{\alpha/2}$. Otherwise, do not reject H_0.

The Mann-Whitney test is designed to determine whether two random samples have been drawn from the same or different populations. If we assume that any difference between the two population distributions is due only to the difference in location of the two distributions, then the Mann-Whitney test is equivalent to testing whether or not the two population means are equal. This is similar to the two independent sample t-test, where we must assume that the two population variances are equal.

The Mann-Whitney test is based on the notion that if the two independent random samples have been drawn from the same population, then the average of the sample ranks $r(x_i)$ and $r(y_j)$ should be approximately equal. If the average of the $r(x_i)$ is much greater or smaller than the average of the $r(y_j)$, then this indicates that the two samples likely came from different populations.

Example 20.2 New employees of the ABC Corporation are given a training program to acquaint them with business procedures and principles. Two groups of ten each are selected randomly from a large set of new employees. The first group is trained using method A, and the second group is trained using method B. At the end of the training period, each group is given the same test to determine how much information has been assimilated. The data are given in the following table.

Method A		Method B	
55	81	50	88
70	72	91	84
70	58	90	78
65	67	62	82
62	50	75	80

At the $\alpha = 0.05$ level, do these samples come from the same population?

Solution The ordered combined sample with ranks is shown in the table.

Measurement	Rank	Measurement	Rank
50, 50	1.5	72	11
		75	12
55	3	78	13
58	4	80	14
62, 62	5.5	81	15
		82	16
65	7	84	17
67	8	88	18
70, 70	9.5	90	19
		91	20

The shaded observations are from method A. Notice the treatment of ties. If two observations tie, the rank given to each is the average of the two relevant ranks. If three numbers tie, say for ranks 8, 9, and 10, then each number is assigned the rank of 9—the average of the three ranks. Now,

$$s = \sum_{i=1}^{10} r(x_i) = 1.5 + 3 + 4 + 5.5 + 7 + 8 + 9.5 + 9.5 + 11 + 15 = 74$$

$$t = s - \frac{n(n+1)}{2} = 74 - \frac{(10)(11)}{2} = 19$$

From Table B.11 in Appendix B, $W_{0.025} = 24$ and $W_{0.975} = (10)(10) - 24 = 76$. Since $t = 19 < W_{0.025} = 24$, we reject the null hypothesis that the two samples come from the same population. Since $t = 19$ falls between $W_{0.005}$ and $W_{0.01}$, we can conclude that the p-value lies between $2(0.005) = 0.010$ and $2(0.01) = 0.02$.

If we can assume that the only difference between the two populations is due to location alone, then we may conclude based on the Mann-Whitney test that the two population means differ. Is that a reasonable assumption based on these sample data (see Problem 20.3 at the end of this chapter)?

■ 20.5 TWO MATCHED SAMPLES: THE WILCOXON SIGNED RANK TEST

The Wilcoxon signed rank test is the nonparametric analogue of the parametric paired t-test for matched samples.

Definition 20.4
Wilcoxon signed rank test

Assumptions:

The data consist of n matched pairs, (X_i, Y_i). Let D_i denote the difference between X_i and Y_i for the ith pair. Then,

1. Each D_i must be a continuous random variable,
2. The distribution of each D_i must be symmetric,
3. The pairs, (X_i, Y_i), $i = 1, 2, \ldots, n$, represent a random sample from a bivariate distribution,
4. The measurement scale for the X and Y values is at least interval.

Hypotheses:

H_0: $\mu_x = \mu_y$
H_A: $\mu_x \neq \mu_y$

Test statistic value:

Determine the differences $d_i = x_i - y_i$, $i = 1, 2, \ldots, n$. If any $d_i = 0$, drop it from the set and decrease n by one. Rank the absolute values, $|d_i|$. If ties occur, average the ranks of the items involved in the tie and use the average as the rank of each tied item. Each rank is suffixed with the sign of the difference d_i corresponding to it. Let r^+ be the total of the positive ranks and r^- be the total of the negative ranks. The test statistic value is the smaller of r^+ and r^-; call this number r.

Decision rule:

Reject H_0 at the α significance level if r exceeds $W_{1-\alpha/2}$ or r is less than $W_{\alpha/2}$, where $W_{\alpha/2}$ is given in Table B.12 in Appendix B and $W_{1-\alpha/2} = [n(n + 1)/2] - W_{\alpha/2}$. Otherwise, do not reject H_0.

Example 20.3 Ten employees of a company are randomly selected to determine whether a speed reading program can improve their reading rates. The individuals are given standardized speed reading exams before and after the program. The data in words per minute are shown in the table.

Individual	Before	After
1	200	225
2	150	375
3	300	650
4	400	380
5	120	100
6	250	250
7	320	410
8	175	180
9	100	130
10	500	600

At the $\alpha = 0.05$ significance level, test the null hypothesis that the two population means are equal, using the Wilcoxon signed rank test.

Solution   The differences, ranks of $|d_i|$, r^+, r^-, and r are given in the table.

Individual	d_i	Rank of $|d_i|$	r^+	r^-
1	-25	4		4
2	-225	8		8
3	-350	9		9
4	$+20$⎤		2.5	
5	$+20$⎦	2.5	2.5	
6	0			
7	-90	6		6
8	-5	1		1
9	-30	5		5
10	-100	7		7
		Totals:	$r = 5.0$	40

Eliminate: $d_6 = 0$, $r^+ = 2.5 + 2.5 = 5.0$, and $r^- = 40.0$; $r = 5.0$
· From Table B.12 in Appendix B, $W_{0.025} = 6$ with $n = 9$. $W_{0.975} = [n(n + 1)/2] - W_{0.025} = [9(10)/2] - 6 = 39$. Since $r < W_{0.025}$, reject H_0; the speed reading program appears to be effective. Since $r = 5$ is between $W_{0.01}$ and $W_{0.025}$, we may conclude that the p-value lies between $2(0.01) = 0.02$ and $2(0.025) = 0.05$.

■ 20.6   SEVERAL INDEPENDENT SAMPLES: THE KRUSKAL-WALLIS TEST

The Kruskal-Wallis test is an extension of the Mann-Whitney test when there are more than two populations. It is the nonparametric analogue of the parametric single-factor, completely randomized analysis of variance design discussed in Chapter 12. The data for the Kruskal-Wallis test must be in the following form:

Sample 1	Sample 2	$\cdots$	Sample k
x_{11}	x_{21}		x_{k1}
x_{12}	x_{22}		x_{k2}
.	.		.
.	.		.
.	.		.
x_{1,n_1}	x_{2,n_2}		x_{k,n_k}

The total number of observations is given by $N = \Sigma_{i=1}^{k} n_i$. The test depends on ranks and is similar to the Mann-Whitney test. Assign ranks 1 to N to all N observations when they have been ordered from the smallest to the largest, disregarding from which of the k samples the observations came. Let r_i be the sum of the ranks assigned to the ith sample.

$$r_i = \sum_{j=1}^{n_i} r(x_{ij}); \qquad \bar{r}_i = \frac{r_i}{n_i}$$

where $r(x_{ij})$ = rank assigned to x_{ij}. If the $\bar{r}_i$ values are approximately the same, it supports the null hypothesis that the k samples came from the same population. If the $\bar{r}_i$ values are not the same, then it indicates that one or more populations are likely composed of different values than the rest.

If they occur, ties are treated as in the Mann-Whitney test.

Definition 20.5
Kruskal-Wallis test

Assumptions:
1. The k random samples are mutually independent.
2. All random variables X_{ij} are continuous.
3. The measurement scale is at least ordinal.

Hypotheses:
H_0: The k population distributions are equal.
H_A: At least one population tends to yield different observations than the rest.

Test statistic value:

$$t = \frac{12}{N(N+1)} \sum_{i=1}^{k} \frac{[r_i - (\frac{1}{2})n_i(N+1)]^2}{n_i}$$

Definition 20.5 (continued)

where

$$n_i = i\text{th sample size}, \ N = \sum_{i=1}^{k} n_i$$

$$r_i = \sum_{j=1}^{n_i} r(x_{ij}), \text{ where } i = 1, 2, \ldots, k$$

$$r(x_{ij}) = \text{Rank assigned to observation } x_{ij}$$

Decision rule:

Table B.13 in Appendix B gives critical T-values at exact significance levels α for $k = 3$ and samples up to and including a sample of size five. If $k > 3$ and/or $n_i > 5$ for at least one sample, the χ^2 distribution with $df = k - 1$ may be used to find the approximate critical T value. Reject the null hypothesis if t is greater than the critical value in Table B.13 or from the χ^2 distribution. Otherwise, do not reject H_0. Note that this is a one-sided test.

We will work an example for which Table B.13 in Appendix B is appropriate. The chi-square approximation for the critical T-value appears to be good even if k and n_i are only slightly larger than 3 and 5, respectively. If, for example, $k = 6$ and we set $\alpha = 0.05$, the approximate critical T-value is $\chi^2_{\alpha=0.05:\,k-1=5} = 11.1$, from Table B.6 in Appendix B. In this case, if $t > 11.1$, we reject the null hypothesis.

If the population distributions differ, but only in location, then the null hypothesis in the Kruskal-Wallis test is equivalent to testing the equality of the k population means, $\mu_1, \mu_2, \ldots, \mu_k$.

Example 20.4 A manager wishes to study the production output of three machines, A, B, and C. The hourly output of each machine is measured for five randomly selected hours of operation.

Observation \ Machine	A	B	C
1	25	18	26
2	22	23	28
3	31	21	24
4	26	*	25
5	20	24	32

* Observation lost due to machine failure.

Test the null hypothesis at the $\alpha = 0.05$ significance level that the three population distributions are equal by using the Kruskal-Wallis test.

Solution   The ordered data are shown in the table.

Rank	A	B	C
1..............		18	
2..............	20		
3..............		21	
4..............	22		
5..............		23	
6.5		24	24
8.5	25		25
10.5	26		26
12..............			28
13..............	31		
14..............			32

$r_1 = 2 + 4 + 8.5 + 10.5 + 13 = 38$

$r_2 = 1 + 3 + 5 + 6.5 = 15.5$

$r_3 = 6.5 + 8.5 + 10.5 + 12 + 14 = 51.5$

Thus,

$$t = \frac{12}{14(15)} \left\{ \frac{[38 - (\frac{1}{2})(5)(15)]^2}{5} + \frac{[15.5 - (\frac{1}{2})(4)(15)]^2}{4} + \frac{[51.5 - (\frac{1}{2})(5)(15)]^2}{5} \right\}$$
$$= (0.057)[(0.05) + (52.5625) + (39.2)] = 5.233$$

From Table B.13 in Appendix B, the critical T-value is 5.6429. Notice that this value corresponds to $n_1 = 5$, $n_2 = 5$, and $n_3 = 4$ in the table. But the order of the sample sizes does not affect the critical value. Since $t = 5.233$ is not greater than 5.6429, we do not reject the null hypothesis that the three population distributions are equal. Since $t = 5.233$ lies between the 0.05 significance value (5.6429) and the 0.10 significance value (4.5229), we may conclude that the p-value for the test is between 0.05 and 0.10.

20.7   RANK CORRELATION: SPEARMAN'S RHO

Spearman's rho is the nonparametric equivalent of the Pearson product moment correlation coefficient given in Chapter 13.

Definition 20.6
Spearman's rho coefficient

Assumptions:
1. The n pairs, (X_i, Y_i), represent a random sample drawn from a bivariate population distribution of continuous random variables X and Y.
2. The measurement scale is at least ordinal.

Hypotheses:
H_0: The X_i and Y_i values are uncorrelated.
H_A: Either there is a tendency for *larger* values of X to be paired with larger values of Y, or there is a tendency for *smaller* values of X to be paired with larger values of Y.

Definition 20.6 (continued)

Test statistic value:

Let $r(x_i)$ be the ranks of the **X** values and $r(y_i)$ be the ranks of the **Y** values. Ties are handled as usual—assign to each tied value the average of the ranks that would have been assigned had there been no ties. Spearman's rho—the correlation measure and the test statistic value—is given by

$$\text{rho} = 1 - 6 \sum_{i=1}^{n} \frac{[r(x_i) - r(y_i)]^2}{n(n^2 - 1)}$$

Decision rule:

Reject H_0 if rho is greater than $\rho_{1-\alpha/2}$ or if rho is less than $\rho_{\alpha/2}$, where $\rho_{\alpha/2}$ and $\rho_{1-\alpha/2}$ are given in Table B.14 in Appendix B. Otherwise, do not reject H_0.

If there are no ties, Spearman's rho can be calculated from Pearson's product-moment correlation coefficient,

$$r = \frac{\sum_{i=1}^{n} (x_i - \bar{x})(y_i - \bar{y})}{\left[\sum_{i=1}^{n} (x_i - \bar{x})^2 \sum_{i=1}^{n} (y_i - \bar{y})^2 \right]^{1/2}}$$

by replacing the x_i and y_i values by their ranks.

Example 20.5   In a target marketing population, ten married couples are randomly selected to participate in an experiment to determine the relationship between the ratings husbands and wives express for a product. Each couple is shown the company's product, and each individual is then asked to rate the product on a scale from 0 (terrible product) to 100 (terrific product). The ten pairs of ratings are shown in the table.

Couple	Husband x_i	Wife y_i
1	90	70
2	100	60
3	75	60
4	80	80
5	60	75
6	75	90
7	85	100
8	40	75
9	95	85
10	65	65

Determine Spearman's rho and test the null hypothesis at the $\alpha = 0.05$ level that **X** and **Y** are uncorrelated.

Solution The ranks of the x_i and y_i values are given in the table.

i	$r(x_i)$	$r(y_i)$
1.............	8	4
2.............	10	1.5
3.............	4.5	1.5
4.............	6	7
5.............	2	5.5
6.............	4.5	9
7.............	7	10
8.............	1	5.5
9.............	9	8
10.............	3	3

The value of rho is:

$$\text{rho} = 1 - \frac{6[(8 - 4)^2 + (10 - 1.5)^2 + (4.5 - 1.5)^2 + \cdots + (3 - 3)^2]}{10(10^2 - 1)}$$

$$= 1 - \frac{(6)(161)}{990} = 1 - 0.976 = 0.024$$

To test the null hypothesis, from Table B.14 in Appendix B, $\rho_{0.975} = 0.6364$ and $\rho_{0.025} = -\rho_{0.975} = -0.6364$. Since rho $= 0.024$ is between $\rho_{0.025}$ and $\rho_{0.975}$, we cannot reject the null hypothesis that **X** and **Y** are uncorrelated. Since the value of the test statistic (rho $= 0.024$) is less than 0.4424, the *p*-value for the test is greater than $2(0.100) = 0.200$.

■ 20.8 GOODNESS OF FIT: THE KOLMOGOROV-SMIRNOV AND LILLIEFORS TESTS

In this section, we will study an exact test for goodness of fit when the population random variable is continuous and the population distribution has been completely specified—the Kolmogorov-Smirnov test. Another test of this type—the Lilliefors test—will be given for the special case of testing the goodness of fit of a set of sample observations to a normal distribution with unspecified population mean and variance.

Before describing these tests, we will review the definition of a cumulative distribution function and define a sample cumulative distribution.

Definition 20.7*

Cumulative distribution function (CDF)

Given a random variable **X**, its *cumulative distribution function* is specified by

$$F(x) = P(\mathbf{X} \le x) \qquad -\infty < x < +\infty$$

* See also definitions 5.5 and 5.9.

Example 20.6   Find the cumulative distribution function for the uniform distribution specified by,

$$f(x) = \begin{cases} 1 & 0 \le x \le 1 \\ 0 & \text{elsewhere} \end{cases}$$

Solution The graph of the uniform density function is illustrated in Figure 20.1. To find its CDF, we must evaluate the function $F(x) = P(X \le x)$. This function "accumulates" the probability to the left of x. For example, if $x = 0.3$, from Figure 20.1, $F(0.3) = P(X \le 0.3) = 0.3$. By selecting more points x, it becomes apparent that $F(x)$ takes the form illustrated in Figure 20.2.[2]

FIGURE 20.1   Uniform distribution

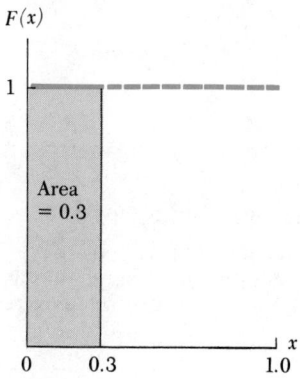

FIGURE 20.2   CDF of the uniform distribution

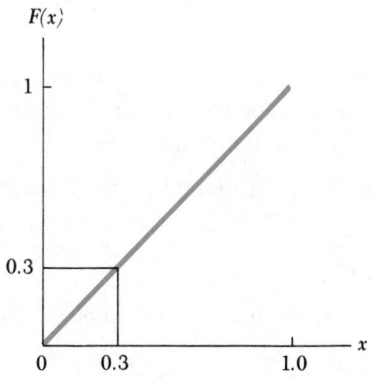

---

[2] The CDF may alternatively be formed directly by integration: $F(x) = P(X \le x) = \int_{-\infty}^{x} f(x)\, dx = \int_{-\infty}^{0} (0)\, dx + \int_{0}^{x} (1)\, dx = 0 + x|_{0}^{x} = (x - 0) = x$, for $0 \le x \le 1$. If $x > 1$, $F(x) = 1.0$. If $x < 0$, $F(x) = 0.0$.

FIGURE 20.3 Sample cumulative distribution function

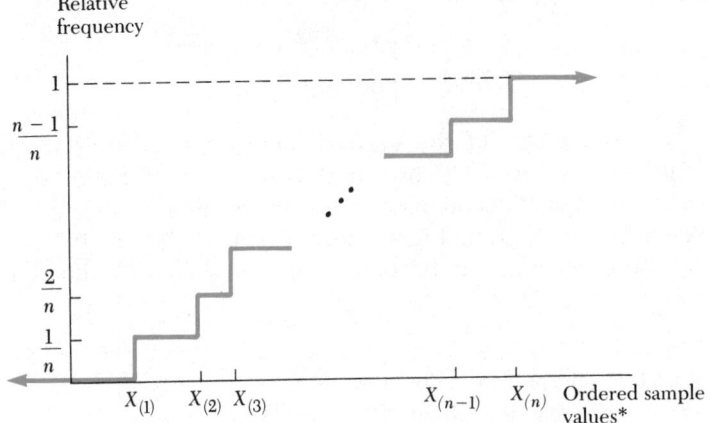

* The parentheses on the subscripts of the x_i values indicate that the $x_{(i)}$ values are the ordered observations, from the smallest to the largest.

Definition 20.8
Sample cumulative distribution function (SCDF)

The *sample cumulative distribution function,* denoted by $S(x)$ for a specific value of **X**, is formed by ordering a set of sample observations from the smallest to the largest, and then plotting the cumulative relative frequencies as illustrated in Figure 20.3.

The SCDF is often called a "jump function"—it jumps (in relative frequency) by $1/n$ each time an observation is encountered as we move along the x-axis from left to right. If the random variable is discrete, it is possible that two or more observations are equal in value. In this case, the SCDF jumps $1/n$ times the number of observations equal to the specific value.

Example 20.7   A random sample, supposedly drawn from the standardized uniform distribution, produced the following ten numbers:

$$0.36 \quad 0.14 \quad 0.60 \quad 0.20 \quad 0.43 \quad 0.81 \quad 0.12 \quad 0.02 \quad 0.55 \quad 0.25$$

Form the SCDF for these ten numbers.

Solution   The ordered numbers are:

$$0.02 \quad 0.12 \quad 0.14 \quad 0.20 \quad 0.25 \quad 0.36 \quad 0.43 \quad 0.55 \quad 0.60 \quad 0.81$$

The SCDF is illustrated in Figure 20.4.

In Figure 20.4, the cumulative distribution function is also shown as the straight line from (0, 0) to (1, 1). Notice that the SCDF is above the CDF almost everywhere. This may suggest that the ten numbers in Example 20.7 do not come from the standardized uniform distribution. In fact, the

FIGURE 20.4 Sample CDF for Example 20.7

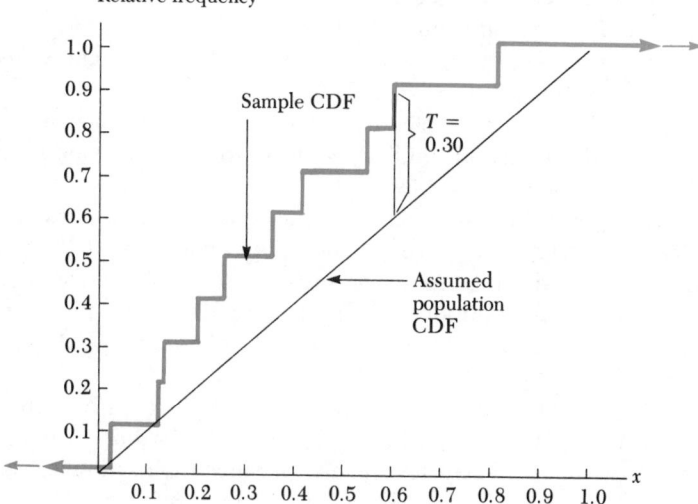

Kolmogorov-Smirnov goodness-of-fit test is based on the single greatest difference between the sample and assumed population CDFs.

Definition 20.9
Kolmogorov-Smirnov goodness-of-fit test

Assumptions:

    1. The sample $X_1, X_2, \ldots, X_n$ is a random sample.

    2. The hypothesized population CDF, denoted by $F_0(x)$, is continuous and completely known (if it has parameters, their values are known).

Hypotheses:

    H_0: $F(x) = F_0(x)$

    H_A: $F(x) \neq F_0(x)$ for at least some x, where $F(x)$ is the unknown CDF of $X_1, X_2, \ldots, X_n$.

Test statistic value:

    Let $S(x)$ be the sample CDF based on the ordered sample observations, $x_{(1)}, x_{(2)}, \ldots, x_{(n)}$. The test statistic value is:

$$t = \sup_X |F_0(x) - S(x)|$$

The symbol, sup, means the greatest difference, and the test statistic value t is
     X

therefore defined as the greatest absolute difference between $F_0(x)$, the hypothesized CDF, and $S(x)$, the sample CDF.

Decision rule:

    Reject the null hypothesis if t is greater than $W_{1-\alpha}$ given in Table B.15 in Appendix B. Otherwise, do not reject H_0.

Example 20.8 Using the data in Example 20.7, test the hypothesis at the $\alpha = 0.05$ significance level that these ten data do represent a random sample from the standardized uniform distribution.

Solution From Figure 20.4, it is apparent that the greatest difference occurs when $x = 0.60$; $F_0(0.60) - S(0.60) = 0.60 - 0.90 = -0.30$. Therefore, $t = |-0.30| = 0.30$. From Table B.15 in Appendix B, $W_{0.95} = 0.409$. Since $t < 0.409$, do not reject the null hypothesis that the data come from the standardized uniform distribution.

The Kolmogorov-Smirnov test can be used for any hypothesized continuous CDF and is typically a very good approximate test if the population random variable is discrete. While this test is applicable in many situations, it is most widely used in goodness of fit for normality, since most parametric tests require the population random variable to be normally distributed, or approximately so.

Example 20.9 The Scholastic Aptitude Test scores are thought to be normally distributed with a mean of 500 and a standard deviation of 100. A random sample of ten recent test results produced the following scores:

$$450 \quad 420 \quad 500 \quad 530 \quad 440 \quad 475 \quad 445 \quad 520 \quad 460 \quad 480$$

Use the Kolmogorov-Smirnov test to determine whether it is likely that these data have been drawn from the specified normal distribution (use $\alpha = 0.10$).

Solution To simplify the calculation of the value of the test statistic, we first standardize the ten scores using

$$z_i = \frac{x_i - \mu}{\sigma}$$

These z-scores are:

$$-0.5 \quad -0.8 \quad 0.0 \quad 0.3 \quad -0.6 \quad -0.25 \quad -0.55 \quad 0.20 \quad -0.40 \quad -0.20$$

Denote by $z_{(i)}$ the ordered z-scores:

$$-0.80 \quad -0.60 \quad -0.55 \quad -0.50 \quad -0.40 \quad -0.25 \quad -0.20 \quad 0.00 \quad 0.20 \quad 0.30$$

The cumulative distribution function for the standardized normal distribution may be plotted by using Table B.3 in Appendix B. This CDF and the sample CDF $S(z)$ are illustrated in Figure 20.5. From Figure 20.5, it is apparent that the greatest difference between $F_0(z)$ and $S(z)$ occurs when $z = 0.30$; $F_0(0.30) = 0.50000 + 0.1179 = 0.6179$, and $S(0.30) = 1.00$. Thus, $t = |0.6179 - 1.0000| = 0.3821$. From Table B.15, $W_{0.90} = 0.369$. Since $t > 0.369$, reject the null hypothesis. These data do not appear to come from a normal distribution with mean 500 and variance 100 at the $\alpha = 0.10$ significance level. Since $t = 0.3821$ lies between $W_{0.10}$ and $W_{0.05}$, we may conclude that the p-value for this test is between 0.10 and 0.05.

Often when we are interested in determining whether a random sample fits well to the normal distribution, we will not know the values of the population parameters μ and σ. But the Kolmogorov-Smirnov test requires that the

FIGURE 20.5 $F_0(z)$ and $S(z)$ functions for Example 20.9

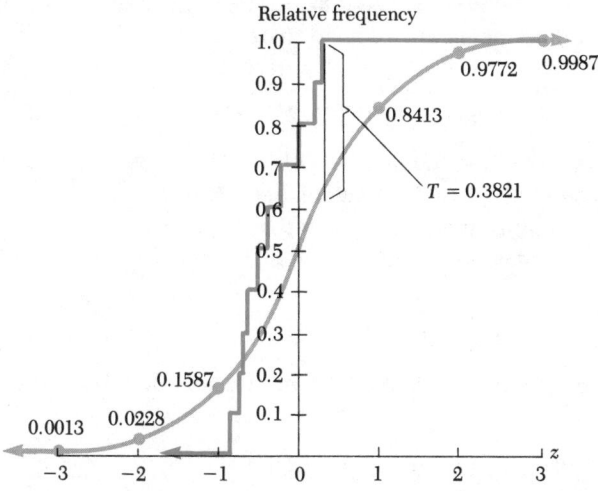

hypothesized CDF, $F_0(x)$, be completely specified. Lilliefors modified the Kolmogorov-Smirnov critical tables so that the value of the test statistic **t** can be based on the sample mean $\bar{x}$ and the sample standard deviation s in place of μ and σ, respectively.

Definition 20.10
Lilliefors test for normality

Assumptions:
1.   The sample $X_1, X_2, \ldots, X_n$ is a random sample.
2.   The hypothesized population CDF, denoted by $F_0(x)$, is continuous.

Hypotheses:
H_0:   The random sample came from a normal distribution, with unspecified mean and standard deviation.
H_A: The cumulative distribution function is not normal.

Test statistic value:
Standardize the x_i values by using

$$z_i = \frac{x_i - \bar{x}}{s}, \text{ where } i = 1, 2, \ldots, n$$

where $\bar{x}$ is the sample mean and s is the sample standard deviation. Let $S(z)$ denote the sample CDF of the z_i values. The test statistic value is:

$$t = \sup_z |F_0(z) - S(z)|$$

that is, it is the greatest difference between $F_0(z)$ and $S(z)$.

Decision rule:
Reject H_0 if t is greater than $W_{1-\alpha}$ given in Table B.16 in Appendix B. Otherwise, do not reject H_0.

Example 20.10 The lifetime of a certain electronics component is thought to be normally distributed. A random sample of eight components is selected, and the components are allowed to operate until failure. The failure times in hours are:

$$1,000 \quad 1,500 \quad 1,600 \quad 1,450 \quad 1,725 \quad 1,560 \quad 1,650 \quad 1,700$$

Using the Lilliefors test, test the null hypothesis that these data came from a normal distribution at the $\alpha = 0.05$ significance level.

Solution The sample mean and sample standard deviation are $\bar{x} = 1,523.125$ and $s = 231.439$, respectively. Using these values, the standardized variables are:

$$-2.26 \quad -0.10 \quad 0.33 \quad -0.32 \quad 0.87 \quad 0.16 \quad 0.55 \quad 0.76$$

The graphs of the standard normal CDF and the sample CDF are illustrated in Figure 20.6.

FIGURE 20.6 Population CDF and sample CDF for Example 20.10

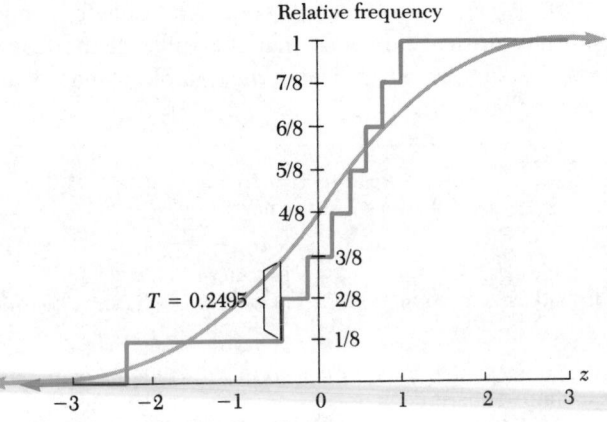

From Figure 20.6, the greatest difference between $F_0(z)$ and $S(z)$ occurs for a value of Z that is approaching $z_{(2)} = -0.32$ from the left, say $z^* = -0.31999$. Then $F(z^*) = 0.3745$, $S(z^*) = 0.125$, and $t = |0.3745 - 0.125| = 0.2495$. From Table B.16 in Appendix B, $W_{0.95} = 0.285$. Since $t < W_{0.95}$, we fail to reject the null hypothesis at the $\alpha = 0.05$ significance level. There is insufficient evidence to reject the sample as not coming from a normal distribution. Since $t = 0.2495$ lies between $W_{0.15}$ and $W_{0.10}$, we may conclude that the p-value for this test is between 0.15 and 0.10.

20.9 RANDOMNESS TEST: THE RUNS TEST

A very simple test has been devised to test for randomness of a set of observations. The test is based on the number of "runs" in a sequence of items, where a run is defined to be an unbroken row of like items that are preceded and followed by items of a different type and includes as many items as possible. Consider the sequence $(- - - + + +)$. This sequence has two runs, while the sequence $(- + - + - +)$ has six runs. The maximum number of runs among n items is n, while the minimum number of runs is one (all items are alike). We will now see how such a simple notion leads to a test for randomness among a set of observations.

Definition 20.11
Runs Test

Assumption:

The observations are listed in the order obtained. Calculate the median of the set of observations. For all those observations *below* the median, assign a $-$ sign, and for all those *above*, assign a $+$ sign. This procedure generates a sequence of $+$'s and $-$'s.

Hypotheses:

H_0: The process that generates the sequence is a random process.

H_A: The process is not random—some items in the sequence are dependent on others in the sequence or some items are distributed differently from others.

Test statistic value:

t = total number of runs of like items in the sequence.

Decision rule:

Reject H_0 if $t > W_{1-\alpha/2}$ or $t < W_{\alpha/2}$, where $W_{\alpha/2}$ and $W_{1-\alpha/2}$ are given in Table B.17 in Appendix B. Otherwise, do not reject H_0.

Example 20.11 The following sequence is purported to be a set of random integers from 0 to 99. Use the runs test to test the hypothesis of randomness at the $\alpha = 0.05$ significance level. The sequence is:

$$28 \quad 4 \quad 23 \quad 98 \quad 44 \quad 10 \quad 6 \quad 25 \quad 54 \quad 81 \quad 12$$
$$6 \quad 4 \quad 33 \quad 67 \quad 55 \quad 71 \quad 66 \quad 22 \quad 18 \quad 49 \quad 85$$

Solution The median is $30.5[(33 + 28) \div 2]$. The sequence of $+$'s and $-$'s is, therefore,

$$-, -, -, +, +, -, -, -, +, +, -, -, -, +, +, +, +, +, -, -, +, +$$

The total number of runs is $t = 8$. From Table B.17 in Appendix B, let N_1 = number of $-$'s and N_2 = number of $+$'s. Then, $N_1 = N_2 = 11$ and $W_{0.025} = 8$, $W_{0.975} = 16$. Since $t = W_{0.025}$ we cannot reject the null hypothesis of randomness at the $\alpha = 0.05$ significance level. Since $W_{0.025} = 8$ and $W_{0.05} = 8$ and $t = 8$, we may conclude that the p-value is between $2(0.025) = 0.05$ and $2(0.05) = 0.10$.

The runs test may be used to check the assumption in analysis of variance and regression that the residuals are independent. See Problem 20.14 at the end of this chapter, for example.

■ 20.10 SUMMARY

In this chapter, we have presented a number of methods that may be appropriate in analyzing data when the standard parametric procedures may not be appropriate. Nonparametric tests are typically easy to compute, but are almost always less powerful than a parametric test if one is appropriate for the data set. Most nonparametric tests require that the data be at least ordinally measurable, while most parametric tests require interval or ratio scale measurements.

Nonparametric tests may be applied to nominally scaled measurements, although we did not consider any of these tests in this chapter. The most widely used nonparametric statistic for the analysis of nominal data is the chi-square statistic introduced in Chapter 19. The chi-square statistic is used to analyze contingency tables, where most often the categories in the table are ordinally measured. The chi-square statistic is, therefore, commonly thought of as a nonparametric statistic, although the chi-square distribution is also used for constructing decision rules for parametric tests (e.g., tests concerning the population variance σ^2—see Chapter 10).

There are, of course, many more nonparametric tests than we have described in this text. An excellent source of the most widely used tests is the Conover reference cited at the end of this chapter.

■ REFERENCES

Conover, W. J. *Practical Nonparametric Statistics*. New York: John Wiley & Sons, Inc., 1971.

Ostle, B., and Mensing, R. W. *Statistics in Research*. 3d ed. Ames: Iowa State University Press, 1976.

Snedecor, G., and Cochran, W. *Statistical Methods*. 7th ed. Ames: Iowa State University Press, 1980.

■ PROBLEMS

20.1. Assuming the distribution of miles per gallon is symmetric so that the mean and median are equivalent, test the hypothesis that the population mean is 30 miles per gallon, using $\alpha = 0.10$ and the data in Example 20.1, by using the t-test in Chapter 10. What is the p-value for this test?

20.2. A marketing research group randomly samples 15 potential buyers of a product to determine their accep-

tance of the product. The measurement scale used is:

0	1	2	3	4
Strongly dislike	Dislike	Neutral	Like	Strongly like

The 15 responses were: 0, 2, 2, 1, 0, 1, 3, 2, 2, 0, 1, 4, 3, 2, 0. Using the median test, test the hypothesis that the population median is 2. (Use $\alpha = 0.1$.) What is the p-value for this test?

20.3. Consider the data in Example 20.2.

a. Is it reasonable to assume that if the two populations differ, they differ only in location? Using the F-test in Chapter 10, test the hypothesis at the $\alpha = 0.5$ significance level that the two population variances σ_1^2 and σ_2^2 are equal.

b. If the null hypothesis in part a, H_0: $\sigma_1^2 = \sigma_2^2$, is not rejected, test the hypothesis that $\mu_1 = \mu_2$ at the $\alpha = 0.05$ level, pooling the two sample variances. Compare this result to one using the Mann-Whitney test in Example 20.2. What is the p-value for this test?

20.4. Two fertilizers, brands A and B, are used on two identical plots of land planted with corn. Each plot is divided into eight equal sections. At the end of the experiment, the yields per section for the two fertilizers are measured. The data are shown in the table.

Fertilizer A	Fertilizer B
80.5	95.4
76.4	84.7
93.2	88.1
90.6	98.2
84.7	101.6
81.2	88.6
78.5	96.4
82.0	97.3

Using the Mann-Whitney test and $\alpha = 0.05$, test the null hypothesis that the two samples came from the same population. What is the p-value for this test?

20.5. Use the paired t-test given in Chapter 11 to test the equivalence of the two population means in Example 20.3 (use $\alpha = 0.05$). Compare your result

with that in Example 20.3, which is based on the Wilcoxon signed rank test. What additional assumptions are required to use the paired t-test? What is the p-value for this test?

20.6. A drug company is interested in determining whether a chemical treatment for a specific form of cancer changes body temperature. Ten patients with the disease are selected at random from a set of patients under experimental control. Their temperatures are measured before and after taking the treatment. The data, given in degrees Fahrenheit, are listed in the table.

Patient	Before	After
1.........	98.2	99.4
2.........	98.4	101.2
3.........	98.0	97.6
4.........	99.0	99.8
5.........	98.6	98.0
6.........	97.5	98.4
7.........	98.4	98.4
8.........	100.0	102.3
9.........	99.2	101.6
10.........	98.6	98.8

Test the null hypothesis that the two population means are equal at the $\alpha = 0.01$ level by using the Wilcoxon signed rank test. What is the p-value for this test?

20.7. Use the analysis of variance method described in Chapter 12 to test the null hypothesis of the equality of the three population means in Example 20.4 ($\alpha = 0.05$). Compare your result with the Kruskal-Wallis test result in Example 20.4. What further assumptions must you make to use the analysis of variance F-test on these data? What is the p-value for this test?

20.8. The breaking strengths of three kinds of wire cord are compared by using four samples of each and measuring the force required to break the cord. The data are shown in this matrix.

Observation \ Type	A	B	C
1	225	315	260
2	260	255	232
3	243	262	256
4	231	287	247

Test the null hypothesis of equal population distributions at the α = approximately 0.05 level, using the Kruskal-Wallis test. What is the p-value for this test?

20.9. A large consulting firm hires a West Coast university to provide an MBA program for its employees. The basic statistics course is taught at four locations of the firm. After completion of the course, standardized tests are given to the participating employees at each location. The results are given in the table.

Observation	Location A	B	C	D
1	96	65	66	60
2	88	74	90	72
3	92	77	88	66
4	75	82	73	75
5	81	70	85	78
6	62	78	87	68
7	86	84	94	
8	86		74	
9	90			
10	85			

Using the Kruskal-Wallis test, test the null hypothesis of equal population distributions. Calculate the p-value for this test. Should the null hypothesis be rejected based on the p-value? Explain.

20.10. A production manager suspects that the level of production among a specific class of workers in the firm is related to their hourly pay. The following data are collected on eight randomly selected workers.

Worker	Hourly pay (x_i)	Production (y_i)*
1	$3.10	50
2	2.50	20
3	4.45	62
4	2.75	30
5	5.00	75
6	5.00	60
7	2.90	42
8	4.75	60

* In units per hour.

a. Calculate Spearman's rho for these data.

b. Calculate Pearson's product-moment correlation coefficient for these data. What assumptions are necessary to compute Pearson's r? What does it measure?

c. Using Spearman's rho, test the hypothesis at the α = 0.10 level that X and Y are uncorrelated. What is the p-value for this test?

20.11. A company believes that its color television tubes have an average lifetime of 18 continuous months of play and that these lifetimes are exponentially distributed. The company has its quality control department randomly sample 12 tubes, which are run to failure. The 12 lifetimes (in months) are:

5.4 21.6 16.2 12.6 19.8
23.4 19.8 16.2 21.6
23.4 16.2 25.2

Use the Kolmogorov-Smirnov test to determine whether it is reasonable to assume that these data came from the hypothesized exponential distribution. Use $\alpha = 0.05$. (Hint: Form the variable $Y = X/\lambda, \lambda = 18$. Use Table B.4 in Appendix B to plot the population CDF.) What is the p-value for this test?

20.12. A major automobile subsidiary believes that its standard batteries have capacities that are normally distributed, with a mean of 100 ampere hours and a standard deviation of 20 ampere hours. Fifteen batteries are tested with the following results:

$$
\begin{array}{cccccc}
120 & 80 & 60 & 90 & 100 & 100 \\
180 & 120 & 160 & 150 & 160 \\
& 140 & 140 & 110 & 150
\end{array}
$$

Test the null hypothesis that these data came from a normal distribution with mean 100 and standard deviation 20, using the Kolmogorov-Smirnov test with $\alpha = 0.05$. What is the p-value for this test?

20.13. The sales research division of a large corporation is conducting research on sales methods for selling one of the products of the corporation. The division has designed a completely randomized one-factor analysis of variance model to investigate the efficacy of three sales methods. The responses are measured in units of $100 in sales. The data are shown in the table.

	Sales method		
Response	A	B	C
1........	20	13	31
2........	23	18	28
3........	22	16	48
4........	22	29	28
5........	36	14	30
6........	45		25
7........	21		
8........	26		

a. Use Lilliefors test to determine whether each treatment sample comes from a normal population. Use $\alpha = 0.05$.

b. What other assumptions must be checked before the parametric analysis of variance F-test, may be used on these data?

20.14. A regression analysis using the model $Y_i = \beta_0 + \beta_1 x_i + \epsilon_i, i = 1, 2, \ldots, 10$, $y_i =$ amount of sales (in $000); and $x_i =$ amount of advertising expenditure (in $000), was performed by a marketing research group to assess the effect on sales of the amount of advertising budget. After fitting the model, the group calculates the following ten residuals to check the aptness of the regression model:

$$
\begin{aligned}
e_1 &= 2.42, e_2 = 2.02, e_3 = -1.08, \\
e_4 &= -0.54, e_5 = -0.98, \\
e_6 &= -0.20, e_7 = -1.60, \\
e_8 &= -0.44, e_9 = -0.80, e_{10} = 1.20
\end{aligned}
$$

a. Use the runs test to determine whether this is a random sequence of residuals. That is, are the residuals independent? Use $\alpha = 0.05$.

b. For statistical hypothesis-testing purposes, the residuals in regression (and in analysis of variance) are supposed to be normally distributed with a mean of 0 and standard deviation σ. Use the Lilliefors test to check this assumption at the $\alpha = 0.05$ significance level for these data.

20.15. A machine plant is interested in acquiring a new stamping machine and has narrowed its selection down to two machines. In order to rate the machines, each machine is set up to produce ten units, and each unit is given a score based on an index that includes such factors as time required to produce the unit, number of

Machine A	50	45	60	60	75	40	46	58	62	72
Machine B	72	61	60	78	56	77	81	75	74	82

flaws, and appearance of the unit. The higher the score, the better the outcome. The index scores are given in the table above. At the $\alpha = 0.05$ significance level, do these machines appear to be producing parts with the same index scores? What is the p value for this test?

20.16. In a production process, it is suspected that the number of defectives produced per hour is related to a skills test score recorded by a worker. The units being produced require a high degree of manual dexterity, and the skills test is designed to test manual dexterity. Ten workers are selected and each is asked to put together 25 units in one hour. Units that are not completed in the hour period are considered to be defective; those that are completed are inspected for defects. The data are given in the table below. Using Spearman's rho, determine whether the correlation between test score and number of defects is significantly different from 0 at the $\alpha = 0.05$ level of significance. What is the p-value for this test?

Worker	1	2	3	4	5	6	7	8	9	10
Test score	90	75	60	65	50	80	75	95	80	70
Number of defectives	2	4	7	7	8	4	3	1	3	5

Statistical decision theory: Decision making under uncertainty

21

In this chapter we begin a discussion of statistical decision theory. This discussion will continue for two more chapters, at which time we will have completed our introductory exposition of this topic.

The material discussed in Chapters 21–23 will mark a distinct deviation from what has been discussed previously. The preceding material is characterized as classical statistics or as classical statistical techniques or methods, whereas the material in these three chapters is characterized as Bayesian statistics, as Bayesian decision theory, or, in abbreviated form, as just decision theory. We will learn as we proceed through this material that there are differences between the two approaches or schools of thought. In Chapter 23, we attempt to summarize the major differences between these two approaches for using statistical data to arrive at a decision, so the reader can make an informed choice as to which method he or she chooses to follow.

In this chapter, we will learn how to characterize a statistical decision problem and will investigate four separate approaches for making decisions in situations in which it is impossible or extremely difficult to assess a probability distribution for a random variable about which we must make a decision. Then, in Chapter 22, we will see how Bayesian decision theory prescribes a course of action to follow when a probability distribution for the random variable of interest exists and is known, but there is no opportunity to obtain any additional information. Then, in Chapter 23, we will see how the Bayesian approach prescribes a decision rule to follow for selecting the sample size and the rejection or critical value of a test statistic

in making a decision where the possibility for obtaining additional information exists (i.e., sampling).

Following the convention in previous chapters (notably Chapters 9 and 10),
we will use the Greek symbol theta in boldface $\boldsymbol{\theta}$ to represent a random
variable or population characteristic of interest (mean, proportion, etc.),
and we will use nonboldfaced theta θ to represent possible values of the
random variable $\boldsymbol{\theta}$. However, it will greatly simplify the discussion of
decision theory if we also refer to the random variable $\boldsymbol{\theta}$ as being a
particular state of nature. Thus, for our purposes, the following two
statements will be equivalent: "The probability that the random variable
$\boldsymbol{\theta}$ is equal to θ_j is denoted by $P(\boldsymbol{\theta} = \theta_j)$" and "The probability that we
are in state θ_j or that state of nature θ_j occurred is denoted by $P(\boldsymbol{\theta} =
\theta_j)$."

We begin our discussion of statistical decision theory or Bayesian decision
theory by an examination of alternate means available for characterizing a
decision problem.

■ 21.2 STRUCTURING A STATISTICAL DECISION PROBLEM

Several alternative methods exist for giving structure to a statistical
decision problem. In this section, we will describe three methods: the payoff
(or loss) matrix, the loss function, and decision trees. Each of these methods
possesses a distinct format which is often helpful in analyzing certain
decision problems.

☐ 21.2.1 Payoff and regret tables

Payoff matrices are a convenient means of organizing and displaying the
various ingredients of a decision problem when the number of possible
states of nature and the number of possible actions are finite. Table 21.1
depicts the basic format of a payoff matrix, where Ω_{ij} represents the payoff
(or loss, for Ω_{ij} negative) that results when action a_i is taken and the
particular state of nature is θ_j. Table 21.1 is often referred to also as a *value
matrix*—denoting the "value" of each act for each state of the world or state
of nature that occurs.

It is customary to denote the various states of nature across the top of the
table (columns) and the various possible acts (alternatives) on the left side of
the matrix (rows) as is shown. The intersection of the act chosen and the
state of nature existing thus denotes the payoff from the particular act-state of
nature combination. In Table 21.1 we have indicated an act labeled a_0 and a
state of nature denoted θ_0. These represent, respectively, the act "do
nothing" (e.g., stock 0 items) and "nothing occurs" or the value of the
random variable is equal to 0 (e.g., no demand, or demand equals 0). Not all
payoff (or loss) tables have this 0 designation, but many do. From Table 21.1,
we see that since we have allowed for the act-state of nature combination
(a_0, θ_0), there are $(m + 1)$ possible actions that can be taken (which includes

TABLE 21.1 General form of a payoff matrix

Act \ State of nature	θ_0	θ_1	θ_2	$\cdots$	θ_j	$\cdots$	θ_n
a_0	Ω_{00}	Ω_{01}	$\cdots$	$\cdots$	$\cdots$	$\cdots$	Ω_{0n}
a_1	Ω_{10}						$\cdot$
a_2	$\cdot$				$\cdot$		$\cdot$
$\cdot$	$\cdot$				$\cdot$		$\cdot$
a_i	$\cdots$	$\cdots$	$\cdots$		Ω_{ij}	$\cdots$	$\cdots$
$\cdot$	$\cdot$				$\cdot$		$\cdot$
$\cdot$	$\cdot$				$\cdot$		$\cdot$
a_m	Ω_{m0}	$\cdots$	$\cdots$	$\cdots$	$\cdots$	$\cdots$	Ω_{mn}

the act "do nothing") and there are $N = (n + 1)$ values the random variable θ can assume.

Also, from Table 21.1, it can be seen that in a statistical decision problem, the payoff from selecting a particular act a_i and having a particular state of nature θ_j occur is known in advance of the decision. There is assumed to be no ambiguity regarding the payoff to be received when the particular act is selected and the state of nature occurs. Hence, we assume that the results of a particular action can be determined without *error* when provided with the state of nature that has or is about to occur. That is, the results in terms of payoffs of our actions are certain; knowledge concerning the state of nature may or may not be certain. Thus, the *stochastic* or the *probabilistic* elements involved in a statistical decision concern states of nature; the *deterministic* or *certain* elements involve the payoffs to be received from acts to be selected when the particular stochastic elements occur.

Use of the payoff matrix in characterizing the ingredients of a decision problem is illustrated in the following example.

Example 21.1 The George Thomas Travel Agency regularly schedules tours from Rochester, Minnesota, to a vacation resort in southern Florida (private resort area). These tours normally take place in the winter months, and George reserves seats on the regularly scheduled airlines by depositing $100 three months in advance of the scheduled departure date. George then sells the tour package to residents of Rochester for $150 (airfare only). George has been given the right to purchase up to six seats in advance for each tour that he schedules.[1] If George has not sold the tour to as many

---

[1] George can actually reserve many more seats than this and, in fact, usually does. However, it will greatly simplify the computational effort involved without our losing an understanding of the subject matter if we restrict the number of alternatives we have to examine in this problem to a more manageable number. This is why we have chosen to limit our discussion of the number of seats available to six.

people as he has reserved seats by one week before the scheduled departure date, he must notify the airline that he is relinquishing his right to the unsold seats, in which case the airline refunds him $80 of his $100 deposit. Thus, George must reserve the seats well in advance of when he will know what the demand for the seats will be. For each seat reserved but unsold by one week before the scheduled departure date, he loses $20.

Characterize this decision problem both in terms of the various states of nature (*demands* for seats on the tour) and in terms of the various alternatives (*supplies* of reserved seats) open to George in the form of a payoff matrix.

Solution Table 21.2 is the payoff matrix for the problem. In this form, the payoff matrix might also be called a *cash-flow matrix,* since the quantities displayed represent the actual cash George will receive ($\Omega_{ij} \geq 0$) or payout ($\Omega_{ij} \leq 0$) if he books a_i seats and θ_j are demanded. Thus, for example, if George reserves five seats and sells only one, his net payoff is a negative $30 (loss).

TABLE 21.2 Payoff matrix for Example 21.1*

Act, a_i (supply) \ State of nature, θ_j (demand)	0	1	2	3	4	5	6
0	0	0	0	0	0	0	0
1	−20	50	50	50	50	50	50
2	−40	30	100	100	100	100	100
3	−60	10	80	150	150	150	150
4	−80	−10	60	130	200	200	200
5	−100	−30	40	110	180	250	250
6	−120	−50	20	90	160	230	300

* Quantities are in dollars.
Source: Data in Example 21.1.

The advantage of displaying the act-state of nature combinations in matrix or tabular form is that the consequence of every act-state of nature pair is readily available from the table. The payoff matrix itself, however, is often criticized because it gives the appearance that certain acts are equally "profitable" for various states of nature. For example, if a decision is made to reserve two seats and there is a demand for six, the cash flow or profit is $100, the same as it would be if the demand is for two. Obviously, the *potential for profit* is greater if the demand is six rather than if the demand is only two. Yet the payoff matrix does not reflect the differences in profit that would result *for a fixed act* when the number of units demanded is greater than the number of units stocked.

For this reason, many decision problems are resolved using an *opportunity loss* or a *regret* matrix to express the consequences of various act-state of nature combinations.

Definition 21.1
Opportinity loss

The *opportunity loss* of selecting a particular act a_i is the difference between the cost or profit that was actually realized for that act and the cost or profit that would have resulted if the decision had been the best one for the state of nature that actually occurred.

The actual computation of the opportunity loss for a given act is very straightforward: subtract each entry in the payoff or the value matrix from the largest entry in its column.

Opportunity loss

$$l(a_i, \theta_j) = (\text{Max } \Omega_{ij}) - \Omega_{ij} \qquad (21.1)$$
$$\underset{i}{}$$

The opportunity loss, then, represents the difference between what was realized under a given act and what could have been realized if the state of nature that actually occurred had been known in advance. Since the opportunity loss is conditional upon the value of θ that occurs, it is also called a *conditional loss* or a *conditional opportunity loss*.

Example 21.2 Compute the opportunity loss or the regret matrix for the data given in Example 21.1.

Solution Table 21.3 gives the opportunity loss or the regret matrix for the data in Example 21.1. The opportunity losses for the entries in the matrix for $\theta = 3$ (three seats are demanded) are computed as follows:

$$\text{Max}(\Omega_{i3}) = \Omega_{33} = 150 \qquad i = 0, 1, 2, \ldots, 6$$
$$\phantom{\text{Max}}\underset{i}{}$$

$$l(a_i, \theta_3) = 150 - \Omega_{i3}$$

Or, for $\theta = 3$, $l(a_i, \theta_3)$ is

$$
\begin{array}{ll}
150 - 0 = 150 & 150 - 130 = 20 \\
150 - 50 = 100 & 150 - 110 = 40 \\
150 - 100 = 50 & 150 - 90 = 60 \\
150 - 150 = 0 &
\end{array}
$$

Note that with the opportunity loss matrix, the consequence or the opportunity loss suffered from a decision to reserve two seats when the demand is six is $200.

The opportunity loss table offers at least one other advantage over the payoff or the value matrix. Namely, the entries in the opportunity loss table are nonnegative (i.e., always either positive or equal to 0) quantities, and the best or the optimal act(s) under a given state of nature can always be found by simply locating those rows (acts) with a 0 entry for that column.

TABLE 21.3 Opportunity loss matrix for Example 21.1*

Act, a_i (supply) \ State of nature, θ_j (demand)	0	1	2	3	4	5	6
0	0	50	100	150	200	250	300
1	20	0	50	100	150	200	250
2	40	20	0	50	100	150	200
3	60	40	20	0	50	100	150
4	80	60	40	20	0	50	100
5	100	80	60	40	20	0	50
6	120	100	80	60	40	20	0

* Quantities are in dollars.
Source: Data from Example 21.1.

☐ **21.2.2 Opportunity loss or payoff functions**

Frequently, the most convenient means for expressing the consequences of various acts for various states of nature is the functional form. While it may or may not always be possible to indicate the consequences of act-state of nature combinations in this fashion, this form is necessary when the number of possible states of nature is infinite (e.g., the diameter of bearings produced on a certain machine) and is often quite helpful when the number of various acts or states of nature is quite large, so that the construction of a payoff or a loss table would be quite cumbersome. For the problem of Example 21.1, the loss function is:

$$l(a_i,\theta_j) = \begin{cases} 50(\theta_j - a_i) & \text{if } \theta_j > a_i \\ 0 & \text{if } \theta_j = a_i \\ 20(a_i - \theta_j) & \text{if } \theta_j < a_i \end{cases}$$

This loss function was derived by noting that no loss occurs (or the loss is 0) when the number of seats reserved is equal to the number of seats demanded ($\theta_j = a_i$), a $20 loss occurs for every seat that is reserved and is not demanded ($\theta_j < a_i$), and a $50 loss (opportunity loss) occurs for every seat that is demanded, but has not been reserved ($\theta_j > a_i$). The quantities ($\theta_j - a_i$) and ($a_i - \theta_j$) denote, respectively, the number of seats demanded but not available, and the number of seats reserved, but not demanded. These quantities are multiplied in the loss function by the per unit costs of being greater or less than the estimate of the quantity demanded.

Figure 21.1 is a graph of the conditional opportunity loss function for the acts reserve 3 and reserve 4. Several features of this graph are worth noting at this time. First, the graph forms a characteristic asymmetric V, with a

FIGURE 21.1 Conditional opportunity losses for acts reserve 3 and reserve 4 in Example 21.1

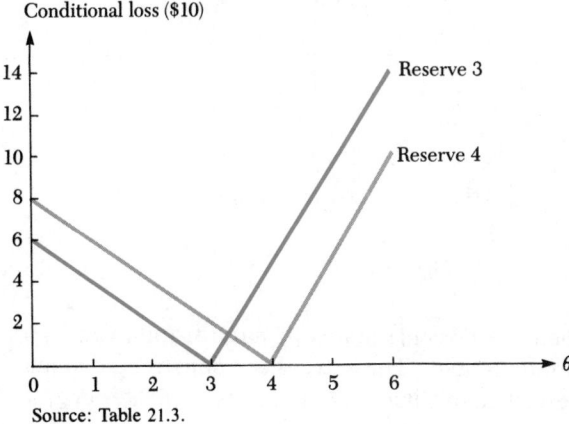

Conditional loss ($10)

Source: Table 21.3.

vertex at the point where conditional losses are 0. Note from the graph that conditional losses are 0 when the number of seats reserved is equal to the number demanded (or the greatest profit act is selected), and conditional losses are positive elsewhere. Second, the slopes of the sides of the V give the rate of decrease in potential profits resulting when the nonoptimal or the best act for a given demand is not selected. The slopes of the two sides are usually different, reflecting the difference in profit forgone (opportunity loss) from an overage as opposed to an underage of an item. *The majority of conditional loss functions exhibit graphs similar to the one shown in Figure 21.1.*

☐ **21.2.3 Decision trees**

A third alternative for structuring pertinent information in a decision problem is the *decision tree*. The decision tree portrays *in sequence* the act

694

FIGURE 21.2 Payoff decision tree for tour problem of Example 21.1

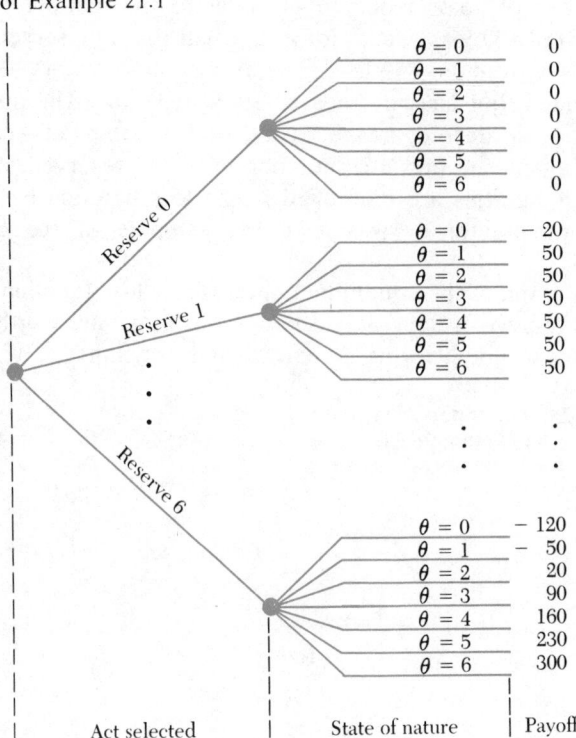

Act selected	State of nature	Payoff

alternatives and then the possible states of nature, followed by the consequences of these combinations. These trees are constructed similarly to the probability trees described in Chapter 4 and, in fact, in later chapters we will associate probabilities with each of the various branches, and conditional losses and profits associated with each.

Example 21.3   Construct a decision tree for the data in Table 21.2.

Solution   The decision tree constructed from the data in Table 21.2 is given in Figure 21.2. Note that the first set of branches emanating from the leftmost node corresponds to the *possible actions* to be taken, and the next set of branches corresponds to the various possible *states of nature* (values of θ). Finally, we indicate in the rightmost column of this figure the consequences (payoffs) of each branch of the decision tree. Note that this format for displaying the combinations or alternatives in a decision problem can become quite burdensome when the number of act-state of nature combinations is very large.

■ 21.3   DECISION MAKING UNDER UNCERTAINTY

In any statistical decision problem, three assumptions can be made regarding knowledge of the true state of nature, θ. First, we can assume that

θ is known *with certainty,* in which case the determination of the act to follow is straightforward. That is, suppose it is known that $\theta = 2$ in the Example 21.1 problem. Then, from the decision tree in Figure 21.2, reserve 2 is the best act to follow.

Second, we can assume total ignorance of the value of θ, in which case we are said to be making a decision *under uncertainty.* Instances in which decisions are made under uncertainty, while few in number, do exist. These instances usually occur whenever something is being done for the first time, although even in these instances we may be able to say something about the probability of various values of θ based on a recent experience of a like occurrence.

And third, we can assess a probability distribution for various values of θ occurring, in which case we are making decisions *under risk.* The majority of both personal and business decisions are made with imperfect knowledge of the value of θ, and hence decision making under risk characterizes most decision situations. While other assumptions are often made concerning knowledge of θ, a close examination of many situations reveals that decisions are made with somewhat less than perfect information, or are actually made under *risk.*

In this section, we will describe four strategies for resolving decisions under uncertainty: maximin profits, maximax profits, minimax regret, and Laplace, or the strategy of insufficient information. Chapters 22 and 23 will then discuss decision making under risk.

☐ 21.3.1 Maximin profits

The maximin profits strategy dictates that we select the best of the worst—that is, that we maximize the worst that can happen to us. It is derived by selecting the minimum payoff for each act and maximizing the minimum of these payoffs.

Maximin profits

$$\underset{i}{\text{Max}}\,[\,\underset{j}{\text{Min}}\,(\Omega_{ij})]$$
(21.2)

Example 21.4 Compute the maximin profits strategy for the data given in Example 21.1.

Solution For the tour problem of Example 21.1, the maximin profits strategy dictates that George schedule no seats in advance:

$$\underset{i}{\text{Max}}(0, -20, -40, -60, -80, -100, -120) = 0$$

With this strategy, George is *guaranteed* of losing no more than $0 irrespective of the number of seats or tours actually demanded. The maximin profits strategy is also depicted in Table 21.4.

The maximin strategy is *ultraconservative* since it hedges against the *worst* that can possibly happen. Its use appears to be based on the premise that forces that control the state of nature are malevolent, that forces are "out to get" the decision maker. It frequently leads to inaction (e.g., reserve 0) since almost every decision *might* lead to losing money (a negative payoff), whereas doing nothing incurs no monetary losses. Still, the strategy does have applicability in certain one-on-one competitive situations as well as others.

TABLE 21.4 Determination of maximin profits strategy for tour problem of Example 21.1*

Act, a_i \\ State of nature, θ_j	0	1	2	3	4	5	6	
0	0	0	0	0	0	0	0	←Maximum = 0 at a_0
1	−20	50	50	50	50	50	50	
2	−40	30	100	100	100	100	100	
3	−60	10	80	150	150	150	150	Minimum payoff of each act (shaded area)
4	−80	−10	60	130	200	200	200	
5	−100	−30	40	110	180	250	250	
6	−120	−50	20	90	160	230	300	

* Quantities are in dollars.
Source: Example 21.1.

☐ 21.3.2 Maximax profits

The decision maker who uses the maximax profits strategy has the opposite view of the world as does the maximin strategist. He views the world as being composed of "friendly forces," and hence chooses to maximize the best consequence of all possible acts. He is termed *ultra-optimistic*.

$$\text{Maximax profits}$$
$$\underset{i}{\text{Max}} \left[\underset{j}{\text{Max}}(\Omega_{ij}) \right] \tag{21.3}$$

Example 21.5 Using the data in Example 21.1, compute the maximax profits strategy.

Solution For the tour problem of Example 21.1, the maximax profits value is $300:

$$\text{Max}_i(0, 50, 100, 150, 200, 250, 300) = \$300$$

The act that corresponds to this maximax profit strategy is "reserve six seats."

This strategy poses a set of disadvantages similar to the maximin strategy. Just as the world is rarely judged to be malevolent, so too is it rarely judged as being composed entirely of friendly forces. This strategy can be appropriate when it is necessary to "go for broke," such as when a major discovery or innovation is required to prevent a firm from going under.

☐ 21.3.3 Minimax loss (minimax opportunity loss, minimax regret)

The minimax loss strategy operates on the opportunity loss matrix in reverse of the way the maximin profits strategy operates on the payoff matrix. First, the maximum opportunity loss for each act is computed. Then the strategy corresponding to the minimum of these maximum losses is selected.

$$\begin{array}{c} \text{Minimax opportunity loss} \\[6pt] \underset{i}{\text{Min}}[\underset{j}{\text{Max}}\, l(a_i, \theta_j)] \end{array} \qquad (21.4)$$

Example 21.6 Using the data in Example 21.2, compute the minimax loss strategy for George Thomas.

Solution From the opportunity loss matrix for the tour problem (Table 21.3), the minimax loss value is:

$$\text{Min}_i(300, 250, 200, 150, 100, 100, 120) = \$100$$

The strategies corresponding to the minimax opportunity losses of $100 are to reserve four *or* five seats for the tour.

Under the minimax loss strategy, George should reserve either four or five seats and minimize the maximum opportunity loss he would suffer. Note that in general, the maximin profit and the minimax loss strategies are not equivalent, as shown in this example.

The minimax loss strategy has the desirable property that it focuses on the analysis of the opportunity costs that arise in decision problems. Thus, for example, the opportunity loss matrix shows more of a loss when two seats are reserved and five versus four are demanded, whereas the payoff matrix indicates the same profit (reward) for both demands for the one reserve level. Still, it is a pessimistic strategy, since it examines only the worst regrets for each act (alternative) and ignores all others.

☐ 21.3.4 Principle of insufficient information or Laplace method

The principle of insufficient information strategy bridges the gap in a sense between decision making under uncertainty and decision making under risk. Under it, we assume that each state of nature is equally likely to occur. The Laplace value of a strategy then equals the arithmetic average of the values for that strategy in the payoff matrix.

$$
\text{Laplace value} \\
\underset{i}{\text{Max}} \left[\frac{1}{N} \sum_j \Omega_{ij} \right] \tag{21.5}
$$

In the formula, N is equal to the number of possible states of nature ($N = 7$ for the tour problem of Example 21.1).

Example 21.7 Compute the Laplace values for each of the acts in Example 21.1, and determine the act to follow that maximizes the Laplace value for this problem.

Solution The Laplace values for the tour problem of Example 21.1 are:

Act (supply of seats reserved)	0	1	2	3	4	5	6
Laplace value	0	40	70	90	100	100	90

Hence, George should reserve either four or five seats and maximize the Laplace value of 100.

The Laplace method does offer certain advantages over the other methods described thus far. First, it uses all the information available in the payoff matrix—it does not examine only the "extremes." Second, it forces the decision maker to think in probabilistic terms; if he has no reason to suspect that one event is more likely to occur than another, then he should be willing to weight them equally in arriving at a decision.

Still, caution should be used when applying the Laplace strategy or the principle of insufficient information whenever we suspect that events are not equally likely to occur. Whenever $P(\theta_0) \neq P(\theta_1) \neq . . \neq P(\theta_n)$, the Laplace method is not appropriate. Too, we should be cautious that the decision maker is not using the Laplace method as a means for avoiding the sometimes difficult task of assessing probabilities for the various states of nature.

■ 21.4 SUMMARY

This chapter has addressed statistical decision making under uncertainty. We use the theory provided whenever knowledge does not exist concerning probabilities of various states of nature occurring, and the opportunity does not exist to obtain additional information.

Different alternatives or methodologies were described for structuring a decision problem: the payoff or loss matrix, the functional form, and decision trees. Payoff or loss matrices offer the advantage of portraying the payoff or loss of every act-state of nature combination when the number of possible states of nature and acts is finite and small. The functional form is a concise means of representing the effects of the interaction of act-state of nature pairs when the number of possible pairs is large or even infinite (in the case of a continuous probability distribution for θ). Decision trees are particularly helpful when it is useful to examine the results of various decisions (act selections) at different decision points.

Decision making, or act selection, was then categorized as being made under certainty, under uncertainty, or under risk, depending on knowledge concerning the state of nature θ. Four different strategies were then described for making decisions under uncertainty: maximin profits, maximax profits, minimax loss, and the principle of insufficient information. The maximin profits strategist believes the world to be malevolent and out to get him, whereas the maximax profits strategist believes the world to be benevolent and composed exclusively of friendly forces. The minimax loss

TABLE 21.5 Strategies for making decisions under uncertainty

The set of acts: $a_i \in A$ Payoff or profit: $P(a_i,\theta_j) = \Omega_{ij}$	The set of possible states of nature: $\theta_j \in \theta$ Opportunity loss: $l(a_i,\theta_j) = (\underset{i}{\text{Max }} \Omega_{ij}) - \Omega_{ij} \geq 0$

Strategy	Derivation	Formula number	Assumptions about the world
Maximin profits	$\underset{i}{\text{Max}}[\underset{j}{\text{Min}}(\Omega_{ij})]$	(21.2)	Malevolent; forces are out to get the decision maker; strategy is ultraconservative
Maximax profits	$\underset{i}{\text{Max}}[\underset{j}{\text{Max}}(\Omega_{ij})]$	(21.3)	Benevolent; world is composed of friendly forces; strategy is ultraoptimistic
Minimax loss	$\underset{i}{\text{Min}}[\underset{j}{\text{Max}} \, l(a_i,\theta_j)]$	(21.4)	Malevolent; the worst will probably happen; strategy is conservative
Laplace (principle of insufficient information)	$\underset{i}{\text{Max}} \dfrac{\sum\limits_{j=1}^{N} \Omega_{ij}}{N}$	(21.5)	All states are equally likely to occur; strategy is neither optimistic nor pessimistic

strategist is still a pessimist, in that he believes in examining only the worst consequences for each act. This strategist does, however, make an attempt at incorporating profit foregone (opportunity loss) into his decision. And finally, the person using the principle of insufficient information would like to incorporate all consequences of each act into his decision (not just the best or the worst) and, lacking information on the various states of nature, is willing to assume that each state of nature is equally likely to occur. This person thus attempts to bridge the gap between decision making under uncertianty and decision making under risk. Table 21.5 is a summary of these four strategies for making decisions under uncertainty.

In Chapter 22, we will describe decision making under uncertainty, where a probability distribution for the possible states of nature either exists or can be readily determined. Then, in Chapter 23, we will again examine decision making under uncertainty, but for the special case in which there is an opportunity to obtain additional (sample) information regarding the various states of nature. We will see in Chapter 23 that statistical decision theory provides us with a model for determining how much information should be obtained and how much we should be willing to pay for it!

▉ REFERENCES

Bierman, H., Jr.; Bonini, C.P.; and Hausman, W. H. *Quantitative Analysis for Business Decisions.* 5th ed. Homewood, Ill.: Richard D. Irwin, Inc., 1977.

Raiffa, H., and Schlaifer, R. *Applied Statistical Decision Theory.* Boston: Division of Research, Harvard University, 1961.

Schlaifer, R. *Probability and Statistics for Business Decisions.* New York: McGraw-Hill Book Company, Inc., 1959.

Willis, R. E., and Chervany, H. L. *Statistical Analysis and Modeling for Management Decision Making.* Belmont, Calif.: Wadsworth Publishing Co., Inc., 1974.

Winkler, R. L. *An Introduction to Bayesian Inference and Decision.* New York: Holt, Rinehart and Winston, Inc., 1972.

▉ PROBLEMS

21.1. Explain the difference between decision making under risk and decision making under uncertainty. Why is the Laplace method considered to be a part of decision making under uncertainty as opposed to decision making under risk?

21.2. Describe a situation you encountered today in which you had to make a choice between alternatives without the benefit of obtaining additional information. Is it possible that you could have reduced the risk in the situation by obtaining additional information if circumstances had been different?

21.3. What is the payoff function (functional form) or the profit function for the tour problem of Example 21.1?

21.4. Given the following payoff table, construct the corresponding loss table.

State of nature / Act	I	II	III	IV	V
1	−10	12	4	7	−3
2	14	8	6	−4	−4
3	3	10	−10	8	4
4	7	0	1	4	9
5	8	4	11	−6	6

21.5. Given the payoff and loss table of Problem 21.4, what is the appropriate act to follow under each of the following strategies?

a. Maximin profits.
b. Maximax profits.
c. Minimax loss.
d. Laplace method.

21.6. Given the following payoff table, can any of the acts (with their associated payoffs) be removed from the table without affecting the optimal act to follow? If so, such acts are said to be

State of nature / Act	I	II	III	IV	V
1	7	−3	2	12	5
2	−2	−4	10	−4	−6
3	2	4	9	6	3
4	6	10	3	11	−2
5	−2	−2	11	4	−4
6	1	9	7	2	8
7	4	−4	0	−1	4

inadmissible or *dominated* acts and can be dropped from further analysis.

21.7. Describe situations (and amounts involved) in which you might use the maximin profits strategy in making a decision (choosing between possible acts), the maximax profits strategy, and the minimax loss strategy.

21.8. Roger Clark obtains long-stem roses from Fred's Flowers each Friday for sale to the public. Fred generally sells Roger flowers that, if left over the weekend, would likely be unsalable the next week. Hence, he is willing to give Roger a good price on his remaining long-stem roses and to take back any roses that are not sold over the weekend and give Roger partial credit for them. Roses are sold to Roger in quantities or units of one dozen each and at a price of $10 per dozen. Roger sells the roses for $15 per dozen, and any remaining unsold over the weekend can be returned to Fred at a credit of $8 per dozen. Although the number of roses remaining to be sold at the end of the week varies, Fred is willing to let Roger have as many as six dozen each weekend and supply him with fresh flowers at the same price if Roger desires more than the quantity of "old" roses that Fred has. Characterize this decision problem both in terms of the various states of nature (demands) and in terms of the various acts (supplies) open to Roger. Construct the payoff and loss matrices for this problem.

21.9. Using the data given in Problem 21.8, compute the appropriate act to follow under each of the following strategies.

a. Maximin profits.
b. Maximax profits.

c. Minimax loss.

d. Laplace method.

21.10. What is the functional form of the profit function for the inventory problem of Problem 21.8?

21.11. Through inheritance, a trust has been setup for a teenager to attend college. Conditions of the trust are such that the funds can be invested in one of four "portfolios" (alternatives), each of which matures in the next three years when the individual will graduate from high school and begin attending college. Depending on economic conditions over the next three years, the net gain or loss in the amount of the trust is as shown in the payoff matrix at the bottom of the page. Compute the appropriate act to follow under each of the following strategies.

a. Maximin profits.

b. Maximax profits

c. Laplace method.

21.12. Construct an opportunity loss matrix for the data in Problem 21.11, and compute the appropriate act to follow under the minimax loss strategy. What type of investment strategy would you guess is being followed under act 3?

21.13. Freddie Flicker is the proprietor of Frieda Flicker's Wholesale Produce Shop. Each week, Freddie must decide how many cases of lettuce to order for sale to retail outlets. In the past, the demand for cases has varied between zero and four per week, with no apparent "pattern" to the demand quantities. For each case sold, Freddie earns a nice profit of $30. For each case unsold at the end of the week, Freddie loses $10. Give the functional form of payoff for this problem.

21.14. Using the data given in Problem 21.13, construct the payoff matrix and compute the appropriate act to follow under the following strategies.

a. Maximin profits.

b. Maximax profits.

c. Laplace method.

21.15. Compute the opportunity loss matrix for the data in Problem 21.13, and compute the optimal act to follow under the minimax loss strategy.

21.16. Compute the opportunity loss function for the data in Problem 21.13, and construct a graph similar to Figure 21.1 for displaying these conditional losses.

21.17. Construct a payoff decision tree for the data in Problem 21.11.

Act	State of nature (the economy) I	II	III
1	1,100	5,600	−2,500
2	6,500	−3,500	1,350
3	2,250	2,250	2,250
4	−4,000	1,590	7,200

Statistical decision theory: Decision making under risk (terminal acts)

22

■ 22.1 INTRODUCTION

In this chapter, we will introduce the topic of statistical decision theory for the important case of decision making when there are not opportunities to obtain additional information, but information does exist on the probability distribution of θ, $P(\theta)$ [or $f(\theta)$]. These instances arise frequently in our daily lives as we are confronted with uncertain situations, but must make an immediate decision. For example, we may look outside to decide whether or not to carry an umbrella on a specific day. We could, if time allowed, listen to the weather report to get additional information about the weather, but in so doing, we run the risk of missing our means of transportation, being late, or missing an appointment. Funds also can be a limiting factor in determining whether additional information can be obtained. For example, we may have spent all that we feel we can in searching out employment opportunities, and we must now arrive at a terminal decision on where we will work.

The case in point is that, for whatever reason, each of us daily faces uncertain situations in which a decision must be made *without benefit of additional information*. Statistical decision making under risk and with terminal acts serves as a medium for analyzing a decision situation and then prescribing a course of action to follow that satisfies a given measure of performance.

In this chapter, we will often refer to the probability distribution of the random variable θ as the *prior distribution* or *prior probability distribution* of θ. For our purposes, the prior probability distribution of θ will simply consist of the probability distribution of θ that currently exists. In Chapter

703

23, we will learn how to revise this distribution in the light of new information, and in fact, will assess just how much effort (dollars) should be expended in obtaining additional information on the value of the random variable. For the moment, however, we will assume that the opportunity does *not* exist to obtain additional information and, hence, will work with the prior (existing) probability distribution of the random variable θ.

■ 22.2 DECISION MAKING UNDER RISK—DISCRETE PROBABILITIES

□ 22.2.1 Finite action decision problems with three or more alternatives (acts)

When making decisions under risk, we inherently assume that a probability distribution can be assessed for the various states of nature, θ_j. This distribution can be assessed using *any* of the methods described in Chapter 3, as well as can be based on sample results. In the majority of decision situations, it is possible to assess some probability distribution for the various states of nature (e.g., the probability of passing this course), even though this assessment may be a difficult task! In situations in which probabilities can be assessed for the various states of nature, the *equivalent* strategies of maximizing expected monetary value (EMV) (or payoff) and minimizing expected opportunity loss (EOL) may be appropriate.

Expected value decision making thus incorporates the likelihood or the probability of certain events occurring in the analysis of a decision. Unlike maximin profits, maximax profits, or minimax loss (or regret) from Chapter 21, expected value decision making takes into account *all* the values in the payoff or the loss matrix, as well as the relative likelihood of events occurring. And, unlike the Laplace method, probabilities of events occurring are determined by objective and subjective estimates, rather than by the belief that all events are equally likely.

□ 22.2.2 Bayes criterion and expected monetary value

In Chapter 5, we defined an expected value of a discrete random variable to be the arithmetic mean or average of a random variable. That is, in computing an expected value, we weight each value of the random variable by its relative frequency or its probability of occurrence. Since an expectation determined in statistical decision theory usually involves monetary units, such an expectation is referred to as an *expected monetary value* (EMV), or possibly as an *expected opportunity loss* (EOL).

The expected profit (payoff) maximization *criterion* in statistical decision theory states that the expected profit of each act should be computed, and that act (course of action) selected that yields the greatest expected profit.

> Maximize expected monetary value (EMV) (profit)
>
> $$\text{Max}_i [E(\Omega_{ij})] = \text{Max}_i \sum_j \Omega_{ij} P(\theta_j) \qquad (22.1)$$

In the equation, Ω_{ij} = profit if act a_i is selected and state of nature θ_j occurs, and $P(\theta_j)$ = probability that state of nature θ_j occurs.

The Bayes criterion or the minimize expected opportunity loss criterion is statistical decision making states that the expected opportunity loss of each act should be computed, and that act selected with the minimum expected opportunity loss.

> Bayes criterion (minimize EOL)
>
> $$\text{Min}_i [E\{l(a_i,\theta_j)\}] = \text{Min}_i \left[\sum_j l(a_i,\theta_j) \cdot P(\theta_j) \right] \qquad (22.2)$$

We will illustrate the use of these two decision-making criteria in the following example.

Example 22.1 Assume that George in Example 21.1 has kept track of the number of seats sold on each tour and has assessed probabilities of various sales amounts occurring according to the long-run relative frequency interpretation of probability as given in Table 22.1.

TABLE 22.1 Probabilities (long-run relative frequency) of various sales levels (demands)

State of nature, θ_j (sales)	0	1	2	3	4	5	6
Long-run relative frequency	0.01	0.08	0.32	0.31	0.19	0.06	0.03

Source: George's 6-year travel agency records.

Compute the expected payoff (profit) and the expected opportunity loss of each act. Which act (number of seats to reserve) should George select?

Solution For the tour problem, the act "reserve 2" yields the following expected profit:

$$-40(0.01) + 30(0.08) + 100(0.32) + 100(0.31) + 100(0.19)$$
$$+ 100(0.06) + 100(0.03) = \$93.00$$

Expected profits for each of the other possible acts are given in Table 22.2, where a stick diagram has been constructed depicting the probabilities of the various states of nature. Based on historical frequencies, George should

TABLE 22.2 Calculation of expected monetary value of all acts for tour problem—empirical discrete distribution

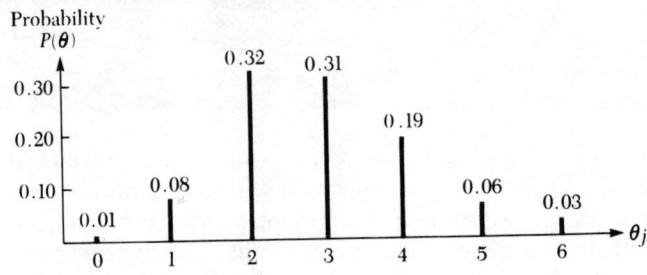

Act, a_i (supply) \ State of nature, θ_j (demand)	0	1	2	3	4	5	6	Expected monetary value of act
0	0	0	0	0	0	0	0	$ 0.00
1	−20	50	50	50	50	50	50	49.30
2	−40	30	100	100	100	100	100	93.00
3	−60	10	80	150	150	150	150	114.30 ←
4	−80	−10	60	130	200	200	200	113.90
5	−100	−30	40	110	180	250	250	100.20
6	−120	−50	20	90	160	230	300	82.30

Source: Examples 21.1 and 21.2.

reserve three seats on each tour and maximize his expected profit of that act of $114.30 as indicated in Table 22.2. Any other act would reduce his expected profit below $114.30.

For the same problem, the act "reserve 2" yields an expected opportunity loss of $51.50 as follows:

$$40(0.01) + 20(0.08) + 0(0.32) + 50(0.31) + 100(0.19)$$
$$+ 150(0.06) + 200(0.03) = \$51.50$$

The opportunity losses for each of the other acts are given in Table 22.3. Using the Bayes criterion, George should reserve three seats on each tour, yielding a minimum expected opportunity loss of $30.20. Note that this is the same act selected using the profit (EMV) maximization criterion, and in fact the two methods are equivalent in the sense that they will always lead to the same act. (A proof of this relationship is left to Problem 22.1 at the end of the chapter.)

TABLE 22.3 Calculation of expected opportunity loss of all acts for tour problem—empirical discrete distribution

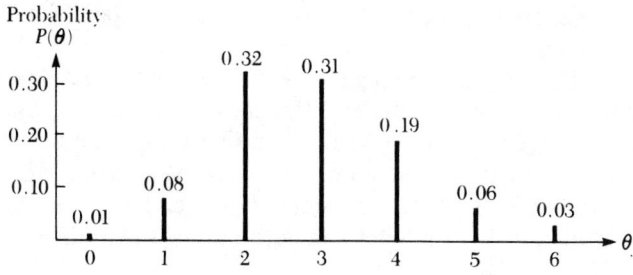

State of nature, θ_j (demand) / Act, a_i (supply)	0	1	2	3	4	5	6	Expected opportunity loss
0	0	50	100	150	200	250	300	$144.50
1	20	0	50	100	150	200	250	95.20
2	40	20	0	50	100	150	200	51.50
3	60	40	20	0	50	100	150	30.20 ←
4	80	60	40	20	0	50	100	30.60
5	100	80	60	40	20	0	50	44.30
6	120	100	80	60	40	20	0	62.20

Source: Examples 21.1 and 21.2.

It is also possible to determine the optimal act to follow by computing the $[C_u/(C_o + C_u)]$th fractile of the probability distribution of θ, where C_o represents the opportunity cost of overbooking one seat, and C_u represents the opportunity cost of underbooking one seat.[1]

From Table 22.1, we can compute the cumulative mass function for θ [i.e., $F(\theta)$] by simply accumulating the probabilities of the various values of θ. Figure 22.1 gives the cumulative distribution function of θ in graphical form for the probability mass function in Table 22.1.

For the data in Example 22.1, $C_u = \$50$ and $C_o = \$20$. Hence,

$$\frac{C_u}{C_o + C_u} = \frac{50}{20 + 50} = \frac{50}{70} = 0.71$$

---

[1] A proof of this relationship involves (in the continuous case) calculus, and hence is omitted here.

By locating 0.71 on the cumulative probability axis and constructing a horizontal line to the point where it intersects the cumulative probability stick as shown in Figure 22.1, the optimal number of units to stock can be determined.

The difficulty in applying this technique to discrete data is that probability is massed only at the integers. Hence, it becomes necessary to determine at which integer (stock level) the cumulative probability is greater than the $[C_u/(C_o + C_u)]$th fractile, and is less than the $[C_u/(C_o + C_u)]$th fractile for the previous admissible level of demand (value of θ). A graph of the cumulative mass function is often helpful in determining the optimal value in this regard.

FIGURE 22.1 Cumulative mass function for data in Table 22.1

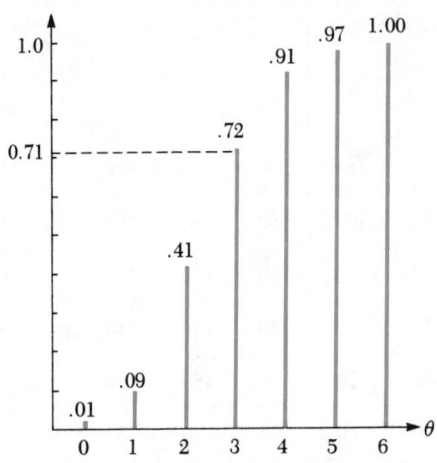

22.2.3 Expected profit under certainty

From Table 22.2, note that if demand is 1, profit is maximized if one seat is reserved; if demand is 2, profit is maximized if two seats are reserved; etc. Thus, the *maximum* profits to be realized from alternate acts *conditional upon the value of* θ are given in the principal diagonal of Table 22.2. The sum of the products of the maximum payoff for each value of θ and the probability of θ occurring have a special meaning in statistical decision making called the *expected profit of a perfect predictor* (EPPP). It represents the expected profit that would be realized if one knew beforehand the value of θ that will occur; or, in other words, what expected profit would be if one could predict precisely the state of nature that might occur.

This quantity, the *expected profit of a perfect predictor*, is also called the *expected profit under certainty* and is given by equation (22.3).

Expected profit under certainty

$$\text{EPPP} = \sum_j P(\theta_j)\,(\underset{i}{\text{Max}}\ \Omega_{ij}) \qquad (22.3)$$

Note how we should interpret the expected profit of a perfect predictor, EPPP. If we knew beforehand that there would be a demand for three seats, we would advise George to reserve three seats, and he would realize a profit of 3($50) = $150. On what percentage of the tours will three seats be demanded? Using the long-run relative frequency interpretation of probability, on 31 percent of the tours, three seats will be demanded. The product (0.31)($150) represents the profit that can be realized if we know in advance what demand will be, multiplied by the relative frequency or the probability with which this demand or profit potential will occur. When we continue to multiply these maximum profits by the probability that (or the frequency with which) they occur, and then add the corresponding products, we obtain the *expected profit* from reserving the optimal number of seats *each time* a tour is selected. That is, we are simply weighting each profit figure by the frequency with which it will occur in the long run.

☐ 22.2.4 Expected value of perfect information

Using equation (22.1), we compute the expected profit to be realized by selecting the optimal act (alternative) under the prior probability distribution of θ. The difference between the profit to be realized with perfect knowledge of the various states of nature and the best we can do (maximize EMV criterion) *without this information* represents an upper bound for the amount we should be willing to pay for this information. That is, the difference between EPPP and the maximum EMV represents the *expected value of perfect information* delineated in equation (22.4).

Expected value of perfect information

$$\text{EVPI} = \left[\sum_j P(\theta_j)(\underset{i}{\text{Max}}\ \Omega_{ij})\right] - \left\{\underset{i}{\text{Max}}\left[\sum_j P(\theta_j)\Omega_{ij}\right]\right\} \qquad (22.4)$$

Example 22.2 Given the information provided in Example 22.1, compute the expected value of perfect information.

Solution First, we must compute the expected profit under certainty, or the expected profit of a perfect predictor of the number of seats demanded on a tour. Using equation (22.3), we compute this quantity as shown in the table.

Expected profit of a perfect predictor

(1) Demand (event)	(2) Probability of event	(3) Conditional (maximum) profit*	(4) (2) · (3)
0	0.01	0	0.00
1	0.08	50	4.00
2	0.32	100	32.00
3	0.31	150	46.50
4	0.19	200	38.00
5	0.06	250	15.00
6	0.03	300	9.00
	1.00		$144.50 = EPPP

* Quantities are in dollars.

The expected profit or payoff for the best act (reserve 3) *without benefit* of a perfect predictor of demand is $114.30. The difference between the expected profit with perfect information and the expected profit with the best act is: EVPI = $144.50 − $114.30 = $30.20. Thus, if George could somehow obtain *perfect* information regarding the number of seats demanded on his tours, he would, over the long run, expect to make $30.20 more on each tour than if he used the act that maximizes expected monetary value—reserve three seats each time. This quantity, the expected value of perfect information, sets an upper limit on the amount George should be willing to pay for such information.

☐ 22.2.5 Conditional value of perfect information

Recall that a conditional opportunity loss is defined to be the difference between the profit or payoff that could have been realized and what was realized from following a certain course of action. The expected value of perfect information can be expressed as the expectation (with respect to θ) of the conditional losses that result from selecting *that particular act that is optimal.* Thus, the expected value of perfect information can be calculated by using either equation (22.2) or (22.4). Because the expected opportunity losses may have already been computed in determining the optimal act, equation (22.2) may be preferred in those cases. Note that whatever formula (22.2 or 22.4) is used to compute EVPI, EVPI can be computed only *after* the optimal act to follow has been selected, and it is computed using the *conditional losses for the optimal act.*

The conditional losses (conditional opportunity losses) for the optimal act are also referred to as the *conditional value of perfect information* (CVPI). That is, *after* the optimal act to follow has been determined, the conditional value of perfect information is equal to the opportunity loss for values of θ

that differ from the value that is optimal under the maximize EMV or the minimize EOL criterion. For example, referring to Table 22.3, the optimal act is to reserve three seats. The *conditional* value of perfect information that θ is equal to 0 (no seats are demanded) is equal to $60. That is, if demand is 0, the decision maker will save $60 if he reserves no seats, as opposed to reserving the optimal number of seats. The decision maker should be willing to spend up to $60 for information that θ will equal 0, because this is the amount that will be saved by reserving zero as opposed to three seats.

The "conditional" in the conditional value of perfect information thus refers to the value of perfect information being conditional on the value of θ that occurs. For example, again referring to Table 22.3, what should the decision maker be willing to spend for information that the number of seats demanded on a tour is five? Since profit can be *increased* by $100 if five as opposed to three seats are reserved, the conditional value of perfect information that $\theta = 5$ in this instance is $100, and the decision maker should be willing to spend up to $100 for this information. Thus, the conditional value of perfect information varies as the value of θ deviates from its optimal value determined by the EMV or EOL criterion. The conditional value of perfect information is delineated in Definition 22.1.

Definition 22.1
Conditional value of perfect information (CVPI)

The *conditional value of perfect information* is equal to the conditional opportunity loss of the optimal act under the probability distribution of θ.

Given the definition of CVPI, the expected value of perfect information can alternately be computed using equation (22.5).

Expected value of perfect information (*EVPI*)

$$\text{EVPI} = \sum_j P(\theta_j)\text{CVPI}_\theta \qquad (22.5)$$

If one examines carefully the sum of the expected opportunity loss and the expected payoff *for each act* in Examples 22.1 and 22.2, an interesting relationship appears. The sum of these two quantities is the same ($144.50) for each act and, in fact, is equal to the expected payoff of a perfect predictor or the expected profit under certainty. This identity, EPPP = EMV + EOL, for each act is often used in checking the calculations performed in solving for the expected payoff or the expected value and the expected opportunity loss of each act.

☐ 22.2.6 Two-action decision problems

Frequently, a statistical decision problem involves the choice between *two* alternatives (acts): for example, whether to lease or buy a piece of equipment; whether to purchase machine A or machine B; whether or not to stay in business; or whether to accept job 1 or job 2. This is to be contrasted to the tour problem of the previous section where the choice was from among several (seven) alternatives (reserve 0, 1, 2, 3, 4, 5, or 6 seats). Frequently, whenever the choice between alternatives admits only two possibilities, the choice between alternatives involves the simple comparison of the expected value of θ, $E(\theta)$, with a break-even value of θ (to be defined). Also, in many decision problems, the computational effort can be additionally simplified whenever the loss or payoff function is linear in θ. We will examine two-action decision problems with linear payoffs and losses in this section. Although the ideas and techniques presented can be generalized to include the type of problem treated in the previous section, these generalizations or extensions are amply treated in the advanced references at the end of this chapter, and so are left to these more detailed works.

We will denote the *costs* associated with the first act (alternative) by:

$$C_1 = B_1 + b_1\theta$$

where B_1 represents a "fixed cost" associated with choosing alternative 1, and $b_1\theta$ represents a cost that varies as a function of the value of the random variable θ (i.e., a constant times the value of the random variable). Similarly, the costs associated with the second alternative are:

$$C_2 = B_2 + b_2\theta$$

First, note that the expressions for C_1 and C_2 are *linear functions* of θ (these expressions involve no squares, logarithms, etc.). Second, note that it is possible for B_1, B_2, b_1, or b_2 to be equal to 0, and the cost expressions will still be linear functions. And finally, note that we could have just as easily chosen to work with profit as opposed to cost figures to derive an optimal plan.

Let us compute the expected value of each of the cost expressions C_1 and C_2. Recall from Chapter 5 that the expected value of a constant is equal to the value of the constant, and that the expected value of a constant times a random variable is equal to the constant times the expected value of the random variable. Thus,

$$E(C_1) = E(B_1 + b_1\theta) = E(B_1) + E(b_1\theta) = B_1 + b_1E(\theta)$$

Similarly,

$$E(C_2) = E(B_2 + b_2\theta) = E(B_2) + E(b_2\theta) = B_2 + b_2E(\theta)$$

We would prefer alternative 1 to alternative 2 if $E(C_1) < E(C_2)$. Substituting into this expression, we obtain:

$$B_1 + b_1 E(\theta) < B_2 + b_2 E(\theta) \qquad \text{or} \qquad B_1 - B_2 < (b_2 - b_1)E(\theta)$$

If we now assume that $b_2 > b_1$ (if b_2 is not greater than b_1, we can simply renumber our acts such that the second alternative always has the larger value of b), we can divide both sides of the preceding inequality by $(b_2 - b_1)$ and not change its direction. Thus,

$$\frac{B_1 - B_2}{b_2 - b_1} < E(\theta) \qquad \text{or} \qquad E(\theta) > \frac{B_1 - B_2}{b_2 - b_1}$$

If $E(\theta) > (B_1 - B_2)/(b_2 - b_1)$, we prefer alternative 1. If $E(\theta) < (B_1 - B_2)/(b_2 - b_1)$, we prefer alternative 2.

The quantity $(B_1 - B_2)/(b_2 - b_1)$ is called the *break-even value of θ, θ_b*, and is given in equation (22.6).

Break-even value of θ, θ_b

$$\theta_b = \frac{B_1 - B_2}{b_2 - b_1} \qquad\qquad (22.6)$$

The use of equation (22.6) in determining the optimal act to follow is illustrated in the following example.

Example 22.3   Burt Calkins, Jr., a confused Christmas tree farmer, is faced with the following situation. Annually, he plants 10,000 trees. The trees that remain healthy are ready for market after eight years. The trees are subject, however, to a virus that slows their growth and adds two years to the time required for them to reach a marketable stage. Mr. Calkins estimates that it costs $0.18 per year per tree to care for them for the additional two years if it is required. Hence, for those trees infected with the virus, Burt spends an additional $0.36 each.

From previous growing experience, Mr. Calkins estimates the probability distribution shown in Table 22.4 for the number of trees stunted by the virus in each lot of 10,000 trees planted. The entire lot of 10,000 trees can be sprayed at a cost of $0.10 per tree to control the virus so that the trees will reach marketable size in eight years.

Given the information provided, what is the break-even value of the number of trees infected that makes spraying versus nonspraying equally costly? What is the optimal (using the Bayes criterion) act to follow? Construct a decision tree for this problem.

Solution   The break-even proportion of trees can be determined using equation (22.6). Let C_S represent the cost of spraying the trees and let C_{NS} represent the cost of caring for the trees the extra two years. Because the entire lot of trees is either sprayed or is not sprayed,

$$C_S = 0.10(10,000) + 0.0\theta = \$1,000 \qquad \text{and} \qquad C_{NS} = 0.0 + 0.36\theta$$

TABLE 22.4 Probability distribution of number and proportion of trees stunted by the virus (in lots of 10,000 trees)

Number of trees infected per 10,000	Proportion of trees infected	Long-run relative frequency (probability)
x	π	$P(\theta)$
2,000....................	0.2	0.50
3,000....................	0.3	0.30
4,000....................	0.4	0.10
5,000....................	0.5	0.05
6,000....................	0.6	0.05
		1.00

Source: Burt's records.

Thus, as a function of θ, the random variable of interest, the break-even number of trees is

$$\theta_b = \frac{B_S - B_{NS}}{b_{NS} - b_S} = \frac{1,000 - 0}{0.36 - 0} \approx 2,778$$

Because costs are linear in θ, the optimal act to follow can be determined by comparing $E(\theta)$ with θ_b. The expected value of θ is:

θ	$P(\theta)$	$\theta P(\theta)$
2,000	0.50	1,000
3,000	0.30	900
4,000	0.10	400
5,000	0.05	250
6,000	0.05	300
	$E(\theta) =$	2,850

Since the expected number of trees infected $[E(\theta) = 2,850]$ is greater than the break-even number ($\theta_b = 2,778$), the optimal act is to spray the trees at a cost of \$1,000. A decision tree delineating the terminal costs of each act is given in Figure 22.2.

FIGURE 22.2 Decision tree for terminal act in Example 22.3

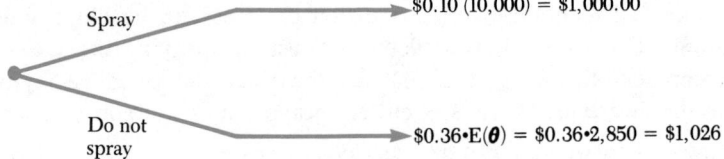

Spray → \$0.10 (10,000) = \$1,000.00

Do not spray → \$0.36•E(θ) = \$0.36•2,850 = \$1,026

FIGURE 22.3   Costs of spraying and not spraying for Example 22.3 data

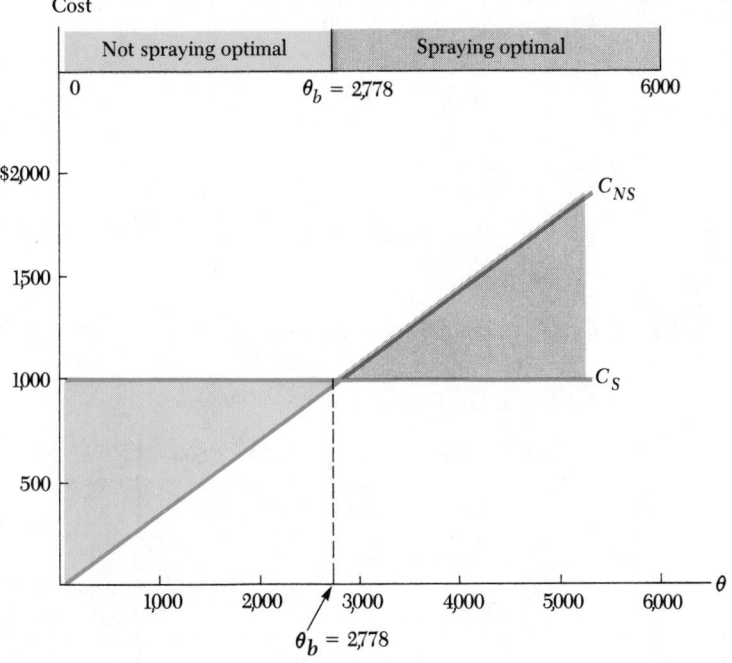

In Figure 22.3 is a graph of the costs of spraying and the costs of not spraying for this example. As can be seen, if the number of trees infected is less than 2,778, the break-even number of trees, not spraying is the optimal (least costly) decision; if the number of trees infected is greater than 2,778, then spraying is the least costly (optimal) act. Since $E(\theta) = 2,850 > 2,778$, spraying is the optimal act under the prior probability distribution of θ.

Example 22.4   Determine the conditional value of perfect information (CVPI) and the expected value of perfect information (EVPI) in Example 22.3.

Solution   The costs of both acts as a function of the number of trees infected are given in Table 22.5. Also shown in Table 22.5 are the conditional values of perfect information for the case in which $E(\theta) < \theta_b$ *and the* case in which $E(\theta) > \theta_b$. Since $E(\theta) > \theta_b$ in this example, CVPI is equal to:

θ	2,000	3,000	4,000	5,000	6,000
CVPI	280	0	0	0	0

Since the optimal act under the prior distribution is to spray, EVPI is equal to:

θ	$P(\theta)$	CVPI	$P(\theta)$ CVPI
2,000	0.50	280	140
3,000	0.30	0	0
4,000	0.10	0	0
5,000	0.05	0	0
6,000	0.05	0	0
	1.00		
			EVPI = 140

TABLE 22.5 Calculation of CVPI for Example 22.4

Act \ θ	2,000	3,000	4,000	5,000	6,000	
Spray	1,000	1,000	1,000	1,000	1,000	Costs
Do not spray	720	1,080	1,440	1,800	2,160	
CVPI if $E(\theta) < \theta_b$	0	80	440	800	1,160	Conditional opportunity losses
CVPI if $E(\theta) > \theta_b$	280	0	0	0	0	

From Example 22.4, we see that the conditional value of perfect information is greater than 0 only if the value of the random variable would lead us to select a different alternative than was suggested by comparing $E(\theta)$ with θ_b. In two-action decision problems, CVPI will be greater than 0 only for those values of θ less than (greater than) the break-even value of θ when $E(\theta)$ is greater than (less than) θ_b. Thus, in computing EVPI, we need to multiply CVPI by $P(\theta)$ only for those values of θ that would cause us to *change our act* from that selected using the maximize EMV criteria. Table 22.5 illustrates that CVPI = 0 for all values of θ for which the expectation of θ leads to the optimal act, and that CVPI > 0 for all other values of θ.

Example 22.5 Assume in Example 22.3 that the expected value of θ, $E(\theta)$, is less than the break-even value of θ, θ_b. What is the expected value of perfect information (EVPI)?

Solution Since the information would now lead us not to spray, CVPI > 0 for values of θ greater than θ_b. EVPI thus becomes:

EVPI = (0.30)(80) + (0.10)(440) + (0.05)(800) + (0.05)(1,160) = \$166

■ 22.3 DECISION MAKING UNDER RISK—CONTINUOUS
PROBABILITIES

In this section, we will select from a set of acts when the number of states
of nature is infinite! In the first section, both the number of states of nature
and the number of possible acts are (or are assumed to be) infinite. Then we
will treat the situation where the number of states of nature is infinite, but the
decision maker must choose between *two* alternatives (acts). As in the
previous section, this latter situation is stressed because so many decision
problems involve the choice between two alternatives in the final analysis.

☐ 22.3.1 **Decision making with infinite states of nature and
infinite alternatives (normal distribution)**

Since the random variable is now assumed to be continuous, its probability distribution is specified by its density function, $f(\theta)$. We will assume in
this section again that losses (opportunity losses) are linear in the random
variable θ, and we will refer to losses of overstocking and understocking a
demand item. These losses occur whenever we underestimate or overestimate the value of the random variable θ. Furthermore, we will assume that
the probability distribution governing demand for the item is normal.

Let the cost of overstocking an item be represented by C_o and the cost of
understocking an item be represented by C_u. Furthermore, assume a_i units
are stocked to meet demand. Then a loss from overstocking occurs
whenever $a_i > \theta$, the demand, and a loss from understocking occurs
whenever $a_i < \theta$. No loss occurs whenever $a_i = \theta$. This type of loss function
is given in equation (22.7). Figure 22.4 is a graph of the corresponding loss
function superimposed on probability distribution of demand.

FIGURE 22.4 Loss function
superimposed on the probability
distribution of demand

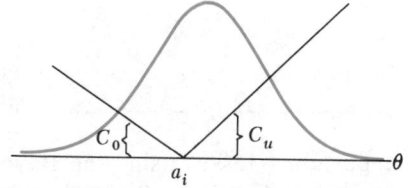

Linear losses of overstocking and understocking a demand item (continuous
distribution)

$$l(a_i, \theta) = \begin{cases} 0 & \text{if } a_i = \theta \\ C_o(a_i - \theta) & \text{if } a_i > \theta \\ C_u(\theta - a_i) & \text{if } a_i < \theta \end{cases} \qquad (22.7)$$

Shifting the quantity stocked from a_i to a_j, where $j \neq i$, shifts the vertex of the "loss function" from a_i to a_j, as shown in Figure 22.5. Thus, in a graphical sense, the solution of the statistical decision problem consists of determining where the vertex of the loss function should be placed on the θ-axis.

FIGURE 22.5 Opportunity loss functions (conditional value of perfect information) for several stocking points a_i

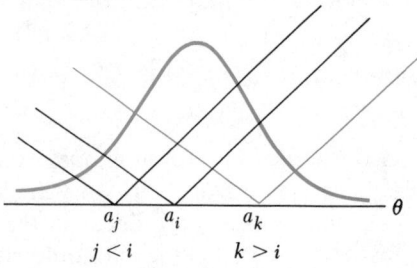

$$a_j \quad a_i \quad a_k$$
$$j < i \qquad k > i$$

The Bayes criterion for solving this problem consists of minimizing the expected losses (opportunity losses) associated with overstocking or under-stocking the demand item. In the discrete case, this loss was given by the sum of the products of the conditional losses and the probability of these losses occurring. In the continuous case, the expected loss can be determined by multiplying each of the losses by the value of the probability density function, $f(\theta)$, where the loss occurs. Since we are dealing with a continuous distribution, this involves an infinite number of multiplications!

Fortunately, it can be shown that the optimal act is to stock a_i units, where the cumulative density function of θ, $F(a)$, is equal to the following:

Optimal fractile of demand distribution to minimize opportunity losses

$$F(a_i) = \frac{C_u}{C_o + C_u} \qquad (22.8)$$

Hence, to minimize expected opportunity losses using the Bayes criterion, the decision maker should stock the $[C_u/(C_o + C_u)]$th fractile of his demand distribution. For example, if $C_u = C_o$, he should stock

$$\frac{C_u}{2C_u} = 0.50 \text{ fractile} = \text{Median}$$

If $3C_u = C_o$, he should stock

$$\frac{C_u}{4C_u} = 0.25 \text{ fractile}.$$

If $C_u = 3C_o$, he should stock

$$\frac{C_u}{4/3\,C_u} = 0.75 \text{ fractile.}$$

The use of equation (22.8) in determining the optimal number of units to stock is illustrated in the following example.

Example 22.6   A baker each morning must decide how many doughnuts to make to meet that day's demand. Each doughnut sold yields a net profit of 3 cents. Doughnuts left unsold at the end of the day are sold at a loss of 2 cents each. The baker feels that the demand distribution for doughnuts can be *approximated* by a normal distribution with an average demand of 1,200 doughnuts and a standard deviation of 100. How many doughnuts should the baker stock to minimize his expected losses (expected opportunity losses)?

Solution   Since $C_u = 3¢$ and $C_o = 2¢$, the baker should stock the $3/(2 + 3)$ $= 3/5 = 0.60$ fractile of his demand distribution. Since the baker approximates his actual demand distribution by a normal distribution with mean $\mu = 1,200$ and standard deviation $\sigma = 100$, he should stock doughnuts such that:

$$P(\theta \leq 1,200 + z \cdot 100) = 0.60$$

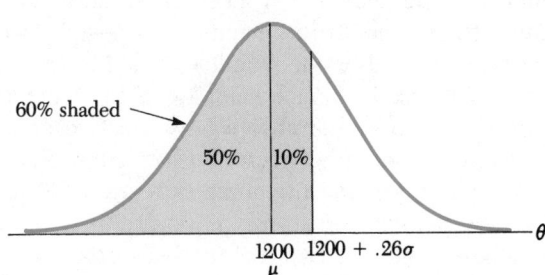

From Table B.3, Appendix B, the normal distribution table, $z \doteq 0.26$. Hence, the baker should stock 1,226 doughnuts per day to minimize his expected losses (expected opportunity losses).

Example 22.7   Assume the same facts as in Example 22.6 except that $C_o = 3¢$ and $C_u = 2¢$. Superimpose the baker's linear loss function on his assumed distribution of demand for the optimal act.

Solution

$$F(a_i) = \frac{2}{2 + 3} = \frac{2}{5} = 0.40, \qquad a_i = 1,200 - z \cdot 100$$

where

$$P(\theta \leq 1,200 - z \cdot 100) = 0.40$$

From the normal distribution table, $z \doteq 0.26$. Hence, he should stock $a_i = 1,200 - 0.26 \cdot 100 = 1,174$ doughnuts.

The application of equation (22.8) is appropriate whenever the decision maker can choose from an infinite number of acts (or assumed infinite number), the number of states of nature is infinite, and he wishes to base his

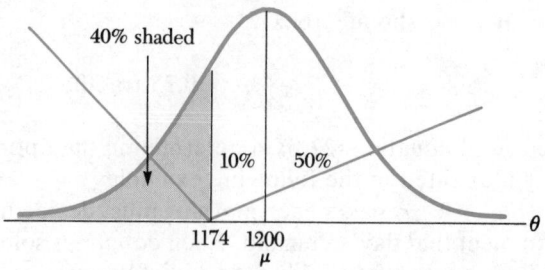

choice on minimizing his expected losses. The reason for using the normal approximation whenever the number of possible alternatives is finite (but large!) is that a table of normal probabilities is frequently available and this greatly simplifies the computation of the decision quantity as described in Chapter 7.

☐ 22.3.2 Decision making with infinite states of nature and the choice of two alternatives (normal distribution)

Frequently, the decision on whether to automate involves the choice of two alternatives. Should the machinery or whatever be purchased for the purpose of saving labor, or would the firm be just as well off because of uncertainty or risk in demand not to make a capital expenditure? Other problems also involve the choice between two alternatives—for example, when a firm is deciding between two particular pieces of equipment or two production processes and the number of states of nature is infinite.

When the number of states of nature is infinite and the number of alternatives is limited to two, the analysis proceeds similarly to the case in which the number of states of nature is finite and the number of actions is limited to two. That is, we compute a break-even value of the random variable θ and compare $E(\theta)$ with it. This comparison leads to one of the two actions being selected.

A difficulty encountered when the number of states of nature is infinite that was not encountered when the number of states of nature is finite, however, is shown in Figure 22.6. To compute the expected value of perfect

FIGURE 22.6 Conditional value of perfect informa-tion—continuous example

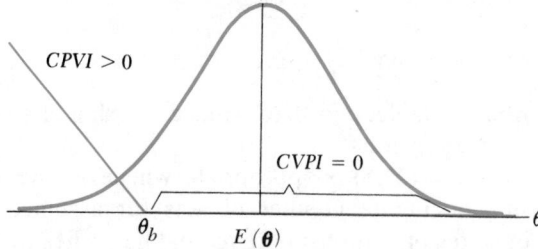

information (EVPI), we need to multiply the conditional value of perfect information by the probability density function of demand to the left or to the right of the break-even value an infinite number of times! Fortunately, tables have been provided for accomplishing this multiplication (integration) in the case of a continuous distribution. Table B.10 in Appendix B is such a table giving the required summation of products for the case in which the demand distribution is normal. That is, $N(D)$ from Table B.10, when substituted into the expression for EVPI in equation (22.9), gives the expected value of perfect information.

Expected value of perfect information—normal distribution

$$\text{EVPI} = C \cdot \sigma_\theta \cdot N(D) \qquad (22.9)$$

where

C = absolute value of the slope of the conditional value or opportunity loss line

σ_θ = standard deviation of the prior distribution

$D = \dfrac{|\theta_b - E(\theta)|}{\sigma_\theta}$

$N(D)$ is found in Table B.10 in Appendix B.

The use of Table B.10 and equation (22.9) in computing EVPI when the number of states of nature is infinite and the number of actions is limited to two is provided in the following example.

Example 22.8 Assume that a firm has the opportunity to invest in a certain piece of capital equipment that is marketed as having a rather large laborsaving potential. The cost to the firm of the particular piece of equipment is $10,000 per year (capital investment, depreciation, maintenance, etc.). The firm presently pays $5 an hour for labor. Hence, if the machine can save the company more than 2,000 labor hours per year, it will prove beneficial to purchase the equipment (i.e., $\theta_b = 2,000$ hours).

An in-house analysis is made of the potential number of labor hours saved based on previous experience with automated equipment, demand for the finished product, and other relevant information. The number of labor hours saved each year is estimated to be $E(\theta) = 2,250$, with a standard deviation of $\sigma_\theta = 150$. The firm also feels that the probability distribution of hours saved can be approximated by a normal distribution.

What is the optimal act under the prior distribution? What is the conditional value of perfect information? The expected value of perfect information?

Solution Let θ represent the number of labor hours saved. Then *profits* from the purchase of the piece of equipment can be represented by the linear function

$$\mathbf{P}_E = -10,000 + 5\theta$$

To maximize expected profits, we should compute the expectation of $\mathbf{P}_E$, $E(\mathbf{P}_E)$. Recall, however, that because the profit equation is a linear function of θ, the expected profit from use of the machine is given by

$$E(\mathbf{P}_E) = -10,000 + 5E(\theta) = -10,000 + 5 \cdot 2,250 = \$1,250$$

Since the expected profit with the machine alternative is greater than 0, the optimal act is to purchase the piece of equipment. Note that we could have derived the same conclusion by computing $\theta_b = 10,000/5 = 2,000$ hours and compared it with $E(\theta)$. since $E(\theta) > \theta_b$ ($2,250 > 2,000$), the optimal act is to purchase the piece of equipment. (Note that we are dealing with *profits* in this example and not costs. Hence, the optimizing criteria is to maximize EMV, and the optimal strategy is therefore to purchase the equipment.)

Since the optimal act under the prior distribution is to purchase the piece of equipment, CVPI is positive only for values of θ less than θ_b. We lose (or profits are decreased) $5 an hour for each hour less than the $\theta_b = 2,000$ hours that are actually saved. Hence, CVPI can be expressed as:

$$\text{CVPI} = \begin{cases} 0 & \text{if } \theta \geq \theta_b \\ 5(\theta_b - \theta) & \text{if } \theta < \theta_b \end{cases}$$

This functional relationship is superimposed on a graph of the probability density function of demand in Figure 22.7.

FIGURE 22.7 Conditional value of perfect information—automation example

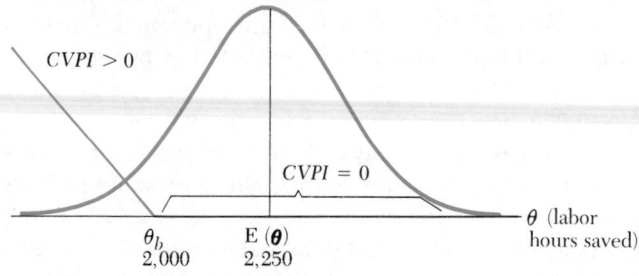

The expected value of perfect information is given by the sum of the products of the normal probability density function, $f(\theta)$, and the conditional value of perfect information to the "left" of θ_b, the break-even value.

For the current example,

$$\text{EVPI} = 5 \cdot 150 \cdot N(D) = 5 \cdot 150 \cdot N\left(\frac{|2,000 - 2,250|}{150}\right)$$

$$= 750 \cdot N(1.67) = 750 \cdot 0.01967 = \$14.75$$

Hence, the expected value of perfect information is $14.75. The reason, of course, for the relatively small value of EVPI is the distance of θ_b from $E(\theta)$ and the relatively small standard deviation, σ_θ, in relation to the mean, $E(\theta)$.

In the next section, we will examine decision situations in which the maximization of expected monetary value or the equivalent minimization of expected opportunity loss may *not* be the appropriate decision criterion when making decisions under risk.

■ **22.4 UTILITY—WHEN MAXIMIZING EXPECTED MONETARY VALUE MAY NOT BE AN APPROPRIATE GUIDE TO ACTION**

Previous sections have presented the maximization of EMV or the minimization of EOL (Bayes criterion) as the appropriate decision strategy for making decisions under risk. To see why either of these equivalent procedures *may not* be appropriate, consider the following two examples.

Example 22.9 Assume you are offered either of the following alternate courses of action:

A1: You receive $250,000.

A2: A fair coin is tossed in the air. If heads, you receive $750,000. If tails, you receive $0.

Which alternative, A1 or A2, would you select?

Solution The expected value (profit) of the second act (A2) is $375,000 [0.5($750,000) + 0.5($0)], and yet most of us would rather have $250,000 *for sure* than a 50–50 chance of winning $750,000. So, although the expected payoff is higher with A2, most of us would prefer to select A1 and guarantee ourselves a payoff of $250,000. The reason we would prefer A1 (if indeed you do) is that the *utility* of $250,000 for sure is greater than the utility of a 50–50 chance of $750,000 (or else $0!). That is, the *relative value* to us of $250,000 is greater than the relative value of a 50–50 chance of $750,000. Consider also the following example.

Example 22.10 You have an opportunity to play in a game (or not) in which a fair coin is to be tossed in the air *once*. If it lands heads up, you receive $1; if it lands tails up, you lose 60 cents (−60 cents payoff). Would you play this game? Would you play the game if the payoffs are $10 and −$6, respectively; if the payoffs are $10,000 and −$6,000, respectively?

Solution The expected payoff of not participating in any game is, of course, $0. The expected value of participating in the three games is (respectively) 20¢, $2, and $2,000.

Depending on one's attitude toward gambling, most of us would be willing to participate in the game when the expected payoff is 20 cents, since the loss of 60 cents will not materially affect our financial well-being. At an expected payoff of $2, a few of us would probably withdraw from the game since the potential loss of $6 is serious enough to make the bet undesirable. And with an expected payoff of $2,000, few of us would be left in the game since the loss of $6,000 may materially affect our financial well-being. We would prefer not to play in the game with the higher stakes even though the expected payoff is greater than the payoff of not participating.

The reason we might not choose to participate in the game with the larger stakes, or that we may in general not select the act with the highest expected payoff, is that there are factors involved *other than the monetary payoff* represented. For example, a loss of $6,000 may affect your ability to finance an automobile or a home, or you may simply face the additional harassment from your spouse for gambling with such high stakes! Similarly, most of us would prefer $250,000 for sure because it might give us an advantageous start in life, as opposed to the chance of winding up with nothing. These considerations and the St. Petersburg paradox (Problem 22.13 at the end of the chapter) led Daniel Bernoulli to investigate the maximization of expected *utility* as opposed to the maximization of expected monetary value (payoff) as a criterion for decision making under risk.

We will not describe in detail how one might assess utilities for various consequences—the actual assessment of a utility function or a utility curve is treated adequately in the references at the end of the chapter. However, whenever it is felt that factors other than those involved with the monetary payoff of an action will influence the choice of alternatives, the utility or the relative worth of each alternative should be assessed. Given the utilities of various actions, the appropriate strategy to follow is the maximization of expected utility (MEU). This is accomplished by substituting "utilities" for "payoffs" and then using the same procedures described in the text.

■ 22.5 SUMMARY

In this chapter, we examined decision making under risk where the probability distribution of the random variable either currently exists or can be readily assessed. The cases of decision making under risk included decision making with both a finite and an infinite number of states of nature and possible acts. When making decisions under risk, we inherently assume that a probability distribution can be assessed for the various states of nature, θ. By using the information provided by the probability mass function (discrete case) or the probability density function (continuous case), it becomes possible to maximize the expected monetary value or minimize the expected opportunity loss (Bayes criterion) of any decision. Hence, the equivalent strategies of maximizing EMV or minimizing EOL were put forth as the desired objective to be reached in decision making under risk.

It was then shown that if payoffs (and consequently losses) are linear in θ (i.e., of the form $P = a + b\theta$), then all that is required to compute the profit or the loss associated with each act is to compute the expectation or the mean value of θ, and substitute the value in the profit or cost equation. In these instances, the mean or expectation is said to be a *certainty equivalent*. Table

TABLE 22.6 Expected value decision making

Criterion	Payoff or loss matrix representation	Functional form, payoffs linear in θ
Maximize expected payoff	$\underset{i}{\text{Max}}\left[\sum_j \Omega_{ij}P(\theta_j)\right]$ (22.1)	$\underset{i}{\text{Max}}\ [P_i = B_i + b_i E(\theta)],$ $P_i = \text{Payoff of act } i$
Minimize expected loss	$\underset{i}{\text{Min}}\left[\sum_j l(a_i,\theta_j)P(\theta_j)\right]$ (22.2)	$\underset{i}{\text{Min}}\ [L_i = B_i' + b_i' E(\theta)],$ $L_i = \text{Loss of act } i$

22.6 is a summary of the computations required to determine the optimal act for both the payoff matrix and the functional forms.

For problems involving only two possible acts (actually the analysis can be extended to any problem involving a finite number of acts, but these cases were not treated in the chapter) and in which payoffs are linear in θ, it is possible to compute a break-even value of θ, θ_b, which gives the point at which the two acts are equally attractive. Let act 1 be denoted by a_1 with payoff $P_1 = B_1 + b_1\theta$, and let act 2 be denoted by a_2 with payoff $P_2 = B_2 + b_2\theta$, where $b_2 > b_1$. Then $\theta_b = (B_1 - B_2)/(b_2 - b_1)$, and if $E(\theta) < \theta_b$, act a_1 is optimal, and if $E(\theta) > \theta_b$, act a_2 is optimal: hence, the reduction in computations involved when one uses the expectation of θ as a certainty equivalent in these two-action problems.

The expected profit of a perfect predictor (EPPP) and the expected value of perfect information were then described. These measures denote, respectively, the expected profit to be realized if one could predict the state of nature perfectly, and the expected value of perfect information *if it could be obtained* delineating the true state of nature. The determination of the expected value of perfect information can be a difficult task depending on the probability distribution of θ, $P(\theta)$ or $f(\theta)$. Fortunately, tables of linear loss integrals such as Table B.10 in Appendix B have been derived to simplify the computations in many instances. A summary table with the computation of EVPI under a variety of conditions is given in Table 22.7. Note that EVPI is the sum of the products of the conditional value of perfect information and the probability of θ.

The case of decision making with an infinite number of states of nature and possible acts was then described. To minimize losses or maximize profits, a decision maker should "stock" the $[C_u/(C_o + C_u)]$th fractile of $f(\theta)$, where C_u represents the cost of being understocked and C_o represents the cost of being overstocked a "unit" θ.

Finally, the maximization of expected utility was put forth as an alternative to the maximization of expected monetary value whenever the range of

TABLE 22.7 Calculation of expected value of perfect information (EVPI)

Payoff matrix form, finite number of states of nature and acts	Functional form, finite number of states of nature, two acts	Functional form, infinite number of states of nature, two acts, normal distribution
$\sum_{j} P(\theta_j)$ $\left(\underset{i}{\text{Max }} \Omega_{ij} \right) -$ $\underset{i}{\text{Max}} \left(\sum_{j} P(\theta_j)\Omega_{ij} \right)$	$\displaystyle\sum_{\theta=\theta_b+1}^{\theta_{\max}} P(\theta) \cdot \text{CVPI} \quad \text{if } E(\boldsymbol{\theta}) < \theta_b$ $\displaystyle\sum_{\theta=\theta_{\min}}^{\theta_b-1} P(\theta) \cdot \text{CVPI} \quad \text{if } E(\boldsymbol{\theta}) > \theta_b$ CVPI is the slope of the opportunity loss line	$C \cdot \sigma_\theta \cdot N(D)$ where: C is the absolute value of the slope of the opportunity loss line σ_θ = standard deviation of the prior distribution $D = \dfrac{\lvert \theta_b - E(\boldsymbol{\theta}) \rvert}{\sigma_\theta}$ $N(D)$ is found in Table B.10

potential payoffs (losses) is so large that factors other than monetary ones need be considered in the evaluation of a decision.

■ REFERENCES

Bierman, H., Jr.; Bonini, C. P.; and Hausman, W. H. *Quantitative Analysis for Business Decisions*. 5th ed. Homewood, Ill.: Richard D. Irwin, Inc., 1977.

Raiffa, H., and Schlaifer, R. *Applied Statistical Decision Theory*. Boston: Division of Research, Harvard University, 1961.

Schlaifer, R. *Probability and Statistics for Business Decisions*. New York: McGraw-Hill Book Company, 1959.

Willis, R. E., and Chervany, H. L. *Statistical Analysis and Modeling for Management Decision Making*. Belmont, Calif.: Wadsworth Publishing Co., Inc., 1974.

Winkler, R. L. *An Introduction to Bayesian Inference and Decision*. New York: Holt, Rinehart and Winston, Inc., 1972.

■ PROBLEMS

22.1. Show that maximization of expected monetary value (expected payoff) is equivalent to minimization of expected opportunity loss for the case of a finite number of states of nature and possible acts. Equivalency should be established by showing that both will lead to the same action.

22.2. Every "football" Saturday, Joe Raisins wheels out his hot dog stand and sells hot dogs to attendees of the football game. Hot dogs cost Joe 20 cents each to sell, plus a 5-cent kickback per hot dog to the local syndicate for the "privilege" of operating the stand. Joe sells hot dogs for 40 cents each, and hot dogs unsold at the end of the game (plus buns, etc.) are sold back to his distributor for redistribution to college dormitory cafeterias for 10 cents each (and you always wondered why the "dorms" have hot dogs on football weekends!). For unsold hot dogs, Joe does *not* have to pay the required 5 cents to the syndicate. Joe has built up the following sales history over his last nine years in college.

Hot dogs sold (in dozens)	Frequency
10	4
11	5
12	7
13	9
14	8
15	8
16	4

Construct payoff and loss matrices similar to those in Tables 22.2 and 22.3 for this problem. What is the optimal act for Joe to follow? What is the expected value of perfect information?

22.3. Write the functional form of the payoff and loss quantities in Problem 22.2. Then graph the profit equations for each act, stock 10 dozen, stock 11 dozen, . . . , stock 16 dozen. What information does the graph provide?

22.4. Joe (in Problem 22.2) feels that his demand distribution can be represented by a binomial distribution. Using the data provided (in dozens), fit a binomial distribution to model Joe's empirical demand history. What is the revised EVPI using the binomial distribution as the demand distribution?

22.5. For the inventory (flower) problem (21.8) of Chapter 21, assume Roger is able to assess the following probability distribution of demand (θ) based on sales over past weekends:

θ	0	1	2	3	4	5	6
$P(\theta)$	0.05	0.05	0.10	0.10	0.30	0.30	0.10

Construct payoff matrices and opportunity loss matrices similar to Tables 22.2 and 22.3 for this problem. What is the conditional value of perfect information? The expected value of perfect information? How many dozens of roses should Roger stock to maximize EMV? To minimize EOL?

22.6. What is the functional form of the payoff and loss quantities in Problem 22.5?

22.7. Bruce Calkins, son of fabled midwestern Christmas tree grower Burton Calkins, Jr., regularly plants ten Christmas trees a year for sale to retail sellers to supplement their regular stocks. (Due to acreage limitations, Bruce can harvest no more than ten trees a year.) The harvesting of these trees involves cutting each tree and wrapping or binding it in steel twine for delivery. There are two methods available to Bruce to harvest these trees for market. Under the first alternative, Bruce can rent a combination chain saw-baler for $4 per day plus 50 cents per tree for the twine (all trees can be harvested in one day). Alternatively, Bruce can hire his good neighbor from Koleen, Dave Montgomery, to harvest the trees for him for delivery at a cost of $1 per tree. Bruce has collected statistics on the number of trees sold in the past 20 years and assesses the probability distribution of demand (prior) based on long-run relative frequency as given in the table at the top of the following page. Trees sell for $2.50 each and are ordered from Bruce on the first Saturday of December each year and must be delivered the following day. However, the harvesting machine

Number of trees demanded	Long-run relative frequency
0	0.00
1	0.00
2	0.00
3	0.00
4	0.05
5	0.05
6	0.05
7	0.10
8	0.30
9	0.30
10	0.15
	1.00

must be *reserved* for the December cutting by Thanksgiving of each year in order to give the owner of the harvesting unit an opportunity to rent the equipment to someone else if Bruce decides not to rent it. Hence, Bruce must rent the machine before he knows what the demand for trees will actually be. What is the expected cost of harvesting the trees using each alternative? What is the expected value of perfect information regarding demand? What is the conditional value of perfect information? What is the optimal act under the prior probability distribution assessed by Bruce?

22.8. Construct a decision tree delineating the courses of action open to Bruce, the possible states of nature, and the expected opportunity cost of each act for Problem 22.7.

22.9. What is the functional form of the conditional value of perfect information in Problem 22.7?

22.10. Referring to the equipment investment problem (Example 22.8), assume that (*a*) the standard deviation of labor hours saved is 350, (*b*) the cost of labor is $10 (changing, of course, θ_b), and (*c*) both the standard

deviation and labor costs are changed as in (*a*) and (*b*). What is the revised expected value of perfect information in each of these three instances? What does this tell you about the relationship between the standard deviation and EVPI, the "distance" between θ_b and $E(\theta)$ and EVPI, and the interaction of the "distance" between $E(\theta)$ and θ_b and the standard deviation with EVPI?

22.11. Again referring to Example 22.8, what would be the optimal decision if $E(\theta) = 1,500$ and the other quantities remain unchanged? What is the EVPI in this instance? What is the functional form of CVPI?

22.12. A firm is considering making a capital improvement to one of its production processes which will increase the quality of the end product. If the improvement is made, the firm feels it may be able to charge $5 more for the end product, although there is a 50-50 chance the improvement will lead to less than a $4 or greater than a $6 increase in what the firm can charge for the product. The cost of the improvement is $20,000 and the number of units of expected sales over the life of the improvement (ignore present value considerations) is 3,600. Fit a normal distribution to this situation and calculate the expected profit of the best act and the expected value of perfect information. Should the investment be made?

22.13. St. Petersburg paradox: a coin is to be tossed into the air k times until a head appears (i.e., a head first appears on the kth trial, $k = 1, 2, \ldots$). Your reward is to be 2^k. Thus, if a head appears on the first toss, you win $2^1 = 2; if a head does not appear until the second toss (i.e., the sequence TH results), you win $2^2 = 4; etc. What is the expected payoff of this game? Given this expected payoff, what would you be willing to pay to play the game once? What does this tell you about your utility for money?

22.14. Explain in your own words why it is "consistent" for an individual to purchase insurance and also make small bets. What roles does the variability or the dispersion of payoffs (losses) play in the determination of this type of behavior?

22.15. For the data in Problem 21.11, assume the trust officer is able to obtain a probability distribution for the various states of the economy over the next three years from his economics instructor in an evening course he is taking. The probabilities of the three states are as given in the table.

State of the economy	Probability
I	0.2
II	0.5
III	0.3
	1.0

Determine the expected monetary value and the expected opportunity loss of each act, and determine the optimal act to follow under the maxi-

mize EMV or minimize EOL criterion. What is the expected value of perfect information regarding the state of the economy? What is the conditional value of perfect information?

22.16. Peter the paperboy sells on the average 100 newspapers per day, with a standard deviation of 21 newspapers. Papers cost Peter 15 cents each and are sold for 20 cents each. Papers remaining at the end of the day can be sold back to the publisher for 11 cents each. Assuming the demand for newspapers is normally distributed, determine the number of newspapers to stock each day to minimize the expected opportunity losses of the decision.

22.17. If papers can be returned to the publisher for 12 cents net return each, how does this affect the number of papers Peter should stock in Problem 22.16?

22.18. Given the payoff table and probability distribution for the various states of nature shown below, determine the following.
  a. The maximum expected payoff.
  b. The minimum expected loss.
  c. The expected profit with a perfect predictor.
  d. The expected value of perfect information.

Act \ State of nature	I	II	III	IV	V
1	− 10	12	4	7	− 3
2	14	8	6	− 4	− 4
3	3	10	− 10	8	4
4	7	0	1	4	9
5	8	4	11	− 6	6

State of nature	I	II	III	IV	V
Probability	0.21	0.09	0.13	0.36	0.21

22.19. A firm has the opportunity to select from two alternatives for manufacturing its product: a machine alternative (M) and a labor alternative (L). The relevant costs as a function of the quantity produced, θ, are:

Machine alternative: $C_M = 5.5 + 1.8\theta$
Labor alternative: $C_L = 1.0 + 2.5\theta$

The prior probability distribution for θ is:

θ	4	5	6	7	8	9
$P(\theta)$	$1/6$	$1/6$	$1/6$	$1/6$	$1/6$	$1/6$

a. Under the prior distribution, the optimal (minimum cost) alternative is to select which method?
b. What is the conditional value of perfect information?
c. What is the value of EVPI?

22.20. If we *assume* in Problem 22.19 that $E(\theta) = 7.0$, then the CVPI that $\theta = 5$ is equal to what number?

22.21. Assume $E(\theta) = 6.0$ in Problem 22.19.
a. What is the CVPI that $\theta = 5$?
b. What is EVPI equal to?

22.22. The table at the bottom of this column shows the probability distribution of demand for a certain product.
a. Assume the cost of an overage is $6 and the cost of an underage is $4. How many units should be stocked?
b. Assume the cost of an overage is $1 and the cost of an underage is $4. How many units should be stocked?
c. Assume $C_0 = \$4$ and $C_u = \$1$. How many units should be stocked?

22.23. A firm has an opportunity to invest in a piece of equipment, the expected savings of which is 2,500 hours a year with a standard deviation of 500 hours (normally distributed). The cost of the equipment is $15,000 a year and labor costs are $6.10 an hour. What course should the firm follow, and what is the expected value of perfect information?

Demand	0	1	2	3	4	5
Probability	0.20	0.20	0.40	0.10	0.05	0.05

Statistical decision theory: The decision to sample

23

23.1 INTRODUCTION

In Chapter 22, we introduced the topic of statistical decision making under risk, where the opportunity did not exist to obtain additional information—a terminal decision had to be made. The criterion for arriving at a decision involved the equivalent strategies of maximizing expected monetary value (EMV) (payoff) and minimizing expected opportunity loss (Bayes criterion). In this chapter, we will continue the line of reasoning developed in the previous chapter in making decisions under risk, but we will include the provision for obtaining imperfect—or sample—information. In making our decision, we will continue to use the Bayes criterion and minimize expected opportunity losses (or, simply, total expected losses). Because we are in essence asking the question, "What should we do before we make a terminal decision?," the type of analysis described in the chapter is often called *preposterior analysis* or preposterior decision making.

The notion of obtaining additional information should not be new to us. Indeed, Chapter 10 and others were devoted entirely to obtaining additional information (through a sample) and evaluating various decisions based on this additional information. What will be new in this chapter will be the incorporation of the consequence of making an incorrect decision into the construction of a decision rule. We will learn that the decision to sample can be regarded much in the same light as the setting of a "critical value" or an "action limit" in a test of a hypothesis (see Chapter 10). Both these decisions have an important bearing on the choices to be made in arriving at a decision that maximizes some measure of performance (or minimizes losses). Because ours is an introductory exposition, we will of

731

necessity have to limit our treatment of the subject. We will consider only two-action or the choice between two alternative types of problems, and we will limit our discussion of these to include only the cases in which we have a discrete prior distribution and binomial sampling, or a normal prior distribution and normal sampling. More advanced and in-depth coverage of the decision to buy imperfect (sample) information can be found in the references at the end of the chapter.

23.2 DECISION MAKING WITH A FIXED SAMPLE SIZE—BINOMIAL SAMPLING

Prior to making a terminal decision, it is often possible to obtain additional information through a sampling process. When sampling is possible, a problem arises in statistical decision theory that has not appeared thus far—namely, using the Bayes criterion, how is a sampling decision rule designed to minimize expected opportunity losses? As we shall see shortly, the Bayes criterion will lead to the determination of an optimal sample size, n, and an optimal critical (rejection) number, x. In this section, however, we will assume that if a sample is to be taken, it must consist of a fixed sample size, and we will compare the opportunity losses of various decision rules with the minimum opportunity loss of a *terminal decision*. In the next section, we will specify the optimal sampling plan, where we are free to choose a value for the sample size, n.

We introduce the topic of statistical decision making with sample information in the following example.

Example 23.1   Before making a decision to spray the entire lot of 10,000 trees, Burt Calkins, Jr. (see Example 22.3), can send sample branch cuttings to his state university to determine whether the sampled trees contain virus in sufficient quantities to slow their growth. The university charges 15 cents to test each cutting, plus a fixed fee of $3.00. Determine the *maximum* amount Burt should be willing to spend on a sample. Express the opportunity losses in this problem of spraying and of not spraying in functional form (as a function of π, the proportion of trees that are infected), and graph the resulting opportunity loss functions.

Solution   If Burt could predict in advance the number of trees that are infected, he would, of course, select the act corresponding to the minimum cost for the specified level of infection. Given this information, we can compute the expected *costs* with a prefect predictor (ECPP) similar to the calculation of EPPP in Chapter 22. Relevant computations are shown in the table.

Under the prior distribution assessed by Burt, he can *expect* (1) to spend $860 even if he can predict with certainty the proportion or the number of trees infected with the virus each year or (2) spend $1,000 by spraying (see Example 22.3). The difference between these quantities ($1,000 − 860 = $140), which is always equal to EVPI in a two-action decision problem, is the

(1)	(2)	(3) Optimal (minimum) cost	(4) ECPP (2) · (3)
θ	P(θ)		
2,000	0.50	$ 720	$360
3,000	0.30	1,000	300
4,000	0.10	1,000	100
5,000	0.05	1,000	50
6,000	0.05	1,000	50
	1.00		

$$\text{ECPP} = \$860$$

cost of uncertainty. Thus, Burt should not be willing to spend more than $140 in sampling costs to determine the optimal act to follow.

The determination of the functional form of the conditional value of perfect information (opportunity loss) initially involves the computation of the break-even value of the random variable, π. The break-even proportion of trees infected is:

$$\pi_b = \frac{2,778}{10,000} = 0.2778$$

An error (opportunity loss) occurs if we spray the trees and the proportion infected is less than 0.2778. Since there are 10,000 trees in each lot, the opportunity loss of spraying is

$$0.36 \cdot 10,000(0.2778 - \pi), \qquad \pi < 0.2778$$

or

Opportunity loss of spraying

$$\text{Loss} = \begin{cases} \$3,600(0.2778 - \pi) & \text{if } \pi < 0.2778 \\ \$0 & \text{if } \pi \geq 0.2778 \end{cases}$$

Similarly, an opportunity loss occurs if we decide *not* to spray and more than 27.78 percent of the trees are infected. Hence, the opportunity loss of not spraying is:

Opportunity loss of not spraying

$$\text{Loss} = \begin{cases} \$0 & \text{if } \pi \leq 0.2778 \\ \$3,600(\pi - 0.2778) & \text{if } \pi > 0.2778 \end{cases}$$

These losses are graphed in Figure 23.1.

FIGURE 23.1 Opportunity losses for Example 23.1

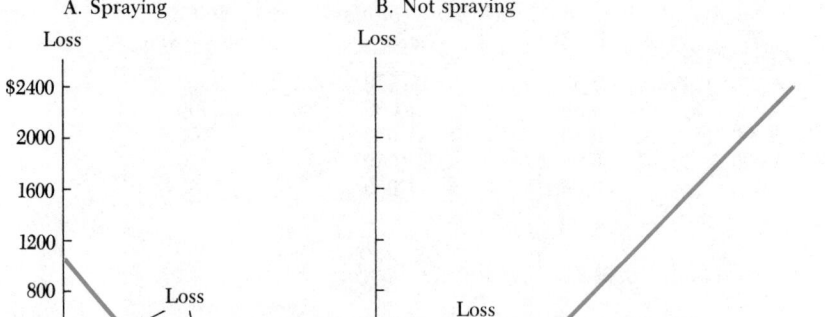

Figure 23.2 gives a decision tree for the problem of Example 23.1 where Burt must decide whether to spray his lot of 10,000 Christmas trees or else inspect a sample of the trees for the presence of virus where $n = 10$ has been stipulated as the number of units (trees) that might be sampled. As can be seen in this diagram, it is necessary to compare the opportunity losses of all acts and select that act with the minimum expected opportunity loss (cost). There are 13 possible acts delineated in Figure 23.2—2 involve choices

FIGURE 23.2 Decision tree for Example 23.1 and a sample of ten units

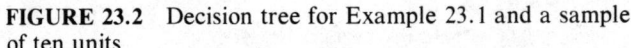

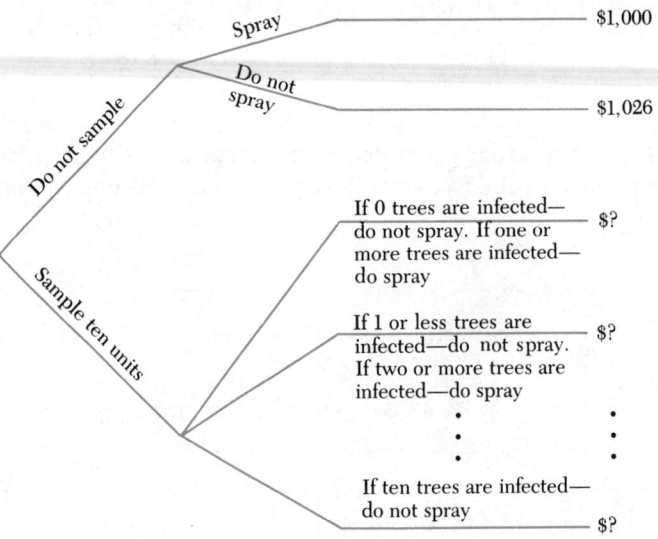

based on an initial decision to not sample, and eleven involve selecting a sample of size 10 and spraying or not spraying depending on the number of trees (11 choices: 0, 1, 2, . . . , 10) found to be infected with the virus.

We will initially evaluate the following decision rule.

Bayesian decision rule

Take a sample of size $n = 10$ trees. If fewer than one are infected, conclude that the proportion that are infected is less than the break-even proportion, $\pi_b = 0.2778$, and do not spray. Otherwise spray.

This decision rule is similar to the classic hypothesis-testing decision rule discussed in Chapter 10 in that the act not to spray is based on failure to reject the null hypothesis H_0: $\pi \leq 0.2778$, and the decision to spray is based on rejection of the null hypothesis (the alternate hypothesis is H_A: $\pi > 0.2778$). Note, though, that the critical value of π, $\hat{\pi}_c = 0.2778$, is based on the *break-even value of the random variable* π, the proportion of trees that are infected, and that nothing has been specified regarding the maximum level of α (probability of a Type I error)! We will now compare the economic consequences of this decision rule with the economic consequences of other decision rules with the same sample size for the Example 23.1 data.

First, we will weight the consequences of making an error with the probability that an error will occur for every permissible value of the random variable π. To do this, we must first evaluate the probabilities of making an error with the decision rule $(n, x) = (10, 1)$.[1] The probability of making an error is

$$P(\text{error}/\pi) = \begin{cases} P(X \geq 1/n = 10, \pi) & \text{if } \pi < 0.2778 \\ P(X < 1/n = 10, \pi) & \text{if } \pi > 0.2778 \end{cases}$$

since an error occurs if we conclude that the proportion of infected trees is greater than π_b and it is not (Type I error), or else we conclude that the proportion of infected trees is less than π_b and it is not (Type II error). Table 23.1 gives the error probabilities for the decision rule (10, 1).

The probabilities given in Table 23.1 have been determined using Table B.1 in Appendix B. For example, the probability that one or more of the ten sample cuttings will be infected when the true proportion of infected trees in the population is equal to 0.20 is:

$$P(X \geq 1/10, 0.2) = 1.0 - P(X = 0/10, 0.2) = 1.0 - 0.107 = 0.893$$

---

[1] Hereafter, the notation (n, x) will refer to the following decision rule: Take a sample of size n. If fewer than x defectives are found, do not reject the null hypothesis H_0: $\pi \leq \pi_c = \pi_b$. Otherwise, reject the null hypothesis.

TABLE 23.1 Error probabilities of the decision rule (10, 1)

Notation	π	$P(error/\pi)$
$P(X \geq 1/10, 0.2)$	0.2	0.893
$P(X < 1/10, 0.3)$	0.3	0.028
$P(X < 1/10, 0.4)$	0.4	0.006
$P(X < 1/10, 0.5)$	0.5	0.001
$P(X < 1/10, 0.6)$	0.6	0.000

The economic consequences (conditional losses) of making an error are given in the bottom two rows of Table 22.4. That is, the conditional value of perfect information is equal to the conditional loss when the wrong act is selected (an error is made).

In order to evaluate the decision rule (10, 1), we must multiply each of the conditional opportunity losses by its respective probability of occurrence. Table 23.2 gives these products, and in Figure 23.3 we have graphed the

TABLE 23.2 Conditional expected opportunity losses (CEOL) of the decision rule (10, 1)

π	$P(error)$ (Table 23.1)	Conditional loss (Table 22.4)	$P(error) \times$ Conditional loss
0.2.............	0.893	280	250.040
0.3.............	0.028	80	2.240
0.4.............	0.006	440	2.640
0.5.............	0.001	800	0.800
0.6.............	0.000	1,160	0.000

respective opportunity losses under the decision rule (10, 1). The quantities given in the last column of Table 23.2 and the graph of Figure 23.3 are called *conditional expected opportunity losses* (CEOL)—conditional on π, but expected as regards the critical number in our decision rule, x.

Definition 23.1
Conditional expected opportunity loss (CEOL) of the decision rule (n, x)

The *conditional expected opportunity loss* of a decision rule (n, x) for a *specific value of the random variable* is the conditional opportunity loss for that value of the random variable multiplied by the probability that an error will be made (the loss will occur).

FIGURE 23.3 Conditional expected opportunity loss (CEOL) of the decision rule (10, 1)

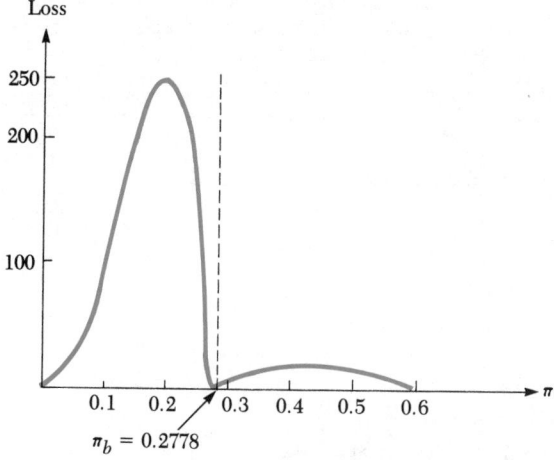

If the quantities in the last column of Table 23.2 are multiplied by their probabilities of occurrence and *these* products are summed, the result is the *unconditional expected opportunity loss* of the decision rule (10, 1). The necessary computations for determining the UEOL of the decision rule (10, 1) are given in Table 23.3. If we add to this expected opportunity loss the cost of sampling ($3.00 + 0.15 \cdot 10 = $4.50), we derive the *unconditional expected total loss* (UETL) of the decision rule (10, 1). Figure 23.4 summarizes the determination of the unconditional expected total loss (UETL) for the decision rule (10, 1).

TABLE 23.3 UEOL of the decision rule (10, 1)

π	Conditional expected opportunity loss (Table 23.2)	Prior probability distribution, $P(\pi)$	Unconditional expected opportunity loss
0.2............	$250.04	0.50	$125.02
0.3............	2.24	0.30	0.67
0.4............	2.64	0.10	0.26
0.5............	0.80	0.05	0.04
0.6............	0.00	0.05	0.00
		1.00	UEOL = $125.99

Definition 23.2
Unconditional expected opportunity loss of a decision rule (n, x)

$$\text{UEOL} = \sum_{\theta} P(\theta) \cdot P(\text{error}/\theta) \cdot \text{CVPI}_\theta$$

Definition 23.3
Unconditional expected total loss (UETL) of a decision rule (n, x)

UETL = UEOL + Cost of sampling

FIGURE 23.4 Unconditional expected total loss (UETL) for the decision rule (10, 1)

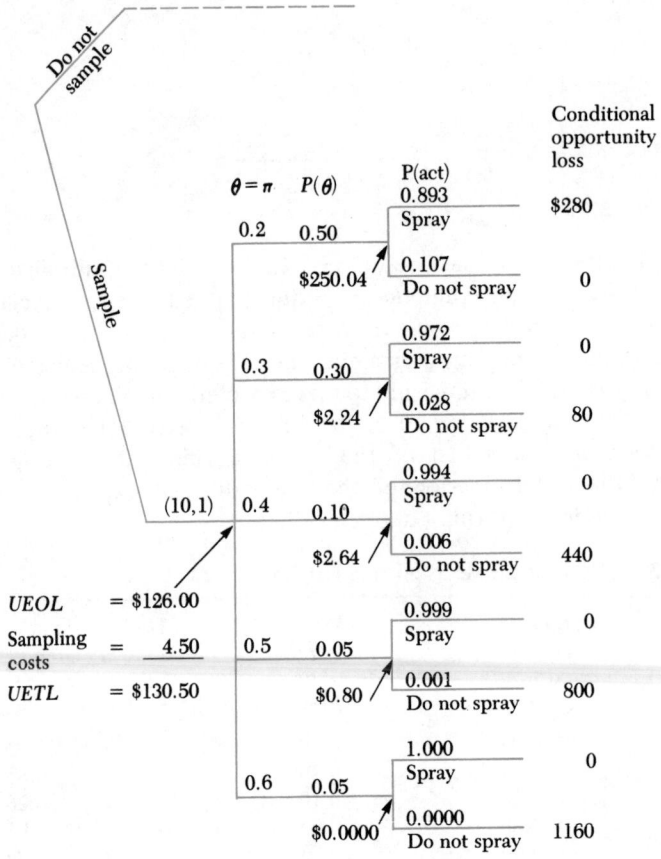

It may be well to recapitulate the operations that have occurred up to this point in evaluating the decision rule (10, 1).

1. The conditional opportunity losses (CVPI) of making an incorrect decision were determined. These losses represent the difference in costs between the most costly act for a given value of the random variable and the least costly act (two alternative decision problems).

2. The probabilities of making an error (an incorrect choice) under a

particular decision rule were computed, yielding an error characteristic curve for that decision rule.

(3.) The conditional losses were multiplied by the probability of making an error *for each value of the random variable,* yielding conditional expected opportunity losses (CEOL).

(4.) Unconditional expected opportunity losses (UEOL) were computed by multiplying the results obtained in step 3 by the probability that the random variable assumes each particular value (given by the prior distribution), and summing these products.

(5.) The cost of sampling was added to the UEOL determined in step 4 to derive the UETL of the decision rule.

A worksheet for accomplishing the operations in steps 1 through 5 is given in Table 23.4.

The development of the UETL for the decision rule of the form (n, x) in Example 23.1 can be reasoned as follows. The prior distribution, $P(\theta)$, gives the prior probability that a certain proportion of trees are infected. For each

TABLE 23.4   Computation worksheet for determining the unconditional expected total loss of a decision rule of the form (n, x)—discrete prior, binomial sampling, two-action problems

(1)	(2)	(3)	(4)	(5)	(6)
θ	Prior $P(\theta)$	P(error)	Joint probability $(2) \cdot (3)$	Conditional loss	UEOL $(4) \cdot (5)$
θ_b					
				UEOL	
				+ Cost of sample	_____
				= UETL	_____

level or proportion of infection, there exists the possibility that the actual sample selected will lead to an incorrect decision. That is, we will have incurred more expense than necessary because we chose a more costly alternative based on the sample results. The probability that we will make such an error is a conditional one—conditional on the actual proportion of infected trees, π (state of nature). If we multiply the probability of there being a certain proportion of trees infected by the conditional probability of an error *given* the proportion of infected trees, we derive the *joint probability* of being in a particular state (level of infection) *and* making an incorrect choice. If we then multiply these joint probabilities by the *conditional* opportunity losses of making an error and sum these products, we derive the UEOL of the sampling plan.

　Example 23.2　Determine the unconditional expected total loss of the decision rule (10, 1) using a table similar to Table 23.4.

　Solution　The necessary computations for determining UETL for the decision rule (10, 1) are given in Table 23.5.

TABLE 23.5　Computation of the unconditional expected total loss for decision rule (10, 1)

	(1) π	(2) Prior $P(\pi)$	(3) Conditional $P(error/\pi)$	(4) Joint probability $(2) \cdot (3)$	(5) Conditional loss	(6) $(4) \cdot (5)$
	0.2	0.50	0.893	0.4465	280	125.02
π_b						
	0.3	0.30	0.028	0.0084	80	0.67
	0.4	0.10	0.006	0.0006	440	0.26
	0.5	0.05	0.001	0.00005	800	0.04
	0.6	0.05	0.000	0.0000	1,160	0.00
		1.00			UEOL	$125.99
					+ Cost of sample	4.50
					= UETL	$130.49

　　Thus far we have evaluated the UETL of the decision rule (10, 1). In order to compare this decision rule with others of the form (10, x), we must increment x at one-unit intervals to determine the particular value of x that yields the minimum UETL of the decision rule (10, x). The UETL values of the decision rules (10, 2), (10, 3), (10, 4), and (10, 5) are given in the table below. For each of the values listed, it is necessary to prepare a table similar to Table 23.5 to compute the UETL of the decision rule given.

Decision rule	UETL
(10, 1)	$130.49
(10, 2)	98.02
(10, 3)	69.02
(10, 4)	63.92
(10, 5)	82.08

The sixth step in the evaluation of a decision rule for a *fixed sample size* is:

6. Determine the UETL of the decision rules $(n, 1)$, $(n, 2)$, $(n, 3)$, ... The decision rule with the lowest UETL is the optimal rule for the given sample size.

It can be shown that in problems of the general type described, a graph of UETL as a function of the rejection or critical number x and for a fixed sample size n has at most one low point or dip (a convex function). Hence, when the UETL for $x = 4$ was less than it was for $x = 3$ and $x = 5$, the calculation of the optimal critical number for the fixed sample size was determined.

Before considering taking a sample of *any* size, the cost of uncertainty was determined to be $140. Taking a sample of size ten and not spraying if fewer than four trees are found to be infected (otherwise spray) results in an unconditional expected total loss of $63.92. The difference between the cost of uncertainty (EVPI) and the UETL of a particular sample plan is called the *expected net gain of sampling* (ENGS). For a decision rule of the form (10, 4):

$$\text{ENGS} = \text{EVPI} - \text{UETL} = \$140.00 - 63.92 = \$76.08$$

On an *expected basis,* we will gain $76.08 by taking a sample of size ten [using the decision rule (10, 4)], as opposed to not taking a sample and making a terminal decision.

Definition 23.4
Expected net gain of sampling (ENGS)

Let the minimum UETL for a decision rule of the form (n, x) be denoted by TL. Then the *expected net gain of sampling* is equal to:

$$\text{ENGS}(n) = \text{EVPI} - \text{TL}$$

The difference between the cost of uncertainty, EVPI, before a sample is selected and the unconditional expected opportunity loss of a particular sampling plan of the form (n, x), UEOL, is called the *expected value of sample information* (EVSI). EVSI can thus be determined by the identity:

$$\text{EVSI} = \text{EVPI} - \text{UEOL}$$

Hence, EVSI represents the expected reduction in uncertainty (EVPI) provided by the sample (UEOL). The ENGS is simply the EVSI minus the cost of that sample:

$$\text{ENGS} = \text{EVSI} - \text{Cost of sampling}$$

Definition 23.5
Expected value of sample information (EVSI)

The *expected value of sample information* is the expected reduction in the cost of uncertainty (EVPI) provided by sample information:

$$\text{EVSI} = \text{EVPI} - \text{UEOL}$$

From the Example 23.1 data,

$$\text{EVSI} = \text{EVPI} - \text{UEOL} = \$140.00 - 59.42 = \$80.58$$

Thus, taking a sample of size $n = 10$ and spraying if four or more infected trees are found results in an expected reduction in the cost of uncertainty of $80.58. When we subtract from this amount the cost of taking the sample ($4.50), we obtain the expected *net* gain of sampling of $76.08 determined previously.

23.3 DETERMINATION OF OPTIMAL SAMPLE SIZE n AND CRITICAL NUMBER x—BERNOULLI PROCESS AND BINOMIAL SAMPLING

In the previous section, we saw how taking a sample of size $n = 10$ and spraying if four or more trees were found to be infected resulted in an expected gain (ENGS) of $76.08 over taking action without benefit of sampling. If taking a sample of size ten reduced the expected cost of uncertainty (EVPI) from $140.00 to $59.42, then it may be that taking a sample larger than $n = 10$ will reduce the cost of uncertainty even further. Indeed, this will always be the case. But, unfortunately, larger samples usually cost more money! Hence, we must seek a balance between the cost of a particular sample size and the information (reduction in uncertainty) the sample provides. We can evaluate various sampling plans by comparing the UETL or the ENGS for the given sample size using the optimal critical number for each sample size. The sampling plan with the lowest UETL or the highest ENGS is then the optimal plan. Thus, when the n of our decision rule is not fixed as it was in the previous section, we must evaluate the ENGS or the UETL of a decision rule for each sample size and select that rule with the (n, x) pair yielding the minimum value of UETL or the maximum ENGS.

When the sample size is not fixed, we must add to steps 1–6 given in the previous section the additional step:

(7.) Compute the optimal critical value x for sampling plans $(1, x)$ $(2, x)$, $(3, x)$, ... The optimal sampling plan of the form (n, x) is the plan that yields the maximum ENGS or the minimum UETL as defined previously.

The relationship between EVSI and EVPI is illustrated in Figure 23.5. Since EVSI = EVPI − UEOL (of a particular sampling plan), EVSI will equal EVPI only when UEOL = 0, and this occurs when every unit in a population is examined (a census).

FIGURE 23.5 Relationship between EVPI and EVSI

EVSI

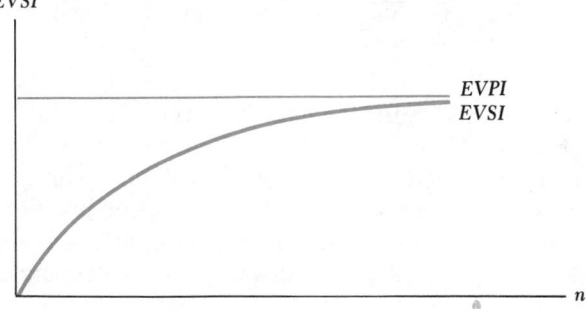

The relationship of UEOL, the cost of sampling, UETL, and ENGS is shown in Figure 23.6. Note that the value of n that minimizes UETL also maximizes ENGS.

FIGURE 23.6 Relationship of UEOL, the cost of sampling, UETL, and ENGS
Cost

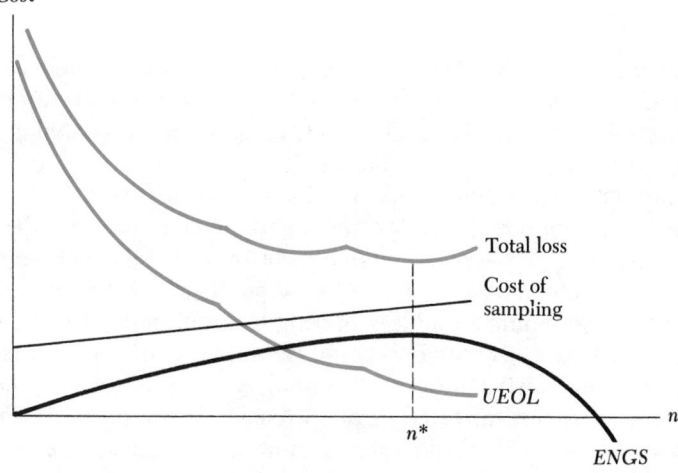

FIGURE 23.7 Graph of ENGS for Example 23.1 data

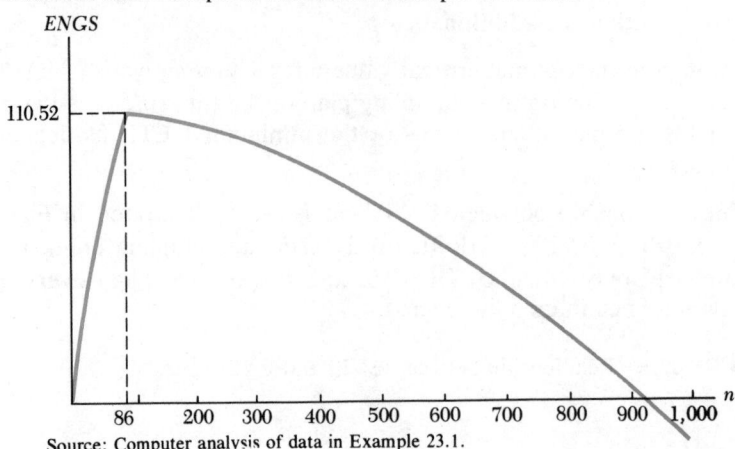

Source: Computer analysis of data in Example 23.1.

Figure 23.7 gives the ENGS for the data in Example 23.1 for values of *n* equal to 0 through 1,000. These ENGS values were determined by selecting a value of *n* and computing ENGS for the optimal sampling plan for the value of *n* selected, similar to the computation of ENGS for the decision rule (10, 4) in the previous section. Fortunately, the majority of calculations required in determining an optimal sampling plan this way can be accomplished on a computer!

If Burt of Example 23.1 wants to minimize his unconditional expected total loss, he should send *n* = 86 sample cuttings to his state university, and if 23 or more infected trees are found in the sample, he should spray the entire lot of 10,000 trees, Otherwise, he should not spray.

23.4 REVISING PROBABILITIES: BAYES THEOREM FOR DISCRETE PRIOR AND BINOMIAL SAMPLING

In Chapter 4, Bayes theorem was presented as a method for revising probabilities for discrete *events*. Bayes theorem was shown to be a simple restatement of conditional probability. This theorem is repeated as Theorem 23.1 *for a discrete random variable*.

The probability revision process is illustrated in Figure 23.8. The prior probability distribution [$f(\theta)$ for the continuous case, and $P(\theta)$ for the discrete case] can be based on *symmetry* or the *long-run relative frequency* definition of probability, or it can be a purely "subjective" probability distribution of a random variable θ. Additional information [$f(x/\theta)$ or $P(x/\theta)$] is usually provided by some type of sampling process as described in Chapter 8. This distribution—the sampling distribution or the "likelihood"—received the majority of attention in Chapters 9–12 when we in essence asked: What is the probability of obtaining the sample results *given*

FIGURE 23.8 Probability revision process

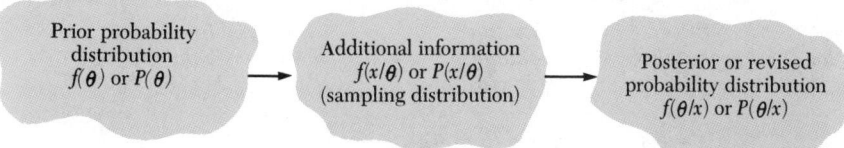

the value of the population parameter? Indeed, it is with the sampling distribution that the classical statistician focuses his attention in the inference or the hypothesis-testing process. And finally, the posterior or the revised distribution [$f(\theta/x)$ or $P(\theta/x)$] is the end result of the probability revision process.

Theorem 23.1

Bayes theorem for discrete random variables

Let the random variable θ correspond to a discrete random with probability mass function given by $P(\theta)$. Additional information on the value of θ is provided by the observed result $\mathbf{X} = x$. The revised probability that θ assumes a particular value, θ_i, where $i = 1, 2, \ldots , n$, is given by

$$P(\theta = \theta_i/\mathbf{X} = x) = \frac{P(\theta = \theta_i)P(\mathbf{X} = x/\theta = \theta_i)}{\displaystyle\sum_{i=1}^{n} P(\theta = \theta_i)P(\mathbf{X} = x/\theta = \theta_i)}$$

Table 23.6 gives the general format for revising probabilities for a discrete *random variable*. This table differs from Table 4.2 in the following ways. First, events are now being identified as random variables (first column in Table 23.6). Moreover, the "events" are now *parameters* of specific distributions. This is a subtle point but is worth noting at this time; we will come back to it at the end of the chapter. Second, titles have been added to each of the columns in Table 23.6 to emphasize the role that each distribution plays in the probability revision process. The quantity $P(x)$ in Table 23.6 gives the marginal probability of the observed sample result; that is, it gives the probability of observing the sample result irrespective of the value of the population parameter, θ. Each of the remaining columns in Table 23.6 has been identified as prior, sample, joint, or revised probability. We illustrate the use of a computation worksheet like Table 23.6 in the following example.

Example 23.3 Unknown to Burt in Example 23.1, his foreman, Barbara, has mailed 20 sample cuttings from his Christmas tree patch to the local university. Burt discovers that a sample has been taken when he receives a letter from the university indicating that of the 20 cuttings mailed in, only 1 contains virus in sufficient quantities to stunt the growth of the trees. What is

TABLE 23.6 Computation worksheet for revising discrete probabilities for qualitative and quantitative random variables using Bayes theorem

Value of random variable, θ	Prior probability, $P(\theta)$	×	Sample probability, $P(x/\theta)$	=	Joint probability, $P(x, \theta)$	Revised probability, $P(\theta/x)$
θ_1	$P(\theta = \theta_1)$		$P(X = x/\theta = \theta_1)$			$\dfrac{P(x, \theta_1)}{P(x)}$
θ_2	$P(\theta = \theta_2)$		$P(X = x/\theta = \theta_2)$		$P(\theta = \theta_i) \times P(X = x/\theta = \theta_i)$ $i = 1, 2, \ldots, n$	$\dfrac{P(x, \theta_2)}{P(x)}$
.	.		.			.
.	.		.			.
.	.		.			.
θ_n	$P(\theta = \theta_n)$		$P(X = x/\theta = \theta_n)$			$\dfrac{P(x, \theta_n)}{P(x)}$
Total	1.000				$P(x) = \sum\limits_{i=1}^{n} P(\theta = \theta_i)P(X = x/\theta = \theta_i)$ Marginal probability	1.00

Burt's revised probability distribution of the proportion of trees that are infected with the virus?

Solution A computation worksheet similar to Table 23.6 is given in Table 23.7 for revising the prior probabilities of the proportion of trees that are infected with the virus. The sample probabilities are determined through the use of Table B.1 in Appendix B, and represent the likelihood or the probability of observing the sample result of 1 infected tree out of 20 given the value of π. For example, the probability that the sample will yield 1 infected cutting out of 20 when in fact the true proportion of infected trees is $\pi = 0.3 = 3{,}000/10{,}000$ is:

$$P(X = 1/n = 20, \pi = 0.3) = P(X \le 1/n = 20, \pi = 0.3)$$
$$- P(X = 0/n = 20, \pi = 0.3)$$
$$= 0.008 - 0.001 = 0.007$$

The remaining sample likelihoods (probabilities) are determined similarly.

Frequently, as was the case in this example, a sample is selected without regard to an "optimal" sample plan like the one developed in the preceding section. Indeed, in many instances, the statistician is called in *after the fact* and asked: How do I best put to use the information I have received from the sample? While it is always better to design the sampling plan and *then* actually select the sample, the question that must now be addressed is: Just how is the additional information provided by the sample put to use?

When a sample has been selected without first designing some type of sampling decision rule, the information in the sample should be combined with the prior information to obtain a posterior distribution of the random variable. Using the posterior distribution, the expected opportunity loss of each act should be computed and the act selected with the minimum expected opportunity loss (Bayes criterion). This process is illustrated in the following example.

Example 23.4 Using the posterior probability distribution of the proportion of trees infected given in Table 23.7, construct a decision tree similar to Figure 22.2 using the posterior probabilities to determine the optimal act to follow.

TABLE 23.7 Revision of probabilities in Example 23.3

π	Prior probability, $P(\pi)$	Sample probability, $\times\, P(X = 1/20,\ \pi)$	Joint probability $= P(X = 1/20,\ \pi)P(\pi)$	Revised probability, $P(\pi/X = 1)$
0.2	0.50	0.057	0.0285	0.9314
0.3	0.30	0.007	0.0020	0.0654
0.4	0.10	0.001	0.0001	0.0032
0.5	0.05	0.000	0.0000	0.0000
0.6	0.05	0.000	0.0000	0.0000
	1.00			1.000
			$P(X = 1) = 0.0306$	

Solution A decision tree delineating the expected cost of each act is given in Figure 23.9. Given this revised information, Burt should *not spray* his trees and minimize his expected losses (costs) of $745.20. Note that we could have obtained this same result by computing the posterior expectation

FIGURE 23.9 Decision tree for terminal act in Example 23.4

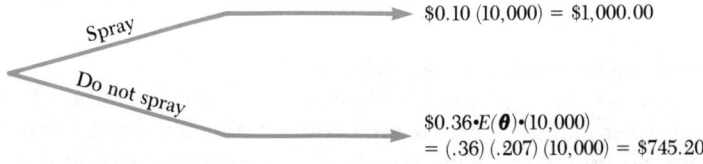

Spray → $0.10 (10,000) = $1,000.00

Do not spray → $0.36·E($\boldsymbol{\theta}$)·(10,000)
= (.36) (.207) (10,000) = $745.20

$E(\boldsymbol{\theta}) = 0.207$ and comparing this quantity with $\theta_b = 0.2778$. In this example, the information provided by the sample, when combined with prior knowledge on the proportion of infected trees, leads us to select a different act than was optimal under the prior distribution. Expected costs are now minimized by not spraying the lot of 10,000 trees.

The question now might be raised as to how one might proceed if additional evidence becomes available after probabilities are revised. For example, if other sample evidence becomes available, how might we revise our "revised" probabilities? In this case, the *revised* probabilities from the first sample become the *prior* probabilities to be revised by the results of the second sample. In this manner, probabilities can be revised as often as additional information becomes available. The effects, say, of combining *two* samples of size $n = 20$ with a prior distribution (two probability revisions) would be the same as revising the prior distribution with the information provided by the larger combined sample of size $n = 40$.

■ 23.5 PREPOSTERIOR DECISION MAKING WITH A NORMAL PRIOR DISTRIBUTION AND NORMAL SAMPLING

In Chapter 22, it was shown that the computation of EVPI is greatly simplified using Table B.10 in Appendix B if the probability distribution of the random variable is one of the theoretical probability distribution models, namely, the normal distribution. It turns out that the numerical calculations required in determining EVSI and ENGS are *also* greatly simplified over the methods just discussed if the prior distribution is normal (as is sampling). EVSI, like EVPI, can be determined through the use of Table B.10. We will discuss the use of Table B.10 in determining EVSI and an optimal sampling plan in the following example.

Example 23.5 The Calcit Company is considering selling an electronic calculator with programming features in each of its 2,000 franchised outlets. The investment required to manufacture and promote the model under consideration is $400,000 over the two-year expected life of the unit (before a newer model must be introduced to stay competitive). The profit to the firm on sales of the calculator is $20 per unit.

The firm has been marketing similar (but nonprogrammable) calculators for several years and estimates demand to be 14 units per store, with a standard deviation of 4 units on the model under consideration. The standard deviation of the distribution of sales by each of the 2,000 franchised outlets is felt to be known and to be equal to 12. This quantity, σ_X, is an indication of the dispersion from the average sales of all stores.

The firm has the opportunity to sample its franchised outlets to estimate retail sales per outlet. The estimated cost of sampling a retail outlet is $130 per store because of the wide geographic dispersion of the outlets.

Given the information provided, should the firm undertake marketing the new product? What is the EVPI? The CVPI? What are EVSI and ENGS for sample sizes of $n = 20, 25, 30,$ and 40?

Solution First, we must compute the break-even quantity of sales per store. Since the required investment is $400,000, each of 2,000 retail outlets must sell

$$\theta_b = \mu_b = \frac{\$400,000}{2,000 \cdot \$20} = 10 \text{ units}$$

for the firm to break even on its investment. Since the prior expected level of

FIGURE 23.10 CVPI for Example 23.5

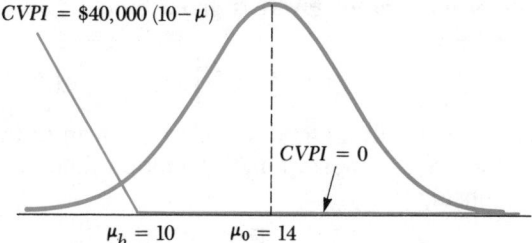

$CVPI = \$40,000\ (10-\mu)$

$CVPI = 0$

$\mu_b = 10$ $\mu_0 = 14$

sales per store is $E(\theta) = \mu_0 = 14$ units, $E(\theta) > \theta_b$ ($14 > 10$),[2] and hence, the firm would make an expected profit if it marketed the programmable unit.

The CVPI for this example is shown in Figure 23.10. That is, since any additional information that sales per store will be greater than $\theta_b = \mu_b = 10$ units leads to the same act (introduce the product), CVPI $= 0$ for values of $\mu \geq 10$. For every unit decrease in mean sales per store below 10 units, however, the firm will incur an opportunity loss of ($\$20$)($2,000$)($1$), or $\$40,000$. Hence, the functional form of CVPI is:

$$\text{CVPI} = \begin{cases} \$40,000(10 - \mu) & \text{for } \mu < 10 \\ 0 & \text{for } \mu \geq 10 \end{cases}$$

The expected value of perfect information (EVPI) under the prior distribution is:

$$\text{EVPI} = C\sigma_0 N(D) = \$40,000 \cdot 4 \cdot N\left(\frac{|10 - 14|}{4}\right) \quad \text{[equation (22.9)]}$$

From Table B.10, $N(1) = 0.08332$. Hence,

$$\text{EVPI} = \$40,000 \cdot 4 \cdot 0.08332 = \$13,331.20$$

The EVPI sets an upper limit on the amount to spend for sample (imperfect) information.

Since it costs $\$130$ to sample each outlet, we would never consider taking a sample larger than $n = (13,331.20/130) \doteq 102$ outlets. Let's assume for the moment that a decision is made to take a sample of 25 outlets. The cost of sampling is thus $\$130.00 \cdot 25 = \$3,250$.

We must now compute the variance of the sampling distribution. Since the producer estimates the standard deviation of the population to be 12,

$$\sigma_{\bar{x}} = \frac{\sigma_x}{\sqrt{n}} = \frac{12}{\sqrt{25}} = 2.4 \quad \text{and} \quad \sigma_{\bar{x}}^2 = (2.4)^2 = 5.76$$

---

[2] For the normal distribution, we will subscript the parameters of the prior distribution with 0 and subscript the parameters of the revised distribution with 1. Hence, μ_0 and σ_0^2 will denote, respectively, the mean and the variance of the prior distribution.

Before the sample is actually selected, the estimate of the standard deviation used in the determination of EVSI is given by:

$$\sigma_* = \sqrt{\sigma_0^2\left(\frac{\sigma_0^2}{\sigma_0^2 + \sigma_{\bar{X}}^2}\right)} \quad \text{or} \quad \sigma_* = \sqrt{4^2\left[\frac{4^2}{4^2 + (2.4)^2}\right]} = 3.43$$

This quantity, σ_*, represents the "amount" of revision in the standard deviation from the prior to the posterior distribution and is derived through use of the relationship

$$\sigma_* = \sqrt{\sigma_0^2 - \sigma_1^2}$$

Given σ_*, EVSI for a normal prior and normal sampling is as defined here.

The expected value of sample information—
normal prior distribution and normal sampling

$$\text{EVSI} = C\sigma_* N(D_*) \tag{23.1}$$

where $D_* = |\mu_b - \mu_0|/\sigma_*$ and $N(D_*)$ is found in Table B.10 in Appendix B.

For the data in Example 23.5,

$$\text{EVSI} = \$40,000 \cdot 3.43 \cdot N\left(\frac{|10 - 14|}{3.43}\right) = \$40,000 \cdot 3.43 \cdot N(1.17)$$
$$= \$40,000 \cdot 3.43 \cdot 0.05964 = \$8,182.61$$

Since a sample of size $n = 25$ costs $3,250,

$$\text{ENGS(25)} = \$8,182.61 - \$3,250 = \$4,932.61$$

Hence, total losses or total opportunity losses should be reduced by $4,932.61 by taking a sample of 25 outlets.

The ENGS for sample sizes of 20, 25, 30, and 40 are given in Table 23.8. From this table, it is apparent that the optimal sample size is between 25 and 40. If we vary n by units of one starting at $n = 30$ and calculate ENGS(n), we will determine the optimal sample size. These results are also given in Table

TABLE 23.8 Determination of optimal sample size for data in Example 23.5

Sample size, n	$\sigma_{\bar{X}} = \dfrac{12}{\sqrt{n}}$	σ_*	D_*	$N(D_*)$	EVSI	Cost of sampling	ENGS
20.......	2.68	3.32	1.20	0.05610	$7,450.08	$2,600	$4,850.08
25.......	2.40	3.43	1.17	0.05964	8,182.61	3,250	4,932.61
29.......	2.23	3.49	1.45	0.06273[a]	8,757.11	3,770	4,987.11
30.......	2.19	3.51	1.14	0.06336	8,895.75	3,900	4,995.75
31.......	2.15	3.52	1.14	0.06336	8,921.09	4,030	4,891.09
40.......	1.90	3.61	1.11	0.06727	9,713.79	5,200	4,513.79

[a] Determined by interpolation.

23.8. From this table, we should take a sample of size $n = 30$ and compute the posterior mean, $E(\boldsymbol{\theta})$. If $E(\boldsymbol{\theta}) < \theta_b = \mu_b$, we should not manufacture the product. If $E(\boldsymbol{\theta}) > \theta_b$, we should manufacture the product. This decision rule will minimize the unconditional expected total losses of the decision.

■ 23.6 REVISING PROBABILITIES: BAYES THEOREM FOR A NORMAL PRIOR DISTRIBUTION AND NORMAL SAMPLING

In this section, we will describe the revision of probability for a random variable that is normally distributed. Only one important case of the normal distribution will be covered in this chapter; namely, the case in which the variance of the sampling distribution of the mean, $\sigma_{\bar{x}}^2$, is known and the mean (average) is unknown. The cases where either the variance of the sampling distribution is unknown and the mean is known, or where both the sample mean and the sampling distribution variance are unknown require the knowledge of probability distributions beyond the scope of this text. In most practical situations in which n is large ($n > 30$), we can treat the variance of the sampling distribution of the mean as being known through use of the relationship

$$\sigma_{\bar{x}}^2 = \frac{\sigma_{\bar{x}}^2}{n} \doteq \frac{S^2}{n}$$

as discussed in Chapter 8.

Before presenting the formulas necessary for revising normal probabilities as well as an example of their use, it will be useful to introduce the following notation. I represents the amount of information contained in a distribution, and it is equal to the reciprocal of the variance. For the prior distribution, for example, I_0 is:

$$I_0 = \frac{1}{\sigma_0^2}$$

The notion of setting the amount of information in a distribution equal to the reciprocal of the variance has great intuitive appeal, for the larger the value of σ^2, the less certain we are about the value of μ, and the smaller is the value of I, the amount of information available. This is illustrated in Figure 23.11. Obviously, the amount of information available in Figure 23.11A is less than that available in Figure 23.11B.

Once again, stressing that it is assumed that *the value of the variance of the sampling distribution of the mean is known,* the results for revising normal probabilities can now be given.

The mean of the posterior distribution, μ_1, is a weighted average of the prior mean, μ_0, and sample mean, $\bar{x}$; the weights are determined by the amount of information contained in the prior distribution, I_0, and in the sampling distribution, $I_{\bar{x}}$. The standard deviation of the revised distribution, σ_1, reflects the pooling of information in the prior distribution and in the sample according to the relationship $I_1 = I_0 + I_{\bar{x}}$. The amount of information

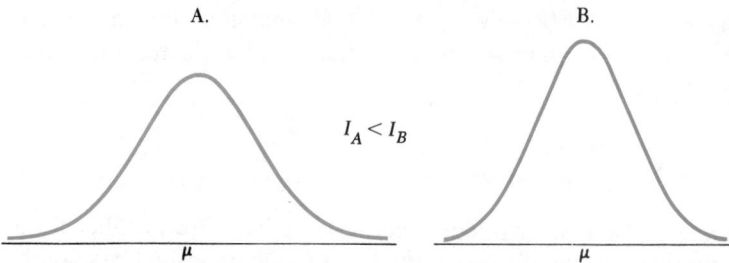

FIGURE 23.11 Two normal distributions with equal means and unequal variances (unequal amounts of information)

A.

B.

$$I_A < I_B$$

μ

μ

Theorem 23.2

Revision of normal prior probabilities with normal sampling (variance of sampling distribution known)[3]

If the prior probability distribution of a population mean is normal with parameters μ_0 and σ_0, and if the sampling distribution is normal with known variance $\sigma_{\bar{X}}^2$ (or $\sigma_{\bar{X}}^2$ can be estimated from the sample variance for n large, $n \geq 30$), and a sample of size n is taken yielding an estimate $\bar{x}$ of μ, then the posterior distribution of $\theta = \mu$ is normal with parameters μ_1 and σ_1 given by:

$$I_0 = \frac{1}{\sigma_0^2}$$

$$I_{\bar{X}} = \frac{1}{\sigma_{\bar{X}}^2} \left(\text{or } I_{\bar{X}} = \frac{n}{s^2} \quad \begin{array}{l} \text{if } \sigma_{\bar{X}}^2 \text{ is unknown, } n \geq 30 \\ \text{and } s^2 \text{ is the sample variance} \end{array} \right)$$

$$\mu_1 = \frac{I_0 \mu_0 + I_{\bar{X}} \bar{x}}{I_0 + I_{\bar{X}}}$$

$$\sigma_1 = \sqrt{\frac{\sigma_0^2 \cdot \sigma_{\bar{X}}^2}{\sigma_0^2 + \sigma_{\bar{X}}^2}}$$

---

in the revised distribution is thus the sum of the information contained in the prior distribution and in the sample. The use of the formulas in Theorem 23.2 for revising normal probabilities is illustrated in the following example.

Example 23.6 A consumer products firm purchases bearings in lots of 500 units for installation on roller skates, which are subsequently marketed to the public. Each lot of bearings is sampled and if the average diameter is within a "tolerable level," the whole lot of bearings is accepted. Because of the low value of these bearings in relation to the total product, a small sample ($n = 10$) is usually selected and the information provided by the sample is combined with the prior information on the behavior of the process

---

[3] The interested reader who desires more information on the development of these formulas as well as on the explicit form of the sampling and marginal distributions is referred to the Jedamus and Frame reference at the end of the chapter.

responsible for producing these bearings to arrive at a decision on whether or not to accept the shipment.

The prior distribution (derived from recent experience) on the average diameter of bearings produced is normal with parameters $\mu_0 = 1.50$ inches and standard deviation $\sigma_0 = 0.01$ inch. The standard deviation of bearings produced by the machine is known to be equal to $\sigma_X = 0.04472$ inch. A sample of ten bearings yields a mean $\bar{x} = 1.48$.

If the revised probability of the average diameter of bearings in the lot being within ±0.02 inch of 1.50 inches is not at least 0.90, the whole lot is rejected; otherwise, it is accepted.

Using the criteria outlined by the firm, should the current shipment of bearings be accepted or rejected?

Solution First, we must compute the parameters of the revised, normal probability distribution. They are:

$$I_0 = \frac{1}{\sigma_0^2} = \frac{1}{0.0001} = 10,000$$

$$\sigma_{\bar{X}}^2 = \frac{\sigma_X^2}{n} = \frac{(0.04472)^2}{10} = 0.0002$$

$$I_{\bar{X}} = \frac{1}{\sigma_{\bar{X}}^2} = \frac{1}{0.0002} = 5,000$$

$$\mu_1 = \frac{10,000 \cdot 1.50 + 5,000 \cdot 1.48}{10,000 + 5,000} = 1.4933$$

$$\sigma_1 = \sqrt{\frac{(0.0001)(0.0002)}{0.0001 + 0.0002}} = \sqrt{0.000066} = 0.0082$$

The revised distribution is a normal distribution with parameters $\mu_1 = 1.4933$ and $\sigma_1 = 0.0082$. Calculation of the required probability is shown below. Since the revised probability that the average diameter of bearings in the lot is within 1.48–1.52 exceeds the firm's requirements (0.9473 > 0.9000), the lot of bearings should be accepted.

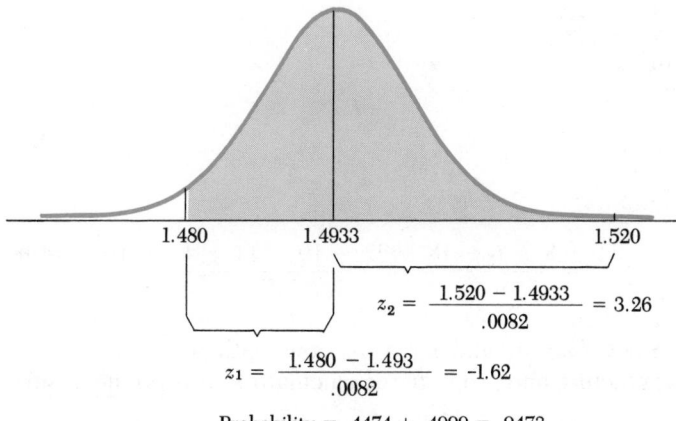

$$z_2 = \frac{1.520 - 1.4933}{.0082} = 3.26$$

$$z_1 = \frac{1.480 - 1.493}{.0082} = -1.62$$

Probability = .4474 + .4999 = .9473

Notice that the amount of information in the prior distribution, I_0, is twice as large as the amount of information in the sample, $I_{\bar{x}}$. This is reflected in the value of the posterior mean, which is closer in value to the prior mean than to the sample mean.

■ 23.7 TERMINAL DECISION MAKING WHEN A SAMPLE HAS
         ALREADY BEEN SELECTED—NORMAL PROCESS

As noted in Section 23.4, the statistician is frequently called in *after* a sample has already been selected and is requested to use the information that has been made available. If both the prior mean and the sample mean are less than or both are greater than μ_b (i.e., $\mu_0 > \mu_b$ and $\bar{x} > \mu_b$; or $\mu_0 < \mu_b$ and $\bar{x} < \mu_b$), then the act that was optimal under the prior is still optimal under the posterior and we do not need to determine a posterior mean.

If, however, $\mu_0 < \mu_b$ and $\bar{x} > \mu_b$, or if $\mu_0 > \mu_b$ and $\bar{x} < \mu_b$, then the posterior mean *may not* bear the same inequality relationship to $\theta_b = \mu_b$. If this inequality relationship changes, then the optimal act may change also. This is illustrated in the following two examples.

Example 23.7 Assume that a sample of ten stores selected in Example 23.5 yields the results given in the table. Does the optimal act to follow change under the posterior distribution?

Store	Calculators sold
1	8
2	6
3	18
4	22
5	10
6	11
7	12
8	10
9	14
10	9

Solution

$$\bar{x} = \frac{8 + 6 + 18 + 22 + 10 + 11 + 12 + 10 + 14 + 9}{10} = 12$$

Since $\bar{x} > \mu_b$ and $\mu_0 > \mu_b$, the optimal act to follow will still be to manufacture and promote the calculators. It is not necessary to compute μ_1.

Example 23.8 Assume now that a sample of ten outlets is selected and yields the results given in the table. What is the optimal act under the revised distribution?

Store	Calculators sold
1	8
2	11
3	13
4	5
5	4
6	10
7	8
8	7
9	6
10	8

Solution

$$\bar{x} = \frac{8 + 11 + 13 + 5 + 4 + 10 + 8 + 7 + 6 + 8}{10} = 8$$

Since $\bar{x} < \mu_b$ and $\mu_0 > \mu_b$, it is necessary to calculate μ_1 using Theorem 23.2. Since σ_X is known to be 12, $\sigma_{\bar{x}} = 12/\sqrt{10} = 3.80$, and

$$I_0 = \frac{1}{4^2} = 0.0625$$

$$I_{\bar{x}} = \frac{1}{(3.80)^2} = 0.06925$$

$$\mu_1 = \frac{0.0625 \cdot 14 + 0.06925 \cdot 8}{0.0625 + 0.06925} = \frac{0.875 + 0.554}{0.13175} = 10.85$$

Because $\mu_1 > \mu_b$ (10.85 > 10.00), the optimal act is still to produce and market the programmable calculators.

■ **23.8 A COMPARISON OF BAYESIAN AND CLASSICAL APPROACHES**

In this section we characterize the major differences between the approaches taken by classical and Bayesian statisticians in arriving at a decision under risk. These differences include whether to admit a subjective probability distribution into the analysis of a problem, and concern statements that can be made regarding the interpretation of results using the different approaches.

☐ 23.8.1 Use of subjective probabilities

The major point of deviation between Bayesian and classical statisticians concerns the admittance of subjective probabilities (reflecting degrees of belief) into the analysis of a problem. The Bayesian, believing that *all* information pertinent to a decision should be incorporated into the decision process, is perfectly willing to use—and indeed, often encourages the use of—subjective probability distributions in arriving at a decision. The classical statistician, on the other hand, will not use a subjective probability distribution to construct, say, a test of a hypothesis or a confidence interval. He bases his judgments solely on the sampling distribution or on the likelihood function described in Chapters 8–11. As such, the classical statistician is often called a "frequentist," basing all probabilities on the long-run frequency interpretation of probability.

☐ 23.8.2 Estimating the value of a population parameter

The second point of departure between the two "camps" of statistics concerns the "value" of a population parameter. The classical statistician does not allow probability statements to be made regarding possible values of a population parameter. The population parameter either is equal to some value or is not, and it is "incorrect" to speak of the probability of the parameter being some value. The Bayesian, on the other hand, views an unknown population parameter as a random variable, and thus is perfectly willing to make probability statements concerning it.

We show a graph of this distinction in Figure 23.12. Figure 23.12A represents the *posterior distribution* of the random variable μ, the population

FIGURE 23.12 Posterior and sampling distribution

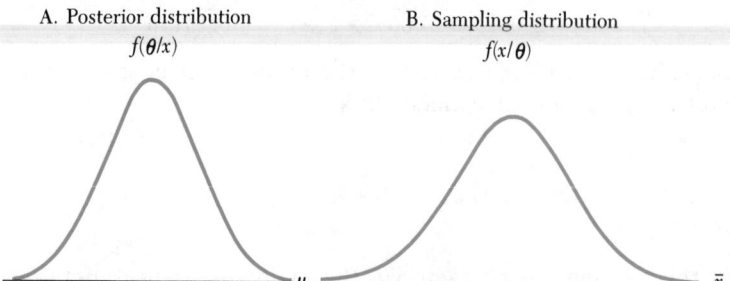

A. Posterior distribution

$f(\theta/x)$

B. Sampling distribution

$f(x/\theta)$

mean or expected value, and Figure 23.12B represents the *sampling distribution* of the random variable $\overline{X}$. The Bayesian statistician would use the posterior distribution of the *random variable* μ to make probability statements regarding the population mean, whereas the classical statistician would use the sampling distribution of the mean to base his statements on the population mean value. Note the difference in the two distributions: $P(\theta/x)$ $[f(\theta/x)]$ represents the probability distribution of the random variable θ given the sample result x, and $P(x/\theta)$ $[f(x/\theta)]$ represents a probability distribution of sample outcomes, conditional on values of the random variable θ. The use of these two distributions leads to some interesting interpretations of probability statements in estimation and hypothesis testing.

Let's take first the case of confidence interval estimation and consider a $(1 - \alpha)$ percent confidence interval on the mean of some process. The Bayesian statistician would state, "The probability is $(1 - \alpha)$ that the mean of the process lies in the stated interval," whereas the classical statistician would say, "$(1 - \alpha)$ percent of the confidence intervals so constructed contain the true mean of the process." The Bayesian is making statements about the interval at hand, while the classicist is making statements about the long run. To differentiate between the two interval estimates, the classical probability interval is called a *confidence interval*, and the Bayesian interval is called a *credible interval*.

Now let's consider the case of hypothesis testing (for discussion purposes, we'll consider a one-tail test of the form:

$$H_0: \quad \mu \le \mu_0$$
$$H_A: \quad \mu > \mu_0,$$

but the same distinction holds for either a one-tail or a two-tail test of a hypothesis). The classical statistician will either reject or not reject the null hypothesis H_0. The Bayesian statistician, on the other hand, will speak of the probability that H_0 is true *and* the probability that H_A is true. The classical statistician specifies the distribution under H_0, but not under H_A because of his concern with the probability of making a Type I error, α. The Bayesian statistician takes into account the distribution under both H_0 *and* H_A in making probability statements regarding both hypotheses.

☐ **23.8.3   Formal versus informal consideration of losses of an incorrect decision**

The classical statistician, in performing a test of a hypothesis, usually sets α (the probability of making a Type I error) equal to a predetermined value of 0.1, 0.05, or 0.01. The actual value of α (of the three) is selected to "reflect" the economic consequences of rejecting the null hypothesis when it is true. Thus, we say that the determination of α is done somewhat informally in regard to the incorporation of losses into the decision. Given a value of α, a test is selected that gives the smallest value of β, the probability of making a Type II error.

The Bayesian statistician bases the determination of α *and* β on the economic consequences that might result if the decision is incorrect. Since the maximization of EMV or the minimization of EOL *formally* incorporates the costs or the losses of making an incorrect decison into a decision rule, we say that the Bayesian statistician formally incorporates the consequences of making both types of errors into his decision rule. It is safe to say that a Bayesian statistician can "tolerate" a much higher probability of making a Type I error (α) than can his classical counterpart!

There are other differences between Bayesian and classical statisticians than those mentioned above, to be sure. We have attempted in the preceding discussion to summarize a few of the major differences between the two approaches so that the reader can make an informed choice between them. References by Schlaifer and by Winkler at the end of the chapter are *excellent* treatments of advanced topics in statistical decision theory or in Bayesian decision theory, which the reader who is interested in more on this topical area may want to consider.

■ REFERENCES

Bierman, H., Jr.; Bonini, C. P.; and Hausman, W. H. *Quantitative Analysis for Business Decisions.* 5th ed. Homewood, Ill.: Richard D. Irwin, Inc., 1977.

Jedamus, Paul, and Frame, Robert J. *Business Decision Theory.* New York: McGraw-Hill Book Company, 1969.

Luce, Duncan R., and Raiffa, Howard. *Games and Decisions.* New York: John Wiley & Sons, Inc., 1957.

Pratt, J. W.; Raiffa, H.; and Schlaifer, R. *Introduction to Statistical Decision Theory.* Preliminary edition. New York: McGraw-Hill Book Company, 1965.

Raiffa, Howard. *Decision Analysis: Introductory Lectures on Choices under Uncertainty.* Reading, Mass.: Addison-Wesley Publishing Co., Inc., 1968.

Raiffa, Howard, and Schlaifer, R. *Applied Statistical Decision Theory.* Boston: Division of Research, Harvard University, 1961.

Schlaifer, Robert. *Analysis of Decisions under Uncertainty.* New York: McGraw-Hill Book Company, 1969.

————. *Computer Programs for Elementary Decision Analysis.* Boston: Division of Research, Harvard University, 1971.

————. *Introduction to Statistics for Business Decisions.* New York: McGraw-Hill Book Company, 1961.

————. *Probability and Statistics for Business Decisions.* New York: McGraw-Hill Book Company, 1959.

Winkler, Robert L. *An Introduction to Bayesian Inference and Decision.* New York: Holt, Rinehart and Winston, Inc., 1972.

■ PROBLEMS

23.1. Assuming you are willing to let subjective judgments enter into a determination of a prior probability distribution, state formally the steps you would take and the factors you would consider in assessing a probability distribution for the outcome of your college's first football game.

23.2. Why would it be difficult (really impossible) to derive a composite probability distribution for the proportion of games won in a season by considering each game as a simple Bernoulli experiment? Why isn't the process generating wins (versus losses or ties) a Bernoulli process?

23.3. Referring to Theorem 23.1, what (in your words) is the correspondence among $P(A/B)$, $P(B/A)$, $P(A)$, $P(B/A)P(A)$, and $P(B/\bar{A})P(\bar{A})$ and $f(\theta)$, $f(x)$, $f(x/\theta)$, $f(x, \theta)$, and $f(\theta/x)$?

23.4. Verify the quantities given in Section 23.2 for the unconditional expected total losses of the decision rules (10, 2), (10, 3), (10, 4), and (10, 5).

23.5. In Example 23.1, assume that the cost of caring for a tree the extra two years is 28 cents. What is the value of θ_b? Give CVPI in tabular and in functional form where θ is expressed as a proportion. What is EVPI? What is the optimal act under the prior distribution?

23.6. What is the optimal decision rule for a sample of size $n = 10$ in Problem 23.5? What is the value of ENGS? Of EVSI? What is the value of UEOL in the optimal sampling plan?

23.7. Construct a graph of the CEOL of the optimal decision rule determined in Problem 23.6. Call the rule (10, x^*). Then determine the CEOL of the decision rule (20, $2x^*$) and graph this decision rule on the same graph. What is the effect in terms of CEOL of doubling the sample size n and the critical number x? What is α, the probability of a Type I error, in the above two decision rules?

23.8. What is the maximum Burt would be willing to spend to take a sample in Problem 23.5? Why?

23.9. Suppose in Example 23.3 two sam-

ple cuttings contained virus in sufficient quantity to stunt the growth of the trees. What is the optimal act under the posterior distribution? What is the minimum expected total cost?

23.10. Referring to the data of Example 22.1, assume George has the opportunity to obtain sample information on a specific tour. Based on a sample of ten individuals, two say they will purchase a tour. (Assume that 50 people are approached per tour to see if they are interested in purchasing it.) Treating George's long-run relative frequency distribution as the prior distribution, what is his revised distribution of tour sales? Using this revised distribution, what is the optimal act (number of seats to reserve)? What is the expected payoff of a perfect predictor? What is the expected value of perfect information?

23.11. A certain machine produces bearings in 500-unit lots. The frequency distribution for the number of defectives in a lot based on recent experience is given in the table.

Number of defectives	Frequency
5	0.10
10	0.18
15	0.43
20	0.15
25	0.14
	1.00

A sample of 25 bearings from one 500-unit lot is taken and one unit is found to be defective. What is the revised probability distribution for defectives in the lot?

23.12. A corporation has undertaken a training program to introduce members of its accounting staff to new computer languages. After the program is completed, a test is administered to evaluate training efficiency. From previous experience, it is known that the probability of passing the test after training is 70 percent, whereas the probability of failing the test without the training is 50 percent. (Employees may, if they wish, take the examination, and if they pass, they are not expected to enroll in the training program.) Ten percent of the accounting staff has participated in the training program. An employee (accountant) selected at random is known to have passed the test. What is the probability that he participated in the training program? That he did not? What probability would you give to passing the test if you know nothing about whether the individual had the training or not?

23.13. A manufacturer of electronic calculators relies on mailed circulars to generate sales of its products. Before any new item is put into full production, a sample mailing is made and the proportion of sampled individuals purchasing the product is recorded. If the proportion of favorable responses (purchases) is greater than what the firm considers to be its break-even point, the item is put into full production in anticipation of forthcoming sales. If the proportion is less than the break-even point, the firm usually decides not to market the product or else to redesign the product so that it will be salable. In either case, the market is such that if the product is released late, competition will severely reduce sales, thus making the item unprofitable. If the item is put into production when there is an insufficient demand for it, large

sums of money will be lost in additional circulars advertising the product at a reduced price. The firm has built up the sales experience shown in the table below in marketing calculators similar to the one now under consideration for production, the "High Performance IV." One hundred circulars advertising the new calculator are mailed, and they result in 5 purchases and 95 negative responses. What is the *revised probability distribution* of sales for the High Performance IV?

Proportion, π	Prior probability (long-run relative frequency from previous sales)
0.02	0.21
0.03	0.23
0.04	0.45
0.05	0.09
0.06	0.01
0.07	0.01
	1.00

Note: To be able to answer this question, you will have to work directly with the probability mass function of a binomial random variable, or you will have to obtain a table of binomial probabilities from your college or university library for $n = 100$.

23.14. Given the information in Problem 23.13 for sales of electronic calculators, assume that the firm in question decided to market the product based on the prior distribution and that 10,000 circulars were then mailed to prospective customers. How many calculators can the firm expect to sell? Justify your result, and comment on the probability of selling at least half again as many as you originally estimated.

23.15. Assume the same facts in Problem 23.14, except that the firm chooses to work with the posterior distribution after probabilities have been revised (see Problem 23.13). How many calculators can the firm now expect to sell? What is the probability of selling half again as many as originally estimated or more now that the probability distribution has been revised?

23.16. Beginning with the relationship

$$I_1 = I_0 + I_{\bar{x}}$$

show that

$$\sigma_1 = \sqrt{\frac{\sigma_0^2 \cdot \sigma_{\bar{x}}^2}{\sigma_0^2 + \sigma_{\bar{x}}^2}}$$

23.17. A competitor of the electronic calculator company of Problem 23.13 markets its calculators principally through retail department stores. In the past, it has chosen to market only the larger, nonportable, non-LSI circuitry models. Recently, however, the firm has been "persuaded" that the market for calculators is shifting to the smaller units, so it has decided to test-market a new model conforming in size to a portable one. One hundred stores from the 10,000 retail outlets it currently has under contract are provided with the new units, resulting in an average sale per store of 200 units and a standard deviation of 25 units. The firm's prior distribution on sales of the new units (based on sales histories of the older models) is assumed normal with mean $\mu_0 = 120$ units and standard deviation $\sigma_0 = 40$ units. What is the firm's posterior distribution for sales of its new calculator?

23.18. A decision maker assesses a prior distribution of sales for a new product as being normal with mean $\mu_0 =$ 190 units per store and standard deviation $\sigma_0 = 30$ units per store. A sample of 100 stores is selected at random for testing sales of the new product, yielding a sample mean $\bar{x} = 225$ units per store and a sample standard deviation $s = 25$ units. What is the posterior distribution of sales for the new product?

23.19. Assume in Problem 23.18 that the sample consists of only 20 stores, but that the statistics remain the same. What is the revised probability distribution?

23.20. In what way does the sample size n affect the amount of information present in a normal sampling distribution? Describe the exact relationship.

23.21. In Example 23.8, assume the following sample results were obtained.

Store	Calculators sold
1	8
2	5
3	7
4	5
5	2
6	8
7	8
8	7
9	6
10	8

What is the optimal act under the posterior distribution of sales?

23.22. Assume the same facts as in Example 23.8 and that $\sigma_X = 9$. What are EVSI and ENGS for the optimal decision rule (sample size)?

23.23. Herman, as his main source of income, inspects garbage disposals at the local plumbing fixtures factory. A lot of 10,000 disposals is awaiting shipment to retail outlets. Herman's boss, Fort Worth Annie, has instructed Herman to inspect 10 of the 10,000 garbage disposals, and at the 5 percent significance level (i.e., $\alpha = 0.05$) either conclude that 10 percent or fewer are defective and ship the batch of 10,000 disposals, or else conclude that more than 10 percent of the disposals are defective and inspect the entire lot of disposals before shipment. Construct a "classical" decision rule for Herman so that he can decide whether to ship the disposals as is or else inspect the entire lot and repair the defective ones. Comment on the choice of $\alpha = 0.05$, and the consequences that might result by setting α at this level (alternative losses).

APPENDIX A

Special topics: A calculus-based study of continuous random variables

The student who has had calculus can appreciate the relationship that exists between the area under a curve [say $f(x)$—a probability density function] between two limits c and d, and the probability that a continuous random variable $\mathbf{X}$ assumes a value between c and d, i.e., $P(c \leq \mathbf{X} \leq d)$. That is, if $\mathbf{X}$ is a continuous random variable with probability density function $f(x)$ defined between the limits a and b, then the probability that the random variable $\mathbf{X}$ assumes a value between c and d, where $c \geq a$, $d \leq b$, and $d > c$, is given by:

$$P(c \leq \mathbf{X} \leq d) = \int_c^d f(x)\, dx$$

In what follows, we present the continuous analogue of many of the definitions for a discrete random variable given in Chapter 5.

The integration symbol $\int$ for a continuous random variable is analogous to the summation symbol for a discrete random variable. In the former case, we are "summing" areas, whereas in the latter case, we are summing over discrete values. Table A.1 summarizes the similarities and differences between the two types of random variables. This table is similar to Table 5.13, except that here we have indicated the integration required to determine the needed probabilities.

The student who has had calculus should begin to recognize some familiar notions at this point. First, the area under a continuous curve, say $f(x)$, can be determined by integration. Indeed, the second condition in Definition 5.8 [the area under the curve $f(x)$ from $x = a$ to $x = b$ must be 1] can be stated concisely as:

**Definition 5.8
(Addendum)**

$$\int_a^b f(x)\, dx = 1$$

For example, consider the density function

$$f(x) = \begin{cases} \dfrac{1}{b - a} & a \leq x \leq b \\ 0 & \text{all other } x \end{cases}$$

developed in Chapter 7 (continuous uniform probability density function). Definition 5.8 (addendum) requires that we show that $\int_a^b [1/(b - a)]\, dx = 1$ for this density function. Since $\int_a^b k\, dx = k \cdot (b - a)$,

$$\int_a^b \left(\frac{1}{b - a}\right) dx = \frac{1}{b - a}(b - a) = 1, \qquad \boxed{k = \frac{1}{b - a}}$$

TABLE A.1 Comparison of properties of discrete and continuous random variables

Property	Discrete random variable	Continuous random variable
Number of possible values of the random variable	Finite *or* countably infinite	Infinite (not countable)
$P(c \leq X \leq d)$	$\displaystyle\sum_{x=c}^{x=d} P(x)$	Area under the curve $f(x)$ from c to d = $\int_c^d f(x)\, dx$
$P(X = e)$.............	$P(e)$	Always zero, $P(X = e) = \int_e^e f(x)\, dx \equiv 0$

In order to show that the probability a continuous random variable assumes a particular value, say $X = e$, is *always equal to 0* using calculus, note the following. The notation $\int_a^b f(x)\, dx$ represents the area under the curve $f(x)$ from $x = a$ to $x = b$. By using the second property in Table A.1, we may write:

$$P(X = e) = P(e \leq X \leq e) = \int_e^e f(x)\, dx = 0$$

since the integral defined over the limits e to e is always equal to 0! Hence, using calculus, it is relatively simple to argue that the probability that a continuous random variable assumes a particular value indeed is always 0.

The expected value of a continuous random variable defined between the limits a and b involves the integration of the product of x and $f(x)$ over the interval (a, b); that is, it is the area under the curve of the function $xf(x)$ between a and b. We give the definition of the expected value of a continuous random variable in Definition A.1.

Definition A.1
Expectation of a random variable (continuous case)

Let X be a continuous random variable defined over the interval (a, b) with probability density function $f(x)$. The *mean* or the *expected value* of the random variable X, denoted by $E(X)$, is given by

$$E(X) = \int_a^b xf(x)\, dx$$

Again considering the continuous uniform random variable **X** described previously,

$$E(X) = \int_a^b x \left(\frac{1}{b-a}\right) dx = \left(\frac{1}{b-a}\right) \int_a^b x \, dx = \left(\frac{1}{b-a}\right)\left(\frac{x^2}{2}\right)\Big|_a^b$$

$$= \frac{1}{b-a}\left(\frac{b^2}{2} - \frac{a^2}{2}\right) = \frac{1}{2}\left(\frac{b^2-a^2}{b-a}\right) = \frac{1}{2}\left[\frac{(b-a)(b+a)}{b-a}\right] = \frac{b+a}{2}$$

Definition A.2 gives the formula for determining the variance of a continuous random variable. Integration is almost always required to determine the variance of a continuous random variable, as was the case for the expected value of a continuous random variable.

Definition A.2
Variance of a random variable (continuous case)

Let **X** be a continuous random variable defined over the interval (a, b) with probability density function $f(x)$. The *variance* of the random variable **X**, denoted by $V(X)$, is given by

$$V(X) = \int_a^b [x - E(X)]^2 f(x) \, dx = \int_a^b x^2 f(x) \, dx - [E(X)]^2$$

For the uniform continuous random variable introduced, we have:

$$V(X) = \int_a^b x^2 f(x) \, dx - [E(X)]^2 = \int_a^b x^2 \left(\frac{1}{b-a}\right) dx - [E(X)]^2$$

$$= \frac{1}{b-a} \int_a^b x^2 \, dx - \left(\frac{b+a}{2}\right)^2 = \frac{1}{b-a}\left(\frac{x^3}{3}\right)\Big|_a^b - \left(\frac{b+a}{2}\right)^2$$

$$= \left(\frac{1}{b-a}\right) \cdot \left(\frac{b^3}{3} - \frac{a^3}{3}\right) - \left(\frac{b+a}{2}\right)^2$$

which simplifies to,

$$V(X) = \frac{(b-a)^2}{12}$$

It is interesting to note the similarity in the computation of the expected value and the variance of a discrete and a continuous random variable as illustrated in Table A.2. The analogy to summing in the discrete case (Σ) is, of course, integrating in the continuous case ($\int$).

TABLE A.2 Comparison of the expected value and variance formulas for discrete and continuous random variables

Property	Discrete random variable	Continuous random variable
Expected value, $E(X)$	$\Sigma x P(x)$	$\int x f(x) \, dx$
Variance, $V(X)$	$\Sigma [x - E(X)]^2 P(x)$	$\int [x - E(X)]^2 f(x) \, dx$

Finally, we present the definition of the expectation of a function of a continuous random variable. This definition is similar to Definition 5.10 for the expectation of a function of a discrete random variable.

> **Definition A.3**
> Expectation of a function of a random variable (continuous case)
>
> Let **X** be a continuous random variable defined between the limits a and b (a < b), with probability density function $f(x)$ and consider a function of **X**, denoted by $h(\mathbf{X})$. The *average* or the *expected value* of $h(\mathbf{X})$ is given by:
>
> $$E[h(\mathbf{X})] = \int_a^b h(x)f(x)\, dx$$

We conclude this section by deriving the expected value and the variance of an exponential random variable, a random variable introduced in Chapter 7.

An exponential random variable has the following probability density function:

$$f(x) = \lambda e^{-\lambda x}; \qquad 0 \le x \le \infty, \quad \lambda > 0$$
$$E(\mathbf{X}) = \int_0^\infty x\lambda e^{-\lambda x}\, dx$$

Using integration by parts,

$$E(\mathbf{X}) = (-xe^{-\lambda x})\big|_0^\infty - \int_0^\infty (-e^{-\lambda x})\, dx$$

$$= (-xe^{-\lambda x})\big|_0^\infty - \left(\frac{1}{\lambda}e^{-\lambda x}\right)\bigg|_0^\infty = \frac{1}{\lambda}$$

$$E(\mathbf{X}^2) = \int_0^\infty x^2\lambda e^{-\lambda x}\, dx = (-x^2 e^{-\lambda x})\big|_0^\infty - \int_0^\infty (-2xe^{-\lambda x})\, dx$$

$$= (-x^2 e^{-\lambda x})\big|_0^\infty + 2\left(-\frac{1}{\lambda}xe^{-\lambda x}\right)\bigg|_0^\infty + \int_0^\infty \frac{1}{\lambda}e^{-\lambda x}\, dx$$

$$= (-x^2 e^{-\lambda x})\big|_0^\infty + 2\left(-\frac{1}{\lambda}xe^{-\lambda x}\right)\bigg|_0^\infty - 2\left(\frac{1}{\lambda^2}e^{-\lambda x}\right)\bigg|_0^\infty = \frac{2}{\lambda^2}$$

$$\therefore V(\mathbf{X}) = E(\mathbf{X}^2) - [E(\mathbf{X})]^2 = \frac{2}{\lambda^2} - \left(\frac{1}{\lambda}\right)^2 = \frac{1}{\lambda^2}$$

Table interpolation illustration—standard normal distribution:

1. $P(Z \geq 1.364) = 0.50 - P(0 \leq Z \leq 1.364)$:

z	$P(0 \leq Z \leq z)$

$$0.010 \begin{bmatrix} 1.360 \\ 1.364 \\ 1.370 \end{bmatrix} 0.004 \qquad 0.0016 \begin{bmatrix} 0.4131 \\ * \\ 0.4147 \end{bmatrix} \begin{matrix} 0.004 \\ 0.010 \end{matrix} (0.0016) = 0.00064$$

$$P(0 \leq Z \leq 1.364) = 0.4131 + 0.00064 = 0.4137 = *$$
$$P(Z \geq 1.364) = 0.5000 - 0.4137 = 0.0863$$

2. $P(0 \leq Z \leq z) = 0.1977$, find z.

z	$P(0 \leq Z \leq z)$

$$\frac{0.0027}{0.0035} (0.010) = 0.0077 \begin{bmatrix} 0.510 \\ z \\ 0.520 \end{bmatrix} 0.010 \qquad \begin{bmatrix} 0.1950 \\ 0.1977 \\ 0.1985 \end{bmatrix} \begin{matrix} 0.0027 \end{matrix} 0.0035$$

$$z = 0.5100 + 0.0077 = 0.5177; \quad P(0 \leq Z \leq 0.5177) = 0.1977$$

TABLE B.1 Binomial probability tables
Tabulated values give the probability that the random variable X assumes a value less than or equal to $x = a$. Tabulated values are

$$P(X \leq x = a/n,\pi) = \sum_{x=0}^{a} P(x) = P(0) + P(1) + \cdots + P(a)$$

$n = 1$

a＼π	.010	.050	.100	.200	.300	.400	.500	.600	.700	.800	.900	.950	.990
0	.990	.950	.900	.800	.700	.600	.500	.400	.300	.200	.100	.050	.010

$n = 2$

a＼π	.010	.050	.100	.200	.300	.400	.500	.600	.700	.800	.900	.950	.990
0	.980	.903	.810	.640	.490	.360	.250	.160	.090	.040	.010	.003	.000
1	1.000	.998	.990	.960	.910	.840	.750	.640	.510	.360	.190	.097	.020

$n = 3$

a＼π	.010	.050	.100	.200	.300	.400	.500	.600	.700	.800	.900	.950	.990
0	.970	.857	.729	.512	.343	.216	.125	.064	.027	.008	.001	.000	.000
1	1.000	.993	.972	.896	.784	.648	.500	.352	.216	.104	.028	.007	.000
2	1.000	1.000	.999	.992	.973	.936	.875	.784	.657	.488	.271	.143	.030

TABLE B.1 (*continued*)

$n = 4$

a \ π	.010	.050	.100	.200	.300	.400	.500	.600	.700	.800	.900	.950	.990
0	.961	.815	.656	.410	.240	.130	.063	.026	.008	.002	.000	.000	.000
1	.999	.986	.948	.819	.652	.475	.313	.179	.084	.027	.004	.000	.000
2	1.000	1.000	.996	.973	.916	.821	.688	.525	.348	.181	.052	.014	.001
3	1.000	1.000	1.000	.998	.992	.974	.938	.870	.760	.590	.344	.185	.039

$n = 5$

a \ π	.010	.050	.100	.200	.300	.400	.500	.600	.700	.800	.900	.950	.990
0	.951	.774	.590	.328	.168	.078	.031	.010	.002	.000	.000	.000	.000
1	.999	.977	.919	.737	.528	.337	.188	.087	.031	.007	.000	.000	.000
2	1.000	.999	.991	.942	.837	.683	.500	.317	.163	.058	.009	.001	.000
3	1.000	1.000	1.000	.993	.969	.913	.813	.663	.472	.263	.081	.023	.001
4	1.000	1.000	1.000	1.000	.998	.990	.969	.922	.832	.672	.410	.226	.049

$n = 6$

a \ π	.010	.050	.100	.200	.300	.400	.500	.600	.700	.800	.900	.950	.990
0	.941	.735	.531	.262	.118	.047	.016	.004	.001	.000	.000	.000	.000
1	.999	.967	.886	.655	.420	.233	.109	.041	.011	.002	.000	.000	.000
2	1.000	.998	.984	.901	.744	.544	.344	.179	.070	.017	.001	.000	.000
3	1.000	1.000	.999	.983	.930	.821	.656	.456	.256	.099	.016	.002	.000
4	1.000	1.000	1.000	.998	.989	.959	.891	.767	.580	.345	.114	.033	.001
5	1.000	1.000	1.000	1.000	.999	.996	.984	.953	.882	.738	.469	.265	.059

$n = 7$

a \ π	.010	.050	.100	.200	.300	.400	.500	.600	.700	.800	.900	.950	.990
0	.932	.698	.478	.210	.082	.028	.008	.002	.000	.000	.000	.000	.000
1	.998	.956	.850	.577	.329	.159	.063	.019	.004	.000	.000	.000	.000
2	1.000	.996	.974	.852	.647	.420	.227	.096	.029	.005	.000	.000	.000
3	1.000	1.000	.997	.967	.874	.710	.500	.290	.126	.033	.003	.000	.000
4	1.000	1.000	1.000	.995	.971	.904	.773	.580	.353	.148	.026	.004	.000
5	1.000	1.000	1.000	1.000	.996	.981	.938	.841	.671	.423	.150	.044	.002
6	1.000	1.000	1.000	1.000	1.000	.998	.992	.972	.918	.790	.522	.302	.068

$n = 8$

a \ π	.010	.050	.100	.200	.300	.400	.500	.600	.700	.800	.900	.950	.990
0	.923	.663	.430	.168	.058	.017	.004	.001	.000	.000	.000	.000	.000
1	.997	.943	.813	.503	.255	.106	.035	.009	.001	.000	.000	.000	.000
2	1.000	.994	.962	.797	.552	.315	.145	.050	.011	.001	.000	.000	.000
3	1.000	1.000	.995	.944	.806	.594	.363	.174	.058	.010	.000	.000	.000
4	1.000	1.000	1.000	.990	.942	.826	.637	.406	.194	.056	.005	.000	.000
5	1.000	1.000	1.000	.999	.989	.950	.855	.685	.448	.203	.038	.006	.000
6	1.000	1.000	1.000	1.000	.999	.991	.965	.894	.745	.497	.187	.057	.003
7	1.000	1.000	1.000	1.000	1.000	.999	.996	.983	.942	.832	.570	.337	.077

$n = 9$

a \ π	.010	.050	.100	.200	.300	.400	.500	.600	.700	.800	.900	.950	.990
0	.914	.630	.387	.134	.040	.010	.002	.000	.000	.000	.000	.000	.000
1	.997	.929	.775	.436	.196	.071	.020	.004	.000	.000	.000	.000	.000
2	1.000	.992	.947	.738	.463	.232	.090	.025	.004	.000	.000	.000	.000
3	1.000	.999	.992	.914	.730	.483	.254	.099	.025	.003	.000	.000	.000
4	1.000	1.000	.999	.980	.901	.733	.500	.267	.099	.020	.001	.000	.000
5	1.000	1.000	1.000	.997	.975	.901	.746	.517	.270	.086	.008	.001	.000
6	1.000	1.000	1.000	1.000	.996	.975	.910	.768	.537	.262	.053	.008	.000
7	1.000	1.000	1.000	1.000	1.000	.996	.980	.929	.804	.564	.225	.071	.003
8	1.000	1.000	1.000	1.000	1.000	1.000	.998	.990	.960	.866	.613	.370	.086

TABLE B.1 (*continued*)

n = 10

a \ π	.010	.050	.100	.200	.300	.400	.500	.600	.700	.800	.900	.950	.990
0	.904	.599	.349	.107	.028	.006	.001	.000	.000	.000	.000	.000	.000
1	.996	.914	.736	.376	.149	.046	.011	.002	.000	.000	.000	.000	.000
2	1.000	.988	.930	.678	.383	.167	.055	.012	.002	.000	.000	.000	.000
3	1.000	.999	.987	.879	.650	.382	.172	.055	.011	.001	.000	.000	.000
4	1.000	1.000	.998	.967	.850	.633	.377	.166	.047	.006	.000	.000	.000
5	1.000	1.000	1.000	.994	.953	.834	.623	.367	.150	.033	.002	.000	.000
6	1.000	1.000	1.000	.999	.989	.945	.828	.618	.350	.121	.013	.001	.000
7	1.000	1.000	1.000	1.000	.998	.988	.945	.833	.617	.322	.070	.012	.000
8	1.000	1.000	1.000	1.000	1.000	.998	.989	.954	.851	.624	.264	.086	.004
9	1.000	1.000	1.000	1.000	1.000	1.000	.999	.994	.972	.893	.651	.401	.096

n = 11

a \ π	.010	.050	.100	.200	.300	.400	.500	.600	.700	.800	.900	.950	.990
0	.895	.569	.314	.086	.020	.004	.000	.000	.000	.000	.000	.000	.000
1	.995	.898	.697	.322	.113	.030	.006	.001	.000	.000	.000	.000	.000
2	1.000	.985	.910	.617	.313	.119	.033	.006	.001	.000	.000	.000	.000
3	1.000	.998	.981	.839	.570	.296	.113	.029	.004	.000	.000	.000	.000
4	1.000	1.000	.997	.950	.790	.533	.274	.099	.022	.002	.000	.000	.000
5	1.000	1.000	1.000	.988	.922	.753	.500	.247	.078	.012	.000	.000	.000
6	1.000	1.000	1.000	.998	.978	.901	.726	.467	.210	.050	.003	.000	.000
7	1.000	1.000	1.000	1.000	.996	.971	.887	.704	.430	.161	.019	.002	.000
8	1.000	1.000	1.000	1.000	.999	.994	.967	.881	.687	.383	.090	.015	.000
9	1.000	1.000	1.000	1.000	1.000	.999	.994	.970	.887	.678	.303	.102	.005
10	1.000	1.000	1.000	1.000	1.000	1.000	1.000	.996	.980	.914	.686	.431	.105

n = 12

a \ π	.010	.050	.100	.200	.300	.400	.500	.600	.700	.800	.900	.950	.990
0	.886	.540	.282	.069	.014	.002	.000	.000	.000	.000	.000	.000	.000
1	.994	.882	.659	.275	.085	.020	.003	.000	.000	.000	.000	.000	.000
2	1.000	.980	.889	.558	.253	.083	.019	.003	.000	.000	.000	.000	.000
3	1.000	.998	.974	.795	.493	.225	.073	.015	.002	.000	.000	.000	.000
4	1.000	1.000	.996	.927	.724	.438	.194	.057	.009	.001	.000	.000	.000
5	1.000	1.000	.999	.981	.882	.665	.387	.158	.039	.004	.000	.000	.000
6	1.000	1.000	1.000	.996	.961	.842	.613	.335	.118	.019	.001	.000	.000
7	1.000	1.000	1.000	.999	.991	.943	.806	.562	.276	.073	.004	.000	.000
8	1.000	1.000	1.000	1.000	.998	.985	.927	.775	.507	.205	.026	.002	.000
9	1.000	1.000	1.000	1.000	1.000	.997	.981	.917	.747	.442	.111	.020	.000
10	1.000	1.000	1.000	1.000	1.000	1.000	.997	.980	.915	.725	.341	.118	.006
11	1.000	1.000	1.000	1.000	1.000	1.000	1.000	.998	.986	.931	.718	.460	.114

n = 13

a \ π	.010	.050	.100	.200	.300	.400	.500	.600	.700	.800	.900	.950	.990
0	.878	.513	.254	.055	.010	.001	.000	.000	.000	.000	.000	.000	.000
1	.993	.865	.621	.234	.064	.013	.002	.000	.000	.000	.000	.000	.000
2	1.000	.975	.866	.502	.202	.058	.011	.001	.000	.000	.000	.000	.000
3	1.000	.997	.966	.747	.421	.169	.046	.008	.001	.000	.000	.000	.000
4	1.000	1.000	.994	.901	.654	.353	.133	.032	.004	.000	.000	.000	.000
5	1.000	1.000	.999	.970	.835	.574	.291	.098	.018	.001	.000	.000	.000
6	1.000	1.000	1.000	.993	.938	.771	.500	.229	.062	.007	.000	.000	.000
7	1.000	1.000	1.000	.999	.982	.902	.709	.426	.165	.030	.001	.000	.000
8	1.000	1.000	1.000	1.000	.996	.968	.867	.647	.346	.099	.006	.000	.000
9	1.000	1.000	1.000	1.000	.999	.992	.954	.831	.579	.253	.034	.003	.000
10	1.000	1.000	1.000	1.000	1.000	.999	.989	.942	.798	.498	.134	.025	.000
11	1.000	1.000	1.000	1.000	1.000	1.000	.998	.987	.936	.766	.379	.135	.007
12	1.000	1.000	1.000	1.000	1.000	1.000	1.000	.999	.990	.945	.746	.487	.122

TABLE B.1 (continued)

$n = 14$

a \ π	.010	.050	.100	.200	.300	.400	.500	.600	.700	.800	.900	.950	.990
0	.869	.488	.229	.044	.007	.001	.000	.000	.000	.000	.000	.000	.000
1	.992	.847	.585	.198	.047	.008	.001	.000	.000	.000	.000	.000	.000
2	1.000	.970	.842	.448	.161	.040	.006	.001	.000	.000	.000	.000	.000
3	1.000	.996	.956	.698	.355	.124	.029	.004	.000	.000	.000	.000	.000
4	1.000	1.000	.991	.870	.584	.279	.090	.018	.002	.000	.000	.000	.000
5	1.000	1.000	.999	.956	.781	.486	.212	.058	.008	.000	.000	.000	.000
6	1.000	1.000	1.000	.988	.907	.692	.395	.150	.031	.002	.000	.000	.000
7	1.000	1.000	1.000	.998	.969	.850	.605	.308	.093	.012	.000	.000	.000
8	1.000	1.000	1.000	1.000	.992	.942	.788	.514	.219	.044	.001	.000	.000
9	1.000	1.000	1.000	1.000	.998	.982	.910	.721	.416	.130	.009	.000	.000
10	1.000	1.000	1.000	1.000	1.000	.996	.971	.876	.645	.302	.044	.004	.000
11	1.000	1.000	1.000	1.000	1.000	.999	.994	.960	.839	.552	.158	.030	.000
12	1.000	1.000	1.000	1.000	1.000	1.000	.999	.992	.953	.802	.415	.153	.008
13	1.000	1.000	1.000	1.000	1.000	1.000	1.000	.999	.993	.956	.771	.512	.131

$n = 15$

a \ π	.010	.050	.100	.200	.300	.400	.500	.600	.700	.800	.900	.950	.990
0	.860	.463	.206	.035	.005	.000	.000	.000	.000	.000	.000	.000	.000
1	.990	.829	.549	.167	.035	.005	.000	.000	.000	.000	.000	.000	.000
2	1.000	.964	.816	.398	.127	.027	.004	.000	.000	.000	.000	.000	.000
3	1.000	.995	.944	.648	.297	.091	.018	.002	.000	.000	.000	.000	.000
4	1.000	.999	.987	.836	.515	.217	.059	.009	.001	.000	.000	.000	.000
5	1.000	1.000	.998	.939	.722	.403	.151	.034	.004	.000	.000	.000	.000
6	1.000	1.000	1.000	.982	.869	.610	.304	.095	.015	.001	.000	.000	.000
7	1.000	1.000	1.000	.996	.950	.787	.500	.213	.050	.004	.000	.000	.000
8	1.000	1.000	1.000	.999	.985	.905	.696	.390	.131	.018	.000	.000	.000
9	1.000	1.000	1.000	1.000	.996	.966	.849	.597	.278	.061	.002	.000	.000
10	1.000	1.000	1.000	1.000	.999	.991	.941	.783	.485	.164	.013	.001	.000
11	1.000	1.000	1.000	1.000	1.000	.998	.982	.909	.703	.352	.056	.005	.000
12	1.000	1.000	1.000	1.000	1.000	1.000	.996	.973	.873	.602	.184	.036	.000
13	1.000	1.000	1.000	1.000	1.000	1.000	1.000	.995	.965	.833	.451	.171	.010
14	1.000	1.000	1.000	1.000	1.000	1.000	1.000	1.000	.995	.965	.794	.537	.140

$n = 16$

a \ π	.010	.050	.100	.200	.300	.400	.500	.600	.700	.800	.900	.950	.990
0	.851	.440	.185	.028	.003	.000	.000	.000	.000	.000	.000	.000	.000
1	.989	.811	.515	.141	.026	.003	.000	.000	.000	.000	.000	.000	.000
2	.999	.957	.789	.352	.099	.018	.002	.000	.000	.000	.000	.000	.000
3	1.000	.993	.932	.598	.246	.065	.011	.001	.000	.000	.000	.000	.000
4	1.000	.999	.983	.798	.450	.167	.038	.005	.000	.000	.000	.000	.000
5	1.000	1.000	.997	.918	.660	.329	.105	.019	.002	.000	.000	.000	.000
6	1.000	1.000	.999	.973	.825	.527	.227	.058	.007	.000	.000	.000	.000
7	1.000	1.000	1.000	.993	.926	.716	.402	.142	.026	.001	.000	.000	.000
8	1.000	1.000	1.000	.999	.974	.858	.598	.284	.074	.007	.000	.000	.000
9	1.000	1.000	1.000	1.000	.993	.942	.773	.473	.175	.027	.001	.000	.000
10	1.000	1.000	1.000	1.000	.998	.981	.895	.671	.340	.082	.003	.000	.000
11	1.000	1.000	1.000	1.000	1.000	.995	.962	.833	.550	.202	.017	.001	.000
12	1.000	1.000	1.000	1.000	1.000	.999	.989	.935	.754	.402	.068	.007	.000
13	1.000	1.000	1.000	1.000	1.000	1.000	.998	.982	.901	.648	.211	.043	.001
14	1.000	1.000	1.000	1.000	1.000	1.000	1.000	.997	.974	.859	.485	.189	.011
15	1.000	1.000	1.000	1.000	1.000	1.000	1.000	1.000	.997	.972	.815	.560	.149

TABLE B.1   (*continued*)

$n = 17$

a \ π	.010	.050	.100	.200	.300	.400	.500	.600	.700	.800	.900	.950	.990
0	.843	.418	.167	.023	.002	.000	.000	.000	.000	.000	.000	.000	.000
1	.988	.792	.482	.118	.019	.002	.000	.000	.000	.000	.000	.000	.000
2	.999	.950	.762	.310	.077	.012	.001	.000	.000	.000	.000	.000	.000
3	1.000	.991	.917	.549	.202	.046	.006	.000	.000	.000	.000	.000	.000
4	1.000	.999	.978	.758	.389	.126	.025	.003	.000	.000	.000	.000	.000
5	1.000	1.000	.995	.894	.597	.264	.072	.011	.001	.000	.000	.000	.000
6	1.000	1.000	.999	.962	.775	.448	.166	.035	.003	.000	.000	.000	.000
7	1.000	1.000	1.000	.989	.895	.641	.315	.092	.013	.000	.000	.000	.000
8	1.000	1.000	1.000	.997	.960	.801	.500	.199	.040	.003	.000	.000	.000
9	1.000	1.000	1.000	1.000	.987	.908	.685	.359	.105	.011	.000	.000	.000
10	1.000	1.000	1.000	1.000	.997	.965	.834	.552	.225	.038	.001	.000	.000
11	1.000	1.000	1.000	1.000	.999	.989	.928	.736	.403	.106	.005	.000	.000
12	1.000	1.000	1.000	1.000	1.000	.997	.975	.874	.611	.242	.022	.001	.000
13	1.000	1.000	1.000	1.000	1.000	1.000	.994	.954	.798	.451	.083	.009	.000
14	1.000	1.000	1.000	1.000	1.000	1.000	.999	.988	.923	.690	.238	.050	.001
15	1.000	1.000	1.000	1.000	1.000	1.000	1.000	.998	.981	.882	.518	.208	.012
16	1.000	1.000	1.000	1.000	1.000	1.000	1.000	1.000	.998	.977	.833	.582	.157

$n = 18$

a \ π	.010	.050	.100	.200	.300	.400	.500	.600	.700	.800	.900	.950	.990
0	.835	.397	.150	.018	.002	.000	.000	.000	.000	.000	.000	.000	.000
1	.986	.774	.450	.099	.014	.001	.000	.000	.000	.000	.000	.000	.000
2	.999	.942	.734	.271	.060	.008	.001	.000	.000	.000	.000	.000	.000
3	1.000	.989	.902	.501	.165	.033	.004	.000	.000	.000	.000	.000	.000
4	1.000	.998	.972	.716	.333	.094	.015	.001	.000	.000	.000	.000	.000
5	1.000	1.000	.994	.867	.534	.209	.048	.006	.000	.000	.000	.000	.000
6	1.000	1.000	.999	.949	.722	.374	.119	.020	.001	.000	.000	.000	.000
7	1.000	1.000	1.000	.984	.859	.563	.240	.058	.006	.000	.000	.000	.000
8	1.000	1.000	1.000	.996	.940	.737	.407	.135	.021	.001	.000	.000	.000
9	1.000	1.000	1.000	.999	.979	.865	.593	.263	.060	.004	.000	.000	.000
10	1.000	1.000	1.000	1.000	.994	.942	.760	.437	.141	.016	.000	.000	.000
11	1.000	1.000	1.000	1.000	.999	.980	.881	.626	.278	.051	.001	.000	.000
12	1.000	1.000	1.000	1.000	1.000	.994	.952	.791	.466	.133	.006	.000	.000
13	1.000	1.000	1.000	1.000	1.000	.999	.985	.906	.667	.284	.028	.002	.000
14	1.000	1.000	1.000	1.000	1.000	1.000	.996	.967	.835	.499	.098	.011	.000
15	1.000	1.000	1.000	1.000	1.000	1.000	.999	.992	.940	.729	.266	.058	.001
16	1.000	1.000	1.000	1.000	1.000	1.000	1.000	.999	.986	.901	.550	.226	.014
17	1.000	1.000	1.000	1.000	1.000	1.000	1.000	1.000	.998	.982	.850	.603	.165

$n = 19$

a \ π	.010	.050	.100	.200	.300	.400	.500	.600	.700	.800	.900	.950	.990
0	.826	.377	.135	.014	.001	.000	.000	.000	.000	.000	.000	.000	.000
1	.985	.755	.420	.083	.010	.001	.000	.000	.000	.000	.000	.000	.000
2	.999	.933	.705	.237	.046	.005	.000	.000	.000	.000	.000	.000	.000
3	1.000	.987	.885	.455	.133	.023	.002	.000	.000	.000	.000	.000	.000
4	1.000	.998	.965	.673	.282	.070	.010	.001	.000	.000	.000	.000	.000
5	1.000	1.000	.991	.837	.474	.163	.032	.003	.000	.000	.000	.000	.000
6	1.000	1.000	.998	.932	.666	.308	.084	.012	.001	.000	.000	.000	.000
7	1.000	1.000	1.000	.977	.818	.488	.180	.035	.003	.000	.000	.000	.000
8	1.000	1.000	1.000	.993	.916	.667	.324	.088	.011	.000	.000	.000	.000
9	1.000	1.000	1.000	.998	.967	.814	.500	.186	.033	.002	.000	.000	.000
10	1.000	1.000	1.000	1.000	.989	.912	.676	.333	.084	.007	.000	.000	.000
11	1.000	1.000	1.000	1.000	.997	.965	.820	.512	.182	.023	.000	.000	.000
12	1.000	1.000	1.000	1.000	.999	.988	.916	.692	.334	.068	.002	.000	.000
13	1.000	1.000	1.000	1.000	1.000	.997	.968	.837	.526	.163	.009	.000	.000
14	1.000	1.000	1.000	1.000	1.000	.999	.990	.930	.718	.327	.035	.002	.000
15	1.000	1.000	1.000	1.000	1.000	1.000	.998	.977	.867	.545	.115	.013	.000
16	1.000	1.000	1.000	1.000	1.000	1.000	1.000	.995	.954	.763	.295	.067	.001
17	1.000	1.000	1.000	1.000	1.000	1.000	1.000	.999	.990	.917	.580	.245	.015
18	1.000	1.000	1.000	1.000	1.000	1.000	1.000	1.000	.999	.986	.865	.623	.174

TABLE B.1 (*continued*)

n = 20

a \ π	.010	.050	.100	.200	.300	.400	.500	.600	.700	.800	.900	.950	.990
0	.818	.358	.122	.012	.001	.000	.000	.000	.000	.000	.000	.000	.000
1	.983	.736	.392	.069	.008	.001	.000	.000	.000	.000	.000	.000	.000
2	.999	.925	.677	.206	.035	.004	.000	.000	.000	.000	.000	.000	.000
3	1.000	.984	.867	.411	.107	.016	.001	.000	.000	.000	.000	.000	.000
4	1.000	.997	.957	.630	.238	.051	.006	.000	.000	.000	.000	.000	.000
5	1.000	1.000	.989	.804	.416	.126	.021	.002	.000	.000	.000	.000	.000
6	1.000	1.000	.998	.913	.608	.250	.058	.006	.000	.000	.000	.000	.000
7	1.000	1.000	1.000	.968	.772	.416	.132	.021	.001	.000	.000	.000	.000
8	1.000	1.000	1.000	.990	.887	.596	.252	.057	.005	.000	.000	.000	.000
9	1.000	1.000	1.000	.997	.952	.755	.412	.128	.017	.001	.000	.000	.000
10	1.000	1.000	1.000	.999	.983	.872	.588	.245	.048	.003	.000	.000	.000
11	1.000	1.000	1.000	1.000	.995	.943	.748	.404	.113	.010	.000	.000	.000
12	1.000	1.000	1.000	1.000	.999	.979	.868	.584	.228	.032	.000	.000	.000
13	1.000	1.000	1.000	1.000	1.000	.994	.942	.750	.392	.087	.002	.000	.000
14	1.000	1.000	1.000	1.000	1.000	.998	.979	.874	.584	.196	.011	.000	.000
15	1.000	1.000	1.000	1.000	1.000	1.000	.994	.949	.762	.370	.043	.003	.000
16	1.000	1.000	1.000	1.000	1.000	1.000	.999	.984	.893	.589	.133	.016	.000
17	1.000	1.000	1.000	1.000	1.000	1.000	1.000	.996	.965	.794	.323	.075	.001
18	1.000	1.000	1.000	1.000	1.000	1.000	1.000	.999	.992	.931	.608	.264	.017
19	1.000	1.000	1.000	1.000	1.000	1.000	1.000	1.000	.999	.988	.878	.642	.182

n = 21

a \ π	.010	.050	.100	.200	.300	.400	.500	.600	.700	.800	.900	.950	.990
0	.810	.341	.109	.009	.001	.000	.000	.000	.000	.000	.000	.000	.000
1	.981	.717	.365	.058	.006	.000	.000	.000	.000	.000	.000	.000	.000
2	.999	.915	.648	.179	.027	.002	.000	.000	.000	.000	.000	.000	.000
3	1.000	.981	.848	.370	.086	.011	.001	.000	.000	.000	.000	.000	.000
4	1.000	.997	.948	.586	.198	.037	.004	.000	.000	.000	.000	.000	.000
5	1.000	1.000	.986	.769	.363	.096	.013	.001	.000	.000	.000	.000	.000
6	1.000	1.000	.997	.891	.551	.200	.039	.004	.000	.000	.000	.000	.000
7	1.000	1.000	.999	.957	.723	.350	.095	.012	.001	.000	.000	.000	.000
8	1.000	1.000	1.000	.986	.852	.524	.192	.035	.002	.000	.000	.000	.000
9	1.000	1.000	1.000	.996	.932	.691	.332	.085	.009	.000	.000	.000	.000
10	1.000	1.000	1.000	.999	.974	.826	.500	.174	.026	.001	.000	.000	.000
11	1.000	1.000	1.000	1.000	.991	.915	.668	.309	.068	.004	.000	.000	.000
12	1.000	1.000	1.000	1.000	.998	.965	.808	.476	.148	.014	.000	.000	.000
13	1.000	1.000	1.000	1.000	.999	.988	.905	.650	.277	.043	.001	.000	.000
14	1.000	1.000	1.000	1.000	1.000	.996	.961	.800	.449	.109	.003	.000	.000
15	1.000	1.000	1.000	1.000	1.000	.999	.987	.904	.637	.231	.014	.000	.000
16	1.000	1.000	1.000	1.000	1.000	1.000	.996	.963	.802	.414	.052	.003	.000
17	1.000	1.000	1.000	1.000	1.000	1.000	.999	.989	.914	.630	.152	.019	.000
18	1.000	1.000	1.000	1.000	1.000	1.000	1.000	.998	.973	.821	.352	.085	.001
19	1.000	1.000	1.000	1.000	1.000	1.000	1.000	1.000	.994	.942	.635	.283	.019
20	1.000	1.000	1.000	1.000	1.000	1.000	1.000	1.000	.999	.991	.891	.659	.190

TABLE B.1 (*continued*)

n = 22

a \ π	.010	.050	.100	.200	.300	.400	.500	.600	.700	.800	.900	.950	.990
0	.802	.324	.098	.007	.000	.000	.000	.000	.000	.000	.000	.000	.000
1	.980	.698	.339	.048	.004	.000	.000	.000	.000	.000	.000	.000	.000
2	.999	.905	.620	.154	.021	.002	.000	.000	.000	.000	.000	.000	.000
3	1.000	.978	.828	.332	.068	.008	.000	.000	.000	.000	.000	.000	.000
4	1.000	.996	.938	.543	.165	.027	.002	.000	.000	.000	.000	.000	.000
5	1.000	.999	.982	.733	.313	.072	.008	.000	.000	.000	.000	.000	.000
6	1.000	1.000	.996	.867	.494	.158	.026	.002	.000	.000	.000	.000	.000
7	1.000	1.000	.999	.944	.671	.290	.067	.007	.000	.000	.000	.000	.000
8	1.000	1.000	1.000	.980	.814	.454	.143	.021	.001	.000	.000	.000	.000
9	1.000	1.000	1.000	.994	.908	.624	.262	.055	.004	.000	.000	.000	.000
10	1.000	1.000	1.000	.998	.961	.772	.416	.121	.014	.000	.000	.000	.000
11	1.000	1.000	1.000	1.000	.986	.879	.584	.228	.039	.002	.000	.000	.000
12	1.000	1.000	1.000	1.000	.996	.945	.738	.376	.092	.006	.000	.000	.000
13	1.000	1.000	1.000	1.000	.999	.979	.857	.546	.186	.020	.000	.000	.000
14	1.000	1.000	1.000	1.000	1.000	.993	.933	.710	.329	.056	.001	.000	.000
15	1.000	1.000	1.000	1.000	1.000	.998	.974	.842	.506	.133	.004	.000	.000
16	1.000	1.000	1.000	1.000	1.000	1.000	.992	.928	.687	.267	.018	.001	.000
17	1.000	1.000	1.000	1.000	1.000	1.000	.998	.973	.835	.457	.062	.004	.000
18	1.000	1.000	1.000	1.000	1.000	1.000	1.000	.992	.932	.668	.172	.022	.000
19	1.000	1.000	1.000	1.000	1.000	1.000	1.000	.998	.979	.846	.380	.095	.001
20	1.000	1.000	1.000	1.000	1.000	1.000	1.000	1.000	.996	.952	.661	.302	.020
21	1.000	1.000	1.000	1.000	1.000	1.000	1.000	1.000	1.000	.993	.902	.676	.198

n = 23

a \ π	.010	.050	.100	.200	.300	.400	.500	.600	.700	.800	.900	.950	.990
0	.794	.307	.089	.006	.000	.000	.000	.000	.000	.000	.000	.000	.000
1	.978	.679	.315	.040	.003	.000	.000	.000	.000	.000	.000	.000	.000
2	.998	.895	.592	.133	.016	.001	.000	.000	.000	.000	.000	.000	.000
3	1.000	.974	.807	.297	.054	.005	.000	.000	.000	.000	.000	.000	.000
4	1.000	.995	.927	.501	.136	.019	.001	.000	.000	.000	.000	.000	.000
5	1.000	.999	.977	.695	.269	.054	.005	.000	.000	.000	.000	.000	.000
6	1.000	1.000	.994	.840	.440	.124	.017	.001	.000	.000	.000	.000	.000
7	1.000	1.000	.999	.928	.618	.237	.047	.004	.000	.000	.000	.000	.000
8	1.000	1.000	1.000	.973	.771	.388	.105	.013	.001	.000	.000	.000	.000
9	1.000	1.000	1.000	.991	.880	.556	.202	.035	.002	.000	.000	.000	.000
10	1.000	1.000	1.000	.997	.945	.713	.339	.081	.007	.000	.000	.000	.000
11	1.000	1.000	1.000	.999	.979	.836	.500	.164	.021	.001	.000	.000	.000
12	1.000	1.000	1.000	1.000	.993	.919	.661	.287	.055	.003	.000	.000	.000
13	1.000	1.000	1.000	1.000	.998	.965	.798	.444	.120	.009	.000	.000	.000
14	1.000	1.000	1.000	1.000	.999	.987	.895	.612	.229	.027	.000	.000	.000
15	1.000	1.000	1.000	1.000	1.000	.996	.953	.763	.382	.072	.001	.000	.000
16	1.000	1.000	1.000	1.000	1.000	.999	.983	.876	.560	.160	.006	.000	.000
17	1.000	1.000	1.000	1.000	1.000	1.000	.995	.946	.731	.305	.023	.001	.000
18	1.000	1.000	1.000	1.000	1.000	1.000	.999	.981	.864	.499	.073	.005	.000
19	1.000	1.000	1.000	1.000	1.000	1.000	1.000	.995	.946	.703	.193	.026	.000
20	1.000	1.000	1.000	1.000	1.000	1.000	1.000	.999	.984	.867	.408	.105	.002
21	1.000	1.000	1.000	1.000	1.000	1.000	1.000	1.000	.997	.960	.685	.321	.022
22	1.000	1.000	1.000	1.000	1.000	1.000	1.000	1.000	1.000	.994	.911	.693	.206

TABLE B.1 (*continued*)

n = 24

a \ π	.010	.050	.100	.200	.300	.400	.500	.600	.700	.800	.900	.950	.990
0	.786	.292	.080	.005	.000	.000	.000	.000	.000	.000	.000	.000	.000
1	.976	.661	.292	.033	.002	.000	.000	.000	.000	.000	.000	.000	.000
2	.998	.884	.564	.115	.012	.001	.000	.000	.000	.000	.000	.000	.000
3	1.000	.970	.786	.264	.042	.004	.000	.000	.000	.000	.000	.000	.000
4	1.000	.994	.915	.460	.111	.013	.001	.000	.000	.000	.000	.000	.000
5	1.000	.999	.972	.656	.229	.040	.003	.000	.000	.000	.000	.000	.000
6	1.000	1.000	.993	.811	.389	.096	.011	.001	.000	.000	.000	.000	.000
7	1.000	1.000	.998	.911	.565	.192	.032	.002	.000	.000	.000	.000	.000
8	1.000	1.000	1.000	.964	.725	.328	.076	.008	.000	.000	.000	.000	.000
9	1.000	1.000	1.000	.987	.847	.489	.154	.022	.001	.000	.000	.000	.000
10	1.000	1.000	1.000	.996	.926	.650	.271	.053	.004	.000	.000	.000	.000
11	1.000	1.000	1.000	.999	.969	.787	.419	.114	.012	.000	.000	.000	.000
12	1.000	1.000	1.000	1.000	.988	.886	.581	.213	.031	.001	.000	.000	.000
13	1.000	1.000	1.000	1.000	.996	.947	.729	.350	.074	.004	.000	.000	.000
14	1.000	1.000	1.000	1.000	.999	.978	.846	.511	.153	.013	.000	.000	.000
15	1.000	1.000	1.000	1.000	1.000	.992	.924	.672	.275	.036	.000	.000	.000
16	1.000	1.000	1.000	1.000	1.000	.998	.968	.808	.435	.089	.002	.000	.000
17	1.000	1.000	1.000	1.000	1.000	.999	.989	.904	.611	.189	.007	.000	.000
18	1.000	1.000	1.000	1.000	1.000	1.000	.997	.960	.771	.344	.028	.001	.000
19	1.000	1.000	1.000	1.000	1.000	1.000	.999	.987	.889	.540	.085	.006	.000
20	1.000	1.000	1.000	1.000	1.000	1.000	1.000	.996	.958	.736	.214	.030	.000
21	1.000	1.000	1.000	1.000	1.000	1.000	1.000	.999	.988	.885	.436	.116	.002
22	1.000	1.000	1.000	1.000	1.000	1.000	1.000	1.000	.998	.967	.708	.339	.024
23	1.000	1.000	1.000	1.000	1.000	1.000	1.000	1.000	1.000	.995	.920	.708	.214

n = 25

a \ π	.010	.050	.100	.200	.300	.400	.500	.600	.700	.800	.900	.950	.990
0	.778	.277	.072	.004	.000	.000	.000	.000	.000	.000	.000	.000	.000
1	.974	.642	.271	.027	.002	.000	.000	.000	.000	.000	.000	.000	.000
2	.998	.873	.537	.098	.009	.000	.000	.000	.000	.000	.000	.000	.000
3	1.000	.966	.764	.234	.033	.002	.000	.000	.000	.000	.000	.000	.000
4	1.000	.993	.902	.421	.090	.009	.000	.000	.000	.000	.000	.000	.000
5	1.000	.999	.967	.617	.193	.029	.002	.000	.000	.000	.000	.000	.000
6	1.000	1.000	.991	.780	.341	.074	.007	.000	.000	.000	.000	.000	.000
7	1.000	1.000	.998	.891	.512	.154	.022	.001	.000	.000	.000	.000	.000
8	1.000	1.000	1.000	.953	.677	.274	.054	.004	.000	.000	.000	.000	.000
9	1.000	1.000	1.000	.983	.811	.425	.115	.013	.000	.000	.000	.000	.000
10	1.000	1.000	1.000	.994	.902	.586	.212	.034	.002	.000	.000	.000	.000
11	1.000	1.000	1.000	.998	.956	.732	.345	.078	.006	.000	.000	.000	.000
12	1.000	1.000	1.000	1.000	.983	.846	.500	.154	.017	.000	.000	.000	.000
13	1.000	1.000	1.000	1.000	.994	.922	.655	.268	.044	.002	.000	.000	.000
14	1.000	1.000	1.000	1.000	.998	.966	.788	.414	.098	.006	.000	.000	.000
15	1.000	1.000	1.000	1.000	1.000	.987	.885	.575	.189	.017	.000	.000	.000
16	1.000	1.000	1.000	1.000	1.000	.996	.946	.726	.323	.047	.000	.000	.000
17	1.000	1.000	1.000	1.000	1.000	.999	.978	.846	.488	.109	.002	.000	.000
18	1.000	1.000	1.000	1.000	1.000	1.000	.993	.926	.659	.220	.009	.000	.000
19	1.000	1.000	1.000	1.000	1.000	1.000	.998	.971	.807	.383	.033	.001	.000
20	1.000	1.000	1.000	1.000	1.000	1.000	1.000	.991	.910	.579	.098	.007	.000
21	1.000	1.000	1.000	1.000	1.000	1.000	1.000	.998	.967	.766	.236	.034	.000
22	1.000	1.000	1.000	1.000	1.000	1.000	1.000	1.000	.991	.902	.463	.127	.002
23	1.000	1.000	1.000	1.000	1.000	1.000	1.000	1.000	.998	.973	.729	.358	.026
24	1.000	1.000	1.000	1.000	1.000	1.000	1.000	1.000	1.000	.996	.928	.723	.222

TABLE B.1 (*concluded*)

$n = 50$

a \ π	.010	.050	.100	.200	.300	.400	.500	.600	.700	.800	.900	.950	.990
0	.605	.077	.005	.000	.000	.000	.000	.000	.000	.000	.000	.000	.000
1	.911	.279	.034	.000	.000	.000	.000	.000	.000	.000	.000	.000	.000
2	.986	.541	.112	.001	.000	.000	.000	.000	.000	.000	.000	.000	.000
3	.998	.760	.250	.006	.000	.000	.000	.000	.000	.000	.000	.000	.000
4	1.000	.896	.431	.018	.000	.000	.000	.000	.000	.000	.000	.000	.000
5	1.000	.962	.616	.048	.001	.000	.000	.000	.000	.000	.000	.000	.000
6	1.000	.988	.770	.103	.002	.000	.000	.000	.000	.000	.000	.000	.000
7	1.000	.997	.878	.190	.007	.000	.000	.000	.000	.000	.000	.000	.000
8	1.000	.999	.942	.307	.018	.000	.000	.000	.000	.000	.000	.000	.000
9	1.000	1.000	.975	.444	.040	.001	.000	.000	.000	.000	.000	.000	.000
10	1.000	1.000	.991	.584	.079	.002	.000	.000	.000	.000	.000	.000	.000
11	1.000	1.000	.997	.711	.139	.006	.000	.000	.000	.000	.000	.000	.000
12	1.000	1.000	.999	.814	.223	.013	.000	.000	.000	.000	.000	.000	.000
13	1.000	1.000	1.000	.889	.328	.028	.000	.000	.000	.000	.000	.000	.000
14	1.000	1.000	1.000	.939	.447	.054	.001	.000	.000	.000	.000	.000	.000
15	1.000	1.000	1.000	.969	.569	.096	.003	.000	.000	.000	.000	.000	.000
16	1.000	1.000	1.000	.986	.684	.156	.008	.000	.000	.000	.000	.000	.000
17	1.000	1.000	1.000	.994	.782	.237	.016	.000	.000	.000	.000	.000	.000
18	1.000	1.000	1.000	.997	.859	.336	.032	.001	.000	.000	.000	.000	.000
19	1.000	1.000	1.000	.999	.915	.446	.059	.001	.000	.000	.000	.000	.000
20	1.000	1.000	1.000	1.000	.952	.561	.101	.003	.000	.000	.000	.000	.000
21	1.000	1.000	1.000	1.000	.975	.670	.161	.008	.000	.000	.000	.000	.000
22	1.000	1.000	1.000	1.000	.988	.766	.240	.016	.000	.000	.000	.000	.000
23	1.000	1.000	1.000	1.000	.994	.844	.336	.031	.000	.000	.000	.000	.000
24	1.000	1.000	1.000	1.000	.998	.902	.444	.057	.001	.000	.000	.000	.000

$n = 50$ (continued)

a \ π	.010	.050	.100	.200	.300	.400	.500	.600	.700	.800	.900	.950	.990
25	1.000	1.000	1.000	1.000	.999	.943	.556	.098	.002	.000	.000	.000	.000
26	1.000	1.000	1.000	1.000	1.000	.969	.664	.156	.006	.000	.000	.000	.000
27	1.000	1.000	1.000	1.000	1.000	.984	.760	.234	.012	.000	.000	.000	.000
28	1.000	1.000	1.000	1.000	1.000	.992	.839	.330	.025	.000	.000	.000	.000
29	1.000	1.000	1.000	1.000	1.000	.997	.899	.439	.048	.000	.000	.000	.000
30	1.000	1.000	1.000	1.000	1.000	.999	.941	.554	.085	.001	.000	.000	.000
31	1.000	1.000	1.000	1.000	1.000	.999	.968	.664	.141	.003	.000	.000	.000
32	1.000	1.000	1.000	1.000	1.000	1.000	.984	.763	.218	.006	.000	.000	.000
33	1.000	1.000	1.000	1.000	1.000	1.000	.992	.844	.316	.014	.000	.000	.000
34	1.000	1.000	1.000	1.000	1.000	1.000	.997	.904	.431	.031	.000	.000	.000
35	1.000	1.000	1.000	1.000	1.000	1.000	.999	.946	.553	.061	.000	.000	.000
36	1.000	1.000	1.000	1.000	1.000	1.000	1.000	.972	.672	.111	.000	.000	.000
37	1.000	1.000	1.000	1.000	1.000	1.000	1.000	.987	.777	.186	.001	.000	.000
38	1.000	1.000	1.000	1.000	1.000	1.000	1.000	.994	.861	.289	.003	.000	.000
39	1.000	1.000	1.000	1.000	1.000	1.000	1.000	.998	.921	.416	.009	.000	.000
40	1.000	1.000	1.000	1.000	1.000	1.000	1.000	.999	.960	.556	.025	.000	.000
41	1.000	1.000	1.000	1.000	1.000	1.000	1.000	1.000	.982	.693	.058	.001	.000
42	1.000	1.000	1.000	1.000	1.000	1.000	1.000	1.000	.993	.810	.122	.003	.000
43	1.000	1.000	1.000	1.000	1.000	1.000	1.000	1.000	.998	.897	.230	.012	.000
44	1.000	1.000	1.000	1.000	1.000	1.000	1.000	1.000	.999	.952	.384	.038	.000
45	1.000	1.000	1.000	1.000	1.000	1.000	1.000	1.000	1.000	.982	.569	.104	.000
46	1.000	1.000	1.000	1.000	1.000	1.000	1.000	1.000	1.000	.994	.750	.240	.002
47	1.000	1.000	1.000	1.000	1.000	1.000	1.000	1.000	1.000	.999	.888	.459	.014
48	1.000	1.000	1.000	1.000	1.000	1.000	1.000	1.000	1.000	1.000	.966	.721	.089
49	1.000	1.000	1.000	1.000	1.000	1.000	1.000	1.000	1.000	1.000	.995	.923	.395

TABLE B.2 Poisson probabilities

					λ					
X	0.1	0.2	0.3	0.4	0.5	0.6	0.7	0.8	0.9	1.0
0	.9048	.8187	.7408	.6703	.6065	.5488	.4966	.4493	.4066	.3679
1	.0905	.1637	.2222	.2681	.3033	.3293	.3476	.3595	.3659	.3679
2	.0045	.0164	.0333	.0536	.0758	.0988	.1217	.1438	.1647	.1839
3	.0002	.0011	.0033	.0072	.0126	.0198	.0284	.0383	.0494	.0613
4		.0001	.0002	.0007	.0016	.0030 ·	.0050	.0077	.0111	.0153
5				.0001	.0002	.0004	.0007	.0012	.0020	.0031
6							.0001	.0002	.0003	.0005
7										.0001

					λ					
X	1.5	2.0	2.5	3.0	3.5	4.0	4.5	5.0	6.0	7.0
0	.2231	.1353	.0821	.0498	.0302	.0183	.0111	.0067	.0025	.0009
1	.3347	.2707	.2052	.1494	.1057	.0733	.0500	.0337	.0149	.0064
2	.2510	.2707	.2565	.2240	.1850	.1465	.1125	.0842	.0446	.0223
3	.1255	.1804	.2138	.2240	.2158	.1954	.1687	.1404	.0892	.0521
4	.0471	.0902	.1336	.1680	.1888	.1954	.1898	.1755	.1339	.0912
5	.0141	.0361	.0668	.1008	.1322	.1563	.1708	.1755	.1606	.1277
6	.0035	.0120	.0278	.0504	.0771	.1042	.1281	.1462	.1606	.1490
7	.0008	.0034	.0099	.0216	.0385	.0595	.0824	.1044	.1377	.1490
8	.0001	.0009	.0031	.0081	.0169	.0298	.0463	.0653	.1033	.1304
9		.0002	.0009	.0027	.0066	.0132	.0232	.0363	.0688	.1014
10			.0002	.0008	.0023	.0053	.0104	.0181	.0413	.0710
11				.0002	.0007	.0019	.0043	.0082	.0225	.0452
12				.0001	.0002	.0006	.0016	.0034	.0113	.0264
13					.0001	.0002	.0006	.0013	.0052	.0142
14						.0001	.0002	.0005	.0022	.0071
15							.0001	.0002	.0009	.0033
16									.0003	.0014
17									.0001	.0006
18										.0002
19										.0001

Table entries are P($\dot{X} = x/\lambda$).

Source: John Neter, William Wasserman, and G. A. Whitmore, *Fundamental Statistics for Business and Economics*, 4th ed. (Boston: Allyn & Bacon, Inc., 1972 ©).

TABLE B.3 Standard normal distribution areas

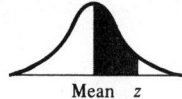

Mean z

z	.00	.01	.02	.03	.04	.05	.06	.07	.08	.09
0.0	.0000	.0040	.0080	.0120	.0160	.0199	.0239	.0279	.0319	.0359
0.1	.0398	.0438	.0478	.0517	.0557	.0596	.0636	.0675	.0714	.0753
0.2	.0793	.0832	.0871	.0910	.0948	.0987	.1026	.1064	.1103	.1141
0.3	.1179	.1217	.1255	.1293	.1331	.1368	.1406	.1443	.1480	.1517
0.4	.1554	.1591	.1628	.1664	.1700	.1736	.1772	.1808	.1844	.1879
0.5	.1915	.1950	.1985	.2019	.2054	.2088	.2123	.2157	.2190	.2224
0.6	.2257	.2291	.2324	.2357	.2389	.2422	.2454	.2486	.2518	.2549
0.7	.2580	.2612	.2642	.2673	.2704	.2734	.2764	.2794	.2823	.2852
0.8	.2881	.2910	.2939	.2967	.2995	.3023	.3051	.3078	.3106	.3133
0.9	.3159	.3186	.3212	.3238	.3264	.3289	.3315	.3340	.3365	.3389
1.0	.3413	.3438	.3461	.3485	.3508	.3531	.3554	.3577	.3599	.3621
1.1	.3643	.3665	.3686	.3708	.3729	.3749	.3770	.3790	.3810	.3830
1.2	.3849	.3869	.3888	.3907	3925	.3944	.3962	.3980	.3997	.4015
1.3	.4032	.4049	.4066	.4082	.4099	.4115	.4131	.4147	.4162	.4177
1.4	.4192	.4207	.4222	.4236	.4251	.4265	.4279	.4292	.4306	.4319
1.5	.4332	.4345	.4357	.4370	.4382	.4394	.4406	.4418	.4429	.4441
1.6	.4452	.4463	.4474	.4484	.4495	.4505	.4515	.4525	.4535	.4545
1.7	.4554	.4564	.4573	.4582	.4591	.4599	.4608	.4616	.4625	.4633
1.8	.4641	.4649	.4656	.4664	.4671	.4678	.4686	.4693	.4699	.4706
1.9	.4713	.4719	.4726	.4732	.4738	.4744	.4750	.4756	.4761	.4767
2.0	.4772	.4778	.4783	.4788	.4793	.4798	.4803	.4808	.4812	.4817
2.1	.4821	.4826	.4830	.4834	.4838	.4842	.4846	.4850	.4854	.4857
2.2	.4861	.4864	.4868	.4871	.4875	.4878	.4881	.4884	.4887	.4890
2.3	.4893	.4896	.4898	.4901	.4904	.4906	.4909	.4911	.4913	.4916
2.4	.4918	.4920	.4922	.4925	.4927	.4929	.4931	.4932	.4934	.4936
2.5	.4938	.4940	.4941	.4943	.4945	.4946	.4948	.4949	.4951	.4952
2.6	.4953	.4955	.4956	.4957	.4959	.4960	.4961	.4962	.4963	.4964
2.7	.4965	.4966	.4967	.4968	.4969	.4970	.4971	.4972	.4973	.4974
2.8	.4974	.4975	.4976	.4977	.4977	.4978	.4979	.4979	.4980	.4981
2.9	.4981	.4982	.4982	.4983	.4984	.4984	.4985	.4985	.4986	.4986
3.0	.49865	.4987	.4987	.4988	.4988	.4989	.4989	.4989	.4990	.4990
4.0	.4999683									

Source: John Neter, William Wasserman, and G. A. Whitmore, *Fundamental Statistics for Business and Economics*, 4th ed. (Boston: Allyn & Bacon, Inc., 1972 ©).

TABLE B.4 Values of $F(\lambda x)$ where **X** has the exponential distribution $f(x) = \lambda e^{-\lambda x}$

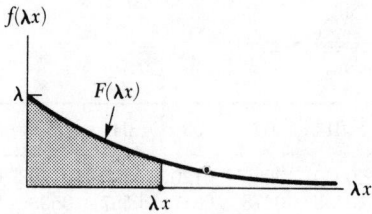

λx	$F(\lambda x)$	λx	$F(\lambda x)$	λx	$F(\lambda x)$	λx	$F(\lambda x)$
0.0	0.000	2.5	0.918	5.0	0.9933	7.5	0.99945
0.1	0.095	2.6	0.926	5.1	0.9939	7.6	0.99950
0.2	0.181	2.7	0.933	5.2	0.9945	7.7	0.99955
0.3	0.259	2.8	0.939	5.3	0.9950	7.8	0.99959
0.4	0.330	2.9	0.945	5.4	0.9955	7.9	0.99963
0.5	0.393	3.0	0.950	5.5	0.9959	8.0	0.99966
0.6	0.451	3.1	0.955	5.6	0.9963	8.1	0.99970
0.7	0.503	3.2	0.959	5.7	0.9967	8.2	0.99972
0.8	0.551	3.3	0.963	5.8	0.9970	8.3	0.99975
0.9	0.593	3.4	0.967	5.9	0.9973	8.4	0.99978
1.0	0.632	3.5	0.970	6.0	0.9975	8.5	0.99980
1.1	0.667	3.6	0.973	6.1	0.9978	8.6	0.99982
1.2	0.699	3.7	0.975	6.2	0.9980	8.7	0.99983
1.3	0.727	3.8	0.978	6.3	0.9982	8.8	0.99985
1.4	0.753	3.9	0.980	6.4	0.9983	8.9	0.99986
1.5	0.777	4.0	0.982	6.5	0.9985	9.0	0.99989
1.6	0.798	4.1	0.983	6.6	0.9986	9.1	0.99989
1.7	0.817	4.2	0.985	6.7	0.9988	9.2	0.99990
1.8	0.835	4.3	0.986	6.8	0.9989	9.3	0.99991
1.9	0.850	4.4	0.988	6.9	0.9990	9.4	0.99992
2.0	0.865	4.5	0.989	7.0	0.9991	9.5	0.99992
2.1	0.878	4.6	0.990	7.1	0.9992	9.6	0.99993
2.2	0.889	4.7	0.991	7.2	0.9993	9.7	0.99994
2.3	0.900	4.8	0.992	7.3	0.9993	9.8	0.99994
2.4	0.909	4.9	0.993	7.4	0.9993	9.9	0.99995

Example: If $\lambda = \frac{1}{4}$, the probability of observing a value of **X** less than or equal to 2 is $F(\lambda x) = F[(\frac{1}{4})(2)] = F(\frac{1}{2}) = 0.393$.

TABLE B.5   The *t*-distribution

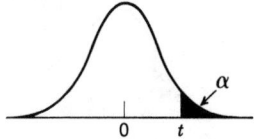

α d.f.	.10	.05	.025	.01	.005
1	3.078	6.314	12.706	31.821	63.657
2	1.886	2.920	4.303	6.965	9.925
3	1.638	2.353	3.182	4.541	5.841
4	1.533	2.132	2.776	3.747	4.604
5	1.476	2.015	2.571	3.365	4.032
6	1.440	1.943	2.447	3.143	3.707
7	1.415	1.895	2.365	2.998	3.499
8	1.397	1.860	2.306	2.896	3.355
9	1.383	1.833	2.262	2.821	3.250
10	1.372	1.812	2.228	2.764	3.169
11	1.363	1.796	2.201	2.718	3.106
12	1.356	1.782	2.179	2.681	3.055
13	1.350	1.771	2.160	2.650	3.012
14	1.345	1.761	2.145	2.624	2.977
15	1.341	1.753	2.131	2.602	2.947
16	1.337	1.746	2.120	2.583	2.921
17	1.333	1.740	2.110	2.567	2.898
18	1.330	1.734	2.101	2.552	2.878
19	1.328	1.729	2.093	2.539	2.861
20	1.325	1.725	2.086	2.528	2.845
21	1.323	1.721	2.080	2.518	2.831
22	1.321	1.717	2.074	2.508	2.819
23	1.319	1.714	2.069	2.500	2.807
24	1.318	1.711	2.064	2.492	2.797
25	1.316	1.708	2.060	2.485	2.787
26	1.315	1.706	2.056	2.479	2.779
27	1.314	1.703	2.052	2.473	2.771
28	1.313	1.701	2.048	2.467	2.763
29	1.311	1.699	2.045	2.462	2.756
30	1.310	1.697	2.042	2.457	2.750
40	1.303	1.684	2.021	2.423	2.704
60	1.296	1.671	2.000	2.390	2.660
120	1.289	1.658	1.980	2.358	2.617
∞	1.282	1.645	1.960	2.326	2.576

Source: P. Hoel, *Elementary Statistics*, 3d ed. (New York: John Wiley & Sons, Inc., 1971 ©).

TABLE B.6   The χ^2-distribution

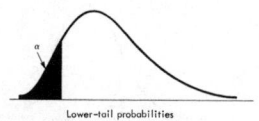

Lower-tail probabilities

df \ α	.001	.005	.010	.025	.050	.100
1	.000	.000	.000	.001	.004	.016
2	.002	.010	.020	.051	.103	.211
3	.024	.072	.115	.216	.352	.584
4	.091	.207	.297	.484	.711	1.06
5	.210	.412	.554	.831	1.15	1.61
6	.381	.676	.872	1.24	1.64	2.20
7	.598	.989	1.24	1.69	2.17	2.83
8	.857	1.34	1.65	2.18	2.73	3.49
9	1.15	1.73	2.09	2.70	3.33	4.17
10	1.48	2.16	2.56	3.25	3.94	4.87
11	1.83	2.60	3.05	3.82	4.57	5.58
12	2.21	3.07	3.57	4.40	5.23	6.30
13	2.62	3.57	4.11	5.01	5.89	7.04
14	3.04	4.07	4.66	5.63	6.57	7.79
15	3.48	4.60	5.23	6.26	7.26	8.55
16	3.94	5.14	5.81	6.91	7.96	9.31
17	4.42	5.70	6.41	7.56	8.67	10.1
18	4.90	6.26	7.01	8.23	9.39	10.9
19	5.41	6.84	7.63	8.91	10.1	11.7
20	5.92	7.43	8.26	9.59	10.9	12.4
21	6.45	8.03	8.90	10.3	11.6	13.2
22	6.98	8.64	9.54	11.0	12.3	14.0
23	7.53	9.26	10.2	11.7	13.1	14.8
24	8.08	9.89	10.9	12.4	13.8	15.7
25	8.65	10.5	11.5	13.1	14.6	16.5
26	9.22	11.2	12.2	13.8	15.4	17.3
27	9.80	11.8	12.9	14.6	16.2	18.1
28	10.4	12.5	13.6	15.3	16.9	18.9
29	11.0	13.1	14.3	16.0	17.7	19.8
30	11.6	13.8	15.0	16.8	18.5	20.6
35	14.7	17.2	18.5	20.6	22.5	24.8
40	17.9	20.7	22.2	24.4	26.5	29.1
45	21.3	24.3	25.9	28.4	30.6	33.4
50	24.7	28.0	29.7	32.4	34.8	37.7
55	28.2	31.7	33.6	36.4	39.0	42.1
60	31.7	35.5	37.5	40.5	43.2	46.5
65	35.4	39.4	41.4	44.6	47.4	50.9
70	39.0	43.3	45.4	48.8	51.7	55.3
75	42.8	47.2	49.5	52.9	56.1	59.8
80	46.5	51.2	53.5	57.2	60.4	64.3
85	50.3	55.2	57.6	61.4	64.7	68.8
90	54.2	59.2	61.8	65.6	69.1	73.3
95	58.0	63.2	65.9	69.9	73.5	77.8
100	61.9	67.3	70.1	74.2	77.9	82.4

TABLE B.6 (*continued*)

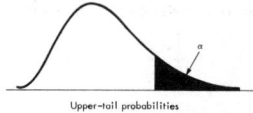

Upper-tail probabilities

df \ α	.100	.050	.025	.010	.005	.001
1	2.71	3.84	5.02	6.63	7.88	10.8
2	4.61	5.99	7.38	9.21	10.6	13.8
3	6.25	7.81	9.35	11.3	12.8	16.3
4	7.78	9.49	11.1	13.3	14.9	18.5
5	9.24	11.1	12.8	15.1	16.7	20.5
6	10.6	12.6	14.4	16.8	18.5	22.5
7	12.0	14.1	16.0	18.5	20.3	24.3
8	13.4	15.5	17.5	20.1	22.0	26.1
9	14.7	16.9	19.0	21.7	23.6	27.9
10	16.0	18.3	20.5	23.2	25.2	29.6
11	17.3	19.7	21.9	24.7	26.8	31.3
12	18.5	21.0	23.3	26.2	28.3	32.9
13	19.8	22.4	24.7	27.7	29.8	34.5
14	21.1	23.7	26.1	29.1	31.3	36.1
15	22.3	25.0	27.5	30.6	32.8	37.7
16	23.5	26.3	28.8	32.0	34.3	39.3
17	24.8	27.6	30.2	33.4	35.7	40.8
18	26.0	28.9	31.5	34.8	37.2	42.3
19	27.2	30.1	32.9	36.2	38.6	43.8
20	28.4	31.4	34.2	37.6	40.0	45.3
21	29.6	32.7	35.5	38.9	41.4	46.8
22	30.8	33.9	36.8	40.3	42.8	48.3
23	32.0	35.2	38.1	41.6	44.2	49.7
24	33.2	36.4	39.4	43.0	45.6	51.2
25	34.4	37.7	40.6	44.3	46.9	52.6
26	35.6	38.9	41.9	45.6	48.3	54.1
27	36.7	40.1	43.2	47.0	49.6	55.5
28	37.9	41.3	44.5	48.3	51.0	56.9
29	39.1	42.6	45.7	49.6	52.3	58.3
30	40.3	43.8	47.0	50.9	53.7	59.7
35	46.1	49.8	53.2	57.3	60.3	66.6
40	51.8	55.8	59.3	63.7	66.8	73.4
45	57.5	61.7	65.4	70.0	73.2	80.1
50	63.2	67.5	71.4	76.2	79.5	86.7
55	68.8	73.3	77.4	82.3	85.7	93.2
60	74.4	79.1	83.3	88.4	92.0	99.6
65	80.0	84.8	89.2	94.4	98.1	106.0
70	85.5	90.5	95.0	100.4	104.2	112.3
75	91.1	96.2	100.8	106.4	110.3	118.6
80	96.6	101.9	106.6	112.3	116.3	124.8
85	102.1	107.5	112.4	118.2	122.3	131.0
90	107.6	113.1	118.1	124.1	128.3	137.2
95	113.0	118.8	123.9	130.0	134.2	143.3
100	118.5	124.3	129.6	135.8	140.2	149.4

TABLE B.7   The F-distribution

Percentiles of the F-distribution: entry is $F(1 - \alpha; r_1, r_2)$ where $P[\mathbf{F} \leq F(1 - \alpha; r_1, r_2)]$ $= 1 - \alpha$.

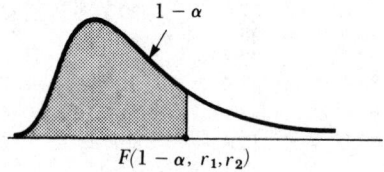

TABLE B.7 (*continued*)

r_2	$1-\alpha$	r_1: 1	2	3	4	5	6	7	8	9
1	.50	1.00	1.50	1.71	1.82	1.89	1.94	1.98	2.00	2.03
	.90	39.9	49.5	53.6	55.8	57.2	58.2	58.9	59.4	59.9
	.95	161	200	216	225	230	234	237	239	241
	.975	648	800	864	900	922	937	948	957	963
	.99	4,052	5,000	5,403	5,625	5,764	5,859	5,928	5,981	6,022
	.995	16,211	20,000	21,615	22,500	23,056	23,437	23,715	23,925	24,091
	.999	405,280	500,000	540,380	562,500	576,400	585,940	592,870	598,140	602,280
2	.50	0.667	1.00	1.13	1.21	1.25	1.28	1.30	1.32	1.33
	.90	8.53	9.00	9.16	9.24	9.29	9.33	9.35	9.37	9.38
	.95	18.5	19.0	19.2	19.2	19.3	19.3	19.4	19.4	19.4
	.975	38.5	39.0	39.2	39.2	39.3	39.3	39.4	39.4	39.4
	.99	98.5	99.0	99.2	99.2	99.3	99.3	99.4	99.4	99.4
	.995	199	199	199	199	199	199	199	199	199
	.999	998.5	999.0	999.2	999.2	999.3	999.3	999.4	999.4	999.4
3	.50	0.585	0.881	1.00	1.06	1.10	1.13	1.15	1.16	1.17
	.90	5.54	5.46	5.39	5.34	5.31	5.28	5.27	5.25	5.24
	.95	10.1	9.55	9.28	9.12	9.01	8.94	8.89	8.85	8.81
	.975	17.4	16.0	15.4	15.1	14.9	14.7	14.6	14.5	14.5
	.99	34.1	30.8	29.5	28.7	28.2	27.9	27.7	27.5	27.3
	.995	55.6	49.8	47.5	46.2	45.4	44.8	44.4	44.1	43.9
	.999	167.0	148.5	141.1	137.1	134.6	132.8	131.6	130.6	129.9
4	.50	0.549	0.828	0.941	1.00	1.04	1.06	1.08	1.09	1.10
	.90	4.54	4.32	4.19	4.11	4.05	4.01	3.98	3.95	3.94
	.95	7.71	6.94	6.59	6.39	6.26	6.16	6.09	6.04	6.00
	.975	12.2	10.6	9.98	9.60	9.36	9.20	9.07	8.98	8.90
	.99	21.2	18.0	16.7	16.0	15.5	15.2	15.0	14.8	14.7
	.995	31.3	26.3	24.3	23.2	22.5	22.0	21.6	21.4	21.1
	.999	74.1	61.2	56.2	53.4	51.7	50.5	49.7	49.0	48.5
5	.50	0.528	0.799	0.907	0.965	1.00	1.02	1.04	1.05	1.06
	.90	4.06	3.78	3.62	3.52	3.45	3.40	3.37	3.34	3.32
	.95	6.61	5.79	5.41	5.19	5.05	4.95	4.88	4.82	4.77
	.975	10.0	8.43	7.76	7.39	7.15	6.98	6.85	6.76	6.68
	.99	16.3	13.3	12.1	11.4	11.0	10.7	10.5	10.3	10.2
	.995	22.8	18.3	16.5	15.6	14.9	14.5	14.2	14.0	13.8
	.999	47.2	37.1	33.2	31.1	29.8	28.8	28.2	27.6	27.2
6	.50	0.515	0.780	0.886	0.942	0.977	1.00	1.02	1.03	1.04
	.90	3.78	3.46	3.29	3.18	3.11	3.05	3.01	2.98	2.96
	.95	5.99	5.14	4.76	4.53	4.39	4.28	4.21	4.15	4.10
	.975	8.81	7.26	6.60	6.23	5.99	5.82	5.70	5.60	5.52
	.99	13.7	10.9	9.78	9.15	8.75	8.47	8.26	8.10	7.98
	.995	18.6	14.5	12.9	12.0	11.5	11.1	10.8	10.6	10.4
	.999	35.5	27.0	23.7	21.9	20.8	20.0	19.5	19.0	18.7
7	.50	0.506	0.767	0.871	0.926	0.960	0.983	1.00	1.01	1.02
	.90	3.59	3.26	3.07	2.96	2.88	2.83	2.78	2.75	2.72
	.95	5.59	4.74	4.35	4.12	3.97	3.87	3.79	3.73	3.68
	.975	8.07	6.54	5.89	5.52	5.29	5.12	4.99	4.90	4.82
	.99	12.2	9.55	8.45	7.85	7.46	7.19	6.99	6.84	6.72
	.995	16.2	12.4	10.9	10.1	9.52	9.16	8.89	8.68	8.51
	.999	29.2	21.7	18.8	17.2	16.2	15.5	15.0	14.6	14.3

TABLE B.7 (*continued*)

					r_1					
r_2	$1-\alpha$	10	12	15	20	24	30	60	120	∞
1	.50	2.04	2.07	2.09	2.12	2.13	2.15	2.17	2.18	2.20
	.90	60.2	60.7	61.2	61.7	62.0	62.3	62.8	63.1	63.3
	.95	242	244	246	248	249	250	252	253	254
	.975	969	977	985	993	997	1,001	1,010	1,014	1,018
	.99	6,056	6,106	6,157	6,209	6,235	6,261	6,313	6,339	6,366
	.995	24,224	24,426	24,630	24,836	24,940	25,044	25,253	25,359	25,464
	.999	605,620	610,670	615,760	620,910	623,500	626,100	631,340	633,970	636,620
2	.50	1.34	1.36	1.38	1.39	1.40	1.41	1.43	1.43	1.44
	.90	9.39	9.41	9.42	9.44	9.45	9.46	9.47	9.48	9.49
	.95	19.4	19.4	19.4	19.4	19.5	19.5	19.5	19.5	19.5
	.975	39.4	39.4	39.4	39.4	39.5	39.5	39.5	39.5	39.5
	.99	99.4	99.4	99.4	99.4	99.5	99.5	99.5	99.5	99.5
	.995	199	199	199	199	199	199	199	199	200
	.999	999.4	999.4	999.4	999.4	999.5	999.5	999.5	999.5	999.5
3	.50	1.18	1.20	1.21	1.23	1.23	1.24	1.25	1.26	1.27
	.90	5.23	5.22	5.20	5.18	5.18	5.17	5.15	5.14	5.13
	.95	8.79	8.74	8.70	8.66	8.64	8.62	8.57	8.55	8.53
	.975	14.4	14.3	14.3	14.2	14.1	14.1	14.0	13.9	13.9
	.99	27.2	27.1	26.9	26.7	26.6	26.5	26.3	26.2	26.1
	.995	43.7	43.4	43.1	42.8	42.6	42.5	42.1	42.0	41.8
	.999	129.2	128.3	127.4	126.4	125.9	125.4	124.5	124.0	123.5
4	.50	1.11	1.13	1.14	1.15	1.16	1.16	1.18	1.18	1.19
	.90	3.92	3.90	3.87	3.84	3.83	3.82	3.79	3.78	3.76
	.95	5.96	5.91	5.86	5.80	5.77	5.75	5.69	5.66	5.63
	.975	8.84	8.75	8.66	8.56	8.51	8.46	8.36	8.31	8.26
	.99	14.5	14.4	14.2	14.0	13.9	13.8	13.7	13.6	13.5
	.995	21.0	20.7	20.4	20.2	20.0	19.9	19.6	19.5	19.3
	.999	48.1	47.4	46.8	46.1	45.8	45.4	44.7	44.4	44.1
5	.50	1.07	1.09	1.10	1.11	1.12	1.12	1.14	1.14	1.15
	.90	3.30	3.27	3.24	3.21	3.19	3.17	3.14	3.12	3.11
	.95	4.74	4.68	4.62	4.56	4.53	4.50	4.43	4.40	4.37
	.975	6.62	6.52	6.43	6.33	6.28	6.23	6.12	6.07	6.02
	.99	10.1	9.89	9.72	9.55	9.47	9.38	9.20	9.11	9.02
	.995	13.6	13.4	13.1	12.9	12.8	12.7	12.4	12.3	12.1
	.999	26.9	26.4	25.9	25.4	25.1	24.9	24.3	24.1	23.8
6	.50	1.05	1.06	1.07	1.08	1.09	1.10	1.11	1.12	1.12
	.90	2.94	2.90	2.87	2.84	2.82	2.80	2.76	2.74	2.72
	.95	4.06	4.00	3.94	3.87	3.84	3.81	3.74	3.70	3.67
	.975	5.46	5.37	5.27	5.17	5.12	5.07	4.96	4.90	4.85
	.99	7.87	7.72	7.56	7.40	7.31	7.23	7.06	6.97	6.88
	.995	10.2	10.0	9.81	9.59	9.47	9.36	9.12	9.00	8.88
	.999	18.4	18.0	17.6	17.1	16.9	16.7	16.2	16.0	15.7
7	.50	1.03	1.04	1.05	1.07	1.07	1.08	1.09	1.10	1.10
	.90	2.70	2.67	2.63	2.59	2.58	2.56	2.51	2.49	2.47
	.95	3.64	3.57	3.51	3.44	3.41	3.38	3.30	3.27	3.23
	.975	4.76	4.67	4.57	4.47	4.42	4.36	4.25	4.20	4.14
	.99	6.62	6.47	6.31	6.16	6.07	5.99	5.82	5.74	5.65
	.995	8.38	8.18	7.97	7.75	7.65	7.53	7.31	7.19	7.08
	.999	14.1	13.7	13.3	12.9	12.7	12.5	12.1	11.9	11.7

TABLE B.7 (*continued*)

r_2	$1-\alpha$	\multicolumn{9}{c}{r_1}
		1	2	3	4	5	6	7	8	9
8	.50	0.499	0.757	0.860	0.915	0.948	0.971	0.988	1.00	1.01
	.90	3.46	3.11	2.92	2.81	2.73	2.67	2.62	2.59	2.56
	.95	5.32	4.46	4.07	3.84	3.69	3.58	3.50	3.44	3.39
	.975	7.57	6.06	5.42	5.05	4.82	4.65	4.53	4.43	4.36
	.99	11.3	8.65	7.59	7.01	6.63	6.37	6.18	6.03	5.91
	.995	14.7	11.0	9.60	8.81	8.30	7.95	7.69	7.50	7.34
	.999	25.4	18.5	15.8	14.4	13.5	12.9	12.4	12.0	11.8
9	.50	0.494	0.749	0.852	0.906	0.939	0.962	0.978	0.990	1.00
	.90	3.36	3.01	2.81	2.69	2.61	2.55	2.51	2.47	2.44
	.95	5.12	4.26	3.86	3.63	3.48	3.37	3.29	3.23	3.18
	.975	7.21	5.71	5.08	4.72	4.48	4.32	4.20	4.10	4.03
	.99	10.6	8.02	6.99	6.42	6.06	5.80	5.61	5.47	5.35
	.995	13.6	10.1	8.72	7.96	7.47	7.13	6.88	6.69	6.54
	.999	22.9	16.4	13.9	12.6	11.7	11.1	10.7	10.4	10.1
10	.50	0.490	0.743	0.845	0.899	0.932	0.954	0.971	0.983	0.992
	.90	3.29	2.92	2.73	2.61	2.52	2.46	2.41	2.38	2.35
	.95	4.96	4.10	3.71	3.48	3.33	3.22	3.14	3.07	3.02
	.975	6.94	5.46	4.83	4.47	4.24	4.07	3.95	3.85	3.78
	.99	10.0	7.56	6.55	5.99	5.64	5.39	5.20	5.06	4.94
	.995	12.8	9.43	8.08	7.34	6.87	6.54	6.30	6.12	5.97
	.999	21.0	14.9	12.6	11.3	10.5	9.93	9.52	9.20	8.96
12	.50	0.484	0.735	0.835	0.888	0.921	0.943	0.959	0.972	0.981
	.90	3.18	2.81	2.61	2.48	2.39	2.33	2.28	2.24	2.21
	.95	4.75	3.89	3.49	3.26	3.11	3.00	2.91	2.85	2.80
	.975	6.55	5.10	4.47	4.12	3.89	3.73	3.61	3.51	3.44
	.99	9.33	6.93	5.95	5.41	5.06	4.82	4.64	4.50	4.39
	.995	11.8	8.51	7.23	6.52	6.07	5.76	5.52	5.35	5.20
	.999	18.6	13.0	10.8	9.63	8.89	8.38	8.00	7.71	7.48
15	.50	0.478	0.726	0.826	0.878	0.911	0.933	0.949	0.960	0.970
	.90	3.07	2.70	2.49	2.36	2.27	2.21	2.16	2.12	2.09
	.95	4.54	3.68	3.29	3.06	2.90	2.79	2.71	2.64	2.59
	.975	6.20	4.77	4.15	3.80	3.58	3.41	3.29	3.20	3.12
	.99	8.68	6.36	5.42	4.89	4.56	4.32	4.14	4.00	3.89
	.995	10.8	7.70	6.48	5.80	5.37	5.07	4.85	4.67	4.54
	.999	16.6	11.3	9.34	8.25	7.57	7.09	6.74	6.47	6.26
20	.50	0.472	0.718	0.816	0.868	0.900	0.922	0.938	0.950	0.959
	.90	2.97	2.59	2.38	2.25	2.16	2.09	2.04	2.00	1.96
	.95	4.35	3.49	3.10	2.87	2.71	2.60	2.51	2.45	2.39
	.975	5.87	4.46	3.86	3.51	3.29	3.13	3.01	2.91	2.84
	.99	8.10	5.85	4.94	4.43	4.10	3.87	3.70	3.56	3.46
	.995	9.94	6.99	5.82	5.17	4.76	4.47	4.26	4.09	3.96
	.999	14.8	9.95	8.10	7.10	6.46	6.02	5.69	5.44	5.24
24	.50	0.469	0.714	0.812	0.863	0.895	0.917	0.932	0.944	0.953
	.90	2.93	2.54	2.33	2.19	2.10	2.04	1.98	1.94	1.91
	.95	4.26	3.40	3.01	2.78	2.62	2.51	2.42	2.36	2.30
	.975	5.72	4.32	3.72	3.38	3.15	2.99	2.87	2.78	2.70
	.99	7.82	5.61	4.72	4.22	3.90	3.67	3.50	3.36	3.26
	.995	9.55	6.66	5.52	4.89	4.49	4.20	3.99	3.83	3.69
	.999	14.0	9.34	7.55	6.59	5.98	5.55	5.23	4.99	4.80

TABLE B.7 (*continued*)

r_2	$1-\alpha$	10	12	15	20	24	30	60	120	∞
8	.50	1.02	1.03	1.04	1.05	1.06	1.07	1.08	1.08	1.09
	.90	2.54	2.50	2.46	2.42	2.40	2.38	2.34	2.32	2.29
	.95	3.35	3.28	3.22	3.15	3.12	3.08	3.01	2.97	2.93
	.975	4.30	4.20	4.10	4.00	3.95	3.89	3.78	3.73	3.67
	.99	5.81	5.67	5.52	5.36	5.28	5.20	5.03	4.95	4.86
	.995	7.21	7.01	6.81	6.61	6.50	6.40	6.18	6.06	5.95
	.999	11.5	11.2	10.8	10.5	10.3	10.1	9.73	9.53	9.33
9	.50	1.01	1.02	1.03	1.04	1.05	1.05	1.07	1.07	1.08
	.90	2.42	2.38	2.34	2.30	2.28	2.25	2.21	2.18	2.16
	.95	3.14	3.07	3.01	2.94	2.90	2.86	2.79	2.75	2.71
	.975	3.96	3.87	3.77	3.67	3.61	3.56	3.45	3.39	3.33
	.99	5.26	5.11	4.96	4.81	4.73	4.65	4.48	4.40	4.31
	.995	6.42	6.23	6.03	5.83	5.73	5.62	5.41	5.30	5.19
	.999	9.89	9.57	9.24	8.90	8.72	8.55	8.19	8.00	7.81
10	.50	1.00	1.01	1.02	1.03	1.04	1.05	1.06	1.06	1.07
	.90	2.32	2.28	2.24	2.20	2.18	2.16	2.11	2.08	2.06
	.95	2.98	2.91	2.84	2.77	2.74	2.70	2.62	2.58	2.54
	.975	3.72	3.62	3.52	3.42	3.37	3.31	3.20	3.14	3.08
	.99	4.85	4.71	4.56	4.41	4.33	4.25	4.08	4.00	3.91
	.995	5.85	5.66	5.47	5.27	5.17	5.07	4.86	4.75	4.64
	.999	8.75	8.45	8.13	7.80	7.64	7.47	7.12	6.94	6.76
12	.50	0.989	1.00	1.01	1.02	1.03	1.03	1.05	1.05	1.06
	.90	2.19	2.15	2.10	2.06	2.04	2.01	1.96	1.93	1.90
	.95	2.75	2.69	2.62	2.54	2.51	2.47	2.38	2.34	2.30
	.975	3.37	3.28	3.18	3.07	3.02	2.96	2.85	2.79	2.72
	.99	4.30	4.16	4.01	3.86	3.78	3.70	3.54	3.45	3.36
	.995	5.09	4.91	4.72	4.53	4.43	4.33	4.12	4.01	3.90
	.999	7.29	7.00	6.71	6.40	6.25	6.09	5.76	5.59	5.42
15	.50	0.977	0.989	1.00	1.01	1.02	1.02	1.03	1.04	1.05
	.90	2.06	2.02	1.97	1.92	1.90	1.87	1.82	1.79	1.76
	.95	2.54	2.48	2.40	2.33	2.29	2.25	2.16	2.11	2.07
	.975	3.06	2.96	2.86	2.76	2.70	2.64	2.52	2.46	2.40
	.99	3.80	3.67	3.52	3.37	3.29	3.21	3.05	2.96	2.87
	.995	4.42	4.25	4.07	3.88	3.79	3.69	3.48	3.37	3.26
	.999	6.08	5.81	5.54	5.25	5.10	4.95	4.64	4.48	4.31
20	.50	0.966	0.977	0.989	1.00	1.01	1.01	1.02	1.03	1.03
	.90	1.94	1.89	1.84	1.79	1.77	1.74	1.68	1.64	1.61
	.95	2.35	2.28	2.20	2.12	2.08	2.04	1.95	1.90	1.84
	.975	2.77	2.68	2.57	2.46	2.41	2.35	2.22	2.16	2.09
	.99	3.37	3.23	3.09	2.94	2.86	2.78	2.61	2.52	2.42
	.995	3.85	3.68	3.50	3.32	3.22	3.12	2.92	2.81	2.69
	.999	5.08	4.82	4.56	4.29	4.15	4.00	3.70	3.54	3.38
24	.50	0.961	0.972	0.983	0.994	1.00	1.01	1.02	1.02	1.03
	.90	1.88	1.83	1.78	1.73	1.70	1.67	1.61	1.57	1.53
	.95	2.25	2.18	2.11	2.03	1.98	1.94	1.84	1.79	1.73
	.975	2.64	2.54	2.44	2.33	2.27	2.21	2.08	2.01	1.94
	.99	3.17	3.03	2.89	2.74	2.66	2.58	2.40	2.31	2.21
	.995	3.59	3.42	3.25	3.06	2.97	2.87	2.66	2.55	2.43
	.999	4.64	4.39	4.14	3.87	3.74	3.59	3.29	3.14	2.97

TABLE B.7 (*continued*)

r_2 $1-\alpha$		1	2	3	4	5	6	7	8	9
						r_1				
30	.50	0.466	0.709	0.807	0.858	0.890	0.912	0.927	0.939	0.948
	.90	2.88	2.49	2.28	2.14	2.05	1.98	1.93	1.88	1.85
	.95	4.17	3.32	2.92	2.69	2.53	2.42	2.33	2.27	2.21
	.975	5.57	4.18	3.59	3.25	3.03	2.87	2.75	2.65	2.57
	.99	7.56	5.39	4.51	4.02	3.70	3.47	3.30	3.17	3.07
	.995	9.18	6.35	5.24	4.62	4.23	3.95	3.74	3.58	3.45
	.999	13.3	8.77	7.05	6.12	5.53	5.12	4.82	4.58	4.39
60	.50	0.461	0.701	0.798	0.849	0.880	0.901	0.917	0.928	0.937
	.90	2.79	2.39	2.18	2.04	1.95	1.87	1.82	1.77	1.74
	.95	4.00	3.15	2.76	2.53	2.37	2.25	2.17	2.10	2.04
	.975	5.29	3.93	3.34	3.01	2.79	2.63	2.51	2.41	2.33
	.99	7.08	4.98	4.13	3.65	3.34	3.12	2.95	2.82	2.72
	.995	8.49	5.80	4.73	4.14	3.76	3.49	3.29	3.13	3.01
	.999	12.0	7.77	6.17	5.31	4.76	4.37	4.09	3.86	3.69
120	.50	0.458	0.697	0.793	0.844	0.875	0.896	0.912	0.923	0.932
	.90	2.75	2.35	2.13	1.99	1.90	1.82	1.77	1.72	1.68
	.95	3.92	3.07	2.68	2.45	2.29	2.18	2.09	2.02	1.96
	.975	5.15	3.80	3.23	2.89	2.67	2.52	2.39	2.30	2.22
	.99	6.85	4.79	3.95	3.48	3.17	2.96	2.79	2.66	2.56
	.995	8.18	5.54	4.50	3.92	3.55	3.28	3.09	2.93	2.81
	.999	11.4	7.32	5.78	4.95	4.42	4.04	3.77	3.55	3.38
∞	.50	0.455	0.693	0.789	0.839	0.870	0.891	0.907	0.918	0.927
	.90	2.71	2.30	2.08	1.94	1.85	1.77	1.72	1.67	1.63
	.95	3.84	3.00	2.60	2.37	2.21	2.10	2.01	1.94	1.88
	.975	5.02	3.69	3.12	2.79	2.57	2.41	2.29	2.19	2.11
	.99	6.63	4.61	3.78	3.32	3.02	2.80	2.64	2.51	2.41
	.995	7.88	5.30	4.28	3.72	3.35	3.09	2.90	2.74	2.62
	.999	10.8	6.91	5.42	4.62	4.10	3.74	3.47	3.27	3.10

TABLE B.7 (*concluded*)

r_2 $1-\alpha$		10	12	15	20	24	30	60	120	∞
30	.50	0.955	0.966	0.978	0.989	0.994	1.00	1.01	1.02	1.02
	.90	1.82	1.77	1.72	1.67	1.64	1.61	1.54	1.50	1.46
	.95	2.16	2.09	2.01	1.93	1.89	1.84	1.74	1.68	1.62
	.975	2.51	2.41	2.31	2.20	2.14	2.07	1.94	1.87	1.79
	.99	2.98	2.84	2.70	2.55	2.47	2.39	2.21	2.11	2.01
	.995	3.34	3.18	3.01	2.82	2.73	2.63	2.42	2.30	2.18
	.999	4.24	4.00	3.75	3.49	3.36	3.22	2.92	2.76	2.59
60	.50	0.945	0.956	0.967	0.978	0.983	0.989	1.00	1.01	1.01
	.90	1.71	1.66	1.60	1.54	1.51	1.48	1.40	1.35	1.29
	.95	1.99	1.92	1.84	1.75	1.70	1.65	1.53	1.47	1.39
	.975	2.27	2.17	2.06	1.94	1.88	1.82	1.67	1.58	1.48
	.99	2.63	2.50	2.35	2.20	2.12	2.03	1.84	1.73	1.60
	.995	2.90	2.74	2.57	2.39	2.29	2.19	1.96	1.83	1.69
	.999	3.54	3.32	3.08	2.83	2.69	2.55	2.25	2.08	1.89
120	.50	0.939	0.950	0.961	0.972	0.978	0.983	0.994	1.00	1.01
	.90	1.65	1.60	1.55	1.48	1.45	1.41	1.32	1.26	1.19
	.95	1.91	1.83	1.75	1.66	1.61	1.55	1.43	1.35	1.25
	.975	2.16	2.05	1.95	1.82	1.76	1.69	1.53	1.43	1.31
	.99	2.47	2.34	2.19	2.03	1.95	1.86	1.66	1.53	1.38
	.995	2.71	2.54	2.37	2.19	2.09	1.98	1.75	1.61	1.43
	.999	3.24	3.02	2.78	2.53	2.40	2.26	1.95	1.77	1.54
∞	.50	0.934	0.945	0.956	0.967	0.972	0.978	0.989	0.994	1.00
	.90	1.60	1.55	1.49	1.42	1.38	1.34	1.24	1.17	1.00
	.95	1.83	1.75	1.67	1.57	1.52	1.46	1.32	1.22	1.00
	.975	2.05	1.94	1.83	1.71	1.64	1.57	1.39	1.27	1.00
	.99	2.32	2.18	2.04	1.88	1.79	1.70	1.47	1.32	1.00
	.995	2.52	2.36	2.19	2.00	1.90	1.79	1.53	1.36	1.00
	.999	2.96	2.74	2.51	2.27	2.13	1.99	1.66	1.45	1.00

Source: Adapted from table 5 of Pearson and Hartley, *Biometrika Tables for Statisticians*, volume 2, 1972, published by the Cambridge University Press, on behalf of The Biometrika Society, by permission of the authors and publishers.

TABLE B.8 Five-place logarithms

N	0	1	2	3	4	5	6	7	8	9
0	$-\infty$	00000	30103	47712	60206	69897	77815	84510	90309	95424
10	00000	00432	00860	01284	01703	02119	02531	02938	03342	03743
11	04139	04532	04922	05308	05690	06070	06446	06819	07188	07555
12	07918	08279	08636	08991	09342	09691	10037	10380	10721	11059
13	11394	11727	12057	12385	12710	13033	13354	13672	13988	14301
14	14613	14922	15229	15534	15836	16137	16435	16732	17026	17319
15	17609	17898	18184	18469	18752	19033	19312	19590	19866	20140
16	20412	20683	20952	21219	21484	21748	22011	22272	22531	22789
17	23045	23300	23533	23805	24055	24304	24551	24797	25042	25285
18	25527	25768	26007	26245	26482	26717	26951	27184	27416	27646
19	27875	28103	28330	28556	28780	29003	29226	29447	29667	29885
20	30103	30320	30535	30750	30963	31175	31387	31597	31806	32015
21	32222	32428	32634	32838	33041	33244	33445	33646	33846	34044
22	34242	34439	34635	34830	35025	35218	35411	35603	35793	35984
23	36173	36361	36549	36736	36922	37107	37291	37475	37658	37840
24	38021	38202	38382	38561	38739	38917	39094	39270	39445	39620
25	39794	39967	40140	40312	40483	40654	40824	40993	41162	41330
26	41497	41664	41830	41996	42160	42325	42488	42651	42813	42975
27	43136	43297	43457	43616	43775	43933	44091	44248	44404	44560
28	44716	44871	45025	45179	45332	45484	45637	45788	45939	46090
29	46240	46389	46538	46687	46835	46982	47129	47276	47422	47567
30	47712	47857	48001	48144	48287	48430	48572	48714	48855	48996
31	49136	49276	49415	49554	49693	49831	49969	50106	50243	50379
32	50515	50651	50786	50920	51055	51188	51322	51455	51587	51720
33	51851	51983	52114	52244	52375	52504	52634	52763	52892	53020
34	53148	53275	53403	53529	53656	53782	53908	54033	54158	54283
35	54407	54531	54654	54777	54900	55023	55145	55267	55388	55509
36	55630	55751	55871	55991	56110	56229	56348	56467	56585	56703
37	56820	56937	57054	57171	57287	57403	57519	57634	57749	57864
38	57978	58092	58206	58320	58433	58546	58659	58771	58883	58995
39	59106	59218	59329	59439	59550	59660	59770	59879	59988	60097
40	60206	60314	60423	60531	60638	60746	60853	60959	61066	61172
41	61278	61384	61490	61595	61700	61805	61909	62014	62118	62221
42	62325	62428	62531	62634	62737	62839	62941	63043	63144	63246
43	63347	63448	63548	63649	63749	63849	63949	64048	64147	64246
44	64345	64444	64542	64640	64738	64836	64933	65031	65128	65225
45	65321	65418	65514	65610	65706	65801	65896	65992	66087	66181
46	66276	66370	66464	66558	66652	66745	66839	66932	67025	67117
47	67210	67302	67394	67486	67578	67669	67761	67852	67943	68034
48	68124	68215	68305	68395	68485	68574	68664	68753	68842	68931
49	69020	69108	69197	69285	69373	69461	69548	69636	69723	69810

TABLE B.8 (*continued*)

N	0	1	2	3	4	5	6	7	8	9
50	69897	69984	70070	70157	70243	70329	70415	70501	70586	70672
51	70757	70842	70927	71012	71096	71181	71265	71349	71433	71517
52	71600	71684	71767	71850	71933	72016	72099	72181	72263	72346
53	72428	72509	72591	72673	72754	72835	72916	72997	73078	73159
54	73239	73320	73400	73480	73560	73640	73719	73799	73878	73957
55	74036	74115	74194	74273	74351	74429	74507	74586	74663	74741
56	74819	74896	74974	75051	75128	75205	75282	75358	75435	75511
57	75587	75664	75740	75815	75891	75967	76042	76118	76193	76268
58	76343	76418	76492	76567	76641	76716	76790	76864	76938	77012
59	77085	77159	77232	77305	77379	77452	77525	77597	77670	77743
60	77815	77887	77960	78032	78104	78176	78247	78319	78390	78462
61	78533	78604	78675	78746	78817	78888	78958	79029	79099	79169
62	79239	79309	79379	79449	79518	79588	79657	79727	79796	79865
63	79934	80003	80072	80140	80209	80277	80346	80414	80482	80550
64	80618	80686	80754	80821	80889	80956	81023	81090	81158	81224
65	81291	81358	81425	81491	81558	81624	81690	81757	81823	81889
66	81954	82020	82086	82151	82217	82282	82347	82413	82478	82543
67	82607	82672	82737	82802	82866	82930	82995	83059	83123	83187
68	83251	83315	83378	83442	83506	83569	83632	83696	83759	83822
69	83885	83948	84011	84073	84136	84198	84261	84323	84386	84448
70	84510	84572	84634	84696	84757	84819	84880	84942	85003	85065
71	85126	85187	85248	85309	85370	85431	85491	85552	85612	85673
72	85733	85794	85854	85914	85974	86034	86094	86153	86213	86273
73	86332	86392	86451	86510	86570	86629	86688	86747	86806	86864
74	86923	86982	87040	87099	87157	87216	87274	87332	87390	87448
75	87506	87564	87622	87679	87737	87795	87852	87910	87967	88024
76	88081	88138	88195	88252	88309	88366	88423	88480	88536	88593
77	88649	88705	88762	88818	88874	88930	88986	89042	89098	89154
78	89209	89265	89321	89376	89432	89487	89542	89597	89653	89708
79	89763	89818	89873	89927	89982	90037	90091	90146	90200	90255
80	90309	90363	90417	90472	90526	90580	90634	90687	90741	90795
81	90849	90902	90956	91009	91062	91116	91169	91222	91275	91328
82	91381	91434	91487	91540	91593	91645	91698	91751	91803	91855
83	91908	91960	92012	92065	92117	92169	92221	92273	92324	92376
84	92428	92480	92531	92583	92634	92686	92737	92788	92840	92891
85	92942	92993	93044	93095	93146	93197	93247	93298	93349	93399
86	93450	93500	93551	93601	93651	93702	93752	93802	93852	93902
87	93952	94002	94052	94101	94151	94201	94250	94300	94349	94399
88	94448	94498	94547	94596	94645	94694	94743	94792	94841	94890
89	94939	94988	95036	95085	95134	95182	95231	95279	95328	95376
90	95424	95472	95521	95569	95617	95665	95713	95761	95809	95856
91	95904	95952	95999	96047	96095	96142	96190	96237	96284	96332
92	96379	96426	96473	96520	96567	96614	96661	96708	96755	96802
93	96848	96895	96942	96988	97035	97081	97128	97174	97220	97267
94	97313	97359	97405	97451	97497	97543	97589	97635	97681	97727
95	97772	97818	97864	97909	97955	98000	98046	98091	98137	98182
96	98227	98272	98318	98363	98408	98453	98498	98543	98588	98632
97	98677	98722	98767	98811	98856	98900	98945	98989	99034	99078
98	99123	99167	99211	99255	99300	99344	99388	99432	99476	99520
99	99564	99607	99651	99695	99739	99782	99826	99870	99913	99957

Source: John Neter, William Wasserman, and G. A. Whitmore, *Fundamental Statistics for Business and Economics*, 4th ed. (Boston: Allyn & Bacon, Inc., 1972 ©).

TABLE B.9　　Critical values for Hartley's H-statistic

$$1 - \alpha = .95$$

n	2	3	4	5	6	7	8	9	10	11	12
3	39.0	87.5	142	202	266	333	403	475	550	626	704
4	15.4	27.8	39.2	50.7	62.0	72.9	83.5	93.9	104	114	124
5	9.60	15.5	20.6	25.2	29.5	33.6	37.5	41.1	44.6	48.0	51.4
6	7.15	10.8	13.7	16.3	18.7	20.8	22.9	24.7	26.5	28.2	29.9
7	5.82	8.38	10.4	12.1	13.7	15.0	16.3	17.5	18.6	19.7	20.7
8	4.99	6.94	8.44	9.70	10.8	11.8	12.7	13.5	14.3	15.1	15.8
9	4.43	6.00	7.18	8.12	9.03	9.78	10.5	11.1	11.7	12.2	12.7
10	4.03	5.34	6.31	7.11	7.80	8.41	8.95	9.45	9.91	10.3	10.7
11	3.72	4.85	5.67	6.34	6.92	7.42	7.87	8.28	8.66	9.01	9.34
13	3.28	4.16	4.79	5.30	5.72	6.09	6.42	6.72	7.00	7.25	7.48
16	2.86	3.54	4.01	4.37	4.68	4.95	5.19	5.40	5.59	5.77	5.93
21	2.46	2.95	3.29	3.54	3.76	3.94	4.10	4.24	4.37	4.49	4.59
31	2.07	2.40	2.61	2.78	2.91	3.02	3.12	3.21	3.29	3.36	3.39
61	1.67	1.85	1.96	2.04	2.11	2.17	2.22	2.26	2.30	2.33	2.36
∞	1.00	1.00	1.00	1.00	1.00	1.00	1.00	1.00	1.00	1.00	1.00

$$1 - \alpha = .99$$

n	2	3	4	5	6	7	8	9	10	11	12
3	199	448	729	1,036	1,362	1,705	2,063	2,432	2,813	3,204	3,605
4	47.5	85	120	151	184	216	249	281	310	337	361
5	23.2	37	49	59	69	79	89	97	106	113	120
6	14.9	22	28	33	38	42	46	50	54	57	60
7	11.1	15.5	19.1	22	25	27	30	32	34	36	37
8	8.89	12.1	14.5	16.5	18.4	20	22	23	24	26	27
9	7.50	9.9	11.7	13.2	14.5	15.8	16.9	17.9	18.9	19.8	21
10	6.54	8.5	9.9	11.1	12.1	13.1	13.9	14.7	15.3	16.0	16.6
11	5.85	7.4	8.6	9.6	10.4	11.1	11.8	12.4	12.9	13.4	13.9
13	4.91	6.1	6.9	7.6	8.2	8.7	9.1	9.5	9.9	10.2	10.6
16	4.07	4.9	5.5	6.0	6.4	6.7	7.1	7.3	7.5	7.8	8.0
21	3.32	3.8	4.3	4.6	4.9	5.1	5.3	5.5	5.6	5.8	5.9
31	2.63	3.0	3.3	3.4	3.6	3.7	3.8	3.9	4.0	4.1	4.2
61	1.96	2.2	2.3	2.4	2.4	2.5	2.5	2.6	2.6	2.7	2.7
∞	1.00	1.0	1.0	1.0	1.0	1.0	1.0	1.0	1.0	1.0	1.0

Source: Reprinted, with permission, from H. A. David, "Upper 5 and 1% Points of the Maximum F-Ratio," *Biometrika*, vol. 39 (1952), pp. 422–24.

TABLE B.10 Unit normal losses

D	.00	.01	.02	.03	.04	.05	.06	.07	.08	.09
.0	.3989	.3940	.3890	.3841	.3793	.3744	.3697	.3649	.3602	.3556
.1	.3509	.3464	.3418	.3373	.3328	.3284	.3240	.3197	.3154	.3111
.2	.3069	.3027	.2986	.2944	.2904	.2863	.2824	.2784	.2745	.2706
.3	.2668	.2630	.2592	.2555	.2518	.2481	.2445	.2409	.2374	.2339
.4	.2304	.2270	.2236	.2203	.2169	.2137	.2104	.2072	.2040	.2009
.5	.1978	.1947	.1917	.1887	.1857	.1828	.1799	.1771	.1742	.1714
.6	.1687	.1659	.1633	.1606	.1580	.1554	.1528	.1503	.1478	.1453
.7	.1429	.1405	.1381	.1358	.1334	.1312	.1289	.1267	.1245	.1223
.8	.1202	.1181	.1160	.1140	.1120	.1100	.1080	.1061	.1042	.1023
.9	.1004	.09860	.09680	.09503	.09328	.09156	.08986	.08819	.08654	.08491
1.0	.08332	.08174	.08019	.07866	.07716	.07568	.07422	.07279	.07138	.06999
1.1	.06862	.06727	.06595	.06465	.06336	.06210	.06086	.05964	.05844	.05726
1.2	.05610	.05496	.05384	.05274	.05165	.05059	.04954	.04851	.04750	.04650
1.3	.04553	.04457	.04363	.04270	.04179	.04090	.04002	.03916	.03831	.03748
1.4	.03667	.03587	.03508	.03431	.03356	.03281	.03208	.03137	.03067	.02998
1.5	.02931	.02865	.02800	.02736	.02674	.02612	.02552	.02494	.02436	.02380
1.6	.02324	.02270	.02217	.02165	.02114	.02064	.02015	.01967	.01920	.01874
1.7	.01829	.01785	.01742	.01699	.01658	.01617	.01578	.01539	.01501	.01464
1.8	.01428	.01392	.01357	.01323	.01290	.01257	.01226	.01195	.01164	.01134
1.9	.01105	.01077	.01049	.01022	$.0^2 9957$	$.0^2 9698$	$.0^2 9445$	$.0^2 9198$	$.0^2 8957$	$.0^2 8721$
2.0	$.0^2 8491$	$.0^2 8266$	$.0^2 8046$	$.0^2 7832$	$.0^2 7623$	$.0^2 7418$	$.0^2 7219$	$.0^2 7024$	$.0^2 6835$	$.0^2 6649$
2.1	$.0^2 6468$	$.0^2 6292$	$.0^2 6120$	$.0^2 5952$	$.0^2 5788$	$.0^2 5628$	$.0^2 5472$	$.0^2 5320$	$.0^2 5172$	$.0^2 5028$
2.2	$.0^2 4887$	$.0^2 4750$	$.0^2 4616$	$.0^2 4486$	$.0^2 4358$	$.0^2 4235$	$.0^2 4114$	$.0^2 3996$	$.0^2 3882$	$.0^2 3770$
2.3	$.0^2 3662$	$.0^2 3556$	$.0^2 3453$	$.0^2 3352$	$.0^2 3255$	$.0^2 3159$	$.0^2 3067$	$.0^2 2977$	$.0^2 2889$	$.0^2 2804$
2.4	$.0^2 2720$	$.0^2 2640$	$.0^2 2561$	$.0^2 2484$	$.0^2 2410$	$.0^2 2337$	$.0^2 2267$	$.0^2 2199$	$.0^2 2132$	$.0^2 2067$
2.5	$.0^2 2004$	$.0^2 1943$	$.0^2 1883$	$.0^2 1826$	$.0^2 1769$	$.0^2 1715$	$.0^2 1662$	$.0^2 1610$	$.0^2 1560$	$.0^2 1511$
3.0	$.0^3 3822$	$.0^3 3689$	$.0^3 3560$	$.0^3 3436$	$.0^3 3316$	$.0^3 3199$	$.0^3 3087$	$.0^3 2978$	$.0^3 2873$	$.0^3 2771$
3.5	$.0^4 5848$	$.0^4 5620$	$.0^4 5400$	$.0^4 5188$	$.0^4 4984$	$.0^4 4788$	$.0^4 4599$	$.0^4 4417$	$.0^4 4242$	$.0^4 4073$
4.0	$.0^5 7145$	$.0^5 6835$	$.0^5 6538$	$.0^5 6253$	$.0^5 5980$	$.0^5 5718$	$.0^5 5468$	$.0^5 5227$	$.0^5 4997$	$.0^5 4777$

Illustration: The value of $L(D)$ for $D = 3.01$ is $0.0^3 3689 = 0.0003689$.

Source: Reproduced from Robert Schlaifer, *Introduction to Statistics for Business Decisions*, published by McGraw-Hill Book Company, 1961, by permission from the copyright holder, the President and Fellows of Harvard College.

TABLE B.11 Critical values of the Mann-Whitney test statistic

n	α	m = 2	3	4	5	6	7	8	9	10	11	12	13	14	15	16	17	18	19	20
2	.001	0	0	0	0	0	0	0	0	0	0	0	0	0	0	0	0	0	0	0
	.005	0	0	0	0	0	0	0	0	0	0	0	0	0	0	0	0	0	1	1
	.01	0	0	0	0	0	0	0	0	0	0	0	1	1	1	1	1	1	2	2
	.025	0	0	0	0	0	0	0	1	1	1	2	2	2	2	3	3	3	3	3
	.05	0	0	0	0	1	1	1	2	2	2	3	3	4	4	4	4	5	5	5
	.10	0	0	1	1	2	2	3	3	4	4	5	5	5	6	6	7	7	8	8
3	.001	0	0	0	0	0	0	0	0	0	0	0	0	0	0	0	0	1	1	1
	.005	0	0	0	0	0	0	0	1	1	1	2	2	2	3	3	3	3	4	4
	.01	0	0	0	0	0	1	1	2	2	2	3	3	3	4	4	5	5	5	6
	.025	0	0	0	1	2	2	3	3	4	4	5	5	6	6	7	7	8	8	9
	.05	0	1	1	2	3	3	4	5	5	6	6	7	8	8	9	10	10	11	12
	.10	1	2	2	3	4	5	6	6	7	8	9	10	11	11	12	13	14	15	16
4	.001	0	0	0	0	0	0	0	0	1	1	1	2	2	2	3	3	4	4	4
	.005	0	0	0	0	1	1	2	2	3	3	4	4	5	6	6	7	7	8	9
	.01	0	0	0	1	2	2	3	4	4	5	6	6	7	8	9	9	10	10	11
	.025	0	0	1	2	3	4	5	5	6	7	8	9	10	11	12	12	13	14	15
	.05	0	1	2	3	4	5	6	7	8	9	10	11	12	13	15	16	17	18	19
	.10	1	2	4	5	6	7	8	10	11	12	13	14	16	17	18	19	21	22	23
5	.001	0	0	0	0	0	0	1	2	2	3	3	4	4	5	6	6	7	8	8
	.005	0	0	0	1	2	2	3	4	5	6	7	8	8	9	10	11	12	13	14
	.01	0	0	1	2	3	4	5	6	7	8	9	10	11	12	13	14	15	16	17
	.025	0	1	2	3	4	6	7	8	9	10	12	13	14	15	16	18	19	20	21
	.05	1	2	3	5	6	7	9	10	12	13	14	16	17	19	20	21	23	24	26
	.10	2	3	5	6	8	9	11	13	14	16	18	19	21	23	24	26	28	29	31
6	.001	0	0	0	0	0	0	2	3	4	5	5	6	7	8	9	10	11	12	13
	.005	0	0	1	2	3	4	5	6	7	8	10	11	12	13	14	16	17	18	19
	.01	0	0	2	3	4	5	7	8	9	10	12	13	14	16	17	19	20	21	23
	.025	1	2	3	4	6	7	9	11	12	14	15	17	18	20	22	23	25	26	28
	.05	1	3	4	6	8	9	11	13	15	17	18	20	22	24	26	27	29	31	33
	.10	2	4	6	8	10	12	14	16	18	20	22	24	26	28	30	32	35	37	39

Source: Adapted from L. R. Verdooren, "Extended Tables of Critical Values for Wilcoxon's Test Statistic," *Biometrika*, vol. 50 (1963), pp. 177–86.

TABLE B.11 *(continued)*

n	α	m = 2	3	4	5	6	7	8	9	10	11	12	13	14	15	16	17	18	19	20
7	.001	0	0	0	0	1	2	3	4	6	7	8	9	10	11	12	14	15	16	17
	.005	0	0	1	2	4	5	7	8	10	11	13	14	16	17	19	20	22	23	25
	.01	0	1	2	4	5	7	8	10	12	13	15	17	18	20	22	24	25	27	29
	.025	1	2	4	6	7	9	11	13	15	17	19	21	23	25	27	29	31	33	35
	.05	1	3	5	7	9	12	14	16	18	20	22	25	27	29	31	34	36	38	40
	.10	2	5	7	9	12	14	17	19	22	24	27	29	32	34	37	39	42	44	47
8	.001	0	0	0	1	2	3	5	6	7	9	10	12	13	15	16	18	19	21	22
	.005	0	0	2	3	5	7	8	10	12	14	16	18	19	21	23	25	27	29	31
	.01	0	1	3	5	7	8	10	12	14	16	18	21	23	25	27	29	31	33	35
	.025	1	3	5	7	9	11	14	16	18	20	23	25	27	30	32	35	37	39	42
	.05	2	4	6	9	11	14	16	19	21	24	27	29	32	34	37	40	42	45	48
	.10	3	6	8	11	14	17	20	23	25	28	31	34	37	40	43	46	49	52	55
9	.001	0	0	0	2	3	4	6	8	9	11	13	15	16	18	20	22	24	26	27
	.005	0	1	2	4	6	8	10	12	14	17	19	21	23	25	28	30	32	34	37
	.01	0	2	4	6	8	10	12	15	17	19	22	24	27	29	32	34	37	39	41
	.025	1	3	6	8	10	13	16	18	21	24	27	29	32	35	38	40	43	46	49
	.05	2	5	8	10	13	16	19	22	25	28	31	34	37	40	43	46	49	52	55
	.10	3	6	9	13	16	19	23	26	29	32	36	39	42	46	49	53	56	59	63
10	.001	0	0	0	2	4	6	7	9	11	13	15	18	20	22	24	26	28	30	33
	.005	0	1	2	5	7	10	12	14	17	19	22	25	27	30	32	35	38	40	43
	.01	0	2	4	6	9	12	14	17	20	23	25	28	31	34	37	39	42	45	48
	.025	1	4	7	9	12	15	18	21	24	27	30	34	37	40	43	46	49	53	56
	.05	2	5	8	12	15	18	21	25	28	32	35	38	42	45	49	52	56	59	63
	.10	4	8	11	14	18	22	25	29	33	37	40	44	48	52	55	59	63	67	71
11	.001	0	0	0	3	5	7	9	11	13	16	18	21	23	25	28	30	33	35	38
	.005	0	1	2	6	8	11	14	17	19	22	25	28	31	34	37	40	43	46	49
	.01	0	2	5	8	10	13	16	19	23	26	29	32	35	38	42	45	48	51	54
	.025	1	4	7	10	14	17	20	24	27	31	34	38	41	45	48	52	56	59	63
	.05	2	6	9	13	17	20	24	28	32	35	39	43	47	51	55	58	62	66	70
	.10	4	8	12	16	20	24	28	32	37	41	45	49	53	58	62	66	70	74	79

TABLE B.11 (continued)

n	α	m = 2	3	4	5	6	7	8	9	10	11	12	13	14	15	16	17	18	19	20
12	.001	0	0	1	3	5	8	10	13	15	18	21	24	26	29	32	35	38	41	43
	.005	0	2	4	7	10	13	16	19	22	25	28	32	35	38	42	45	48	52	55
	.01	0	3	6	9	12	15	18	22	25	29	32	36	39	43	47	50	54	57	61
	.025	2	5	8	12	15	19	23	27	30	34	38	42	46	50	54	58	62	66	70
	.05	3	6	10	14	18	22	27	31	35	39	43	48	52	56	61	65	69	73	78
	.10	5	9	13	18	22	27	31	36	40	45	50	54	59	64	68	73	78	82	87
13	.001	0	0	2	4	6	9	12	15	18	21	24	27	30	33	36	39	43	46	49
	.005	0	2	4	8	11	14	18	21	25	28	32	35	39	43	46	50	54	58	61
	.01	1	3	6	10	13	17	21	24	28	32	36	40	44	48	52	56	60	64	68
	.025	2	5	9	13	17	21	25	29	34	38	42	46	51	55	60	64	68	73	77
	.05	3	7	11	16	20	25	29	34	38	43	48	52	57	62	66	71	76	81	85
	.10	5	10	14	19	24	29	34	39	44	49	54	59	64	69	75	80	85	90	95
14	.001	0	0	2	4	7	10	13	16	20	23	26	30	33	37	40	44	47	51	55
	.005	0	2	5	8	12	16	19	23	27	31	35	39	43	47	51	55	59	64	68
	.01	1	3	7	11	14	18	23	27	31	35	39	44	48	52	57	61	66	70	74
	.025	2	6	10	14	18	23	27	32	37	41	46	51	56	60	65	70	75	79	84
	.05	4	8	12	17	22	27	32	37	42	47	52	57	62	67	72	78	83	88	93
	.10	5	11	16	21	26	32	37	42	48	53	59	64	70	75	81	86	92	98	103
15	.001	0	0	2	5	8	11	15	18	22	25	29	33	37	41	44	48	52	56	60
	.005	0	3	6	9	13	17	21	25	30	34	38	43	47	52	56	61	65	70	74
	.01	1	4	8	12	16	20	25	29	34	38	43	48	52	57	62	67	71	76	81
	.025	2	6	11	15	20	25	30	35	40	45	50	55	60	65	71	76	81	86	91
	.05	4	8	13	19	24	29	34	40	45	51	56	62	67	73	78	84	89	95	101
	.10	6	11	17	23	28	34	40	46	52	58	64	69	75	81	87	93	99	105	111
16	.001	0	0	3	6	9	12	16	20	24	28	32	36	40	44	49	53	57	61	66
	.005	0	3	6	10	14	19	23	28	32	37	42	46	51	56	61	66	71	75	80
	.01	1	4	8	13	17	22	27	32	37	42	47	52	57	62	67	72	77	83	88
	.025	2	7	12	16	22	27	32	38	43	48	54	60	65	71	76	82	87	93	99
	.05	4	9	15	20	26	31	37	43	49	55	61	66	72	78	84	90	96	102	108
	.10	6	12	18	24	30	37	43	49	55	62	68	75	81	87	94	100	107	113	120

TABLE B.11 (concluded)

n	α	m = 2	3	4	5	6	7	8	9	10	11	12	13	14	15	16	17	18	19	20
17	.001	0	1	3	6	10	14	18	22	26	30	35	39	44	48	53	58	62	67	71
	.005	0	3	7	11	16	20	25	30	35	40	45	50	55	61	66	71	76	82	87
	.01	1	5	9	14	19	24	29	34	39	45	50	56	61	67	72	78	83	89	94
	.025	3	7	12	18	23	29	35	40	46	52	58	64	70	76	82	88	94	100	106
	.05	4	10	16	21	27	34	40	46	52	58	65	71	78	84	90	97	103	110	116
	.10	7	13	19	26	32	39	46	53	59	66	73	80	86	93	100	107	114	121	128
18	.001	0	1	4	7	11	15	19	24	28	33	38	43	47	52	57	62	67	72	77
	.005	0	3	7	12	17	22	27	32	38	43	48	54	59	65	71	76	82	88	93
	.01	1	5	10	15	20	25	31	37	42	48	54	60	66	71	77	83	89	95	101
	.025	3	8	13	19	25	31	37	43	49	56	62	68	75	81	87	94	100	107	113
	.05	5	10	17	23	29	36	42	49	56	62	69	76	83	89	96	103	110	117	124
	.10	7	14	21	28	35	42	49	56	63	70	78	85	92	99	107	114	121	129	136
19	.001	0	1	4	8	12	16	21	26	30	35	41	46	51	56	61	67	72	78	83
	.005	1	4	8	13	18	23	29	34	40	46	52	58	64	70	75	82	88	94	100
	.01	2	5	10	16	21	27	33	39	45	51	57	64	70	76	83	89	95	102	108
	.025	3	8	14	20	26	33	39	46	53	59	66	73	79	86	93	100	107	114	120
	.05	5	11	18	24	31	38	45	52	59	66	73	81	88	95	102	110	117	124	131
	.10	8	15	22	29	37	44	52	59	67	74	82	90	98	105	113	121	129	136	144
20	.001	0	1	4	8	13	17	22	27	33	38	43	49	55	60	66	71	77	83	89
	.005	1	4	9	14	19	25	31	37	43	49	55	61	68	74	80	87	93	100	106
	.001	2	6	11	17	23	29	35	41	48	54	61	68	74	81	88	94	101	108	115
	.025	3	9	15	21	28	35	42	49	56	63	70	77	84	91	99	106	113	120	128
	.05	5	12	19	26	33	40	48	55	63	70	78	85	93	101	108	116	124	131	139
	.10	8	16	23	31	39	47	55	63	71	79	87	95	103	111	120	128	136	144	152

Table entries are upper-tail critical values W_α such that $P(W \geq W_\alpha) = \alpha$.

For n or m greater than 20, the critical value W_α may be approximated by

$$W_\alpha = \frac{nm}{2} + z_\alpha \sqrt{\frac{nm(n+m+1)}{12}}$$

where z_α is the standard normal variate such that a proportion α of the area is to the right of z_α.

TABLE B.12 Critical values of the Wilcoxon signed rank test statistic

Sample size n	$W_{.005}$	$W_{.01}$	$W_{.025}$	$W_{.05}$	$W_{.10}$	$W_{.20}$	$n(n+1)/2$
4	0	0	0	0	1	3	10
5	0	0	0	1	3	4	15
6	0	0	1	3	4	6	21
7	0	1	3	4	6	9	28
8	1	2	4	6	9	12	36
9	2	4	6	9	11	15	45
10	4	6	9	11	15	19	55
11	6	8	11	14	18	23	66
12	8	10	14	18	22	28	78
13	10	13	18	22	27	33	91
14	13	16	22	26	32	39	105
15	16	20	26	31	37	45	120
16	20	24	30	36	43	51	136
17	24	28	35	42	49	58	153
18	28	33	41	48	56	66	171
19	33	38	47	54	63	74	190
20	38	44	53	61	70	82	210

Table entries are lower-tail critical values: W_α such that $P(W \leq W_\alpha) = \alpha$.
The upper-tail critical values may be determined by using the relationship:

$$W_\alpha = [n(n+1)/2] - W_{1-\alpha}, \qquad \alpha > 0.50$$

For example, if $n = 10$, $\alpha = 0.05$, and the test is two-tailed, $W_{\alpha/2} = W_{0.025} = 9$ from the table;

$$W_{1-\alpha/2} = W_{0.975} = [10(10 + 1)/2] - W_{0.025} = 55 - 9 = 46$$

For $n > 20$, the critical value W_α may be approximated from:

$$W_\alpha = [n(n+1)/4] + z_\alpha \sqrt{n(n+1)(2n+1)/24}$$

where z_α is the standard normal variate such that a proportion α of the area is to the right of z_α.

Source: Adapted from R. L. McCornack, "Extended Tables of the Wilcoxon Matched Pairs Signed Rank Statistics," *Journal of the American Statistical Association*, vol. 60 (1965), pp. 864–71.

TABLE B.13   Critical values of the Kruskal-Wallis test statistic for three samples and small sample size.

\multicolumn{3}{c}{Sample sizes}	Critical value	α	\multicolumn{3}{c}{Sample sizes}	Critical value	α
n_1	n_2	n_3			n_1	n_2	n_3		
2	1	1	2.7000	.500	4	3	2	6.4444	.009
2	2	1	3.6000	.267				6.3000	.011
2	2	2	4.5714	.067				5.4444	.046
			3.7143	.200				5.4000	.051
								4.5111	.098
3	1	1	3.2000	.300				4.4444	.102
3	2	1	4.2857	.100	4	3	3	6.7455	.010
			3.8571	.133				6.7091	.013
3	2	2	5.3572	.029				5.7909	.046
			4.7143	.048				5.7273	.050
			4.5000	.067				4.7091	.092
			4.4643	.105				4.7000	.101
3	3	1	5.1429	.043	4	4	1	6.6667	.010
			4.5714	.100				6.1667	.022
			4.0000	.129				4.9667	.048
3	3	2	6.2500	.011				4.8667	.054
			5.3611	.032				4.1667	.082
			5.1389	.061				4.0667	.102
			4.5556	.100	4	4	2	7.0364	.006
			4.2500	.121				6.8727	.011
3	3	3	7.2000	.004				5.4545	.046
			6.4889	.001				5.2364	.052
			5.6889	.029				4.5545	.098
			5.6000	.050				4.4455	.103
			5.0667	.086	4	4	3	7.1439	.010
			4.6222	.100				7.1364	.011
4	1	1	3.5714	.200				5.5985	.049
								5.5758	.051
4	2	1	4.8214	.057				4.5455	.099
			4.5000	.076				4.4773	.102
			4.0179	.114	4	4	4	7.6538	.008
4	2	2	6.0000	.014				7.5385	.011
			5.3333	.033				5.6923	.049
			5.1250	.052				5.6538	.054
			4.3750	.100				4.6539	.097
			4.1667	.105				4.5001	.104
4	3	1	5.8333	.021	5	1	1	3.8571	.143
			5.2083	.050	5	2	1	5.2500	.036
			5.0000	.057				5.0000	.048
			4.0556	.093				4.4500	.071
			3.8889	.129				4.2000	.095
								4.0500	.119

Source: Adapted from W. H. Kruskal and W. A. Wallis, "Use of Ranks on One-Criterion Variance Analysis," *Journal of the American Statistical Association,* vol. 47 (1952), pp. 583–621.

TABLE B.13 (*continued*)

n_1	n_2	n_3	Critical value	α	n_1	n_2	n_3	Critical value	α
5	2	2	6.5333	.008	5	4	4	7.7604	.009
			6.1333	.013				7.7440	.011
			5.1600	.034				5.6571	.049
			5.0400	.056				5.6176	.050
			4.3733	.090				4.6187	.100
			4.2933	.112				4.5527	.102
5	3	1	6.4000	.012	5	5	1	7.3091	.009
			4.9600	.048				6.8364	.011
			4.8711	.052				5.1273	.046
			4.0178	.095				4.9091	.053
			3.8400	.123				4.1091	.086
5	3	2	6.9091	.009				4.0364	.105
			6.8281	.010	5	5	2	7.3385	.010
			5.2509	.049				7.2692	.010
			5.1055	.052				5.3385	.047
			4.6509	.091				5.2462	.051
			4.4945	.101				4.6231	.097
5	3	3	7.0788	.009				4.5077	.100
			6.9818	.011	5	5	3	7.5780	.010
			5.6485	.049				7.5429	.010
			5.5152	.051				5.7055	.046
			4.5333	.097				5.6264	.051
			4.4121	.109				4.5451	.100
5	4	1	6.9545	.008				4.5363	.102
			6.8400	.011	5	5	4	7.8229	.010
			4.9855	.044				7.7914	.010
			4.8600	.056				5.6657	.049
			3.9873	.098				5.6429	.050
			3.9600	.102				4.5229	.100
5	4	2	7.2045	.009				4.5200	.101
			7.1182	.010	5	5	5	8.0000	.009
			5.2727	.049				7.9800	.010
			5.2682	.050				5.7800	.049
			4.5409	.098				5.6600	.051
			4.5182	.101				4.5600	.100
5	4	3	7.4449	.010				4.5000	.102
			7.3949	.011					
			5.6564	.049					
			5.6308	.050					
			4.5487	.099					
			4.5231	.103					

TABLE B.14 Critical values of the Spearman test statistic

n	$\alpha = .100$	.050	.025	.010	.005	.001
4	.8000	.8000				
5	.7000	.8000	.9000	.9000		
6	.6000	.7714	.8286	.8857	.9429	
7	.5357	.6786	.7450	.8571	.8929	.9643
8	.5000	.6190	.7143	.8095	.8571	.9286
9	.4667	.5833	.6833	.7667	.8167	.9000
10	.4424	.5515	.6364	.7333	.7818	.8667
11	.4182	.5273	.6091	.7000	.7455	.8364
12	.3986	.4965	.5804	.6713	.7273	.8182
13	.3791	.4780	.5549	.6429	.6978	.7912
14	.3626	.4593	.5341	.6220	.6747	.7670
15	.3500	.4429	.5179	.6000	.6536	.7464
16	.3382	.4265	.5000	.5824	.6324	.7265
17	.3260	.4118	.4853	.5637	.6152	.7083
18	.3148	.3994	.4716	.5480	.5975	.6904
19	.3070	.3895	.4579	.5333	.5825	.6737
20	.2977	.3789	.4451	.5203	.5684	.6586
21	.2909	.3688	.4351	.5078	.5545	.6455
22	.2829	.3597	.4241	.4963	.5426	.6318
23	.2767	.3518	.4150	.4852	.5306	.6186
24	.2704	.3435	.4061	.4748	.5200	.6070
25	.2646	.3362	.3977	.4654	.5100	.5962
26	.2588	.3299	.3894	.4564	.5002	.5856
27	.2540	.3236	.3822	.4481	.4915	.5757
28	.2490	.3175	.3749	.4401	.4828	.5660
29	.2443	.3113	.3685	.4320	.4744	.5567
30	.2400	.3059	.3620	.4251	.4665	.5479

Table entries are critical values ρ_α such that the $P(\rho \geq \rho_\alpha) = \alpha$.
If $n > 30$, the approximate critical values ρ_α may be obtained from

$$\rho_\alpha = \frac{z_\alpha}{\sqrt{n-1}}$$

where z_α is the standard normal variate such that a proportion α of the area is to the right of z_α. The lower percentiles may be obtained from the equation,

$$\rho_\alpha = -\rho_{1-\alpha}, \qquad \rho < 0.50$$

Source: Adapted from G. J. Glasser and R. F. Winter, "Critical Values of the Coefficient of Rank Correlation for Testing the Hypothesis of Independence," *Biometrika,* vol. 48 (1961), pp. 444–48.

TABLE B.15 Critical values of the Kolmogorov test statistic: Two-sided test

	α = .20	.10	.05	.02	.01		α = .20	.10	.05	.02	.01
n = 1	.900	.950	.975	.990	.995	n = 21	.226	.259	.287	.321	.344
2	.684	.776	.842	.900	.929	22	.221	.253	.281	.314	.337
3	.565	.636	.708	.785	.829	23	.216	.247	.275	.307	.330
4	.493	.565	.624	.689	.734	24	.212	.242	.269	.301	.323
5	.447	.509	.563	.627	.669	25	.208	.238	.264	.295	.317
6	.410	.468	.519	.577	.617	26	.204	.233	.259	.290	.311
7	.381	.436	.483	.538	.576	27	.200	.229	.254	.284	.305
8	.358	.410	.454	.507	.542	28	.197	.225	.250	.279	.300
9	.339	.387	.430	.480	.513	29	.193	.221	.246	.275	.295
10	.323	.369	.409	.457	.489	30	.190	.218	.242	.270	.290
11	.308	.352	.391	.437	.468	31	.187	.214	.238	.266	.285
12	.296	.338	.375	.419	.449	32	.184	.211	.234	.262	.281
13	.285	.325	.361	.404	.432	33	.182	.208	.231	.258	.277
14	.275	.314	.349	.390	.418	34	.179	.205	.227	.254	.273
15	.266	.304	.338	.377	.404	35	.177	.202	.224	.251	.269
16	.258	.295	.327	.366	.392	36	.174	.199	.221	.247	.265
17	.250	.286	.318	.355	.381	37	.172	.196	.218	.244	.262
18	.244	.279	.309	.346	.371	38	.170	.194	.215	.241	.258
19	.237	.271	.301	.337	.361	39	.168	.191	.213	.238	.255
20	.232	.265	.294	.329	.352	40	.165	.189	.210	.235	.252
				Approximation for n > 40			$\dfrac{1.07}{\sqrt{n}}$	$\dfrac{1.22}{\sqrt{n}}$	$\dfrac{1.36}{\sqrt{n}}$	$\dfrac{1.52}{\sqrt{n}}$	$\dfrac{1.63}{\sqrt{n}}$

Table entries are critical values W_α such that $P(W \geq W_\alpha) = \alpha$.
Source: Adapted from L. H. Miller, "Tables of Percentage Points of Kolmogorov Statistic," *Journal of the American Statistical Association*, vol. 51 (1956), pp. 111–21.

TABLE B.16 Critical values of the Lilliefors test statistic

	α = .20	.15	.10	.05	.01
Sample size n = 4	.300	.319	.352	.381	.417
5	.285	.299	.315	.337	.405
6	.265	.277	.294	.319	.364
7	.247	.258	.276	.300	.348
8	.233	.244	.261	.285	.331
9	.223	.233	.249	.271	.311
10	.215	.224	.239	.258	.294
11	.206	.217	.230	.249	.284
12	.199	.212	.223	.242	.275
13	.190	.202	.214	.234	.268
14	.183	.194	.207	.227	.261
15	.177	.187	.201	.220	.257
16	.173	.182	.195	.213	.250
17	.169	.177	.189	.206	.245
18	.166	.173	.184	.200	.239
19	.163	.169	.179	.195	.235
20	.160	.166	.174	.190	.231
25	.149	.153	.165	.180	.203
30	.131	.136	.144	.161	.187
Over 30	$\dfrac{.736}{\sqrt{n}}$	$\dfrac{.768}{\sqrt{n}}$	$\dfrac{.805}{\sqrt{n}}$	$\dfrac{.886}{\sqrt{n}}$	$\dfrac{1.031}{\sqrt{n}}$

Table entries are critical values W_α such that $P(W \geq W_\alpha) = \alpha$.
Source: Adapted from H. W. Lilliefors, "On the Kolmogorov-Smirnov Test for Normality with Mean and Variance Unknown," *Journal of the American Statistical Association*, vol. 62 (1967), pp. 399–402.

TABLE B.17 Critical values of the runs statistic

		W_α					$W_{1-\alpha}$				
N_1	N_2	$W_{.005}$	$W_{.01}$	$W_{.025}$	$W_{.05}$	$W_{.10}$	$W_{.90}$	$W_{.95}$	$W_{.975}$	$W_{.99}$	$W_{.995}$
2	5	—	—	—	—	3	—	—	—	—	—
	8	—	—	—	3	3	—	—	—	—	—
	11	—	—	—	3	3	—	—	—	—	—
	14	—	—	3	3	3	—	—	—	—	—
	17	—	—	3	3	3	—	—	—	—	—
	20	—	3	3	3	4	—	—	—	—	—
5	5	—	3	3	4	4	8	8	9	9	—
	8	3	3	4	4	5	9	10	10	—	—
	11	4	4	5	5	6	10	—	—	—	—
	14	4	4	5	6	6	—	—	—	—	—
	17	4	5	5	6	7	—	—	—	—	—
	20	5	5	6	6	7	—	—	—	—	—
8	8	4	5	5	6	6	12	12	13	13	14
	11	5	6	6	7	8	13	14	14	15	15
	14	6	6	7	8	8	14	15	15	16	16
	17	6	7	8	8	9	15	15	16	—	—
	20	7	7	8	9	10	15	16	16	—	—
11	11	6	7	8	8	9	15	16	16	17	18
	14	7	8	9	9	10	16	17	18	19	19
	17	8	9	10	10	11	17	18	19	20	21
	20	9	9	10	11	12	18	19	20	21	21
14	14	8	9	10	11	12	18	19	20	21	22
	17	9	10	11	12	13	20	21	22	23	23
	20	10	11	12	13	14	21	22	23	24	24
17	17	11	11	12	13	14	22	23	24	25	25
	20	12	12	14	14	16	23	24	25	26	27
20	20	13	14	15	16	17	25	26	27	28	29

Table entries are critical values W_p such that $P(W \le W_p) = p$, $p = \alpha$ if $\alpha \le 0.10$ or $p = 1 - \alpha$ if $\alpha \ge 0.90$.

For n or m greater than 20, the critical value $W_p(p = \alpha$ or $p = 1 - \alpha)$ may be approximated by:

$$W_p = \frac{2nm}{n + m} + 1 + z_p\sqrt{\frac{2nm(2nm - n - m)}{(n + m)^2(n + m + 1)}}$$

where z_p is the standard normal variate such that a proportion p of the area is to the right of z_p.

To use the table: Let N_1 be the smaller sample size and N_2 the larger. If the exact values of N_1 and N_2 are not listed, use the nearest values given as an approximation. Reject H_0 if T is less than W_α (or greater than $W_{1-\alpha}$) for the one-tailed test at the significance level. For the two-tailed test, reject H_0 if either $T > W_{1-\alpha/2}$ or $T < W_{\alpha/2}$ at the α significance level.

Source: Adapted from F. S. Swed and C. Eisenhart, "Tables for Testing Randomness of Grouping in a Sequence of Alternatives," *The Annals of Mathematical Statistics*, vol. 14 (1943), pp. 66–87.

TABLE B.18 Relationship between z and r

z	.00	.01	.02	.03	.04	.05	.06	.07	.08	.09
.0	.0000	.0100	.0200	.0300	.0400	.0500	.0599	.0699	.0798	.0898
.1	.0997	.1096	.1194	.1293	.1391	.1489	.1587	.1684	.1781	.1878
.2	.1974	.2070	.2165	.2260	.2355	.2449	.2543	.2636	.2729	.2821
.3	.2913	.3004	.3095	.3185	.3275	.3364	.3452	.3540	.3627	.3714
.4	.3800	.3885	.3969	.4053	.4136	.4219	.4301	.4382	.4462	.4542
.5	.4621	.4700	.4777	.4854	.4930	.5005	.5080	.5154	.5227	.5299
.6	.5370	.5441	.5511	.5581	.5649	.5717	.5784	.5850	.5915	.5980
.7	.6044	.6107	.6169	.6231	.6291	.6352	.6411	.6469	.6527	.6584
.8	.6640	.6696	.6751	.6805	.6858	.6911	.6963	.7014	.7064	.7114
.9	.7163	.7211	.7259	.7306	.7352	.7398	.7443	.7487	.7531	.7574
1.0	.7616	.7658	.7699	.7739	.7779	.7818	.7857	.7895	.7932	.7969
1.1	.8005	.8041	.8076	.8110	.8144	.8178	.8210	.8243	.8275	.8306
1.2	.8337	.8367	.8397	.8426	.8455	.8483	.8511	.8538	.8565	.8591
1.3	.8617	.8643	.8668	.8693	.8717	.8741	.8764	.8787	.8810	.8832
1.4	.8854	.8875	.8896	.8917	.8937	.8957	.8977	.8996	.9015	.9033
1.5	.9052	.9069	.9087	.9104	.9121	.9138	.9154	.9170	.9186	.9202
1.6	.9217	.9232	.9246	.9261	.9275	.9289	.9302	.9316	.9329	.9342
1.7	.9354	.9367	.9379	.9391	.9402	.9414	.9425	.9436	.9447	.9458
1.8	.9468	.9478	.9498	.9488	.9508	.9518	.9527	.9536	.9545	.9554
1.9	.9562	.9571	.9579	.9587	.9595	.9603	.9611	.9619	.9626	.9633
2.0	.9640	.9647	.9654	.9661	.9668	.9674	.9680	.9687	.9693	.9699
2.1	.9705	.9710	.9716	.9722	.9727	.9732	.9738	.9743	.9748	.9753
2.2	.9757	.9762	.9767	.9771	.9776	.9780	.9785	.9789	.9793	.9797
2.3	.9801	.9805	.9809	.9812	.9816	.9820	.9823	.9827	.9830	.9834
2.4	.9837	.9840	.9843	.9846	.9849	.9852	.9855	.9858	.9861	.9863
2.5	.9866	.9869	.9871	.9874	.9876	.9879	.9881	.9884	.9886	.9888
2.6	.9890	.9892	.9895	.9897	.9899	.9901	.9903	.9905	.9906	.9908
2.7	.9910	.9912	.9914	.9915	.9917	.9919	.9920	.9922	.9923	.9925
2.8	.9926	.9928	.9929	.9931	.9932	.9933	.9935	.9936	.9937	.9938
2.9	.9940	.9941	.9942	.9943	.9944	.9945	.9946	.9947	.9949	.9950
3.0	.9951									
4.0	.9993									
5.0	.9999									

The z-values appear in the scales at the left and above the table; the r-values appear in the body of the table.

Source: Huntsberger, Bellingsley, and Croft, *Statistical Inference for Management and Economics* (Boston: Allyn & Bacon, Inc., 1975©).

TABLE B.19 Random digits

Line	(1)–(5)	(6)–(10)	(11)–(15)	(16)–(20)	(21)–(25)	(26)–(30)	(31)–(35)
101	13284	16834	74151	92027	24670	36665	00770
102	21224	00370	30420	03883	94648	89428	41583
103	99052	47887	81085	64933	66279	80432	65793
104	00199	50993	98603	38452	87890	94624	69721
105	60578	06483	28733	37867	07936	98710	98539
106	91240	18312	17441	01929	18163	69201	31211
107	97458	14229	12063	59611	32249	90466	33216
108	35249	38646	34475	72417	60514	69257	12489
109	38980	46600	11759	11900	46743	27860	77940
110	10750	52745	38749	87365	58959	53731	89295
111	36247	27850	73958	20673	37800	63835	71051
112	70994	66986	99744	72438	01174	42159	11392
113	99638	94702	11463	18148	81386	80431	90628
114	72055	15774	43857	99805	10419	76939	25993
115	24038	65541	85788	55835	38835	59399	13790
116	74976	14631	35908	28221	39470	91548	12854
117	35553	71628	70189	26436	63407	91178	90348
118	35676	12797	51434	82976	42010	26344	92920
119	74815	67523	72985	23183	02446	63594	98924
120	45246	88048	65173	50989	91060	89894	36036
121	76509	47069	86378	41797	11910	49672	88575
122	19689	90332	04315	21358	97248	11188	39062
123	42751	35318	97513	61537	54955	08159	00337
124	11946	22681	45045	13964	57517	59419	58045
125	96518	48688	20996	11090	48396	57177	83867
126	35726	58643	76869	84622	39098	36083	72505
127	39737	42750	48968	70536	84864	64952	38404
128	97025	66492	56177	04049	80312	48028	26408
129	62814	08075	09788	56350	76787	51591	54509
130	25578	22950	15227	83291	41737	59599	96191
131	68763	69576	88991	49662	46704	63362	56625
132	17900	00813	64361	60725	88974	61005	99709
133	71944	60227	63551	71109	05624	43836	58254
134	54684	93691	85132	64399	29182	44324	14491
135	25946	27623	11258	65204	52832	50880	22273
136	01353	39318	44961	44972	91766	90262	56073
137	99083	88191	27662	99113	57174	35571	99884
138	52021	45406	37945	75234	24327	86978	22644
139	78755	47744	43776	83098	03225	14281	83637
140	25282	69106	59180	16257	22810	43609	12224
141	11959	94202	02743	86847	79725	51811	12998
142	11644	13792	98190	01424	30078	28197	55583
143	06307	97912	68110	59812	95448	43244	31262
144	76285	75714	89585	99296	52640	46518	55486
145	55322	07598	39600	60866	63007	20007	66819
146	78017	90928	90220	92503	83375	26986	74399
147	44768	43342	20696	26331	43140	69744	82928
148	25100	19336	14605	86603	51680	97678	24261
149	83612	46623	62876	85197	07824	91392	58317
150	41347	81666	82961	60413	71020	83658	02415

Source: *Table of 105,000 Random Decimal Digits,* Interstate Commerce Commission, Bureau of Transport Economics and Statistics, 1949.

APPENDIX C
Answers to selected problems

1.2. *Schedule*—a questionnaire filled out by an interviewer.
Questionnaire—a printed form containing a set of questions.
A schedule contains directions for the interviewer, whereas a questionnaire contains directions for the respondent. A schedule is used in a personal interview, while a questionnaire is used in a survey or in an experiment.

1.4. *Primary data*—data obtained from the organization that collected them.
Secondary data—data obtained from a source other than the organization that collected them.
Primary data are usually more reliable. The more times the data are presented, the greater the chance for errors (typographical errors, misinterpretations, and so forth). In addition, the primary source will often clearly state the nature of the survey or experiment that produced the data, while the secondary sources typically will not.

1.6. *Design*—selecting the kinds of questions to be asked (dichotomous, multiple choice, or free answer), the order of the questions, the degree of directness, the length of the questionnaire or schedule, and so forth.
Pretest—testing the questionnaire or schedule on a small number of respondents to determine its readiness for distribution.
Editing—checking the completed questionnaire or schedule for errors.

1.10. *a.* Qualitative; *b.* quantitative; *c.* qualitative; *d.* quantitative.

1.12. *a.* If the manufacturer and supplier both selected random samples fairly, then the difference between the two sample results could be ascribed to sampling error.
b. Yes—by arguing that the manufacturer's sample was not representative of the population; that is, the sample was subject to sampling error.

1.14. *a.* The target population is the set of *potential* customers of the drugstore.
b. Yes, the respondents are those who *currently* use the drugstore, and this sample excludes those who have never used the store or who used it at one time but are presently not customers.
c. By household interview or mail survey of residents located within the market area of the drugstore.

1.16. *a.* Census; *b.* sample; *c.* sample (although computerization of tax returns may make a census possible).

1.18. *a.* It is a leading question—it asks for agreement. "The amount of money spent on national defense is (*a*) about right, (*b*) too much, (*c*) too little."
b. Few people will be able to recall the number over an entire year. The length of the period should be shortened to one month or two weeks.
c. This statement may "place" the name in some people's minds. "List the five (or so) brand names that come to mind when hi-fi equipment is mentioned."
d. It is a leading question. "The money allocated to send men into space is (*a*) well spent, (*b*) not well spent."

1.20. *a.* An experiment—the factors that affect gas mileage (precision of recording equipment, type of tires, type of drivers, and so forth) need to be closely controlled.

b. A survey—there is no reason to control factors that may affect the number of sick days taken per month.

c. A survey—again, we are not interested in controlling factors that may affect the interest rate.

1.22. a. The residents of Seal Point and, perhaps, surrounding communities.

b. Residents with telephones, a possible source of bias.

c. 183 yes votes, 307 no votes, and 24 no-opinion votes.

d. Ratio—120 votes is "twice" as many as 60.

e. 183, 307, 24.

1.24. Any examples are suitable that use descriptive statistics for describing data sets and inferential statistics to induce a result about a population from a sample.

1.26. a. Nothing is said about the percentage of students who contracted the flu but did not receive the shots. It is a faulty conclusion.

b. Not necessarily. They may be misclassified or incorrectly recorded, or the numbers or answers may later be subject to nonsampling error, such as mistabulation, keypunching errors, and so on.

c. Yes, this is generally the case.

Chapter 2

2.2.

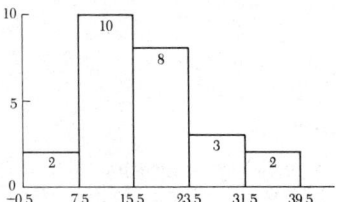

2.4. Mean = 1.3; median = 3; mode = 0, 4 (bimodal); range = 20; variance = 34.41.

2.6. $\mu = 1.3$, $\sigma = \sqrt{34.41} = 5.87$
$\mu \pm \sigma$: -4.57 to 7.17, $(7/10)(100) = 70$ percent
$\mu \pm 2\sigma$: -10.44 to 13.04, $(10/10)(100) = 100$ percent
$\mu \pm 3\sigma$: -16.31 to 18.91, $(10/10)(100) = 100$ percent
Chebychef: $\mu \pm \sigma$, at least 0 percent; $\mu \pm 2\sigma$, at least 75 percent; $\mu \pm 3\sigma$, at least 89 percent.

2.8. $\gamma = 3(1.3 - 3)/5.87 = -0.87$; $\gamma < 0$, thus, the distribution of numbers is skewed left.

2.10. $\mu \pm (3/4)\sigma$: Chebychef does not apply (k must be greater than 1).
$\mu \pm (5/2)\sigma$: at least $^{21}/_{25}$.
$\mu \pm 4\sigma$: at least $^{15}/_{16}$.

2.12. Arithmetic mean = 234; geometric mean = 54.93; median = 50; mode = 10.
Either the geometric mean or the median is probably most representative, but the determination of which measure is most representative depends on the nature of the process that produced the data.

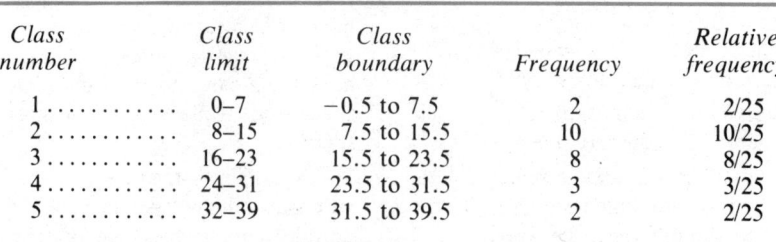

Class number	Class limit	Class boundary	Frequency	Relative frequency
1	0–7	−0.5 to 7.5	2	2/25
2	8–15	7.5 to 15.5	10	10/25
3	16–23	15.5 to 23.5	8	8/25
4	24–31	23.5 to 31.5	3	3/25
5	32–39	31.5 to 39.5	2	2/25

Net return	Tally	Frequency	Cumulative relative frequency
3.50–3.99.........	II	2	2/60 = 0.033
4.00–4.49.........	JHf I	6	8/60 = 0.133
4.50–4.99.........	JHf IIII	9	17/60 = 0.283
5.00–5.49.........	JHf IIII	9	26/60 = 0.433
5.50–5.99.........	JHf JHf JHf III	18	44/60 = 0.733
6.00–6.49.........	JHf JHf JHf	15	59/60 = 0.983
6.50–6.99.........	I	1	60/60 = 1.000
		60	

2.14. When the distribution is moderately skewed. In this case, the center of gravity will be "pulled" toward the skewed end of the distribution and will therefore, "favor" those observations in the tail of the distribution.

2.16. *a.* See table at top of page.

b.

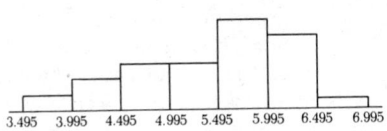

c.

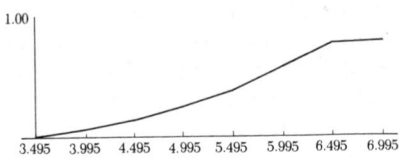

2.18. $\mu \pm 2\sigma$: at least 75 percent; -69.21 to 57.21; $9/10$.

$\mu \pm 3\sigma$: at least 89 percent; -100.82 to 88.82; $10/10$.

2.20. $\mu = \$20,000$; $\sigma = \$2,000$; $0.84 = [1 - (1/k^2)]$, $k = 2.5$.
$\mu \pm 2.5\sigma \rightarrow \$15,000$ to $\$25,000$.

2.22.

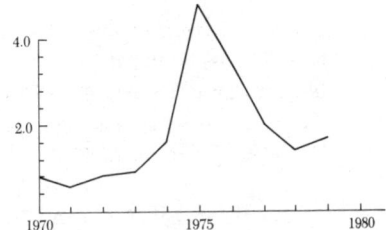

2.24. *a.* mean = 7.364; standard deviation = 0.233; median = 7.3; mode = 7.2.

b. See table at bottom of page.

c.

$$\mu \doteq \frac{\Sigma f_i m_i}{\Sigma f_i} = 185.775/25 = 7.431$$

Class	m_i Class mark	Tally	f_i Frequency	$f_i m_i$	$f_i m_i^2$
7.00–7.19.........	7.095	IIII	4	28.380	201.356
7.20–7.39.........	7.295	JHf IIII	9	65.655	478.953
7.40–7.59.........	7.495	JHf	5	37.475	280.875
7.60–7.79.........	7.695	JHf	5	38.475	296.065
7.80–7.99.........	7.895	II	2	15.790	124.662
			25	185.775	1,381.911

$$\sigma \doteq \sqrt{\frac{\Sigma f_i m_i^2 - \frac{(\Sigma f_i m_i)^2}{n}}{n}}$$

$$= \sqrt{\frac{1{,}381.911 - \frac{(185.775)^2}{25}}{25}} = 0.238$$

2.26. *a.* $\mu = 18{,}642;\quad \sigma = 3{,}613;\quad M = 16{,}021;$

$$\gamma = \frac{3(\mu - M)}{\sigma} =$$
$$\frac{3(18{,}642 - 16{,}021)}{3613} = 2.176$$

skewed right

b. $\mu \pm 2\sigma$ contains at least 75 percent → 11,416 to 25,868.
$\mu \pm 3\sigma$ contains at least 89 percent → 7,803 to 29,481.

Chapter 3

3.2. Let M_1 = younger man, M_2 = older man, F_1 = younger woman, F_2 = older woman. (See table below.)

3.4. Event A: probability = $^3/_{12}$; event B: $^6/_{12}$; event C: $^6/_{12}$; event D: $^2/_{12}$: event E: $^8/_{12}$.

3.6. *a.* Sample space: $3! = 6$ equally likely points: $E_1(1, 2, 3)$; $E_2(1, 3, 2)$; $E_3(2, 1, 3)$; $E_4(2, 3, 1)$; $E_5(3, 1, 2)$; $E_6(3, 2, 1)$.

b. A: E_1, E_2 $P(A) = P(E_1) + P(E_2) = ^1/_6 + ^1/_6 = ^1/_3$.

c. B: E_1, E_2, E_5, E_6 $P(B) = ^4/_6$.

d. C: E_1, E_2 $P(C) = ^2/_6$.

3.8. Product rule: $(26)(26)(26)(10)(10)(10) = 17{,}576{,}000$.

3.10. *a.* Product rule: $(6)(5)(4)(3)(2)(1) = 720$ (or $P_6^6 = 720$).

b. Product rule: $(A)(5)(4)(3)(2)(1) = 120$.

3.12. *a.* Product rule: $(5)(5)(5)(5) = 625$.

b. $(3)(5)(5)(5) = 375$.

c. $(1)(1)(1)(5) = 5$, if he knows in which two courses he will receive an A.

3.14. $C_7^{10} = C_3^{10} = 120$.

3.16. *a.* Subjective.

b. Objective or subjective.

c. Objective (if all have been tested).

d. Subjective.

e. Subjective.

3.18. *a.* There are m groups of objects. We select one object from the first group of k_1 distinguishable objects, one object from the second group of k_2 distinguishable objects and so on. We wish to know how many sets of m objects may be formed in this manner.

b. We select r objects from n distinguishable objects and wish to know how many different *groups* of r objects may be selected.

c. We select r objects from n distin-

a. Simple events	Chairman	Secretary
E_1	M_1	M_2
E_2	M_1	F_1
E_3	M_1	F_2
E_4	M_2	M_1
E_5	M_2	F_1
E_6	M_2	F_2
E_7	F_1	M_1
E_8	F_1	M_2
E_9	F_1	F_2
E_{10}..........	F_2	M_1
E_{11}..........	F_2	M_2
E_{12}..........	F_2	F_1

b. A: E_1, E_2, E_3
c. B: $E_1, E_2, E_3, E_4, E_5, E_6$
d. C: $E_2, E_3, E_5, E_6, E_9, E_{12}$
e. D: E_2, E_3
f. E: $E_1, E_2, E_3, E_4, E_5, E_6, E_9, E_{12}$

guishable objects and wish to know how many different groups and orderings of r objects there are.

d. We select r_1 objects from n_1 of the first type and r_2 objects from n_2 of the second type and wish to know how many groups of the $r = r_1 + r_2$ objects are possible.

3.20. *a.* Sample space: $3! = 6$ outcomes.
e_1: ABC e_3: BAC e_5: CAB
e_2: ACB e_4: BCA e_6: CBA
b. $P(e_i) = \frac{1}{6}$, where $i = 1, 2, \ldots 6$.
$P(A \text{ or } B \text{ first}) = P(e_1) + P(e_2) + P(e_3) + P(e_4) = \frac{4}{6}$.

3.22. Hypergeometric: $C_3^6 C_2^4 = \dfrac{6!}{3!3!} \dfrac{4!}{2!2!}$
$= (20)(6) = 120.$

3.24. *a.* $P_6^6 = 6! = 720;$

$\boxed{6}\boxed{5}\boxed{4}\boxed{3}\boxed{2}\boxed{1} = 720.$

b. $\boxed{}\boxed{}\ \boxed{}\boxed{}\ \boxed{}\boxed{}$
  $\;\;C_1\quad\; C_2\quad\; C_3$
$3! = 6$ orderings of couples;
within couples, MF or FM,

$\boxed{2}\boxed{2}\boxed{2} \times 6$ couple orderings $=$
$2^3 \cdot 6 = 48.$

3.26. Total points in sample space: $P_5^5 = 5!$
$= 120.$
Total points in event space:

$\boxtimes\boxtimes\boxtimes\boxed{2}\boxed{1} = 2.$

Probability (event) $= \frac{2}{120} = \frac{1}{60}.$

Chapter 4

4.2. *a.* $\frac{10}{50}.$
b. $\frac{3}{13}.$
c. $(0 + 2)/(3 + 13) = \frac{2}{16}.$

d. $(2 + 8 + 10)/(3 + 13 + 15) = \frac{20}{31}.$
e. $(30 + 15 - 10)/50 = \frac{35}{50}.$
f. $P(\text{office worker/plant worker}) =$
  $0;$
  $P(\text{office worker}) = \frac{15}{50}; 0 \neq \frac{15}{50}:$
  no.
g. $P(\text{office worker/favors change}) =$
  $\frac{10}{30};$
  $P(\text{office worker}) = \frac{15}{50}; \frac{10}{30} \neq$
  $\frac{15}{50}; no.$
h. $P(\text{executive} \cap \text{plant worker}) = 0;$
  yes.
i. $P(\text{executive} \cap \text{no opinion}) = 0;$
  yes.

4.4. *a.* $P(B_3) = \frac{30}{200} = 0.15.$
b. $P(A_3 \cap B_2) = \frac{15}{200} = 0.075.$
c. $P[(A_1 \cup A_2)/B_3] = (10 + 10)/30 =$
  $0.6667.$
d. $P[(A_1 \cup A_3)/B_2] = (15 + 15)/70 =$
  $0.4286.$
e. $P[A_1/(B_1 \cup B_2)] = (25 + 15)/(100$
  $+ 70) = \frac{40}{170} = 0.2353.$
f. $P(A_1/B_3) = \frac{10}{30} = 0.3333.$
g. $P(A_1/B_1) = \frac{25}{100} = 0.25; P(A_1) =$
  $\frac{50}{200} = 0.25; 0.25 = 0.25; yes.$

4.6.

	B_1	B_2	
A_1	0.25	0.25	0.50
A_2	0.15	0.35	0.50
	0.40	0.60	1.00

a. $(A_1, A_2), (B_1, B_2).$
b. $(A_1, A_2), (A_1, B_1), (A_1, B_2), (A_2, B_1), (A_2, B_2), (B_1, B_2).$
c. $P(B_2/A_1) = 0.25/0.50 = 0.50.$
d. $P(A_1 \cap B_2) + P(A_2 \cap B_1) = 0.25 + 0.15 = 0.40.$

4.8. *a.* *Assuming* that each candidate is taking the exam for the first time,

1	2	3	
P	P	P	$(0.8)^3 = 0.512$
P	P	$\bar{P}$	$(0.8)^2(0.2) = 0.128$
P	$\bar{P}$	P	0.128
$\bar{P}$	P	P	0.128
$\bar{P}$	$\bar{P}$	P	$(0.8)(0.2)^2 = 0.032$
$\bar{P}$	P	$\bar{P}$	0.032
P	$\bar{P}$	$\bar{P}$	0.032
$\bar{P}$	$\bar{P}$	$\bar{P}$	$(0.2)^3 = 0.008$

P(two pass) $= 0.384$

P(at least one passes)
$= 1 - 0.008 = 0.992$

b. $P(\bar{P}_1 \cap \bar{P}_2) = P(\bar{P}_1)P(\bar{P}_2/\bar{P}_1) =$
$(0.20)(0.10) = 0.02.$

c. $P(\text{pass}) = P(P_1) + P(P_2/\bar{P}_1)P(\bar{P}_1) =$
$0.80 + (0.90)(0.20) = 0.98.$

4.10.

Hats (outcomes) \ Men	1	2	3
1	1	2	3
2	1	3	2
3	2	1	3
4	3	1	2
5	3	2	1
6	2	3	1

Number of points in sample space $= P_3^3 = 3!$
$= 6.$

a. P(no right hats) $= {}^2/_6$ (outcomes 4 and 6).
b. P(exactly one gets right hat) $= {}^3/_6$ (outcomes 2, 3, and 5)
c. P(exactly two get right hat) $= 0$ (impossible—if two get right hats, all three must get right hats)

4.12. Let $S_i =$ plane shot down at ith battery.

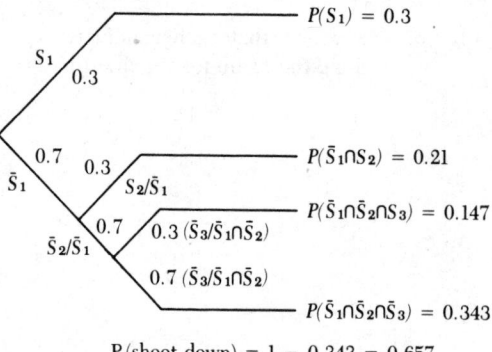

$P(S_1) = 0.3$

$P(\bar{S}_1 \cap S_2) = 0.21$

$P(\bar{S}_1 \cap \bar{S}_2 \cap S_3) = 0.147$

$P(\bar{S}_1 \cap \bar{S}_2 \cap \bar{S}_3) = 0.343$

P(shoot down) $= 1 - 0.343 = 0.657.$

4.14. Let $D_i = i$th item is defective, where $i = 1, 2, 3.$
a. $P(D_1 \cap D_2 \cap D_3) =$
$P(D_1)P(D_2)P(D_3) = (0.10)^3 = 0.001.$
b. P(at least one defective) $= 1 - P$(none defective) $= 1 - (0.90)^3 = 0.271$

4.16. *a.* P(tranquilizer pills) $= {}^1/_3.$
b.

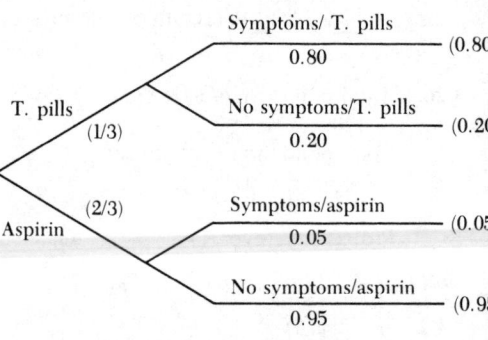

$$P(\text{T. pills/symptoms}) = \frac{P(\text{T. pills/symptoms})}{P(\text{symptoms})}$$
$$= \frac{(0.80)({}^1/_3)}{(0.80)({}^1/_3) + (0.05)({}^2/_3)} = \frac{0.80}{0.80 + .10} = \frac{8}{9}.$$
c. P(aspirin/symptoms) $= 1 - {}^8/_9 = {}^1/_9.$

4.18.

$$P(A_1 \cap A_2) = \frac{90}{2,450} = 0.037.$$

$$A_2/A_1 \quad A_1 \cap A_2 : \left(\frac{10}{50}\right)\left(\frac{9}{49}\right) = \frac{90}{2,450}$$

$$A_1 \quad 9/49$$
$$10/50 \quad 40/49$$

$$\bar{A}_2/A_1 \quad A_1 \cap \bar{A}_2 : \left(\frac{10}{50}\right)\left(\frac{40}{49}\right) = \frac{400}{2,450}$$

$$40/50 \quad A_2/\bar{A}_1 \quad \bar{A}_1 \cap A_2 : \left(\frac{40}{50}\right)\left(\frac{10}{49}\right) = \frac{400}{2,450}$$

$$\bar{A}_1 \quad 10/49$$

$$39/49$$
$$\bar{A}_2/\bar{A}_1 \quad \bar{A}_1 \cap \bar{A}_2 : \left(\frac{40}{50}\right)\left(\frac{39}{49}\right) = \frac{1,560}{2,450}$$
$$1.000$$

$$P(A_1 \cap A_2) = \frac{90}{2,450} = 0.037.$$

4.20.

$$D/A \; 0.01 \quad P(A \cap D) = (0.2)(0.01) = 0.002$$

$$\bar{D}/A \; 0.99 \quad P(A \cap \bar{D}) = (0.2)(0.99) = 0.198$$

$$A \; 0.20$$

$$B \; 0.50 \quad D/B \; 0.02 \quad P(B \cap D) = (0.5)(0.02) = 0.010$$

$$\bar{D}/B \; 0.98 \quad P(B \cap \bar{D}) = (0.5)(0.98) = 0.490$$

$$C \; 0.30 \quad D/C \; 0.03 \quad P(C \cap D) = (0.3)(0.03) = 0.009$$

$$\bar{D}/C \; 0.97 \quad P(C \cap \bar{D}) = (0.3)(0.97) = \underline{0.291}$$
$$1.000$$

$$P(B/D) = \frac{P(B \cap D)}{P(D)} = \frac{0.010}{0.002 + 0.010 + 0.009} = \frac{10}{21} = 0.476.$$

Note: Bayes' theorem could have been used.

4.22. No, if $P(B) = 0$, event B cannot occur. It makes no sense to condition an event (A) on an event that cannot occur.

4.24.

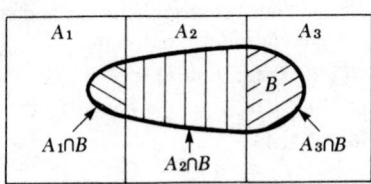

$P(B/A_i)P(A_i) = P(A_i \cap B)$ (multiplicative law). Thus, $P(B) = P(A_1 \cap B) + P(A_2 \cap B) + P(A_3 \cap B)$.

4.26. $P(A \cup B) = P(A) + P(B) - P(A \cap B)$.
$P(A \cup B)$ will be less when $P(A \cap B) > 0$.

4.28. $P(A) = 1/3; P(B) = 3/4; P(A \cap B) = 1/6$.
a. $P(A \cup B) = P(A) + P(B) - P(A \cap B) = 1/3 + 3/4 - 1/6 = 4/12 + 9/12 - 2/12 = 11/12$.
b. $P(B/A) = \dfrac{P(A \cap B)}{P(A)} = \dfrac{1/6}{1/3} = 1/2$.
c. $P(B/A) = 1/2; P(B) = 3/4$; since $P(B/A) \neq P(B)$, no.

4.30. a. $(1/2)^5 = 1/32$.
b. $1/2$.

4.32. a. $P(\text{no pages}) = (0.99)^6 = 0.94148$.

P(at least one page) $= 1 - P$(no pages) $= 1 - 0.94148 = 0.0585$.

b. P(three pages) $= C_3^6(0.01)^3(0.99)^3 = 0.000019406$.

Chapter 5

5.4. The PMF gives the probability that the random variable assumes a value x; the CMF gives the probability that the random variable assumes a value less than or equal to x.

5.6. 0.0611; 0.2894; 0.4379; 0.2116.

5.8. $F(1) = 0.30; F(2) = 0.90; F(3) = 1.00$.

5.10. x = 0, 1, 2, or 3; P(x) = 0.1667, 0.5, 0.3, or 0.0333.

5.12. 0.8; 0.64.

5.14. No—sum is not equal to 1.0.

5.16. 1.8; 0.7054.

5.18. a. 1.00; b. 0.60; c. 10; d. 11; e. 0.60; f. 15; g. 2.4.

5.20. $74.25.

5.22. $10.

5.24. a. $36.50; b. $3,650,000; c. slow production.

5.26. a. 16.05; b. 0.40; c. 0.20; d. 0.40; e. −0.12.

5.28. a. 7.60, 3.00, 4.64, 0; b. 22.80; c. 0; yes.

Chapter 6

6.2. 0.17 to 5.83; 100 percent.

6.4. $^{21}/_6$; $^{91}/_6 - (^{21}/_6)^2$.

6.6. 400.

6.8. a. successes and failures only, with independent trials; b. the population of successes and failures is divided between two subpopulations, and trials are not independent; c. independence versus nonindependence.

6.10. a. 0.2424; b. 2.

6.12. 0.0044.

6.14. 0.432.

6.16. a. 0.969; b. 0.938.

6.18. a. 0.048; b. 0.140; c. 0.031.

6.20. a. 0.098; b. 0.266; c. 2.5.

6.22. a. 1; b. 0.774; c. 0.109; d. 0.349; e. 0.736; f. 0.001; g. 0.019; h. 0.098; i. 0.677; j. 0.001.

6.24. 0.1316.

6.26. 0.128.

6.28. a. 0.608; b. 0.594; c. binomial.

6.30. a. 0.1804; b. 0.3233; c. yes.

6.32. 0.9084; Poisson distribution requirements are met.

6.34. 0.406.

Chapter 7

7.2. a. 0.3849; b. 0.3159; c. 0.4265; d. 0.4251; e. 0.3227; f. 0.0224; g. 0.9516; h. 0.5070; i. 0.0062; j. 0.7257; k. 0.1112; l. 0.9515.

7.4. a. 0.0228; b. 0.0668; c. 0.8944.

7.6. 0.0228.

7.8. 78.32 inches.

7.10. 0.8399.

7.12. $32,800.

7.14. a. $38,840; b. $37,010; c. $32,990.

7.16. a. 0.50; b. 0.7123; c. 0.4246.

7.18. 0.0228.

7.20. a. 0.030 to 0.034; b. 0.9544; c. Chebychef sets a *lower* bound.

7.22. a. 0.578; b. 0.5752.

7.24. 0.9946.

7.26. 0.6757.

7.28. 0.2877.

7.30. a. It assigns some probability to the negative portion of the real line; b. no, because the mean and standard deviation are not equal.

7.32. a. 0.22; b. 0.777; c. 0.472.

7.34. a. 0.1429; b. 0.135; c. 0.097.

7.36. 0.181.

Chapter 8

8.2. a. $\overline{X}$ is normally distributed with mean μ and standard deviation $\sigma/\sqrt{n}$.

b. $\overline{X}$ is approximately normally distributed with mean μ and standard deviation $\sigma/\sqrt{n}$ for large n.

8.4. μ = average of all population values; $\overline{X}$ = average of all sample values; $\mu_{\overline{X}}$ = average of $\overline{X}$ values from repeated samples of a fixed size $[E(\overline{X}) = \mu_{\overline{X}}]$; μ and $\overline{X}$ should be of approximately the same magnitude; $\mu_{\overline{X}} = \mu$.

8.6. Binomial distribution; the normal distribution can be used as a model if n is "large."

8.8. $\sqrt{(N - n)/(N - 1)}$; when sampling from a finite population.

8.10. *a.* 0.1056; *b.* 0.9392.

8.12. *a.* 0.1587; *b.* 0.8664.

8.14. *a.* 0.0228; *b.* no—sample size is too small for the central limit theorem to apply.

8.16. Note: Use finite population correction factor.
a. 0.5962 (without interpolation); *b.* 0.1118 (without interpolation); *c.* 0 ($\sigma_{\overline{X}} = 0$).

8.18. Using the finite population correction factor, $\sigma_P = 0.0217$; 0.0838 (without interpolation). This is an approximate answer, using normal approximation to binomial with $n = 100$ and $\pi = 0.05$.

8.24. *a.* $C_3^7 = 35$; *e.* sample mean.

Chapter 9

9.2. *Estimator*—a random variable that has a sampling distribution (e.g., $\overline{X}$ is an estimator of μ).
Estimate—a value of an estimator in a specific sample (e.g., $\bar{x}$ is an estimate of μ).

9.4. Point estimation and interval estimation.

9.6. A confidence interval provides more information than does a point estimate. It gives an interval within which the value of the population parameter

is likely to occur rather than a single (point) number guess.

9.8. *95 percent confidence*—in repeated samples of a fixed size, 95 percent of *all* samples will produce confidence intervals that contain the value of the population parameter and 5 percent of the samples will not.

9.10. Sample size is directly related to the precision of the estimation process.

9.12. If n is doubled, width decreases by a factor of $\sqrt{2}$.

9.14. $p = 0.40$; 0.40 ± 0.098.

9.16. *a.* 44.67 to 55.33.

9.18. *a.* 3.57 to 4.43; *b.* 3.55 to 4.45; *c.* interval in part *b*.

9.20. 8.92 to 11.27.

9.22. 2,401.

9.24. 148.5 to 160.1.

9.26. 16,577, using $z = 2.575$.

9.28. *a.* $p = 0.42$, $0.42 + 1.28(0.049) = 0.483$; *b.* no.

Chapter 10

10.2.

H_0: $\theta_1 = \theta_2$	H_0: $\theta_1 \geq \theta_2$	H_0: $\theta_1 \leq \theta_2$
H_A: $\theta_1 \neq \theta_2$	H_A: $\theta_1 < \theta_2$	H_A: $\theta_1 > \theta_2$

10.4. Type I error: rejecting H_0 when H_0 is true.
Type II error: not rejecting H_0 when H_0 is false.

10.6. Given a fixed level of confidence (and, therefore, significance level), there is a one-to-one correspondence between hypothesis testing and confidence intervals; if the confidence interval contains the hypothesized value, H_0 is not rejected; if the confidence interval does not contain the hypothesized value, H_0 is rejected.

10.8. Test statistic is the sample statistic whose sampling distribution is used to decide whether or not to reject the null hypothesis.

10.10. If the sample size is large, the approximate sampling distribution of a test statistic is often taken to be the normal distribution via the central limit theorem.

10.12. *a.* 0.0228; *b.* 0.0228.

10.14. *a.* 0.0124; *b.* 0.6915 (*Note:* area to left of $z = -4.5$ is negligible).

10.16. H_0: $\mu \geq 54$; H_A: $\mu < 54$; $z = -3$; p-value < 0.0013; reject H_0.

10.18. $t = (\bar{x} - \mu)/(s/\sqrt{n}) = 3.6$; $p = $ value < 0.005; reject H_0.

10.20. *a.* $z = 2.21$; p-value $= 0.9864$; do not reject H_0.

10.22. H_0: $\mu \leq 10$; H_A: $\mu > 10$; $n = 36$; $\bar{x} = 12$; $s = 2$; $t = 6.00$; p-value < 0.005; reject H_0.

10.24. *a.* 0.0526 to 0.0624; *b.* $z = -1$; p-value $= 2(0.1587) = 0.3174$.

10.26. $p = 0.13$.

10.28. At least 2963 invoices.

10.30. At least 174 calls.

Chapter 11

11.2. 35.

11.4. *a.* 0.18 to 0.40; *b.* $z = 5.13$, reject H_0, *c.* p-value much less than 0.001.

11.6. *a.* $\sigma_1^2 = \sigma_2^2$; each sample is drawn from a normal distribution.
b. The differences, $d_i = x_{1i} - x_{2i}$, are normally distributed.

c. The sample sizes are sufficiently large for the central limit theorem to apply to the difference between the sample proportions.
d. Each sample is drawn from a normal population.

11.8. *a.* Paired; two observations on the same salesperson.
b. $t = 1.76$; approximate p-value $= 2(0.05) = 0.10$; do not reject H_0.

11.10. More efficient.

11.12. *a.* -19.50 to -10.50 ($t = 2.013$ by interpolation); yes.
b. $f = 4$; $F(1 - \alpha/2; n_1 - 1, n_2 - 1) = F(0.95; 24,24) = 1.98$; reject H_0 since $f = 4 > 1.98$; p-value < 0.001.

11.14. $f = 2.78$; by using the $F(1 - \alpha; 24,20)$ table (actually, $r_1 = 23, r_2 = 19$), $0.01 < p$-value < 0.025.

11.16. H_0: $\mu_A \geq \mu_B$; H_A: $\mu_A < \mu_B$; $t = -4.24$; $df = 59$; p-value < 0.005; reject H_0; small chance a Type I error will be committed by rejecting H_0.

11.18. H_0: $\pi_{NE} \leq \pi_S$; H_A: $\pi_{NE} > \pi_S$; $p_1 = 0.667$; $p_2 = 0.500$; $z = 2.93$; p-value $= 0.0017$; reject H_0; very small chance of committing a Type I error by rejecting H_0.

11.20. -38.44 to -21.56, assuming $\sigma_1^2 = \sigma_2^2$.

11.22. -0.54 to -0.16.

Chapter 12

12.2. Testing the hypothesis of equal population means in t populations; identifying the sources of variation in a response (dependent) variable.

12.4. In the randomized block design, the blocking factor is considered to be fixed and a "treatment" is one level of the single factor. In the two-factor experiment, the experimental "treatments" are all combinations of the levels of both factors.

12.6. Explore differences among the treatment means.

12.8. *a.* SST = 4374.95; SSTR = 2,668.05; $f = 28.14$; reject H_0 at $\alpha = 0.10$ significance level; p-value < 0.001.
b. $S_p = 9.74$; $t = -5.30$; reject H_0.
c. $t^2 = f$.

12.10. SSTR = 50,138.3; SSE = 23,661.7; $f = 9.54$; reject H_0 at $\alpha = 0.05$ significance level; $0.01 < p$-value < 0.005.

12.12. SSTR = 1.20; SSE = 0.60; f = 12.5; reject H_0 at α = 0.05 significance level; p-value < 0.001.

12.14. $\hat{\sigma}_L^2 = \text{MSE} \sum_{i=1}^{t} \frac{C_i^2}{b} = (0.67) \left[\frac{(1)^2}{3} + \frac{(-1)^2}{3} + \frac{(0)^2}{3} + \frac{(0)}{3} \right]$ (for $\mu_1 - \mu_2$)

$$= (0.67)^2; \hat{\sigma}_L = 0.67$$

$s^2 = (t - 1) F[(1 - \alpha); (t - 1), (t - 1)(b - 1)] = (3) F(0.95; 3, 6) = 14.28.$
$s = 3.78; \bar{x}_1 = 17, \bar{x}_2 = 21, \bar{x}_3 = 18, \bar{x}_4 = 15.$

$\mu_1 - \mu_2$:	-4 ± 2.52	$\mu_2 - \mu_3$:	3 ± 2.52
$\mu_1 - \mu_3$:	-1 ± 2.52	$\mu_2 - \mu_4$:	6 ± 2.52
$\mu_1 - \mu_4$:	2 ± 2.52	$\mu_3 - \mu_4$:	3 ± 2.52

12.16. $\hat{\sigma}_L^2 = 2.744$; $s^2 = 11.48$; $s\,\hat{\sigma}_L = 5.62$,
$\bar{x}_1 = 41, \bar{x}_2 = 42.2, \bar{x}_3 = 31.2, \bar{x}_4 = 38.2, \bar{x}_5 = 30.8.$

$\mu_1 - \mu_2$:	-1.2 ± 5.62	$\mu_2 - \mu_4$:	4.0 ± 5.62
$\mu_1 - \mu_3$:	9.8 ± 5.62	$\mu_2 - \mu_5$:	11.4 ± 5.62
$\mu_1 - \mu_4$:	2.8 ± 5.62	$\mu_3 - \mu_4$:	-7.0 ± 5.62
$\mu_1 - \mu_5$:	10.8 ± 5.62	$\mu_3 - \mu_5$:	0.4 ± 5.62
$\mu_2 - \mu_3$:	11.0 ± 5.62	$\mu_4 - \mu_5$:	7.4 ± 5.62

12.18. *a.* Randomized block design, area = block.
b. SSTR = 1,288.25; SSB = 3,868.17; SSE = 552.50;
p-value = 0.053 (by interpolation).
Reasonable evidence to reject; roughly equivalent to α = 0.05 significance level.
c. $\hat{\sigma}_L = 7.835$; $s = 3.78$, $\bar{x}_A = 77, \bar{x}_B = 72, \bar{x}_C = 95.67, \bar{x}_D = 69.$

$\mu_A - \mu_B$:	5 ± 29.61	$\mu_B - \mu_C$:	-23.67 ± 29.61
$\mu_A - \mu_C$:	-18.67 ± 29.61	$\mu_B - \mu_D$:	3 ± 29.61
$\mu_A - \mu_D$:	8 ± 29.61	$\mu_C - \mu_D$:	26.67 ± 29.61

12.20. *a.* $X_{ij} = \mu + \tau_i + B_j + e_{ij}$, where i = 1, 2, 3, 4, 5; j = 1, 2, 3, 4, 5.
b. f = 25; $F(0.95; 4, 16)$ = 3.02 (by interpolation).
Since f = 25 > 3.02, reject H_0 at the α = 0.05 significance level; p-value much less than 0.001.
c. $\hat{\sigma}_L = 2$, $s = 3.476$.

$\mu_0 - \mu_A$:	-5 ± 6.95	$\mu_A - \mu_C$:	-10 ± 6.95
$\mu_0 - \mu_B$:	-12 ± 6.95	$\mu_A - \mu_D$:	-14 ± 6.95
$\mu_0 - \mu_C$:	-15 ± 6.95	$\mu_B - \mu_C$:	-3 ± 6.95
$\mu_0 - \mu_D$:	-19 ± 6.95	$\mu_B - \mu_D$:	-7 ± 6.95
$\mu_A - \mu_B$:	-7 ± 6.95	$\mu_C - \mu_D$:	-4 ± 6.95

12.22. *a.* f = 5.
b. $0.005 < p$-value < 0.01; $F(0.99; 4, 15)$ = 4.89, $F(0.995; 4, 15)$ = 5.80.
c. s = 3.50
   i. $\hat{\sigma}_L = 0.5$; -0.5 ± 1.75.
   ii. $\hat{\sigma}_L = 0.707$; 3.0 ± 2.47.
   iii. $\hat{\sigma}_L = 0.866$; 1.5 ± 3.03.

Chapter 13

13.2. To assess whether the relationship is linear.

13.4. The normal equations are derived from the calculus and, when solved simultaneously, produce least squares estimates of the population slope and intercept quantities.

13.6. It is the ratio of the explained to the total variation.

13.8. The estimators are unbiased and are

minimum variance among all linear estimators.

13.10. The coefficient of determination is the square of the coefficient of correlation.

13.12. In correlation, X is a random variable, whereas in regression, x is fixed.

13.14. Interval; no.

13.16. $r = (s_X/s_Y) \cdot b_1$.

13.18. a. $y = 100 + 3x$; b. linear statistical; c. $y = 100 + x^2$; d. $y = (x - 1)^2$.

13.20. b. Linear, with positive slope; c. $\hat{y} = 0.16 + 0.011x$; d. the residuals appear to be highly correlated.

13.22. a. 0.75; b. 90 percent.

13.24. a. $\hat{y} = -24{,}451.96 + 90.462x$; b. 0.916, 6,909.165; c. \$65,919.58; d. the residuals may be correlated.

13.26. b. 35.906, 1.293; c. 12.856, 0.7245; d. yes, yes.

13.28. a. $\hat{y} = 33.06 - 0.76x$; b. 0.681; c. 25.46, 6.45, 32.30, 33.06; d. 5.07; e. yes.

13.30. b. 0.8312.

13.32. b. 0.38; c. we would prefer regression if we desire to estimate the form of the relationship.

13.34. b. 0.993; c. a very strong linear relationship.

Chapter 14

14.2. A forecast interval is for an individual value, whereas a confidence interval is constructed for a mean value.

14.4. b_1 is a known sample quantity.

14.6. a. Yes; b. $\hat{y} = 221.86 + 10.508x$; c. \$6,758; d. \$6,521 to \$6,995; e. \$6,216 to \$7,300; f. $t = 19.13$, reject H_0; g. $f = 366.35$, reject H_0; h. yes, since the tests are equivalent.

14.8. $t = 5.16$; hence, reject H_0: $\rho = 0$.

14.10. Since $96.4 > 8.40$, reject H_0.

14.12. Since $t = 2.45 > 2.064$, reject H_0.

14.14. Since $t = 10.05 > 2.120$, reject H_0.

14.16. 0.0281.

14.18. a. $\hat{y} = 18.21 + 0.974x$; b. 81.5 percent; c. since $t = 5.56 > 1.895$, reject H_0; d. 85.52 to 143.13; e. 79.62 to 100.73.

14.20. Since $t = 19.1 > 2.365$, reject H_0; p-value is less than 0.01; average daily temperature is significant in predicting fuel use.

14.22. a. Reject H_0 since $t = 4.58 > 1.96$; b. 0.2449 to 0.5717.

14.24. a. Yes, reject H_0: $\beta_1 = 0$ beyond the 0.01 significance level; b. \$8.37 to \$16.22.

14.26. a. 0.8809; b. since $t = 6.45 > 3.055$, reject H_0.

14.28. a. 0.585 to 0.790; b. a strong relationship does not exist.

14.30. Since $t = 0.4136 < 2.776$, do *not* reject H_0.

14.32. a. Since $-1.96 < t = 1.90 < 1.96$, do not reject H_0; b. -0.01 to 0.2355; c. these variables are not significantly linearly related.

14.34. Since $t = 1.43 < 1.96$, do not reject H_0.

Chapter 15

15.4. The predictive ability and the explanatory properties of the model may improve if more than one independent variable is included.

15.6. If all three beta-values are simultaneously equal to 0, there is a 8.76 percent chance of getting an F-value as large as or larger than the one obtained in the sample.

15.8. x_2.

15.10. a. For GPA, $\hat{y} = -13.41 + 15.15$GPA, GPA is significant for all alpha-values greater than 0.0128; for HOURS, $\hat{y} = 18.777 + 1.224$HOURS, HOURS is significant for all alpha-values greater than 0.0046; for GPA and HOURS, $\hat{y} = -3.302 + 8.837$GPA $+ 0.8744$HOURS, the model is

Adam acknowledged it. The man was afraid. *Of me? Of having become too intimate, if it goes against me?*

His frigate, *Anemone,* had turned to face a vastly superior enemy, out-gunned and out-manned, with many of his company sent away as prize crews. He had not acted out of arrogance, or reckless pride, but to save the convoy of three heavily laden merchantmen he had been escorting to the Bermudas. *Anemone's* challenge had given the convoy time to escape, to find safety when darkness came. He remembered *Unity's* impressive commander, Nathan Beer, who had had him moved to his own quarters, and had come to visit him as he was treated by the surgeon. Even through the mists of agony and delirium, Adam had sensed the big American's presence and concern. Beer had spoken to him more like a father to his son than like a fellow captain, and an enemy.

And now Beer was dead. Adam's uncle, Sir Richard Bolitho, had met and engaged the Americans in a brief and bloody encounter, and it had been Bolitho's turn to give comfort to his dying adversary. Bolitho believed they had been fated to meet: neither had been surprised by the conflict or its ferocity.

Adam had been given another frigate, *Zest,* whose captain had been killed while engaging an unknown vessel. He had been the only casualty, just as Adam had been the only survivor from *Anemone* apart from a twelve-year-old ship's boy. The others had been killed, drowned, or taken prisoner.

The only verbal evidence submitted this morning had been his own. There had been one other source of information. When *Unity* had been captured and taken into Halifax, they had found the log which Nathan Beer had been keeping at the time of *Anemone's* attack. The court had been as silent as the falling snow as the senior clerk read aloud Beer's comments concerning the fierce engagement, and the explosion aboard *Anemone* which had

significant for alpha-values greater than 0.0055; *b.* combined: $32,139, GPA: $33,858, HOURS: $29,743; the combined model is probably the best; *c.* $r_{x_1x_2} = 0.5343$.

15.12. Since $f = 120.98 > 3.03$, reject H_0.

15.14. $r_{x_1x_2} = 0.89$.

15.16. *a.* 0.8683; *b.* 0.260.

15.18. Since $f = 6.84 > 2.40$, reject H_0.

15.20. *a.* 5.09; *b.* 5.17.

15.22. *a.* $\hat{y} = 3.95 + 2.06\text{EDUCATION} + 0.010\text{SERVICE}$; *b.* $12.25; *c.* $16.35; *d.* with EDUCATION equal to x_1, since $t = 5.88 > 3.499$, reject H_0: $\beta_1 = 0$, and with SERVICE equal to x_2, since $0.09 < 3.499$, do not reject H_0: $\beta_2 = 0$; *e.* years of postsecondary education likely contribute to the expected wage rate, whereas length of service does not; *f.* Since $f = 25.64 > 4.74$, reject H_0; *g.* there appears to be no strong violation of the underlying assumptions.

15.24. *a.* $\hat{y} = 6.93 + 1.18\text{SERVICE} + 1.19\text{SEX}$; *b.* $f = 1.997$; hence, do not reject H_0: $\beta_2 = 0$ at the 0.1, 0.05, or 0.01 significance level; *c.* no, unless it can be shown that hiring practices in the past have been biased against hiring women, so that they have shorter tenures with the company.

Chapter 16

16.4. 14.2687.

16.6. $\hat{y} = 100.770 + 10.270t$, $t = 0$ at 1977, t is in years.

16.8. *a.* Company A: $\hat{y} = 155.764 + 4.8260t$, $t = 0$ at 1969, t is in years, y is in $000, or $\log \hat{y} = 2.196 + 0.012t$, $t = 0$ at 1969, t is in years, y is in $000; Company B: $\hat{y} = 98.803 + 3.465t$ (same conditions), or $\log \hat{y} = 1.999 + 0.013t$ (same conditions); *b.* based on the semilogarithmic model for both firms, the cyclical pattern is present in both series, and they are

similar; *c.* using the semilogarithmic model for both firms, estimated sales are 221.50 and 147.23 (in $000), respectively.

16.10. $\hat{y} = 2,579.57 + 21.32t + 6.75t^2$, $t = 0$ at 1974, t is in years; the model appears to be appropriate; a very pronounced cyclical component is present.

16.12. 17.623.

16.14. $\hat{y} = 27.709 + 9.989t$, $t = 0$ at 1969, t is in years; a nonlinear function may be more appropriate.

16.16. $\log \hat{y} = 1.720 + 0.069t$, $t = 0$ at 1969, t is in years, y is in $ billions: $\hat{y}_{1980} = 301.30 billion; the model developed appears to be a good one.

16.18. 2.1, -2.6, -4.0, 19.6, 3.8, 10.6, -10.5, -3.9, 0.3, 0.7, -20.3; -1.5, -0.9, -5.2, 8.6, -4.2, 13.8, 2.6, 6.5, -2.5, 6.2, -13.9.

16.20. Based on the median of specific relatives: 81.86, 118.74, 106.33, and 93.07.

16.22. 456.12.

16.24. 45,708 units; 41,088 units; the first estimate should be the more accurate.

Chapter 17

17.2. 187.40.

17.4. 12; 1975.

17.6. 4.

17.8. $\hat{y}_t = 0.945 + 1.058y_{t-2}$.

17.10. $\hat{y} = 44.069 + 2.087t$ (starting with $t = 0$).

17.12. 44.07; 46.84; 49.19; 49.07; 52.27; 54.11; 56.97; 58.60; 61.58; 61.76; 65.19; 66.37; 68.32; 71.26.

17.14. Harmonic terms may not be significant.

17.16. $z_t = y_t - y_{t-1}$; mean is not changing over time.

17.18. 197.467; 263.741; all appear to be significant.

17.20. $\hat{y}_t = 1,846.12 + 0.0672y_{t-1}$.

State of nature / Act	1	2	3	4	5
1	24	0	7	1	12
2	0	4	5	12	13
3	11	2	21	0	5
4	7	12	10	4	0
5	6	8	0	14	3

H_0 at the $\alpha = 0.05$ significance level; p-value < 0.001.

21.6. Yes, act 1 dominates act 7 and act 5 dominates act 2.

21.8.

Payoff matrix:

Act, a_i (supply) \ State of nature, θ_j (demand)	0	1	2	3	4	5	6
0	0	0	0	0	0	0	0
1	−2	5	5	5	5	5	5
2	−4	3	10	10	10	10	10
3	−6	1	8	15	15	15	15
4	−8	−1	6	13	20	20	20
5	−10	−3	4	11	18	25	25
6	−12	−5	2	9	16	23	30

Quantities are in dollars; supply and demand quantities are in dozens.

Opportunity loss matrix:

Act, a_i (supply) \ State of nature, θ_j (demand)	0	1	2	3	4	5	6
0	0	5	10	15	20	25	30
1	2	0	5	10	15	20	25
2	4	2	0	5	10	15	20
3	6	4	2	0	5	10	15
4	8	6	4	2	0	5	10
5	10	8	6	4	2	0	5
6	12	10	8	6	4	2	0

Quantities are in dollars.

21.10

$$\Omega_{ij} = \begin{cases} \$5\,a_i & \text{if } \theta_j \geq a_i \\ \$7\theta_j + \$2a_i & \text{if } \theta_j < a_i \end{cases}$$

21.12. $\min_{i} \{9{,}700;\ 9{,}100;\ 4{,}950;\ 10{,}500\} =$ 4,950;
select portfolio 3; a bank deposit.

21.14. 0; 4; 3.

21.16.
$$l(a_i,\ \theta_j) = \begin{cases} 30(\theta_j - a_i) & \text{if } \theta_j > a_i \\ 0 & \text{if } \theta_j = a_i \\ 10(a_i - \theta_j) & \text{if } \theta_j < a_i \end{cases}$$

Chapter 22

22.2. Stock 14 dozen; EVPI = $2.08.

22.4. $n = 17$; $\pi = 0.77$; EVPI = $2.01.

22.6.
$$l(a_i,\ \theta_j) = \begin{cases} 5(\theta_j - a_i) & \text{if } \theta_j > a_i \\ 0 & \text{if } \theta_j = a_i \\ 2(a_i - \theta_j) & \text{if } \theta_j < a_i \end{cases}$$

$$\Omega_{ij} = \begin{cases} 5a_i & \text{if } \theta_j \geq a_i \\ 7\theta_j - 2a_i & \text{if } \theta_j < a_i \end{cases}$$

22.8. $C_m = \$8.025$; $C_k = \$8.05$.

22.10. $245.88; $0.00; $0.15.

22.12. $\sigma_\theta = 1.50$, $\theta_b = \$5.56$; do not purchase; EVPI = $1,300.86.

22.14. Consideration of nonmonetary consequences: utility is not linear in dollars.

22.16. 103.

22.18. *a.* 4.93; *b.* 5.29; *c.* 10.22; *d.* 5.29.

22.20. 1.0.

22.22. *a.* 1; *b.* 2; *c.* 0.

Chapter 23

23.2. Outcomes (trials) are probably not independent.

23.4. Construct tables similar to Table 23.5.

23.6. (10, 5); $30.76; $26.26; $35.24.

23.8. EVPI = $66.

23.10.

θ	$P(\theta)$
0.00	0.00
0.02	0.01
0.04	0.18
0.06	0.32
0.08	0.30
0.10	0.12
0.12	0.07
	1.00

4; $177.50; $27.20.

23.12. $P(\text{T/P}) = 0.135$; $P(\bar{\text{T}}/\text{P}) = 0.865$; $P(\text{P}) = 0.52$.

23.14. 349; highly unlikely.

23.16. Substitute $1/\sigma_1^2$ for I_1, etc; solve for σ_1 and simplify.

23.18. $\mu_1 = 224.76$, $\sigma_1 = 2.49$ (normal).

23.20. As n increases, the amount of information increases by $1/s^2$, or by the reciprocal of the variance.

23.22. $n = 21$; EVSI = $9,659.97; ENGS = $6,929.97.

Index

This book has been set VIP in 10 and 9 point Times Roman, leaded 2 points. Chapter numbers are Avant Garde Bold oversize, and chapter titles are 24 point Helvetica. The size of the type page is 31 by 47 picas.

The *t*-distribution

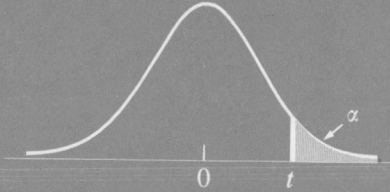

d.f.	.10	.05	.025	.01	.005
1	3.078	6.314	12.706	31.821	63.657
2	1.886	2.920	4.303	6.965	9.925
3	1.638	2.353	3.182	4.541	5.841
4	1.533	2.132	2.776	3.747	4.604
5	1.476	2.015	2.571	3.365	4.032
6	1.440	1.943	2.447	3.143	3.707
7	1.415	1.895	2.365	2.998	3.499
8	1.397	1.860	2.306	2.896	3.355
9	1.383	1.833	2.262	2.821	3.250
10	1.372	1.812	2.228	2.764	3.169
11	1.363	1.796	2.201	2.718	3.106
12	1.356	1.782	2.179	2.681	3.055
13	1.350	1.771	2.160	2.650	3.012
14	1.345	1.761	2.145	2.624	2.977
15	1.341	1.753	2.131	2.602	2.947
16	1.337	1.746	2.120	2.583	2.921
17	1.333	1.740	2.110	2.567	2.898
18	1.330	1.734	2.101	2.552	2.878
19	1.328	1.729	2.093	2.539	2.861
20	1.325	1.725	2.086	2.528	2.845
21	1.323	1.721	2.080	2.518	2.831
22	1.321	1.717	2.074	2.508	2.819
23	1.319	1.714	2.069	2.500	2.807
24	1.318	1.711	2.064	2.492	2.797
25	1.316	1.708	2.060	2.485	2.787
26	1.315	1.706	2.056	2.479	2.779
27	1.314	1.703	2.052	2.473	2.771
28	1.313	1.701	2.048	2.467	2.763
29	1.311	1.699	2.045	2.462	2.756
30	1.310	1.697	2.042	2.457	2.750
40	1.303	1.684	2.021	2.423	2.704
60	1.296	1.671	2.000	2.390	2.660
120	1.289	1.658	1.980	2.358	2.617
∞	1.282	1.645	1.960	2.326	2.576

Source: Hoel, *Elementary Statistics*, 3d ed. (New York: John Wiley & Sons, ©1971).